Spray Drying of Vaccines

Cordin Arpagaus

Spray Drying of Vaccines

From Laboratory Research to Industrial Applications

Springer

Cordin Arpagaus
Institute for Energy Systems
Eastern Switzerland University of Applied
Sciences
Buchs, Switzerland

ISBN 978-3-031-24322-6 ISBN 978-3-031-24323-3 (eBook)
https://doi.org/10.1007/978-3-031-24323-3

This Springer imprint is published by the registered company Springer Nature Switzerland AG
The registered company address is: Gewerbestrasse 11, 6330 Cham, Switzerland

Declaration: The author declares no conflicts of interest. The author is solely responsible for the content and writing of the book. The opinions, findings, and conclusions expressed in the book are personal views of the author and do not constitute an endorsement by any institution or authority. The author has no relevant affiliations or financial interests with any organization or entity that may have a financial interest in the subject discussed in the book. Names of specific vendors, manufacturers, or products are included for public interest and informational purposes.

Preface

The Global Race for the COVID-19 Vaccine

As of September 28, 2022, more than 12.7 billion COVID-19 vaccine doses have been administered in 184 countries since early 2021, making it the largest mass vaccination program in human history. Since then, more than 60% of the world's population has received at least one dose of more than 20 different COVID-19 vaccines. Thus, vaccination is widely recognized as the linchpin of medicine to end the COVID-19 pandemic. Indeed, preventing infection is a much better option than treating it. However, producing billions of vaccine doses is a global challenge for controlling the pandemic and returning lives to normal. Effective vaccines must reach people in low- and middle-income countries. At a cost of $10 to $30 per dose (including administration) and a vaccination coverage rate of 60% of the population with a 2-dose series, this translates into a lucrative market in the range of $150 to $450 billion worldwide.

Spray Drying Technology as an Option for Future Thermostable Vaccines

Aseptic spray drying technology could produce a large share of vaccines in stable powder form against many diseases, overcome the cooling challenges, and better match production with demand. Spray-dried powder is relatively easy, fast, cheap to produce, and stable at room temperature with suitable excipients and glass stabilizers, unlike most traditional vaccines that must be stored cold. The cold chain would be relieved, saving millions of dollars and providing effective vaccines for the world's most vulnerable people in low- and middle-income countries. Significant interest in spray-dried vaccines is also based on the capability of particle engineering. Thus, new administration routes of powder vaccines like pulmonary, intranasal, oral, or skin routes are possible, besides today's reconstitution and injections requiring needles and syringes. Thermostable vaccines are currently at crossroads. The traditional drying technology to convert a liquid vaccine into a stable dry solid form is freeze drying (lyophilization), but spray drying would be an economical alternative.

My Motivation for Writing this Book

It took me precisely two and a half years to write this book. The first lockdown of the COVID-19 pandemic in February 2020 was the main trigger to start writing. After that, I felt an inner urge to review the state of the art of spray drying vaccines. After a short time, I had the framework for this book and received the Go from the publisher Springer Nature to start writing this book. Immediately, I jumped into the literature research and contacted various researchers and manufacturers who published on this topic to get reliable background information.

Web of Science, Scopus, Google Scholar, PubMed, Sci-Hub, and Espacenet provided an essential basis for searching and retrieving original research and review articles from the last 30 years. The search terms used were mainly "spray drying," "vaccines," and "powder." All published studies found are included in the book as cited references. In addition, the language translation programs DeepL and Grammarly provided valuable help in writing sound text in English.

Another motivational boost came in the early 2021 when the first mRNA COVID-19 vaccines were introduced. There was a lot of news about the ultra-cold storage requirements with an enormous logistical challenge, especially in low- and middle-income countries where such cooling infrastructure is not always available. In addition, the COVID-19 pandemic has definitely spurred interest in potent and thermostable SARS-CoV-2 vaccines, which further prompted me to look at the thermostability of vaccines.

Overall, I may say that in the course of writing, I have also broadened my horizons and learned a lot about vaccines. I sincerely hope that a reader will appreciate the book and possibly make it at least selectively useful for his application. The thematic scope meant that many aspects could only be touched on the surface. However, in certain sections, there is more depth. I have tried to illustrate the relevant facts with many figures to enhance understanding and draw various conclusions supported by scientific references. The result is a comprehensive book with seven chapters, 241 figures, 88 tables, and over 1,250 references for further reading.

The Target Audience of this Book

The book is intended to provide a reference source for those involved in the vaccine formulation and the biopharmaceutical processing industry. The book's target audience is academics, researchers, students, vaccine developers, process engineers, industry professionals, and possibly funders, including government agencies, who are active in the field of vaccines.

Special Thanks and Dedication

My special thanks go to numerous experts listed in the acknowledgments for reviewing various text sections and providing permission for images to be published in this book.

Further, thanks go to the Institute for Energy Systems (IES) at the Eastern Switzerland University of Applied Sciences (OST), where I work as Senior Research Engineer on industrial high-temperature heat pumps and other thermal energy systems, which allowed me to work flexible hours in my home office.

Finally, the book is dedicated to my family for their patience and support and for giving me the freedom to write it, especially in the unique situations during the COVID-19 pandemic.

My special thanks go to my lovely cat, Chima, who supported me during coffee breaks!

Domat/Ems, Switzerland
September 2022

Cordin Arpagaus

Acknowledgments

I cordially thank the following experts for their generous, voluntary, and constructive support and assistance in helping me bring this book into the world.

- Dr. Sune Klint Andersen, Principal Scientist Particle Engineering Oral Solids Development at Janssen Pharmaceutica N. V. (Belgium), for the advice and expertise in reviewing text sections on aseptic spray drying technology, Design of Experiment (DOE) studies, scale-up, and comparison of drying technologies.
- Dr. Vu Truong-Le, Founder and CEO, and Dr. Phillip M. Lovalanti, Director R&D, at Aridis Pharmaceuticals (USA) for assistance in reviewing text sections on spray-dried influenza vaccines, measles vaccines, aseptic spray drying technology, Design of Experiment (DOE) studies, scale-up, and comparison of drying technologies.
- Dr. Ahmad M. Abdul-Fattah, Senior Manager at CSL Behring (Germany), for assistance in reviewing text sections on drying technologies for vaccines, aseptic spray drying technology, and vaccine formulation in a dry state.
- Prof. Pavan Muttil from the University of New Mexico (USA) for assistance in reviewing text sections on drying technologies for vaccines, aseptic spray drying technology, and vaccine formulation in a dry state, and for permission of using the picture of the Mini Spray Dryer B-290 in the biosafety cabinet.
- Dr. Keith Masters, Book Author and Spray Drying Expert from SprayDryConsult (Denmark), for assistance in reviewing text sections on drying technologies for vaccines, aseptic spray drying technology, and producing vaccines in powder form.
- Debra Kristensen, Director of Vaccine Technology Strategy and Policy from PATH (USA), for the discussion on the opportunities and challenges of spray drying of vaccines and reviewing text sections on thermostability and CTC-approved vaccines.
- Dr. Manjari Lal, Program Advisor Formulation Technologies for Vaccines and Pharmaceuticals at PATH (USA), for the discussion on the opportunities and challenges of spray drying of vaccines.
- Prof. Dr. Regina Scherließ from the Kiel University (Germany) for assistance in reviewing text sections on spray-dried influenza vaccines.
- Dr. David McAdams, Scientific Consultant at Dave McAdams Consulting (USA), for assistance in reviewing text sections on spray-dried measles vaccines

and the Carbon Dioxide-Assisted Nebulization with a Bubble Dryer® (CAN-BD) technology.

- Robert Turok, Technology Manager at SPX Flow Inc. Anhydro Dryer Group (USA), for advice and expertise on aseptic spray drying and the permission to use the image of the aseptic spray dryer.
- Dr. Bruce Roser, Chairman and Founder of Stablepharma Ltd. (UK), for assistance in reviewing text sections on the patented technologies StableVax™, HydRIS™ (Hypodermic Rehydration Injection System, and VitRIS™ (Vitrified Readily Injectable Suspension) and for the rights to use the images of the StableVax™.
- Dr. Sam de Costa, Stabilization Projects Manager at Nova Laboratories Ltd. (UK), for the discussion on aseptic spray drying options and permission to use the images of the aseptic spray dryer within a sterile environment.
- PD Dr. Steve Pascolo, Cofounder of CureVac and former Chief Scientific Officer, today immunologist at the University Hospital of Zurich (Switzerland), for assistance in reviewing the text section on mRNA vaccines.
- Dr. Mario Amacker (Mymetics SA, Switzerland) for assistance in reviewing the text section on HIV vaccines.
- Dr. Dennis Christensen, Head of Vaccine Delivery and Formulation Research Section at Statens Serum Institut (Denmark), for the discussion on opportunities and challenges of spray drying of vaccines and reviewing text sections on spray-dried influenza vaccines.
- Dr. Nitesh Kunda, Assistant Professor from the College of Pharmacy and Health Sciences at St. John's University (USA), for the assistance in reviewing text sections on spray-dried human papillomavirus.
- Dr. Jeffrey Breit, Senior Director of Pharmaceutical Sciences and Development at Rezolute Ltd. (USA), for assistance in reviewing the text section on spray-dried anthrax vaccines.
- Daniel A. LeClair from the McMaster University (Canada) for assistance in reviewing text sections on the Design of Experiment (DOE) studies.
- Varsha Singh, Ph.D. candidate from McMaster University (Canada), for assistance in reviewing text sections on the Design of Experiment (DOE) studies.
- Dr. Gaurav Kanojia, Project Manager at Intravacc (The Netherlands), for discussing opportunities and challenges of spray drying vaccines and assistance in reviewing text sections on the Design of Experiment (DOE) studies and scale-up.
- Dan Dobry, Director and Head of Commercial Development Drug Product at Lonza (USA), for assistance in reviewing text sections on the Design of Experiment (DOE) studies and scale-up.
- David Vodak, Executive Director and Head of Small Molecules R&D at Lonza (USA), for assistance in reviewing the text section on spray-dried norovirus vaccines.
- Fabrice Rose, Senior Formulation Specialist at Ziccum AB (Sweden), for assistance in reviewing text sections on alternative drying technologies for vaccines, spray-dried tuberculosis vaccines and permission to use the images of the LaminarPace™ spray dryer.

- Roger Lassing, Vice President of Business Development at Iconovo (Sweden), for assistance in reviewing text sections on single-use disposable dry powder inhalers for pulmonary vaccine delivery and permission to use the image of the ICOone® inhaler.
- Marco Meuri, Business Area Manager Spray Dryer and Freeze Drying at BÜCHI Labortechnik AG (Switzerland), for permission to use the images of the Mini Spray Dryer B-290 with accessory options.
- Debra Repko, Product Manager Spray Dryer and Encapsulation at Buchi Corporation (USA), for the discussion on laboratory spray dryers.
- Marco Laackmann, Sales Director Inhalation Technologies at Harro Höfliger (Germany), for assistance in reviewing text sections on aseptic powder filling and permission to use images of the dosing and filling systems.
- Steven Vineis, Marketing Director of Flow Sciences Inc. (USA), for permission to use images of the enclosures for the Mini Spray Dryer B-290 to process active pharmaceutical ingredients (APIs).
- Søren Terp Madsen, Product Manager at GEA Process Engineering A/S (Denmark), for permission to use the images of the PHARMA-SD® spray dryers.
- Andrew Birkmire, Process Development Manager at GEA Systems North America LLC (USA), for the discussion on aseptic spray drying technology.
- Christian Dunne, Head of Sterile Solutions of ChargePoint Technology Ltd. (UK), for permission to use the images of the split butterfly valve and the spray dryer discharge process.
- Gregory Cocchiaro, Sales and Marketing Administration at The Baker Company (USA), for permission to use the image of the BioPROTEC® cabinet.
- Isabel Pina, Director External Communications at Hovione (Portugal), for permission to use the images of the spray dryers on a laboratory and commercial scale.
- Maria Grazia Lolli, Marketing Specialist at IMA Life (Italy), for permission to use the images of the LYnfinity spray freeze drying system and the SMART PWD aseptic powder dosing and filling systems.
- Carlo Cattenati, Head of Product Management at Romaco s.r.l. (Italy), for permission to use the images of the automatic MicroMaxX series for sterile dosing of powders.
- André Hermsen, Manager Team Pharma of Hosokawa Micron BV (The Netherlands), for permission to use the images of the Active Freeze Dryer products.
- Dr. Richard Johnson, Founder and CEO of Upperton Pharma Solutions (UK), for permission to use the images of the spray dryers on a laboratory and commercial scale.
- Dr. Yuh-Fun Maa, Senior Principal Engineer at Genentech (USA), for permission to use the images of the laboratory spray dryers designs.
- Dr. Bernhard Luy, CEO of Meridion Technologies GmbH (Germany), for permission to use the images of the SprayCon™ and Lyo-Motion™ technologies.

About This Book

This book addresses the stabilization of vaccine powders by spray drying and provides an overview of the current state of the art on a laboratory and industrial scale. The book aims to familiarize readers with the advances in vaccine spray drying technology to understand its application potential better. In particular, the book addresses the design of aseptic spray dryers, parameters affecting the spray drying process, sterile powder processing, cleaning procedures, and powder filling. In addition, different drying technologies for the production of dry powder vaccines are compared to discuss the unique capabilities of spray drying as a particle engineering technology for vaccines. Special attention is given to research studies on spray-dried vaccines published over the past 30 years, with key findings from laboratory research to clinical trials. Potential applications of spray-dried vaccines and routes of administration are presented in detail. Finally, an outlook is given on how close the aseptic spray drying of vaccines is to the market and the challenges that need to be overcome to be commercially successful. The book's target audience is academics, researchers, vaccine developers, industry experts, students, and possibly funders, including government agencies, who are active in the field. In addition, the book is a reference source for those involved in the vaccine formulation and biopharmaceutical processing industry.

Contents

About the Author

Dr. Cordin Arpagaus is a process engineer and received his master's degree from the Swiss Federal Institute of Technology Zurich (ETH Zurich) in 2001. In 2005, he completed his doctorate (Dr. sc. techn.) at the Institute of Process Engineering at ETH Zurich in the field of plasma treatment of polymer powders. He also holds a master's degree in business administration from ETH Zurich.

Dr. Arpagaus then worked for several years in industry in management positions of Swiss companies active in laboratory and process technology. Between 2006 and 2011, he was a global product group manager for laboratory spray dryers and encapsulation systems at BUCHI Labortechnik AG (Flawil, Switzerland). His primary focus was developing new nano spray drying technologies and analyzing recent market trends in the pharmaceutical, materials, and food industries. From 2011 to 2015, he was the team leader of technical support for sensors and laboratory products at Hamilton Bonaduz AG (Switzerland).

Since April 2015, he has been back in academia and is working as a Senior Research Engineer at the Institute of Energy Systems (IES) at the Eastern Switzerland University of Applied Sciences in Buchs (OST—Ostschweizer Fachhochschule). His research activities focus on thermal energy systems and energy efficiency in industrial processes, focusing on multi-temperature and high-temperature heat pumps, and heat pumps with turbo-compressors.

He has extensive professional experience in process engineering in research and industry with particular interest in spray drying, heat pumps, and plasma processes. So far, Dr. Arpagaus has published more than

50 research articles and reviews papers in international journals, authored a book on *High-Temperature Heat Pumps*, written several book chapters on nano spray drying, and given numerous presentations at international conferences.

Dr. Arpagaus is an active researcher in the International Drying Symposium series, a member of the Swiss Society of Process and Chemical Engineers (SGVC), the International Institute of Refrigeration (IIR), and a reviewer of several international journals, including *International Journal of Refrigeration*, *Applied Thermal Energy*, *Applied Energy*, *Renewable and Sustained Energy Reviews*, *Energy*, *Trends in Food Science and Technology*, *Food and Bioproducts Processing*, *Food Hydrocolloids*, *Powder Technology*, *Additive Manufacturing*, *MDPI Materials*, and *MDPI Pharmaceuticals*.

Nomenclature

AAL	Aleuria aurantia lectin
Ad5	Adenovirus serotype 5
AdCh68	Chimpanzee adenovirus serotype 68
AdHu5	Human adenovirus serotype 5
Ag85A	Mycobacterium tuberculosis antigen 85A
AIDS	Acquired Immunodeficiency Syndrome
ALD	Atomic layer deposition
API	Active pharmaceutical ingredient
APMV-1	Avian paramyxovirus serotype 1
B-190	Mini Spray Dryer B-190
B-191	Mini Spray Dryer B-191
B-290	Mini Spray Dryer B-290
B-296	Dehumidifier B-296
B-90	Nano Spray Dryer B-90
BALB	Bagg and Albino
BCG	Bacille Calmette-Guérin
BMGF	Bill & Melinda Gates Foundation
BSA	Bovine serum albumin
CAN-BD	Carbon Dioxide-Assisted Nebulization with a Bubble Dryer®
CCID50	50% cell culture infectious dose
CD	Circular Dichroism
CDC	Centers for Disease Control and Prevention
CDMO	Contract Development and Manufacturing Organization
CFD	Computational fluid dynamics
COPD	Chronic obstructive pulmonary disease
COVID-19	Coronavirus disease 2019
CPAD	Compact prefilled auto-disable devices
CPS	Capsular polysaccharides
CQA	Critical quality attribute
CTC	Controlled Temperture Chain
Culp	Cutinase-like proteins
DCM	Dichloromethane
dD	Droplet size
DDA	Dimethyl-dioctadecyl ammonium bromide

DLS	Dynamic light scattering
DNA	Deoxyribonucleic acid
DOE	Design of experiments
DPPC	1, 2-dipalmitoyl-sn-glycero-3-phosphatidylcholine
DSC	Differential scanning calorimetry
DSPC	1, 2 distearoyl-sn-glycero-3-phosphocholine
DT	Diphtheria-tetanus (vaccine)
DTP	Diphtheria-tetanus-pertussis (vaccine)
EDTA	Ethylenediaminetetraacetic acid
EF	Edema factor
EF	Emitted fraction
ELISA	Enzyme-linked Immunosorbent Assay
EMA	European Medicines Agency
EPI	Expanded Programme on Immunization
F	Feed flow rate
FDA	Food and Drug Administration
FFA	Fluorescent focus assay
FFU	Fluorescent forming units
FPF	Fine particle fraction < 5 μm
G	Gas flow rate
GAVI	Global Alliance for Vaccines and Immunizations, the Vaccine Alliance
GLA	Glucopyranosyl lipid A
GMP	Good manufacturing practice
GOI	Antigen of interest
GRAS	Generally Regarded As Safe (excipients)
GTFCC	Global Task Force on Cholera Control
HA	Hemagglutinin
HAI	Haemagglutination inhibition assay
HBS	HEPES buffered saline
HBsAg	Purified hepatitis B virus surface protein
HBV	Hepatitis B virus
HepB	Hepatitis B vaccine
HepB-BD	Hepatitis B vaccine for birth dose
HEV	Hepatitis E virus
Hib	Haemophilus influenzae type b
HIC	High-income countries
HIT	Herd immunity threshold
HIV	Human immunodeficiency virus
HN	Hemagglutinin-neuraminidase protein
HP	High-performance
HPMC	Hydroxypropyl methylcellulose
HPMCAS	Hydroxypropyl methylcellulose acetate succinate
HPV	Human papilloma virus
HSV	Herpes simplex virus

ICG	International Coordinating Group
ICH	International Conference on Harmonisation
ID	Intradermal
IgG	Immunoglobulin G
IglC	Intracellular growth locus subunit C
IM	Intramuscular
IPV	Inactivated polio vaccine
IVI	International Vaccine Institute
ivt	In vitro transcribed (ivt mRNA)
KPO_4	Phosphate buffer
LF	Lethal factor
Lm	Listeria monocytogenes
LMICs	Low- and middle-income countries
LMP	Lipid-based nanoparticle
LVS	Live vaccine strain
mAb	Monoclonal antibody
MACIVIVA	Manufacturing process for Cold-chain Independent Virosome-based Vaccines
MAP	Microarray patch
MCV1	Measles-containing vaccine first dose
MCV2	Measles-containing vaccine second dose
MenA	Neisseria meningitidis A
MenAfriVac	Serogroup A meningococcal conjugate vaccine
MHRA	UK's Medicines and Healthcare products Regulatory Agency
MMAD	Mass median aerodynamic diameter
MMR	Measles-mumps-rubella vaccine
MMRV	Measles-mumps-rubella-varicella vaccine
MR	Measles-rubella vaccine
mRNA	Messenger ribonucleic acid
MSF	Médecins Sans Frontières
MV	Membrane vesicles
n.a.	Not available
NA	Neuraminidase
NaH_2PO_4	Sodium dihydrogen phosphate
NCMP	Nanocomposite microparticle
ND	Newcastle disease
NDV	Newcastle disease virus
nsPs	Nonstructural proteins
OCV	Oral cholera vaccine
ODF	Orally disintegrating film
OIE	World Organization for Animal Health (formerly Office International des Epizooties)
OMV	Outer membrane vesicles
OPV	Oral polio vaccine
PA	Protective antigen

PATH	Program for Appropriate Technology in Health
PBS	Phosphate buffered saline
PCV	Pneumococcal conjugate vaccine
PDPs	Product Development Partnerships
Pe	Peclet number
PEG	Polyethylene glycol
PfTrx	Pyrococcus furious thioredoxin
PGA-co-PDL	Poly(glycerol adipate-co-ω-pentadecalactone
PLA	Polylactic acid
PLGA	Poly(lactic-co-glycolic) acid
PNAP	Porous nanoparticle-aggregate particle
Polio	Poliomyelitis
PPSV23	23-valent pneumococcal polysaccharide vaccine
PSD	PHARMA-SD, Particle size distribution
PspA	Pneumococcal surface protein A
PT	Pertussis
PVA	Polyvinyl alcohol
PVP	Polyvinylpyrrolidone
QbD	Quality by design
R0	Baseline reproduction number
RF	Respirable fraction
RH	Relative humidity
RMC	Residual moisture content
RNA	Ribonucleic acid
rPA	Recombinant protective antigen
RSA	Rat serum albumin
RSV	Respiratory syncytial virus
RT	Room temperature
SARS-CoV-2	Severe acute respiratory syndrome coronavirus 2
SC	Subcutaneous
SDS-PAGE	Sodium dodecyl sulfate-polyacrylamide gel electrophoresis
SEC	Size exclusion chromatography
SEM	Scanning Electron Microscopy
SERI	Swiss State Secretariat for Education, Research, and Innovation
sIPV	Sabin inactivated polio vaccine
siRNA	Small interfering RNA
SLS	Sodium lauryl sulfate
SPS	Stabilization and protection solution
T	Titre log loss
TB	Tuberculosis
$TCID_{50}$	50% tissue culture infectious dose
Td	Tetanus-diphtheria (reduced component) toxoid (vaccine)
TDP	Tetanus-diphtheria-pertussis (vaccine)
Tg	Glas transition temperature
T_{in}	Inlet drying gas temperature

Tin	Inlet drying gas temperature
Tm	Phase transition temperature
TMA	Trimethylaluminum
TNA	Toxin neutralizing antibody
T_{out}	Outlet drying gas temperature
TT	Tetanus toxoid (vaccine)
T_{WB}	Wet-bulb temperature
UNICEF	United Nations International Children's Emergency Fund
USSR	Union of Soviet Socialist Republics
VLP	Virus-like particle
VSV	Vesicular stomatitis virus
VVM	Vaccine Vial Monitor
WHO	World Health Organization
XPX	X-ray spectroscopy
XRD	X-ray Diffraction
YF	Yellow fever (vaccine)

A Short Introduction to Vaccines

1

Abstract

This chapter provides a brief introduction to the subject of vaccines. First, the most important terms are explained to help the reader get started. Next, it gives an overview of the vaccines currently on the market, their types and achievements, global vaccination coverage, the concept of herd immunity, market size, expected growth, and the vaccine development process for routine and COVID-19 vaccines. Finally, the importance of the vaccine market to key market players, cooperation partners, and society is highlighted.

1.1 Vaccines Safe Lives

Vaccines are without question one of the greatest achievements of modern medicine for humanity, providing an efficacious and cost-effective intervention to prevent deadly diseases (Ulmer et al. 2006; Bregu et al. 2011; Nabel 2013; Delany et al. 2014; Greenwood 2014; GAVI 2021a). Vaccines will also play a major part in safeguarding people's health in the twenty-first century (Rappuoli 2011; Rappuoli et al. 2011, 2021; Delany et al. 2014). Today, more than 70 vaccines are approved by the WHO to prevent at least 25 different bacterial and viral infectious diseases (Nabel 2013; WHO 2021). Figure 1.1 shows a compilation of the vaccines, including COVID-19 (against SARS-CoV-2 virus). In addition, a global pipeline of more than 120 new vaccine products is under development, 60 of which are of particular importance for low-income countries (Kaddar 2013; D'Amore and Yang 2019; GEA 2019).

Since the development of the smallpox vaccine in 1798 by Edward Jenner (Jenner 1798), vaccination has controlled major diseases in the world, including smallpox, rabies, diphtheria, tetanus, yellow fever, pertussis, Haemophilus influenzae type b, poliomyelitis, measles, mumps, rubella, typhoid, rotavirus,

C. Arpagaus, *Spray Drying of Vaccines*,
https://doi.org/10.1007/978-3-031-24323-3_1

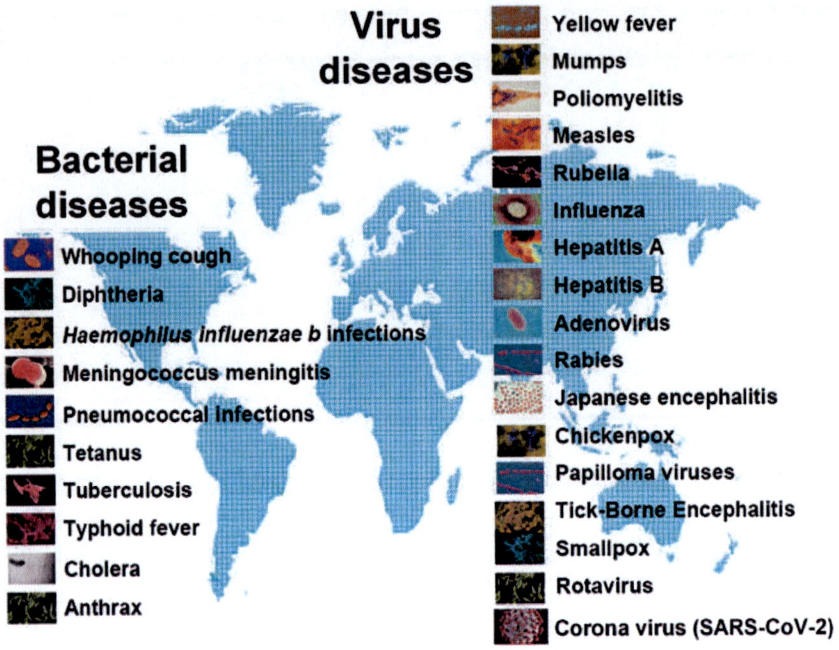

Fig. 1.1 Vaccines for humans against bacterial and viral infectious diseases (Extended from Kaddar (2013))

hepatitis B, and others (Plotkin and Plotkin 2018). As a result, some of the world's deadliest and most disabling diseases are virtually eliminated through routine childhood vaccination in many countries.

In 1980, smallpox was certified as eradicated by the WHO (1980), and cases of polio, measles, rubella, yellow fever, mumps, and influenza have declined significantly around the world (Plotkin and Plotkin 2018). For example, cases from wild polioviruses have reduced by over 99% since 1988, from an estimated 350,000 to only 33 reported cases in 2018 (WHO 2019a). Another success is the MenAfriVac vaccine, which has effectively reduced the number of meningitis A cases in sub-Saharan Africa because it can withstand high temperatures during storage and transport (Lydon et al. 2014).

Over 100 million infants are vaccinated each year, and the WHO estimated that vaccination saves about 2 to 3 million lives per year worldwide (WHO/UNICEF/World Bank 2009; Delany et al. 2014; WHO 2020a). However, this figure is a low estimate, with accurate calculation being virtually impossible (Vanderslott et al. 2019). An example that makes this clear is the impact of smallpox vaccination. If the vaccine had never been developed, (Hinman 1998) estimated that about 5 million people would die from smallpox each year without the vaccine, implying that about 150 to 200 million lives were saved between 1980 and 2018

(Vanderslott et al. 2019). Similar calculations could be made for other diseases, such as polio, measles, diphtheria, pertussis, or tetanus.

For COVID-19, a recent study from (Gupta et al. 2021) estimated that the vaccination campaigns in the US prevented nearly 140,000 deaths during the first 5 months of 2021 (December 14, 2020, to May 9, 2021). In addition, it may have prevented nearly 3 million cases of the disease. Galvani et al. (2021) estimated that by the end of June 2021, the COVID-19 vaccination program in the US averted approx. 279,000 additional deaths and up to 1.25 million additional hospitalizations. If the US had achieved only 50% of the actual vaccination rate, there would have been nearly 121,000 additional deaths and more than 450,000 additional hospitalizations during that period. Figure 1.2 illustrates these findings. Most of the

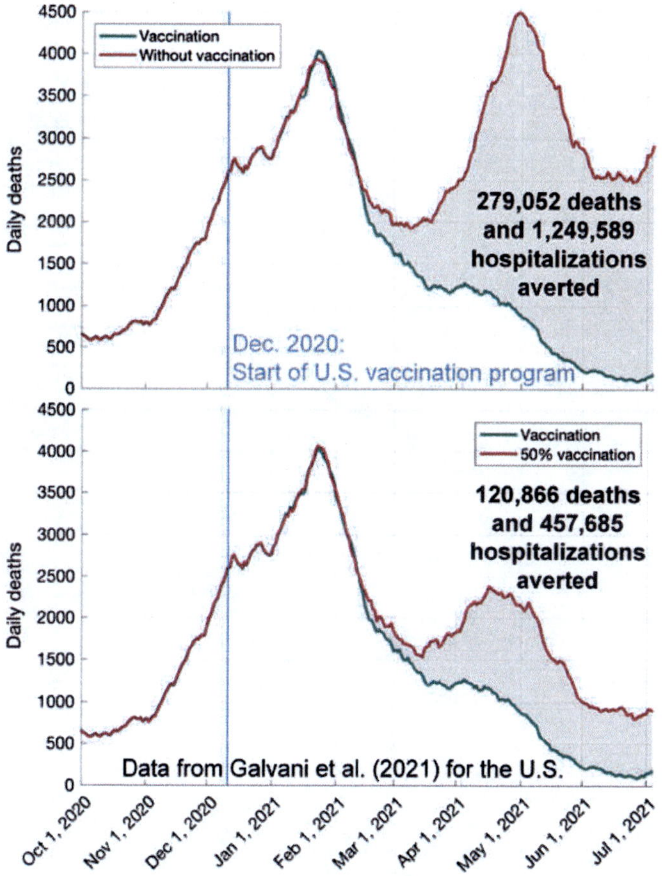

Fig. 1.2 Estimated U.S. seven-day rolling average of daily deaths with and without vaccination and if the vaccination rate had been 50% lower than the actually achieved rate (Data and figures adapted from Galvani et al. (2021))

additional deaths would be due to the increase and spread of the transmissible Alpha variant during the second wave in 2021 (the shaded area in the figure).

1.2 Global Vaccination Coverage

Figure 1.3 illustrates the global vaccination coverage rates of the main vaccines from 1980 to 2019 based on the immunization report of WHO/UNICEF (2020). The summary table shows coverage rates by WHO geographic regions, such as Europe, Southeast Asia, Western Pacific, Americas, Eastern Mediterranean, and Africa. For many essential vaccines, coverage is now well over 80%. However, as indicated by the green, yellow, and red traffic lights, Africa has the greatest need to catch up.

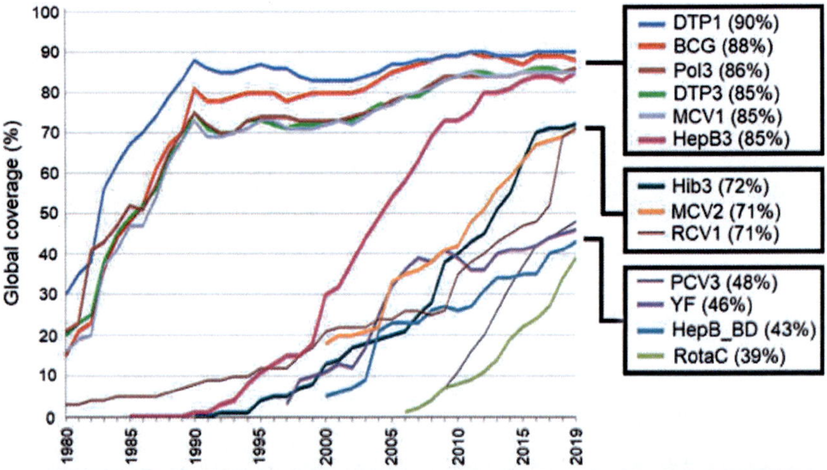

Region Vaccin	DTP1	BCG	Pol3	DTP3	MCV1	HepB3	IPV1	Hib3	MCV2	RCV1	MCV2	PCV3	YFV	HepB_BD	RotaC
Global coverage (%) (2019	90	88	86	85	85	85	82	72	71	71	57	48	46	43	39
European	97	92	95	95	96	92	95	79	91	96	14	80	0	41	25
South East Asian	94	93	90	91	94	91	83	89	83	93	80	23	0	54	37
Western Pacific	95	96	94	94	94	94	90	24	91	94	85	14	0	84	2
American	90	83	87	84	88	81	86	85	75	88	25	83	60	55	74
Eastern Mediterranean	89	87	83	82	82	82	83	82	75	45	75	52	0	34	49
African	81	80	74	74	69	73	71	73	33	33	33	70	45	6	50

Fig. 1.3 Global vaccination coverage estimates from 1980 to 2019 of 194 WHO Member States. The table is structured by region and vaccine for 2019, green: > 90% coverage, yellow: between 50 and 90%, red: < 50% (Data source: WHO/UNICEF (2020)). Abbreviations: *DTP1* First dose of diphtheria-tetanus-toxoid and pertussis vaccine, *BCG* Bacillus Calmette-Guérin, *Pol3* Third dose of polio vaccine, *DTP3* Third dose of diphtheria-tetanus-toxoid and pertussis vaccine, *MCV1* First dose of Measles-containing vaccine, *HepB3* Third dose of Hepatitis B vaccine, *Hib3* Third dose of Haemophilus influenzae type B vaccine, *MCV2* Second dose of Measles-containing vaccine, *RCV1* First dose of Rubella-containing vaccine, *PCV3* Third dose of Pneumococcal conjugate vaccine, *YF* Yellow fever vaccine, *HepB_BD* Hepatitis B birth dose, and *RotaC* Rotavirus last dose

In 2019, about 85% of young children (about 116 million) received 3 doses of the diphtheria-tetanus-pertussis vaccine (DTP3), protecting them from infectious diseases that can cause severe illness and disability or be fatal. By 2019, 125 member states have achieved at least 90% coverage of the DTP3 vaccine. The chart shows that the proportion receiving the recommended vaccines has remained stable in recent years.

In 2019, 14 million infants did not receive their first dose of the DTP1 vaccine, indicating a lack of access to immunization and other health services, and another 5.7 million were only partially immunized (WHO/UNICEF 2020). More than 60% of these children live in the 10 countries of Angola, Brazil, the Democratic Republic of Congo, Ethiopia, India, Indonesia, Mexico, Nigeria, Pakistan, and the Philippines. WHO's response to this situation is to work closely with these countries and collaborate with partners to improve global vaccination coverage. Overall, the goal is to push to reach 90% coverage in the following years.

As shown in Fig. 1.4, the leading infectious cause of mortality is tuberculosis, with an estimated 1.18 million vaccine-preventable deaths per year in 2017. In addition, there was a significant decrease in measles (−86%) and tetanus deaths (−88%) in 2017 compared to 1990 (Vanderslott et al. 2019).

However, despite successes in global vaccination, the WHO and the GAVI Alliance (formerly the Global Alliance for Vaccines and Immunization) estimate that more than 1.5 million people (3 per minute) still die annually from vaccine-preventable diseases (WHO 2020b; GAVI 2021a).

Many vaccine-preventable deaths occur in low- and middle-income countries (LMICs) where vaccination programs are in place but cold chain infrastructure for vaccine storage and transport is inadequate (Cheyne 1989; Galazka et al. 1999; Kaufmann et al. 2011; Price et al. 2020). This unfortunate finding is directly related to the difficulty of distributing life-saving vaccines to people in poorly served

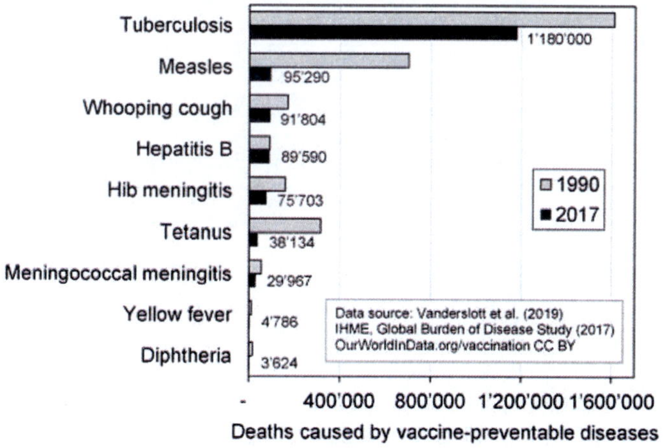

Fig. 1.4 Deaths caused by vaccine-preventable diseases in 1990 and 2017 (Data adapted from Vanderslott et al. (2019))

remote villages, fragile states, and strife-torn regions (WHO/UNICEF/World Bank 2009), as well as to the limited long-term storage stability, i.e., thermostability (Rexroad et al. 2002; Dumpa et al. 2019).

1.3 Herd Immunity

The primary mechanism by which vaccines work is quite simple. Vaccines generate immunity in an individual, and there is a collective social benefit when vaccination coverage is high (Vanderslott et al. 2019). Herd immunity is a key concept for epidemic control and occurs when a high percentage of the population is immune, making it less likely that infectious disease will spread (Randolph and Barreiro 2020; Fontanet and Cauchemez 2020). When the number of immune individuals is reached, a particular disease no longer exponentially grows or declines is called the herd immunity threshold (HIT). This threshold can be calculated from the basic reproduction number (R_0).

Figure 1.5 shows the HIT for various diseases. Measles and pertussis are highly contagious airborne diseases, and much of the population needs to be immunized to stop transmission. For COVID-19 (SARS-CoV-2 virus), most estimates of R_0 are in

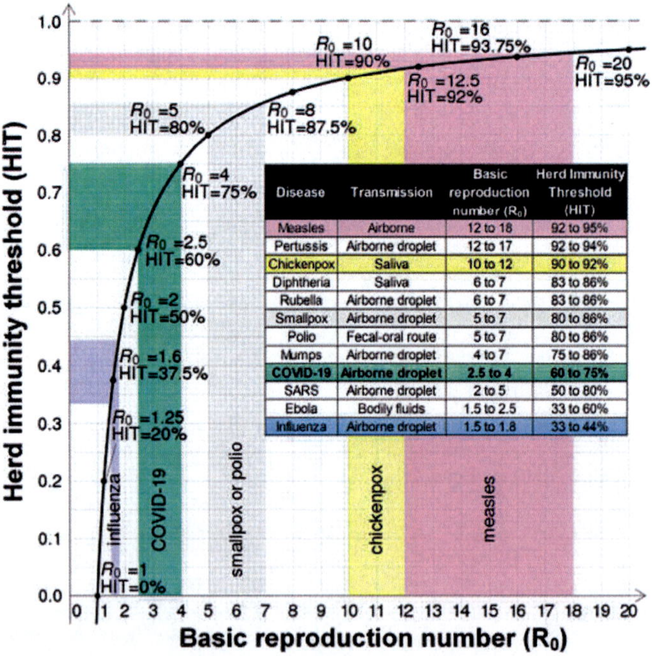

Disease	Transmission	Basic reproduction number (R_0)	Herd Immunity Threshold (HIT)
Measles	Airborne	12 to 18	92 to 95%
Pertussis	Airborne droplet	12 to 17	92 to 94%
Chickenpox	Saliva	10 to 12	90 to 92%
Diphtheria	Saliva	6 to 7	83 to 86%
Rubella	Airborne droplet	6 to 7	83 to 86%
Smallpox	Airborne droplet	5 to 7	80 to 86%
Polio	Fecal-oral route	5 to 7	80 to 86%
Mumps	Airborne droplet	4 to 7	75 to 86%
COVID-19	Airborne droplet	2.5 to 4	60 to 75%
SARS	Airborne droplet	2 to 5	50 to 80%
Ebola	Bodily fluids	1.5 to 2.5	33 to 60%
Influenza	Airborne droplet	1.5 to 1.8	33 to 44%

Fig. 1.5 Herd immunity thresholds (HIT) of vaccine-preventable diseases (Adapted from Vanderslott et al. (2019), Randolph and Barreiro (2020), Fontanet and Cauchemez (2020) and Wikipedia CC BY-SA 4.0)

the range of 2.5–4, corresponding to a theoretical HIT of 60–75% (Randolph and Barreiro 2020; Fontanet and Cauchemez 2020). One assumption in these calculations is that populations are homogeneous or well-mixed. High vaccination rates benefit everyone by reducing the spread of infection. In addition, high rates in one country help other countries; and increased rates in one generation benefit the next generation (Ehreth 2003).

1.4 Types of Vaccines

The word "vaccine" originates from the Latin Variolae Vaccinae (smallpox of the cow), which Edward Jenner demonstrated in 1798 could prevent smallpox in humans (Baxby 2011; Delany et al. 2014; Plotkin and Plotkin 2018). The WHO defines vaccines as "a heterogeneous class of medicinal products containing immunogenic substances capable of inducing specific, active, and protective host immunity against infectious disease" (WHO 2006).

In other words, a vaccine trains the body's immune system to produce the correct type of neutralizing antibody to fight a viral or bacterial disease it has not previously encountered. Therefore, an effective vaccine provides a vaccinated person with long-term protection.

Viruses are ubiquitous in the biological environment, and despite many powerful protection methods, they sometimes gain the upper hand and threaten life (Goodsell 2009). As shown schematically in Fig. 1.6, most viruses range in size from tens to hundreds of nanometers. This size diversity, and consequently superficial and internal structures, presents a major challenge in vaccine formulation and stabilization.

Scientific research has led to the development of various types of vaccines that elicit safe immune responses and protect against infections, and new vaccine strategies continue to be investigated to prevent existing and emerging diseases. Each type of vaccine is designed to teach the immune system how to fight off certain types of pathogens and the serious diseases they cause.

Figure 1.7 shows several vaccine platforms designed to protect against viral diseases and approved for human use (McGill COVID19 Vaccine Tracker Team 2021). The platforms can be divided into component vaccines and whole vaccines, including:

- live-attenuated vaccines (viral or bacterial) that use a weakened form of the pathogen,
- inactivated vaccines, in which the pathogen is killed and cannot self-replicate,
- protein subunit vaccines, from the surface of the pathogen,
- virus-like particles (VLPs) that mimic the pathogen itself,
- DNA- or RNA-based vaccines containing genetic material from the pathogen only, and viral vector vaccines (non-replicating, replicating) that use another virus that cannot copy itself but carries the pathogen's genes.

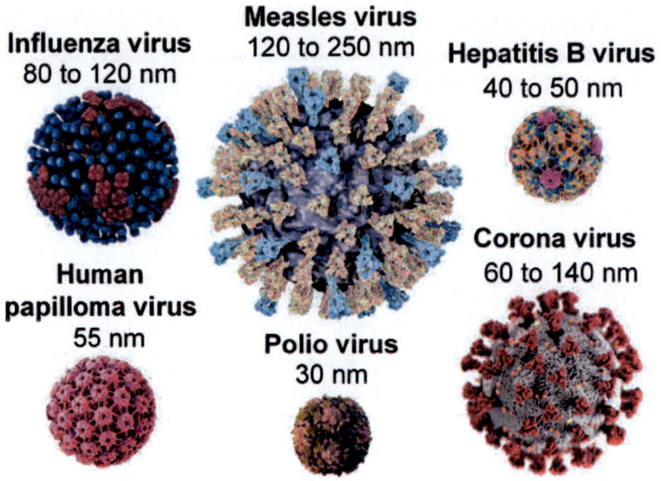

Fig. 1.6 3D structure and size range of some viruses and illustration of proteins exposed on their surface. Photo credits: Influenza virus (Centers for Disease Control and Prevention, National Center for Immunization and Respiratory Diseases, CC BY 4.0), Measles virus (Shutterstock, ID#: 505,972,186), Hepatitis B virus (Shutterstock, ID#: 1,103,225,849), Human papilloma virus (HPV) (Shutterstock, ID#: 533,717,263), Polio virus (Centers for Disease Control and Prevention), Corona virus (Centers for Disease Control and Prevention, ID#: 23,311)

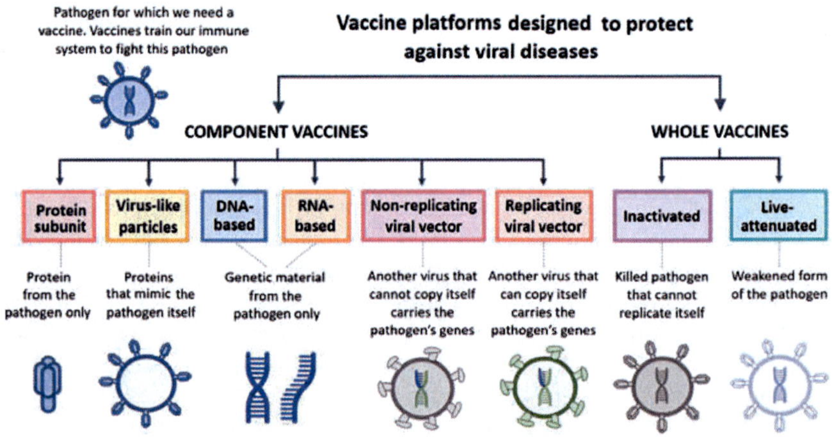

Fig. 1.7 Types of vaccine platforms designed to protect against viral diseases (Adapted from McGill COVID19 Vaccine Tracker Team (2021))

The major types of vaccines approved for human use include live-attenuated, killed, or inactivated organisms and subunit vaccines (e.g., protein and carbohydrate antigens) based on partially purified components. Table 1.1 reports a list of the main vaccines approved for immunization in the US by the D (FDA 2021).

Live-attenuated vaccines (i.e., bacteria or viruses) use a weakened form of the pathogen through physical or chemical treatment. Because these vaccines are similar to natural infections, they can produce a strong and long-lasting immune response with 1 or 2 doses (HHS 2021). Adjuvants are not required to boost the immune response, but live-attenuated vaccines are more sensitive to loss of efficacy during storage, especially at elevated temperatures. This means they have to be stored and transported in a refrigerated environment. Live-attenuated vaccines are used, for example, to protect against measles, mumps, rubella (MMR combination vaccine), rotavirus, smallpox, chickenpox, and yellow fever (Ulmer et al. 2006; Rosini et al. 2011; Delany et al. 2014; FDA 2021).

The second group of whole vaccines is **inactivated or killed pathogens** by chemicals (e.g., formaldehyde), heat, or irradiation with low-energy electrons, while leaving their proteins intact so that the immune system recognizes the vaccine as a whole and produces the appropriate antibodies (Ulmer et al. 2006). Inactivated vaccines generally do not provide as strong immunity and protection as live-attenuated vaccines and therefore require multiple doses over time (e.g., booster vaccinations) to maintain long-lasting immunity to the disease. Inactivated vaccines are used to protect against, for example, hepatitis A, influenza, polio, or rabies (HHS 2021).

Instead of the whole pathogen, **subunit vaccines** contain only the components (antigens) of the infectious agent that best stimulate the immune system (NIAID 2019). **Protein subunit vaccines** include only the proteins from the surface of the pathogen. Although this design makes vaccines safer by minimizing side effects

Table 1.1 Vaccines licensed for immunization in the US by the FDA

Live-attenuated vaccines	Inactivated vaccines	Subunit vaccines
Cholera		
Chickenpox		
Herpes zoster		Anthrax
Influenza		Diphtheria
Measles	Hepatitis A	Haemophilus influenzae B
Mumps	Influenza	Hepatitis B
Rotavirus	Japanese encephalitis	Human papillomavirus
Rubella	Plague	Influenza
Smallpox	Pertussis	Pertussis
Typhoid	Polio	Meningococcus
Tuberculosis	Rabies	Pneumococcal
Varicella		Tetanus
Yellow fever		Typhoid
Zoster		

Summarized based on Ulmer et al. (2006), Rosini et al. (2011), Delany et al. (2014), FDA (2021), HHS (2021)

and easier to manufacture, adjuvants must often be added to enhance the vaccine's protective immune response and subsequent clinical efficacy because the antigens alone are not sufficient to induce adequate long-term immunity.

Some vaccines consist of two or more antigens combined in the final formulation stage or mixed immediately before administration. Such vaccines are intended to protect against more than one disease or a disease caused by different strains or serotypes of the same organism. Some examples of combination vaccines for children are (PATH 2012; CDC 2014; Kumru et al. 2014):

- Comvax (Merck) combines Hib and Hep B
- Twinrix (GSK) combines Hep A and Hep B
- Pediarix (GSK) combines DTaP, Hep B, and IPV (polio)
- ProQuad (Merck) combines MMR and varicella (chickenpox)
- Kinrix (GSK) combines DTaP and IPV (polio)
- Pentacel (Sanofi Pasteur) combines DTaP, IPV (polio), and Hib.

Some vaccines to prevent invasive bacterial diseases are based on polysaccharides or sugars that form the outer shell of many bacteria. Today, there are also conjugated vaccines in which a protein antigen is covalently bound to a carrier protein or bacterial polysaccharide to enhance the immunogenicity of the antigen, e.g., against Hib, pneumococcal, and meningococcal infections (NIAID 2019). They can also be used on people with weakened immune systems and long-term health problems. However, one limitation of these vaccines is that booster shots are required in order to get ongoing protection against the diseases (HHS 2021).

Virus-like particles (VLPs) are virus-derived proteins and associated molecules that can self-assemble and form a particulate structure that mimics the shape and size of a virus particle but lacks the genetic material required for the virus to replicate in the host cells (Al-Barwani et al. 2014; NIAID 2019; Nooraei et al. 2021). This makes them safer for vaccine recipients and operators in contact with the vaccine (Fuenmayor et al. 2017). VLP-based vaccines have the safety profile of a subunit vaccine but with efficacy and vaccination outcomes comparable to killed or live-attenuated vaccines (Donaldson et al. 2018). To date, several VLP-based vaccines are on the market with good results, including.

- Recombivax HB and Engerix-B (GSK) for hepatitis B virus (HBV),
- Gardasil (Merck), Cervarix (GSK), and Gardasil-9 (Merck) for human papillomavirus (HPV), and
- Hecolin (Xiamen Innovax Biotech Co. Ltd) to prevent hepatitis E virus (HEV) infection (Lua et al. 2014; Al-Barwani et al. 2014; Nooraei et al. 2021).

Several other VLP-based vaccines are also in various design, production, and approval stages.

For decades, scientists studied **viral vector vaccines**, which use a modified version of a different virus as a vector to deliver protection, including influenza virus, vesicular stomatitis virus (VSV), measles virus, and adenovirus (Ad) (Holman et al. 2009; Ura et al. 2014; Lauer et al. 2017). For example, human Ad serotype 5 (Ad5) has been widely investigated as a gene delivery vector because it can be easily produced in high titers (Ura et al. 2014). Adenoviruses are also used as a viral vector to produce COVID-19 vaccines such as ChAdOx1 nCov-19 (AZD1222, Oxford University and AstraZeneca), Sputnik V (Gam-COVID-Vac, Gamaleya Research Institute), Ad5-nCoV (Convidecia, CanSino Biologics), and Ad26.COV2.S (Janssen, Johnson & Johnson) (AboulFotouh et al. 2021).

Another platform is **DNA- and RNA-based vaccines** that contain only genetic material from the pathogen, not the virus itself, i.e., DNA (as plasmids) or RNA (as messenger RNA) that is taken up by cells and translated into proteins to elicit an immune response (Leitner et al. 1999; Callaway 2020). Currently, the field of mRNA vaccines is experiencing a burst in R&D with dozens of preclinical and clinical reports demonstrating the efficacy of these platforms (Pardi et al. 2018; Kowalski et al. 2019; Samaridou et al. 2020). For example, CureVac, BioNTech, and Moderna run multiple preclinical and clinical trials with mRNA vaccines, and large pharmaceutical companies such as Genentech, Amgen, Merck, Pfizer, and Lonza have entered this field (Kowalski et al. 2019).

The **mRNA vaccines** platform offers high potency, capacity for rapid development, the potential for cost-effective manufacturing (i.e., scale-up), and safe administration (Pardi et al. 2018; Rosa et al. 2021). Moreover, the genetic code of mRNA vaccines can be rapidly modified to address any new virus mutations (AboulFotouh et al. 2021). Compared to plasmid DNA and viral vector vaccines that have to reach the nucleus, mRNA vaccines are active in the cell's cytoplasm and do not integrate the genome (Delany et al. 2014; Steinle et al. 2017). Therefore, the translation of the genetic information encoded on the mRNA to produce the desired protein starts instantly (Guan and Rosenecker 2017). Finally, the mRNA is rapidly degraded by cellular processes in the cytoplasm so that no traces can be found after 2–3 days (Steinle et al. 2017).

Efficient intracellular delivery is the key and is achieved by encasing mRNA vaccines in carrier molecules that protect against enzymatic degradation and allow rapid uptake and expression in the cytoplasm. **Lipid nanoparticles (LNPs)** have become the most attractive and widely used mRNA vaccine delivery systems, as discussed in several recent reviews (Whitehead et al. 2009; Midoux and Pichon 2015; Reichmuth et al. 2016; Guan and Rosenecker 2017; Pardi et al. 2018; Samaridou et al. 2020; Rosa et al. 2021). Figure 1.8 shows that mRNA must cross the cell membrane to reach the cytoplasm among the many functional barriers. A positively charged LNP protects the mRNA from enzymatic degradation, favors the mRNA localization at the negatively charged cell membrane, facilitates endocytosis into the cytoplasm, and allows the mRNA release from the endosome. The mRNA interacts with ribosomes to produce the proteins. LNP formulations are typically composed of (Reichmuth et al. 2016; Pardi et al. 2018; Kowalski et al. 2019):

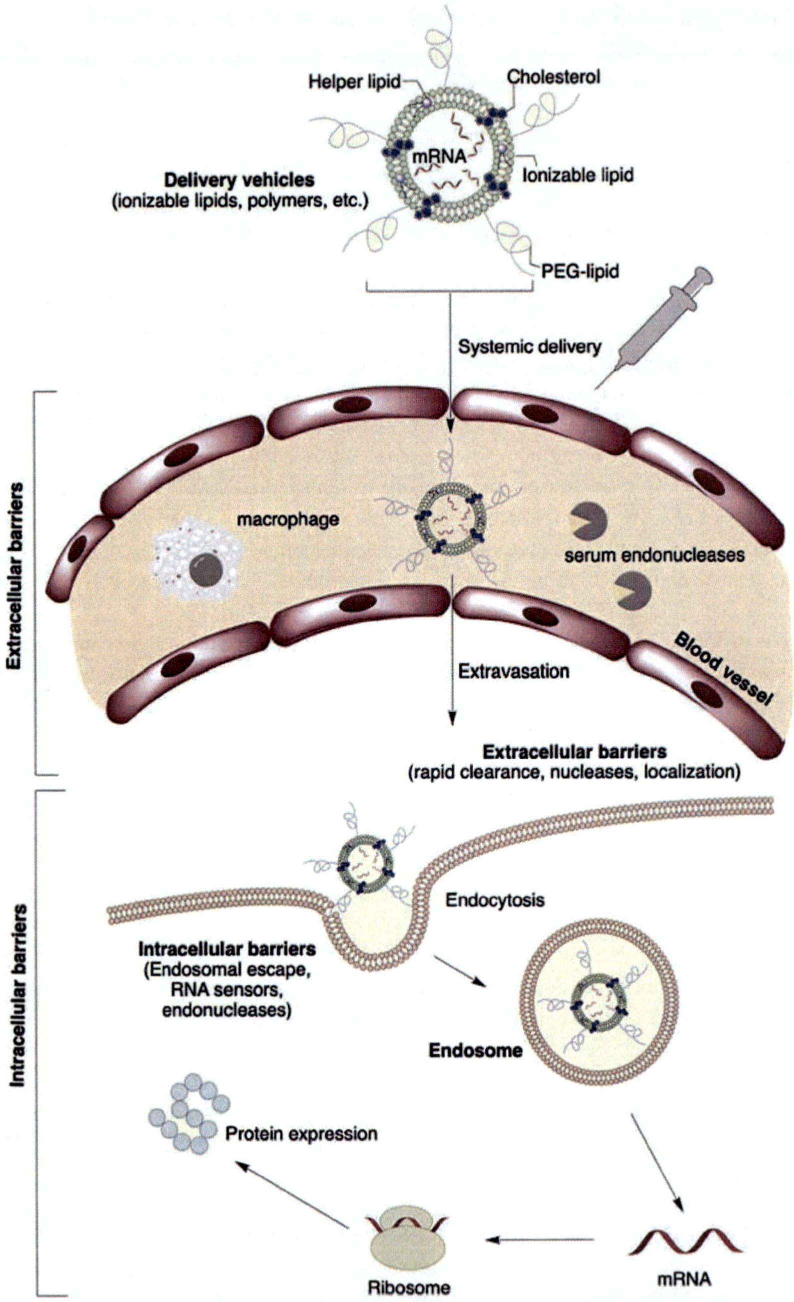

Fig. 1.8 Schematic representation of extra- and intracellular barriers for mRNA delivery based on lipid nanoparticles (Adapted from Kowalski et al. (2019), open access under CC BY-NC-ND 4.0 license)

- an ionizable cationic lipid or polymeric material bearing tertiary or quaternary amines to encapsulate the polyanionic mRNA,
- a zwitterionic lipid (e.g., 1,2-dioleoyl-sn-glycero-3-phosphoethanolamine, DOPE) that resembles the lipids in the cell membrane,
- cholesterol to stabilize the lipid bilayer of the LNP, and
- a polyethylene glycol (PEG)-lipid to lend the nanoparticle a hydrating layer, improve colloidal stability, and reduce protein absorption.

Table 1.2 compares the mRNA-COVID-19 vaccines from Pfizer-BioNTech (BNT162b2/Comirnaty) and Moderna (mRNA-1273) based on several reviews (Crommelin et al. 2021; Kim et al. 2021; Schoenmaker et al. 2021; Verbeke et al. 2021).

Both mRNA vaccines use LNPs for delivery carriers, encode the spike (S) protein of SARS-CoV-2, and show similar efficacy (approximately 95%) with the two-dose regimen (prime, booster). However, their mRNA doses (30 μg for BNT162b2 versus 100 μg for mRNA-1273), dosing intervals, formulation characteristics, and storage conditions are different. A drawback of the current mRNA COVID-19 vaccines is that they require storage conditions at (ultra)low temperatures, limiting their distribution to locations where cold storage is scarce or unavailable (Schoenmaker et al. 2021).

1.5 Vaccine Development Process

mRNA vaccines were the first vaccines approved in both the US and the EU as a prophylactic treatment to combat the COVID-19 pandemic (Rosa et al. 2021). There has been unprecedented speed in vaccine development (Corbett et al. 2020; Mulligan et al. 2020; Rappuoli et al. 2021; Verbeke et al. 2021). Figure 1.9 illustrates the timeline for COVID-19 vaccine development, from genome code submission over the internet to synthetic gene identification and first human clinical trials (Rappuoli et al. 2021).

Within a week after the genomic sequence of the coronavirus SARS-CoV-2 became available (GenBank ID: MN908947.3, January 10, 2020) (Wu et al. 2020), the vaccine developer Moderna was able to produce mRNA encoding SARS-CoV-2 prefusion-stabilized spike protein and formulated in LNPs as a delivery vehicle (Rappuoli et al. 2021). This led to a first-in-human phase I clinical trial starting on March 16, 66 days after the viral sequence was posted (Jackson et al. 2020), and a phase II trial 74 days later on May 29 (Corbett et al. 2020), and a Phase III trial on July 27 (Verbeke et al. 2021) (see Fig. 1.10).

For Pfizer-BioNTech with BNT162b2, the pace was similar. Phase I clinical trials began with 4 vaccine candidates on April 23 in Germany and May 4 in the US. Phase II/III trials started on July 27. Finally, the fast-track designations by the FDA (2020a, b) and EMA (2020, 2021) resulted in their approval to the market by both regulatory agencies by the end of December 2020 for emergency use. This

Table 1.2 Comparison of mRNA vaccines from Pfizer-BioNTech and Moderna against COVID-19 (dosage profiles, formulation compositions, storage requirements)

Category	Pfizer-BioNTech	Moderna
Product name	BNT162b2 (Comirnaty)	mRNA-1273
mRNA dose (volume)	30 μg (0.3 mL)	100 μg (0.5 mL)
Dosing interval	2 doses with a 21-day interval	2 doses with a 28-day interval
Route of administration	Intramuscular	Intramuscular
Lipid nanoparticle (LNP) components	0.43 mg ALC-0315 (Acuitas) = (4-hydroxybutyl) azanediyl)bis(hexane-6,1-diyl)bis(2-hexyldecanoate) 0.05 mg ALC-0159 = 2-[(polyethylene glycol)-2000]-N,N ditetradecylacetamide 0.09 mg 1,2-distearoyl-sn-glycero-3-phosphocholine (DSPC) 0.2 mg Cholesterol	SM-102 = Heptadecan-9-yl 8-((2-hydroxyethyl)(8-nonyloxy)-8-oxooctyl) amino)octanoate PEG2000-DMG = 1-monomethoxypolyethyleneglycol-2,3-dimyristylglycerol with polyethylene glycol of average molecular weight 2000 1,2-distearoyl-sn-glycero-3 phosphocholine (DSPC) Cholesterol
Molar lipid ratios (mol %) Ionizable cationic lipid/neutral lipid/cholesterol/PEG-ylated lipid	46.3/9.4/42.7/1.6	50/10/38.5/1.5
Buffer	0.01 mg Potassium dihydrogen phosphate 0.07 mg Disodium hydrogen phosphate dihydrate pH 7–8	Tris (tromethamine) pH 7–8
Other excipients	0.01 mg Potassium chloride 0.36 mg Sodium chloride 6 mg Sucrose Water for injection	Sodium acetate Sucrose Water for injection
Efficacy	95%	94.5%
Storage requirements	−60 to −80 °C (6 months) protected from light 2–8 °C (5 days) protected from light Room temperature (2 h)	−15 to −25 °C (6 months) protected from light 2–8 °C (30 days) protected from light 8–25 °C (12 h)

Crommelin et al. (2021), Kim et al. (2021), Schoenmaker et al. (2021), Verbeke et al. (2021)

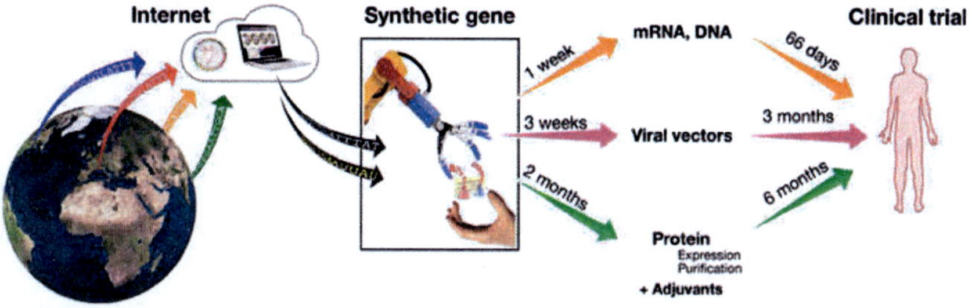

Fig. 1.9 Timeline for COVID-19 vaccine development by internet submission of the genome code to synthetic gene identification and human clinical trials (Reproduced from Rappuoli et al. (2021), open access under CC BY-NC-ND 4.0 license)

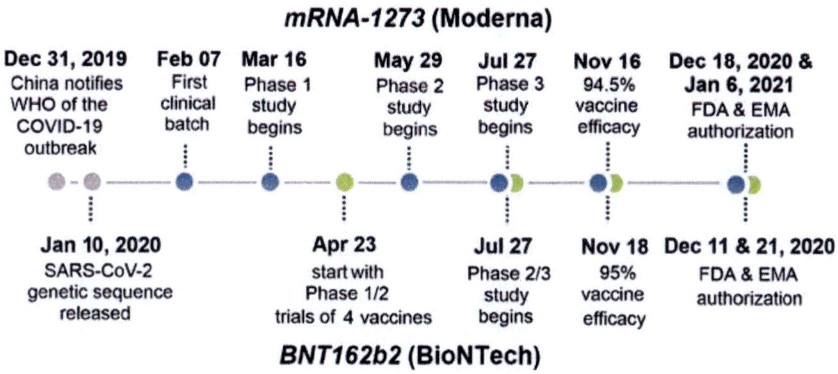

Fig. 1.10 Timeline for the development of mRNA vaccines against COVID-19 from BioNTech (BNT162b2) and Moderna (mRNA-1273) (Adapted from Verbeke et al. (2021))

demonstrates ways to accelerate the vaccine development timeline with mRNA vaccines in a pandemic. Promising immunogenicity and efficacy data from phase III trials in humans may be produced in about 10 months (Sahin et al. 2020; Anderson et al. 2020; Rappuoli et al. 2021).

The remarkable speed of COVID-19 vaccine development was possible because scientists combined for the first time the advances of three decades in independent fields in an unprecedented global effort, including the following technologies (Rappuoli et al. 2021):

- Reverse vaccinology (technology to identify novel vaccine antigens from the genome sequence of pathogens using bioinformatics approaches) (Ulmer et al. 2006; Rosini et al. 2011),
- Structural vaccinology (or structure-based antigen design) (Delany et al. 2014),

- Synthetic biology (ability to use synthetic genes for vaccines or cancer therapies) (Dolgin 2021; Tan et al. 2021), and
- Vaccine adjuvants (substances added to increase vaccine effectiveness) (Egli et al. 2014).

In general, however, vaccine development is lengthy, complex, costly, and involves a combination of public and private involvement (Plotkin et al. 2017; Douglas and Samant 2018). Table 1.3 outlines the key stages for vaccine development and objectives (Plotkin et al. 2017).

For example, an analysis of 352 vaccine development projects over 11 years (2003 to 2013) found that only 21 vaccine candidates, or 6%, achieved market entry (Stephens 2014; Halliday 2017). However, another study revealed a 16.2% success rate from Phase 1 to license approval (Thomas et al. 2016) (see Table 1.4).

Table 1.5 lists vaccine development timelines of 10–30 years for some of the currently approved vaccines, a time frame that most consider too long (Douglas and Samant 2018; Artaud et al. 2019).

Table 1.3 Key vaccine development stages and objectives

Stage	Objectives
Preclinical	• Assess safety and immunogenicity of a target antigen or cell in cell culture or animal disease models • Assess a safe starting dose for human clinical studies
Clinical trial authorization application	• Apply for approval to conduct human clinical studies
Phase 1 clinical trials	• Assess the safety of the candidate vaccine • Determine the type and extent of immune response that the vaccine provokes
Phase 2 clinical trials	• Assess candidate vaccine safety, immunogenicity, dose–response, schedule of immunizations, and method of delivery
Phase 3 clinical trials	• Assess the candidate vaccine in the target populations for safety and rare adverse events • Vaccine efficacy is estimated • Vaccine manufacturing consistency is confirmed • Concomitant testing with other prescribed vaccines may be required
Approval and licensure	• Submit and gain approval of the Biological Product Application
Post-Licensure monitoring	• Confirm filed use of the vaccine is consistent with expectations from the clinical studies and finalized manufacturing and release process
License amendments	• Confirm any changes to the intended use of the vaccine in different populations or/and modifications to the manufacturing process (seeds, raw material sources, process steps, release steps, equipment, facilities, etc.) do not adversely affect the product purity, safety, or effectiveness

Based on Plotkin et al. (2017)

Table 1.4 Overall success rates for vaccines from Phase 1 to license approval over 10 years (2006–2015)

Phase 1 to 2	Phase 2 to 3	Phase 3 to BLA	BLA to Approval	Overall
66.3%	32.9%	74.3%	100%	16.2%

BLA Biologics License Agreement
Data source Thomas et al. (2016)

Table 1.5 Vaccine development timelines

Vaccines	Years to approval
Varicella (chickenpox)	25 to 30
FluMist (influenza)	25 to 30
Human papillomavirus (HPV)	14 to 16
Rotavirus	14 to 16
Pediatric combination vaccines	10 to 12

Data source Douglas and Samant (2018)

Figure 1.11 shows a typical vaccine development timeline that takes about 5–10 years, sometimes longer, and in contrast, the accelerated timeline for vaccine development during a pandemic of about 1–2 years only. Several opportunities were used to accelerate vaccine development in the COVID-19 pandemic, including (John Hopkins University & Medicine 2021):

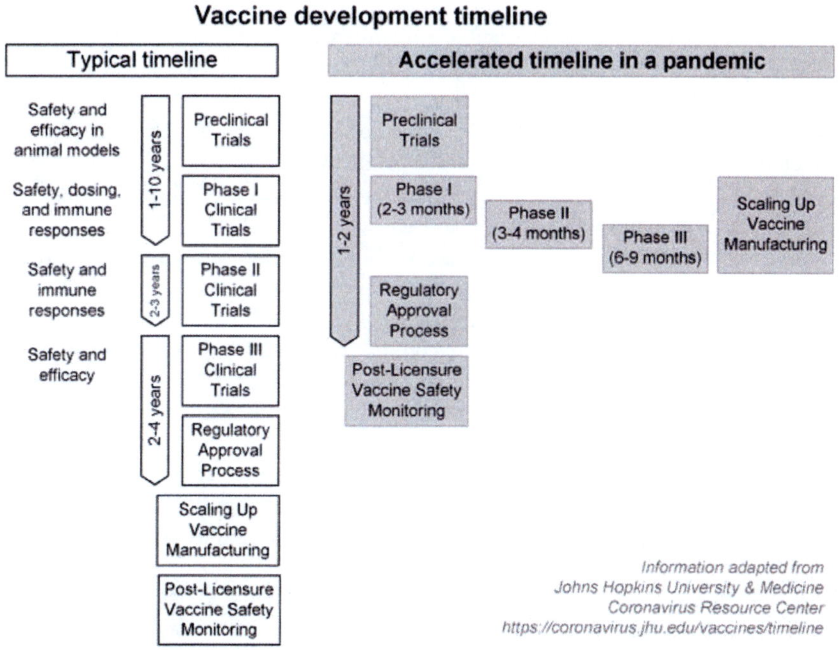

Fig. 1.11 Timeline for vaccine development. Left: Typical timeline of 5–10 years, sometimes longer. Right: Accelerated timeline during a pandemic (Adapted from John Hopkins University & Medicine (2021), Coronavirus Resource Center)

- Clinical trial phases were combined to evaluate the safety and immune responses.
- Differences in disease risk between those who received the viral vaccine and those who received a placebo or a comparator vaccine were measured more quickly.
- Fast track designations by the FDA and EMA accelerated the start of clinical trials and regulatory review for market approval.
- Governments and others invested heavily in manufacturing capacity to produce large numbers of vaccine doses before Phase III trial results were available.

1.6 Vaccine Market and Growth

With a market share of about 3.4% in 2017 (Evaluate 2018), the vaccine industry is relatively small compared to the whole pharmaceutical sector but rapidly growing (Douglas and Samant 2018; D'Amore and Yang 2019). Indeed, vaccines are becoming a driver for the pharmaceutical industry.

Table 1.6 and Fig. 1.12 highlight the projected development of the vaccine market size in billions of US$ from 2000 to 2028 based on several market reports (Evaluate 2018; Globe Newswire 2018; PR Newswire 2018; Research and Markets 2018; Markets and Markets 2020; Research and Markets 2020).

- Evaluate (2018) reported a total worldwide vaccine market revenue of US$ 27.7 billion in 2017 and growth to US$ 44.6 billion until 2024 at a compound annual growth rate (CAGR) of 7.0%.
- Research and Markets (2018) accounted for US$ 38.4 billion in 2018 and forecasted US$ 81.3 billion by 2027 (CAGR of 8.7%).
- Research and Markets (2020) valued the market at US$ 29.6 billion in 2018 and expected a growth rate of 10.3% to reach US$ 43.8 billion in 2022.
- PR Newswire (2018) reported a US$ 33.7 billion market in 2018 and expected it to reach US$ 57.5 billion by 2025, growing at 7.9%.
- Markets and Markets (2020) projected the global market to reach US$ 58.4 billion by 2024 from US$ 41.7 billion in 2019 (CAGR of 7.0%).
- Globe Newswire (2018) estimated the market to reach US$ 103.6 billion by 2028, growing at 11.0% in the forecast period 2018 to 2028.

The WHO also projects that the global vaccine market could be worth US$ 100 billion by 2025 (GEA 2019), but there are a considerable volume and value divide between the high-income countries and the low-to-middle income countries. High-income countries account for approximately 82% of global vaccine sales, but only 20% of the annual volume of vaccines used worldwide (GEA 2019).

The growth of the vaccine market is majorly attributed to the high prevalence of infectious diseases, increasing support of the key pharmaceutical players to enhance vaccine R&D, growing governmental initiatives towards immunization, rising

Table 1.6 Vaccine market trends from 2000 to 2028 in billion US $ based on various market reports

Year	Vaccine market (Billion US$)	References
2000	4; 5	D'Amore and Yang (2019), GEA (2019)
2005	10	D'Amore and Yang (2019)
2013	24; 25.6	Douglas and Samant (2018), D'Amore and Yang (2019)
2017	27.7	Douglas and Samant (2018)
2018	29.6; 33.7; 38.4	Research and Markets (2018, 2020), D'Amore and Yang (2019)
2019	41.7	PR Newswire (2018), Markets and Markets (2020)
Estimates		
2022	43.8	Research and Markets (2020)
2024	44.6; 58.4	Evaluate (2018), PR Newswire (2018), Markets and Markets (2020)
2025	57.5; 100	D'Amore and Yang (2019), GEA (2019)
2027	81.3	Research and Markets (2018)
2028	100; 103.6	Globe Newswire (2018), D'Amore and Yang (2019)

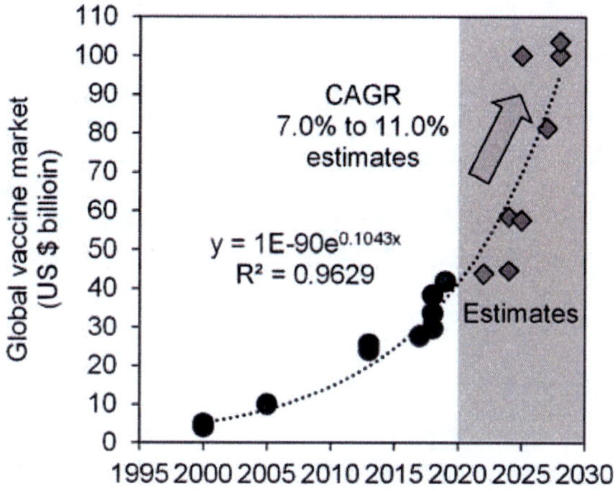

Fig. 1.12 Vaccine market trends from 2000 to 2028 in billion US$ based on various market reports (Douglas and Samant 2018; Evaluate 2018; Globe Newswire (2018); PR Newswire (2018); Research and Markets 2018; D'Amore and Yang 2019; GEA 2019; Markets and Markets 2020; Research and Markets 2020). The global vaccine market is estimated to reach US$ 100 billion by 2028, with an average CAGR of 8.7 ± 1.7% from 2020 to 2028

adoption of vaccines, technological advancements in the vaccine industry, and a strong pipeline for vaccines (PR Newswire 2018). As a result, more than 120 new products are in development, 60 of which are particularly important for low-income countries (D'Amore and Yang 2019). Besides, increasing epidemic potentials and a growing focus on therapeutic vaccines are expected to offer significant growth opportunities for players operating in the vaccine market. However, the substantial capital investments required for developing vaccines may restrain the market growth. Vaccines are also seen to play a crucial role in preventing widespread antimicrobial resistance, predicted to cause an estimated 10 million deaths by 2050 (The Review on Antimicrobial Resistance 2016). By preventing infections that would otherwise require antibiotic treatment, vaccination can help to combat the overuse of antibiotics and help curb the spread of resistant strains. They also provide a crucial defense against diseases that can no longer be treated with antibiotics.

The global vaccine market is dominated by subunit and conjugate vaccines (Globe Newswire 2018). Subunit vaccines can comprise multiple antigens using recombinant DNA technology and are available as protein-based subunit vaccines, polysaccharide vaccines, and subunit conjugate vaccines. However, in terms of the route of administration, the global vaccine market is dominated by intramuscular, with US$18 billion in sales in 2017 (Globe Newswire 2018).

According to GlobalData (2021), the size of the COVID-19 vaccine market across the seven major markets (US, Germany, France, Italy, Spain, UK, and Japan) is expected to increase from about US$ 13.6 billion in 2021 to US$ 19.5 billion in 2026 at a CAGR of 7.6% should annual vaccination be required (Fig. 1.13). COVID-19 vaccines based on mRNA technology dominate these countries with a market share of 95.4% in 2021 and US$ 9.5 billion in sales for Pfizer/BioNTech and US$ 3.5 billion for Moderna. With introducing a nanoparticle vaccine from Novavax and a recombinant protein vaccine from Sanofi and GSK, the market share of mRNA vaccines is expected to decline from 95.4% to 84.7% by 2026. Pfizer/BioNTech and Moderna vaccines will continue to dominate.

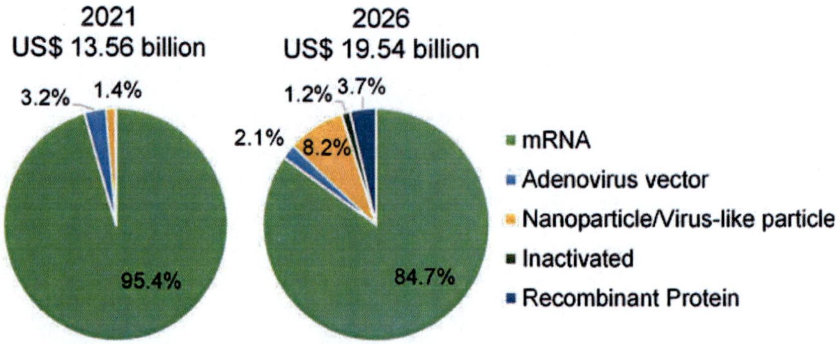

Fig. 1.13 Global sales forecast by class for COVID-19 vaccines in 2021 and 2026 across the seven major markets (US, Germany, France, Italy, Spain, UK, and Japan) (Data adapted from GlobalData (2021))

1.7 Vaccine Suppliers and Sales Numbers

According to the market study of Evaluate (2018), the top four vaccine suppliers in the global vaccine market are GlaxoSmithKline (UK), Merck & Co. (US), Pfizer (US), and Sanofi Pasteur (France), with a dominant share of the vaccine business of approximately 90% of the sales (Table 1.7).

Other prominent key players in this market include Novavax (US), Emergent BioSolutions (US), Inovio Pharmaceuticals (US), CSL (Australia), Bavarian Nordic (Denmark), and Mitsubishi Tanabe Pharma Corporation (Japan), as well as Johnson & Johnson (US) and AstraZeneca (UK), which were both not listed in the study of Evaluate (2018). The remainder comes from regional vaccine companies, the largest located in middle-income countries such as India, China, and Brazil (Douglas and Samant 2018). These regional companies are gradually growing their market share on a dose basis. On the other hand, the top four companies are slowly losing

Table 1.7 Companies and worldwide vaccine sales 2017–2024 with compound annual growth rate (CAGR) and market shares

Rank	Company	Sales 2017 (US$ billion)	Sales 2024 (US$ billion)	CAGR 2017 to 2024 (%)	Market share 2017 (%)	Market share 2024 (%)
1	GlaxoSmithKline (GSK) (UK)	6.7	10.7	+7	24.0	24.1
2	Merck & Co. (USA)	6.5	9.4	+5	23.6	21.1
3	Sanofi Pasteur (France)	5.8	8.1	+5	20.8	18.2
4	Pfizer (USA)	6.0	7.3	+3	21.7	16.3
5	Novavax (USA)	–	2.7	n/a	–	5.9
6	Emergent BioSolutions (USA)	0.3	1.1	+21	1.0	2.5
7	CSL (Australia)	0.8	1.1	+4	3.0	2.4
8	Inovio Pharmaceuticals (USA)	–	0.7	n/a	–	1.5
9	Bavarian Nordic (Denmark)	0.1	0.5	+84	0.0	1.2
10	Mitsubishi Tanabe Pharma (Japan)	0.4	0.5	+3	1.5	1.1
	Top 10	26.5	42.1	+7	95.7	94.3
	Others	1.2	2.5	+12	4.3	5.7
	Total industry	27.7	44.6	+7	100.0	100.0

Data adapted from Evaluate (2018), May 2018

market share in doses to the Developing Countries Vaccine Manufacturers Network (DCVMN) sourced doses (Douglas and Samant 2018). The key players contributing significantly to the global vaccine market are Serum Institute of India (SII) (India), Daiichi Sankyo Company (Japan), Panacea Biotec (India), Biological E (India), Bharat Biotech (India), Sinovac Biotech (China), Incepta Vaccine (Bangladesh), Valneva SE (France), VBI Vaccines (US), and PT Bio Farma (Brazil). Emergin manufacturers are playing an active role.

Serum Institute of India (SII) is a more than 50 years old private company that produces the highest number of vaccine doses globally. It makes 1.5 billion doses annually and employs over 7000 workers (Thomas 2020). The company had a large freeze-drying capacity of about 30 million doses per year in 2011, which could be increased to 40 million doses in an emergency (Dhere et al. 2011). It mainly operates from its two facilities in Pune, although it has two other small plants in the Netherlands and the Czech Republic. It exports 80% of its products, around 20 vaccines to 165 countries, selling them at an average of 50 cents a dose, which makes them among the most affordable in the world (Thomas 2020).

Evaluate (2018) expects GSK to lead the vaccines market in 2024 with a market share of 24.1%. This reflects the diverse breadth of GSK's vaccines portfolio. Merck & Co. is forecast to take second place with sales of US$ 9.4 billion. The company's top-selling vaccine, Gardasil, is expected to grow by US$ 0.9 billion. Emergent Biosciences is expected to grow 21% owing to its R&D asset, NuThrax, an anthrax vaccine. Finally, Novavax has the highest value R&D vaccine asset with its pre-clinical combination respiratory vaccine, expected to break into the top 5 vaccine sales by 2024.

However, these market analyses were conducted before the Corona pandemic. The most extensive vaccination campaign in history is well underway. Figure 1.14 provides an overview of publicly disclosed COVID-19 vaccine contracts between countries and manufacturers (Bloomberg 2021a). These statistics show that around 9.6 billion COVID-19 vaccine doses were under contract as of March 9, 2021. That would be enough to immunize more than half the world's population (most vaccines require two doses) if the vaccines were evenly distributed. However, that is not the case. Rich countries have signed extensive supply contracts. However, the vast majority of these doses have yet to be produced and delivered.

The AstraZeneca/Oxford vaccine is by far the most in-demand COVID-19 vaccine, primarily because it can be stored at normal refrigerator temperatures (2–8 °C), while other vaccines require ultra-cold storage. As of March 9, 2021, India has secured a total of about 2.2 billion COVID-19 vaccine doses through contracts with AstraZeneca/Oxford, Novovax, and Gamaleya (Bloomberg 2021a). In addition, the EU-27 has secured a total of nearly 1.84 billion doses with various manufacturers, and the U.S. has secured 1.21 billion doses.

As of September 22, 2021, more than 6.03 billion doses have been administered worldwide in 184 countries (Bloomberg 2021b). Globally, the latest rate averaged 31.5 million doses per day. At this rate, it will take another 6 months to cover 75% of the population (Bloomberg 2021b).

Volume of COVID-19 vaccine contracts between countries and manufacturers* (in million doses, as of March 9, 2021) Data source: Bloomberg (2021), COVID-19 Vaccine Deals Tracker															
Vaccine Supplier / Country	AstraZeneca/Oxford	Novavax	BioNTech/Pfizer	Moderna	Gamaleya	Sanofi/GSK	Johnson & Johnson	Sinovac Biotech	CureVac	Sinopharm	UBI Group	Fosun Pharma	Valneva	CanSino Biologics	Medicago	Total
India	1'000	1'000	-	-	200	-	-	-	-	-	-	-	-	-	-	2'200
EU (27)	300	-	500	310	-	300	200	-	225	-	-	-	-	-	-	1'835
USA	300	110	300	300	-	100	100	-	-	-	-	-	-	-	-	1'210
Covax**	300	-	-	-	-	200	-	-	-	-	200	-	-	-	-	700
UK	100	60	40	17	-	60	30	-	50	-	-	-	60	-	-	417
Japan	120	-	144	50	-	-	-	-	-	-	-	-	-	-	-	314
Indonesia	50	50	-	-	-	-	-	126	-	60	-	-	-	15	-	301
Brazil	100	-	-	-	100	-	-	100	-	-	-	-	-	-	-	300
China	200	-	-	-	-	-	-	-	-	-	-	100	-	-	-	300
Canada	20	52	40	44	-	52	10	-	-	-	-	-	-	-	20	238
Mexico	77	-	34	-	64	-	-	-	-	-	-	-	-	35	-	210
Russia	-	-	-	-	160	-	-	-	-	-	-	-	-	-	-	160
Latin America***	150	-	-	-	-	-	-	-	-	-	-	-	-	-	-	150
South Korea	20	40	26	40	-	-	4	-	-	-	-	-	-	-	-	130
Australia	54	51	20	-	-	-	-	-	-	-	-	-	-	-	-	125
Turkey	-	-	5	-	-	-	-	100	-	-	-	-	-	-	-	105
Pakistan	-	-	-	-	-	-	-	-	-	88	-	-	-	-	-	88
Malaysia	13	-	32	-	13	-	-	14	-	-	-	-	-	4	-	76
Middle East****	-	-	-	-	-	-	-	-	-	75	-	-	-	-	-	75
Uzbekistan	-	-	-	-	70	-	-	-	-	-	-	-	-	-	-	70
Philippines	17	30	-	20	-	-	-	-	-	-	-	-	-	-	-	67
Thailand	61	-	-	-	-	-	-	2	-	-	-	-	-	-	-	63
Egypt	-	-	-	-	50	-	-	-	-	-	-	-	-	-	-	50
Nepal	-	-	-	-	50	-	-	-	-	-	-	-	-	-	-	50
Others	127	11	79	35	59	-	24	26	-	7	2	-	-	-	-	370
Total	3'009	1'404	1'220	816	766	712	368	368	275	230	202	100	60	54	20	9'604

* This is a selection of the best-known vaccines that have been approved or are about to be approved.

** Covax covers most countries in the world, including many in Africa.

*** Latin America includes all countries in the region except Brazil.

**** Middle East includes the United Arab Emirates, Egypt, Bahrain, and Jordan.

Fig. 1.14 Countries and regions with the largest COVID-19 vaccine contracts as of March 9, 2021 (in million doses). Data source: Bloomberg (2021a), based on 137 publicly disclosed vaccine agreements between countries and vaccine suppliers

1.8 Cooperations to Improve Vaccines Accessibility

To improve the accessibility of routine vaccines in low-income countries, global health organizations such as The Vaccine Alliance (GAVI), the United Nations International Children's Emergency Fund (UNICEF), the World Bank, the World Health Organization (WHO), the Bill and Melinda Gates Foundation (BMGF), and various donors, including governments, the European Commission, private sector foundations, organizations, and corporate partners work cooperatively in a massive undertaking to purchase and distribute vaccines (Hill et al. 2016).

Figure 1.15 illustrates the cooperation of those philanthropic organizations to implement vaccination programs (Hill et al. 2016). Created in 2000, GAVI is an international organization bringing together public and private sectors with the shared goal of saving lives and protecting people's health by increasing the equitable and sustainable use of vaccines (GAVI 2021a, b).

The BMGF's initial pledge of US$ 750 million in 1999 provided the seed money to launch GAVI. In 2015, GAVI spent US$ 1.5 billion to fund vaccination programs. In addition, donors demonstrated commitments to GAVI of US$ 2 billion in 2019, bringing the total to US$ 19.1 billion since 2000 (GAVI 2021c).

GAVI then provides funding to other partner organizations, such as UNICEF, to purchase vaccines and ensure their distribution in low-income countries (Hill et al. 2016). UNICEF contracts with vaccine manufacturers to buy vaccines at prices far below the usual retail value. In this way, UNICEF plays a vital role as the world's largest buyer and supplier of vaccines to low-income countries where those in need cannot afford them. The World Bank brings the expertise of the world's largest source of development assistance.

BMGF also funds research-oriented institutions such as the Program for Appropriate Technology in Health (PATH) and other product development partnerships to identify promising technologies and support their development. This supports developing vaccines for which there is little market incentive and helps improve vaccine technology (e.g., overcoming temperature stability limitations) (Hill et al. 2016).

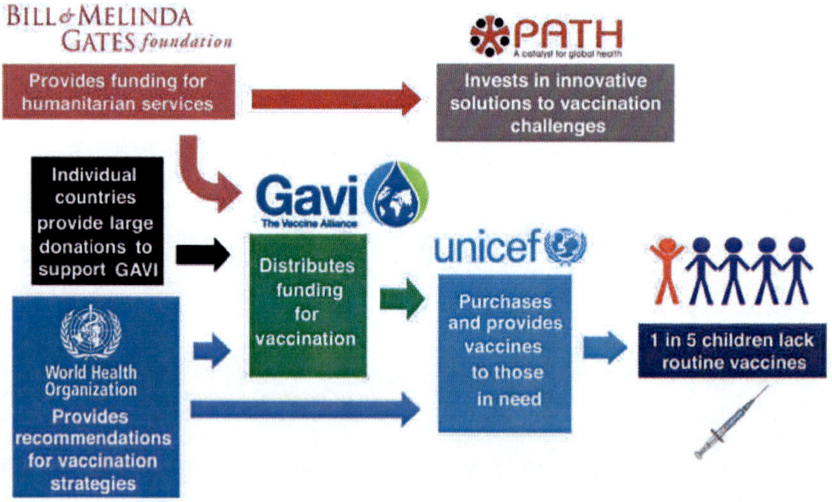

Fig. 1.15 Cooperation of the main philanthropic organizations to implement vaccination programs (Adapted from Hill et al. (2016) with permission from Elsevier)

1.9 Vaccine Prices

Countries primarily self-procure vaccines (60% of doses) (WHO 2019b). The remaining procurement occurs through UNICEF SD (36% of doses) and PAHO (Pan American Health Organization) Revolving Fund (4% of doses). Prices sold to UNICEF are substantially lower than prices for equivalent vaccines sold in the private sector in high-income countries. Table 1.8 compares some vaccine prices sold to UNICEF (2021) with private sector prices (CDC 2021) in the US and shows price reductions ranging from 78 to 99%.

Vaccine prices vary widely by vaccine type. Figure 1.16a shows the median price per dose and the average multiplier factor between the lowest and highest price by vaccine type for "non-GAVI, non-PAHO middle-income countries" (WHO 2018). These countries are neither eligible for GAVI support nor have access to the PAHO Revolving Fund, i.e., outside the Americas' Region.

Tetanus diphtheria (Td), Tetanus Toxoid (TT), and Measles-Mumps-Rubella (MMR) contributed most significantly to the high price variance with multiplying factors of about 212, 102, and 93, respectively. However, these vaccine types had relatively low median prices (US$ 3.53 for MMR, US$ 0.16 for Td, and US$ 0.15 for TT).

Figure 1.16b illustrates the range of vaccine prices in the U.S. public and private sectors (CDC 2021). Private sector prices are reported annually to CDC by vaccine manufacturers. Public sector prices are from vaccine contracts, i.e., state health departments, certain urban immunization projects, and certain current and former U.S. territories.

1.10 Summary of the Introduction

This brief introduction to vaccines clearly shows that vaccines play a central role in ensuring human health in the twenty-first century. The extent of lives saved could be illustrated by examples of numerous bacterial and viral diseases, from smallpox to COVID-19 (SARS-CoV-2 virus).

Vaccines are becoming a driving force for the pharmaceutical industry. The global vaccine market of around US$ 35 billion today is expected to reach US$ 100 billion by 2028, with an expected annual growth rate of around 9%. The market is divided between a few major global players accounting for about 90% of the revenue. Regional companies are gradually growing their market share. Key market players and collaborators have been highlighted. Analysis of vaccine prices shows substantial variations between vaccine types.

Global vaccination coverage has risen sharply since 2000, thanks to the continued efforts of GAVI, The Vaccine Alliance, and the active funding support of international health organizations and donors to improve access to vaccines in low-income countries and enable significant price reductions.

Table 1.8 Vaccine prices per dose (in US$) in the private sector compared to those sold to UNICEF

Vaccines	Private sector in the US[a]	UNICEF[b] (median)	Price reduction
Bacille calmette guerin (BCG)	n.a	0.12	–
Diphtheria and tetanus (DT)	n.a	0.17	–
Diphtheria, tetanus and pertussis (DTP)	26.3	0.18	99.3%
DTP-HepB-Hib (Pentavalent)	n.a	1.00	–
Ebola	n.a	98.6	–
Hepatitis A (HepA)	34.8	7.55	78.3%
Hepatitis B (HepB)	24.9	0.40	98.4%
Human papillomavirus (HPV)	239.3	11.4	95.2%
Inactivated polio vaccine (IPV)	36.9	2.8	92.4%
Japanese encephalitis	n.a	0.45	–
Measles	n.a	0.32	–
Measles and rubella (MR)	n.a	0.66	–
Measles, mumps, rubella (MMR)	82.5	2.15	97.4%
Meningococcal meningitis	134.8	6.03	95.5%
Oral cholera vaccine (OCV)	n.a	2.22	–
Oral polio vaccine (OPV)	n.a	0.15	–
Rabies	n.a	9.0	–
Rotavirus vaccine (RV)	87.9	1.6	98.2%
Pneumococcal conjugate vaccine (PCV)	110.5	4.0	96.4%
Tetanus, diphteria (Td)	25.9	0.11	99.6%
Typhoid conjugate	n.a	1.5	–
Yellow Fever (YF)	n.a	1.08	–

[a]Private sector prices were obtained from CDC (2021), the lowest available price is indicated
[b]UNICEP prices (median) from UNICEF (2021)

The most common cause of death from vaccine-preventable diseases is still tuberculosis. Herd immunity is achieved at different vaccination coverage rates depending on the disease. Estimates for COVID-19 suggest that 60% to 75% is required to return to normal life.

An important lesson learned from the COVID-19 pandemic is that the timeline for vaccine development could be drastically reduced to about 1 to 2 years. This remarkable speed was made possible by an unprecedented global effort combining advances from the independent fields of reverse and structural vaccines, synthetic biology, and vaccine adjuvants.

Of the vaccine platforms, mRNA vaccines, viral vectors, and protein subunit vaccines have been the most popular. mRNA vaccines in lipid nanoparticles as delivery systems have emerged as an attractive platform predicted to be of great a

Vaccine type	Median price per dose (US$)	Average multiplier factor between the lowest and highest price
DTaP-HepB-Hib-IPV	35.7	3
PCV	23.9	20
DTaP-Hib-IPV	21.7	3
Rabies	16.3	24
Seas. Influenza (adult)	13.8	12
YF	12.1	41
Rotavirus	8.20	4
Hib	5.66	4
MMR	3.53	93
HPV	3.32	11
IPV	2.30	9
MR	1.09	9
Hep B (adult)	1.06	49
DTWP-HepB-Hib	0.82	10
Hep B	0.47	46
DT	0.28	16
Measles	0.28	21
DTwP	0.23	32
bOPV	0.18	27
Td	0.16	212
BCG	0.15	44
TT	0.15	102

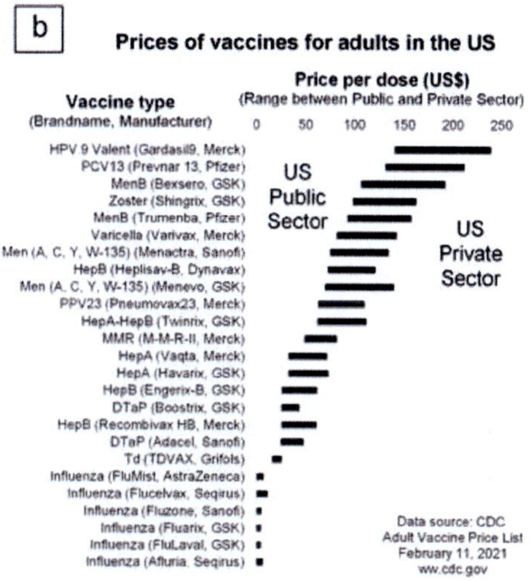

Fig. 1.16 **a** Vaccine prices by vaccine type and average multiplier factor between the lowest and highest price for "non-GAVI, non-PAHO middle-income countries" (are neither eligible for GAVI support nor have access to the PAHO Revolving Fund, i.e., lie outside the Region of the Americas) (Data from WHO (2018)). **b** Vaccine prices by vaccine type for adults in the US comparing the range between public and private sector (Data from CDC (2021))

importance in the future. However, available mRNA vaccines require ultra-cold storage and transport, a major challenge for the cold chain infrastructure.

Improving the thermal stability of vaccines has been a focus area for decades. The development of thermostable vaccines can significantly improve the distribution and accessibility of vaccines worldwide. This will be the main topic of the next chapter.

References

AboulFotouh K, Cui Z, Williams RO (2021) Next-generation COVID-19 vaccines should take efficiency of distribution into consideration. AAPS PharmSciTech 22:126. https://doi.org/10.1208/s12249-021-01974-3

Al-Barwani F, Donaldson B, Pelham SJ et al (2014) Antigen delivery by virus-like particles for immunotherapeutic vaccination. Ther Deliv 5:1223–1240. https://doi.org/10.4155/tde.14.74

Anderson EJ, Rouphael NG, Widge AT et al (2020) Safety and immunogenicity of SARS-CoV-2 mRNA-1273 vaccine in older adults. N Engl J Med 383:2427–2438. https://doi.org/10.1056/NEJMoa2028436

Artaud C, Kara L, Launay O (2019) Vaccine development: from preclinical studies to phase 1/2 clinical trials. In: Methods in molecular biology, pp 165–176

Baxby D (2011) Edward Jenner's role in the introduction of smallpox vaccine. History of vaccine development. Springer, New York, New York, NY, pp 13–19

Bloomberg (2021a) COVID-19 vaccine deals tracker. https://www.bloomberg.com/graphics/covid-vaccine-tracker-global-distribution/contracts-purchasing-agreements.html. Accessed on 23 Sep 2021a

Bloomberg (2021b) COVID-19 vaccine tracker. https://www.bloomberg.com/graphics/covid-vaccine-tracker-global-distribution/. Accessed on 23 Sep 2021b

Bregu M, Draper SJ, Hill AVS, Greenwood BM (2011) Accelerating vaccine development and deployment: report of a Royal Society satellite meeting. Philos Trans R Soc B Biol Sci 366:2841–2849. https://doi.org/10.1098/rstb.2011.0100

Callaway E (2020) The race for coronavirus vaccines: a graphical guide. Nature 580:576–577. https://doi.org/10.1038/d41586-020-01221-y

CDC (2014) Combination vaccines, information for parents. Last updated April 2014. https://www.cdc.gov/vaccines/hcp/patient-ed/conversations/downloads/fs-combo-vac.pdf

CDC (2021) CDC vaccine price list, centers for disease and control and prevention. Accessed on 11 Feb 2021 (Adult Vaccine Price List). https://www.cdc.gov/vaccines/programs/vfc/awardees/vaccine-management/price-list/

Cheyne J (1989) Vaccine Delivery management. Clin Infect Dis 11:S617–S622. https://doi.org/10.1093/clinids/11.Supplement_3.S617

Corbett KS, Edwards DK, Leist SR et al (2020) SARS-CoV-2 mRNA vaccine design enabled by prototype pathogen preparedness. Nature 586:567–571. https://doi.org/10.1038/s41586-020-2622-0

Crommelin DJA, Anchordoquy TJ, Volkin DB et al (2021) Addressing the cold reality of mRNA vaccine stability. J Pharm Sci 110:997–1001. https://doi.org/10.1016/j.xphs.2020.12.006

D'Amore T, Yang Y (2019) Advances and challenges in vaccine development and manufacture. In: www.bioprocessintl.com. https://bioprocessintl.com/manufacturing/vaccines/advances-and-challenges-in-vaccine-development-and-manufacture

Delany I, Rappuoli R, De Gregorio E (2014) Vaccines for the 21st century. EMBO Mol Med 6:708–720. https://doi.org/10.1002/emmm.201403876

Dhere R, Yeolekar L, Kulkarni P et al (2011) A pandemic influenza vaccine in India: from strain to sale within 12 months. Vaccine 29:A16–A21. https://doi.org/10.1016/j.vaccine.2011.04.119

Dolgin E (2021) Synthetic biology speeds vaccine development. In: Nat. Milestones Vaccines. https://media.nature.com/original/magazine-assets/d42859-020-00025-4/d42859-020-00025-4.pdf

Donaldson B, Lateef Z, Walker GF et al (2018) Virus-like particle vaccines: immunology and formulation for clinical translation. Expert Rev Vaccines 17:833–849. https://doi.org/10.1080/14760584.2018.1516552

Douglas RG, Samant VB (2018) The vaccine industry. In: Plotkin's Vaccines. Elsevier, pp 41–50

Dumpa N, Goel K, Guo Y et al (2019) Stability of vaccines. AAPS PharmSciTech 20:42. https://doi.org/10.1208/s12249-018-1254-2

Egli A, Santer DM, Barakat K et al (2014) Vaccine adjuvants—understanding molecular mechanisms to improve vaccines. Swiss Med Wkly 144:1–16. https://doi.org/10.4414/smw.2014.13940

Ehreth J (2003) The global value of vaccination. Vaccine 21:596–600. https://doi.org/10.1016/S0264-410X(02)00623-0

EMA (2020) European Medicines Agency, EMA recommends first COVID-19 vaccine for authorisation in the EU, 21 Dec 2020. https://www.ema.europa.eu/en/news/ema-recommends-first-covid-19-vaccine-authorisation-eu

EMA (2021) EMA recommends COVID-19 vaccine moderna for authorisation in the EU, 6 Jan 2021. https://www.ema.europa.eu/en/news/ema-recommends-covid-19-vaccine-moderna-authorisation-eu

Evaluate (2018) Evaluate pharma world preview 2018, Outlook to 2024

FDA (2020a) FDA briefing document, BNT162b2, Pfizer-BioNTech COVID-19 vaccine, 10 Dec 2020. https://www.fda.gov/media/144245/download

FDA (2020b) FDA briefing document, mRNA-1273, moderna COVID-19 vaccine, 17 Dec 2020. https://www.fda.gov/media/144434/download

FDA (2021) Vaccines licensed for use in the United States, content current as of 12/01/2021. https://www.fda.gov/vaccines-blood-biologics/vaccines/vaccines-licensed-use-united-states

Fontanet A, Cauchemez S (2020) COVID-19 herd immunity: where are we? Nat Rev Immunol 20:583–584. https://doi.org/10.1038/s41577-020-00451-5

Fuenmayor J, Gòdia F, Cervera L (2017) Production of virus-like particles for vaccines. N Biotechnol 39:174–180. https://doi.org/10.1016/j.nbt.2017.07.010

Galazka A, Milstien J, Zaffran M (1999) The thermostability of vaccines, global programme for vaccines and immunization. WHO, Geneva, 1998. WHO/GPV/98.07. https://apps.who.int/iris/bitstream/handle/10665/64980/WHO_GPV_98.07.pdf

Galvani A, Moghadas SM, Schneider EC (2021) Deaths and hospitalizations averted by Rapid U.S. vaccination rollout (Commonwealth Fund, July 2021). https://www.commonwealthfund.org/publications/issue-briefs/2021/jul/deaths-and-hospitalizations-averted-rapid-us-vaccination-rollout

GAVI (2021a) Facts & figures, updated 22 Feb 2021. https://www.gavi.org/sites/default/files/document/2021/Gavi-Facts-and-figures-February-21.pdf

GAVI (2021b) The vaccine alliance (formerly the global alliance for vaccines and immunizations). https://www.gavi.org/our-alliance. Accessed on 24 Sep 2021

GAVI (2021c) Annual progress report, year 4 of our five-year strategy. https://www.gavi.org/sites/default/files/programmes-impact/our-impact/apr/Gavi-Progress-Report-2019_1.pdf

GEA (2019) High quality vaccine production with GEA technology

GlobalData (2021) COVID-19 vaccine market set to reach $19.5 bn by 2026. https://www.globaldata.com/covid-19-vaccine-market-set-reach-19-5bn-2026/. Accessed on 14 Jul 2021

Globe Newswire (2018) Global vaccine market analysis & forecast (2018–2028): a $ 103.57 billion opportunity, growing at a CAGR of 11.02%

Goodsell DS (2009) Viruses. The machinery of life. Springer, New York, New York, NY, pp 126–138

Greenwood B (2014) The contribution of vaccination to global health: past, present and future. Philos Trans R Soc B Biol Sci 369:20130433. https://doi.org/10.1098/rstb.2013.0433

Guan S, Rosenecker J (2017) Nanotechnologies in delivery of mRNA therapeutics using nonviral vector-based delivery systems. Gene Ther 24:133–143. https://doi.org/10.1038/gt.2017.5

Gupta S, Cantor J, Simon KI et al (2021) Vaccinations against COVID-19 May have averted up to 140,000 deaths in the United States. Health Aff 40:1465–1472. https://doi.org/10.1377/hlthaff.2021.00619

Halliday J (2017) Commercial aspects of vaccine development. In: Micro and nanotechnology in vaccine development. Elsevier, pp 411–421. http://dx.doi.org/10.1016/B978-0-323-39981-4/00022-1

HHS (2021) Vaccine types. U.S. Department of Health & Human Services. Accessed on 12 June 2021. https://www.hhs.gov/immunization/basics/types

Hill AB, Kilgore C, McGlynn M, Jones CH (2016) Improving global vaccine accessibility. Curr Opin Biotechnol 42:67–73. https://doi.org/10.1016/j.copbio.2016.03.002

Hinman AR (1998) Global progress in infectious disease control. Vaccine 16:1116–1121. https://doi.org/10.1016/S0264-410X(98)80107-2

Holman DH, Wang D, Woraratanadharm J, Dong JY (2009) Viral vectors. In: Vaccines for biodefense and emerging and neglected diseases. Elsevier, pp 77–91

Jackson LA, Anderson EJ, Rouphael NG et al (2020) An mRNA vaccine against SARS-CoV-2—preliminary report. N Engl J Med 383:1920–1931. https://doi.org/10.1056/NEJMoa2022483

Jenner E (1798) An inquiry into the causes and effects of the variolae vaccinae, a disease discovered in some of the western counties of England, particularly gloucestershire, and known by the name of the Cow Pox. Sampson Low, London

John Hopkins University & Medicine (2021) Vaccine research and development: how can COVID-19 vaccine development be done quickly and safely? https://coronavirus.jhu.edu/vaccines/timeline. Accessed on 24 Sep 2021

Kaddar M (2013) Global vaccine market features and trends. In: WHO, IVB, Geneva. https://www.who.int/immunization/programmes_systems/procurement/market/world_vaccine_market_trends.pdf

Kaufmann JR, Miller R, Cheyne J (2011) Vaccine supply chains need to be better funded and strengthened, or lives will be at risk. Health Aff 30:1113–1121. https://doi.org/10.1377/hlthaff.2011.0368

Kim J, Eygeris Y, Gupta M, Sahay G (2021) Self-assembled mRNA vaccines. Adv Drug Deliv Rev 170:83–112. https://doi.org/10.1016/j.addr.2020.12.014

Kowalski PS, Rudra A, Miao L, Anderson DG (2019) Delivering the messenger: advances in technologies for therapeutic mRNA delivery. Mol Ther 27:710–728. https://doi.org/10.1016/j.ymthe.2019.02.012

Kumru OS, Joshi SB, Smith DE et al (2014) Vaccine instability in the cold chain: mechanisms, analysis and formulation strategies. Biologicals 42:237–259. https://doi.org/10.1016/j.biologicals.2014.05.007

Lauer KB, Borrow R, Blanchard TJ (2017) Multivalent and multipathogen viral vector vaccines. Clin Vaccine Immunol 24. https://doi.org/10.1128/CVI.00298-16

Leitner WW, Ying H, Restifo NP (1999) DNA and RNA-based vaccines: principles, progress and prospects. Vaccine 18:765–777. https://doi.org/10.1016/S0264-410X(99)00271-6

Lua LHL, Connors NK, Sainsbury F et al (2014) Bioengineering virus-like particles as vaccines. Biotechnol Bioeng 111:425–440. https://doi.org/10.1002/bit.25159

Lydon P, Zipursky S, Tevi-Benissan C et al (2014) Economic benefits of keeping vaccines at ambient temperature during mass vaccination: the case of meningitis A vaccine in Chad. Bull World Health Organ 92:86–92. https://doi.org/10.2471/BLT.13.123471

Markets and Markets (2020) Vaccines market—global forecast to 2024, published Jan 2020

McGill COVID19 Vaccine Tracker Team (2021) About vaccines. https://covid19.trackvaccines.org/how-do-vaccines-work/

Midoux P, Pichon C (2015) Lipid-based mRNA vaccine delivery systems. Expert Rev Vaccines 14:221–234. https://doi.org/10.1586/14760584.2015.986104

Mulligan MJ, Lyke KE, Kitchin N et al (2020) Phase I/II study of COVID-19 RNA vaccine BNT162b1 in adults. Nature 586:589–593. https://doi.org/10.1038/s41586-020-2639-4

Nabel GJ (2013) Designing tomorrow's vaccines. N Engl J Med 368:551–560. https://doi.org/10.1056/NEJMra1204186

NIAID (2019) Vaccine types. National Institute of Allergy and Infectious Diseases. https://www.niaid.nih.gov/research/vaccine-types

Nooraei S, Bahrulolum H, Hoseini ZS et al (2021) Virus-like particles: preparation, immunogenicity and their roles as nanovaccines and drug nanocarriers. J Nanobiotechnology 19:59. https://doi.org/10.1186/s12951-021-00806-7

Pardi N, Hogan MJ, Porter FW, Weissman D (2018) mRNA vaccines—a new era in vaccinology. Nat Rev Drug Discov 17:261–279. https://doi.org/10.1038/nrd.2017.243

PATH (2012) Summary of stability data for licensed vaccines, 29 Nov 2012, produced by Working in Tandem, Ltd. for PATH

Plotkin S, Robinson JM, Cunningham G et al (2017) The complexity and cost of vaccine manufacturing—an overview. Vaccine 35:4064–4071. https://doi.org/10.1016/j.vaccine.2017.06.003

Plotkin SL, Plotkin SA (2018) A short history of vaccination. In: Plotkin's Vaccines. Elsevier, pp 1–15, e8

Price DN, Kunda NK, Ellis R, Muttil P (2020) Design and optimization of a temperature-stable dry powder BCG vaccine. Pharm Res 37:11. https://doi.org/10.1007/s11095-019-2739-8

PR Newswire (2018) Global vaccines market forecasts 2018–2025: strong pipeline, rising adoption of vaccination, increasing government focus on immunization programs & technological progress in vaccine administration, 12 Dec 2018

Randolph HE, Barreiro LB (2020) Herd immunity: understanding COVID-19. Immunity 52:737–741. https://doi.org/10.1016/j.immuni.2020.04.012

Rappuoli R (2011) Twenty-first century vaccines. Philos Trans R Soc B Biol Sci 366:2756–2758. https://doi.org/10.1098/rstb.2011.0075

Rappuoli R, Mandl CW, Black S, De Gregorio E (2011) Vaccines for the twenty-first century society. Nat Rev Immunol 11:865–872. https://doi.org/10.1038/nri3085

Rappuoli R, De Gregorio E, Del Giudice G et al (2021) Vaccinology in the post−COVID-19 era. Proc Natl Acad Sci 118:e2020368118. https://doi.org/10.1073/pnas.2020368118

Reichmuth AM, Oberli MA, Jaklenec A et al (2016) mRNA vaccine delivery using lipid nanoparticles. Ther Deliv 7:319–334. https://doi.org/10.4155/tde-2016-0006

Research and Markets (2018) Vaccines—global market outlook (2018–2027)

Research and Markets (2020) Vaccines global market report 2020, December 2019

Rexroad J, Wiethoff CM, Jones LS, Middaugh CR (2002) Lyophilization and the thermostability of vaccines. Cell Preserv Technol 1:91–104. https://doi.org/10.1089/153834402320882593

Rosa SS, Prazeres DMF, Azevedo AM, Marques MPC (2021) mRNA vaccines manufacturing: challenges and bottlenecks. Vaccine 39:2190–2200. https://doi.org/10.1016/j.vaccine.2021.03.038

Rosini R, Barocchi MA, Rappuoli R (2011) Microbial vaccine design: the reverse vaccinology approach. Development of vaccines. John Wiley & Sons Inc., Hoboken, NJ, USA, pp 1–18

Sahin U, Muik A, Derhovanessian E et al (2020) COVID-19 vaccine BNT162b1 elicits human antibody and TH1 T cell responses. Nature 586:594–599. https://doi.org/10.1038/s41586-020-2814-7

Samaridou E, Heyes J, Lutwyche P (2020) Lipid nanoparticles for nucleic acid delivery: current perspectives. Adv Drug Deliv Rev 154–155:37–63. https://doi.org/10.1016/j.addr.2020.06.002

Schoenmaker L, Witzigmann D, Kulkarni JA et al (2021) mRNA-lipid nanoparticle COVID-19 vaccines: structure and stability. Int J Pharm 601:120586. https://doi.org/10.1016/j.ijpharm.2021.120586

Steinle H, Behring A, Schlensak C et al (2017) Concise review: application of in vitro transcribed messenger RNA for cellular engineering and reprogramming: progress and challenges. Stem Cells 35:68–79. https://doi.org/10.1002/stem.2402

Stephens P (2014) Vaccine R&D: past performance is no guide to the future. Vaccine 32:2139–2142. https://doi.org/10.1016/j.vaccine.2014.02.047

Tan X, Letendre JH, Collins JJ, Wong WW (2021) Synthetic biology in the clinic: engineering vaccines, diagnostics, and therapeutics. Cell 184:881–898. https://doi.org/10.1016/j.cell.2021.01.017

The Review on Antimicrobial Resistance (2016) Tackling drug-resistant infections globally: final report and recommendations

Thomas L (2020) India's Serum Institute gears up to make millions of doses of potential COVID-19 vaccine. News-MedicalNet, Apr 28 2020 1–5

Thomas DW, Burns J, Audette J et al (2016) Clinical development success rates 2006–2015, June 2016. BioMedTracker June:1–28

Ulmer JB, Valley U, Rappuoli R (2006) Vaccine manufacturing: challenges and solutions. Nat Biotechnol 24:1377–1383. https://doi.org/10.1038/nbt1261

UNICEF (2021) Vaccines pricing data. https://public.tableau.com/app/profile/supply.division/viz/UNICEFPricedataoverviewforvaccines/Fulldashboard. Accessed on 26 Sep 2021

Ura T, Okuda K, Shimada M (2014) Developments in viral vector-based vaccines. Vaccines 2:624–641. https://doi.org/10.3390/vaccines2030624

Vanderslott S, Dadonaite B, Roser M (2019) Vaccination, published online at OurWorldInData.org. Our World Data. https://ourworldindata.org/vaccination

Verbeke R, Lentacker I, De Smedt SC, Dewitte H (2021) The dawn of mRNA vaccines: the COVID-19 case. J Control Release 333:511–520. https://doi.org/10.1016/j.jconrel.2021.03.043

Whitehead KA, Langer R, Anderson DG (2009) Knocking down barriers: advances in siRNA delivery. Nat Rev Drug Discov 8:129–138. https://doi.org/10.1038/nrd2742

WHO (1980) The global eradication of smallpox, final report of the global commission for the certification of smallpox eradication, Geneva, December 1979. Hist Int Public Heal 4:1–122

WHO (2006) Guidelines on stability evaluation of vaccines, WHO/BS/0.6.2049

WHO (2018) Global vaccine action plan, monitoring, evaluation & accountability, secretariat annual report 2018

WHO (2019a) Poliomyelitis, key facts, 22 July 2019. https://www.who.int/news-room/fact-sheets/detail/poliomyelitis

WHO (2019b) Global vaccine market report, WHO/IVB/19.03. https://www.who.int/immunization/programmes_systems/procurement/mi4a/platform/module2/MI4A_Global_Vaccine_Market_Report.pdf

WHO (2020a) Immunization coverage: key facts, 15 July 2020. https://www.who.int/en/news-room/fact-sheets/detail/immunization-coverage

WHO (2020b) Tuberculosis, key facts, 14 Oct 2020. https://www.who.int/news-room/fact-sheets/detail/tuberculosis

WHO (2021) WHO prequalified vaccines. Online database: https://extranet.who.int/gavi/PQ_Web. Accessed on 20 March 2021. https://extranet.who.int/gavi/PQ_Web

WHO/UNICEF (2020) Progress towards global immunization goals—2019, summary presentation of key indicators, updated July 2020, pp 1–28

WHO/UNICEF/World Bank (2009) State of the world's vaccines and immunization, 3rd ed. Geneva, World Health Organization, 2009. https://apps.who.int/iris/bitstream/handle/10665/44169/9789241563864_eng.pdf

Wu F, Zhao S, Yu B et al (2020) A new coronavirus associated with human respiratory disease in China. Nature 579:265–269. https://doi.org/10.1038/s41586-020-2008-3

Thermostability of Vaccines

2

Abstract

This chapter discusses the recommended storage conditions of prequalified vaccines. First, data from the literature on heat and freeze stability are summarized and explained. Next, an excursus provides current information on the shelf life and storage requirements of COVID-19 vaccines. Then, the challenge of cold storage and its importance in vaccine distribution is discussed. Finally, a vision for an ideal vaccine is outlined, and potential improvements are identified.

2.1 Recommended Storage Conditions and Administration Routes

Vaccines are large and complex 3D molecular structures made up of proteins, nucleic acids, lipids and carbohydrates, inactivated microorganisms, or in some cases live-attenuated microorganisms as active components, as well as stabilizers, adjuvants, preservatives, and other substances which together contribute to overall vaccine efficacy and safety (WHO 2006a). In addition, vaccines are very sensitive to environmental factors, primarily due to their inherent temperature sensitivity.

When a pharmaceutical or biotech company makes a vaccine, it must be kept under controlled temperature conditions during manufacturing, storage, and transportation until administration to the patient. A central reference document on the temperature sensitivity of vaccines is the WHO guideline (WHO 2006b). Table 2.1 shows the recommended storage conditions of WHO prequalified vaccines used in immunization programs at the distribution levels of the cold chain (WHO 2006a, b).

C. Arpagaus, *Spray Drying of Vaccines*,
https://doi.org/10.1007/978-3-031-24323-3_2

Table 2.1 Recommended storage conditions for WHO prequalified vaccines

Vaccine type	Primary vaccine storage	Intermediate vaccine storage		Health center	Health post
		Region	District		
OPV	−15 °C to −25 °C		All vaccines are recommended to be stored at +2 °C to +8 °C		
BCG	WHO no longer recommends that freeze-dried vaccines be stored at −20 °C. Storing them at −20 °C is not harmful but is unnecessary. Instead, these vaccines should be kept in refrigeration and transported at +2 °C to +8 °C				
Measles					
MMR					
MR					
YF					
Hib (freeze-dried)					
Meningococcal A&C					
HepB	+2 °C to +8 °C These vaccines are freeze sensitive and must never be frozen				
IPV					
DT					
DTP					
DTP-HepB					
Hib (liquid)					
Td					
TT					
Diluent vials must NEVER be frozen. If the manufacturer supplies a freeze-dried vaccine packet with its diluent. ALWAYS store the product at between +2 °C and +8 °C. If space permits, diluents supplied separately from the vaccine may safely be stored in the cold chain between +2 °C and +8 °C					

Adapted from WHO (2006b)

OPV Oral polio vaccine, *BCG* Bacille Calmette-Guerin, *MMR* Measles-mumps-rubella, *MR* Measles-rubella, *YF* Yellow fever, *Hib* Haemophilus influenzae type b conjugate vaccine, *HepB* Hepatitis B, *IPV* Inactivated polio vaccine, *DT* Diphtheria-tetanus, *DTP* Diphtheria-tetanus-pertussis, *Td* Tetanus-diphtheria toxoid vaccine, *TT* Tetanus toxoid

Most of the more than 4.7 billion doses of vaccine administered each year worldwide (Cavalier 2020; Cavalier et al. 2021) (before the COVID-19 pandemic) require constant refrigeration and must be stored refrigerated between +2 to +8 °C at all levels of the cold chain (WHO 2006b).

An exception is the oral polio vaccine (OPV) that must be kept frozen between −15 °C and −25 °C at the higher levels of the cold chain, i.e., at the national (primary) and regional or province-level (Table 2.1). Freeze-dried vaccines (i.e., BCG, measles, MMR, MR, YF, and Hib) may also be kept frozen at −15 °C to −25 °C if cold chain space permits, but this is neither essential nor recommended.

At other levels of the cold chain (i.e., intermediate vaccine stores and health facilities), these vaccines should be stored between +2 °C and +8 °C. Some vaccines are sensitive to freezing and should be protected from sub-zero temperatures. These are predominantly liquid formulations containing vaccines against (WHO 2006b, 2015a) cholera, DTP, HepB, Hib (liquid formulation), HPV, IPV) and their combinations, influenza, pneumococcal, and rotavirus (liquid formulation).

Table 2.2 provides a more up-to-date summary of vaccines licensed by the FDA for use in the United States (as of January 31, 2022, including COVID-19 vaccines) (FDA 2022). The vaccines are structured by disease, trade name, vaccine type, dosage form, route of administration, and storage temperature. As shown in Table 2.2, the majority of vaccine types in use are inactivated (approximately 27%), live-attenuated (21%), recombinant subunit (20%), conjugate (11%), toxoid (10%), mRNA, or viral vector.

Figure 2.1 shows the distribution of administration routes for the FDA-approved vaccines summarized in Table 2.2 (FDA 2022). Most vaccines are administered by intramuscular injection (IM, approximately 75%), by subcutaneous injections (SC, 12%) into the fatty layer under the skin (hypodermis) (WHO 2021a). Intradermal injections (ID, 1%) target the dermis, and percutaneous administration (PC, 3%) usually refers to the administration of Bacille Calmette Guérin (BCG) into the dermis and epidermis using multiple punctures or multi-tipped devices with 1-mm needles (PATH 2009). In addition, some live-attenuated vaccines are administered orally (e.g., typhoid, rabies, and rotavirus) or intranasally (IN) (e.g., influenza) to mimic the natural route of infection. Oral administration of vaccines facilitates immunization by eliminating the need for a needle and syringe (WHO 2021a).

The dosage forms are mainly suspensions (approximately 63%), solutions (10%), emulsions (1%), or lyophilized powder (26%). For stability reasons, live attenuated vaccines are often freeze-dried in the presence of excipients and additives to ensure adequate stability during long-term storage. A total of 18 vaccines listed in Table 2.3 are freeze-dried and need to be reconstituted with a suitable diluent before administration. The bulk of FDA-approved vaccines (approximately 87%) requires storage under refrigerated conditions at +2 to +8 °C. Only a few of the listed vaccines can be kept at room temperature (25 °C) for a few days, and none is stable long-term above refrigerated temperatures. Several must be kept frozen at temperatures below −15 °C and even below −60 °C to maintain potency (e.g., mRNA COVID-19 or Ebola vaccines).

2.2 Heat and Freeze Stability of Vaccines

Tables 2.3 and 2.4 provides an overview of some globally representative vaccines against viruses and bacteria, grouped by vaccine type, product, manufacturer, formulation (i.e., liquid or lyophilized form), route of administration, adjuvant, storage temperature, shelf-life, freeze, and heat sensitivity (WHO 2006b, 2021b; PATH 2012; Kumru et al. 2014; FDA 2022).

A general trend can be seen that live-attenuated vaccines do not contain adjuvants but are more sensitive to heat during storage and distribution (Kumru et al. 2014). These vaccines contain weakened, attenuated versions of infectious viruses and bacteria that must replicate in vivo and mimic natural infection.

Table 2.2 Vaccines licensed by the FDA for use in the United States structured by disease, trade name, vaccine type, dosage form, administration route, and storage temperature

Disease	Trade name	Vaccine type	Dosage form	Administration route	Storage temperature
Anthrax	Biothrax	Inactivated	Suspension	IM or SC	2–8 °C (not freeze)
Cholera	Vaxchora	Live-attenuated	Lyo	Oral	2–8 °C (25 °C/72 h, not freeze)
COVID-19	Comirnaty	mRNA	Suspension	IM	−90 to −60 °C (8 °C/30 days)
	Janssen Vaccine	Viral vector			2–8 °C (25 °C/12 h)
	SPIKEVAX	mRNA			−50 to −15 °C (8 °C/30 days, 25 °C/24 h)
Dengue	DENGVAXIA	Live-attenuated	Lyo	SC	2–8 °C (not freeze)
DT	TDVAX, TENIVAC	Toxoid	Suspension	IM	2–8 °C (not freeze)
DTP	Adacel, Boostrix, DAPTACEL, Infanrix, KINRIX	Toxoid	Suspension	IM	2–8 °C (not freeze)
DTP and polio	Quadracel	Toxoid, inactivated	Suspension	IM	2–8 °C (not freeze)
DTP, hepB, and polio	Pediarix				
DTP, polio, and hib	Pentacel	Toxoid, inactivated, conjugated			
DTP, polio, hepB, and hib	VAXELIS				
Ebola virus disease	ERVEBO	Live-attenuated	Suspension	IM	−80 to −60 °C (8 °C/2 weeks)
Hepatitis A	Havrix, VAQTA	Inactivated	Suspension	IM	2–8 °C (not freeze)
Hepatitis A and B	Twinrix	Recombinant			

(continued)

Table 2.2 (continued)

Disease	Trade name	Vaccine type	Dosage form	Administration route	Storage temperature
Hepatitis B	Engerix-B	Recombinant	Suspension	IM	2–8 °C (not freeze)
	PREHEVBRIO				
	Recombivax HB				2–8 °C (25 °C/72 h, not freeze)
	HEPLISAV-B		Solution		2–8 °C (not freeze)
Herpes varicella-zoster virus (shingles)	SHINGRIX	Recombinant	Lyo	IM	2–8 °C (not freeze)
	Zostavax	Live-attenuated		SC	−50 to 8 °C
Hib	ActHIB	Conjugate	Lyo	IM	2–8 °C (not freeze)
	Hiberix				
	PedvaxHIB		Suspension		
HPV	Cervarix	Recombinant	Suspension	IM	2–8 °C (not freeze)
	Gardasil				2–8 °C (25 °C/72 h, not freeze)
	Gardasil 9				
Influenza	Agriflu, Afluria, Fluad, Fluarix, Flucelvax, Flulaval, Fluvirin, Fluzone	Inactivated	Suspension	IM	2–8 °C (not freeze)
	FluMist	Live-attenuated		IN	
	Flublok	Recombinant	Solution	IM	
	AUDENZ	Inactivated	Emulsion		
	Fluzone Intradermal		Suspension	ID	
Japanese encephalitis	Ixiaro	Inactivated	Suspension	IM	2–8 °C (not freeze)
MMR	M-M-R II	Live-attenuated	Lyo	SC	−50 to 8 °C
MMR and varicella	ProQuad				

(continued)

Table 2.2 (continued)

Disease	Trade name	Vaccine type	Dosage form	Administration route	Storage temperature
Meningitis B	BEXSERO	Recombinant	Suspension	IM	2–8 °C (not freeze)
Meningococcal	Menactra	Conjugate	Solution	IM	2–8 °C (not freeze)
	Menveo		Lyo		
	MenQuadfi	Recombinant	Solution		
	TRUMENBA		Suspension		
	Menomune-A/C/Y/W-135		Lyo	SC	
Pneumococcal	Pneumovax 23	Inactivated	Solution	IM or SC	2–8 °C
	Prevnar 13, Prevnar 20	Conjugate	Suspension	IM	2–8 °C (25 °C/96 h)
	VAXNEUVANCE				2–8 °C (not freeze)
Polio	IPOL	Inactivated	Suspension	IM or SC	2–8 °C (not freeze)
Rabies	Imovax	Inactivated	Lyo	IM	2–8 °C (not freeze)
	RabAvert			Oral	2–8 °C
Rotavirus	ROTARIX	Live-attenuated	Lyo	Oral	2–8 °C (not freeze)
	RotaTeq		Solution		
Smallpox	ACAM2000	Live-attenuated	Lyo	PC	−25 to -15 °C
Smallpox and monkeypox	JYNNEOS		Suspension	SC	
Tuberculosis	BCG Vaccine	Live-attenuated	Lyo	PC	2–8 °C (not freeze)
Typhoid	TYPHIM Vi	Inactivated	Solution	IM	2–8 °C (not freeze)
	Vivotif	Live-attenuated	Lyo	Oral	2–8 °C
Varicella (chickenpox)	Varivax	Live-attenuated	Lyo	SC	−50 to −15 °C (8 °C/72 h)
Yellow fever	YF-Vax	Live-attenuated	Lyo	SC	2–8 °C (not freeze)

Lyo Lyophilized powder, *IM* intramuscular, *SC* subcutaneous, *PC* Percutaneous, *ID* intradermal, *IN* intranasal spray, *DT* diphtheria and tetanus, *DTP* diphtheria, tetanus, and pertussis, *hepB* hepatitis B, *MMR* measles, mumps, and rubella. Based on data from FDA (2022), as of Jan. 31, 2002, excluding vaccines without trade names, including 3 emergency use authorized COVID-19 vaccines)

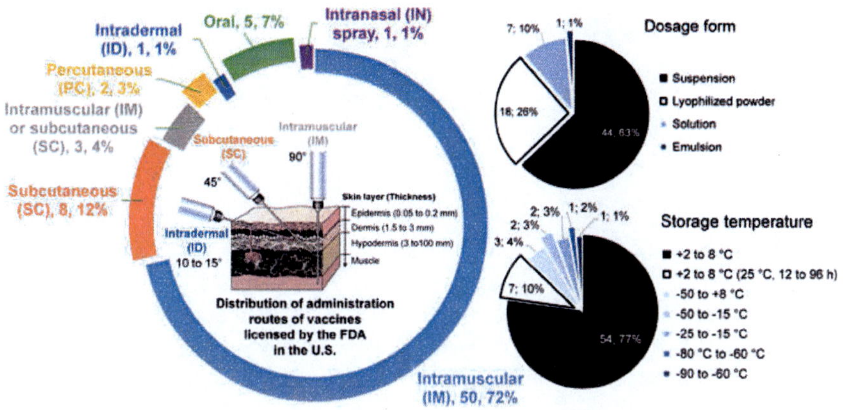

Fig. 2.1 Administration routes, dosage forms, and storage temperatures of 70 vaccines licensed by the FDA for use in the United States (Based on data from FDA (2022) as of Jan. 31, 2022, including 3 emergency use authorized COVID-19 vaccines)

A second trend is that non-replicating vaccines often contain adjuvants (e.g., aluminum) to enhance immune responses and ensure a sufficient level of protective immunity. These vaccines include inactivated viruses and bacteria, purified subunit proteins and carbohydrate antigens, and recombinant subunit protein antigens. Inactivated and subunit vaccines are generally more stable for long-term storage and are usually developed as liquid formulations. However, these vaccines may lose potency if accidentally frozen during storage and distribution (Kumru et al. 2014).

Some of the vaccines used today have some inherent stability in addition to what is currently indicated on their product label (WHO 2006b). This additional stability would allow for limited use of the vaccine outside the cold chain without maintaining the restrictions of 2–8 °C.

Nevertheless, most of the available vaccines on the market require an uninterrupted cold chain throughout the product's shelf-life. This situation limits the reach of vaccines to remote regions in low-and-middle-income countries (Kaddar et al. 2013) that do not have adequate cold chain infrastructure (Kunda et al. 2016).

Chen and Kristensen (2009) have provided further stability data for approved vaccines at various storage temperatures (i.e., 2–8 °C, 25 °C, 37 °C, and $\geq$ 45 °C), as summarized in Table 2.5. The stability data are based on published studies available to the public (WHO 2006b; Chen and Kristensen 2009). For this purpose, reference is made to the manufacturer's product label.

The data show that most existing, commonly used vaccines have a shelf life of 2 years or longer at 2–8 °C. However, the sensitivity of the vaccines to excursions outside this range varies considerably. Lyophilized vaccines are generally only stable before reconstitution. Once reconstituted, the efficacy of live attenuated vaccines, such as vaccines against measles and yellow fever, can rapidly decrease. Therefore, WHO recommendations state that reconstituted vaccines must be kept

Table 2.3 Examples of FDA-approved bacterial vaccines

Vaccine type	Product	Manufacturer	Formulation (admin route)	Adjuvant	Storage temperature	Shelf-life	Freeze sensitivity	Heat sensitivity
Live-attenuated bacteria								
BCG	TICE BCG	Organon	Lyo (ID, PC)	None	2–8 °C	≥ 2 years	No	≥ 9 days (54 °C)
	BCG Vaccine	Organon	Lyo (PC)			N/A		48 h (23 °C)
Typhoid	Vivotif	Berna	Capsule (Oral)	None	2–8 °C	1.5 years	Yes	12 h (37 °C)
Conjugates polysaccharide-carrier								
Meningococcal (Groups ACYW)	Menactra	Sanofi Pasteur	Liquid (IM)	None	2–8 °C	1 year	Yes	N/A
	Nenveo	Novartis	Lyo (IM)	None	2–8 °C	2 years	Yes	N/A
Pneumococcal (7-valent)	Prevnar	Pfizer	Liquid (IM)	Aluminum	2–8 °C	2 years	Yes	N/A
Pneumococcal (13-valent)	Prevnar-13	Pfizer	Liquid (IM)	Aluminum		2 years		4 days (40 °C)
Pneumococcal (23-valent)	Pneumovax 23	Merck	Liquid (IM, SC)	None		N/A		N/A
Hib	Hiberix	GSK	Lyo (IM)	None	2–8 °C	N/A	No	≥ 5 weeks (55 °C)
	ActHIB	Sanofi Pasteur				3 years	Yes	N/A
	PedvaxHIB	Merck	Liquid (IM)	Aluminum	2–8 °C	3 years	Yes	N/A
Subunit, purified bacterial antigens								
Tetanus Toxoid	No trade name	Sanofi Pasteur	Liquid (IM, SC)	None	2–8 °C	N/A	Yes	N/A
Tetanus Toxoid Adsorbed	No trade name	Sanofi Pasteur	Liquid (IM)	Aluminum		N/A		
Anthrax	Biothrax	Emergent	Liquid (IM, SC)	Aluminum		4 years		

(continued)

Table 2.3 (continued)

Vaccine type	Product	Manufacturer	Formulation (admin route)	Adjuvant	Storage temperature	Shelf-life	Freeze sensitivity	Heat sensitivity
Typhoid	Typhim Vi	Sanofi Pasteur	Liquid (IM)	None		3 years		
Meningococcal (Groups ACWY)	Menomune	Sanofi Pasteur	Lyo (SC)	None	2–8 °C	3 years	Yes	6 weeks (60 °C)
Combination vaccines								
DTP w/HepB	Pediarix	GSK	Liquid (IM)	Aluminum	2–8 °C	N/A	Yes	N/A
DTP	Infanrix	GSK				3 years		
DTP	Tripedia	Sanofi Pasteur				N/A		
DTP	Daptacel	Sanofi Pasteur				N/A		
DTP w/IPV	Kinrix	GSK				N/A		
Hib, HepB	Comvax	Merck				N/A		
DT	No trade name	Sanofi Pasteur				N/A		
DT	DecaVac	Sanofi Pasteur				N/A		
DT	TeniVac	Sanofi Pasteur				N/A		
DT	Diphtheria, Tetanus	Mass Biologics				N/A		
TT	Adacel	Sanofi Pasteur				3 years		

(continued)

Table 2.3 (continued)

Vaccine type	Product	Manufacturer	Formulation (admin route)	Adjuvant	Storage temperature	Shelf-life	Freeze sensitivity	Heat sensitivity
Diphtheria and Acellular pertussis	Boostrix	GSK				3 years		
DTP, IPV, Hib	Pentacel	Sanofi Pasteur				N/A		
Meningococcal, Hib (Groups CY-Hib)	Menhibrix	GSK	Lyo (IM)	None	2–8 °C	N/A	Yes	N/A

Data reproduced from Kumru et al. (2014), open access under CC-BY-NC-ND license
IM Intramuscular, *PC* Percutaneous, *SC* Subcutaneous, *Lyo* Lyophilized, *N/A* Not available

Table 2.4 Examples of FDA-approved viral vaccines

Vaccine type and disease	Trade name	Manufacturer	Formulation (admin route)	Adjuvant	Storage temperature	Shelf-life	Freeze sensitive	Heat sensitivity
Live, attenuated virus								
Varicella[a]	Varivax	Merck	Lyo (SC)	None	Frozen	2 years	No	6 h (27 °C)
Zoster[a]	Zostavax	Merck	Lyo (SC)			18 months		N/A
Rotavirus	Rotarix	GSK	Lyo (Oral)		2–8 °C	3 years		7 days (20 °C)
Rotavirus	RotaTeq	Merck	Liquid (Oral)			2 years	N/A	N/A
Influenza[d]	Flumist	MedImmune	Liquid (Nasal)			18 weeks	Yes	
Adenovirus (Type 4 and 7)	[b]	Barr Labs	Tablet (Oral)			N/A		
Smallpox	ACAM2000	Sanofi Pasteur	Lyo (PC)		Frozen	18 months	No	
Yellow Fever	YF-Vax	Sanofi Pasteur	Lyo (SC)		2–8 °C	3 years	Yes	
Inactivated virus								
Hepatitis A	Vaqta	Merck	Liquid (IM)	Aluminum	2–8 °C	3 years	Yes	3 months (28 °C)
	Havrix	GSK						1–3 weeks (37 °C)

(continued)

Table 2.4 (continued)

Vaccine type and disease	Trade name	Manufacturer	Formulation (admin route)	Adjuvant	Storage temperature	Shelf-life	Freeze sensitive	Heat sensitivity
Influenza	Fluarix[d]	GSK	Liquid (IM)	None	2–8 °C	1 year	Yes	12 weeks (20 °C)
	Flulaval[d]	ID Biomed						N/A
	Agriflu	Novartis						
	Fluvirin	Novartis						
	Afluria	CSL Limited						
	Fluzone[d,f]	Sanofi Pasteur				N/A		
	Flucelvax	Novartis						
	H5N1	Sanofi Pasteur						
	[b]2009 (H1N1)	CSL Limited						
	[b]2009 (H1N1)	ID Biomed						
	[b]2009 (H1N1)	Novartis						
	[b]2009 (H1N1)	Sanofi Pasteur						
	H5N1	ID Biomed		AS03[e)]				
	[b]2009 (H1N1)	MedImmune	Liquid (Nasal)	None				
Rabies	RabAvert	Novartis	Lyo (IM)	None	2–8 °C	3 years	No	3 months (37 °C)
	Imovax	Sanofi Pasteur				N/A	Yes	N/A
Polio	IPOL	Sanofi Pasteur	Liquid (IM, SC)	None	2–8 °C	$\geq$ 2 years	Yes	N/A
Japanese Encephalitis	Ixiaro	Intercell	Liquid (IM)	Aluminum	2–8 °C	1 year	Yes	N/A
	JE-Vax	Osaka	Lyo (SC)	None		N/A		1 month (27 °C)

(continued)

Table 2.4 (continued)

Vaccine type and disease	Trade name	Manufacturer	Formulation (admin route)	Adjuvant	Storage temperature	Shelf-life	Freeze sensitive	Heat sensitivity
Recombinant vaccines								
HepB	Engerix B	GSK	Liquid (IM)	Aluminum	2–8 °C	3 years	Yes	72 h (25 °C)
	Recombivax HB	Merck				3 years		N/A
HPV	Gardasil	Merck	Liquid (IM)	Aluminum	2–8 °C	3 years	Yes	3 months (45 °C)
	Cervarix	GSK		AS04[e)]		4 years		7 days (37 °C)
Influenza	FluBlock	Protein Sciences	Liquid (IM)	None	2–8 °C	16 weeks	Yes	N/A
Combination vaccines								
MMR	M-M-R II	Merck	Lyo (SC)	None	Frozen	2 years	No	7 days (37 °C)
DTP w/HepB	Pediarix	GSK	Liquid (IM)	Aluminum	2–8 °C	N/A	Yes	N/A
DTP w/IPVo	Kinrix	GSK	Liquid (IM)	Aluminum	2–8 °C	N/A	Yes	N/A
Hepatitis A and B	Twinrix	GSK	Liquid (IM)	Aluminum	2–8 °C	3 years	Yes	1 week (37 °C)
MMR w/Varicella	ProQuad	Merck	Lyo (SC)	None	Frozen	2 years	No	N/A

Data reproduced from Kumru et al. (2014), open access under CC-BY-NC-ND license
IM Intramuscular, *PC* Percutaneous, *SC* Subcutaneous, *Lyo* Lyophilized, *N/A* Not available
[a]Refrigerated formulation available, [b]No trade name identified, [c]Aluminum hydroxide and lipid A, [d]Quadrivalent formulation available, [e]Squalene, DL-a-tocopherol, and polysorbate 80, [f]High dose intradermal formulation available

Table 2.5 Temperature stability during storage of vaccines commonly used in national immunization programs and structured by formulation (liquid and lyophilized)

Vaccine	Formulation	Adjuvant	Freeze sensitive	Temperature stability during storage			
				2–8 °C	25 °C	37 °C	≥ 45 °C
HPV (Shank-Retzlaff et al. 2006)	Liquid	Aluminum	Yes	> 3 years	> 2 years	1.5 months	14 days
DT	Liquid, usually in combination vaccines	Aluminum	Yes	> 3 years	Months	Months	Unstable above 55 °C
HepB	Liquid	Aluminum	Yes	> 4 years	Months	Weeks	Days at 45 °C
Pertussis (whole cell)	Liquid, always in combination vaccines	Aluminum	Yes	1.5–2 years	2 weeks	1 week	> 10% potency loss per day
Pneumococcal, 7-valent, conjugate	Liquid	Aluminum	Yes	> 2 years	No data	Maybe unstable depending on the presentation	Unstable
Inactivated poliovirus (IPV)	Liquid, can be in combination formulation	No	Yes	1–4 years	Weeks	Weeks	Little data available
OPV	Liquid, live-attenuated	No	No	≤ 1 year (≥ 2 years at −20 °C)	Weeks	2 days	Unstable
Influenza, inactivated	Liquid	No	Possibly	≤ 1 year	Not known	Not known	Not known
Hib	Liquid or lyophilized, in a combination of vaccines	No	Yes (liquid)	> 2 years	> 2 years (lyophilized)	Maybe unstable depending on the presentation	Unstable
Rotavirus	Lyophilized or liquid, live-attenuated	No	No	> 2 years	2 years	Not known	Not known

(continued)

Table 2.5 (continued)

Vaccine	Formulation	Adjuvant	Freeze sensitive	Temperature stability during storage			
				2–8 °C	25 °C	37 °C	≥ 45 °C
Japanese encephalitis B	Lyophilized, live-attenuated	No	No	1.5 years	4 months	7–10 days	Unstable
YF	Lyophilized, live-attenuated	No	No	> 2 years	Months	2 weeks	Unstable
MRR	Lyophilized, live-attenuated	No	No	2 years	> 1 month	> 1 week	Unstable
BCG (mycobacterium tuberculosis)	Lyophilized, live-attenuated	No	No	1–2 years	Months	Loss of > 20% potency after 1 month	Unstable
Meningococcal serogroups ACWY (polysaccharide)	Lyophilized	No	Yes	2 years	Weeks to months	Not known	Not known

Data adapted from WHO (2006b), Chen and Kristensen (2009)

cold during the immunization session, and any unused vaccine from a multi-dose vial must be discarded within 6 h (WHO 2006b). This handling is partly related to reconstituted vaccines' instability and minimizes the risk of bacterial contamination as live vaccines do not contain preservatives.

Based on the Expanded Program on Immunization (EPI) established in 1974 (48 years ago), two types of vaccines are distinguished:

- heat-stable but freeze-sensitive, and
- freeze-stable but heat-sensitive.

Accordingly, (Kartoglu and Milstien 2014) ranked the commonly used EPI vaccines according to their freeze sensitivity and heat stability. Figure 2.2 (left) shows the breakdown of freeze sensitivity plotted versus heat stability for traditional vaccines used in national immunization programs, such as OPV, measles, YF, BCG, TT, DTP, and HepB.

The axes are generated as an ordinal scale based on the vaccines' freezing sensitivity and heat stability (Kartoglu and Milstien 2014). Thus, it can be recognized that there are two main categories among the traditional vaccines, namely live attenuated vaccines (e.g., OPV, measles, BCG, YF), which are not freeze-sensitive and not stable to high temperatures, and inactivated vaccines (e.g., TT, HepB, DTP) which should not be frozen but are more heat-stable (Kartoglu and Milstien 2014).

On the same scale, Fig. 2.2 (right) shows newer vaccines included in the national immunization program since 2000. These are in particular vaccines against (Kartoglu and Milstien 2014):

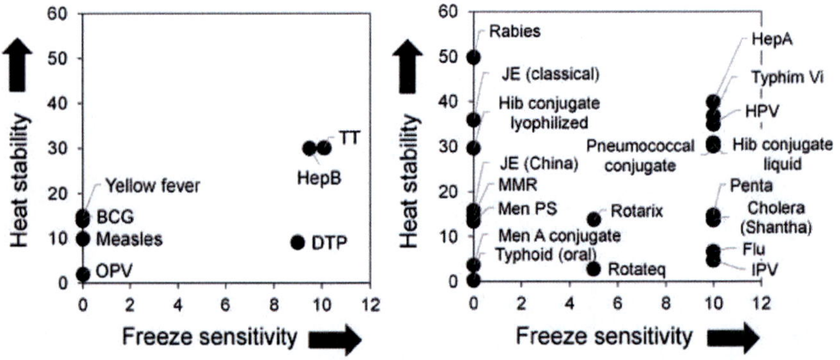

Fig. 2.2 Heat stability and freeze sensitivity of traditional (left) and newer (right) vaccines (Data adapted from Kartoglu and Milstien (2014). *BCG* Bacille Carmen Guerin, *DTP* Diphtheria-tetanus-pertussis, *HepA* Hepatitis A, *HepB* Hepatitis B, *Hib* Haemophilus influenza type b, *HPV* Human papillomavirus, *IPV* Inactivated polio vaccine, *JE* Japanese encephalitis, *Men A* Meningitis A, *Men PS* Meningitis polysaccharide, *MMR* Measles-mumps-rubella, *OPV* Oral poliomyelitis vaccine, *Penta* DTP + HepB + Hib, *Rotarix and Rotateq* Rotavirus vaccine (GSK and Merck), *TT* Tetanus toxoid, *Typhim Vi* Typhoid polysaccharide vaccine (Sanofi Pasteur)

- MMR,
- Hib conjugate (liquid and lyophilized),
- pentavalent vaccine (Penta),
- meningococcus both polysaccharide (Men PS) and conjugate (Men A conjugate),
- rotavirus (Rotarix, Rotateq),
- pneumococcus conjugate,
- IPV,
- HPV,
- rabies,
- influenza (Flu), and
- threshold vaccines against Japanese Encephalitis (JE), Hepatitis A (Hep A), varicella, cholera, and typhoid fever.

As shown in Fig. 2.2 (right), the spectrum of behaviors for newer vaccines is broader. For example, certain vaccines are freeze-stable, but their heat stability varies. Some have a medium freezing stability, and many vaccines should not be frozen according to the manufacturer's instructions. Therefore, today's cold chain must accommodate all of these vaccine products, which is quite challenging.

Figure 2.3 (left) summarizes in another way which vaccines are damaged by freezing and which are not (WHO/PATH 2014; WHO 2015b). Many vaccines can tolerate freezing, including freeze-dried vaccines (shown in bold type) and oral polio vaccine (OPV). However, freezing can irreversibly damage vaccines that contain aluminum-salt adjuvants (e.g., aluminum phosphate or hydroxide). These vaccines precipitated faster than non-frozen samples (Kurzątkowski et al. 2013), adversely affecting the immunological properties. Aluminum adjuvants are included in vaccines containing diphtheria toxoid, tetanus toxoid, hepatitis B antigens, and the diluent for meningitis A conjugate vaccine (WHO 2015b). Some vaccines that do not contain an aluminum adjuvant are also damaged by freezing, including some oral cholera, inactivated polio, and influenza vaccines.

Besides, Fig. 2.3 (right) provides an overview of the relative heat sensitivity of the vaccines divided into six groups (WHO/PATH 2014; WHO 2015b). The vaccines in the upper groups are more heat-sensitive than those in the groups below them. This is because all vaccines lose their potency over time, and heat accelerates this potency loss.

Figure 2.4 relates the relative heat and freeze sensitivity of the vaccines to the formulation type, i.e., freeze-dried (black) and liquid form (blue: no adjuvant, yellow: with aluminum adjuvant) (WHO/PATH 2014). Each bubble represents the overall or representative heat/freeze sensitivity for that type of vaccine. Individual vaccines of the same type but different manufacturers can have different stabilities.

As can be seen, vaccines in liquid form containing an aluminum adjuvant are generally the most sensitive to freezing. On the other end of the spectrum, the freeze-dried vaccines are fully resistant to freezing damage, becoming

	Vaccine
All these vaccines are damaged by freezing	DTaP DTaP-hepatitis B-Hib-IPV (hexavalent) DTwP DTwP-hepatitis B-Hib (pentavalent) Hepatitis A Hepatitis B Human papillomavirus Meningitis C (polysaccharide-protein conjugate) Pneumococcal (polysaccharide-protein conjugate) TT, DT, Td
	Cholera (inactivated) Influenza (inactivated, split) Hib (liquid) Inactivated poliovirus Typhoid polysaccharide
These vaccines are not damaged by freezing	**Meningitis A (polysaccharide-protein conjugate)*** Rotavirus (liquid and freeze dried) Yellow fever
	Bacillus Calmette-Guérin **Hib (freeze dried)** **Japanese encephalitis (live and inactivated)** **Measles** **Measles, mumps, rubella** Oral poliovirus **Rabies** **Rubella** **Varicella-zoster virus**

Heat sensitivity	**Vaccine**
Most-sensitive group	Oral poliovirus **Varicella-zoster virus**
	Influenza (inactivated, split)
	Inactivated poliovirus **Japanese encephalitis (live)** **Measles, mumps, rubella**
	Cholera (inactivated) DTaP DTwP DTaP-hepatitis B-Hib-IPV (hexavalent) DTwP-hepatitis B-Hib (pentavalent) Hib (liquid) **Measles** Rotavirus (liquid and **freeze dried**) **Rubella** **Yellow fever**
	Bacillus Calmette-Guérin Human papillomavirus **Japanese encephalitis (inactivated)** TT, DT, Td
Least-sensitive group	Hepatitis A Hepatitis B **Hib (freeze dried)** **Meningitis A (polysaccharide-protein conjugate)** Meningitis C (polysaccharide-protein conjugate) Pneumococcal (polysaccharide-protein conjugate) **Rabies** Typhoid polysaccharide

Fig. 2.3 Temperature sensitivity of vaccines. The vaccines are arranged alphabetically in groups, not in order of sensitivity to freezing or heat. Freeze-dried vaccines are bold. *DT* diphtheria and tetanus toxoid vaccine, *DTaP* diphtheria, tetanus, and (acellular) pertussis vaccine, *DTwP* diphtheria and tetanus toxoid (whole cell) and pertussis vaccine, *Hib* Haemophilus influenzae type b vaccine, *IPV* inactivated poliovirus vaccine, *Td* tetanus and diphtheria vaccine with reduced diphtheria content for adults, *T* tetanus toxoid. *The diluent for meningitis A vaccine is damaged by freezing (Adapted from WHO/PATH (2014), WHO (2015b))

heat-sensitive after reconstitution (WHO/PATH 2014). Rabies vaccines, pneumococcal conjugate vaccines (PCV), human papillomavirus vaccines (HPV), hepatitis B vaccines, diphtheria, and tetanus toxoid vaccines are among the least sensitive to high temperatures (MSF 2014). Freeze-dried measles, yellow fever, and BCG vaccines occupy the middle position. Oral poliomyelitis vaccine (OPV) and Varicella-zoster are the most temperature-fragile (WHO 2006b).

Overall, vaccines that do not require refrigeration for part of their shelf life enable a significant simplification of storage and delivery logistics to reach even the most remote parts of the world without harm. It will also lead to significant savings in wasted vaccines and logistics, including cold chain equipment. The development of thermostable vaccines is, therefore, a highly desirable strategy.

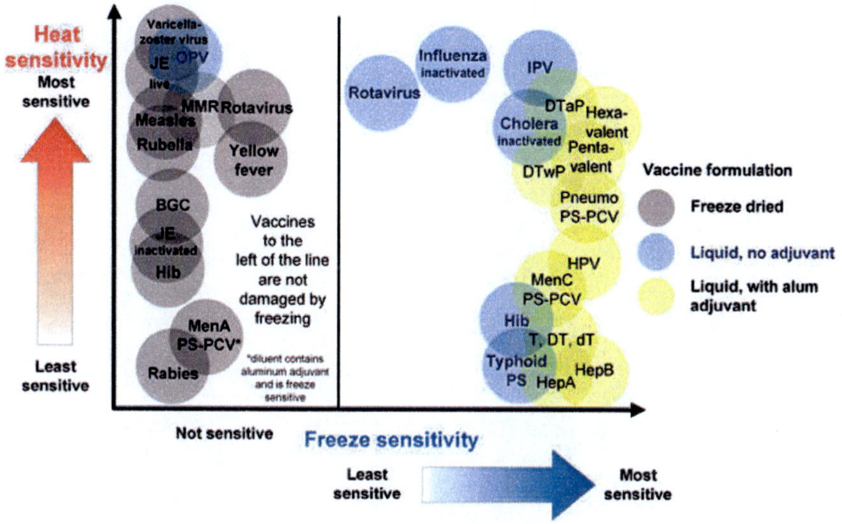

Fig. 2.4 Temperature sensitivity of various vaccines (Adapted from WHO/PATH (2014))

2.3 Excursus: Thermostability of COVID-19 Vaccines

As soon as the necessary information about the SARS-CoV-2 virus that causes COVID-19 disease became available, scientists quickly began to design the mRNA instructions for the cells to translate the unique spike protein into an mRNA vaccine. As a result, mRNA vaccines represent a new era in vaccinology, and interest has grown because this vaccine platform can be developed in the laboratory with readily available materials, meaning that the process can be standardized and easily scaled up, making vaccine development faster than traditional methods (Pardi et al. 2018).

The COVID-19 mRNA-based vaccines made by Pfizer-BioNTech, Moderna, and CureVac are encapsulated in positively charged lipid nanoparticles (LNPs), enabling resistance to RNase-mediated degradation and stable self-assembling particles that can be injected.

However, as shown in Table 2.6, a disadvantage of those mRNA vaccines is that they have to be shipped and stored in a frozen state at ultra-low temperatures of −70 °C, −20 °C, and −60 °C, respectively, to prevent degradation of the constructs (CureVac 2020a, b; Moderna 2020; Pfizer-BioNTech 2020; Cavalier et al. 2021; Crommelin et al. 2021).

The COVID-19 mRNA-based vaccine from Pfizer-BioNTech requires ultra-cold storage at −80 to −60 °C (Pfizer-BioNTech 2020). Newer stability data show that the Pfizer-BioNTech vaccine (Comirnaty®) can be stored at −25 to −15 °C for a total of 2 weeks and support decentralized distribution plans for vaccination at general practitioners' offices across European Union (Pfizer 2021a). After thawing,

Table 2.6 Storage stability and requirements of mRNA COVID-19 vaccines

Manufacturer	Stability in frozen state	Stability at 2–8 °C	Stability at room temperature	2 Doses (injection volume), dosing schedule	References
Pfizer-BioNTech	−80 to −60 °C, up to 6 months	Up to 5 days	Up to 2 h (Up to 6 h after dilution in its original multidose vial)	30 μg (0.3 mL) 3 weeks apart	Pfizer-BioNTech (2020)
Moderna	−25 to −15 °C, up to 6 months	Up to 30 days	Up to 12 h	100 μg (0.5 mL) 4 weeks apart	Moderna (2020)
CureVac	At below −60 °C, at least 3 months	At least 3 months at +5 °C	Up to 24 h	12 μg (no information) 4 weeks apart	CureVac (2020a, b)

Based on Cavalier et al. (2021), Crommelin et al. (2021) and fact sheets for healthcare providers administering the vaccines (CureVac 2020a, b; Moderna 2020; Pfizer-BioNTech 2020)

the undiluted multidose vials (5 doses) can be stored in a refrigerator at +2 to +8 °C for up to 5 days. In comparison, Moderna's vaccine can be stored at −25 to −15 °C (Moderna 2020), facilitating distribution and storage in areas without easy access to dry ice or dedicated ultra-low temperature freezers. Once thawed, the vials (10 doses) can be stored refrigerated at +2 to +8 °C for up to 30 days. In addition, CureVac announced that its mRNA-based vaccine candidate is stable for at least 3 months at a refrigerator temperature of +5 °C and can be used as a ready-to-use vaccine for up to 24 h at room temperature (CureVac 2020a, b).

On October 12, 2021, CureVac announced the strategic decision to focus its COVID-19 vaccine development on second-generation mRNA vaccine candidates (CV2CoV) in collaboration with GSK and to withdraw its first-generation COVID-19 vaccine candidate (CVnCoV) from the ongoing regulatory process at the European Medicines Agency (EMA) (CureVac 2021).

Otherwise, the differences between these high-tech mRNA vaccines are relatively minimal, except that Moderna (100 μg) requires more than 3 times as much substance per dose as Pfizer-BioNTech (30 μg) and more than 8 times as much as CureVac (12 μg) (Crommelin et al. 2020). Thus, this characteristic is relevant in scaling up manufacturing with billions of doses.

An open question is also how long vaccination confers protection against COVID-19 disease. The general opinion is that the duration will certainly be at least 6 months. It is also unclear whether vaccinated persons can transmit the virus; this will only become clear over time and with experience from the field. A booster vaccination is likely necessary due to viral mutations, similar to seasonal influenza vaccination.

First experiences in the use of ultra-cold storage at −70 °C (Cavalier 2020) have already been gained with the Ebola vaccine in several African countries such as Sierra Leone (Jusu et al. 2018) and the Democratic Republic of Congo (Rigby 2020; Uwagbale 2020). However, the logistical challenges and global scale in the COVID-19 pandemic are much more significant than Ebola, which is relatively confined to a specific African region. Furthermore, shipping and storage at −70 °C is a major challenge even for modern hospitals in high-income countries. Therefore, it is expected that these ultra-cold storage vaccines will be used primarily in centralized hospitals.

Recently, (Heida et al. 2021) pointed out that most COVID-19 vaccine candidates in clinical trials are administered by injection. However, such liquid vaccines require a highly regulated cold chain infrastructure, which poses a significant challenge for future global vaccine distribution, particularly in remote and tropical areas.

The question, therefore, remains whether vaccine distribution is effective enough to ensure mass-scale vaccination campaigns while minimizing costs. While pharmaceutical companies hint that they are producing a lyophilized version of their vaccine to increase vaccine stability, these still need to be reconstituted before administration. While this partially resolves the stability issues of liquid formulations, it does not eliminate the downsides of injection. A potential solution may be found in inhalable dry powder vaccines (Heida et al. 2021).

Figure 2.5 shows the relative progress of the different COVID-19 vaccine candidates at various stages of clinical development, including viral vector vaccines, inactivated and live-attenuated virus vaccines, virus-like particles, protein-based vaccines, and RNA or DNA vaccines (Callaway 2020; Jain et al. 2020).

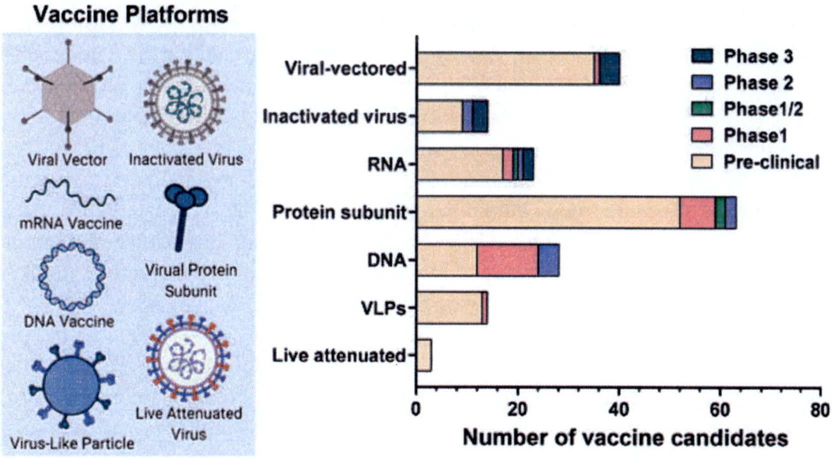

Fig. 2.5 Different vaccine platforms for the COVID-19 vaccine and the number of vaccine candidates in various phases of clinical trials (Adapted from Jain et al. (2020), open access under CC-BY 4.0 license, published on November 3, 2020)

As of December 24, 2021, the COVID-19 vaccine landscape listed 137 vaccine candidates in clinical trials and 194 vaccines in preclinical development (WHO 2021c). About 35% of the vaccine candidates in the clinical phase are based on protein subunit, 17% on RNA, 15% on viral vector (non-replicating), 13% on the inactivated virus, and 11% on DNA platforms.

Table 2.7 lists the leading COVID-19 vaccines in Phase 3 clinical trials, including information on thermostability during storage (McGill COVID19 Vaccine Tracker Team 2021; UNICEF 2021; WHO 2021c; Zimmer et al. 2021; AboulFotouh et al. 2021). Figure 2.6 shows how the COVID-19 vaccine heat sensitivity varies depending on the vaccine platform, and even vaccines based on the same platform have different stability profiles depending on the manufacturer (AboulFotouh et al. 2021).

Almost all licensed inactivated virus-based COVID-19 vaccines require storage at +2 to +8 °C. The Adenovirus vector-based vaccine candidates are more likely to be stored at 2–8 °C. Storing frozen at −25 to −15 °C would significantly increase their shelf-life (AboulFotouh et al. 2021). The mRNA vaccines of Pfizer-BioNTech and Moderna have 6 months of shelf-life when kept frozen at −70 °C at −20 °C, respectively. CureVac's mRNA vaccine is stable for 3 months at 5 °C.

There are only a limited number of COVID-19 vaccines that can be stored at around 25 °C for several hours. SinoVac Biotechnology announced that its aluminum salt adjuvanted inactivated CoronaVac vaccine would be stable for at least 5 months at 2–8 °C, 42 days at 25 °C, and 28 days at 37 °C (AboulFotouh et al. 2021). Zydus Cadila's ZyCoV-D three-dose plasmid DNA vaccine can be stored in liquid form at 30 °C for 3 months (Das 2020). However, so far, there is a lack of dry powder vaccines that could provide thermostability at 40 °C for several months, which is more attractive in the long term.

Broadening vaccine research scope towards inhalable dry powder formulations would offer several benefits, including increased vaccine stability, longer shelf life, facilitation of mass vaccination campaigns, and better protection by eliciting an immune response through the mucous membranes (Heida et al. 2021). However, as shown in Table 2.7, the main route of COVID-19 vaccine administration is by liquid injection into the muscle (skin) via needle and syringe.

As DNA-based vaccines are more stable at room temperature and do not require cold storage, they are more suitable for stockpiling. Vaccine candidates from AnGES/Osaka University/Takara Bio, Zydus Cadila, and Inovio are already in Phase 3 clinical trials. The developed DNA-based vaccines are delivered into skin cells via skin patches or electric pulses from a hand-held device.

The COVID-19 vaccine ChAdOx1 nCov-19 (AZD1222) developed by the University of Oxford and AstraZeneca is based on an attenuated version of an adenovirus that contains the genetic material of the SARS-CoV-2 spike protein. It can be stored and transported at +2 to +8 °C, facilitating delivery and administration in existing healthcare facilities.

The Sputnik V, Convidecia, and Ad26COV2S vaccines are similar to the AstraZeneca vaccine from a technological and thermostability standpoint. However, Sputnik V is a two-dose vaccine based on two recombinant adenovirus types,

Table 2.7 Information on the storage stability and efficacy of COVID-19 vaccines that are approved or in Phase 3 clinical trials

Vaccine manufacturer	Vaccine name	Vaccine platform	Efficacy (doses)	Storage stability	Route of administration
Pfizer/BioNTech	Comirnaty (tozinameran, BNT162)	mRNA (encapsulated in lipid-nanoparticles)	95% (2 doses, 3 weeks apart)	6 months at −70 °C, 5 days at 2–8 °C, 2 h at 25 °C	Muscle injection
Moderna/NIH	mRNA-1273	mRNA (encapsulated in lipid-nanoparticles)	94.5% (2 doses 4 weeks apart)	6 months at −20 °C, 30 days at 2–8 °C, 12 h at 25 °C	Muscle injection
CureVac	CVnCoV	mRNA	Unknown (2 doses, 4 weeks apart)	3 months at 2–8 °C, 2 h at 25 °C	Muscle injection
Oxford University/AstraZeneca	ChAdOx1 nCov-19 (AZD1222)	Adenovirus-based viral vector (Ad5)	62% to 90%, depending on dosage (2 doses, 4 weeks apart)	6 months at 2–8 °C 6 h at 25 °C	Muscle injection
Gamaleya Research Institute	Sputnik V (Gam-COVID-Vac) (Gam-COVID-Vac-Lyo)	Adenovirus-based viral vector (Ad5 and Ad26)	91.6% (2 doses, 3 weeks apart)	Liquid form at −18 °C, Lyophilized (dry) form at 2–8 °C	Muscle injection
CanSino Biologics/Academy of Military Medical Sciences	Convidecia (AD5-nCoV)	Adenovirus-based viral vector (Ad5)	65.3% (1 dose)	Stable in refrigerator (2–8 °C)	Muscle injection
Johnson & Johnson, Janssen Pharmaceutical, Beth Israel Deaconess Medical Center	Ad26.COV2.S	Adenovirus-based viral vector (Ad26)	72% (US), 66% (Latin America), 64% (South Africa) (1 dose)	2 years frozen at −20 °C, 3 months at 2–8 °C	Muscle injection
Vector Institute, BEKTOP	EpiVacCorona	Peptide-based	Unknown (2 doses, 3 weeks apart)	2 years at 2–8 °C	Muscle injection
Novavax	NVX-CoV2373	Protein subunit	96%, 86% (B.1.1.7), 49% (B.1.351) (2 doses, 3 weeks apart)	Stable in refrigerator (2–8 °C)	Muscle injection

(continued)

Table 2.7 (continued)

Vaccine manufacturer	Vaccine name	Vaccine platform	Efficacy (doses)	Storage stability	Route of administration
Anhui Zhifei Longcom, Chinese Academy of Medical Sciences	RBD-Dimer (ZF2001)	Protein subunit	Unknown (3 doses, 4 weeks apart)	Stable in refrigerator (2–8 °C)	Muscle injection
Medicago	CoVLP	Virus-like Particle	Unknown (2 doses, 3 weeks apart)	Stable in refrigerator (2–8 °C)	Muscle injection
Sinopharm, Beijing Institute of Biological Products	BBIBP-CorV	Inactivated virus (Vero cells)	79.3% (2 doses, 3 weeks apart)	Stable in refrigerator (2–8 °C)	Muscle injection
Sinopharm, Wuhan Institute of Biological Products	COVID-19 vaccine (Vero cells)	Inactivated virus (Vero cells)	72.5% (2 doses, 3 weeks apart)	Stable in refrigerator (2–8 °C)	Muscle injection
Bharat Biotech, Indian Council of Medical Research, National Institute of Virology	Covaxin (BBV152 A,B, C)	Inactivated virus	78%, (2 doses, 2 weeks apart)	7 days at 37 °C	Muscle injection
SinoVac Biotech	CoronaVac (PiCoVacc)	Inactivated virus	50.6% (Brazil), 91.3% (Turkey) (2 doses, 2 weeks apart)	5 months at 2–8 °C, 42 day at 25 °C, 28 days at 37 °C	Muscle injection
Inovio	INO-4800	DNA plasmid	Unknown (2 doses, 4 weeks apart)	More than 1 year at room temperature	Skin injection (Surface electroporation)
AnGES, Osaka University, Takara Bio	AG0302-COVID19	DNA plasmid	Unknown (2 doses, 2 weeks apart)	More than 1 year at room temperature	Skin injection
Zydus Cadila	ZyCoV-D	DNA plasmid	Unknown (3 doses, 4 weeks apart)	3 months at 30 °C	Skin injection

Compiled from McGill COVID19 Vaccine Tracker Team (2021), UNICEF (2021), WHO (2021c), Zimmer et al. (2021), Wouters et al. (2021), AboulFotouh et al. (2021)

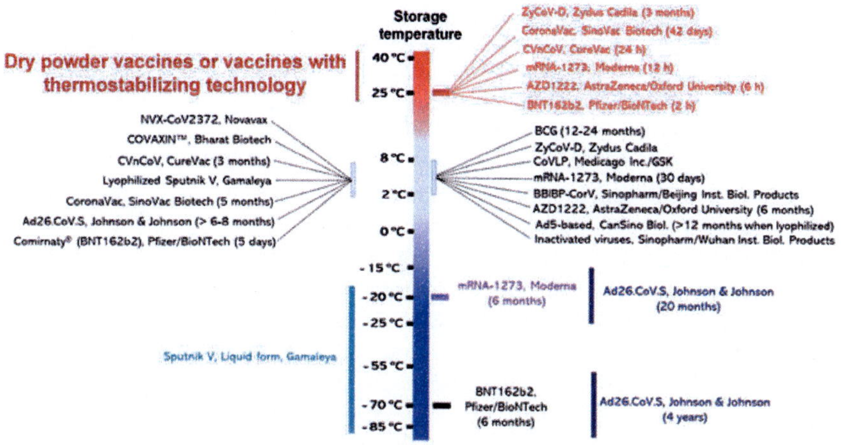

Fig. 2.6 Expected long-term storage temperatures and shelf-lives of COVID-19 vaccine candidates in Phase 3 or 2/3 clinical trials. Thermostable vaccines (25–40 °C) can be developed via thermostabilizing technology or the transformation of liquid vaccines into dry powders (Image adapted from AboulFotouh et al. (2021) with permission from Springer Nature)

rAd26 for prime vaccination and rAd5 for boost vaccination (Logunov et al. 2020). The use of two vectors serves to circumvent possible vector immunity. Clinical trials confirmed the efficacy of 91.4% on day 28 after the first dose and over 95% 42 days after the first dose (21 days after the second dose) (RDIF 2020).

A freeze-dried form of Sputnik V is available (Gam-COVID-Vac-Lyo), which can be stored at +2 to +8 °C, perhaps even at room temperature. It is then diluted before injection. Studies confirmed that the immune response to the lyophilized form is the same as the vaccine's liquid form. Therefore, the switch to the freeze-dried form was expected to occur beginning of 2021 (Ivanova and Sagdiev 2020).

However, Russia is not the only country exploring the freeze-drying of COVID-19 vaccines. In Japan, Daiichi Sankyo Co produces an mRNA-based candidate with an advantage in storage at higher temperatures (Ivanova and Sagdiev 2020).

Pfizer also announced plans to develop a vaccine in dry powder form requiring only standard refrigeration in 2021 (Dunn 2020). In April 2021, a clinical trial evaluating a freeze-dried version of the COVID-19 vaccine began in adults aged 18–55 years in the US (ClinicalTrials.gov Identifier: NCT04816669) (Pfizer 2021b). The Phase 3 trial, which involves about 1100 subjects, will last about 2 months, and researchers aim to determine whether the freeze-dried version is as safe and effective as the already approved version. If successful, the new formulation could be ready for use by early 2022. This achievement would be a big step forward for vaccine formulation technologies (PATH 2020). Having a dried version of the COVID-19 vaccine is in the next round of development. Providing better long-term stability than liquid forms would be immensely important to vaccine manufacturers. Until then, more testing is needed to see if freeze-drying affects the vaccine's efficacy.

As reported to the (WHO 2021d) from national authorities as of 15 September 2021, there have been more than 226 million confirmed infection cases of COVID-19 worldwide, including more than 4.6 million deaths. A total of 5.6 million vaccine doses have been administered. The global attention towards immunization is at an all-time high. However, the lack of a better thermostability of the COVID-19 vaccines is still a significant burden and barrier limiting worldwide distribution (Wu et al. 2020). More temperature-stable vaccines are highly desirable for use in the field and prevent future COVID-19 outbreaks.

To sum up, in the race for a safe and effective vaccine against COVID-19, the vaccine formulation and the route of administration play a critical role. As shown in Fig. 2.7 and Table 2.7, the administration of current COVID-19 vaccine candidates follow traditional ways in liquid form by muscle injection that require refrigerated storage and depend on the cold chain.

However, an ideal COVID-19 vaccine should be in a ready-to-use thermostable form that can be stored at ambient temperatures while maintaining immunogenicity over a long shelf life (Wang et al. 2020). Furthermore, a dried vaccine powder can be reconstituted before injection or used directly for inhalation or intranasal vaccination if it displays good aerosol properties, safety, and efficiency profiles

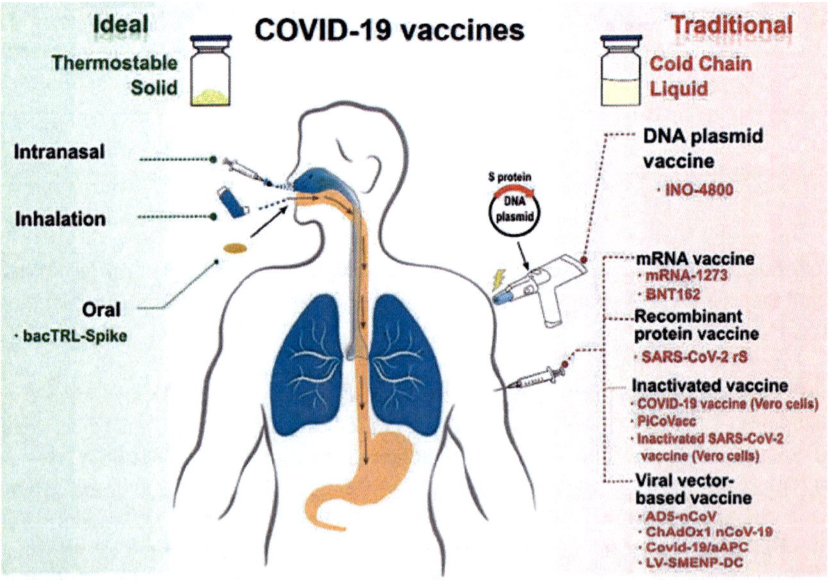

Fig. 2.7 A comparison of routes of administration between the ideal vaccines and the current COVID-19 vaccine candidates (Image from Wang et al. (2020) with permission from Springer Nature)

(Wang et al. 2020). Therefore, in terms of future vaccine formulation, it is crucial to formulate the vaccine in a dosage form that is stable and suitable for large-scale production and that the material can be transported to the patient site with logistical simplicity.

Because COVID-19 is primarily a respiratory disease, needle-free mucosal vaccination via the intranasal pulmonary or oral route may offer advantages to traditional parenteral vaccination for both COVID-19 protection and mass immunization. In addition, formulating vaccines in dry powder form can help improve vaccine thermostability and ultimately reduce immunization costs.

Finally, to conclude this section, reference is made to a recent study by (Taher et al. 2021) where inhalable nano spray-dried powders of remdesivir (an antiviral drug for patients in critical conditions) and hydroxychloroquine (antimalaria drug) using poly(lactic-co-glycolic) acid (PLGA) was proposed as an alternative way to deliver a drug against COVID-19. Dry powder inhalation would greatly benefit patients because the main organ of infection of SARS-CoV-2 is the esophagus and lungs, and delivering the vaccine directly to the site of action would be effective. PLGA is considered a suitable polymer as it is biocompatible and non-cytotoxic. It is most commonly used in drug delivery as a carrier or excipient for optimal drug formulation and distribution (Arpagaus 2019).

Although the "Living Guideline of Therapeutics and COVID-19" of (WHO 2021e) makes a strong and conditional recommendation against the administration of hydroxychloroquine and remdesivir for the treatment of hospitalized patients with COVID-19, this study may encourage other researchers to conduct more clinical trials on the use of recommended therapeutic drugs (e.g., corticosteroids) in the treatment of COVID-19 disease.

In addition, it is suggested that future research should also focus on further commercialization of nano spray drying technology at pilot and industrial scales (Arpagaus et al. 2017, 2018; Arpagaus 2018). Recent advances in particle engineering have enabled the design of nanoparticle-based vaccines with precise control over size, shape, functionality, and surface properties, leading to enhanced antigen presentation and strong immunogenicity (Al-Halifa et al. 2019).

2.4 Cold Chain and the Long Road to Vaccination

The cold chain was principally developed for two types of vaccines; those whose temperature sensitivity is intrinsic to the structure of the vaccine antigen and those whose temperature sensitivity is related to additives and adjuvants (Kartoglu and Milstien 2014).

The cold chain protects vaccines both from heat and cold exposure. It is an integrated system of types of equipment (e.g., cold rooms, shipping containers, refrigerators, vehicles), procedures, records, and activities used to handle, store, transport, distribute and monitor temperature-sensitive products.

As part of a strategy to strengthen vaccine supply chains and achieve better immunization equity and coverage, in 2015, GAVI (The Vaccine Alliance), UNICEF (United Nations Children's Fund), and other partners launched the Cold Chain Equipment Optimization Platform (CCEOP)—a US$ 400 million five-year project to upgrade existing cold chain equipment in 56 countries by 2021 (UNICEF 2019). Strengthening cold chains also supports a rapid response to infectious disease outbreaks, including the COVID-19 pandemic. By the end of 2019, over 20,000 cooling units had been installed in health centers, and nearly 20,000 more units were in the delivery process (UNICEF 2019). The COVID-19 vaccination campaign has revolutionized the vaccine cold chain because of nearly double the usual volumes and the new temperature ranges for storage and transport that require new equipment, new solutions, and new organizations (Cavalier et al. 2021).

Figure 2.8 illustrates the cold chain from the factory down to the health facilities and children in remote villages. Vaccination data are collected, recorded, checked, analyzed, and reported back up the chain to ensure proper monitoring and vaccine forecasting (WHO 2015a).

The availability of a vaccine cold chain might not be so much of a problem in high-income countries, but it can be a real challenge to get vaccines to a village in a pool or township in low-income countries. The cooling infrastructure can be missing or incomplete and presents a significant barrier to effective vaccination schemes, especially when introducing newer and more expensive vaccines (Zaffran et al. 2013). A cold chain is only as strong as its weakest link (Kartoglu and Milstien 2014).

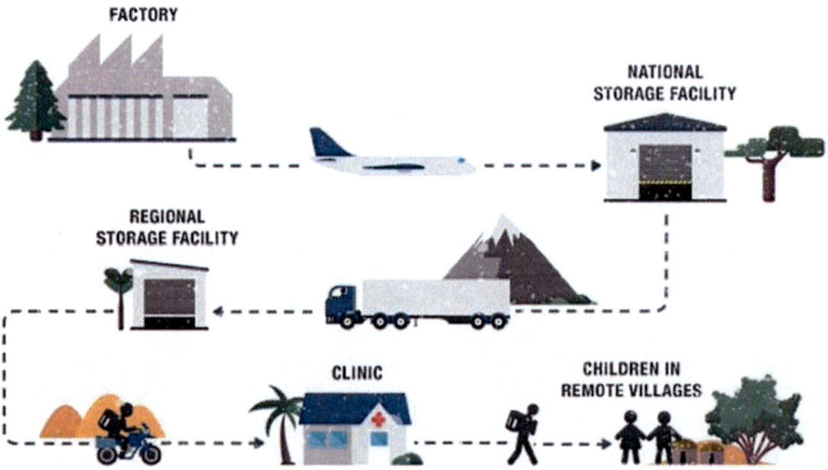

Fig. 2.8 The cold vaccine chain—a long road to vaccination. Vaccines must be kept between +2 °C to +8 °C from the factory to some of the most remote places on earth (Adapted from Médecins Sans Frontières (MSF) published in Jackson et al. (2018))

The future of the cold chain in reducing its dependency by exploiting the stability of each vaccine is discussed. Vaccine distribution without a cold chain would considerably simplify the delivery system and facilitate integration into vaccine distribution in low-income countries (WHO 2006b).

The price of keeping vaccines cold costs countries and health care organizations millions of dollars. Maintaining the cold chain and refrigeration of vaccines from manufacture to delivery costs about $US 200–300 million each year (Lloyd 2000; Das 2004), which increases the cost of vaccination by 14–20% (Lloyd 2000; Manufacturing Chemist 2010). In low-income countries, the cold chain alone can account for up to 80% of the cost of vaccination programs (Kols and Sherris 2000).

2.5 Wastage of Vaccines Due to Heat and Freeze Exposure

Delivering all the vaccines needed for vaccination campaigns anywhere, anytime and on time remains a challenge in the twenty-first century (Cavalier et al. 2021). WHO (2005) estimated that more than 50% of all vaccine doses may be wasted globally every year because of issues related to temperature control, logistics, and shipment (Table 2.8). The main reason for the high wastage is that multi-dose vials of unpreserved vaccines must be discarded within 6 h. Thus, if there are not enough people to vaccinate once a vial is opened, the remaining doses must be discarded when sessions are completed (Lloyd 2000). Temperature-related wastage is less common because a vaccine vial monitor (VVM) monitors heat exposure. However, freeze exposure is likely to go undetected. The direction of freeze-sensitive vaccines to temperatures below recommended ranges in the cold chain may result in loss of vaccine potency (WHO 2006b; Matthias et al. 2007; Hanson et al. 2017). Evidence of vaccine freezing in the cold chain has been provided in a literature review of (PATH 2003) covering over 20 case studies from 1990 to 2003 in Asia, Australasia, Central Asia, Eastern Europe, the UK, and the US. In addition, Matthias et al. (2007) highlighted in a systematic literature review that accidental freezing was

Table 2.8 Reasons for vaccine wastage in unopened and opened vials

Reasons for vaccine wastage	
In unopened vials	In opened vials
• Expiry • VVM indication • Heat exposure • Freezing • Breakage • Missing inventory • Theft • Discarding unused vials returned from an outreach session	In addition to the types listed in the left column: • Discarding remaining doses at the end of a session • Not being able to draw the number of doses indicated on the label of a vial • Poor reconstitution practices • Submergence of opened vials in water • Suspected contamination • Patient reaction requiring more than one dose

Adapted from the guideline for monitoring vaccine wastage at the country level WHO (2005)

pervasive and occurred in all segments of the cold chain. Between 14 and 35% of refrigerators or transport shipments were exposed to freezing temperatures. Studies examining all distribution segments found that 75% to 100% of vaccine shipments were exposed to freezing temperatures.

More recently, Hanson et al. (2017) reported that the percentage of vaccines exposed to temperatures below recommended ranges during storage and shipments was 33% and 38% in high-income countries and 37% and 19% in low-income countries. Other reasons for wastage besides improper temperature control, logistics, and transport-related issues include storage at too high temperatures (i.e., heat exposure) (BIONITY 2004; Murhekar et al. 2013), poor cold chain and warehouse management of unopened vials due to expiry, and interruption of the cold chain, e.g., due to maintenance of refrigerators, and limited access to electricity in the poorest areas (WHO 2005). However, it should also be noted that the wastage in countries is not always documented, so data is sparse.

2.6 Suggested Improvements to Vaccine Products

To find out how vaccine products should be improved to meet better the needs of vaccination programs (Kristensen et al. 2016a) conducted a qualitative survey between October 2011 and March 2012. In semi-structured individual interviews, responses from 158 immunization stakeholders in Brazil, China, India, Peru, the Philippines, and Tanzania (both mainland and Zanzibar) of all levels of the supply chain were collected.

Table 2.9 shows the responses to suggested improvements. The most frequent response was better heat stability (n = 56/155 respondents). There was also a strong interest in pre-filled, single-dose, easy-to-use products (n = 99/155 mentioned one or more of these attributes).

Thermostable vaccines will not entirely remove the need for refrigeration, as it is unlikely that vaccines can be kept outside the cold chain for their entire shelf life. Still, they simplify vaccine distribution logistics and promise to increase vaccination campaigns' efficacy significantly. Even people in the most remote areas could gain access to the benefits of modern medicine. Improved product stability reduces the cold chain requirements and reduces the cost of shipping, handling, and long-term storage. The cost savings in refrigeration, cold boxes, and transport logistics could help finance this higher quality healthcare provision.

Room-temperature stable vaccines offering extended shelf life would facilitate mass immunization programs to unreachable areas without refrigeration. In addition, these types of vaccines allow long-term storage for stockpiling and rapid deployment in public emergencies, including pandemics (Zhu et al. 2014). Ideally, new thermostable vaccines would be approved for shelf-life storage at tropical room temperatures (WHO 2006b).

Table 2.9 Participants' suggested improvements to vaccine products, 155 respondents

Suggested improvements	Number of participants	Percent of participants (%)
Heat stable	56	36
Prefilled	50	32
Single-dose	49	32
Easy to use and understand; standardized	32	21
Low-dose vial (2 or 5 doses)	21	14
Combined antigens (fewer immunizations)	15	10
Small packaging	13	8
Fully liquid presentation (not lyophilized)	12	8
No need to store in the cold chain	11	7
Freeze stable	10	6
Oral presentation/no injection	9	6
Multidose vial (10+)	9	6
Reduced doses per fully immunized child	9	6
Vaccine with preservative	8	5
Vaccine without preservative	4	3
No ampoules	3	2
Vaccine vials with freeze indicators	1	1

Data from Kristensen et al. (2016a) with permission from Elsevier

In an additional study with the same group of immunization stakeholders, (Kristensen et al. 2016b) examined whether thermostable vaccines would help address cold chain challenges. Across all countries, challenges were identified with both heat exposure and freezing of vaccines, with heat exposure being a more significant concern. Overall, most participants (73%, 115/158) were interested in vaccines labeled for short-term use in a CTC. This solution would provide more flexibility during use and in the event of cold chain interruptions. In addition, 41% of respondents (13/32) preferred a hepatitis B vaccine that was freeze-dried and more heat-stable, as a more stable product would reduce risk, cost, and waste.

For public health, the benefits of thermostable vaccines are indisputable (Jackson et al. 2018), but these do not necessarily translate into commercial advantages for vaccine manufacturers (Chen and Kristensen 2009). In addition, there are enormous technical and regulatory challenges. For example, the time and resources needed to prove that a vaccine is still effective after exposure to high temperatures are considerable, and manufacturers need to be satisfied that investing the necessary funds to generate this data is valuable (Jackson et al. 2018).

2.7 Lessons Learned on Vaccine Stabilization

Since 2003, PATH has studied the technical and commercial feasibility of stabilizing vaccines (WHO 2006b; Kristensen and Chen 2010). The work of PATH has facilitated the research of thermostable vaccines in collaboration with commercial partners. Technical research focused mainly on sugar-glass and spray drying-based technology applied to both core and new vaccines of the Expanded Program on Immunization (EPI).

Work by PATH and collaborators in 2010 has yielded a spray-dried hepatitis B vaccine stable for up to 24 months at 37 °C (Chen et al. 2010), and a spray-dried live-attenuated measles vaccine that is stable at 37 °C for 8 weeks (Ohtake et al. 2010). In addition, based on experiences with 32 technical collaborators (Kristensen et al. 2011) to improve the stability of seven different vaccine antigen types with 11 different vaccine manufacturers, Kristensen and Chen (2010) from PATH have summarized key "lessons learned" on vaccine stabilization projects. Table 2.10 summarizes and repeats these lessons learned.

Ideally, vaccine developers should consider vaccine stabilization approaches early in the development process. Many commercially available vaccines could be reformulated to increase stability. For example, vaccines containing an aluminum adjuvant could be protected against frost damage, e.g., propylene glycol, polyethylene glycol 300, or glycerine. However, heat stabilization of vaccines requires a customized approach, and results will vary (Kristensen et al. 2011).

Recent breakthroughs have resulted in spray-dried vaccines and novel vaccine delivery. A heat-stable vaccine could be labeled with a vaccine vial monitor (VVM) and, together with new guidelines, allow controlled storage and transport at ambient temperature. Spray drying could produce stable vaccines in the form of free-flowing powders with controllable particle sizes, enabling new vaccine delivery methods such as aerosols, dry powder jet injection, coated microneedles,

Table 2.10 Stabilization of vaccines: lessons learned by PATH

Lesson	Learning
Lesson 1	Integrate stabilization efforts early in the vaccine development process
Lesson 2	There are circumstances where it makes sense to reformulate existing commercially available vaccines to enhance stability
Lesson 3	Freeze stabilization is possible for vaccines containing aluminum adjuvant when adding propylene glycol, polyethylene glycol 300, or glycerin
Lesson 4	Heat stabilization requires a customized approach, and results will be variable
Lesson 5	The full benefits of heat-stable vaccines will only be realized after changes are made to storage guidelines
Lesson 6	Trade-offs may need to be made between the heat stability and the format characteristics of a vaccine. Some heat-stability improvements result in inferior product formats while others enable new, beneficial formats
Lesson 7	Products with enhanced stability can benefit both vaccine producers and purchasers

Based on Kristensen and Chen (2010)

and biodegradable implants (Chen and Kristensen 2009). Finally, vaccines with enhanced stability could benefit vaccine producers and purchasers (Kristensen and Chen 2010).

2.8 Characteristics of an Ideal Vaccine

Levine (2011) described some fundamental characteristics of an ideal vaccine that would overcome many limitations (Table 2.11). First, an ideal vaccine would require only a single dose and provide long-lasting protection, even in very young children, the elderly, and immunocompromised individuals, and would be well tolerated. Second, the ideal vaccine would be thermostable, thereby eliminating the need for a cold chain. Third, it would be administered without needle and syringe (i.e., by mucosal, transcutaneous, or needle-free parenteral administration) and would be combinable or co-administered with other vaccines. Lastly, the ideal vaccine would leave a marker (e.g., visible or serologic) that would distinguish the immunized person from the unimmunized person or the person who has experienced infection with the wild-type pathogen. Some existing vaccines include several of these features, but none has all (Levine 2011).

2.9 VIPS—The Vaccine Innovation Prioritisation Strategy

In 2017, GAVI, WHO, UNICEF, PATH, and the Bill & Melinda Gates Foundation (BMGF) launched VIPS (The Vaccine Innovation Prioritisation Strategy), intending to develop a single integrated framework to evaluate and prioritize vaccine

Table 2.11 Some fundamental characteristics of an ideal vaccine

Characteristic of an ideal vaccine
• Safe in all ages (including very young infants and elderly) and immunocompromised hosts
• It only requires administration of a single-dose
• Confers a high level of protection in all ages, including young infants, the elderly, and high-risk groups (including immunocompromised hosts)
• Protection begins soon after vaccination and is long-lived (at least 10 years)
• Administered without a needle and syringe
• No cold chain required
• Combinable with other vaccines intended due to be administered at the same time
• Affordable
• Elicits a readily measurable immune response that is a correlate of protection and can be used to differentiate vaccinated (i.e., and protected) subjects from unvaccinated persons or those who have immune responses stimulated by past infection with the wild type pathogen

Adapted from Levine (2011), with permission from Elsevier

product innovations and to drive these innovations forward to better address country needs and support the Alliance's immunization coverage and equity goals (GAVI 2020). The VIPS process involved in-depth research, stakeholder consultations, and the development and application of a methodology that evaluates various technologies at different stages along the product development pipeline. In 2019, the first phase of VIPS resulted in a shortlist of nine innovations, further analyzed in a second phase. In May 2020, the VIPS process concluded with the decision to prioritize the following three innovations for which the Alliance will engage to advance development, policy, and access (VIPS 2020a):

- microarray patches (MAPs),
- heat stable and Controlled Temperature Chain (CTC) qualified vaccines (a combined formulation, regulatory, and novel programmatic approach to vaccine management), and
- barcodes on primary packaging (an implementation/system innovation).

The VIPS partners now focus on defining end-to-end strategies and clear action plans for the three prioritized innovations to accelerate their introduction and impact. The Alliance has prioritized heat-stable and CTC-qualified vaccines, including both liquid and dry formulations (VIPS 2020a). Dry formulations are of interest if they can be coupled with technologies that offer additional benefits, such as eliminating the problems of manual reconstitution, as would be the case with dual-delivery chamber devices, fixed-dose implants, or MAPs. Overall, increased thermostability is a desirable feature of all vaccines to improve vaccine efficacy and, where possible and appropriate, allow storage and transport at elevated temperatures in a CTC.

2.10 The Controlled Temperature Chain (CTC)

By taking advantage of the true heat stability of vaccines, WHO and PATH collaborated in the project Optimize (PATH/WHO 2012) to enable the use of vaccines in a controlled temperature chain (CTC) (WHO 2020), which allows vaccines to be kept at temperatures outside of the traditional cold chain of +2 to +8 °C for a limited time immediately before administration under monitored and controlled conditions. The WHO has established programmatic criteria for a vaccine to be labeled for use in a CTC (WHO 2020):

1. The vaccine should be used in a campaign or special vaccine delivery strategy setting. CTC is not currently recommended for immunization through routine delivery.
2. The vaccine must be able to tolerate ambient temperatures of at least 40 °C for a minimum of three days (just before administration) and should be accompanied by (see Fig. 2.9):

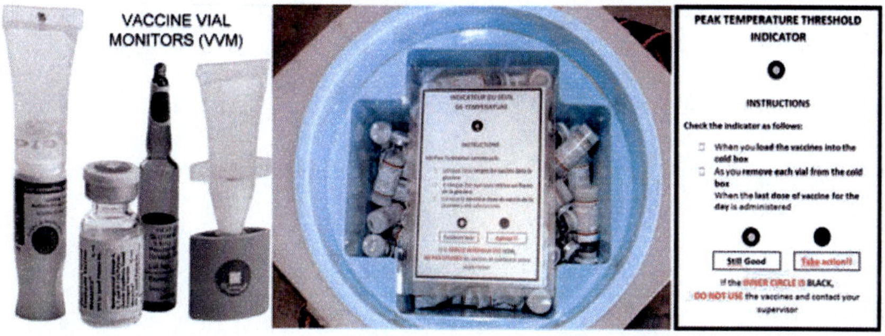

Fig. 2.9 Vaccine Vial Monitors (VVM) placed on vaccine vials and Peak Temperature Threshold Indicator (PTTI) placed in each vaccine carrier delivered by the CTC approach. The PTTI has a temperature-sensitive sticker that changes color when the threshold of 40 °C is reached. Vials found to have been exposed to temperatures above 40 °C would be discarded. (Images adapted from WHO (2002, 2015a) and WHO/PATH (2013))

- a vaccine vial monitor (VVM) on each vial, and
- a peak temperature threshold indicator (PTTI) in each vaccine carrier to indicate exposure to heat.

3. The vaccine must be licensed for use in a CTC by the relevant regulatory authorities, with a label that specifies the conditions.

As a note, removing a vaccine from the cold chain within the context of CTC occurs only once, in the days just before the vaccine is administered. CTC, therefore, does not replace the conventional cold chain but operates in conjunction with it.

In 2016, the WHO established a CTC working group under the Immunization Practices Advisory Committee (IPAC) to promote the development and uptake of thermostable vaccines and provide the essential coordination among stakeholders (Kahn et al. 2017). A key output is the CTC Strategic Roadmap for Priority Vaccines 2017–2020 (WHO 2017), which addresses supply and programmatic issues, aiming for CTC priority vaccines to become licensed for CTC use and facilitating the uptake of the CTC vaccine delivery approach. This roadmap, which is currently being updated with new target vaccines for CTC, focuses primarily on the following four priority CTC vaccines:

(1) Human papilloma virus (HPV)
(2) Oral cholera vaccine (OCV)
(3) Hepatitis B vaccine birth dose (HepB-BD), and
(4) Tetanus toxoid (TT), tetanus-diphtheria low dose (Td), or other tetanus-toxoid-containing vaccines (TT-CVs).

These four vaccine types were selected based on the following criteria (WHO 2017):

- potential in terms of adequate heat stability (i.e., can withstand higher temperatures without losing potency)
- a delivery strategy that would benefit from CTC use/expressed country need, and
- the technical feasibility of CTC licensure.

2.11 Why Consider a Controlled Temperature Chain (CTC)?

The CTC approach complements investments in the supply chain and helps overcome burdens and constraints associated with the last mile of vaccine delivery in a traditional cold chain. Ultimately, CTC can contribute to higher immunization coverage and equity in low-income countries (WHO 2017). As described in a commentary by Kahn et al. (2017), the expanding use of vaccines in a CTC is an effective way to strengthen immunization supply chains and increase vaccine access and coverage, especially among hard-to-reach populations with little to no supply system capacity. More compelling evidence shows that the CTC offers valuable opportunities to maximize supply chain efficiency, protect cold chain investments, and protect more children and families from diseases prevented by vaccination. The various risks and benefits of the CTC have to be weighed up against each other (Fig. 2.10) (Kahn 2019).

Overall, the benefits outweigh the risks because a CTC:

- generates significant costs savings (Lydon et al. 2014; Mvundura et al. 2017) through simplified distribution and logistics when handling vaccines, i.e., specifically (PATH/WHO 2012)

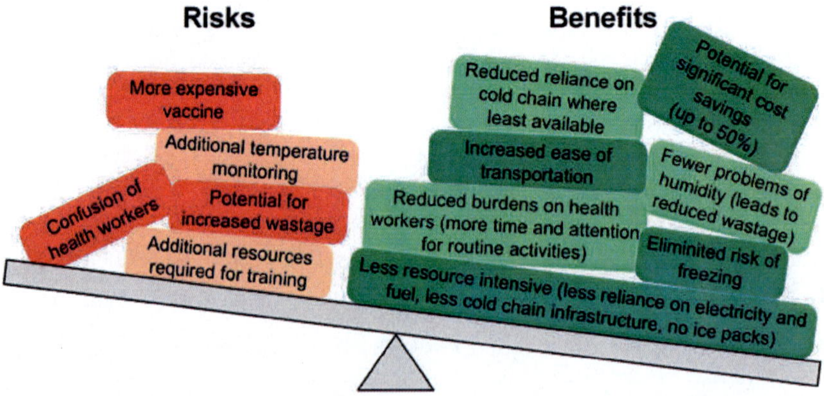

Fig. 2.10 Risks and benefits of a CTC (Adapted from Kahn (2019)

  - removing the need for additional freezers, refrigerators, cold boxes (WHO 2015a)
  - saving time for health workers in logistics and ice pack preparation
  - reducing the required storage space
- reduces the risk of vaccine wastage due to freezing (Matthias et al. 2007),
- expands the geographical reach of vaccines to more people (i.e., hard-to-reach areas and marginalized populations (PATH/WHO 2012), where the cold chain infrastructure is weak (Zipursky et al. 2014), and hence
- increases the potential for higher vaccination coverage (WHO 2013; Mvundura et al. 2017), especially for hepatitis B birth dose to infants, tetanus toxoid vaccine to women, and human papillomavirus vaccine to adolescent girls (PATH/WHO 2012).

CTC also helps secure and enable investments made earlier in the supply chain (Kahn et al. 2017). Other benefits of vaccines delivered with the CTC approach are (WHO 2017):

- improving working conditions for vaccinators by reducing the weight of the vaccine carriers,
- obviating the need to renew ice packs during long journeys,
- avoiding the need to travel to return vaccines to the cold chain after outreach, and
- optimizing staff time by redirecting staff to routine immunization activities typically spent on conditioning ice packs for campaigns.

2.12 Economic Benefits of Thermostable Vaccines in the CTC

So far, only a few studies have evaluated the economic benefits of thermostable vaccines (Otto et al. 1999; Levin et al. 2007; Lee et al. 2012; Lydon et al. 2014; Mvundura et al. 2017).

Otto et al. (1999) investigated vaccination against hepatitis B at birth with a pre-filled, non-reusable vaccination device (UniJect®), which was stored outside the cold chain in a tropical environment for up to one month. The study in the Tabanan district (Bali, Indonesia) demonstrated no statistically significant difference in immunogenicity between standard cold chain storage and storage outside the cold chain.

Levin et al. (2007) evaluated the cost impacts of thermostable vaccines (i.e., single-dose measles, yellow fever, Bacille Calmette-Guerin, and diphtheria-tetanus-pertussis-hepatitis B) in Cambodia, Ghana, and Bangladesh. The introduction of the thermostable vaccines increased the effectiveness of all vaccines, and the incremental costs increased for 3 out of 4 vaccines. Single-dose presentations of thermostable vaccines were concluded to be potentially cost-effective interventions to reduce childhood deaths and disability in low-resource settings in Asia and Africa.

Lee et al. (2012) simulated the impact of making different combinations of six EPI vaccines thermostable in Niger's vaccine supply chain. The most substantial impact came from making the pentavalent (DTP, hepatitis B, and Haemophilus influenza type b) vaccine thermostable, increasing its vaccine availability from 87 to 97%, and vaccine availability of all other remaining non-thermostable EPI vaccines to over 93%. The authors concluded that besides eliminating the need for refrigerators and freezers, thermostable vaccines could free up significant cold storage space for other vaccines and help alleviate supply chain bottlenecks worldwide.

Lydon et al. (2014) explored the economic benefits of CTC use during a mass vaccination campaign of meningitis A vaccine in Chad. The authors concluded that CTC implementation at the most peripheral levels of the supply chain system could result in significant economic benefits and extend vaccine coverage in low-resource settings.

Mvundura et al. (2017) showed during a MenAfriVac™ campaign in Togo in 2014 that the strongest argument for using CTC was reducing logistics costs for remote health centers without cold chain infrastructure. CTC reduced the logistical challenges of vaccine distribution.

2.13 CTC-Approved Vaccines and Pipeline

As of February 2020, two heat-stable liquid vaccines (Gardasil®4 and Shanchol™) and one heat-stable freeze-dried vaccine (MenAfriVac™) were fully licensed, and WHO prequalified for CTC use (VIPS 2019, 2020b), with an additional hepatitis B product under review (WHO 2018). Figure 2.11 presents images of the CTC-qualified vaccines. Moreover, the number of CTC-approved vaccines is tracked by the Indicator SO6.4 of the WHO (2018).

Figure 2.12 visualizes the progress of vaccines licensed for use in a CTC (WHO 2018; Kahn 2019). Since 2012, there has been a noteworthy increase. For example, in October 2012, the lyophilized MenAfriVac™ powder (Men A, administered intramuscularly) was the first vaccine licensed and prequalified by the WHO for use in a CTC at 40 °C for up to 4 days (WHO 2012). MenAfriVac™ is a freeze-dried meningitis A vaccine powder developed by the Serum Institute of India (SII) with support from PATH and the WHO through the Meningitis Vaccine Project. It is made of purified meningococcal A polysaccharide covalently bound to tetanus toxoid (TT) and excipients (WHO 2013). By 2017, more than 4 million people in six countries had received MenAfriVac™ delivered in a CTC (Kahn et al. 2017) (Fig. 2.12).

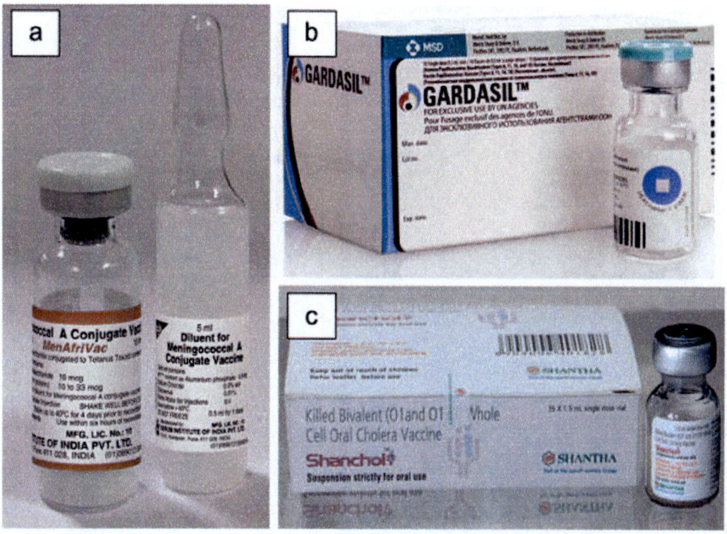

Fig. 2.11 CTC-qualified vaccines as of Feb. 2020. **a** MenAfriVac™ produced by Serum Institute India (meningococcal A conjugate vaccine, lyophilized, administered intramuscular, storage up to 40 °C for 4 days), **b** Merck's Gardasil®4 (quadrivalent human papillomavirus vaccine, liquid, intramuscular, storage up to 42 °C for 3 days), **c** Shantha Biotechnics Shanchol™ (inactivated oral cholera vaccine, liquid, oral, storage up to 40 °C for 14 days) (Images from WHO prequalified vaccines, https://extranet.who.int/pqvdata)

MenAfriVac™ vaccine reduced meningitis cases in sub-Saharan Africa and reduced the costs associated with the storage and transport of the vaccine. For example, during a 10-day meningitis A mass vaccination campaign in three regions of Chad in December 2011 (1.8 million people), Lydon et al. (2014) estimated a cost reduction from US$ 0.24 to US$ 0.12 per person vaccinated if the vaccines were kept at or near ambient temperatures in a CTC for up to 4 days at the district and service delivery levels. Based on the promising results, the WHO (2013) has issued specific guidelines for decision-makers and managers involved in vaccination programs on how to use MenAfriVac™ in a CTC during campaigns.

Zipursky et al. (2014) described the field use of MenAfriVac™ in a CTC during a mass immunization campaign in Benin in December 2012 (155,600 people, 21 supervisors, and 77 vaccinators). The feasibility and acceptability of the practice among vaccinators and supervisors were assessed. Clear benefits from the CTC approach were seen in low-income settings, especially in hard-to-reach areas or where the cold chain was weak. In a survey, 98.7% of supervisors and 100% of vaccinators indicated that, if given a choice, they would prefer to conduct their next campaign in a CTC rather than using the traditional cold chain. Asked which vaccines they would most like to see approved for CTC use, most vaccinators and supervisors named polio (44%), measles (40%), and yellow fever (29%).

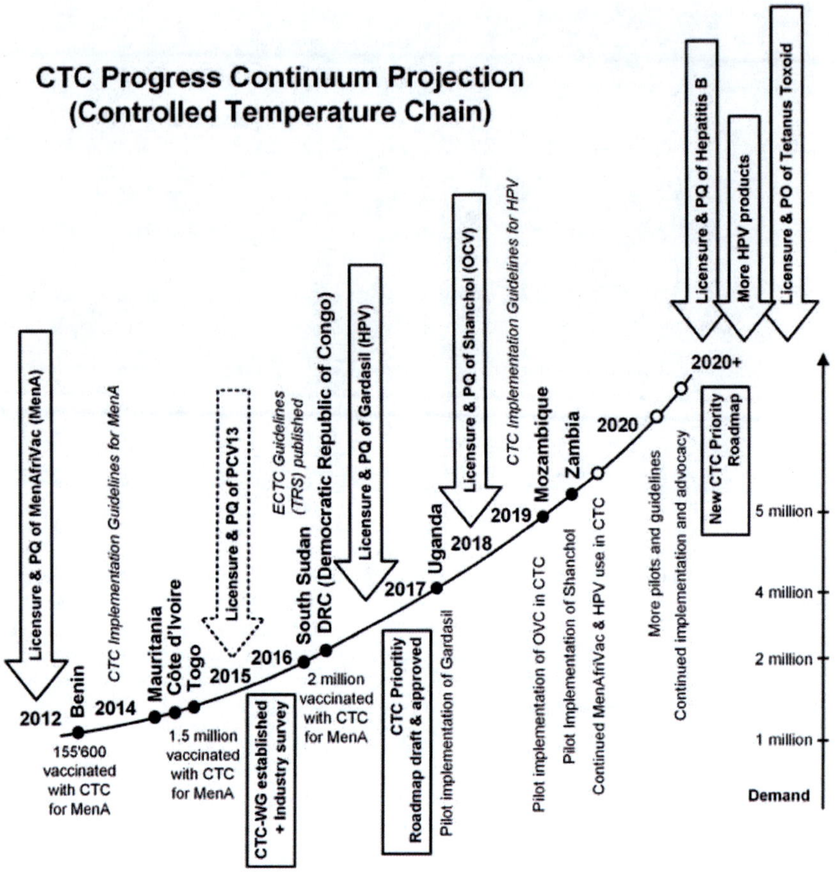

Fig. 2.12 The CTC progress continuum shows a growing pipeline of vaccines qualified for CTC use. *CTC* Controlled Temperature Chain, *ECTC* Extended Controlled Temperature Conditions, *PQ* prequalification (Data adapted from WHO (2018) and Kahn (2019)

From a MenAfriVac™ campaign in Togo in 2014, Mvundura et al. (2017) calculated savings of US$ 0.03–0.107 per dose if the vaccine could have been stored in a CTC from the national store to the health centers. The higher estimates included the savings achieved by not purchasing the additional cold chain equipment for the campaign. The study findings demonstrated that using the CTC approach could especially reduce the costs of conducting campaigns for remote locations without cold chain infrastructure. However, Lydon et al. (2014) also noted that the achievable savings would be smaller in other countries with more efficient cold chain infrastructures. Furthermore, a direct extrapolation of the cost savings to other vaccines was impossible because MenAfriVac™ was a new vaccine launched for a one-off campaign effort in specific geographic zones (WHO 2017). Nevertheless, with the support of GAVI, UNICEF, PATH, WHO, and their partners, along with the commitment of national governments, MenAfriVac™ vaccination campaigns have taken place all over Africa's meningitis belt (WHO 2012). Since then, regulatory and WHO prequalification pathways exist to label vaccines for CTC use (WHO 2015c, 2016a, 2017).

Pfizer's Prevnar 13® (pneumococcal conjugate vaccine, PCV, liquid form) was approved for use at temperatures up to 40 °C for 3 days (WHO 2017). However, this indication was removed in 2016 to allow consistent labeling across Prevnar products (with different presentations and the number of doses per vial). Moreover, the WHO confirmed that this particular product was not a high priority for the CTC as it is typically delivered with other vaccines that still require the traditional cold chain.

In June 2016, Merck's Gardasil®4 (quadrivalent human papillomavirus vaccine, HPV, delivered intramuscularly as liquid suspension) was approved for CTC use at temperatures up to 42 °C for 3 days. Implementation guidelines for using the HPV vaccine in a CTC are available (WHO 2017). In 2017, a successful CTC pilot delivery in Uganda provided a valuable opportunity to document and assess the implementation guidelines for use in a CTC along with lessons learned (WHO 2018).

In January 2018, Shantha Biotechnics (India, a Sanofi company) released Shanchol™ (inactivated oral cholera vaccine, OCV, liquid suspension for oral administration) for use in a CTC at temperatures up to 40 °C for 14 days (WHO 2018). Besides, Euvichol® (OCV, oral liquid suspension) from EuBiologics is currently under review for CTC labeling by the Korean regulatory authority in consultation with the WHO prequalification team. Moreover, the International Vaccine Institute (IVI) recently started working with EuBiologics to obtain a CTC label for Euvichol-Plus®, a vaccine presentation in plastic tubes (Odevall et al. 2018). Dukoral® (OCV) from Valneva Sweden AB is prequalified, but the manufacturer Crucell Sweden AB has yet to seek CTC compatibility (WHO 2017).

Other vaccines such as hepatitis B birth-dose (HepB-BD), tetanus toxoid (TT), and rotavirus vaccines are preparing to seek CTC licensure (Kahn et al. 2017). There exist seven WHO prequalified HepB-BD vaccines (Petit et al. 2017) and six TT vaccines (WHO 2017). However, none of them are licensed for use in a CTC yet. The results of published and manufacturer data from several monovalent HepB-BD vaccines support a recommendation to allow vaccine storage at up to 40 °C for 3 days following CTC (WHO 2016b). In a modeling study, Scott et al. (2018) showed that a CTC strategy for HepB-BD vaccine produced health benefits, improved the timing and coverage of the vaccination, and was cost-saving in the regions of East Asia and Pacific, Latin America and the Caribbean, sub-Saharan Africa, and North Africa and the Middle East.

More recently, Seaman et al. (2020) and (Cui 2020) pointed out that WHO's goal of HepB-BD birth dose coverage of 90% in low-income and middle-income countries (LMICs) (Kaddar et al. 2013) by 2030 is unlikely to be reached under current protocols. Therefore, CTC and compact prefilled auto-disable (CPAD) devices strategies for vaccine delivery are seen as cost-effective solutions to overcome coverage barriers.

Overall, two manufacturers of monovalent HepB-BD are moving forward with stability testing to support an eventual CTC label (WHO 2017). A TT vaccine manufacturer has been discussing the potential for CTC with WHO and partner agencies (WHO 2017). Other vaccine manufacturers are in the process of qualifying their existing and pipeline liquid vaccines for CTC use (VIPS 2020b). Furthermore, dry powder formulations created by other drying methods than freeze-drying are not currently used for commercially available vaccines and are still at the early testing phases of R&D (VIPS 2019). Overall, vaccine manufacturers are encouraged to optimize stability testing and qualify products for CTC use (Patel and Kahn 2018). More vaccines are expected to be licensed for CTC in the coming years.

2.14 Summary

This chapter shows that there are already several thermostable vaccines on the market that are less heat sensitive than others and could be stored outside the cold chain for extended periods without compromising their efficacy. It would be highly recommended to take full advantage of the actual heat stability. Most of the vaccines are administered in liquid form by intramuscular injection. An important finding is that stability efforts should be conducted early in the vaccine development process. There are circumstances in which it is appropriate to reformulate existing, commercially available vaccines to improve stability. An excursus on the current thermostability situation of COVID-19 vaccines revealed a lack of vaccines with thermostability between +25 and +40 °C. Available mRNA vaccines require ultracold storage down to −70 °C. Proposed solutions for an ideal vaccine have been outlined, and dried heat-stable vaccines would provide important innovations for alternative delivery methods than injection. With Controlled Temperature Chain (CTC) qualification and labeling, there is an opportunity to qualify vaccines at higher temperatures, meeting immunization stakeholders' needs. CTC allows for greater flexibility in use, especially in low-income countries.

References

AboulFotouh K, Cui Z, Williams RO (2021) Next-generation COVID-19 vaccines should take efficiency of distribution into consideration. AAPS PharmSciTech 22:126. https://doi.org/10.1208/s12249-021-01974-3

Al-Halifa S, Gauthier L, Arpin D et al (2019) Nanoparticle-based vaccines against respiratory viruses. Front Immunol 10:1–11. https://doi.org/10.3389/fimmu.2019.00022

Arpagaus C (2018) pharmaceutical particle engineering via nano spray drying—process parameters and application examples on the laboratory-scale. Int J Med Nano Res 5. https://doi.org/10.23937/2378-3664.1410026

Arpagaus C (2019) PLA/PLGA nanoparticles prepared by nano spray drying. J Pharm Investig 1–22. https://doi.org/10.1007/s40005-019-00441-3

Arpagaus C, John P, Collenberg A, Rütti D (2017) Nanocapsules formation by nano spray drying. In: Jafari SM (ed) Nanoencapsulation technologies for the food and nutraceutical industries. Elsevier, pp 346–401

Arpagaus C, Collenberg A, Rütti D et al (2018) Nano spray drying for encapsulation of pharmaceuticals. Int J Pharm 546:194–214. https://doi.org/10.1016/j.ijpharm.2018.05.037

BIONITY (2004) Braking the cold chain: a new technology based on processes two billion years old promises to revolutionise the international vaccine programme, 21.10.2004. In: Bionity.-com, News. https://www.bionity.com/en/news/40384/ ©LUMITOS AG

Callaway E (2020) The race for coronavirus vaccines: a graphical guide. Nature 580:576–577. https://doi.org/10.1038/d41586-020-01221-y

Cavalier G (2020) The cold chain: a logistical challenge at the heart of the distribution of the COVID-19 vaccine. https://iifiir.org/en/news, https://iifiir.org/en/news/the-cold-chain-a-logistical-challenge-at-the-heart-of-the-distribution-of-the-covid-19-vaccine

Cavalier G, Fertel C, De Paoli F et al (2021) Cold chain technology brief: vaccines

Chen D, Kristensen D (2009) Opportunities and challenges of developing thermostable vaccines. Expert Rev Vaccines 8:547–557. https://doi.org/10.1586/erv.09.20

Chen D, Kapre S, Goel A et al (2010) Thermostable formulations of a hepatitis B vaccine and a meningitis A polysaccharide conjugate vaccine produced by a spray drying method. Vaccine 28:5093–5099. https://doi.org/10.1016/j.vaccine.2010.04.112

Crommelin DJA, Anchordoquy TJ, Volkin DB et al (2021) Addressing the cold reality of mRNA vaccine stability. J Pharm Sci 110:997–1001. https://doi.org/10.1016/j.xphs.2020.12.006

Cui F (2020) Strategies to increase timely uptake of hepatitis B vaccine birth dose. Lancet Glob Heal 8:e869–e870. https://doi.org/10.1016/S2214-109X(20)30257-6

CureVac (2020a) Clinical trial protocol, COVID-19: a phase 2b/3, Randomized, observer-blinded, placebo-controlled, multicenter clinical study evaluating the efficacy and safety of investigational SARS-CoV-2 mRNA vaccine CVnCoV in adults 18 years of age and older, 11 Nov. 1–152

CureVac (2020b) CureVac's COVID-19 vaccine candidate, CVnCoV, suitable for standard fridge temperature logistics. Press Release, 12 Nov 2020

CureVac (2021) Press Release, 12 Oct 2021, CureVac to Shift Focus of COVID-19 Vaccine Development to Second-Generation mRNA Technology. https://www.curevac.com/en/2021/10/12/curevac-to-shift-focus-of-covid-19-vaccine-development-to-second-generation-mrna-technology/

Das P (2004) Revolutionary vaccine technology breaks the cold chain. Lancet Infect Dis 4:719. https://doi.org/10.1016/S1473-3099(04)01222-8

Das S (2020) ZyCoV-D shot can remain stable at room temperature: Zydus Cadila chairman, 12 Dec 2020. Bus.Stand. https://www.business-standard.com/article/current-affairs/zycov-d-shot-can-remain-stable-at-room-temperature-zydus-cadila-chairman-120121200544_1.html

Dunn A (2020) Pfizer's top scientist tells us the pharma giant is already thinking about a new version of its coronavirus vaccine for 2021 that can overcome one of its biggest limitations, 19 Nov 2020. Bus.Insid. https://www.businessinsider.com/pfizer-is-developing-powder-version-of-its-covid-19-vaccine-2020-11

FDA (2022) Vaccines licensed for use in the United States, Content current as of 31 Jan 2022. https://www.fda.gov/vaccines-blood-biologics/vaccines/vaccines-licensed-use-united-states

GAVI (2020) The vaccine innovation prioritisation strategy. https://www.gavi.org/our-alliance/market-shaping/vaccine-innovation-prioritisation-strategy

Hanson CM, George AM, Sawadogo A, Schreiber B (2017) Is freezing in the vaccine cold chain an ongoing issue? A literature review. Vaccine 35:2127–2133. https://doi.org/10.1016/j.vaccine.2016.09.070

Heida R, Hinrichs WLJ, Frijlink HW (2021) Inhaled vaccine delivery in the combat against respiratory viruses: a 2021 overview of recent developments and implications for COVID-19. Expert Rev Vaccines 00:1–18. https://doi.org/10.1080/14760584.2021.1903878

Ivanova P, Sagdiev R (2020) Exclusive: Russia focuses on freeze-dried vaccine doses as transport fix, 15 Nov 2020. In: Reuters

Jackson T, Peterson G, Naunton M et al (2018) Over half of vaccines are wasted globally for these simple reasons. In: World Econ. Forum. https://www.weforum.org/agenda/2018/07/the-biggest-hurdle-to-universal-vaccination-might-just-be-a-fridge

Jain S, Batra H, Yadav P, Chand S (2020) COVID-19 vaccines currently under preclinical and clinical studies, and associated antiviral immune response. Vaccines 8:649. https://doi.org/10.3390/vaccines8040649

Jusu MO, Glauser G, Seward JF et al (2018) Rapid establishment of a cold chain capacity of—60 °C or colder for the STRIVE ebola vaccine trial during the ebola outbreak in sierra leone. J Infect Dis 217:S48–S55. https://doi.org/10.1093/infdis/jix336

Kaddar M, Schmitt S, Makinen M, Milstien J (2013) Global support for new vaccine implementation in middle-income countries. Vaccine 31:B81–B96. https://doi.org/10.1016/j.vaccine.2012.11.085

Kahn A-L (2019) An update on the WHO work in support of the controlled temperature chain (CTC), September 2019. In: UNICEF Vaccine Ind. Consult. Sept 2019. https://www.unicef.org/supply/media/3216/file/VIC-2019-Session-6-Pres-3-CTCupdate.pdf

Kahn A-L, Kristensen D, Rao R (2017) Extending supply chains and improving immunization coverage and equity through controlled temperature chain use of vaccines. Vaccine 35:2214–2216. https://doi.org/10.1016/j.vaccine.2016.10.091

Kartoglu U, Milstien J (2014) Tools and approaches to ensure quality of vaccines throughout the cold chain. Expert Rev Vaccines 13:843–854. https://doi.org/10.1586/14760584.2014.923761

Kols A, Sherris J (2000) HPV Vaccines: promise and challenges, July 2000. PATH

Kristensen D, Chen D (2010) Stabilization of vaccines: lessons learned. Hum Vaccin 6:227–231. https://doi.org/10.4161/hv.6.3.11618

Kristensen D, Chen D, Cummings R (2011) Vaccine stabilization: research, commercialization, and potential impact. Vaccine 29:7122–7124. https://doi.org/10.1016/j.vaccine.2011.05.070

Kristensen DD, Bartholomew K, Villadiego S, Lorenson K (2016a) What vaccine product attributes do immunization program stakeholders value? Results from interviews in six low- and middle-income countries. Vaccine 34:6236–6242. https://doi.org/10.1016/j.vaccine.2016.10.057

Kristensen DD, Lorenson T, Bartholomew K, Villadiego S (2016b) Can thermostable vaccines help address cold-chain challenges? Results from stakeholder interviews in six low- and middle-income countries. Vaccine 34:899–904. https://doi.org/10.1016/j.vaccine.2016.01.001

Kumru OS, Joshi SB, Smith DE et al (2014) Vaccine instability in the cold chain: mechanisms, analysis and formulation strategies. Biologicals 42:237–259. https://doi.org/10.1016/j.biologicals.2014.05.007

Kunda NK, Wafula D, Tram M et al (2016) A stable live bacterial vaccine. Eur J Pharm Biopharm 103:109–117. https://doi.org/10.1016/j.ejpb.2016.03.027

Kurzątkowski W, Kartoğlu Ü, Staniszewska M et al (2013) Structural damages in adsorbed vaccines affected by freezing. Biologicals 41:71–76. https://doi.org/10.1016/j.biologicals.2011.10.011

Lee BY, Cakouros BE, Assi T-M et al (2012) The impact of making vaccines thermostable in Niger's vaccine supply chain. Vaccine 30:5637–5643. https://doi.org/10.1016/j.vaccine.2012.06.087

Levin A, Levin C, Kristensen D, Matthias D (2007) An economic evaluation of thermostable vaccines in Cambodia, Ghana and Bangladesh. Vaccine 25:6945–6957. https://doi.org/10.1016/j.vaccine.2007.06.065

Levine MM (2011) "IDEAL" vaccines for resource poor settings. Vaccine 29:D116–D125. https://doi.org/10.1016/j.vaccine.2011.11.090

Lloyd J (2000) Technologies for vaccine delivery in the 21st century, WHO/V&B/00.35, December 2000

Logunov DY, Dolzhikova IV, Zubkova OV et al (2020) Safety and immunogenicity of an rAd26 and rAd5 vector-based heterologous prime-boost COVID-19 vaccine in two formulations: two open, non-randomised phase 1/2 studies from Russia. Lancet 396:887–897. https://doi.org/10.1016/S0140-6736(20)31866-3

Lydon P, Zipursky S, Tevi-Benissan C et al (2014) Economic benefits of keeping vaccines at ambient temperature during mass vaccination: the case of meningitis A vaccine in Chad. Bull World Health Organ 92:86–92. https://doi.org/10.2471/BLT.13.123471

Manufacturing Chemist (2010) No need for the cooler, 6 Sept 2010. www.manufacturingchemist.com

Matthias DM, Robertson J, Garrison MM et al (2007) Freezing temperatures in the vaccine cold chain: a systematic literature review. Vaccine 25:3980–3986. https://doi.org/10.1016/j.vaccine.2007.02.052

McGill COVID19 Vaccine Tracker Team (2021) COVID-19 vaccine tracker. Accessed on 5 Jan 2021. https://covid19.trackvaccines.org/types-of-vaccines/

Moderna (2020) Fact sheet for healthcare providers administring vaccine (vaccination providers) emergency use authorization (EUA) of the moderna COVID-19 vaccine to prevent corona virus disease 2019 (COVID-19) in individuals 18 years of age and older, pp 1–5

MSF (2014) Médecins Sans Frontières Access Campaign, Vaccinating Children Beyond The Cold Chain, Issue Brief, 29 April 2019, pp 1–12

Murhekar MV, Dutta S, Kapoor AN et al (2013) Frequent exposure to suboptimal temperatures in vaccine cold-chain system in India: results of temperature monitoring in 10 states. Bull World Health Organ 91:906–913. https://doi.org/10.2471/BLT.13.119974

Mvundura M, Lydon P, Gueye A et al (2017) An economic evaluation of the controlled temperature chain approach for vaccine logistics: evidence from a study conducted during a meningitis A vaccine campaign in Togo. Pan Afr Med J 27:27. https://doi.org/10.11604/pamj.supp.2017.27.3.12087

Odevall L, Hong D, Digilio L et al (2018) The Euvichol story—development and licensure of a safe, effective and affordable oral cholera vaccine through global public private partnerships. Vaccine 36:6606–6614. https://doi.org/10.1016/j.vaccine.2018.09.026

Ohtake S, Martin RA, Yee L et al (2010) Heat-stable measles vaccine produced by spray drying. Vaccine 28:1275–1284. https://doi.org/10.1016/j.vaccine.2009.11.024

Otto BF, Suarnawa IM, Stewart T et al (1999) At-birth immunisation against hepatitis B using a novel pre-filled immunisation device stored outside the cold chain. Vaccine 18:498–502. https://doi.org/10.1016/S0264-410X(99)00242-X

Pardi N, Hogan MJ, Porter FW, Weissman D (2018) mRNA vaccines—a new era in vaccinology. Nat Rev Drug Discov 17:261–279. https://doi.org/10.1038/nrd.2017.243

Patel MK, Kahn AL (2018) Game changing: hepatitis B vaccine in a controlled temperature chain. Lancet Glob Heal 6:e596–e597. https://doi.org/10.1016/S2214-109X(18)30233-X

PATH (2003) Evidence of vaccine freezing in the cold chain, literature review

PATH (2009) Intradermal delivery of vaccines, a review of the literature and the potential for development for use in low- and middle_income countries, 27 Aug 2009. https://www.who.int/immunization/programmes_systems/supply_chain/optimize/Intradermal-delivery-vaccines_report_2009-Sept.pdf

PATH (2012) Summary of stability data for licensed vaccines, 29 Nov 2012, produced by Working in Tandem, Ltd. for PATH

PATH (2020) Beyond the syringe—vaccine formulation, explained, How drops, shots, and powders could make a COVID-19 vaccine available to everyone, 20 Nov 2020. https://www.path.org/articles/beyond-the-syringevaccine-formulation-explained/

PATH/WHO (2012) Fact sheet—optimize—immunization systems and technologies for tomorrow—taking advantage of the true heat stability of vaccines

Petit D, Tevi-Benissan C, Woodring J et al (2017) Countries' interest in a hepatitis B vaccine licensed for the controlled temperature chain; survey results from African and Western Pacific regions. Vaccine 35:6866–6871. https://doi.org/10.1016/j.vaccine.2017.10.025

Pfizer (2021a) EMA approves new storage option for pfizer-biontech vaccine, easing distribution and storage of doses across european union, 26 March 2021. https://www.pfizer.com/news/press-release/press-release-detail/ema-approves-new-storage-option-pfizer-biontech-vaccine

Pfizer (2021b) Study to evaluate the safety, tolerability, and immunogenicity of multiple formulations of BNT162B2 against COVID-19 in healthy adults, ClinicalTrials.gov Identifier: NCT04816669. https://www.pfizer.com/science/find-a-trial/nct04816669

Pfizer-BioNTech (2020) Fact sheet for healthcare providers administring vaccine, emergency use authorization (EUA) of the pfizer-BioNTech COVID-19 Vaccine to prevent coronavirus disease 2019 (COVID-19) in individuals 16 years of age and older, pp 1–6

RDIF (2020) Sputnik V, general information. The Russian Direct Investment Fund (RDIF). https://sputnikvaccine.com/about-vaccine/

Rigby J (2020) How a deep-frozen vaccine reached 400,000 people helped stop Ebola and holds hope for coronavirus, 15 Nov 2020. TheTelegraph. https://www.telegraph.eo.uk/global-health/science-and-disease/deep-frozen-vaccine-reached-400000-people-helped-stop-ebola/

Scott N, Palmer A, Morgan C et al (2018) Cost-effectiveness of the controlled temperature chain for the hepatitis B virus birth dose vaccine in various global settings: a modelling study. Lancet Glob Heal 6:e659–e667. https://doi.org/10.1016/S2214-109X(18)30219-5

Seaman CP, Morgan C, Howell J et al (2020) Use of controlled temperature chain and compact prefilled auto-disable devices to reach 2030 hepatitis B birth dose vaccination targets in LMICs: a modelling and cost-optimisation study. Lancet Glob Heal 8:e931–e941. https://doi.org/10.1016/S2214-109X(20)30231-X

Shank-Retzlaff ML, Zhao Q, Anderson C et al (2006) Evaluation of the thermal stability of Gardasil®. Hum Vaccin 2:147–154. https://doi.org/10.4161/hv.2.4.2989

Taher M, Shaari SS, Susanti D (2021) Potential nanospray inhalation of remdesivir and hydroxychloroquine using poly (lactic-co-glycolic) acid as fast delivery for Covid-19 treatment. J Pharm 1:34–44. https://doi.org/10.31436/jop.v1i1.50

UNICEF (2019) Supply annual report 2019, pp 1–65

UNICEF (2021) COVID-19 vaccine market dashboard. https://www.unicef.org/supply/covid-19-vaccine-market-dashboard

Uwagbale E-E (2020) A successful Ebola vaccine delivery shows how a Covid-19 vaccine would work in Africa, 23 Nov 2020. QUARTZ AFRICA. https://qz.com/africa/1936803/ebola-vaccine-delivery-shows-how-covid-vaccine-can-work-in-africa/

VIPS (2019) VIPS Phase I executive summary: heat-stable/controlled temperature chain (CTC) qualified dry formulations (vaccine innovation prioritisation strategy), pp 1–7

VIPS (2020a) The VIPS prioritisation process: methodology and outcomes (vaccine innovation prioritisation strategy), pp 1–24

VIPS (2020b) VIPS phase II executive summary: heat stable/controlled temperature chain (CTC) qualified liquid formulations (vaccine innovation prioritisation strategy), pp 1–16

Wang J, Peng Y, Xu H et al (2020) The COVID-19 vaccine race: challenges and opportunities in vaccine formulation. AAPS PharmSciTech 21:225. https://doi.org/10.1208/s12249-020-01744-7

WHO (2002) Getting started with vaccine vial monitors. Tech. Sess. Vaccine Vial Monit. 27 March 2002, Geneva, pp 1–17

WHO (2005) Monitoring vaccine wastage at country level, Guidelines for programme managers, WHO/V&B/03.18.Rev.1

WHO (2006a) Guidelines on stability evaluation of vaccines, WHO/BS/0.6.2049

WHO (2006b) Temperature sensitivity of vaccines, WHO/IVB/06.10, August 2006. World Heal Organ Dep Immunization, Vaccines Biol 1–58. WHO/IVB/06.10

WHO (2012) Revolutionary meningitis vaccine breaks another barrier; first to gain approval to travel outside cold chain. Newsroom, 14 Nov 2012, Atalanta. https://www.who.int/immunization/newsroom/menafrivac_20121114/en/

WHO (2013) Use of MenAfriVac™ (meningitis A vaccine) in a controlled temperature chain (CTC) during campaigns, WHO/IVB/13.04

WHO (2015a) Module 2: the vaccine cold chain. WHO 1–46

WHO (2015b) How to monitor temperatures in the vaccine supply chain, WHO/IVB/15.04. WHO Vaccine Manag. Handbook, Modul. VMH-E2-01.1 1–44

WHO (2015c) Assessing the programmatic suitability of vaccine candidates for WHO prequalification (Revision 2014), WHO/IVB/14.10, January 2015

WHO (2016a) Annex 5 guidelines on the stability evaluation of vaccines for use under extended controlled temperature conditions. WHO Tech Rep Ser No 999, 2016. WHO/BS/2015.2268

WHO (2016b) A systematic review of monovalent hepatitis B vaccine thermostability

WHO (2017) Controlled Temperature Chain: Strategic Roadmap for Priority Vaccines 2017–2020, WHO/IVB/17.20 (Orginally distributed as WHO/IVB/17.17)

WHO (2018) Global Vaccine Action Plan, Monitoring, Evaluation & Accountability, Secretariat Annual Report 2018

WHO (2020) Controlled temperature chain page website. http://www.who.int/immunization/programmes_systems/supply_chain/ctc/en

WHO (2021a) Vaccine safety basics e-learing course, module 2: types of vaccines and adverse reactions, route of administration. https://vaccine-safety-training.org/route-of-administration.html

WHO (2021b) WHO prequalified vaccines, online database: https://extranet.who.int/gavi/PQ_Web. Accessed on 20 March 2021. https://extranet.who.int/gavi/PQ_Web

WHO (2021c) Draft landscape of COVID-19 candidate vaccines, 24 Dec 2021. https://www.who.int/publications/m/item/draft-landscape-of-covid-19-candidate-vaccines

WHO (2021d) WHO coronavirus disease (COVID-19) dashboard. Accessed on 13 June 2021. https://covid19.who.int/

WHO (2021e) Therapeutics and COVID-19, living guideline, 31 March 2021e, pp 1–60

WHO/PATH (2013) Delivering MenAfriVAC using the controlled temperature chain approach, project optimize

WHO/PATH (2014) Temperature sensitivity of vaccines, March 2014

Wouters OJ, Shadlen KC, Salcher-Konrad M et al (2021) Challenges in ensuring global access to COVID-19 vaccines: production, affordability, allocation, and deployment. Lancet 397:1023–1034. https://doi.org/10.1016/S0140-6736(21)00306-8

Wu F, Zhao S, Yu B et al (2020) A new coronavirus associated with human respiratory disease in China. Nature 579:265–269. https://doi.org/10.1038/s41586-020-2008-3

Zaffran M, Vandelaer J, Kristensen D et al (2013) The imperative for stronger vaccine supply and logistics systems. Vaccine 31:B73–B80. https://doi.org/10.1016/j.vaccine.2012.11.036

Zhu C, Shoji Y, McCray S et al (2014) Stabilization of HAC1 influenza vaccine by spray drying: formulation development and process scale-up. Pharm Res 31:3006–3018. https://doi.org/10.1007/s11095-014-1394-3

Zimmer C, Corum J, Sui-Lee W (2021) Coronavirus vaccine tracker. Accessed on 5 Jan 2021. New York Times. https://www.nytimes.com/interactive/2020/science/coronavirus-vaccine-tracker.html

Zipursky S, Djingarey MH, Lodjo J-C et al (2014) Benefits of using vaccines out of the cold chain: delivering meningitis A vaccine in a controlled temperature chain during the mass immunization campaign in Benin. Vaccine 32:1431–1435. https://doi.org/10.1016/j.vaccine.2014.01.038

Drying Technologies for Vaccines

3

Abstract

This chapter begins with liquid and solid vaccine formulation approaches and discusses the advantages and limitations. This is followed by a description of drying mechanisms and an overview of the drying technologies most commonly used in the pharmaceutical industry to stabilize vaccines. Next, traditional freeze drying (lyophilization), spray drying, and emerging alternative drying technologies such as spray freeze drying, supercritical fluid drying, and vacuum foam drying are presented, and their respective advantages and disadvantages are compared. Then, an excursus presents newer ways of delivering dried vaccines, such as air-drying vaccines in a sponge, on membrane fibers, microarray patches, or other substrates. A major focus is on the principle of spray drying based on laboratory-scale experience. After that, the chapter is devoted to the question of when spray drying is a suitable alternative to other drying processes for stabilizing vaccines. Numerous case studies and application examples support the chapter for stabilizing vaccines. Facts and figures highlight the main advantages of spray drying for vaccine stabilization. Finally, the main challenges are critically reviewed and discussed. Throughout the chapter, the reader will find relevant literature references that provide further details on the drying process in question.

3.1 Considerations for Dry Powder and Liquid Vaccine Formulations

In the last 20 years, there has been a significant increase in reports of improved stability of dry powder vaccine formulations. Figure 3.1 shows an analysis of the pharmaceutical form of WHO prequalified vaccines based on the WHO database (WHO 2021). As of 17 October 2021, 259 presentations of prequalified vaccines

C. Arpagaus, *Spray Drying of Vaccines*,
https://doi.org/10.1007/978-3-031-24323-3_3

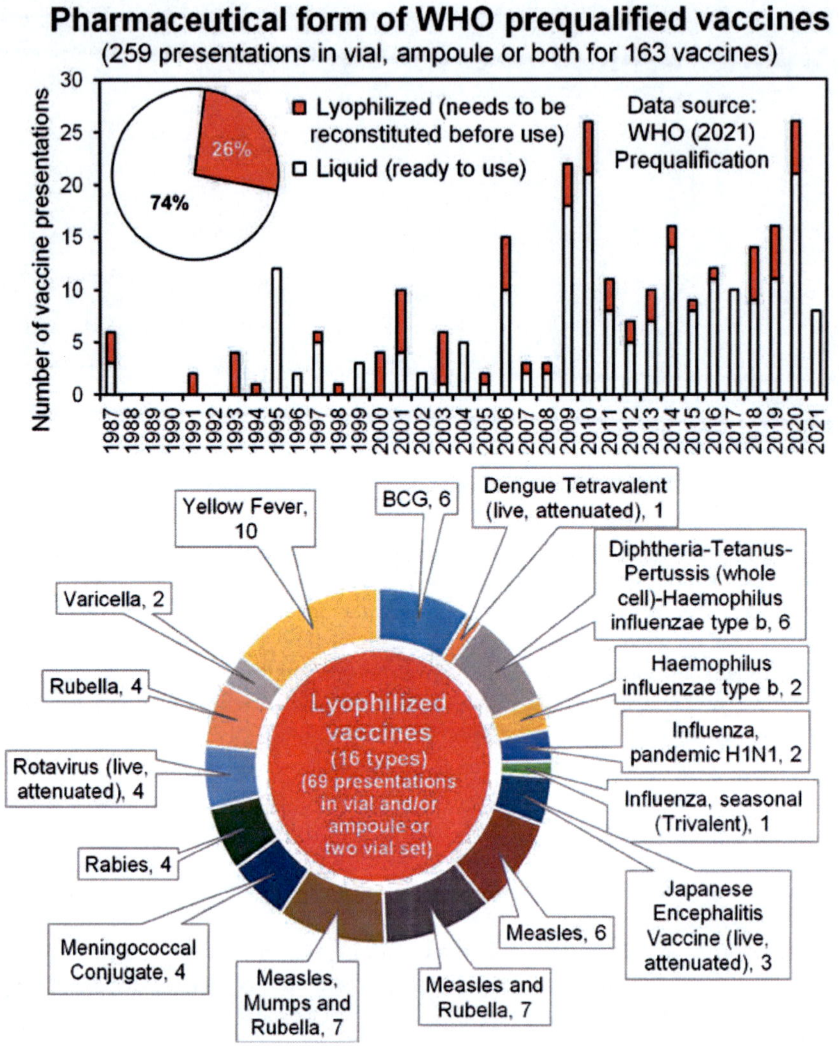

Fig. 3.1 The pharmaceutical form of WHO prequalified vaccines. 26% (69 of 259) is lyophilized and 74% (190 of 252) in liquid form (ready to use) (Data source: WHO (2021), https://extranet.who.int/gavi/PQ_Web, accessed on October 17, 2021)

were listed in the database, of which 74% of the vaccine presentations were in liquid (ready-to-use) and 26% in a lyophilized form (in vials and/or ampoules or two sets of vials) (shown in red).

These 69 lyophilized vaccine presentations cover 16 vaccine types, including (in alphabetic order):

- Bacille Calmette-Guérin (BCG, 6 vaccines),
- Dengue treatment (DT, 1),
- Diphtheria-tetanus-pertussis (DTP, 6),
- Haemophilus influenzae (Hib, 2),
- Influenza (2 pandemics, 1 seasonal),
- Japanese encephalitis (JE, 3),
- Measles (6),
- Measles and rubella (7),
- Measles-mumps-rubella (MMR, 7),
- Meningococcal (Men, 4),
- Rabies (4),
- Rotavirus (4),
- Rubella (4),
- Varicella (2),
- Yellow fever (YF, 10).

Thus, freeze drying has already advanced as the gold standard for vaccine drying that has found its place and to which alternative drying methods for vaccines have to be compared. Since the vials can be sealed in a freeze-drying chamber before removal, it is a reliable aseptic drying process. Before injection to the patient, the lyophilized vaccines have to be reconstituted with excipient diluent (62 vaccine presentations) or liquid active components (7).

To date, no company has launched any spray-dried vaccine on the market, according to the WHO prequalified vaccines database (WHO 2021). On the one hand, spray drying is still considered an emerging technology in the field of vaccines and is thus in the early stages of commercialization, and on the other hand, there are potential barriers to the market introduction of spray-dried vaccines that are elaborated in this book.

As shown in Fig. 3.1, refrigerated liquid vaccine formulations are preferred today over lyophilized formulations, which require manual reconstitution. The reasons for liquid dosage forms of vaccines are mainly the relatively inexpensive and straightforward production, packaging, and fairly straightforward administration to patients through a syringe. However, dry powder vaccines have several potential benefits over liquid vaccines, with one of the most important being the omission of the cold chain, as storage at room temperature or higher is possible.

Table 3.1 summarizes the main reasons for choosing dried over the conventional liquid formulation according to benefits and limitations (Tlaxca et al. 2015; Geng 2018; Bhambhani and Antochshuk 2020).

The thermostability of live-attenuated virus vaccines in liquid form is strongly influenced by pH, ionic strength, osmolarity, and the presence of excipients (Brandau et al. 2003). Compared to dry formulations, aqueous vaccine formulations have a higher risk of physical instability such as denaturation, aggregation, and precipitation (Kunda et al. 2016).

Table 3.1 Potential benefits and limitations of liquid versus dried vaccine products

Formulation	Benefits	Limitations
Liquid	• Ease of manufacturing • Convenient to end-users • Lower cost of goods compared to dried formulation • Established for vaccine administration	• More prone to degradation reactions during storage, agitation, and shear • May suffer from material compatibility issues • A cold chain for storage is necessary
Dried	• Improved heat stability by water removal due to lower mobility and slower degradation kinetics • Less susceptible to stresses that occur after exposure to freezing temperatures • Improved handling, storage, transport, and distribution (storage at room temperature, reduced weight, less dependent on a cold chain) • Potential for stockpiling at room temperature in the event of pandemics • Allows different routes of administration (e.g., pulmonary, intranasal, oral, intradermal)	• Loss of potency during processing, such as by heating, evaporation, vacuum, or freezing with corresponding potential stresses • Requires reconstitution at the patient location if administered by injection • Needs capital and energy-intensive drying equipment • Formulation and new excipients are needed

Summarized from Tlaxca et al. (2015), Geng (2018), Bhambhani and Antochshuk (2020)

Figure 3.2 shows the influence of the pH on the loss of activity for several live-attenuated virus vaccines. It is clear that in most cases, the loss of activity decreases between pH 6 and 8, which is favorable because injectables in this range must be formulated for patient comfort (Brandau et al. 2003).

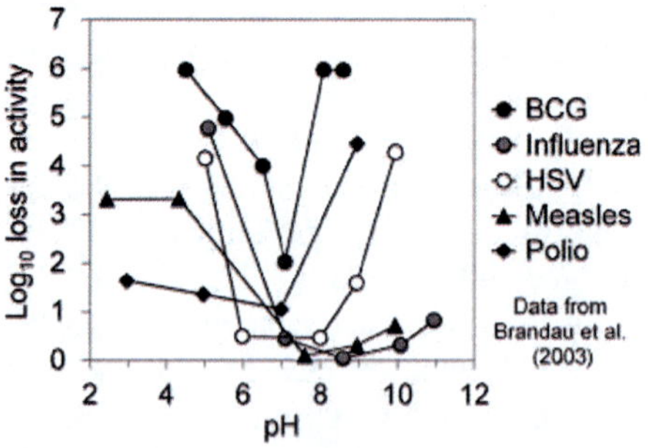

Fig. 3.2 Effect of pH on the loss of activity for several live-attenuated vaccines (Data adapted from Brandau et al. (2003) with permission from Elsevier)

One strategy implemented is to freeze the liquid at very low temperatures (below the glass transition temperature, T_g) to maintain longevity in the frozen state and ease administration as a thawed liquid (Tlaxca et al. 2015). This approach is also evident in the current COVID-19 mRNA vaccines made by Pfizer-BioNTech (2020), Moderna (2020), and CureVac (2020a, b), which require ultracold storage (Cavalier 2020; Crommelin et al. 2021). Although such a strategy can be useful, it presents a more significant challenge in cold chain management. A robust cold chain is necessary to ensure that effective vaccines reach the patients (Cavalier et al. 2021).

If a vaccine is not stable due to pH adjustment and formulation excipients in liquid form, developing a vaccine in dry powder form is an option (Rosini et al. 2011). The choice of drying method will be associated mainly with the economics of drying and the intended route of administration. Drying vaccines can be done in many ways (see Fig. 3.3), such as lyophilization, spray drying, spray freeze drying, foam drying, supercritical fluid drying, air drying, and other approaches. For example, spray-dried powders enable administration via inhalation since the desired control over particle size can be achieved, while freeze drying is commonly applied for injectable products, as sterility and particle-free quality attributes are easier obtained (Abdul-Fattah et al. 2007a).

The removal of water offers advantages such as ease of handling and storage, reduction in refrigeration and transportation costs, and improved stability (Ohtake et al. 2020). In addition, formulating vaccines into a dry form increases their thermal stability due to lower mobility and slower temperature-induced degradation kinetics than the liquid phase (Clapp et al. 2011; Kunda et al. 2016). Besides, a dry solid formulation is more temperature stable and can be stored at higher temperatures, such as room temperature, and is less dependent on the cold chain (Schwartzbach 2011). On the other hand, the introduction of only a few dried

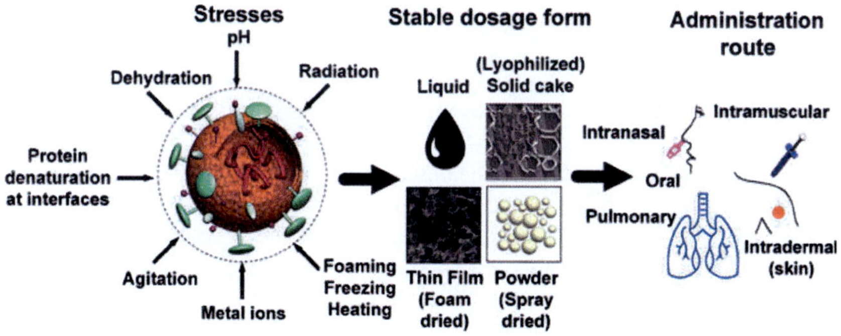

Fig. 3.3 Live-attenuated and inactivated viral vaccines are subjected to various stresses during manufacturing. The typical stabilized dosage forms are liquid, thin film (foam-dried), solid cake (lyophilized), or powder (spray-dried). Besides intramuscular, this opens up new routes of administration such as pulmonary, intranasal, oral, or intradermal (skin) (Adapted from Tlaxca et al. (2015) with permission from Elsevier)

thermostable vaccines is likely to have a relatively negligible impact on the cost of the entire cold chain (Nicholls 2004). However, it would simplify vaccine distribution and impact vaccine availability, especially in remote locations.

Dry powder formulations also enable alternative routes of administration, as shown in Fig. 3.3, including inhalation, intranasal, oral, and intradermal routes. Furthermore, long-term storable dry powder vaccines have great potential for room temperature stockpiling in case of pandemics (e.g., COVID-19) and facilitate mass vaccination programs in areas that cannot be reached without refrigeration (i.e., vaccine distribution to all). New dry vaccine formulations that do not require reconstitution include, for example, microarray patches or spray-dried powders for inhalation and could be preferred over liquid formulations in the future (see Sect. 3.9).

However, there are several limitations to the use of dried vaccine formulations. Most drying methods require either heating or freezing (e.g., ice crystals during lyophilization), which can be problematic for vaccine components. Vaccine degradation kinetics follow the Arrhenius equation as a function of temperature and activation energy, e.g., during storage of spray-dried formulations (Schaefer and Lee 2015; Lovalenti et al. 2016).

Other stresses also include protein denaturation at interfaces, e.g., from foaming or freezing and collapsing during vacuum foam drying, or shear stresses during atomization or agitation, as well as chemical reactions with metal ions or radiation (Tlaxca et al. 2015).

In addition, all freeze-dried vaccine formulations currently in use require reconstitution with a liquid diluent before administration where the factors described above, such as pH, ionic strength, adjuvant stability, etc., become destabilizing again. Depending on the vaccine type, a maximum time interval must be respected between reconstitution and use, ranging from immediate use to 24 h, as indicated in the package insert (IAC 2018).

The following sections will discuss the formulation of dry powder vaccines and the basic concepts of leading drying technologies.

3.2 Dry Powder Vaccine Stabilization Mechanisms

The physical and chemical stabilization of vaccines is fundamental for the long-term storage and maintenance of immunogenicity and efficacy (Gribbon et al. 1996). Drying in a glassy matrix enables the production of thermostable vaccines stored and transported to different climatic zones (e.g., from icy to tropical climates). In addition, sugar-glass-stabilized vaccines are inert and rapidly soluble (Lloyd 2000). It is well documented that sugars and polyols in formulations protect proteins from degradation during drying and storage. The mechanisms of sugar glass drying are well understood for reproducing the process over a wide range of vaccine applications. Sugar-glass drying and stabilization technology is used in numerous pharmaceutical and biological products and many recombinant, bacterial and viral vaccines and with different drying methods.

In general, two accepted hypotheses have been proposed to explain the stabilization of proteins by sugar glasses and excipients in the solid state. One is the water replacement hypothesis describing a thermodynamic stabilization mechanism, and the other is the glass dynamics hypothesis or vitrification, which is based on kinetic consideration. Both theories have been known for more than 40 years and have been discussed extensively in the literature.

3.2.1 Water Replacement Hypothesis

The water replacement hypothesis describes the stabilization of proteins (e.g., biopharmaceuticals, vaccines) from a thermodynamic perspective (Crowe et al. 1985, 1996, 1998b; Carpenter and Crowe 1989; Slade et al. 1991; Carpenter et al. 1994, 2002; Allison et al. 1999). Protein-bound water removed during the final stages of drying contributes to thermodynamic destabilization and loss of activity. When excipients (e.g., sugar molecules, polyols) are present, they can form hydrogen bonds with specific sites on the surface of the proteins, resulting in the preservation of the native structure of the protein. In this theory, the stabilizer preserves the native protein conformation by maintaining the high free energy of unfolding. In other words, the idea is that sugar molecules replace the water with hydrogen bonds, stabilizing the protein structure in the absence of water.

Figure 3.4 illustrates a possible stabilization process according to the water replacement theory. It shows how spray-dried live-attenuated Listeria monocytogenes with non-reducing sugars, such as trehalose, mannitol, and myo-inositol, can stabilize the bacterial cell membrane (Kunda et al. 2016). It is speculated that the phosphate groups in the lipid bilayer (or peptidoglycan layer) of the bacterial membrane form multiple external hydrogen bonds with the hydroxyl groups on the sugar molecules. These hydrogen bonds help maintain the three-dimensional structure of the bacterial cell membrane and thus provide protection for Listeria during spray drying and storage. In addition to the water exchange theory, the glassy formation of sugars in the dry state limits the molecular mobility of these microorganisms, preventing aggregation and increasing stability.

3.2.2 Glass Dynamics or Vitrification Hypothesis

The glass dynamics or vitrification hypothesis proposes that stabilization results from suppression of the molecular mobility of protein molecules (e.g., biopharmaceuticals, vaccines) by the formation of a rigid, glassy matrix (Green and Angell 1989; Cicerone et al. 2006; Yoshioka and Aso 2007; Abdul-Fattah et al. 2007a; Chang and Pikal 2009; Cicerone and Douglas 2012; Cicerone et al. 2015).

This reduced molecular mobility is reflected as α- and β-relaxations of the glassy matrix. α-relaxations, also called global molecular motions, occur mainly due to the

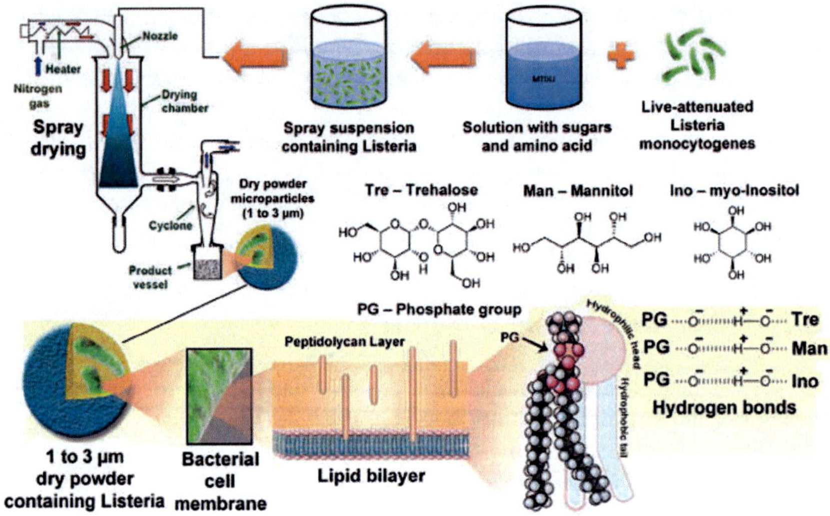

Fig. 3.4 Spray drying of live-attenuated listeria monocytogenes with non-reducing sugars confers structural stability to the bacterial membrane. The phosphate groups in the lipid bilayer (or peptidoglycan layer) form multiple external hydrogen bonds with the hydroxyl groups present on the sugar molecules. These hydrogen bonds stabilize the cell membrane and thus provide protection for Listeria during spray drying and storage (Adapted from Kunda et al. (2016) with permission from Elsevier)

entire molecule's slow translational and rotational motions on a second-to-month scale and strongly influence the diffusion of reactive molecular species. Furthermore, the rotations correlate with the viscosity of the whole system and are thus related to the glass transition.

In contrast, β-relaxations are local molecular motions in spatial proximity that influence motions at a time scale of picoseconds. Suppressed β-relaxations in the matrix by the introduction of certain small-molecule plasticizers (e.g., glycerol, DMSO) to the sugar glasses can enhance stability by acting as shock absorbers and are supposed to slow down the transport of reactive species and thus degradation (Cicerone et al. 2006, 2015; Cicerone and Douglas 2012).

Evidence from the literature suggests that chemical degradation and structural relaxation require some type of motion in the solid-state to correlate storage stability and structural relaxation (Yoshioka and Aso 2007; Abdul-Fattah et al. 2007a; Grasmeijer et al. 2013; Cicerone et al. 2015). Since thermal history affects global relaxation, the drying process impacts stability. For an amorphous sugar, molecular mobility can generally be determined from the difference between measured glass transition temperature and storage temperature according to the Williams-Landel-Ferry relationship (Williams et al. 1955). A rule of thumb is that the T_g should be at least 50 °C above the storage temperature for the translational molecular motion to become insignificant for a typical pharmaceutical product (Hancock et al. 1995).

Figure 3.5 shows an example of vitrification and water replacement's contribution to the stability of the spray-dried protein (alkaline phosphatase) using trehalose and inulin at different T_g and thus different translational molecular motions (Grasmeijer et al. 2013). The researchers measured the enzymatic activity immediately after spray drying and after 12 days of storage at 60 °C. The activity loss became negligible at T_g above 45 °C (close to the spray dryer outlet temperature of about 60 °C). After 12 days of storage at 60 °C, the stability curve shifted to the right, indicating that a higher T_g is required for protein stabilization during storage. Trehalose appeared to stabilize the protein at a slightly lower T_g than inulin, attributed to the larger molecular size of inulin (3908 g/mol vs. 342 g/mol). In general, the results showed that when T_g is below the storage temperature, protein stabilization is dominated by the vitrification mechanism. However, when the T_g is raised significantly above the storage temperature (e.g., +10 to 20 °C), the protein becomes immobile, and the water replacement mechanism becomes dominant for protein stabilization. The example shows that both theories play a role simultaneously and have their justification, but also that neither can fully explain protein stabilization and leaves room for further refinements.

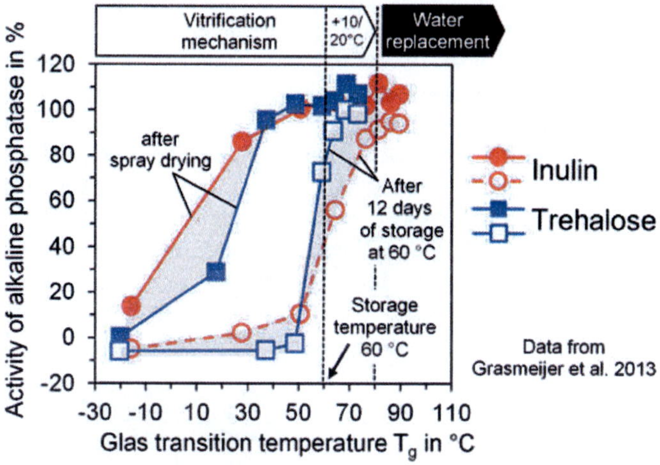

Fig. 3.5 Activity of alkaline phosphatase (protein) incorporated in inulin and trehalose after spray drying and after 12 days of storage at 60 °C. Spray drying conditions: Mini Spray Dryer B-290, T_{in}: 100 °C, T_{out}: 58 to 64 °C, aspirator: 100%, pump speed: 7%, atomizing air: 50 mm). When T_g is below the storage temperature, protein stabilization is dominated by the vitrification mechanism. When T_g is increased above the storage temperature (e.g., +10 to 20 °C), the protein becomes immobile, and the water replacement mechanism becomes dominant (Data reproduced from Grasmeijer et al. (2013) with permission from Elsevier)

3.2.3 Excursus: Sugar Stabilization of the Resurrection Plant and Tardigrades

A scientifically inaccurate but good way to illustrate glass stabilization is preserving insects in amber glass (Fig. 3.6). Thus, their structure is preserved over millions of years. Similarly, tiny vaccine molecules are immobilized in soluble sugar glass (Nicholls 2004), and as soon as these particles come into contact with water (i.e., when reconstituted by diluents for injection or by human body fluids), the sugar glass dissolves, and the vaccines are released.

The evidence of sugars as stabilizers of biological materials comes from cryptobionts, natural organisms that exhibit cryptobiosis, a metabolic state of life in response to adverse environmental conditions such as desiccation, freezing, and oxygen deprivation.

Anhydrobiosis is the best-studied form of cryptobiosis and occurs in situations of extreme desiccation (Crowe et al. 1992, 1998a, b). The term anhydrobiosis comes from the Greek for "life without water," a very ancient process that evolved over 100 million years (Gribbon et al. 1996).

Cryptobionts can completely dry out and regain full metabolic activity when re-exposed to water (Lloyd 2000). These organisms include baker's yeast, invertebrates such as rotifers, tardigrades, brine shrimp, worm-like nematodes, and even larger organisms such as the resurrection plant, which can survive near complete desiccation during dry periods (Weisburd 1988; Crowe et al. 1992). In addition, some plants, animals, and bacteria can survive in a desiccated form for hundreds of years (AskNature 2017).

Figure 3.8a shows a desert resurrection plant that has developed an amazing ability to survive years of drought by replacing the water within the cells with a sugar solution containing about 12.5 wt% dry weight of trehalose and 1.5 wt% sucrose (Richards et al. 2002). Once rehydration occurs, the organism returns to life from a death-like state within a few hours and regains full metabolic activity. Thus, the cycle of desiccation and rehydration can repeatedly occur without harming the

Fig. 3.6 **a** Not scientifically accurate, but an excellent visual illustration of glass stabilization are insects entrapped in amber that have maintained their structure over millions of years (Idea of illustration based on Roser (2006)). **b** Similarly, vaccines are immobilized in sugar glass and rendered temperature stable (Image from Nicholls (2004) with permission from Elsevier)

plant. In addition, when dehydrated, the plant can withstand environmental extremes, including temperatures up to 100 °C, without damage (Roser 1991).

Figure 3.7 illustrates the mechanism of sugar stabilization of phospholipid bilayers in liposomes and cell membranes (Crowe et al. 1985, 1992, 1998a, b). The polar head groups of phospholipids are hydrated, and water molecules spatially separate the head groups.

When the water molecules are removed by dehydration, the packing density of the head groups increases, and the bilayer dries to a gel, which may lead to disturbance of the vesicle (Amorij et al. 2008). According to the water replacement hypothesis, the sugar can stabilize the dry lipid bilayer in trehalose or sucrose and inhibit the phase transition during dehydration. Carbohydrates containing hydroxyl groups can take over the hydrogen bonds of the water and thus squeeze between the lipid head groups. This situation leads to more space between the lipid molecules and a reduction in van der Waals forces resulting in lower phase transition temperatures. Thus, at temperatures where the bilayer would typically be in the gel phase, it is in the liquid crystalline phase and does not undergo a phase transition during rehydration. The structure of the cell membrane is preserved in its dried state.

Another example of cryptobionts is tardigrades, also known as "water bears" (Fig. 3.8b), which live on plants such as the resurrection plant and dry out together with their hosts and revive as soon as they come into contact with water again (Roser 2006). Under normal conditions, a tardigrade that is less than one millimeter long contains 85 wt% water but can survive even with less than 2 wt% of its body weight (Weisburd 1988). However, in the drought-induced dormant phase, the tardigrades harden and can survive extreme heat, liquid nitrogen, strong vacuums, high doses of ionizing radiation, and other damaging conditions of natural and laboratory environments (Weisburd 1988).

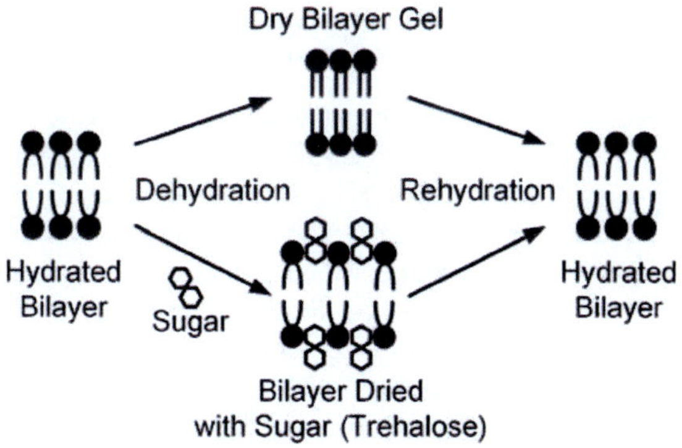

Fig. 3.7 Schematic illustration of the proposed mechanism for stabilizing phospholipid bilayers, liposomes, and membranes of intact cells by disaccharide sugar during dehydration and rehydration. The sugar inhibits the fusion during dehydration (Adapted from Crowe et al. (1985, 1992, 1998a, b) with permission from Elsevier)

Fig. 3.8 Resurrection plant and tardigrades: **a** The resurrection plant can repeatedly cycle between a dry and a hydrated state. Examples have been kept for over 100 years in a museum cupboard in Arizona in dry condition (left) yet revived completely on thorough wetting (right). Their tissues contain large quantities of the inert glass-forming disaccharide trehalose (Picture adapted from Roser (1991, 2006), with permission from Elsevier). **b** SEM images of a tardigrade in the tun (dry) and active (rehydrated) state (Photos from Photo Science Library). Tardigrades ("water bears") live on plants like the resurrection plant, dry out with their hosts, and revive. Tardigrades share the ability to survive extreme heat, ionizing radiation, vacuum, and liquid nitrogen temperatures for decades in the dry form Weisburd (1988), Roser (2006)

The common feature of the biochemistry of these organisms is the accumulation of large amounts of sugars (e.g., trehalose and sucrose) in their cells for desiccation protection at low water content. The sugars essentially replace the hydrogen bonds that would normally be formed by water. The other key mechanism of desiccation protection is creating aqueous glass, which embeds cellular components in a non-crystalline solid matrix and slows down degrading chemical reactions.

The remarkable ability of trehalose to protect cryptobionts from desiccation damage is used on an industrial scale for drying numerous products. The knowledge gained from nature on stabilizing biological materials is well transferred to industrial drying processes. For example, trehalose is added to foods before air drying to protect against denaturation and prevent the loss of aromatic volatiles that give fresh foods their characteristic aromas and flavors (Roser 1991; Colaço and Roser 1994).

As an interesting final comment, trehalose derives its name from "Trehala manna," the edible cocoons of a parasitic beetle of the genus Larinus (Berthelot 1859; Tillequin 2009). This pupa is found on aromatic thorn bushes in the Middle East and contains large amounts of trehalose and protein. It is considered to be the biblical "Manna from Heaven" mentioned in the Old Testament that nutritionally sustained the Israelites in the desert during their journey to the promised land (Colaço and Roser 1994). Today, trehalose is commercially produced from starch by an enzymatic process.

3.2.4 Excipients and Glass Transition Temperature

The selection of suitable glass-forming excipients is a prerequisite for the long-term stability of dried vaccines without cooling. The excipient should have a low number of reducing groups, a high T_g, and a low tendency to crystallize, i.e., form a glassy (amorphous) matrix (Saluja et al. 2010; Sou et al. 2016; Vehring et al. 2020). In general, most excipients probably stabilize by vitrification, and some such as trehalose or inulin also can replace water, although inulin to a lesser extent due to the size of the molecule (Grasmeijer et al. 2013).

For binary amorphous mixture, the T_g can be described by the Gordon-Taylor equation (Gordon and Taylor 1952) (Eq. 3.1)

$$T_g = \left(w_1 T_{g1} + k\, w_2 T_{g2}\right) / \left(w_1 + k\, w_2\right) \tag{3.1}$$

where w_1 and w_2 are the weight fractions and T_{g1} and T_{g2} the glass transition temperatures of the sugar solute and water, respectively (T_g of water is −135 °C) (Roos and Karel 1991; Chen et al. 2000b). The constant k needs to be determined experimentally for different water contents and T_g's.

Moreover, Fig. 3.9 shows the calculated values of k for various carbohydrates as a function of their anhydrous T_g (Roos 1993; Crowe et al. 1996). The points lie on a linear continuum, with trehalose at the top end. Crowe et al. (1996) have empirically determined a k-value of 7.5 ± 0.4 for trehalose, which is close to the estimated value of 7.35 from the retrofitted line.

For determining the T_g of a multi-component formulation, it is referred for further literature (Roos and Karel 1991; Roos 1993; Hancock and Zografi 1994, 1997; Bhandari et al. 1997; Chen et al. 2000b; Katkov and Levine 2004). Experimentally, the glass transition temperature (T_g) of dried products is typically measured by temperature-modulated differential scanning calorimetry (DSC) and is defined as the mid-point deflection of reversing heat flow vs. temperature curve (Jin et al. 2010; Saluja et al. 2010; Sou et al. 2016).

Table 3.2 lists the T_g of some glass-forming excipients studied for the spray drying of vaccine formulations. The list includes non-reducing disaccharides such as sucrose, trehalose, polysaccharides such as inulin and dextran, the polyol mannitol, and synthetic polymer PVP that offers an amorphous matrix with a high T_g. Figure 3.10 illustrates the molecular structures.

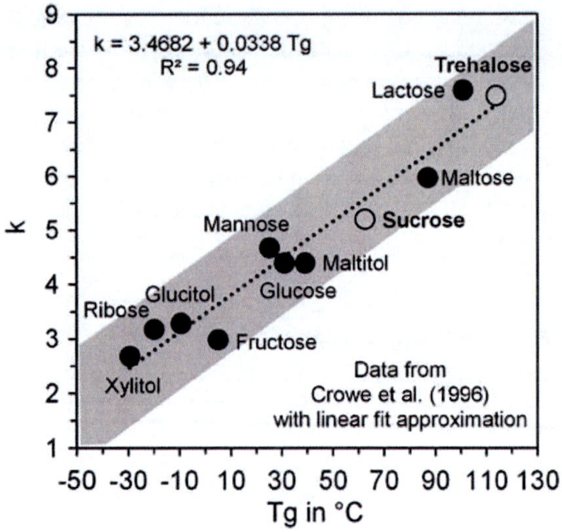

Fig. 3.9 Values of the constant *k* of the Gordon-Taylor equation (Eq. 3.1) for different sugars. For trehalose, the *k*-value is 7.35, for sucrose 5.2. The solid line is a linear regression of the data points (k = 3.4682 + 0.0338 T_g, R^2 = 0.94). The grey area represents a 95% confidence interval (Data reproduced from Crowe et al. (1996) with permission from Elsevier)

A general trend of increasing T_g with the molecular size is observed (Sou et al. 2016) following the order mannitol (T_g from 11 to 13 °C) < trehalose (110–121 °C) < inulin (140–156 °C) < dextran (152–196 °C) < PVP (130–185 °C) < pullulan (261 °C).

Trehalose and sucrose are the most commonly used sugars to dry and stabilize vaccines and other biomolecules, i.e., improve shelf-life and heat stability through hydrogen bonding and vitrification (Ohtake et al. 2010; Ohtake and Wang 2011).

For example, Abdul-Fattah et al. (2007a, b, c) studied a sucrose-based formulation of a live-attenuated virus vaccine of a parainfluenza strain by spray drying and extended the storage stability to 20 weeks at 25 °C and 1 week at 37 °C. Ohtake et al. (2010) used trehalose-sucrose and protein-based spray-drying formulations to form a heat-stable live-attenuated measles vaccine powder with a glassy matrix that remained stable for 8 weeks at 37 °C. Significantly improved long-term stability of a spray-dried stabilized live-attenuated B/harbin influenza vaccine for 14 months at 25 °C was reported by Truong-Le and Pham (2007), Ohtake et al. (2015). Notably, the vaccine formulation contained 5 wt% sucrose and 2 wt% trehalose.

Trehalose has been used as a glass-forming stabilizer in numerous other examples of spray-dried vaccines against influenza (Ajmera and Scherließ 2014; Scherließ et al. 2014; Zhu et al. 2014; Sou et al. 2015; Lovalenti et al. 2016; Kanojia et al. 2016), or measles (LiCalsi et al. 2001; De Swart et al. 2007; Ohtake et al. 2010; Truong-Le et al. 2012, 2014). The reader is referred to Chap. 6 for a more detailed discussion of each application.

Table 3.2 Glass transition temperatures of selected excipients and mixtures used for vaccine stabilization by spray drying

Excipient	T_g (°C)	Molecular weight (kDa, g/mol)	References
Pullulan	261	200–300 kDa	Teekamp et al. (2017)
PVP (polyvinylpyrrolidone)	185	1000 kDa	Hancock and Zografi (1997)
	179	K90: 1250 kDa	Asgreen et al. (2020)
	177	K90: 1250 kDa	Hancock and Zografi (1994)
	176	K90: 1250 kDa	Jameel and Pikal (2010)
	163	K30: 49 kDa	Asgreen et al. (2020)
	161	K30: 40 kDa	Sou et al. (2016)
	140	40 kDa	Drake et al. (2018)
	130	K17: 10 kDa	Alhajj et al. (2021)
Dextran	195	Dex 3.5: 3.7 kDa	Asgreen et al. (2020)
	162	Dex 1.5: 1.6 kDa	Asgreen et al. (2020)
	152	Dex 1: 1.0 kDa	Asgreen et al. (2020)
Inulin	156	4 kD	Amorij et al. (2008)
	154	EXL 608	Hinrichs et al. (2001)
	152	4 kD	Saluja et al. (2010)
	140	RS	Hinrichs et al. (2001)
Trehalose	121	342 g/mol	Hinrichs et al. (2001)
	118		Jameel and Pikal (2010)
	117		Drake et al. (2018)
	117		Miller and de Pablo (2000)
	116		LeClair et al. (2019)
	115		Saleki-Gerhardt and Zografi (1994)
	115		Miller et al. (1997)
	112		Hancock and Zografi (1997)
	110		Crowe et al. (1998b)
Sucrose	75	342 g/mol	Jameel and Pikal (2010)
	74		Hancock and Zografi (1994, 1997)
	73		Miller and de Pablo (2000)
	65		Crowe et al. (1998b)
	62		Roos (1993), Bhandari et al. (1997)
PLGA (75:25)	53	40–100 kDa	Alhajj et al. (2021)
PLGA (50:50)	48		
PLGA (10:90)	45		
Mannitol	13	162 g/mol	Jameel and Pikal (2010)
	11		Sou et al. (2016)

(continued)

Table 3.2 (continued)

Excipient	T_g (°C)	Molecular weight (kDa, g/mol)	References
Trehalose/Leucine (90/10 wt%)	65	342 g/mol/131 g/mol	Sou et al. (2016)
Inulin/Leucine (90/10 wt%)	68	5 kDa/131 g/mol	Sou et al. (2016)
PVP/Leucine (90/10 wt%)	90	40 kDa/131 g/mol	Sou et al. (2016)
Mannitol/PVP (10/0.5 wt%)	86	162 g/mol/8 kDa	Jin et al. (2010)
Mannitol/β-Cyclodextrin/ Trehalose/Dextran (MCTD) (10/0.02/0.2/0.1% wt%)	97	162 g/mol/1135 g/mol/ 342 g/mol/60 to 90 kDa	Jin et al. (2010)
Mannitol/Dextran (67/33 wt%)	134	162 g/mol/40 kDa	Afkhami et al. (2017)
	127	162 g/mol/40 kDa	LeClair et al. (2016a)
Trehalose/PVP (24.5/75.5 wt%)	132	342 g/mol/40 kDa	Drake et al. (2018)
Trehalose/PVP/PBS salt (23.6/69.1/7.3 wt%)	130	342 g/mol/40 kDa/411 g/mol	Drake et al. (2018)
Trileucine	104	358 g/mol	Lechuga-Ballesteros et al. (2008)

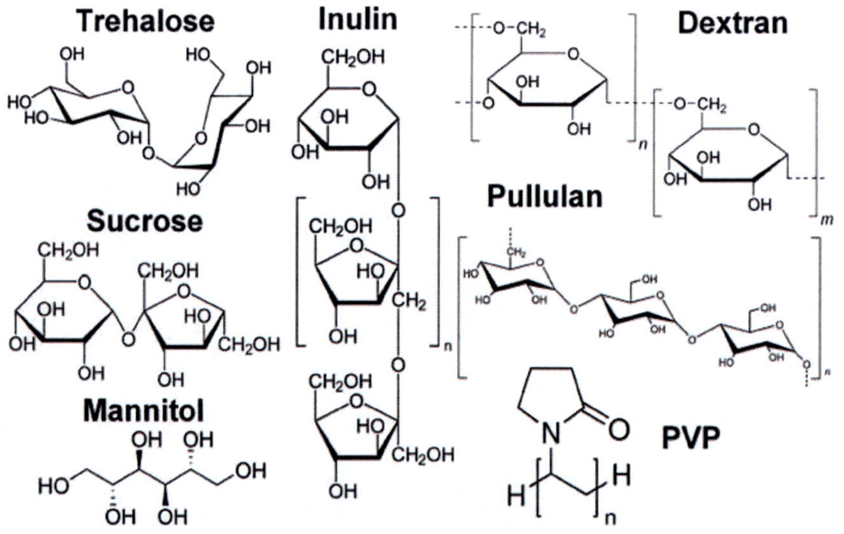

Fig. 3.10 Molecular structures of trehalose, sucrose, mannitol, inulin, dextran, pullulan, and PVP (Images from Wikimedia Commons, the free media repository)

From a molecular structure point of view, trehalose is a simple disaccharide of glucose, also known as α-D-glucopyranosyl-(1-1)-α-D-glucopyranoside (Colaço and Roser 1994; Ohtake and Wang 2011). Their reducing carbons connect the two glucopyranose rings, and the resulting low energy (< 0.1 kcal/mol (Roser 1985) glycosidic bond makes trehalose completely non-reducing and exceptionally stable to hydrolysis (Colaço and Roser 1994). In comparison, the glycosidic bond of sucrose is higher in free energy (27 kcal/mol) (Roser 1991), which makes it more susceptible to hydrolysis. Moreover, trehalose is highly soluble in water at 40.6–69 g per 100 g H_2O at 20 °C, but less soluble than sucrose (200 g per 100 g H_2O) (Ohtake and Wang 2011).

Figure 3.11a shows the effect of the residual moisture on the glass transition temperature (T_g) for trehalose and sucrose (Saleki-Gerhardt and Zografi 1994; Crowe et al. 1998b, a). Trehalose readily dries to an amorphous material with a high T_g in the range of about 110–121 °C (Table 3.2). A linear relationship is observed up to about 0.06 g H_2O/g dry weight (Fig. 3.11b). Depending on the storage temperature, trehalose could turn into a liquid, which affects the stability requirement of a dispersed protein (Jameel and Pikal 2010). The absorption of small amounts of water by sugar glasses strongly lowers their T_g (Slade and Levine 1995); in other words, amorphous solids become plasticized. An increase in water content of 1 wt% decreases T_g by about 10 °C.

The T_g for trehalose is higher than that of sucrose, especially at the lowest water contents. For example, Miller et al. (1997) measured a T_g of 115 °C for anhydrous trehalose, which was 40–50 °C higher than that of sucrose (T_g between 65 and 74 °C) (Saleki-Gerhardt and Zografi 1994).

The higher T_g of trehalose compared to sucrose makes it advantageous for use in situations where it is difficult to achieve low residual moisture in the product. In any case, it must be ensured that the T_g of the product is well above the expected

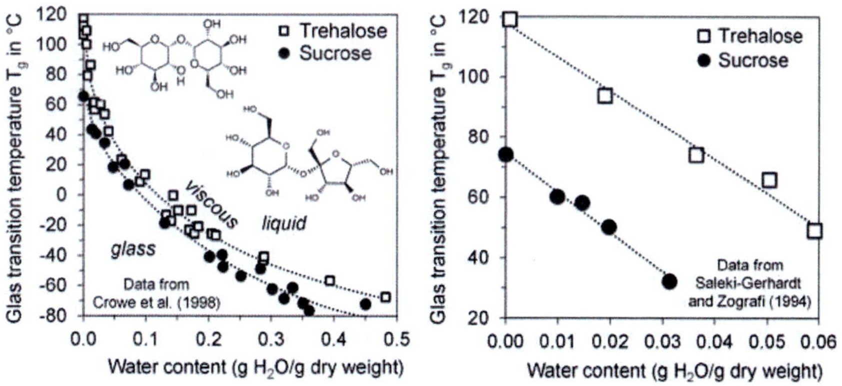

Fig. 3.11 Effect of residual water content (in % w/w) on trehalose and sucrose glass transition temperature (T_g). The T_g for dry trehalose is about 50 °C higher than that for sucrose (Data reproduced from Crowe et al. (1998b, a) and Saleki-Gerhardt and Zografi (1994) with permission from Springer Nature)

storage temperature. Hancock et al. (1995) suggested that for stable storage of an amorphous pharmaceutical product in the dried state, the T_g should be about 50 °C above the storage temperature. In the literature, this point is also referred to as the Kauzmann temperature (Kauzmann 1948), where the potential for molecular mobility is reduced to a minimum. For amorphous solid dispersion using sucrose, the literature recommends a T_g of about 20–30 °C above the storage temperature (Shamblin et al. 1999; Zhou et al. 2002). Hence, for long-term stability at 37 °C, the T_g of a spray-dried trehalose powder should be higher than about 87 °C. According to Fig. 3.11b, at this temperature, the moisture content of the spray-dried trehalose powder should not exceed 2–3 wt%. Based on moisture sorption isotherms of amorphous trehalose at 25–30 °C, this corresponds to relative humidity (RH) of about 10% in the surrounding air (Roe and Labuza 2005; Gomez et al. 2021a).

Mannitol is stable as a powder and resists moisture absorption at relatively high humidities, making it a suitable excipient for encapsulating biopharmaceuticals for inhalation (Jin and Tsao 2013). However, the relatively low T_g of about 11 °C indicates high molecular mobility and rapid crystallization.

Jin et al. (2010) showed that a spray-dried formulation of mannitol, β-cyclodextrin, trehalose, and dextran (10/0.02/0.2/0.1 wt%) increased the T_g to 97 °C (1.45 wt% water content) and was considered suitable for an inhalable tuberculosis vaccine in dry powder form.

Dextran is readily mixed with mannitol to increase glass transition. For example, LeClair et al. (2016a) evaluated a binary mixture of mannitol/dextran (67/33 wt%) and used it to increase T_g to 127 °C for improved thermal stabilization of a human type 5 adenovirus vector by spray drying. Subsequently, Afkhami et al. (2017) spray-dried human adenoviral-vectored tuberculosis vaccines in a mannitol/dextran matrix. Morgan et al. (2020) also stabilized human type 5 adenovirus vector in a solution of mannitol/dextran (3:1 weight ratio).

Inulin has favorable physicochemical properties, such as high and adjustable T_g depending on molecular weight and crystallization resistance (Hinrichs et al. 2001). For example, inulin (4 kD, T_g of 156 °C) has been used as a glass-forming sugar to stabilize subunit influenza vaccines by spray drying (Amorij 2007; Amorij et al. 2008; Saluja et al. 2010). Long-term tests showed complete stabilization of the spray-dried powdered vaccine for at least 3 years at 20 °C (Saluja et al. 2010).

Inulin, dextran, and mixtures of dextran/trehalose were also evaluated in spray freeze drying as protective agents for influenza subunit and whole inactivated virus vaccine (Amorij et al. 2007; Saluja et al. 2010; Audouy et al. 2011; Murugappan et al. 2013) as well as various adjuvants to enhance systemic and particularly mucosal immune responses (Patil et al. 2014, 2015). Further details on the spray freeze dried vaccine applications are described in Sect. 3.8.

Pullulan is a polysaccharide with suitable film-forming properties used in the food industry as an oxygen barrier to extend shelf life (Leathers 2003). It is readily soluble in water and has a high T_g of 261 °C (Mw 200–300 kDa) (Teekamp et al. 2017). Some recent studies have exploited the synergistic effects of pullulan (i.e., high physical stability and surface activity) and trehalose (i.e., close molecular packing with protein) to improve the storage stability of dried vaccines (Teekamp et al. 2017; Leung et al. 2019; Gomez et al. 2021b). Teekamp et al. (2017) improved the storage stability of β-galactosidase (model protein) up to 4 weeks at high moisture conditions (30 °C/56% RH) after freeze-drying. Leung et al. (2019) stabilized live-attenuated herpes simplex virus type 2 (HSV-2) and inactivated influenza A virus vaccines in oral thin films by air-drying und showed that vaccines retained their immunogenicity (in a mice model) for at least 2–3 months of storage at 40 °C. Carrigy et al. (2019) spray-dried various pullulan-trehalose mixtures and demonstrated that a mass fraction of 10 wt% pullulan improved the aerosol properties of the microparticles compared to a trehalose-only formulation. Theoretical modeling predicted Péclet numbers of about 1 for trehalose and 18 for pullulan (Carrigy et al. 2020), suggesting a gradient of pullulan within the drying droplet near the surface, stabilizing bioactive ingredients thanks to high T_g. Gomez et al. (2021b) applied this concept and spray-dried a tuberculosis vaccine candidate for pulmonary delivery (see Chap. 6, Sect. 6.6).

Sou et al. (2016) and Mangal et al. (2015) showed that the addition of L-leucine significantly supported surface enrichment during atomization due to its surface activity and low aqueous solubility and, thus, improved particle formation and processing yield. Lechuga-Ballesteros et al. (2008) added trileucine as a surface-active excipient and minimized protein surface enrichment during spray drying. Trileucine forms an amorphous phase with a high T_g of about 104 °C that allows long-term stability at room temperature. The hydrophobic particle coating also supports the flowability of the powder and enables improved aerosol performance. Carrigy et al. (2020) successfully spray dried amorphous formulations with trileucine as shell-forming excipient and trehalose as bulk glass stabilizer. This combination showed excellent biologic stabilization potential and outperformed the standard leucine-trehalose formulation commonly proposed for stabilizing biologicals by spray drying.

Also listed in Table 3.2 are some formulation mixtures combining, for example, trehalose with PVP (Drake et al. 2018). A combination of trehalose/PVP/PBS salt (23.6/69.1/7.3 wt%) significantly increased the T_g from 83 °C to 130 °C.

3.2.5 Drying Paths in a Sucrose-Water State Diagram

As already emphasized several times, successful drying of vaccines requires a stabilizer, such as sugar, to immobilize the vaccine during drying and subsequent storage. Figure 3.12 shows the principle of sugar glass drying using the state diagram of a binary sucrose/water system (Vehring 2007; Amorij et al. 2008; Lovalenti and Truong-Le 2020). A similar state diagram for trehalose-water, including the glass transition line, is available by Roe and Labuza (2005).

The drying routes for spray drying, freeze drying (same path for spray freeze drying), and vacuum foam drying illustrate a potential vaccine incorporated into the glassy sugar matrix. For simplification, the contributions of the vaccine and buffers

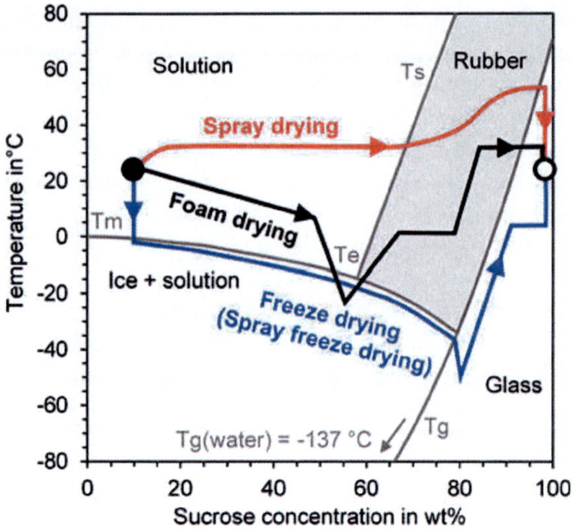

Fig. 3.12 Sucrose-water state diagram with the illustration of the drying paths of spray drying, foam drying, and freeze drying (spray freeze drying is similar to freeze drying but at a much lower temperature) starting from a 10 wt% sucrose solution. T_e eutectic temperature, T_g glass transition curve, T_m melting curve, T_s crystallization curve (solubility limit) (Adapted from Vehring (2007), Amorij et al. (2008), Lovalenti and Truong-Le (2020))

and salts are neglected in the state diagram. In this example, the drying starts with a solution containing 10 wt% solids at room temperature.

In spray drying, water evaporates rapidly from the liquid product droplets so that the temperature is maintained close to ambient conditions by a balance between evaporative cooling and heat input from the warm drying gas (see further details in Sect. 3.5). When enough water has been removed, the dissolved solids reach their solubility limit (T_s) and begin to precipitate until eventually the evaporation rate slows and the temperature of the now solid particles begins to rise, approaching the outlet drying gas temperature of the spray dryer. When the powder is collected, it cools back to room temperature.

In freeze drying, the liquid product is first cooled to its melting temperature (T_m). With continued cooling, separation into two phases occurs, a growing accumulation of ice crystals and an increasingly concentrated product solution containing the sugar. After the liquid product phase has cooled below its glass transition temperature (T_g), primary drying occurs by applying a high vacuum to remove the ice by sublimation. This is followed by a temperature rise for secondary drying, removing water from the solid concentrated product by sublimation, resulting in a denser porous solid or cake (see further details in Sect. 3.4, Figs. 3.32, and 3.42). When the final product returns to room temperature, the glass protects the porous microstructure.

In spray freeze drying, the drying path is similar to conventional freeze drying, but the temperature is much lower, and freezing is much faster. The liquid solution is atomized into a cryogenic medium, e.g., liquid nitrogen, to vitrify the droplets. A key difference is an extremely rapid solidification due to the large surface area of the liquid droplets, as well as the ability to create particles of an adjustable size that can be used for specific drug delivery strategies such as epidermal powder delivery and dry powder inhalation (see further details in Sect. 3.6).

In foam drying, the process starts with a concentration step by applying a low vacuum and temperatures close to ambient. The resulting rapid removal of water by boiling (i.e., minutes to a few hours) produces a sharp drop in product temperature due to latent heat of vaporization, leading to temporary freezing of the product and entrapment in the foam structure. Further moisture removal, or secondary drying (i.e., several hours to a couple of days), may involve a combination of sublimation and evaporation that occurs under a large surface area until the product eventually approaches shelf temperature. An increase in storage temperature above ambient conditions causes further moisture removal. The product then cools to room temperature. Further details on foam-dried vaccine applications are described in Sect. 3.7.

Figure 3.13 shows the phase diagram of water and the thermodynamic equilibria between the ice, liquid, and vapor phases to further illustrate the essential differences between the drying processes at a high level. At the triple point (0.01 °C, 0.0612 bar), solid, gas, and liquid coexist. The different drying regimes with the phase transition processes are indicated by arrows, including sublimation used in freeze drying, vacuum evaporation used in vacuum drying, and evaporation used in spray drying.

Overall, a high research activity demonstrates the feasibility of alternative drying processes. In each case of the drying processes, the goal is to improve the thermostability and shelf-life of vaccines without losing immunogenicity while creating more flexible dosage forms, such as drying the vaccine into a stable powder.

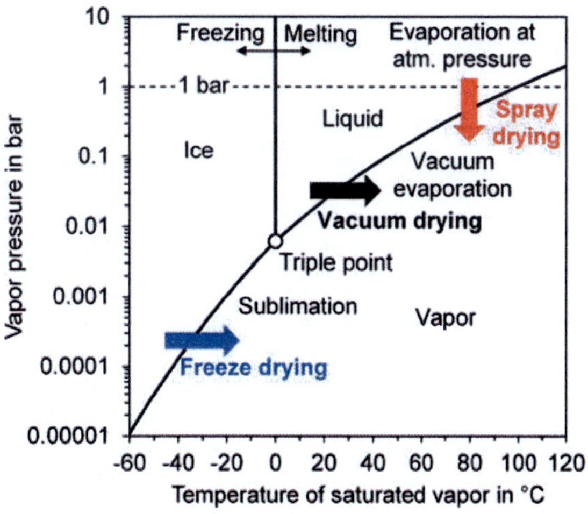

Fig. 3.13 Phase diagram of water with the phase transition processes of sublimation for freeze drying, vacuum evaporation for vacuum drying, and evaporation for spray drying (Adapted from Vass et al. (2019) with permission from Elsevier)

The versatility of the final dry formulation is related to the control of powder properties. In addition, delivery systems for needle-free administration, inhalation, or nasal applications are coming into focus (Bahamondez-Canas and Cui 2018).

Finally, the interested reader is referred to some excellent reviews that give further information about drying technologies to biological materials and vaccines (Amorij et al. 2008; Maltesen and van de Weert 2008; Abdul-Fattah and Truong 2010; Rosini et al. 2011; Walters et al. 2014; Searles and Cherian 2015; Ohtake et al. 2015; Langford et al. 2018b; Emami et al. 2018; Vass et al. 2019; Lovalenti and Truong-Le 2020; Thorat et al. 2020).

Several drying technologies can produce dry powder vaccines based on the formulation excipients and their properties. The most commonly applied drying technologies will be reviewed and discussed in the following sections.

3.3 Freeze Drying of Vaccines

3.3.1 Principle and Benefits of Freeze Drying

Freeze-drying or lyophilization is a multi-stage process that involves freezing, primary drying (by ice sublimation), and secondary drying (by desorption), as visualized in Fig. 3.14 by a representative temperature profile for adjuvanted vaccines (Chan et al. 2017).

First, the vaccine biomolecules and various excipients are mixed to form an appropriate liquid formulation. Next, this liquid is filled into vials, which are half stoppered and loaded onto a temperature-controlled shelf inside a chamber with controlled pressure and temperature. The samples are then frozen at very low temperatures (typically at −40 or −45 °C) (Hansen et al. 2015), followed by lowering the pressure to near-vacuum, and heat is applied by increasing the shelf temperature. This allows sublimation (primary drying) to remove unbound water from the samples and collect it as ice on a condenser. During secondary drying, the temperature increases, and bound water is removed from the samples by desorption. Once the samples are sufficiently dry, atmospheric pressure is restored, and an inert gas may be used to blanket the samples. Finally, the samples are stoppered and ready for use. The result is a fine, microporous structured solid cake as an end product, consisting of a glassy matrix of sugars and others in the formulations. Since the freeze-drying process takes place directly in the vial, the vaccine can be aseptically enclosed at the end of the freeze-drying cycle. The vials are then transported and reconstituted using a diluent to their original form before administration.

Other advantages of freeze drying include high yields of near 100%, even for smaller sample batches of difficult-to-dry products delivered in vials. Equally important, freeze-drying is a scalable technology and ensures consistent product quality across all production scales. In short, traditionally, freeze drying is a kind of gold standard process for drying vaccines.

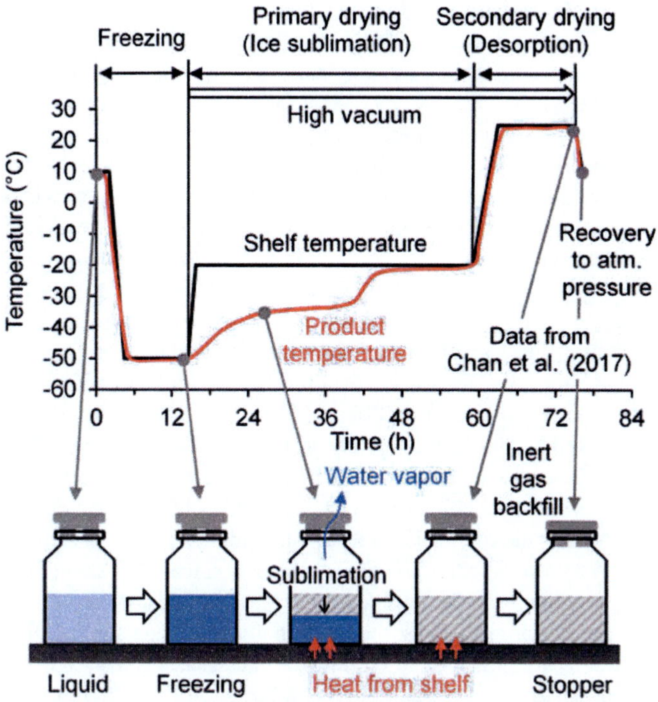

Fig. 3.14 Schematic representation of a typical freeze-drying process for adjuvanted vaccines with freezing, primary drying (ice sublimation), secondary drying (desorption), and stoppering (Shelf temperature profile adapted from Chan et al. (2017) with permission from Springer Nature)

3.3.2 A Short Market Review of Lyophilized Drugs

Freeze drying was invented in the 1930s to preserve blood serum and blood plasma and enabled large-scale penicillin manufacturing in the 1940s (Debus 2017). In the 1950s, onward lyophilization played a critical role in global vaccination against viral diseases such as smallpox. After a heat-stable smallpox vaccine became available through freeze drying, a worldwide vaccination of smallpox was possible (Cross et al. 1957). Even under tropical conditions, the freeze-dried vaccine maintained potency for several weeks at room temperature, which allowed the vaccine's widespread distribution in countries with poor transport to remote regions without a cold chain. It also enabled the maintenance of sufficient reserves in these countries against epidemics. As a consequence, smallpox was declared eradicated by the World Health Assembly in 1980.

Smallpox and other success stories have given freeze drying technology a significant boost to making freeze-drying the most widely used drying technique for removing moisture from biopharmaceutical products and stabilizing sensitive substances such as proteins, peptides, vaccines, oligonucleotides, cell-based products, diagnostic reagents, probiotic cultures, among many others (Costantino 2004; Greb 2009; Hansen et al. 2015; Ohtake et al. 2015; Debus 2017). Many biopharmaceuticals could not be commercially viable without lyophilization. Most CDMOs (Contract Development and Manufacturing Organizations) have freeze drying capabilities, thus contributing to its popularity.

Foods and pharmaceuticals together account for more than 95% of the worldwide market for lyophilized end products, which amounts to approximately US$ 16 billion per year (Alexeenko and Topp 2020). The global market for freeze drying equipment is estimated at US$ 1.7 billion in 2020 and is projected to reach a size of US$ 2.8 billion by 2027, growing at a CAGR of 7.5% (Research and Markets 2020).

A major driver for innovation in freeze drying technologies is the increasing number of newly approved pharmaceutical products manufactured in a freeze-dried dosage form (Alexeenko and Topp 2020). Figure 3.15 shows the development of new FDA-approved lyophilized drugs per year and indications from 1990 to 2014 (Arduini 2016). Since about 2008, a rapid increase of newly approved lyophilized drugs has been evidenced. This growth has been driven mainly by discovered biological drugs such as monoclonal antibodies and antibody–drug conjugates (Debus 2017). Of the 472 lyophilized drugs, FDA-approved since 1990, most new products were in the categories of infectious diseases, oncology, cardiology, and immune, respectively. Furthermore, it was reported that the percentage of all approved injectable/infusible drugs that were lyophilized increased from 12% between 1990 and 1998 to more than half between 2013 and 2015 (Arduini 2016; Langford et al. 2020). For the future, it is expected that approximately 25% of biotechnological and pharmaceutical products will be lyophilized (Teixeira 2005; Wright 2019).

3.3.3 WHO Prequalified Lyophilized Vaccines and Manufacturers

As presented in Sect. 3.1 (Fig. 3.1), 27% (69 presentations) of the 252 prequalified WHO vaccines (WHO 2021) are available in a dried format and are processed by freeze drying. 73% of the vaccines are in liquid (ready-to-use) format and require refrigeration with storage and distribution at 2–8 °C (see Chap. 2).

Table 3.3 provides an overview of the manufacturers producing lyophilized vaccines based on the same WHO database of prequalified vaccines. There are 18 companies from 14 countries, including Belgium, Brazil, Bulgaria, China, Denmark, France, India, Indonesia, Japan, Russia, Senegal, South Korea, Thailand, and the USA. By far, The Serum Institute of India (SII) is leading the market with around 46% of the total lyophilized vaccines (i.e., 32 out of 69).

As an example of the automatic choice of freeze-drying, News10 (2020) reported in December 2020 that everyone in the vaccine industry is now working on a freeze-dried version of a COVID-19 vaccine for the next round (Fig. 3.16). A room temperature stable vaccine was declared necessary for making the vaccine available worldwide, including low-income countries with low refrigeration capabilities.

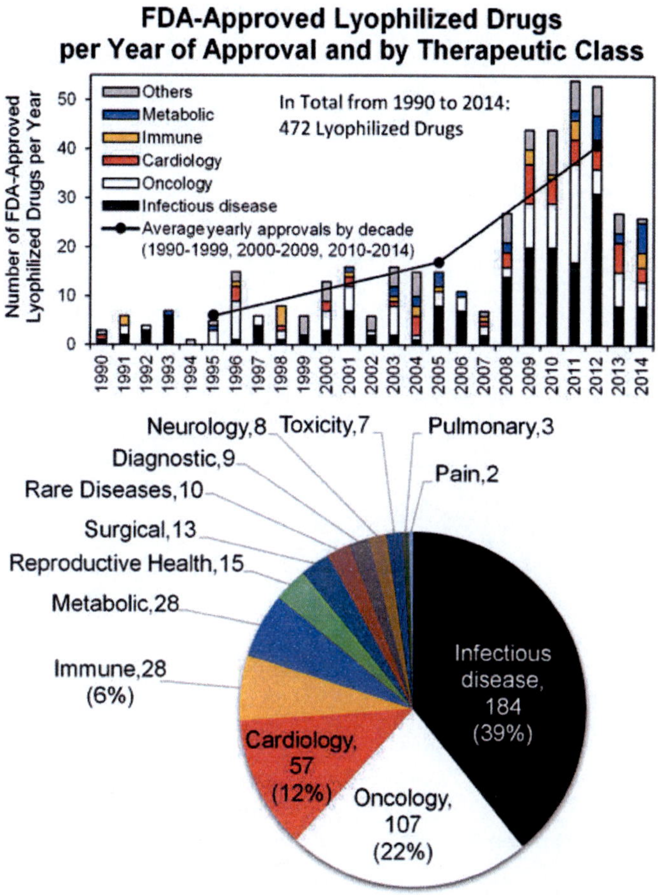

Fig. 3.15 Historical development of the number of FDA-approved lyophilized drugs by year of approval from 1990 to 2014, structured by therapeutic class. Total products: 472 (Data adapted from Arduini (2016) with kind permission from IMA Life, Italy)

3.3.4 Challenges in the Production of Freeze-Dried Vaccines

Although freeze-drying is well established in the vaccine industry and readily available processing equipment, three main challenges exist:

- cost,
- productivity, and
- no particle technology.

Table 3.3 Overview of the 18 vaccine manufacturers that produce 16 types of lyophilized vaccines (69 lyophilized)

Vaccine manufacturer (country)	Vaccine type (number of lyophilized presentations)
Japan BCG Laboratory (Japan)	BCG (2)
BB-NCIPD Ltd. (Bul Bio-National Center of Infectious and Parasitic Diseases Ltd.) (Bulgaria)	BCG (2)
AJ Vaccines A/S (Denmark)	BCG (1)
Serum Institute of India Pvt. Ltd. (India)	BCG (1) Diphtheria-Tetanus-Pertussis (whole cell)-Haemophilus influenzae type b (1) Diphtheria-Tetanus-Pertussis (whole cell)-Hepatitis B-Haemophilus influenzae type b (3) Haemophilus influenzae type b (1) Influenza, pandemic H1N1 (2) Influenza, seasonal (Trivalent) (1) Measles (4) Measles and Rubella (4) Measles, Mumps, and Rubella (4) Meningococcal A Conjugate 10 μg (1) Meningococcal A Conjugate 5 μg (1) Rabies (1) Rotavirus (live, attenuated) (4) Rubella (4)
Sanofi Pasteur SA (France)	Dengue Tetravalent (live, attenuated) (1) Haemophilus influenzae type b (1) Rabies (1) Yellow Fever (1)
Biological E. Limited (India)	Diphtheria-Tetanus-Pertussis (whole cell)-Hepatitis B-Haemophilus influenzae type b (2) Measles and Rubella (3)
Chengdu Institute of Biological Products Co., Ltd. (China)	Japanese Encephalitis Vaccine (live, attenuated) (2)
GPO-MBP Co., Ltd. (Thailand)	Japanese Encephalitis Vaccine (live, attenuated) (1)
PT Bio Farma (Persero) (Indonesia)	Measles (2)
GlaxoSmithKline Biologicals SA (Belgium)	Measles, Mumps, and Rubella (2) Meningococcal ACYW-135 (conjugate vaccine) (1)
Merck Vaccines (USA)	Measles, Mumps, and Rubella (1) Varicella (1)
Pfizer (Belgium)	Meningococcal ACYW-135 (conjugate vaccine) (1)
Chiron Behring Vaccines Private Ltd. (India)	Rabies (1)
Cadila Health Care Ltd (India)	Rabies (1)

(continued)

Table 3.3 (continued)

Vaccine manufacturer (country)	Vaccine type (number of lyophilized presentations)
SK Bioscience Co., Ltd. (South Korea)	Varicella (1)
Institut Pasteur de Dakar (Senegal)	Yellow Fever (3)
Bio-Manguinhos/Fiocruz (Brazil)	Yellow Fever (3)
Federal State Budgetary Scientific Institution, Chumakov Federal Scientific Center for Research & Development of Immune-And Biological Products, Russian Academy of Sciences (Russia)	Yellow Fever (3)

Data summarized from WHO (2021), https://extranet.who.int/gavi/PQ_Web/, accessed on March 20, 2021

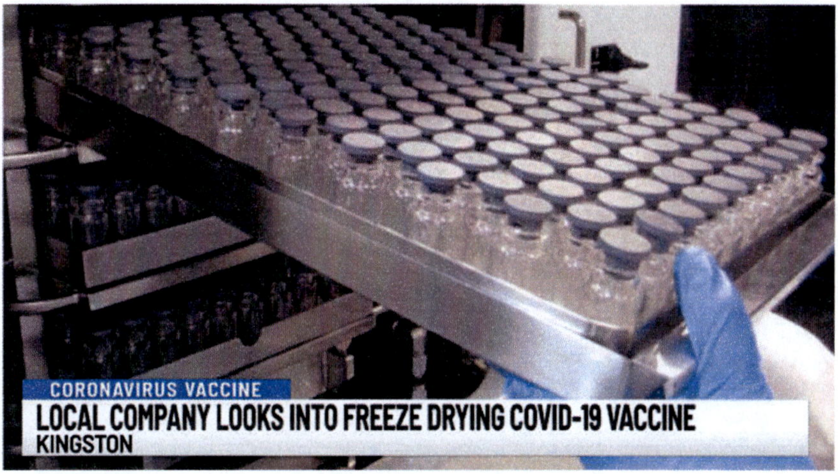

Fig. 3.16 The freeze-drying company Millrock Technology (Kingston, NY, USA) is one step closer to launching a freeze-dried COVID-19 vaccine that can be stored at room temperature (Image from News10 (2020) posted on December 11, 2020)

Freeze drying is among the most time-consuming and expensive unit operations (Roser 1991; Colaço and Roser 1994; Domínguez 2011) with a low energy efficiency of less than 5% in batch mode operation (Alexeenko and Topp 2020).

Table 3.4 shows the fixed and manufacturing costs of different drying methods for foods (Roser 1991; Colaço and Roser 1994; Domínguez 2011) and the costs relative to spray drying.

As can be seen, freeze drying is the most expensive method, while air drying is the most cost-effective. In comparison, the fixed costs for spray drying, including purchase, installation, and engineering, are about 8.3 times lower than freeze-drying and the operating costs 4.9 times lower.

Table 3.4 Costs of different drying methods for foods given as costs per kg of water evaporated per hour and relative to spray drying

	Fixed costs		Manufacturing costs	
Drying method	(US$/kg-h)	Relative to spray drying	(Cents/kg)	Relative to spray drying
Air drying	610	0.4	17	0.9
Fluid bed drying	1000	0.7	17	0.9
Drum drying	1040	0.8	23.4	1.2
Spray drying	1360	1.0	19.4	1.0
Vacuum drying	5860	4.3	49	2.5
Freeze-drying	11,228	8.3	95.2	4.9

The fixed costs include equipment purchase, installation, and engineering costs
Data summarized from Roser (1991), Colaço and Roser (1994), Domínguez (2011)

Another comparison from the flavor industry, where spray drying is the most common method for drying encapsulations, shows that the cost of spray drying is even 30–50 times lower than that of freeze-drying (Desobry et al. 1997).

Unfortunately, there are no publicly published cost data for dry vaccines to make a direct comparison, but it is assumed that the relative cost comparison is plausible and can also be applied in this application field. However, costs depend on many aspects, such as location, utilities, building support requirements, financing exchange rates, and material and labor costs (Masters 2002). Therefore, pricing data should be treated cautiously, as such data may be too imprecise to indicate what the price covers precisely.

Another challenge of freeze drying is the batch-wise operation, which is challenging to adapt and implement within a continuous manufacturing process (Langford et al. 2018c). The complete drying process can take two to three days (GEA 2021). At a fill rate of 600 vials per minute, it takes a few hours to fill a freeze dryer with a capacity for 100,000 vials, but several days to empty the equipment again before it can be refilled (Gladd 2014). Besides, a long processing time carries the risk of losing an entire batch if, for some reason, process conditions are not maintained throughout the freeze-drying cycle. Furthermore, freeze dryer safety systems increase installation and operating costs (Ledet et al. 2015). In addition, freeze-drying equipment takes up a relatively large amount of space within a production facility, which can be a difficult option for some companies (Greb 2009). Regarding continuous manufacturing, efforts are underway to develop a controlled, continuous freeze drying process for individual vials that will allow direct monitoring of the drying history for each vial (De Meyer et al. 2015; Pisano et al. 2019).

Finally, the end product after freeze-drying is a dried cake or foam. Manufacturers need to integrate a grinding step into the process after freeze-drying if a powdered product is desired with specific particle properties such as size distribution and dispersibility (Greb 2009; Gladd 2014).

3.4 Spray Drying

3.4.1 Principle of Spray Drying

Spray drying is a widely applied method to convert aqueous or organic solutions, emulsions, dispersions, and suspensions into dry powder. Applications are found in all major industries, including food, chemicals, pharmaceuticals, and biotechnological materials. Dry milk powder, instant soups, instant coffee, detergents, and dyes are just a few spray-dried products currently available in everyday life (Arpagaus 2007).

A spray dryer atomizes a liquid feed into fine droplets and evaporates the solvent or water using hot drying gas. A complete spray drying process consists of a sequence of four basic steps (Masters 1985; Passos and Mujumdar 2005; Snyder and Lechuga-Ballesteros 2008; Schwartzbach 2011; Vehring et al. 2020):

(1) Feed preparation (solution, suspension, emulsion)
(2) Liquid atomization and atomization of the liquid formulation into a spray, where the droplets are formed
(3) Particle formation and drying by intensive heat and mass transfer between droplets and gas (i.e., (1) shrinkage of liquid droplets by water evaporation from the surface, (2) crust formation on the surface, (3) thickening of the solid crust by vapor diffusion)
(4) Separation of the dried powder from the exhaust drying gas (Fig. 3.17).

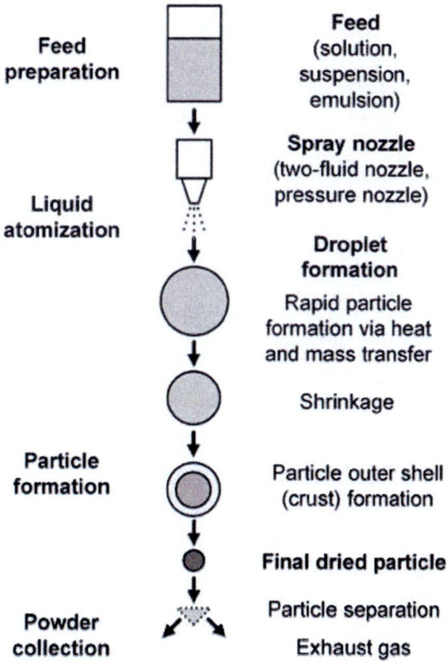

Fig. 3.17 Basic steps of a spray drying process

The fast-drying, short residence time, and predictable temperature exposure result in a gentle drying process that can be adjusted for various process conditions, including drying heat-sensitive materials such as vaccines and complex proteins (Schwartzbach 2011). The process can produce fine, dustless, or agglomerated powders to precise specifications (Masters 1991). The resulting dry particles are collected from the gas stream with a cyclone, filter, or electrostatic particle collector (Li et al. 2010a; Arpagaus 2012; Arpagaus et al. 2018). The solid particles may be further processed, e.g., in tableting or capsule filling, or used directly as a powder, e.g., in an inhalation dosage form. Compared to other drying technologies, spray drying is a very fast drying process and converts a liquid product into a dry powder in a single step, making the process advantageous in terms of cost, scale-up, and process simplification. Besides, spray drying enables particle design of the product by engineering particle shape and structure with different substance properties and drying conditions. Spray drying has also proven to be a viable method to stabilize temperature-sensitive substances such as enzymes, proteins, antibiotics, etc., without much activity loss. This characteristic enables stockpiling and facilitated distribution.

The following sections will highlight the basic principles of spray drying technology. For a discussion on various application examples of spray-dried vaccines, the reader is referred to Chap. 6.

3.4.2 Process Parameters

Laboratory-scale spray dryers are particularly suitable for optimizing process conditions in the context of feasibility studies. Figure 3.18 shows the main adjustable process parameters and formulation variables in a laboratory spray dryer (Mini Spray Dryer B-290 from Büchi Labortechnik AG, Switzerland as an example applied in vaccine research), as well as the depending output parameters. The most critical process parameters using a typical lab-scale spray dryer equipped with a two-fluid nozzle are:

- inlet drying gas temperature,
- drying gas flow rate,
- humidity of the drying gas,
- atomization gas flow rate,
- liquid feed rate,
- feed composition (the type of polymer, excipient, stabilizer, and surfactant concentrations), and
- solids concentration.

Overall, spray drying is a process in which the product properties depend primarily on the correlation and interdependence of the process parameters. Figure 3.19 illustrates the strength of the corresponding influence between the process parameters and their output parameters, which are:

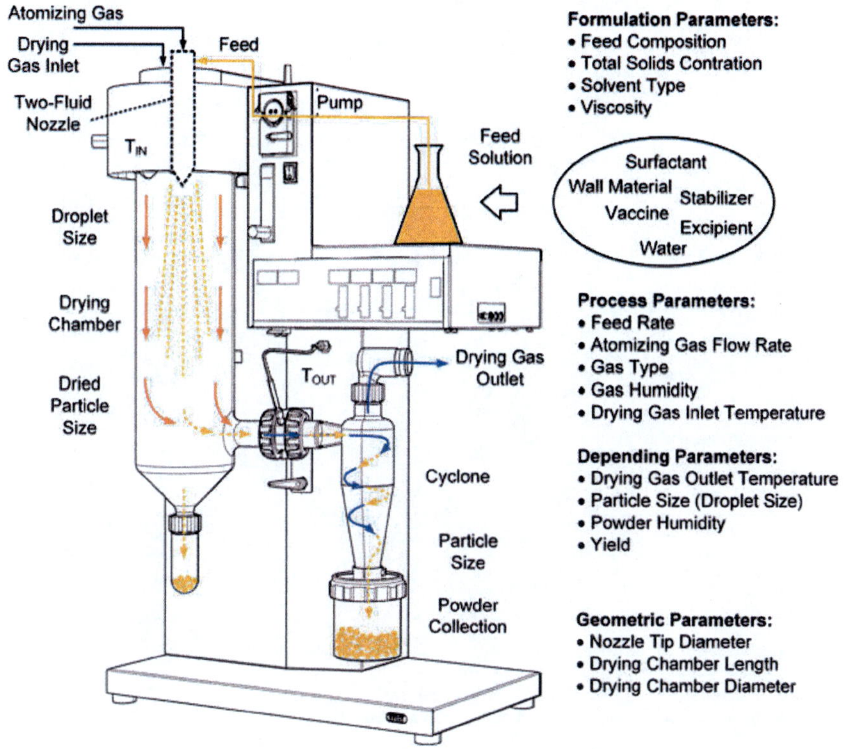

Fig. 3.18 Schematic illustration of the leading spray drying parameters for encapsulation of vaccines in a laboratory spray dryer. Indicated are the most critical formulation and process parameters. An outlet filter is missing in the drawing for simplicity, although it is typically used (Image courtesy by Büchi Labortechnik AG, Switzerland)

- outlet temperature (temperature load)
- particle size (droplet size)
- final powder moisture content (humidity), and
- product yield.

The inlet temperature is understood as the temperature of the heated drying gas. Therefore, it is possible to predict the outlet drying gas temperature from the setting of the inlet temperature. Figure 3.20 shows the interaction between the inlet and outlet temperatures and the water evaporation rate of the Mini Spray Dryer B-290 at a 100% aspirator rate. The lines for the outlet temperature (60, 80, and 100 °C) follow a linear relationship as a first approximation. This information makes it possible to predict first estimates of temperature conditions and water evaporation rates.

The outlet temperature of the drying gas is directly related to the set inlet temperature, the feed rate, the drying gas flow rate, and the atomization gas flow

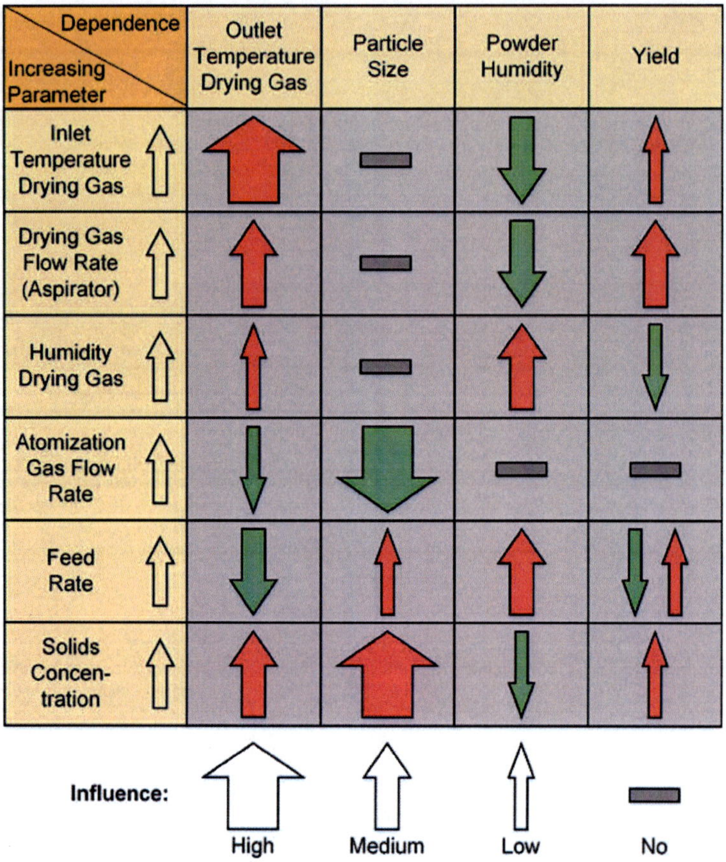

Fig. 3.19 Interplay of process parameters during spray drying and their influence on the product physicochemical properties. High, medium, low, or no influence is indicated by the thickness of the arrows. Green represents inverse and red direct correlation (Based on Arpagaus (2007))

rate. Therefore, changes in any of these parameters will increase or decrease the outlet temperature (see Fig. 3.19).

Figure 3.21a–d visualizes the relationships based on a parameter study with pure water based on data from Maa et al. (1997) in a Mini Spray Dryer B-190 model. Some slopes or gradients are steeper than others, especially the increased outlet temperature at higher inlet temperature.

- A higher inlet temperature causes a proportional increase in the outlet temperature at a fixed feed rate.
- A lower drying gas flow rate leads to a lower outlet temperature at a fixed feed rate.

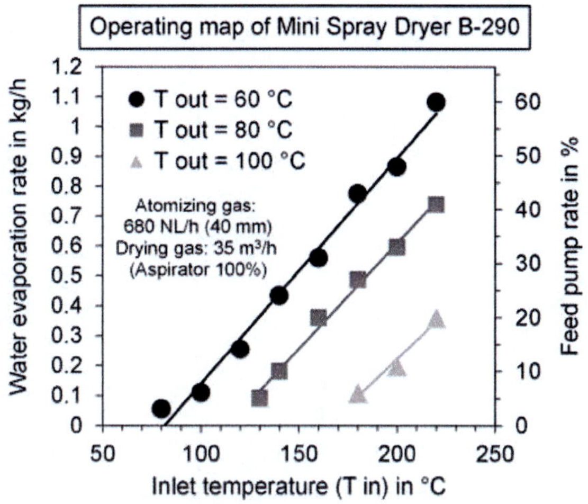

Fig. 3.20 Operating map (design space) of the Büchi Mini Spray Dryer B-290 Advanced. The water evaporation rate is given as a function of the inlet and outlet drying gas temperatures. The spray dryer was set up with a standard cyclone, lamella curtain, and silicone feed tubing. The temperature error is ±3 °C (Data based on Arpagaus (2007))

- A lower liquid feed rate increases the outlet temperature at fixed inlet temperature.
- A higher atomization gas flow rate decreases the outlet temperature slightly at fixed atomization gas temperature and inlet temperature.

An important aspect is the scalability of the water evaporation results of the Mini Spray Dryer B-290 lab-scale model with the former models B-191 (1994 to 2003) and B-190 (sold before 1994 by Büchi Labortechnik AG). Furthermore, the larger spray cylinder of the B-290 (16.5 × 60 cm D × H) results in more significant heat loss and a lower outlet temperature than the B-191 and B-190 models (both 12 × 50 cm D × H).

Figure 3.22 (left) illustrates this relationship by spray drying pure water at a 100% aspirator rate. With the same inlet temperature, drying gas flow rate, and feed rate, the outlet temperature in the B-290 is about 7 °C (at 80 °C T_{in}), 13 °C (at 120 °C), 20 °C (at 160 °C) and 28 °C lower (at 220 °C). On the other hand, the inlet temperature in the B-290 can be increased accordingly to obtain the same outlet temperature as with the B-191 (or B-190).

Figure 3.22 (right) shows the difference in the outlet temperature by using the standard or the narrower high-performance cyclone. When operating with water, the outlet temperature difference (ΔT_{out}) increases from about 6–14 °C at 90 °C and 210 °C inlet temperature because of approximately 35% lower air flow rate through the improved cyclone (Maury et al. 2005). A lower drying air flow rate generally leads to a lower outlet temperature at fixed inlet temperature due to the energy balance.

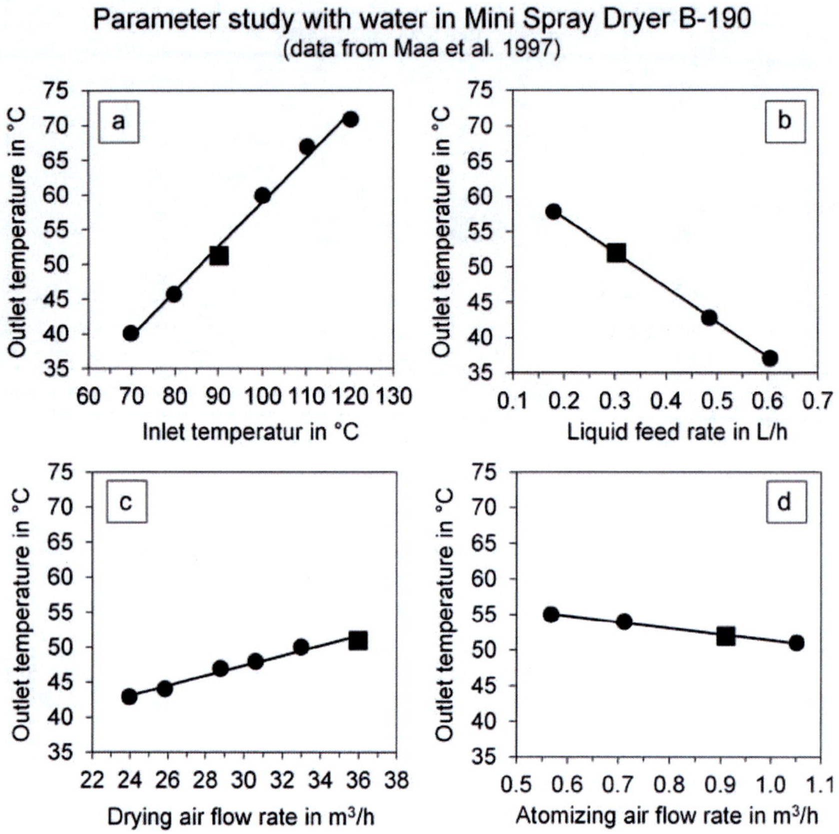

Fig. 3.21 Relationships between the drying air outlet temperature with **a** drying air inlet temperature, **b** liquid feed rate, **c** drying air flow rate, and **d** atomizing airflow rate for spray drying of pure water at standard operating conditions (T_{in}/T_{out} 90 °C/53 °C, 36 m^3/h, 0.9 m^3/h, marked with rectangles) on a Mini Spray Dryer B-190). The four regression lines are linear fit approximations (Data adapted from Maa et al. (1997) with permission from Taylor & Francis)

3.4.3 Temperature Profile of Dry Powder Vaccine During Spray Drying

There is a common belief that the temperatures usually used for the inlet drying gas of a spray dryer would lead directly to product degradation. Indeed, high gas temperatures of around 100 °C or higher are commonly used to spray dry droplets (Tlaxca et al. 2015) rapidly. However, as soon as the droplets of the spray come into contact with the hot drying gas, the temperature at the droplet surface approximates the wet-bulb temperature of the drying gas, a dynamic equilibrium where evaporation takes place on the saturated vapor film on the droplet surface (Masters 1985; Oakley 2004).

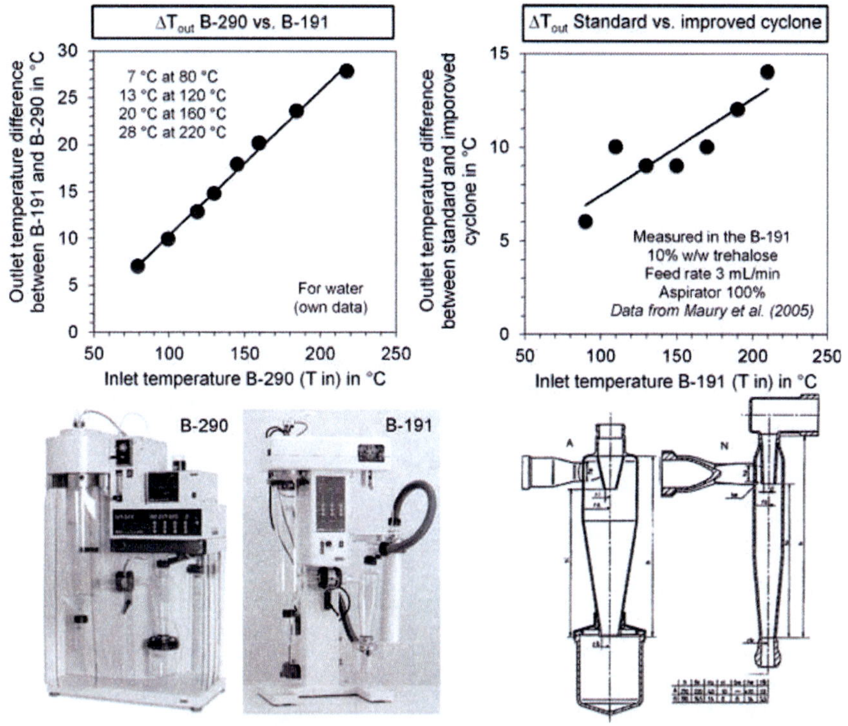

Fig. 3.22 Left: Measured outlet temperature difference (ΔT_{out}) of the Mini Spray dryer B-290 compared to the B-191 model (or B-190) while spray drying water at the same inlet temperature and 100% aspirator rate. Right: Outlet temperature difference (ΔT_{out}) between standard and improved (high-performance, narrower) cyclone measured in the B-191 using 10% w/w trehalose solution at 3 mL/min feed rate and 100% aspirator rate (Data adapted from Maury et al. (2005) with permission from Elsevier)

Figure 3.23 shows the wet-bulb temperature as a function of the gas temperature at normal pressure. For example, a water droplet dried in air at a temperature of 100 °C and relative humidity of 1% will experience only about 34 °C. Higher relative humidity slows the concentration-dependent mass transfer to the gas phase, slowing the particle formation rate along with evaporative cooling of the droplet and increasing the wet-bulb temperature.

That evaporative cooling effect efficiently prevents droplet heating irrespective of its size. As a result, the vaccine inside the droplets does not experience the high drying gas temperatures surrounding the droplet. Instead, it is exposed to a downstream temperature below its glass transition temperature (T_g) that is more amenable to its physical and/or chemical stability (Vehring 2008; Vehring et al. 2020).

Figure 3.24 illustrates an idealized temperature profile of the drying gas and the droplet (product) during spray drying. The gas temperature decreases with time or

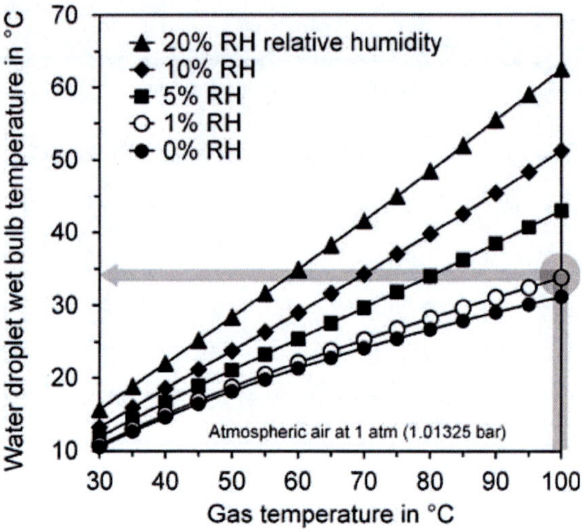

Fig. 3.23 Calculated wet bulb temperature of the water droplet as a function of the drying gas temperature at different relative humidity and normal air pressure of 1 atm (1.01325 bar)

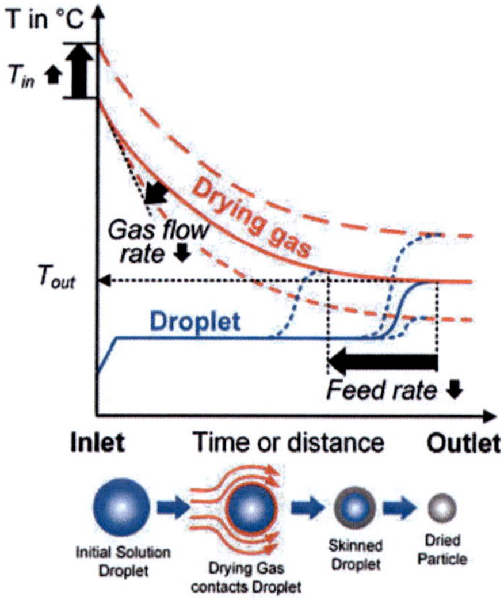

Fig. 3.24 Idealised temperature profiles of the drying gas and the droplets during spray drying in co-current mode. The influence of drying gas temperature, flow rate, and liquid feed rate on droplet temperature is illustrated. T_{in}/T_{out}: inlet/outlet drying gas temperature (Adapted from Arpagaus (2007))

distance in the spray chamber. Due to the high specific surface area and the existing temperature and moisture gradients, intense heat and mass transfer lead to efficient drying. The droplets shrink as water is evaporated. Towards the end of the drying process, evaporation slows, and solidification occurs. The glass particles rapidly equilibrate to the surrounding outlet gas temperature at the bottom outlet of the spray drying chamber, which determines the final outlet temperature (T_{Out}). In co-current spray dryers, the gas temperature is lowest at the bottom exit, enabling the gentle drying of heat-sensitive biological materials like vaccines. The wet-bulb temperature of the droplet and the outlet gas temperature define the thermal gradient experienced by spray-dried particles (Breit and DuBose 2013). As a rule of thumb, the outlet temperature can be regarded as the highest temperature to which the product may be heated, i.e., the upper thermal load on the product.

In spray dryers, the outlet gas temperature cannot be set directly. It is controlled by the applied.

- drying gas flow rate (aspirator),
- inlet drying gas temperature (T_{In}) and
- liquid feed rate.

Furthermore, the outlet temperature is a derivative of the evaporation enthalpy of the solvent, the concentration of the solids in the feed, and heat losses through the drying chamber wall. A higher inlet temperature results in a higher outlet temperature. A lower feed rate increases the outlet temperature because less liquid must be evaporated simultaneously. A lower drying gas flow rate decreases the slope of the gas temperature profile and leads to a lower outlet temperature (Arpagaus 2007).

In addition to the mild temperature exposure, the large surface area of the droplets produced by atomization of the sample enables rapid drying, which leads to a short residence time within the drying gas (Searles and Cherian 2015). The processing time for spray drying involves product travel through the drying chamber (residence time) and downstream powder collection in cyclone and outlet filter of about 45–60 s in a commercial operation with continuous powder discharge. The short residence time minimizes the probability of thermal degradation during processing. These conditions allow spray drying to be applied to thermolabile products, including vaccines (Ohtake et al. 2010; Chen et al. 2010; McAdams et al. 2012; Kunda et al. 2016; Price et al. 2020).

3.4.4 Droplet Size

There are a variety of atomization systems available, which may be classified according to the nozzle design. The energy required is supplied by centrifugal, pressure, kinetic or sonic nozzles (Masters 1985). In rotary atomization, the feed liquid is introduced into the drying chamber using a rotating disc or wheel. With pressure atomization, the liquid is fed into the nozzle under pressure, which causes the liquid to be atomized into droplets as it leaves the nozzle. Finally, the feed liquid and the atomizing gas are fed to the nozzle separately in two-fluid nozzles. The gas causes the feed to be atomized into a spray mist during mixing. Two-fluid nozzles are generally used in laboratory spray dryers.

The atomization gas flow rate is the critical process parameter to adjust the droplet size. A higher gas flow rate leads to smaller droplets and, accordingly, to smaller dried particles. It also results in a slightly lower outlet drying gas temperature due to the gas flow energy balance. It is remarkable that one liter of sprayed liquid forms

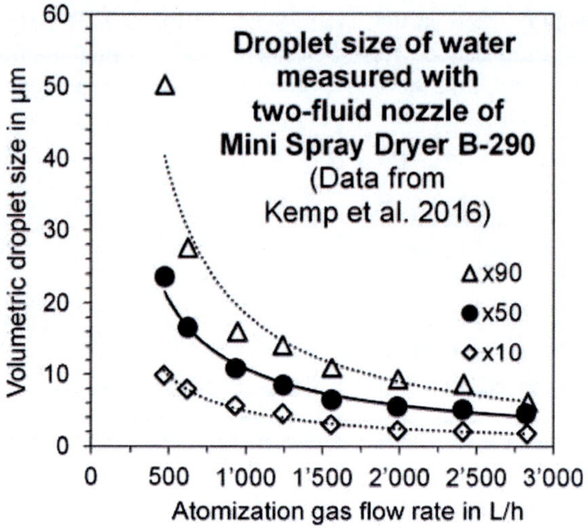

Fig. 3.25 Variation of the volumetric droplet size of water with a higher atomizing gas flow rate. It was measured with Laser Doppler Anemometry at ambient temperature × 10 and × 90 indicate the drop size distributions. A two-fluid nozzle of the Mini Spray Dryer B-290 was used with a 0.7 mm nozzle tip and 1.5 mm cap diameter. The measuring range was 10 mm below the spray nozzle tip (Data adapted from Kemp et al. (2016) with permission from Taylor & Francis)

about 2×10^9 individual droplets of 100 μm size (3.14×10^{-8} m^2 surface area assuming spherical droplets), which corresponds to a total surface area of over 60 m^2.

Figure 3.25 shows the expected decrease in water droplet size with increasing atomizing gas flow rate measured by Laser Doppler Anemometry in a Mini Spray Dryer B-290 with a 0.7 mm two-fluid nozzle (Kemp et al. 2013, 2016). An increasing gas flow rate increases the shear forces for atomization and leads to smaller droplets. The droplet size tends to asymptote at about 5–6 μm, indicating that the atomizing gas flow approaches the speed of sound.

The final dried particle size (d_P) can be estimated from the initial liquid droplet size (d_D) and the solids concentration in the feed (C_S) by a mass balance of the solids fraction. Assuming a one-to-one transformation of droplets to dried particles with no coalescence, agglomeration, or attrition, the relationship can be approximated by Eq. 3.2 (Maa et al. 1998a; Kemp et al. 2013; Vehring et al. 2020), where ρ_P is the particle density.

$$d_P = d_D \cdot (C_S/\rho_P)^{1/3} \qquad (3.2)$$

The formula assumes spherical droplets and solid non-porous particles, negligible volatile content in the particles, and all droplets contain the same solids concentration. Thus, the final particle size is proportional to the droplet size and a lesser degree to the solid concentration (power of 1/3). If the density is the true density of the solid, the equation yields a minimum particle size. However, if the particles are porous or hollow (e.g., because an outer crust has formed), the mean density is lower, and the particle diameter is larger.

Typical droplet lifetimes in a spray dryer, and hence particle formation rates, can occur over a range from milliseconds to many seconds. This time is controlled by the initial liquid droplet size and the evaporation rate (Vehring et al. 2020). A helpful formula for calculating the minimum droplet evaporation time (t) for small particles is shown in Eq. 3.3 (Kemp et al. 2016).

$$t = \rho_D d_D^2 \Delta H_{ev} / 4 Nu \lambda_{\text{gas}} \left(T_{\text{gas}} - T_{wb}\right) \tag{3.3}$$

The required drying time is proportional to the square of the droplet diameter (d_D). Considering a typical case with $T_{\text{gas}} = 80$ °C, $T_{wb} = 30$ °C, $\rho_D = 1000$ kg/m^3, $\Delta H_{ev} = 2500$ kJ/kg, $\lambda_g = 3 \times 10^{-5}$ kJ/mKs, and $Nu = 2$, a drying time of 21 ms for 10 µm particles and 2.1 s for 100 µm particles is obtained (Kemp et al. 2016). This gives a useful indication of the minimum required residence time in a spray dryer. Calculations by Feng et al. (2011) assuming an initial water droplet diameter of 7 µm, a drying temperature of 75 °C, and a drying airflow rate of 100 L/min also resulted in drying times in the order of 10 ms.

3.4.5 Particle Formation in a Spray Dryer

The complete particle formation process in a spray dryer is somewhat complex and can be divided into different stages, such as evaporation, solidification, deformation, and equilibration, as Vehring et al. (2020) described. The evaporation leads to a cooling effect of the gas surrounding the droplets and thus to a small thermal load on the vaccine, which is buffered within the liquid droplets from direct exposure to the high dryer inlet temperature. The internal redistribution of components in the droplets during the evaporation phase is decisive for the particle design (Vehring et al. 2020).

To describe the particle formation process, several authors emphasize the usefulness of the Péclet number as a means to predict the morphology of spray-dried powders (Vehring et al. 2007; Vehring 2008; Gaspar et al. 2014; LeClair et al. 2016b). The dimensionless Péclet number (Pe) given in Eq. 3.4 describes the relative balance of the droplet evaporation rate (κ) (Eq. 3.5) to the rate of diffusion D_i of solutes or suspended components inside the droplet. It is based on the fact that the constant rate of evaporation of a liquid droplet is proportional to its surface area.

$$\text{Pe} = \kappa / 8 D_i \tag{3.4}$$

$$\kappa = hA(T - T_\infty) \tag{3.5}$$

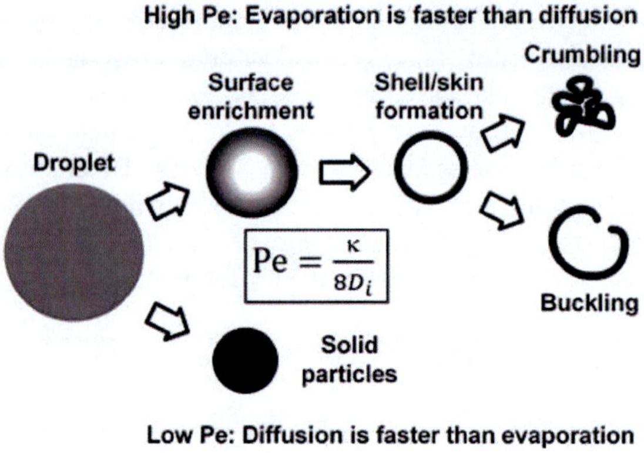

Fig. 3.26 Illustration of a proposed particle formation process and impact of the Péclet number (Pe) (Figure adapted from Vehring (2008), Walters et al. (2014), Vehring et al. (2020) with permission from Springer Nature)

The evaporation rate is increased by a larger temperature gradient between the drying droplet (T) and the air in the spray drying chamber (T_∞) as well as the convective heat transfer coefficient of the process (h) and the surface area of heat transfer (A).

As illustrated in Fig. 3.26, if Pe is high, the evaporation of the solvent takes place faster than the diffusion, and the particle surface becomes enriched in solutes that precipitate. In these cases, an outer layer is formed almost instantaneously at the droplet surface leading to hollow, light, and porous particles. The formation of a shell or skin at high drying temperatures impedes the further shrinking of the droplet so that it is likely that hollow particles with large void volumes are formed. Thus, increasing the inlet temperature of the spray dryer increases the probability of the formation of larger, hollow particles (LeClair et al. 2016b).

On the other hand, for low Pe numbers, the diffusion motion of the solutes is fast compared to the velocity of the decreasing droplet surface, and the droplet is allowed to shrink while solutes migrate to the droplet center (Gaspar et al. 2014). As a result, the solutes distribute evenly inside the evaporating droplet leading to little surface enrichment. At a critical supersaturation level, dense and solid particles are produced with a density close to the true density of the dry components. Typical solid particles are saccharides dried at moderate to low drying gas temperatures (Vehring 2008).

During the deformation stage, partially solid particles may deform to crumbling or buckling structures. The plasticity of pharmaceutical materials (e.g., polymers, proteins, carbohydrates, and others) during spray drying has resulted in a great diversity of particle morphologies (Charlesworth and Marshall 1960; Walton and Mumford 1999; Walton 2000; Gaspar et al. 2014). The morphologies include dense, hollow, porous composite or encapsulated structures with spherical, wrinkled, shriveled, collapsed, or cenospheres (e.g., doughnut-like) shapes (Arpagaus et al. 2017).

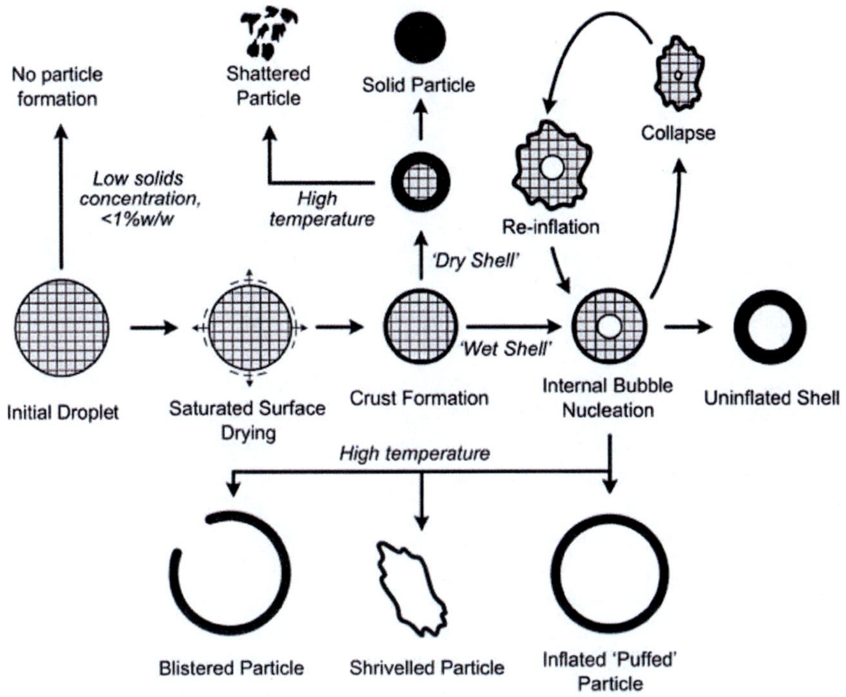

Fig. 3.27 Different particle morphologies that may result when drying droplets containing dissolved or suspended solids (Reproduced from Handscomb et al. (2009) with permission of Elsevier)

As illustrated in Fig. 3.27, numerous particle morphologies may result when drying droplets containing dissolved or suspended solids (Handscomb et al. 2009). Generally, slow-drying leads to more compact particles, whereas fast-drying yields hollow particles (Büchi Labortechnik AG 2020). An increase in solid concentration results in larger and more porous dried particles due to mass balance. A higher feed rate also leads to a larger droplet size, which results in larger dried particles. Of course, the formulation also plays a role in particle characteristics, i.e., type of excipients, total solids content, viscosity of the formulation, etc.

3.4.6 Stickiness, Glass Formation, Residual Moisture, and Reconstitution Time

Stickiness is regarded as a process engineering challenge in spray drying. It leads to product agglomeration, product deposits on the spray chamber wall and raises problems with product caking and agglomeration during packaging of the spray-dried product. In this regard, Table 3.5 shows the qualitative relationship

Table 3.5 Physical properties of some carbohydrates and qualitative stickiness behavior during spray drying

Carbohydrates	Hygroscopicity (relative)	Melting point (°C)	T_g (°C)	Stickiness (relative)
Lactose	+	223	101	+
Maltose	++	165	87	(++)
Sucrose	+++	186	62	+++
Glucose	+++++	146	31	+++++
Fructose	++++++	105	5	++++++

The number of + symbols indicates the relative degree of hygroscopicity and stickiness
Data adapted from Bhandari et al. (1997) with permission from Taylor & Francis

between the physical properties of some carbohydrates used in spray drying of foods and the stickiness behavior based on data from Bhandari et al. (1997).

An optimal formulation for spray drying aims to encapsulate the vaccine in an amorphous excipient matrix that stabilizes the vaccine when dry. As already discussed in Sect. 3.2, for successful drying, a high achievable T_g is a critical factor for the long-term storage stability of vaccines. Consequently, a high T_g also reduces stickiness in the spray dryer.

Since the vaccine concentration is low compared to the excipients, the T_g of the formulation depends mainly on the excipient composition and the residual moisture. Water (T_g of −135 °C) significantly lowers the T_g of the formulation, while trehalose (T_g of 110–121 °C) and high molecular weight components such as dextran (T_g of 152–196 °C) or PVP (140–185 °C) are used to increase the T_g and thus reduce stickiness.

The amino acids L-leucine or trileucine are used as excipients to reduce stickiness (Seville et al. 2007; Mangal et al. 2015; Sou et al. 2016). Due to their surface activity and low water solubility (Gomez et al. 2021b), the amino acids displace both proteins and peptides from the air–liquid interface of the droplets during spray drying, reducing protein aggregation and increasing the dispersity of the dry powder (Lechuga-Ballesteros et al. 2008). Trileucine forms an amorphous phase at room temperature with a high T_g of about 104 °C. Moreover, the addition of L-leucine improves particle formation and spray drying yield (Sou et al. 2016; Price et al. 2020). In particular, the hydrophobic particle coating protects against moisture absorption and improves the aerosol properties of the spray-dried powder, as demonstrated for inhalable vaccines against tuberculosis (Wong et al. 2007; Price et al. 2020; Gomez et al. 2021b), influenza (Sou et al. 2015), or human papillomavirus (HPV) (Saboo et al. 2016).

From an application point of view, the outlet temperature of the spray dryer needs to be lower than the T_g of the excipients to avoid product stickiness and related problems (Breit and DuBose 2013). This situation is shown schematically in Fig. 3.26a with the particle history in a spray dryer assuming an outlet drying gas temperature of 50 °C. The fine spray-dried particles are rapidly dehydrated and experience the spray dryer's approaching outlet gas temperature. For heat-sensitive products such as vaccines, the outlet temperature is usually kept below 50–60 °C.

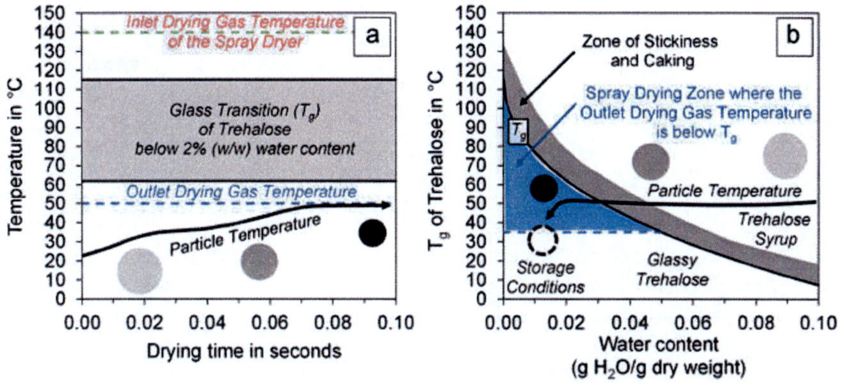

Fig. 3.28 **a** Particle history in a spray dryer with a hypothetical outlet drying gas temperature of 50 °C (Adapted from Breit and DuBose (2013)). **b** The glass transition of spray-dried trehalose particles during water removal (Inspired by Bhandari et al. (1997)). Trehalose is present as syrup at high water contents, forming glassy solid particles at the end of the spray-drying process. The spray drying zone at the outlet is shown as the blue area below the T_g-curve of trehalose up to about 35 °C and up to 5 wt% water content. Trehalose above the T_g at the particle surface can cause stickiness, agglomeration, and caking. The storage conditions of the powder are below T_g and depend on the final water content

Achieving a low residual moisture content in the spray-dried glass is essential for a T_g well above storage temperatures and positively influencing a dried vaccine's chemical and physical stability. Figure 3.28b shows that for successful spray drying of trehalose particles, there is a narrow zone below the T_g curve where the residual water content of the final powder is less than 5 wt% (g H_2O/g dry weight). At higher water contents, trehalose is present as a kind of syrup that forms glassy solid particles at the end of the spray drying process.

If the outlet drying gas temperature is higher than T_g, the particle surfaces will be sticky, leading to agglomeration and caking. For sucrose, for example, the zone of stickiness has been described up to a temperature difference of 20 K between outlet drying gas temperature and T_g (Bhandari et al. 1997). Finally, powder storage requires temperatures below the glass transition (e.g., below 35 °C) to prevent viscous flow.

Since the outlet temperature is a direct function of the inlet temperature, feed rate, solids concentration, and moisture in the drying gas (see discussion in Sect. 3.4.2), a slight change in outlet temperature can significantly affect the structural and storage properties of the product.

As shown in Fig. 3.29 for an aqueous solution, the outlet temperature and residual moisture content are almost linearly dependent on inlet temperature but vary inversely with the liquid feed rate (Abdul-Fattah and Truong 2010). A higher inlet temperature reduces the relative humidity of the drying gas, resulting in a drier and less sticky powder. Therefore, the inlet temperature must be set as high as possible and the temperature difference between inlet and outlet as low as possible

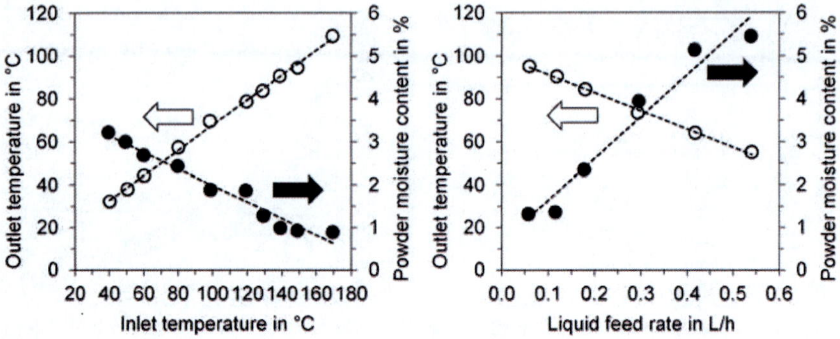

Fig. 3.29 Effect of inlet drying gas temperature (at 1 mL/min feed rate) and liquid feed rate (at 150 °C T_{in}) on outlet temperature and residual moisture content in a spray-dried monoclonal antibody (mAb) powder (Data adapted from Abdul-Fattah and Truong (2010) with permission from John Wiley and Sons)

to minimize the residual moisture content in the final product. Conversely, a higher feed rate lowers the outlet temperature and results in a powder with higher residual moisture content (Maa et al. 1998a). However, too high feed speed can lead to wet and sticky particles that adhere to the spray chamber wall.

Dehumidifying the drying air before entering the spray dryer additionally helps to reduce moisture. However, due to the short residence time in the spray dryer, a small amount of water remains in the spray-dried particles. For example, residual moisture contents of 1.2–4.9% in trehalose (Kanojia et al. 2016), around 3% in trehalose-sucrose (Ohtake et al. 2010), 4.2–6.5% in mannitol (Schüle et al. 2008), and 5.4–8.4% in lactose powder (Maa et al. 1997) were achieved on laboratory spray dryers depending on the application. To further reduce the residual moisture content, a secondary drying step, e.g., in a vacuum dryer or microwave dryer, may be required to meet product quality requirements (Masters 1991; Miller and Gil 2012; Dobry et al. 2015).

A fast reconstitution time for spray-dried vaccine powders in water is essential for administration by injection. As discussed by several authors (Bowen et al. 2013; Batens et al. 2018; Massant et al. 2020), the reconstitution time of a protein powder depends on protein concentration, the solubility of protein and excipients, the type of stabilizers (e.g., amino acids like arginine) and sugars (e.g., sucrose and trehalose), the specific surface area of the dry powder, surface properties, and the procedure of reconstitution, agitation intensity, and medium. As an example, Table 3.6 shows the order of magnitude of reconstitution of lyophilized and spray-dried monoclonal antibodies (mAbs) (human immunoglobulin *G*, IgG) formulations in water for injection (Dani et al. 2007).

Reconstitution of the lyophilized powder (high-dose 200 mg/mL) by the slow swirling of the vial took an average of 13 min (reproducibility from 11 to 20 min). In contrast, spray-dried IgG formulations stabilized with sucrose or trehalose (molar ratio 360:1 sugar:protein) showed a reconstitution time of about 5 min. The faster

Table 3.6 Reconstitution times for lyophilized and spray-dried human immunoglobulin *G* (IgG) formulations stabilized with sucrose or trehalose at high-dose 200 mg/mL monoclonal antibodies (mAbs) concentration in water for injection

Formulation (200 mg mAb per mL of water)	Reconstitution time (min)	
	Average	Range
Lyophilized bulk	~13	11 to 20
Spray-dried IgG + trehalose	~5	4 to 7
Spray-dried IgG + sucrose	<5	4 to 6
Spray-dried IgG + trehalose + Tween 20	~3	2.5 to 3.5

Data reproduced from Dani et al. (2007) with permission from Elsevier

dissolution was attributed to the larger surface area (1 to 4 μm particles). Besides, small amounts of polysorbate (Tween 20) improved the dissolution kinetics to about 3 min.

Another study (Bowen et al. 2013) compared the reconstitution time for spray-dried and freeze-dried mAb powders formulated with a 2:1 and 1:2 mAb: trehalose weight ratio. First, each powder was reconstituted with 1 mL of purified water to prepare a protein solution of 25 mg/mL mAb. Then, the vials were agitated at 450 rpm on a shaker. Most mAb powders were dissolved entirely within 2 min. Overall, the reconstitution time of the spray-dried and freeze-dried powders was comparable and independent of the mAb:trehalose weight ratio.

Overall, in developing an optimal spray-dried vaccine formulation, various excipients such as stabilizers, surfactants, sugars, polymers, buffers are typically investigated. The range of options provides formulation specialists with a whole playing field for improving the handling, flowability, and stability of a spray-dried vaccine powder. Surfactants, such as polysorbates, stabilize the liquid/gas interface and, due to their surface activity, reduce the protein concentration at the droplet surface during the spray drying phase, thus preventing interface-induced denaturation of the protein (Maa et al. 1998b; Adler and Lee 1999; Adler et al. 2000). Nevertheless, the actual effects of buffers, stabilizers, preservatives, surfactants, etc., on the stability of a particular spray-dried vaccine need to be determined on a case-by-case basis.

More details of vaccine case studies supporting these findings are described in Chap. 6 for several vaccine examples, including influenza (Sect. 6.3), measles (Sect. 6.4), hepatitis B (Sect. 6.5), tuberculosis (Sect. 6.6), and others (Sect. 6.7). In addition, the selected excipients and their role are discussed in more detail.

3.4.7 Product Stresses During Spray Drying

Despite the many advantages of spray drying, some challenges need to be considered (Ståhl et al. 2002; Snyder and Lechuga-Ballesteros 2008; Fourie et al. 2008; Greb 2009; Abdul-Fattah and Truong 2010; Walters et al. 2014; Sou et al. 2015; LeClair et al. 2016b; Grasmeijer et al. 2019; Arpagaus 2020; Morgan et al. 2020;

Roth and Scherließ 2021). The main challenges during spray drying which can affect the immunogenicity of the vaccine are

- thermal stress, and
- shear stress during the atomization process.

The transferred thermal energy weakens intramolecular bonds and leads to changes in the native form of a protein. In addition, dehydration stress caused by the removal of water molecules from the surface of the antigen can also destabilize proteins.

The outlet temperature of a spray dryer is considered the most critical operating parameter for determining the residual activity of spray-dried heat-sensitive products. Ideally, the temperature used in the spray drying process should be high enough to ensure efficient drying but not too high that the integrity of the vaccine is compromised to prevent loss of activity and shortening of shelf life. Thus, a compromise must be found between the required outlet temperature for efficient drying and achieving a low moisture content.

The aggregated high molecular weight protein (HMWP) is often used to measure thermal degradation. Figure 3.30a shows a strong correlation between increased HMWP with higher outlet temperature in spray-dried insulin powders (Ståhl et al. 2002). At about 120 °C, HMWP aggregates increased significantly, indicating that the outlet temperature should be kept below 120 °C to avoid thermal degradation. Theoretical optimization of the process resulted in an outlet air temperature of 61 ± 4 °C.

Figure 3.30b shows how spray drying of IgG1 monoclonal antibodies (mAbs) adversely impacted product integrity at inlet temperatures above the unfolding temperature (Abdul-Fattah and Truong 2010). The onset of process-induced mAb aggregation, as measured by DSC, occurred at an inlet drying air temperature of about 140 °C, which corresponded approximately to an outlet temperature of 90 °C. Below that temperature, the loss of mAb aggregation was minimal.

Broadhead et al. (1994) observed that the residual enzymatic activity of spray-dried β-galactosidase without protective excipients was lost at about 50 to 60 °C outlet air temperatures and increased significantly at higher temperatures. Trehalose and arginine enhanced the activity during spray drying and maintained it during storage at 40 °C for one year. Figure 3.30c illustrates the effect on the residual activity of trehalose formulations at inlet temperatures between 105 and 200 °C and clearly shows the ability of trehalose to preserve the enzymatic activity up to about 90–100 °C outlet temperature, which corresponded to an inlet temperature of about 140 °C. The moisture content of the final formulation was typically between 2 and 5%.

In conclusion, even though the inlet drying air temperature may exceed 100 °C in normal spray drying conditions, thermal denaturation of proteins is not generally observed because the temperature of the droplet containing a protein solution hardly exceeds the wet-bulb temperature of water (below 40 °C), and due to the formulation with stabilizers.

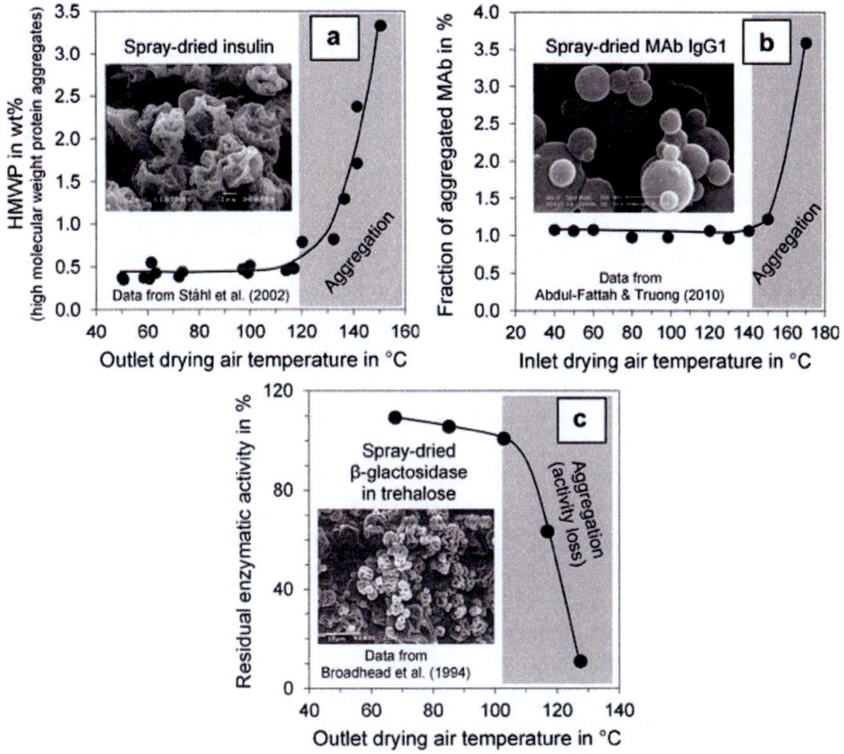

Fig. 3.30 **a** Degradation of spray-dried insulin to high molecular weight protein aggregates (HMWP) as a function of outlet drying air temperature. Spray drying conditions: Mini Spray Dryer B-191, 0.5 mm two-fluid nozzle with cooling water circulation, 500–800 L/h N_2 atomizing gas, 120–300 mL/h feed rate (Data and SEM picture adapted from Ståhl et al. (2002) with permission from Elsevier). **b** Process-induced mAb aggregation of IgG1 monoclonal antibody (mAbs) at an inlet drying air temperature of approximately 140 °C corresponding to an outlet temperature of around 90 °C. Below this temperature, the loss of mAb aggregation was minimal (Data adapted from Abdul-Fattah and Truong (2010) with permission from John Wiley and Sons, SEM picture adapted from Abdul-Fattah et al. (2007b) with permission from Elsevier). **c** The effect of spray dryer outlet temperature on the residual enzymatic activity of trehalose-stabilized β-galactosidase. Spray drying conditions: Mini Spray Dryer B-190, two-fluid nozzle, 2 mL/min feed rate, 500 L/h air atomization rate (Data and SEM picture adapted from Broadhead et al. (1994) with permission from Oxford University Press)

The temperature required for thermal denaturation of proteins is a function of water content and increases sharply with decreasing water content towards the dried solid state. Therefore, dried proteins are relatively stable with denaturation temperatures exceeding 100 °C (Hageman 1988; Snyder and Lechuga-Ballesteros 2008). However, while most of the drying process lasts only a fraction to a few seconds, spray drying can prevent prolonged exposure of the particles to high temperatures (> 100 °C).

Besides thermal stress, the spray-dried vaccine is subjected to additional shear stress and a rapidly increasing air–liquid interface during the atomization step (Maa and Hsu 1997; Maa et al. 1998b; Snyder and Lechuga-Ballesteros 2008; Abdul-Fattah and Truong 2010; Walters et al. 2014; Langford and Ohtake 2019; Morgan et al. 2020). Combining shear stress with the rapid formation of an air–liquid interface can lead to protein unfolding, followed by possible aggregation of protein components and loss of biological activity (Maa and Hsu 1997).

On the other hand, proteins were reported to withstand shear rates on the order of spray atomization (Maa and Hsu 1996; Maa et al. 1998b). However, proteins with high surface activity tend to gravitate more towards the air–water interface of the droplet and are more susceptible to atomization-induced degradation. Human growth hormone (Mumenthaler et al. 1994; Maa and Hsu 1997; Maa et al. 1998b) and bovine serum albumin (Adler et al. 2000) with a distinct amphiphilic nature have been reported to adsorb at the air–water interface and tend to aggregate due to the interaction of the hydrophobic region (Abdul-Fattah and Truong 2010).

It has been demonstrated that the addition of a small amount of surfactant (e.g., 0.1 wt% polysorbate 20 or 80) to the feed solution is effective in preventing proteins from appearing on the surface of spray-dried particles and in reducing surface denaturation and formation of protein aggregates during atomization (Mumenthaler et al. 1994; Adler and Lee 1999; Adler et al. 2000).

However, since the commonly used surfactants (i.e., polysorbates and plurones) are liquid at ambient conditions, they can lead to cohesive particles, limiting their applicability in powder form (Snyder and Lechuga-Ballesteros 2008). Lechuga-Ballesteros et al. (2008) have shown that trileucine, a surface-active tripeptide that is a solid at ambient conditions and has a high T_g of about 104 °C, displaces both proteins and peptides from the air–liquid interface of the droplets, reducing protein aggregation during spray drying. Notably, aerosol performance was also improved of the spray dried powder. Recently, trileucin has been successfully used to develop a formulation platform for a spray-dried inhalable tuberculosis vaccine candidate (Gomez et al. 2021b).

Laboratory studies by Grasmeijer et al. (2019) comparing a Mini Spray Dryer B-290 and a Nano Spray Dryer B-90 showed the droplet generating principle (i.e., nozzle type and configuration) is an important factor in maintaining protein stability. Atomization, heating, ultra-sonication, and recirculation of the feed solution negatively affected the stability of sensitive model protein L-lactic dehydrogenase. In addition, cooling the two-liquid nozzle provided an important advantage over the non-cooled vibrating mesh nozzle.

A previous study by Bürki et al. (2011) demonstrated that the spray mesh size of the Nano Spray Dryer B-90 and the inlet drying temperature influenced the residual activity of model protein β-galactosidase. Nevertheless, the process conditions appeared to be mild and well suited for protein drying. Full enzyme activity was retained when optimized process parameters were applied (80 °C inlet and 36 °C outlet temperature, 4.0 um spray mesh).

More recently, Roth and Scherließ (2021) studied chymotrypsin as a model protein in a Nano Spray Dryer B-90 and identified by fluorescence assays that both the shear forces by ultrasonic vibrations of atomization and the drying process affected the protein structure. However, the enzyme activity did not change. Further experiments would be necessary to clarify the cause of activity loss and to find the best process parameters for a stable product.

In another parametric study using a Mini Spray Dryer B-290, LeClair et al. (2016b) studied the effect of gas flow rate (shear stress) and inlet temperature (thermal stress) on the viral vector activity of a spray-dried recombinant human type 5 adenovirus vaccine encapsulated in mannitol/dextran (weight ratio 2:1, T_g of 130 °C). Figure 3.31 shows the effects on the total viral vector titer loss using a surface plot. A higher gas flow rate increased shear, which atomized the liquid into smaller droplets and increased the air–water interface accordingly. This increased the rate of denaturation for the viral vector proteins and thus the loss of viral activity. Furthermore, a higher inlet temperature also resulted in the denaturation of the labile viral vector. Besides, the titer loss was reduced by increasing the solid concentration of mannitol and dextran, selected as a suitable excipient matrix for thermal stabilization of the viral vector. Overall, the work highlighted the importance of optimal spray drying parameters for maintaining viral vector activity.

A recent follow-up study by Morgan et al. (2020) examined the effect of shear stress on adenovirus activity during spraying at room temperature (25 °C). Adenovirus was dispersed in a 4 or 8 wt% excipient solution of mannitol/dextran (3:1 weight ratio) and sprayed into 50 mL Falcon tubes at various shear rates by adjusting the atomization gas and liquid feed rates using a 0.7 mm two-fluid nozzle. The activity loss due to shear was small (0.16–0.49 log), compared with the total activity loss during the complete spray drying process of approximately 0.5–4 log (i.e., including shear during atomization and heat exposure during spray drying).

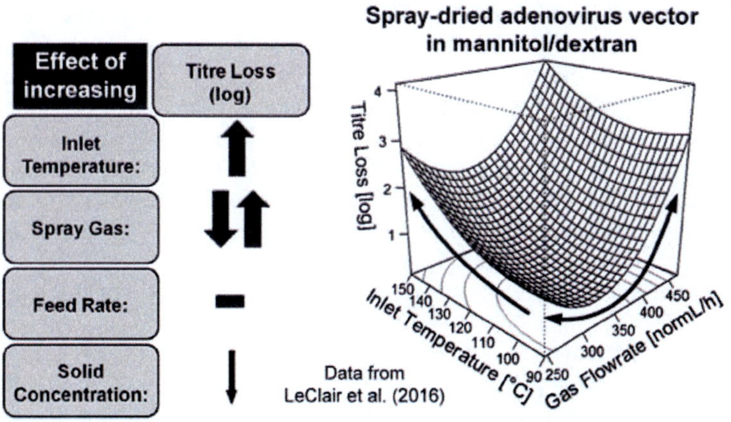

Fig. 3.31 Effect on adenovirus vector titer loss (log) versus inlet temperature (thermal) and gas flow rate (shear) in a Mini Spray Dryer B-290 with 0.7 mm two-fluid nozzle (Data adapted from LeClair et al. (2016b) with permission from Springer Nature)

Overall, the authors concluded that shear stress in the two-fluid nozzle had little effect on the biological agent. Even at the highest shear rate, viral activity was maintained.

Another option to minimize shear stress is to use a two-fluid nozzle with ultrasonic vibration (Truong-Le et al. 2012, 2014; Büchi Labortechnik AG 2014), which can be operated at relatively low pressure and enables the processing of complex biopharmaceuticals such as vaccines, as shown for live-attenuated influenza vaccines (Lovalenti et al. 2016) (see more details in Chap. 6, Sect. 6.3).

3.4.8 Comparison of Spray Drying to Freeze Drying

Table 3.7 compares the main benefits and challenges of spray drying and freeze drying concerning vaccines (Germishuizen et al. 2010; McAdams et al. 2012; Upperton 2015; Vehring et al. 2020). Freeze drying has already established itself as a kind of gold standard for vaccine drying. The drying process is reliable and aseptic in which the vials can be sealed in a freeze-drying chamber before removal. On the other hand, spray drying is the most mature alternative drying technology to lyophilization to stabilize biopharmaceuticals (Walters et al. 2014; Langford et al. 2018b, c). In many cases, spray drying and lyophilization are complementary technologies (Schwartzbach 2011). The end product is a dried biopharmaceutical stabilized in an excipient matrix in both cases. Numerous research studies in the laboratory scale show a lively research activity with spray drying (see especially Chap. 6 with applications). However, no spray-dried vaccine is listed in the WHO database (WHO 2021). In addition, there is a lack of regulatory precedent. Spray drying is still considered an emerging technology in the field of vaccines and is thus in the early stages of commercialization.

While freeze-drying can dry in vials and aseptically enclose samples at the end of the freeze-drying process, spray-dried powders require an additional filling step to dose the vaccine powder into vials, blister packs, cartridges, syringes, bags, or other medical devices. There is an additional risk of material loss (Vehring et al. 2020), and powder filling systems for sterile powder, and effective dosage of 1–2 mg remain challenging. Some gravimetric powder filling systems are available on the market, as presented in Chap. 4 (Sect. 4.6.3). Latest developments in microdosing using vacuum-pressure technology have widened the possibilities of aseptic powder filling to constantly low dosing values of down to 20–30 mg with up to 2% dosing accuracy (ROMACO 2018; Cattenati 2019). Furthermore, there may be difficulties handling hygroscopic and electrostatically charged powders (Walters et al. 2014). A spray-dried powder may adhere to the glass of a vial, making vial inspection difficult (Gladd 2014).

Equally important is that both freeze-drying and spray drying are scalable technologies and maintain consistent product quality across all production scales. Spray drying is a highly reproducible process that can be predictably scaled up to almost any production size (Schwartzbach 2011). For small batches of difficult-to-dry powders to be supplied in vials, lyophilization has an advantage,

Table 3.7 Benefits (+) and challenges (−) of spray drying and freeze-drying for vaccines

Attribute	Freeze-drying (Lyophilization)	Spray drying
Application to vaccines	(+) The gold standard process for drying vaccines	(+) Many research studies on a laboratory scale (−) Lack of commercialized vaccine applications (−) Lack of regulatory precedent
Sterile manufacturing	(+) Established technology (+) Straightforward aseptic processing (+) Equipment widely available	(+) Emerging technology (−) Aseptic processing is more challenging but available (−) Lack of existing aseptic facilities
Final dosage form	(+) Produced in the dosage form in vials (−) The dried cake requires reconstitution before injection	(+) Flowable powder in one step allows diverse delivery options (e.g., pulmonary, intranasal, epidermal) and easy dosage (+) Improved storage and handling of powders (−) Requires additional filling step (+) Versatile vaccine powder filling options (e.g., into vials, blister packs, cartridges, syringes, bags, or other medical devices) (−) Powder requires reconstitution
Scale-up	(−) More challenging to scale up (apparatus)	(+) More straightforward scale-up for the drying (e.g., established for pharmaceuticals) (−) Powder filling of small amounts
Capacity	(−) Smaller volumes (−) Batch process	(+) Higher capacity (+) Continuous process
Yield	(+) Drying takes place directly in a vial (+) Limited process losses (+) Yield almost 100%	(−) It depends on particle size and spray dryer size (−) Limited recovery yield in laboratory scale (50 to > 70%) (+) Higher recovery in pilot and commercial scale (>95%)
Powder properties	(−) Limited control in powder properties (−) Compact solid bulk (cake) (−) Limited product formats and delivery options (−) Additional milling/grinding/micronization steps needed to get powder	(+) Controllable, uniform product format (tunable particle size, distribution, shape, morphology) (+) Particle engineering: powders can be designed for alternative delivery routes (e.g., pulmonary, nasal delivery) (+) Powder stability by encapsulation and for storage
Vacuum	(−) Need for vacuum	(+) No need for a vacuum. It runs at atmospheric pressure
Drying time	(−) Time-consuming method (usually 24 to 48 h)	(+) A quick process, drying in seconds, rapid, instantaneous drying
Heat-sensitive products	(+) Extremely temperature-sensitive process	(+) Suitable for heat-sensitive materials (+) No freezing required

(continued)

Table 3.7 (continued)

Attribute	Freeze-drying (Lyophilization)	Spray drying
	(+) Suitable for heat-sensitive materials (+) Avoids cold storage of dried product (−) Requires freezing step	(−) Thermal stress despite evaporative cooling effect (max. thermal load is the outlet gas temperature) (+) Avoids cold storage of dried product
Shear	(+) No shear stress	(−) Potential shear stress on biologics
Cost	(−) Higher capital/operations costs (−) Higher electrical power requirements	(+) Lower capital/operation costs

whereas spray drying is advantageous if free-flowing powders are required to produce large powder quantities.

A significant advantage of freeze-drying is that the drying process takes place directly in a vial so that a yield of 100% is usually achieved even with smaller sample quantities. However, in laboratory spray dryers, the yields can be significantly lower than in commercial plants and optimized process conditions. Also, yields are not only process-dependent but also formulation-dependent. Typically, yields on a laboratory-scale are around 50–65% (Jin and Tsao 2013; Kemp et al. 2016) but can exceed 70% when using a high-performance cyclone (Maury et al. 2005) and police filter (shown in Chap. 4, Fig. 4.1), or an electrostatic precipitator like in the Nano Spray Dryer B-90 (Schmid et al. 2009, 2011; Arpagaus and Meuri 2010; Li et al. 2010b). This feature is critical to keep in mind to avoid rejection of spray drying technology due to incorrect selection of process conditions or equipment. Pilot spray dryers with kg-scale batch size routinely recover more than 95% of the spray-dried product (Vehring et al. 2020), and industrial-scale spray dryers under optimized conditions even 98–99% (Schwartzbach 2011), however not 100%, which can be an issue considering its implementation for high-cost therapeutics (Walters et al. 2014).

Besides, spray drying provides the flexibility of particle engineering (Vehring et al. 2007, 2020; Vehring 2008; Arpagaus 2018; Langford and Ohtake 2019). A spray-dried product is, by nature, a powder. The product properties such as particle size, shape, morphology, bulk density, dispersibility, flow properties, and so on can be tailored and optimized by formulation, spray atomization, and drying conditions. These properties drive product characteristics such as surface composition, powder flowability, wetting, and rehydration.

On the other hand, freeze-drying offers fewer options for particle property modification. It produces a packaged solid "cake" product inside a vial. Before further processing in a possible powder filling step, a freeze-dried product must be milled or sieved if required.

Figure 3.32a–d shows some SEM pictures examples of both lyophilized/milled and spray-dried powder formulations containing Norovirus vaccine (Vodak et al. 2012),

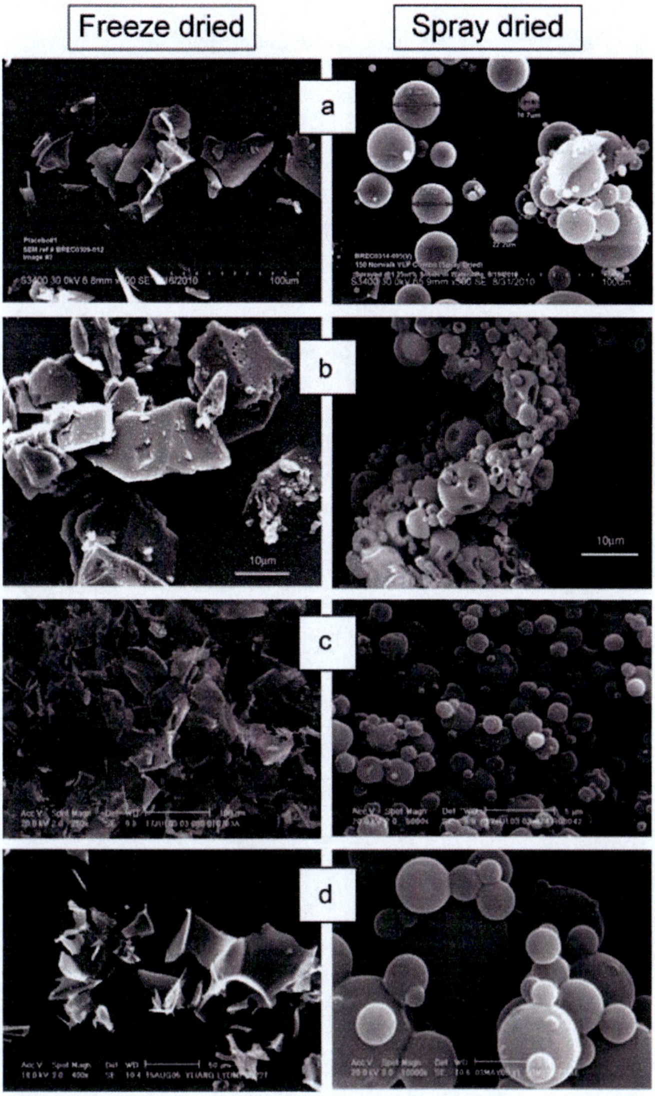

Fig. 3.32 SEM pictures of freeze-dried (left) and spray-dried (right) powder containing: **a** Norovirus vaccine (Vodak et al. 2012), **b** Insulin (Maltesen and van de Weert 2008), c Methionyl human growth hormone in trehalose with Tween-20 (Abdul-Fattah et al. 2008), and **d** Immunoglobulin G (IgG1) antibody with sucrose (Abdul-Fattah et al. 2007b) (Images reproduced with permission from Elsevier and Springer Nature)

insulin (Maltesen and van de Weert 2008), methionyl human growth hormone in trehalose with Tween-20 (Abdul-Fattah et al. 2008), and immunoglobulin G (IgG1) antibody with sucrose (Abdul-Fattah et al. 2007). The particle morphologies of the spray-dried products are more spherical and uniform in shape and size than a jet-milled/lyophilized cake. It should be noted that the particle morphology is process- and formulation-dependent. For example, many protein formulations spray-dried without surfactant have a raisin-like morphology after spray-drying. With the addition of surfactant, the particles are more spherical.

The lyophilized products appear fragmented into irregular shards with varying aspect ratios, while the spray-dried products are spherical. Irregularly shaped particles made by grinding are more likely to aggregate due to charge, thus inhibiting free-flow properties (Roser 2006).

In many instances, it is advantageous to have a dry biopharmaceutical product in a powder form, which facilitates its conversion into capsules, blisters, tablets (Greb 2009; Al-Zoubi et al. 2021) or dispersion in quick dissolving oral thin films (Mao et al. 2010; Ohtake et al. 2011a; Truong-Le et al. 2015). Spray drying eliminates the need for additional milling or sieving, thus reducing manufacturing costs. In that sense, spray drying represents a convenient process that generates powder products with unique particle characteristics (Jin and Tsao 2013). The spray-dried powder can be mixed, compacted, and filled into many desired product formats by suitable aseptic powder treatment and filling processes while maintaining the improved stability of the dried solid product (Vehring et al. 2020).

Furthermore, the encapsulation of vaccines by spray drying in free-flowing powders enables alternative delivery routes that are not feasible with other drying techniques (Langford et al. 2018b, c). Alternative routes of administration could be:

- inhaled powder administered to the lungs by dry powder inhalers (Germishuizen et al. 2010; Vehring et al. 2020),
- powder deposited intranasally (Vodak et al. 2012; Tlaxca et al. 2015; Bahamondez-Canas and Cui 2018),
- enteric-coated (micro)capsules delivered orally to specific regions of the intestinal tract (Brown 1996; Vass et al. 2019), or
- powder to form microneedles on a patch applied on the skin (Gladd 2014).

The adjustable particle sizes from 3 to 5 μm make spray drying particularly attractive for inhaled vaccine therapies with dry powder inhalers. In this context, reference is made to Sect. 4.6.3 in Chap. 4, which presents various delivery devices for spray-dried products, including aseptic powder filling in vials, blisters, cartridges for dry powder inhalers, dual-chamber disposable syringes, and dual-chamber tubes.

While spray drying atomizes a liquid feed into fine droplets and achieves rapid drying continuously in one step, freeze drying is a batch process that can often take 2–3 days before the dried product is available (GEA 2021). The long processing time carries the risk of losing an entire batch if, for some reason, the process conditions are not maintained throughout the freeze-drying cycle. In addition, freeze drying safety systems increase installation and operating costs (Ledet et al. 2015). The preparation time of the vaccine substances and the purification time of the equipment after the experimental runs (i.e., labor costs) are comparable for both technologies (Jin and Tsao 2013). However, freeze-drying requires a vacuum, which is time and energy-consuming for large batch sizes, while spray drying runs at atmospheric pressure, allowing energy savings (Walters et al. 2014). In addition, freeze-dryer loading and unloading require direct human intervention for manual loading or more complex, high-precision automatic loading and unloading equipment increasing cost (Searles and Cherian 2015).

In contrast to lyophilization, where the capacity of the freeze-dryer limits the production, spray drying can be operated continuously, sometimes even over several days, to achieve the required high quantities of powder products (De Costa 2014). This flexibility ultimately helps to reduce the batch costs, hence, driving down the unit cost of a finished biological product. Continuous processing is beneficial for the commercial manufacturing of products required in high quantities, and this is one area where spray drying can be applied. However, powder filling is needed after spray drying, which can dampen the continuous process, as integrating these two elements can be challenging. Examples of aseptic powder filling are presented in Sect. 4.6 of Chap. 4.

As described by Searles and Cherian (2015), the economics and scalability of spray drying are such that a single spray dryer of similar size and cost to a freeze dryer can produce 5–10 times the amount of product. GEA Niro (2017) also stated that a normal-size aseptic spray dryer could replace approximately 6–8 large freeze dryers, significantly reducing capital and operating costs and simplifying plant operations. Geng (2018) reported that one SPX Anhydro MS-1250 aseptic spray dryer could replace approximately 12 freeze dryers with a capacity of 56,000 vials or 6 large freeze dryers with a total of 112,000 vials.

The investment costs for an aseptic spray dryer are process-dependent and product-specific. Aseptic spray dryers are designed to meet customer-specific requirements with CIP, SIP, process validation, documentation effort, and high-level control. Consequently, the costs increase compared to a spray dryer for a pharmaceutical application.

Also, aseptic spray drying of vaccines requires aseptic powder filling, increasing the total investment costs (Andersen 2020). Several aseptic powder filling systems are shown in Chap. 4 (Sect. 4.6.3). Therefore, it can be assumed that the price of a complete aseptic spray drying line will be comparable to that of freeze-drying.

Overall, it can be concluded that spray drying can eliminate bottlenecks in vaccine production. Spray drying is faster (i.e., drying in seconds rather than days for a batch) and can be used for high product throughput (Roser 2009; Schwartzbach 2011; Breit and DuBose 2013; Geng 2018; Andersen 2020; Arpagaus 2020). Furthermore, spray drying offers particle engineering and additional powder dosage forms, although more formulation effort may be required.

3.5 Spray Freeze Drying of Vaccines

This section explains the principle of spray freeze drying and compares the process characteristics with spray drying and freeze drying. In addition, some application examples with vaccines, including pulmonary, intranasal, and epidermal routes of powder delivery, are discussed. In the end, an overview of commercially available equipment is given.

3.5.1 Principle and Comparison to Spray Drying

Spray freeze-drying is an evolving technology that combines spray drying and freeze-drying elements to produce a free-flowing powder. The process steps involved include atomization and generation of droplets, rapid freezing, and drying by sublimation.

In contrast to spray drying, the liquid feed is atomized directly into a cold vapor or liquid serving as a cryogenic medium, such as liquid nitrogen. The large gas–liquid interface of the droplets results in rapid freezing, suitable for thermally unstable compounds, such as vaccines. Freezing concentrates the core compound in the wall matrix of stabilizers by solidifying the solvent, mostly water, into tiny ice crystals. The frozen particles are collected and further sublimated by a freeze-drying step under a vacuum to produce powder in dry form. During the sublimation phase, the ice crystals formed leave pores in the dried particles. The particle size and density are controlled by the solids concentration and chemical composition and the atomization, freezing rate, and temperature of the cryogenic medium.

Figure 3.33 shows the two primary approaches used in spray freeze-drying vaccine powders (Cheow et al. 2011; PowderPro AB 2013; Ledet et al. 2015; Sebastião et al. 2019; Eber et al. 2021). First, the feed is sprayed directly into liquid nitrogen (−196 °C) (Fig. 3.33a). The nitrogen is then removed, and the frozen granules are freeze-dried in separate equipment (Maa et al. 2003; Eber et al. 2021).

In the second approach (Fig. 3.33b), the sample is sprayed into a chilled chamber, collected on a filter, and then freeze-dried with a stream of cold, desiccated gas (Ledet et al. 2015). Alternatively, the frozen granules are fed either directly into a rotating drum or a heated transport system (e.g., a tray of vibrating shelves or a vibrating belt) for lyophilization under vacuum conditions (Bosshammer 2014; Wanning et al. 2015; Pisano et al. 2019; Sebastião et al. 2019; Adali et al. 2020; Meridion 2020). Continuous drum rotation at low speeds (0.5–1 rpm) ensures a gentle circulation of the powder product during the drying process (Bosshammer 2014).

More and more technical solutions have been developed to continuously produce pharmaceutical powders in large quantities by spray freeze-drying in recent years. On this topic, Adali et al. (2020) have provided a comprehensive overview and described advanced equipment concepts, consisting of spray freeze chambers with heated transport systems (e.g., vibratory belts), stirring systems, or cylindrical

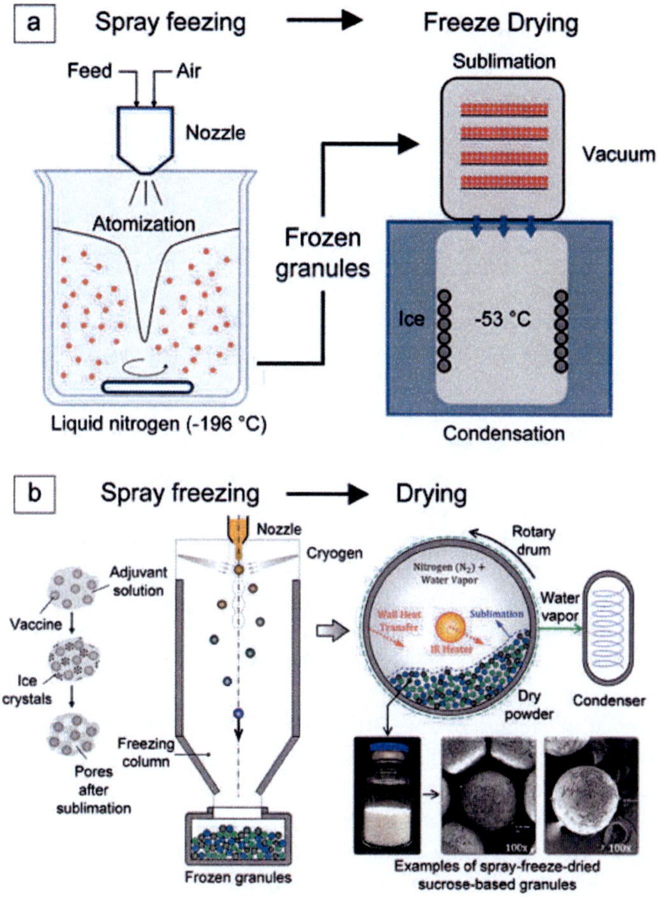

Fig. 3.33 Approaches to spray-freeze drying: **a** Spray-freeze-drying into liquid nitrogen and transfer of frozen granules to freeze-dryer, **b** Atmospheric spray-freeze-drying in freezing column with cryogen. Collection of frozen granules and drying (sublimation) in a rotary drum (Adapted from Cheow et al. (2011), PowderPro AB (2013). Ledet et al. (2015), Sebastião et al. (2019), Eber et al. (2021) with permission from Elsevier)

rotary drums for drying the frozen powders. Finally, the interested reader is referred to (Adali et al. 2020) and (Pisano et al. 2019) for a more general discussion of continuous processes and to other comprehensive reviews of spray freeze drying principles and their applications (Walters et al. 2014; Ledet et al. 2015; Wanning et al. 2015; Langford et al. 2018a, b, c, d; Emami et al. 2018; Sebastião et al. 2019).

Table 3.8 compares some process characteristics of spray drying and spray freeze drying (Maa et al. 1999; Ledet et al. 2015; Wanning et al. 2015; Sebastião et al. 2019). The scalability of the two processes is comparable and straightforward, as the speed of atomization mainly limits the productivity of both. However, spray freeze-drying is more time-consuming and generally more complex to handle with

Table 3.8 Process comparisons between spray drying and spray freeze drying

Factors	Spray drying	Spray freeze drying
Operation	Fast drying process	Fast freezing, more time-consuming drying process due to freezing and sublimation
Scalability	Straightforward	Comparable to spray drying but more complex due to freezing by liquid nitrogen
Operation costs	Lower	Higher
Yield	50 to > 70% (for lab-scale) > 95% (industrial-scale)	> 95%
Aerosolization	Good	Very good (at least 50% better)
Particle size	1 to 200 μm	10 to 800 μm

Based on literature Maa et al. (1999), Ledet et al. (2015), Wanning et al. (2015), Sebastião et al. (2019)

two process steps, i.e., spray freezing and freeze-drying. In addition, the frozen powders must either be transferred to a freeze-drying section or freeze-dried in place. Accordingly, the operating costs are assumed to be higher.

Moreover, spray freeze-drying is a comparatively new method for producing biopharmaceutical powder preparations. A further challenge is developing large-scale equipment with a capacity of over 100 kg of powder per day. Section 3.5.3 shows that industrial equipment is gradually becoming available. In addition, some continuous systems are on the market, which allows easier processing of larger powder quantities.

A significant advantage of spray freeze drying on a laboratory-scale is a high production yield of over 95% (Ledet et al. 2015). Yields in laboratory spray dryers are typically lower, ranging from about 50 to 65% (Jin and Tsao 2013; Kemp et al. 2016), but can exceed 70% when using a high-performance cyclone (Maury et al. 2005). In industrial-scale spray dryers with a clean-wall drying chamber and continuous operation with subsequent powder recovery in a highly efficient cyclone and filter arrangement, yield values above 95% are achieved (Schwartzbach 2011; Vehring et al. 2020).

Figure 3.34 shows some SEM pictures of spray freeze-dried vaccine powders. Compared to spray drying, where the size and shape of the particles are created during the drying process, the particle size in spray freeze-drying is almost equal to the initial droplet size (Wanning et al. 2015). The dried particles are usually spherical, and the internal structure is highly porous. As a result, spray freeze-drying yields larger granules in a wide range of about 10–800 μm (Wanning et al. 2015) that are uniform, free-flowing, dust-free, and dosable into vials with flexible quantities, e.g., even for the personalized medication (Langford et al. 2018d). On the other hand, spray freeze-dried particles may be fragile, affecting powder handling and shipping.

In contrast to conventional freeze-drying in vials, the general sequence of operation is reversed by having the lyophilization process before the vial filling process. As a result, the increased particle surface of the frozen particles leads to high heat transfer during sublimation, and the drying time is shorter (Walters et al. 2014; Langford et al. 2018b, c).

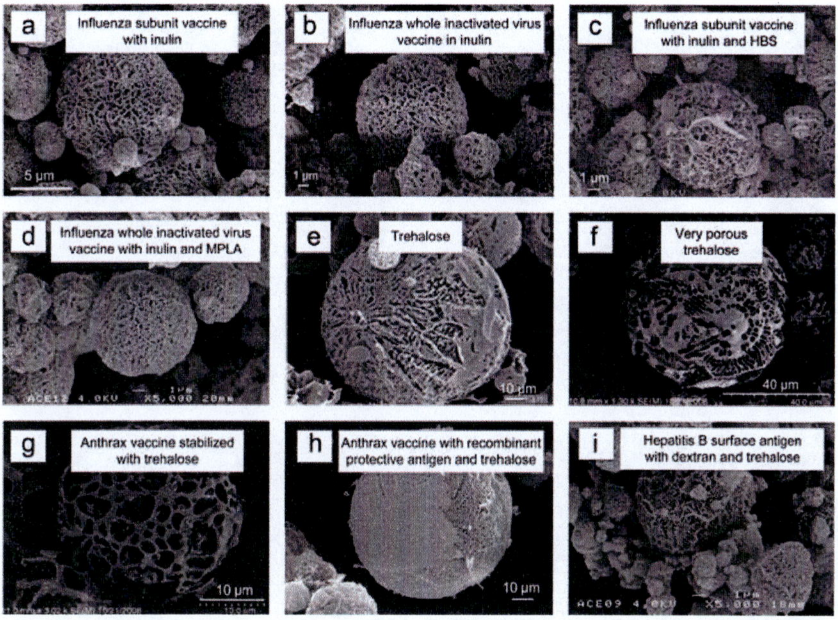

Fig. 3.34 SEM pictures of spray freeze-dried vaccine powder: **a** Spherical and highly porous influenza subunit vaccine powder for pulmonary delivery (Amorij et al. 2007), **b** Highly porous particles of influenza whole inactivated virus vaccine in inulin for pulmonary immunization (Audouy et al. 2011), **c** Influenza subunit vaccine powder with inulin and buffered by HEPES buffer saline (HBS) (Saluja et al. 2010), **d** Whole inactivated virus influenza vaccine powder with inulin and supplemented with monophosphoryl lipid A (MPLA) as a lyoprotectant (Patil et al. 2014), **e** Trehalose (Garmise et al. 2007), **f** Spherical and very porous trehalose with low tapped density (0.08 g/mL) (Wang et al. 2012), **g** Anthrax vaccine powder for nasal administration with hollow porous structure using trehalose as stabilizer (Wang et al. 2012), **h** High-porosity vaccine powder of around 70 µm with Anthrax recombinant protective antigen (rPA) (Mikszta et al. 2005), and **i** Hepatitis B surface antigen vaccine prepared by a combination of dextran and trehalose and after 3 months of storage at 60 °C (Tonnis et al. 2014) (Images reproduced with permission from Elsevier, Springer Nature, and Oxford University Press)

3.5.2 Applications of Spray Freeze-Dried Vaccines

To date, dry powder vaccine formulations produced by spray freeze-drying have drawn attention to pulmonary, nasal, and epidermal applications for their potential advantages over conventional liquid vaccines, including greater stability and longer shelf life, as well as the elimination of the need for refrigeration for storage and distribution, thereby facilitating mass vaccination (Sullivan et al. 2006). Table 3.9 summarizes various research studies based on literature.

Table 3.9 Research studies of spray freeze-dried vaccine powders described in the literature

Delivery route	Vaccine type	Excipients	Particle size	References
Pulmonary	Influenza (subunit)	Inulin	4–30 μm	Amorij et al. (2007)
	Influenza (subunit)	Inulin	4–26 μm	Saluja et al. (2010)
	Influenza (whole inactivated virus)	Inulin	2–11 μm	Audouy et al. (2011)
	Influenza (whole inactivated virus)	Inulin, dextran, dextran/trehalose	3–15 μm	Murugappan et al. (2013)
	Influenza (whole inactivated virus)	Inulin, various adjuvants	1–17 μm	Patil et al. (2014, 2015)
	Hepatitis B	Inulin, dextran/trehalose	1–10 μm	Tonnis et al. (2014)
Nasal	Influenza (whole inactivated virus)	Trehalose	26.9 μm (median)	Garmise et al. (2007)
	Anthrax (antigens)	Trehalose	~ 70 μm (mean)	Mikszta et al. (2005)
	Anthrax (antigens)	Trehalose	~ 70 μm (mean)	Jiang et al. (2006)
	Anthrax (antigens)	Trehalose	8–73 μm (mean ~ 25 μm)	Wang et al. (2012)
	Plague	Mannitol, trehalose, Tween 80	Not available	Huang et al. (2009)
Epidermal	Influenza (subunit)	Trehalose/mannitol/dextran (3:3:4 w/w/w)	31–63 (mean 46 μm)	Chen et al. (2003), Maa et al. (2004), Dean and Chen (2004)
	Influenza (split-virion)	Trehalose/mannitol/dextran (4:4:2 w/w/w)	39–46 μm	Maa et al. (2004)
	Diphtheria and tetanus toxoid	Trehalose/glycine/dextran	38–53 μm	Maa et al. (2003)
	Hepatitis B	Tehalose/mannitol/dextran (3:3:4 w/w/w)	30–60 μm	Maa et al. (2007)

Vaccine powders from influenza subunits and whole inactivated viruses were successfully prepared by spray freeze-drying and investigated via pulmonary immunization in mice (Amorij et al. 2007; Saluja et al. 2010; Audouy et al. 2011; Murugappan et al. 2013; Patil et al. 2014, 2015). Spray freezing was performed by spraying the vaccine formulation in liquid nitrogen with a 0.5 mm two-fluid nozzle of a Mini Spray Dryer B-190 (Büchi Labortechnik AG, Switzerland). The frozen particles were then dried in a freeze dryer for about 11 h.

Inulin, dextran, and mixtures of dextran/trehalose were evaluated as protective agents (Murugappan et al. 2013), as well as various adjuvants to enhance systemic and particularly mucosal immune responses (Patil et al. 2014, 2015). Inulin was used due to its favorable physicochemical properties, such as high and variable T_g depending on molecular weight, as well as crystallization resistance (Hinrichs et al. 2001) (see Table 3.2). High porosity particles were produced (Fig. 3.34d), which enabled them to dissolve within seconds when brought in contact with water. The specific surface areas were found to be about 60–100 m^2/g for the spray freeze-dried powders (Patil et al. 2014, 2015). The powders were reported to be stable at ambient temperature for 3 years (20 °C, 10% relative humidity) (Saluja et al. 2010), and 3 months stored up to 30 °C (Murugappan et al. 2013).

Besides, Tonnis et al. (2014) showed that hepatitis B surface antigen (HBsAg) in an amorphous matrix of inulin or a combination of dextran and trehalose by spray-freeze drying resulted in a formulation that is more stable than the aluminum hydroxide containing formulation. Furthermore, the resulting powder could be stored at room temperature for at least 3 months without a change of antigenicity or conformation of the HBsAg protein.

Research studies on intranasal administration of spray freeze-dried vaccine powder formulations include vaccines against anthrax (Mikszta et al. 2005; Jiang et al. 2006; Sullivan et al. 2006; Wang et al. 2012), influenza (Huang et al. 2004; Garmise et al. 2007), and plague (Huang et al. 2009).

Garmise et al. (2007) reported about intranasal administration of influenza vaccines to rats as possible influenza prophylaxis, as the virus enters the host in this way. Serum IgG antibody titers were comparable to intramuscular administration, and mucosal IgA antibody titers were significantly higher. The dry powder showed improved stability over liquid formulations under various storage conditions.

Studies with rabbits (Mikszta et al. 2005) showed that a spray freeze-dried particle size of around 70 μm was appropriate for deposition in the nasal cavity and an effective route to anthrax vaccination. Based on these findings, Jiang et al. (2006) prepared spray freeze-dried formulations at pH 7–8 using trehalose as a stabilizer. The dry powder maintained the protein integrity at ambient and accelerated aging conditions at 25 and 40 °C for about 1 month, while the liquid formulation showed rapid degradation. Therefore, such powder formulations with the non-invasive intranasal administration were considered an attractive approach for mass biodefense immunization.

In further studies, Wang et al. (2012) showed that the efficacy of spray freeze-dried anthrax recombinant protective antigen powder stored at room temperature for 2.5 years was comparable to a freshly prepared liquid formulation. The produced powders were highly porous with an average size of 25 μm (Fig. 3.34g). In addition, storage stability at 40 °C could be increased up to 2 weeks as long as the powder was exposed to limited humidity.

Furthermore, Huang et al. (2009) demonstrated the preclinical feasibility of reconstituted spray freeze-dried powder vaccine for vaccination against the plague in a mouse model. The intranasal delivery led to 80% protection but was less effective than intramuscular injection or intradermal administration with microneedles. Therefore, an additional dose of vaccine was required to achieve similar protection, attributed to the required absorption through the intranasal mucosal barrier.

In addition to pulmonary and intranasal administration, several vaccine powder formulations for epidermal powder immunization (EPI) have been prepared by spray freeze drying, including antigens against influenza (Chen et al. 2003; Maa et al. 2004; Dean and Chen 2004), diphtheria, tetanus (Maa et al. 2003), and hepatitis B (Tonnis et al. 2014).

EPI provides a needle-free immunization technology designed to deliver antigens to the viable epidermis, the superficial layer of skin containing many Langerhans cells, but lacking blood vessels and sensory nerve endings (Chen et al. 2002; Kim 2017). The Langerhans cells detect, capture, and present antigens to evoke an immune response to invading pathogens. To deliver the powdered vaccines into the epidermal layer of the skin, powder injection methods, such as the PowderJect® ND device, were used to accelerate the particles using high-pressure helium as a propellant (Fig. 3.35). Two different powder formulations have been developed for effective EPI. One uses the precipitation of antigens on 1–2 μm gold particles, and the other is based on the formulation of antigens into larger particles using mono- and disaccharide excipients. The lower density of sugar formulations ($\leq$ 1.5 g/ml) compared to gold (18 g/ml) is compensated by a larger particle size (20–70 μm) to achieve effective skin penetration (Chen et al. 2002; Maa et al. 2003).

Maa et al. (2004) developed high-density sugar particles (tap density $\geq$ 0.5 g/mL) by spray freeze-drying and increasing the solid content of the feed solution to about 35 wt%. An optimized trehalose/mannitol/dextran (10 kDa) formulation was developed in a 3:3:4 ratio with influenza antigen. The flash-frozen and lyophilized particles produced a vaccine powder with appropriate characteristics for EPI. Using ultrasonic nozzle atomization (60 or 120 kHz, Sono-Tek, USA), large particles of about 35–56 μm were obtained at high yields of 82–89%. The powder showed suitable physical properties and stability over 1 year when stored at room temperature (25 °C), obviating the need for an entire cold chain.

Chen et al. (2003) demonstrated in mice and monkey models that QS-21 is a potent adjuvant for augmenting the immunogenicity of co-administered influenza vaccine in non-human primates. Later on, Dean and Chen (2004) conducted a Phase I clinical trial with 36 adults, and the powdered trivalent influenza vaccine was tested with the PowderJect® ND delivery system (Fig. 3.35). The EPI of the influenza vaccine was declared safe and induced a humoral immune response in humans. Site reactions following EPI were primarily mild and self-limiting.

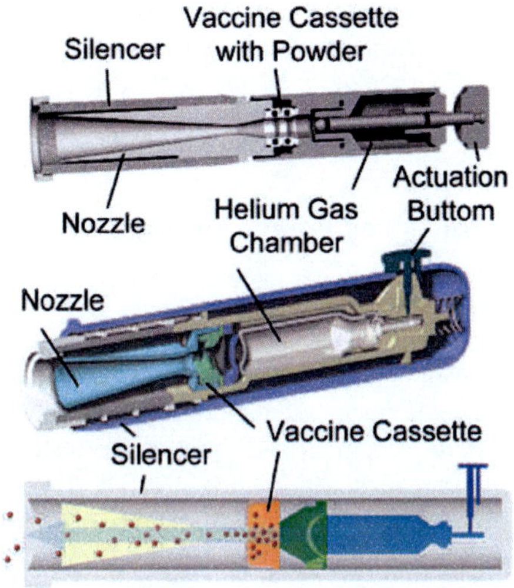

Fig. 3.35 Cross-sectional views of the PowderJect® ND delivery device for epidermal powder immunization (EPI). The device comprises a single-use helium gas cylinder (5 mL volume, 45 bar), a nozzle, a silencer, and a vaccine powder cassette. The cassette contains a polyurethane thermoplastic elastomer washer sealed at both ends with polycarbonate membranes. The device is actuated by depressing the actuation button and breaking the helium microcylinder valve. The helium release from the gas chamber ruptures the cassette membranes and propels the vaccine particles through the nozzle. The particles are accelerated to high velocity with enough force to penetrate the epidermal layer of the skin (Figures adapted from Chen et al. (2000a), Dean and Chen (2004), Kim (2017) with permission from Springer Nature and Elsevier)

Spray freeze drying also offers the advantage of formulating vaccines with aluminum salts, which are often sensitive to slow freezing in conventional freeze-drying. On the other hand, rapid freezing leads to higher nucleation rates and reduced growth of large ice crystals. In this context, Maa et al. (2003) developed stable dry powders of alum-adjuvanted hepatitis B and diphtheria and tetanus toxoid vaccines suitable for EPI. The spray freeze-dried hepatitis B surface antigen (HBsAg) powders were tested in hairless guinea pig models (Maa et al. 2007). The process yield was 82–85%, and a trehalose/mannitol/dextran formulation (weight ratio 3:3:4) showed the best process performance, including particle properties for skin penetration. In addition, long-term efficacy over 16 weeks at 40 °C and maintenance of immunogenicity in guinea pigs was observed.

Overall, it can be stated that several studies showed that spray freeze-drying is an effective process for producing vaccine-containing powders that can be stably stored at room temperature. In addition, rapid freezing and subsequent sublimation allow drying of temperature-sensitive products in highly porous particle morphology and variable size, enabling non-invasive routes of administration such as pulmonary, nasal, and especially epidermal delivery, which has already been demonstrated in Phase I clinical trials. To date, most spray freeze-drying research on vaccines has been conducted using laboratory-made systems. The spray freeze drying methods are still very experimental. Industrial and regulatory aspects such as qualification, process validation, good manufacturing practices have rarely been considered (Searles and Cherian 2015; Wanning et al. 2015). The following section provides an overview of available spray freeze drying equipment.

3.5.3 Equipment for Spray Freeze Drying

There are a few equipment manufacturers that have commercialized spray freeze drying technology. Meridion Technologies GmbH (Germany) has reached this level and is considered the leading technology provider for spray freeze-drying technology on an industrial scale under aseptic conditions. Meridon offers application lab facilities for product development and process verification up to a 20 kg pilot scale and provides project realization on all required scales, covering engineering, project management, equipment design, manufacturing, installation, and start-up and after-sales service (Meridion 2020).

Figure 3.36 illustrates the fully contained sterilizable process line in the production scale. The technology combines an atmospheric spray freezing chamber (SprayCon™ technology) with a rotary drum drying system (Lyo-Motion™ dynamic bulk freeze-drying technology) (Meridion 2020). This patented process (Luy et al. 2012; Plitzko et al. 2012; Struschka et al. 2012) appears to be the first industrial-scale process for the sterile production of spray freeze-dried products (Luy and Stamato 2020).

SprayCon™ uses high-precision frequency-controlled prilling nozzles to break the liquid into individual droplets. The top lid of the spray freezing chamber carries one or more nozzles for droplet generation. The selectable size range of the droplets is between 300 and 1000 μm, depending on nozzle diameter, feed rate, frequency, and viscosity. The spray rates per nozzle range from 10 to 35 g/min (Luy and Stamato 2020).

The droplets flow through a cold gas atmosphere by gravity and freeze into spherical microparticles. Typical drop heights range from 1.5 to 3.5 m with residence times of 1–3 s (Luy and Stamato 2020). The spray freezing chamber is a double-walled cylindrical vessel with the lower end funnel-shaped to guide the frozen microspheres to the exit. The double-wall is cooled with a mixture of liquid and gaseous nitrogen. Typical operating temperatures range from −80 to −150 °C. The freezing step is performed at ambient- or slight overpressure to ensure aseptic processing.

The frozen particles are continuously discharged from the freezing chamber into a pre-cooled drum (around 1.6 m in diameter) of the rotary freeze dryer positioned in a cylindrical vacuum process chamber. The double-walled drum contains cleaning nozzles for steam sterilizing in place (SIP) and cleaning in place (CIP). With the Lyo-Motion™ technology (Bosshammer 2014), the frozen bulk material is gently and continuously freeze-dried by dynamic rotation and mixing under vacuum (e.g., 0.1 mbar or less) (Luy and Stamato 2020). One or more glass-covered thermal infrared radiators provide the required sublimation energy (Meridion 2020).

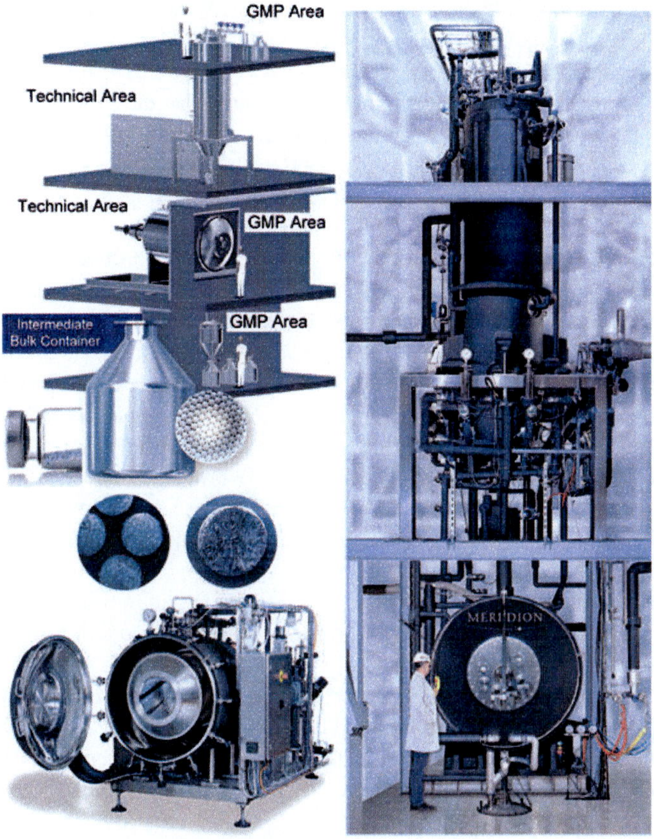

Fig. 3.36 Fully contained sterilizable process line in production scale with spray freezing chamber, dynamic rotating drum freeze dryer, and discharge container. The footprint takes up to 2 or 3 floors (Images courtesy of MERIDION Technologies GmbH, Germany)

Drying times depend on the formulation, particle size, scale and load, and process conditions like drum rotation, pressure, temperatures, and heating rate. A pellet will lose 80–90% of its weight during drying. The resulting microspheres are nearly monodisperse, free-flowing, and dust-free (Luy and Stamato 2020).

By reverse rotation, the bulk material enters integrated discharge blades and is conveyed to the rim of the front opening, where it is collected in a hopper. After drying, the resulting bulk material is loaded into a container coupled to a sterile isolator for powder filling.

Even though this plant configuration is batch-oriented, the technology seems to be the forerunner of new, more flexible, and cost-effective continuous spray freeze-drying processes. In principle, the technology can be operated continuously, with a suitable automated transfer system for the constant discharge of the dried pellets, or as an option, if the spray freeze unit serves two parallel operating drum freeze dryers, which ensure a continuous material flow.

Luy and Stamato (2020) reported that such a process line (Fig. 3.36) was built and successfully tested for a parenteral pharmaceutical product, designed for 100 L (liquid) batch size. With 20% solids content, a processing time of 24 h for lyophilization can be expected, with a total yield of > 97%. The generation of frozen bulk material typically requires 10–20 h, depending on droplet size and spray rate.

Langford et al. (2018a) reported about spray freeze-dried 20% sucrose solution in a laboratory, pilot, and commercial-scale with Meridion's LyoMotion equipment. A 300 μm nozzle was used for the trials, and the pressure in the drying drum was kept between 0.05 and 0.1 mbar. Table 3.10 summarizes the results. The authors reported residual water contents of less than 1% on all scales, and a yield of over 97% was possible on a commercial scale. The highly porous particles promoted rapid wetting and allowed reconstitution in less than 1 min.

In further studies (Langford et al. 2018b, c) processed several formulations containing monoclonal antibodies (mAbs) and sucrose in a weight ratio of 5:2 (mAbs to sucrose) with the spray freeze dryer at Meridion. Free-flowing particles were obtained that were easy to aliquot and re-suspend. Even at a high protein concentration of more than 200 mg/mL, a short reconstitution time of < 7 min was achieved. The results are considered promising for the production of heat-stable vaccine powder formulations.

Spray freeze drying can also be performed continuously, as demonstrated by the aseptic spray freeze drying technology LYnfinity (IMA Life 2019a). IMA Life holds patents for bulk freeze-drying using spray freezing and agitated drying with heating (DeMarco and Renzi 2017; Ganguly et al. 2019).

Table 3.10 Spray freeze-drying results for 20% (w/w) sucrose solution processed at the lab, pilot, and commercial-scale using Meridon's LyoMotion equipment

	Lab-scale (RFD LyoMotion LAB)	Pilot-scale (LyoMotion 30)	Commercial-scale (LyoMotion 200)
Amount processed (kg)	1	6	107
Spray rate per nozzle (g/min)	19 g/min (1 nozzle)	19 g/min (1 nozzle)	26 g/min (3 nozzles)
Droplet size (μm)	550	550	550
Freezing gas temperature (°C)	−150	−150	−120
Drying time (h)	5.5	16	29.25
Yield (%, w/w)	98.6	81.6	97.3
Residual water content (wt%)	≤ 1.0	≤ 1.0	≤ 0.6
Reconstitution time (min)	≤ 1.0	≤ 1.0	≤ 1.0

Data adapted from Langford et al. (2018a)

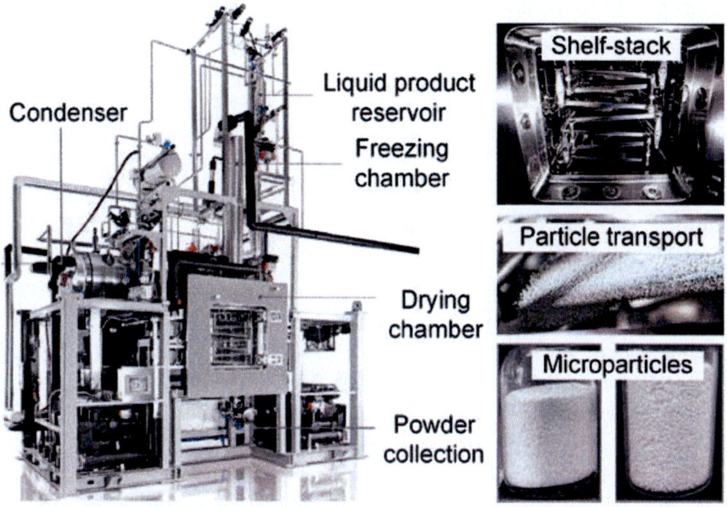

Fig. 3.37 Continuous aseptic spray freeze drying technology LYnfinity with cascading shelf-stack allowing for rapid sublimation and high productivity (Images by courtesy of IMA Life North America, USA)

Figure 3.37 illustrates the LYnfinity technology, characterized by total containment and continuous aseptic production from formulation to powder filling (e.g., into vials, syringes, inhalation systems). The process allows a significant increase in productivity and a reduction in downtime. The intended applications include antibiotics, bacteria, vaccines, and almost 50% of the biopharmaceutical products listed by the FDA and EMA (IMA Life 2020).

The nozzle of LYnfinity operates at a specific frequency and splits a laminar liquid jet into uniform droplets with low shear stress. The droplets enter a cryogenically cooled freezing tower, which is double-walled and cooled by liquid nitrogen and silicone oil. The gas inside the cooling column is kept below −130 °C (IMA Life 2019a, b). The droplet stream falls vertically through the column with the cooled gas. The fast freezing allows maintaining the spherical shape of the droplets. It is estimated that at −150 °C gas temperature, a 500 μm droplet freezes at a distance of less than 1.5 m (IMA Life 2019a, b).

The individual frozen droplets are collected at the bottom of the spray freezing chamber and transferred to the drying chamber. An intermediate chamber, equipped with a special vacuum pump and jacket cooling, allows the transfer between the freezing chamber at atmospheric pressure and the drying chamber under vacuum, without interrupting continuous operation.

The drying chamber is designed for continuous operation. It includes cascaded stacks of vibratory plates, shown in Fig. 3.37, and a dual ice condenser arrangement that allows for condenser change during drying, where one condenser can be defrosted while the other condenser is in operation (IMA Life 2019b). The vibratory movement gently transports the frozen particles on the inclined shelves at a controlled speed to achieve rapid sublimation and prevent agglomeration. Finally, the dried particles are filled into containers for direct powder dosing or storage as bulk material. The system's throughput is up to 4 L/h (wet product, about 100 L in 24 h). Applications with vaccines are not disclosed for confidentiality reasons.

The Japanese company ULVAC Inc. has developed the Micro Powder Dry (μPD series) technology for spray freeze-drying of microparticles (ULVAC 2010a, b). The liquid is sprayed as a liquid column through special nozzles directly into a vacuum chamber (0.05–0.3 mbar). The frozen particles are collected on a heating/cooling belt in the drying chamber. Figure 3.38 shows the different available systems. LyoMoment is a benchtop model suitable for basic research (100 mL batch capacity), μPD400 for small batch production in R&D (5 L batch), and μPD2000 for mass production (150 L) (ULVAC 2010b). Particle sizes from 100 to 400 μm are achieved. Besides, a completely closed design for sterile mass production with powder filling and capping has been patented (Ogata et al. 2013). A cooperation agreement between Azbil Telstar Technologies and ULVAC further develops the concept for a continuous fine-spray freeze-drying process (Bullich 2015; Pisano et al. 2019; Adali et al. 2020). Potential applications are sterile pharmaceuticals such as antibiotics, vaccines, cancer drugs, or blood products in the form of powder, tablets, capsules, or skin patches (ULVAC 2010b).

On a laboratory and small production scale, PowderPro AB (Sweden) provides the freeze granulators LS-2 and LS-6 with a processing capacity of 2–6 L/h (min. 50 mL) of solid suspensions or solutions (PowderPro AB 2013) (Fig. 3.39). The LS

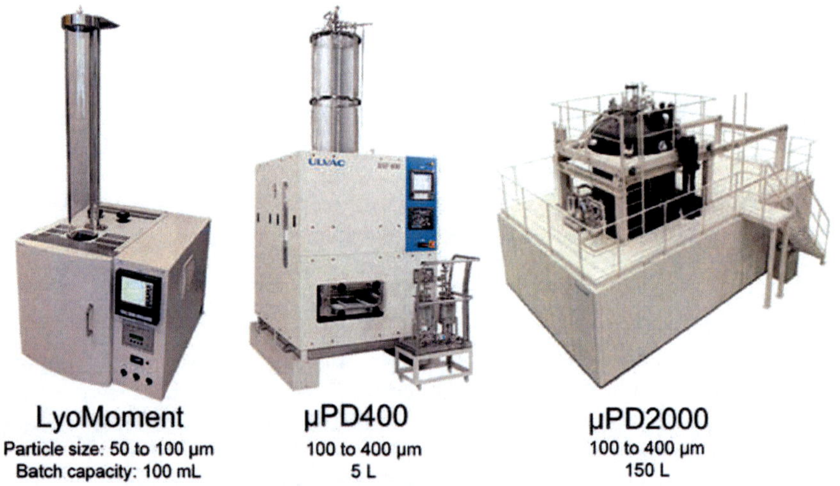

Fig. 3.38 ULVAC micro powder dry (μPD Series) technology for spray freeze-drying of microparticles. LyoMoment (100 mL batch capacity, 50–100 μm particle size), μPD400 (5 L, 100–400 μm), and μPD2000 (150 L, 100–400 μm) equipment (Images by courtesy of ULVAC, Japan)

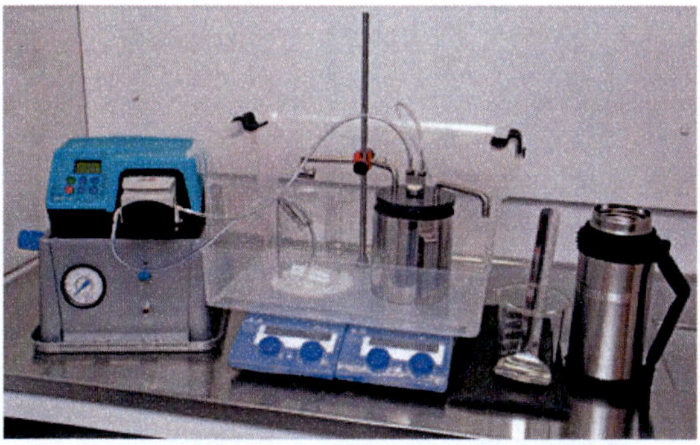

Fig. 3.39 Lab-scale granulator (LS-2) using magnetic or impeller stirring with a capacity of processing small batches of down to 50 mL and up to 2 L/h depending on the material and suspension properties (Image by courtesy of PowderPro AB, Sweden)

models have a feed pump, an air atomizing nozzle, and a spraying chamber with liquid N_2. After spray freezing, the frozen granulates are dried in a separate tray freeze dryer with sufficient capacity.

Finally, the Active Freeze Dryer technology from Hosokawa Micron B.V. is also worth mentioning, which was developed to produce homogeneous batches of sterile material (Van der Wel 2012). The technology is particularly suitable for freeze-drying pharmaceutical applications such as antibiotics, proteins, hormones, viruses, vaccines, bacteria, yeasts, blood serum, liposomes, and transplant materials such as collagen sponges. Solutions, dispersions, pastes, and even wet solids can be processed into a fine, free-flowing, dry powder in a single processing step (Hosokawa Micron B.V. 2019, 2020).

Figure 3.40 shows a schematic of the system components. The technology comprises a jacketed, conical freezing tank with a stirrer and a collecting filter. First, the material is quickly frozen inside the tank. Then, the product is kept in motion and freeze-dried under a vacuum. The energy for sublimation is introduced through the wall and efficiently distributed in the product by the stirrer. The initially coarse granules gradually become smaller by sublimating the connecting ice structure between the frozen material. The SEM images in Fig. 3.40 compare the morphology of a typical active pharmaceutical ingredient (API) from the Active Freeze Dryer with a milled powder dried in a standard tray freeze dryer.

Moreover, it is also a completely closed process designed for aseptic processing, offering Cleaning in Place (CIP) and Sterilization in Place (SIP) by pressurized steam (Hosokawa Micron B.V. 2020). Several batch sizes are available, from a few liters for laboratory work and small-scale production to bulk drying applications involving hundreds of liters. In addition, Active Freeze Dryer models of 1, 5, 60, and 800 L are available for testing at the Hosokawa Micron Test Center in The Netherlands (Hosokawa Micron B.V. 2019, 2020).

In conclusion, spray freeze drying is a promising technology for producing biopharmaceuticals in powder form. However, the availability of commercial spray

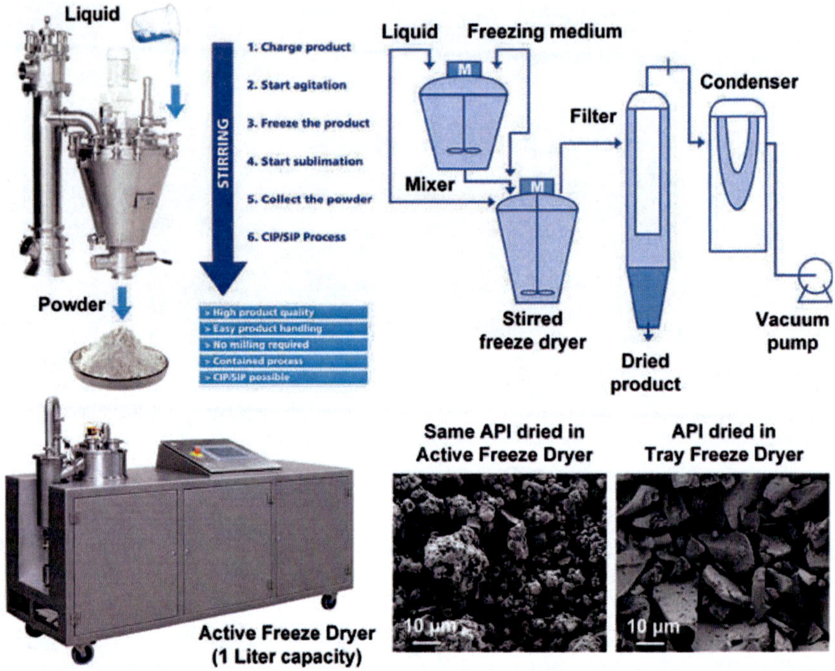

Fig. 3.40 Active freeze dryer technology. Principle drawing, laboratory unit with a 1-L capacity, and SEM pictures of freeze-dried sodium bicarbonate solution (API). Left: Powder product from Active Freeze Dryer, right: Milled API product after conventional tray freeze-drying (Images courtesy of Hosokawa Micron BV, The Netherlands)

freeze-drying equipment on the market is limited, at least according to information from the literature. The most advanced technologies are Meridion's Lyo-Motion™ and SprayCon™, offering bulk freeze-drying with vacuum rotary drum and IMA Life's LYnfinity spray freeze-drying technology enabling a more continuous aseptic operation.

3.6 Vacuum Foam Drying of Vaccines

Vacuum foam drying is an emerging technology for drying sensitive biologics (Vehring 2007; Ohtake et al. 2011b; Langford et al. 2018a, b, c; Lovalenti and Truong-Le 2020). The vaccine-containing liquid solution or suspension is foamed under a vacuum above the freezing point and evaporated to form a foam structure. During foaming, solution bubbles seep upward, burst, and coalesce with neighboring bubbles (cavitation process) (Walters et al. 2014; Ohtake et al. 2015).

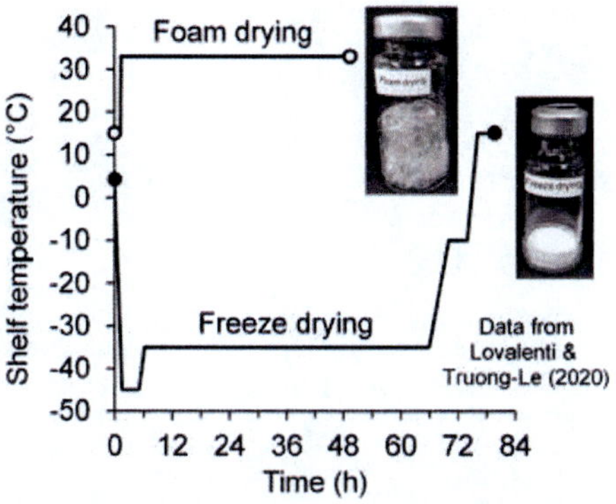

Fig. 3.41 Comparison of representative foam- and freeze-drying shelf temperature profiles in a conventional freeze dryer (Adapted from Lovalenti and Truong-Le (2020) with permission from John Wiley and Sons)

Figure 3.41 shows a typical shelf temperature profile for foam drying compared to a conventional freeze drying cycle (Lovalenti and Truong-Le 2020). Unlike freeze-drying, there is no freezing step so that it can be used with freeze-sensitive vaccines, and the drying process is performed near ambient temperature (e.g., 15–35 °C). After some modification of controls and process-programming software, conventional freeze-dryers can be used to execute an effective foam drying process in stoppered glass vials (Bronshtein 2004). However, the temperature and vacuum must be carefully controlled to avoid freezing due to evaporative cooling.

The resulting dried product is typically a closed-cell foam with macroscopic structure and variable density (Abdul-Fattah and Truong 2010; Ohtake et al. 2015). Figures 3.41 and 3.42a, b, and c show representative samples of foam-dried, freeze-dried, and spray-dried vaccine formulations (Ohtake et al. 2011a; Jangle and Pisal 2012; Lovalenti et al. 2016; Lovalenti and Truong-Le 2020). The foam-dried samples are translucent and brittle and take up a larger volume than a freeze-dried cake or spray-dried powder. Figure 3.42c shows the typical appearance of foam-dried sucrose at different concentrations (Jangle and Pisal 2012). The foam-dried samples have different morphologies and densities depending on the drying cycle and sugar concentration. The foam height and structure can vary from vial to vial within a batch, resulting in lower process robustness and higher reject rates than lyophilization. As the comparison of the morphology of saccharide-rich samples in Fig. 3.42d illustrates, foam-dried structures tend to consist of large thick plates, while freeze-dried samples contain more thin plates and spray-dried particles are spherical (Abdul-Fattah et al. 2007b).

Compared to spray drying, the advantage of foam drying is significantly reducing the air–water interface during drying (Lovalenti and Truong-Le 2020). Process stresses such as surface tension or shear, associated with liquid bubbles and boiling, can be minimized with suitable glass formers, surfactants, and other stabilizers. The reconstitution time for most foam-dried products is generally less than

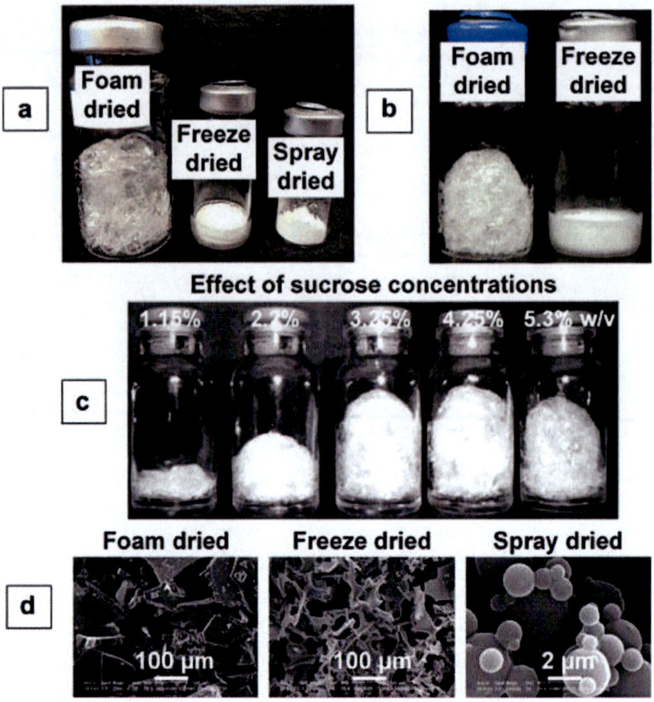

Fig. 3.42 **a** Typical appearance of foam-dried, freeze-dried, and spray-dried vaccine formulations (Images reproduced from Lovalenti et al. (2016) with permission from Springer Nature). **b** Comparison of foam-dried and freeze-dried live-attenuated Salmonella Typhi (Ty21a) bacterial vaccine in 25% sucrose and 25 mM potassium phosphate. The foam-dried structure is translucent and brittle and occupies a larger volume in a vial than a freeze-dried sample of the same composition (Images reproduced from Ohtake et al. (2011a) with permission from Elsevier). **c** Effect of sucrose concentrations on foam-dried product characteristics in various concentrations of 1.15%, 2.2%, 3.25%, 4.25%, and 5.3% w/v (Images reproduced from Jangle and Pisal (2012) with permission from the Indian Pharmaceutical Association). **d** Freeze-dried, foam-dried, and spray-dried morphology of saccharide-rich dried formulations observed by SEM (Images reproduced from Abdul-Fattah et al. (2007b) with permission from Elsevier)

a minute (Lovalenti and Truong-Le 2020). In addition, foam-dried materials typically possess lower specific surface areas than lyophilized and spray-dried materials, leading to stability enhancement (Abdul-Fattah et al. 2007a, b, c; Truong and Abdul-Fattah 2010; Langford et al. 2018c). On the other hand, unlike spray drying, which is very fast, the foam drying process can take about 20–60 h to dry the product as a batch process (see examples in Table 3.11).

Table 3.11 provides an overview of several research studies on foam drying of bacterial and viral vaccines, including:

- Newcastle disease virus (LaSota virus) (Pisal et al. 2006),
- Influenza virus (Truong-Le 2006; Lovalenti et al. 2016),

Table 3.11 A selection of research studies on foam-dried vaccines described in the literature, with details on the excipients used, the cycle of foam drying, and storage stability (time until potency decreased by about 1 log in FFU/ml or $TCID_{50}$/ml)

Vaccine type	Excipients	Foam drying cycle	Storage stability	References
Newcastle disease vaccine (LaSota virus)	Sucrose, PVP, N-Z-amine	10 °C (1 mbar) for 2 h, 28 °C (0.033 mbar) for 18 h	3 weeks at 40 °C	Pisal et al. (2006)
Live-attenuated influenza virus	Sucrose, gelatin, block copolymers	15 °C (0.067 mbar) for 1 h, 33 °C (0.033 mbar) for 48 h	2 years at 25 °C 21 weeks at 37 °C	Truong-Le 2006)
Tularemia (*Francisella tularensis* live vaccine strain)	Trehalose, gelatin, Pluronic F68	15 °C (0.133 mbar) for 40 h, 20 °C (0.133 mbar) for 20 h	12 weeks at 25 °C 1.5 weeks at 37 °C	Ohtake et al. (2011b)
Live-attenuated rabies virus	Sucrose, methylglucoside, phosphate buffer	35 to 45 °C (1 to 4 mbar) for 20 to 24 h	23 months at 22 °C 15 months at 37 °C 3 h at 80 °C	Smith et al. (2015)
Live-attenuated vaccine of porcine reproductive and respiratory syndrome virus	Trehalose, tryptone, other protectants	25 °C (3 mbar) for 1 h, 28 to 30 °C (0.1 mbar) for 12 h, 28 to 30 °C (0.01 mbar) for 18 to 24 h, total process 30 to 42 h	1 year at 25 °C 4 months at 37 °C 25 days at 45 °C	Lv et al. (2016)
Live-attenuated influenza virus	Arginine, gelatin	15 to 45 °C (0.133 mbar) for 72 h	20 weeks at 37 °C	Lovalenti et al. (2016)
Live-attenuated virus vaccine of a parainfluenza strain	Sucrose, Pluronic F-68	15 °C (0.067 mbar) for 1 h, 33 °C (0.033 mbar) for 48 h	20 weeks at 25 °C	Abdul-Fattah et al. (2007a, b, c)

(continued)

Table 3.11 (continued)

Vaccine type	Excipients	Foam drying cycle	Storage stability	References
Salmonella Typhi live bacterial vaccine (Ty21a)	Trehalose, gelatin, methionine	15 °C (0.133 mbar) for 40 h, 20 °C (0.133 mbar) for 20 h	42 weeks at 25 °C 12 weeks at 37 °C	Ohtake et al. (2011a)
Live-attenuated measles virus vaccine (Edmonston-Zagreb)	Arginine, human serum albumin, divalent cations	15 °C (0.067 mbar) for 24 h, 33 °C (0.067 mbar) for 24 h	2 weeks at 37 °C	Ohtake et al. (2010)
Live B/Harbin influenza virus	n.a.	n.a.	18 weeks at 37 °C	Truong and Abdul-Fattah (2010)

- Parainfluenza virus (Abdul-Fattah et al. 2007a, b, c),
- Measles (Ohtake et al. 2010),
- Live B/Harbin Influenza virus (Truong and Abdul-Fattah 2010),
- Salmonella Typhi bacteria (Ohtake et al. 2011a),
- Francisella tularensis bacteria (Ohtake et al. 2011b),
- Rabies virus (Smith et al. 2015), and
- Porcine reproductive and respiratory syndrome virus (Lv et al. 2016).

Table 3.11 also summarizes information on the excipients used, foam drying cycles, and storage stability achieved, indicated as the time required for the vaccine to decrease in potency by about 1 $\log_{10}$ CFU.

Pisal et al. (2006) applied vacuum foam drying to improve the stability of freeze-dried LaSota live virus vaccine for immunization of chickens against Newcastle disease. The foamability improved with sucrose content and by incorporating N-Z-amine and polyvinylpyrrolidone (PVP K30). A combination of sucrose/N-Z-amine/PVP (50/10/3% w/v) provided optimal dehydration and storage stability of 3 weeks at 40 °C. The products were amorphous and showed no devitrification during storage.

In a patent, Truong-Le (2006) reported about foam-dried live-attenuated influenza virus vaccine which could be stored for even 146 days (21 weeks) at 37 °C and at least 2 years at 25 °C (i.e., the time required for a one log order loss in FFU/ml or $TCID_{50}$/ml). The formulation consisted of 40% sucrose, 5% gelatin, and 25 mM 7.2 pH KPO_4 buffer.

In another study, Ohtake et al. (2011b) foam-dried a particularly thermolabile *Francisella tularensis* live vaccine strain, a highly virulent bacterium causing tularemia disease in ticks, rabbits, rodents, and humans. The optimized formulation contained trehalose/gelatin/Pluronic F68 (30/5/0.02% w/v) and enabled storage stability of about 1.5 weeks at 37 °C and 12 weeks at 25 °C. Gelatin decreased the residual water content and increased the T_g.

Smith et al. (2015) formulated live-attenuated rabies virus strain into a stable, dry foam. The formulation contained sucrose and methyl glucoside in phosphate buffer. Virus titer assays showed storage stability for at least 23 months at 22 °C, 15 months at 37 °C, and 3 h at 80 °C. The vaccine effectively induced neutralizing antibodies and protected mice from peripheral rabies infection.

Lv et al. (2016) foam-dried live-attenuated porcine reproductive and respiratory syndrome vaccine. Trehalose, tryptone, and other protectants were used as stabilizers. Storage stability of 12 months at 25 °C, 4 months at 37 °C, and 25 days at 45 °C was achieved, which was sufficient for long-distance transport. Besides, the preserved vaccine was injected into 20-day-old piglets, and the serum titer was at a similar level two weeks after inoculation as with a commercial freeze-dried vaccine.

Several researchers have also investigated the impact of different drying methods on vaccine stability. For example, Table 3.12 provides details on process loss and loss of potency during storage for live-attenuated influenza vaccine (H1N1) (Lovalenti et al. 2016), live-attenuated virus vaccine candidate against human parainfluenza virus (Medi 534) (Abdul-Fattah et al. 2007a, b, c), and Salmonella Typhi live bacterial vaccine (Ty21a) (Ohtake et al. 2011a).

Figure 3.43 illustrates the storage stability of live-attenuated influenza vaccine formulations after foam drying, freeze drying, and spray drying at 37 °C (Lovalenti et al. 2016). The virus titer $TCID_{50}$ (50% infectious dose in tissue culture) of the foam-dried formulations decreased slowly at an average rate of 0.055 log $TCID_{50}$/mL/week, while for the spray-dried formulations, 0.72 log $TCID_{50}$/mL/week and the freeze-dried formulation approximately 0.58 (Nasovac™, commercial freeze-dried product) and 0.66 log $TCID_{50}$/mL/week. Gelatine and arginine were identified as key stabilizers for the foam-dried formulations.

Gelatin was added to increase foaming, confer protection against interfacial stresses, and inhibit aggregation. Arginine provided additional stability and limited mobility. Overall, the foam-dried formulations were about an order of magnitude more stable than the spray-dried or freeze-dried formulations. In addition, the foam-dried vaccines showed complete retention of immunogenicity in a ferret model.

In 2007, Abdul-Fattah et al. (2007a, b, c) compared the stability of a genetically engineered live-attenuated bivalent vaccine candidate against human parainfluenza virus (Medi 534) using foam drying, spray drying, and freeze drying processes with sucrose-based formulations (Table 3.12 b). The order of storage stability at 25 and 37 °C based on square-root-of-time kinetics followed the order:

- foam-dried (with surfactant Pluronic F-68),
- spray-dried (with surfactant),
- foam-dried (without surfactant), and
- spray-dried (without surfactant).

Table 3.12 Comparison of process-associated loss and storage stability for dried vaccine formulations prepared by foam drying, freeze drying, and spray drying

[a]Live-attenuated influenza vaccine (H1N1) (Lovalenti et al. 2016)	Process loss (log $TCID_{50}$/mL)	Rate of titer loss at 37 °C (log $TCID_{50}$/mL/week)
Foam drying	0.4 ± 0.2	0.055 ± 0.011
Spray drying	0.3 ± 0.2	0.72 ± 0.09
Freeze drying	0.5 ± 0.2	0.66 ± 0.12
Nasovac™ (freeze-dried)	n.a.	0.58 ± 0.09
[b]Live-attenuated virus vaccine candidate against human parainfluenza virus (Medi 534) (1:100 vaccine/sucrose by weight) (Abdul-Fattah et al. 2007a, b, c)	Process loss (log $TCID_{50}$/mL)	Rate of potency loss at 37 °C (log $TCID_{50}$/mL/week$^{0.5}$)
Foam drying	0.8	2.31 ± 0.19
Foam drying (with Pluronic F-68)	0.9	1.20 ± 0.19
Spray drying (with Pluronic F-68)	0.4	2.25 ± 0.19
Spray drying	0.8	2.76 ± 0.19
Freeze drying	1.4	2.65 ± 0.30
[c]Salmonella Typhi live bacterial vaccine (Ty21a) (Ohtake et al. 2011a)	Process loss (log_{10} CFU)	Stability loss after 1 week at 25 °C (log_{10} CFU)
Foam drying (25% w/v sucrose, 0.5% w/v methionine, 2.4 wt% DMSO, 25 mM KPO_4, pH 8)	0.3 ± 0.2	0.0 (no loss after 4 weeks at 25 °C)
Spray drying (7% w/v sucrose)	0.7 ± 0.1	1.7 ± 0.1
Freeze drying (28% w/v sucrose, 1% w/v gelatin, 25 mM KPO_4, pH 7)	0.8 ± 0.1	0.5 ± 0.0

[a]Live-attenuated influenza vaccine (H1N1), [b]Live-attenuated virus vaccine candidate against human parainfluenza virus (Medi 534), [c]Salmonella Typhi live bacterial vaccine (Ty21a)
Data reproduced from Abdul-Fattah et al. (2007a, b, c), Ohtake et al. (2011a), Lovalenti et al. (2016) with permission of Elsevier and Springer Nature

The superior storage stability of the foam-dried formulations with surfactant (around 20 weeks at 25 °C) was attributed to the very low specific surface area (SSA) of about 0.1 m^2/g dried product, low vaccine surface coverage during drying, and low molecular mobility within the glassy matrix. These findings were also supported by Truong and Abdul-Fattah (2010) on the storage stability of live influenza B/Harbin vaccine, in which the foam-dried vaccine was the most stable (4.2 months at 37 °C) and the spray-dried powder was the least stable (0.4 months). In addition, the foam-dried powder showed the lowest SSA of 0.06–0.09 m^2/g, while the spray-dried powder exhibited an SSA of 2.6–3.1 m^2/g.

Ohtake et al. (2010) showed that foam-dried measles virus vaccine retained higher storage stability than spray drying or freeze drying. However, foam drying resulted in more significant process losses of virus titers. The optimized formulation tested at 37 °C was stable for approximately 8 weeks (i.e., time for 1 log $TCID_{50}$

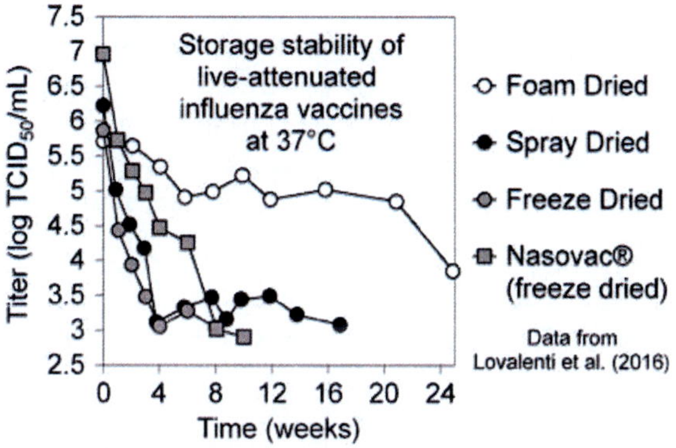

Fig. 3.43 Stability profiles at 37 °C of foam-dried, freeze-dried, and spray-dried formulations of live-attenuated influence vaccine (spray-dried with sucrose, trehalose, and surfactant Pluronic F-68). Nasovac™ is a commercial freeze-dried product marketed by the Serum Institute of India (Data adapted from Lovalenti et al. (2016) with permission of Springer Nature)

loss), exceeding the WHO heat stability requirement of 1 week at 37 °C. The key stabilizers identified during formulation screening were arginine, human serum albumin, and a combination of divalent cations.

Finally, Ohtake et al. (2011a) presented data on an oral live-attenuated vaccine against Salmonella typhi (Ty21a) using trehalose (sucrose), methionine, and gelatine. The optimized foam-dried formulation showed stability of more than 4 weeks at 37 °C and 42 weeks at 25 °C, compared to 12 h and 2 weeks for commercial freeze-dried Vivotif™ vaccine (i.e., potency loss < 0.5 $\log_{10}$ CFU) (Table 3.12 c). In contrast, the spray-dried (7% w/v sucrose) and freeze-dried formulations (28% w/v sucrose, 1% w/v gelatin, 25 mM KPO_4, pH 7) lost 1.7 and 0.5 $\log_{10}$ CFU after 1 week of storage at 25 °C. Moreover, the foam-dried vaccine induced a similar immunogenic response in mice and similar protection after challenge as commercial freeze-dried Vivotif™.

In summary, the foam-dried application examples show improved storage stability compared to lyophilization or spray drying, and the immunogenicity in animal models is comparable to freeze-dried vaccines. The enhanced storage stability can be mainly attributed to the lower specific surface area and the optimal composition of excipients to produce thermostable formulations. Looking ahead, foam drying allows for an increase in preservation time due to improved heat stability and thus the potential for long-term storage, shipping, and distribution, and stockpiling, as well as a reduction in cold chain dependence.

Overall, it can be concluded that vacuum foam drying appears to be a suitable and scalable technology for preserving sensitive biotherapeutics and vaccines in the dry state (Bronshtein 2004; Pisal et al. 2006; Ohtake et al. 2011a). Commercial freeze-dryer systems with particular temperature and pressure control and aseptic processing are available for large-scale processing and scale-up. The main difference between the equipment requirements for foam drying and freeze drying is the condenser moisture load. The moisture load is higher for foam drying due to the high initial rates of moisture losses during the boiling phase and therefore requires a larger condenser capacity. However, challenges remain in process robustness for pilot plant and commercial scale processing (Lovalenti and Truong-Le 2020). To date, no foam-dried vaccine product has made it to market (Lovalenti et al. 2016) and clinical test results are lacking according to the literature reviewed.

3.7 Supercritical Fluid Drying Technology

Supercritical fluid drying is an alternative drying technology for food and biopharmaceutical powders. It is fast and can be performed under mild temperature conditions (Sievers et al. 2001; Jovanović et al. 2004; Amorij et al. 2008; Walters et al. 2014; Emami et al. 2018). The most popular fluid is carbon dioxide (CO_2) since it has a low critical temperature of 31.1 °C (73.8 bar), is nontoxic, nonflammable, and inexpensive. Supercritical CO_2 also has unique solvation properties, polarity, and the potential of recovery and residue-free. The FDA recognizes CO_2 as safe in pharmaceutical processing.

Typically, supercritical fluid methods can be subdivided into two groups, i.e., those that use CO_2 as a solvent and those that use CO_2 as an antisolvent. As a solvent, the process is called Rapid Expansion of Supercritical Solutions (RESS), and as an antisolvent, it is known as gas antisolvent (GAS) or supercritical antisolvent (SAS) (Van Oort and Sacchetti 2007; Kankala et al. 2017).

For the stabilization of drug delivery and related biomedical application examples as dry powder formulations using supercritical fluid technology, the interested reader is referred to various review papers (Kompella and Koushik 2001; Jovanović et al. 2004; Van Oort and Sacchetti 2007; Walters et al. 2014; Kankala et al. 2017).

Examples of laboratory-scale and commercial equipment manufacturers of supercritical fluid systems include Separex SAS (France), Extratex (France), SFE Process (France), NATEX Prozesstechnologie GmbH (Austria), Apeks Supercritical (USA), and AKTIV-DRY LLC (USA) (Walters et al. 2014).

Supercritical fluid drying is a relatively new drying technology with considerable potential for vaccines. Examples include dry powder vaccines for pulmonary delivery, including.

- diphtheria toxoid (Amidi et al. 2007),
- live-attenuated measles virus (Sievers et al. 2007b; Burger et al. 2008; Cape et al. 2010; Kisich et al. 2011; Lin et al. 2011; Agarkhedkar et al. 2014),
- human papillomavirus (HPV) (McAdams 2011; McAdams et al. 2012), and
- Hepatitis B surface antigen (HBsAg) (Sievers et al. 2007b).

Prof. em Robert (Bob) Sievers and his co-workers from the University of Colorado (Boulder, USA) have mainly driven the research on vaccines. The developed supercritical fluid drying technology using Carbon Dioxide-Assisted Nebulization

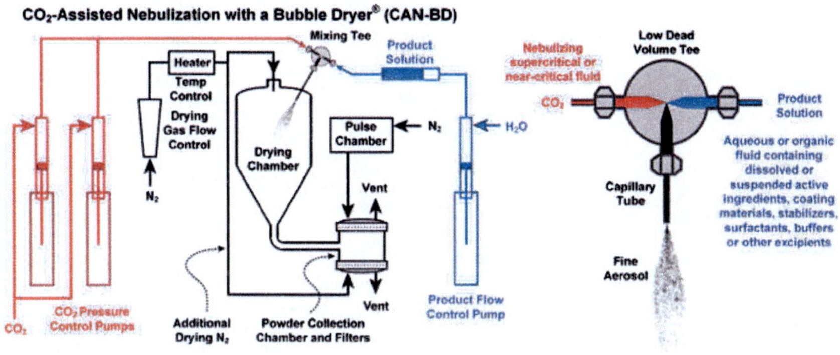

Fig. 3.44 Schematic diagram of the carbon dioxide-assisted nebulization with a bubble dryer® (CAN-BD) process (Adapted from Cape et al. (2010) and Manion et al. (2012) with permission from Taylor & Francis)

with a Bubble Dryer® (CAN-BD) will be discussed in the next section with the main findings related to vaccines. It is a supercritical CO_2 process, patented by Sievers and Karst (1994), Sievers et al. (1999). Figure 3.44a shows a schematic diagram of the CAN-BD process.

An aqueous or organic product solution or suspension (e.g., a vaccine formulation containing 1% to 10% solids) is intimately mixed with supercritical or near-critical CO_2 in a low dead volume mixing tee at room temperature (20 to 35 °C) and usually at 83–103 bar (Cape et al. 2008). CO_2 is used as an aerosolization aid that allows drying at lower temperatures, unlike spray drying, which may favor the drying of thermosensitive vaccines. The pressurized mixture is rapidly expanded through a capillary flow restrictor (usually made of fused silica, stainless steel, or PEEK with an inner diameter of 75–380 μm and a length of about 10 cm) to near atmospheric pressure (Cape et al. 2010).

The rapid expansion of compressed CO_2 from 83 to 1 bar produces a fine aerosol consisting of microdroplets and microbubbles. Further breakup of the microdroplets occurs due to the sudden release of CO_2 dissolved in the liquid. The generated aerosol is fed into a drying chamber and is rapidly dried by preheated nitrogen gas of about 25–65 °C under atmospheric conditions. The particles are collected on a filter with pore sizes between 0.2 and 0.45 μm (Sievers 1999; Sellers et al. 2001; Cape et al. 2008). In this way, particles in a size range of about 3–5 μm can be produced, which is desirable for pulmonary delivery (Burger et al. 2008). In myo-inositol-based formulations, the residual water content can be less than 0.5% (Burger et al. 2008). The average residence time of the droplet/dry particles in a lab-scale 750 ml glass drying chamber has been estimated to be less than 3 s (Sievers et al. 2003). Typical flow rates on a laboratory scale are 0.3–0.6 mL/min of liquid solution and 1–3 mL/min of dense CO_2 (Cape et al. 2008).

Table 3.13 Process conditions for lab-scale production of live-attenuated measles and hepatitis B powder vaccine using CAN-BD

Parameter	Live-attenuated measles vaccine	Hepatitis B vaccine
Product solute	H_2O	H_2O
Nebulizing liquid	CO_2	CO_2
Sample chamber (in-line, downstream of the water pump)	32 mL floating piston column	7 mL reservoir
Drying chamber dimensions	2.4 L glass chamber	3.5 L glass chamber
Mixing tee pressure	1200 psig (83 bar)	1200 psig (83 bar)
Product and nebulizing fluid temperatures	Room temperature	Room temperature
Product flow rate	0.3 to 0.5 mL/min	0.3 to 0.5 mL/min
Nebulizing fluid flow rate	1.5 mL/min	2 mL/min
N2 drying gas flow rate	30 L/min	30 L/min
Drying chamber outlet temperature	50 °C	50 °C
Product composition before drying (aqueous formulations)	4.9 wt% trehalose in reconstituted vaccine which contains Minimal Essential Medium (MEM) culture supernatant + gelatin, sorbitol, L-histidine, L-alanine, tricine, L-arginine hydrochloride, and lactalbumin hydrolysate	27 wt% vaccines suspension + 9 wt% trehalose, 27 wt% vaccine suspension + 5 wt% trehalose + 4 wt% PVP, 40 wt% vaccine suspension + 9 wt% lactose

Data adapted from Sievers et al. (2007b) with permission from Elsevier

Table 3.13 shows some experimental details for lab-scale production of live-attenuated measles and hepatitis B powder vaccines (Sievers et al. 2007b). The CAN-BD process has also been successfully scaled up to process up to 30 mL/min liquid solution (Cape et al. 2008).

Table 3.14 provides an overview of many application examples successfully processed with the CAN-BD drying process, grouped by vaccines, oligonucleotides, antibodies, enzymes, antibiotics, sugar excipient stabilizers, and other components in formulations.

A major highlight achieved with the CAN-BD process includes stabilized dry powder formulations of live-attenuated measles vaccine (Sievers et al. 2007b; Burger et al. 2008; Cape et al. 2010; Kisich et al. 2011; Lin et al. 2011; Agarkhedkar et al. 2014) that were stable for almost 3 years at 2–8 °C, up to about 6 months at 25 °C with less than one log loss in potency, or 7 days at 37 °C with less than one log loss passing the WHO stability criterion (Cape et al. 2008, 2010).

Table 3.14 Overview of applications produced by the CAN-BD drying process

Application	Substances	References
Vaccines	Live-attenuated measles virus (Edmonston-Zagreb), influenza live virus vaccine, hepatitis B surface antigen (HBsAg) vaccine, human papillomavirus (HPV) vaccine	Sievers et al. (2007a, b), Burger et al. (2008), Cape et al. (2008, 2010), McAdams (2011), Kisich et al. (2011), Lin et al. (2011), Agarkhedkar et al. (2014)
Oligonucleotides	Single- and double-stranded siRNA and DNA, aptamers, amidites	Sievers et al. (2007a)
Antibodies	PRIMATIZED® anti-CD4, human IgG, anti-human lambda light chain	Cape et al. (2008)
Enzymes	α1-antitrypsin, trypsinogen, lactate dehydrogenase, lysozyme, insulin, alkaline phosphatase	Sievers et al. (2001), Cape et al. (2008)
Antibiotics	Capreomycin, kanamycin, moxifloxacin hydrochloride, tobramycin sulfate, amoxicillin, doxycycline, cefazolin, ciprofloxacin hydrochloride, amikacin, rifampin	Sievers et al. (2007b), Manion et al. (2012)
Sugar excipient stabilizers	Myo-inositol, trehalose, mannitol, sorbitol, lactose, sucrose	Sellers et al. (2001), Burger et al. (2008), Cape et al. (2010), McAdams (2011), Kisich et al. (2011)
Other components	Phytosterols, PEG, PVP, hydrolyzed gelatin, sodium chloride, DPPC, salbutamol, buffers (tricine, sodium or potassium phosphate, sodium acetate, sodium citrate), surfactants (palmitic acid, stearic acid, Tween 20, Tween 80, Pluronic F68), amino acids (arginine, alanine, histidine, leucine, methionine), and metal chelating agents, such as ethylene diamine tetraacetic acid (EDTA) or diethylene triaminepentaacetic acid (DTPA)	Sievers et al. (2001, 2003), Cape et al. (2010)

List adapted from Cape et al. (2010) and extended with various supporting references

The CAN-BD process also enabled the development of a dried alum-adjuvanted HepB subunit vaccine powder stable at elevated temperatures and retained potency after storage for 43 days at above 66 °C and 4 years at 2–8 °C (Sievers et al. 2007b; Sievers 2015). In addition, unlike liquid formulations of alum-adjuvanted vaccines, the powder was not likely to be damaged by freezing during storage. This is noteworthy because common alum-adjuvanted vaccine products are sensitive to freezing and cannot be dried using standard freeze-drying methods.

Figure 3.45a shows the CAN-BD process at laboratory scale at the University of Colorado (Boulder, USA) for the production of powder vaccines (Sievers 2015), and Fig. 3.45b presents the installation of the pilot-scale CAN-BD system at the Serum Institute of India Ltd. (Pune, India) in a Grade A environment for the GMP production of inhalable dry powder measles vaccine (Cape et al. 2010) for a toxicity study in non-human primates (Lin et al. 2011) and a Phase I clinical trial in humans (ClinicalTrials.gov, NCT01557699) (Agarkhedkar et al. 2014).

Figure 3.46 shows the particle morphology of CAN-BD generated vaccine powders against (a and b) measles (Sievers et al. 2007b; Cape et al. 2010), (c) hepatitis B (Sievers et al. 2007b), and (d) L1 protein-based HPV (McAdams 2011). The powders were prepared from aqueous solutions with various stabilizing ingredients listed in the caption of Fig. 3.45. Most of the particles are respirable. Chap. 6 (Sect. 6.4) will give further details on the studies on measles vaccines.

In summary, the apparent challenge of supercritical fluid drying technology is the scalability of the process to larger powder volumes and a required secondary drying step to further reduce the residual moisture. So far, studies have been limited to a batch size of about 100 mL of liquid vaccine. A highlight is a clinical trial in healthy Indian men with supercritical spray-dried measles vaccine powder administered by inhalation. All subjects tolerated the vaccine well and showed a comparable immune response to an approved measles vaccine administered subcutaneously as a reconstituted lyophilized powder.

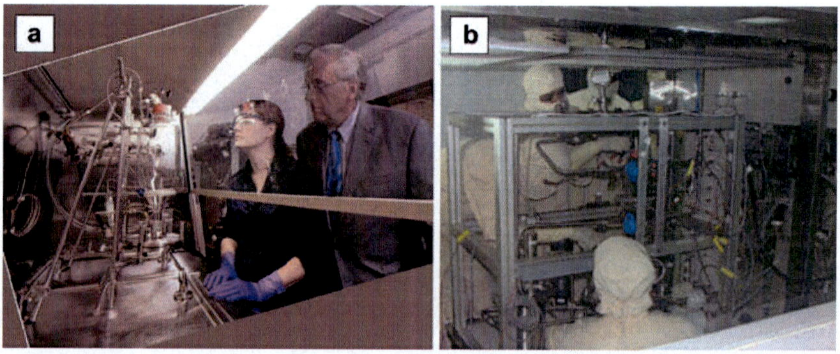

Fig. 3.45 **a** Prof. em. Robert (Bob) Sievers and a colleague look over the CAN-BD process in the laboratory at the University of Colorado (Boulder, USA) for making powder vaccines (Image adapted from Sievers (2015)). **b** Pilot-scale CAN-BD equipment installed at the Serum Institute of India Ltd. (Pune, India) in a Grade A environment for the GMP production of inhalable dry powder measles vaccine for the non-human primate toxicity study (Lin et al. 2011) and Phase I clinical trial (Agarkhedkar et al. 2014) (Image adapted from Cape et al. (2010))

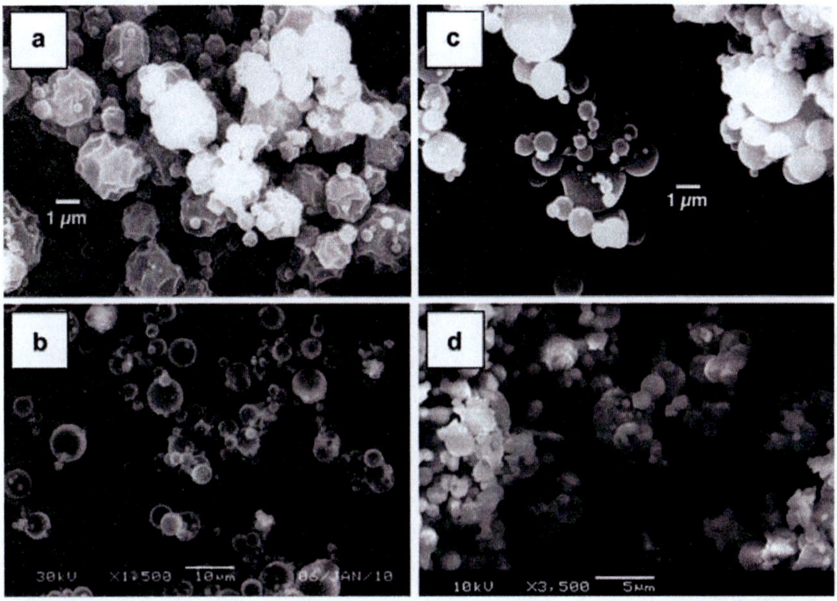

Fig. 3.46 SEM images of vaccine particles produced with the CAN-BD process: **a** Measles vaccine placebo containing 4% trehalose and approximately 0.5% of other placebo ingredients (sorbitol, hydrolyzed gelatin, lactalbumin hydrolysate, tricine buffer, histidine, alanine, arginine HCl, and trace components of Minimum Essential Medium) (MMAD = 1.9 μm, GSD = 1.6, respirable mass fraction < 5 μm = 94%) (Image adapted from Sievers et al. (2007b) with permission from Elsevier). **b** Measles vaccine placebo formulated with myo-inositol (11% total dissolved solids, 50 g/L myo-inositol, 25 g/L gelatin, 16 g/L arginine-HCl, 1 g/L alanine, 2.1 g/L histidine, 3.5 g/L lactalbumin hydrolysate, 3 g/L tricine, pH 6.5–7.0) (Image from Cape et al. (2010)). **c** Hepatitis B vaccine powder with trehalose stabilizer (weight ratio stabilizer/vaccine = 85, mean particle size = 1.6 μm) (Image adapted from Sievers et al. (2007b) with permission from Elsevier). **d** Human Papillomavirus (HPV) L1 protein powder formulated with myo-inositol (11% total dissolved solids, 100.3 g/L myo-inositol, 1.6 g/L L-leucine, 0.9 g/L ammonium acetate, 2.8 g/L NaCl, 2.4 g/L Tween, 1.4 g/L Tris buffer, 0.1 g/L DTT, 0.1 g/L EDTA) (Image from McAdams (2011))

3.8 Overall Comparison of Drying Technologies for Vaccines

As described in the last sections, several drying technologies and formulation approaches can dry vaccine formulations to a stable solid form. In particular, freeze drying, spray drying, spray freeze drying, vacuum foam drying, and supercritical fluid drying and their drying mechanisms were presented and discussed. The primary motivation is typical that dry vaccines offer stability, reduced weight, reduced transportation costs, reduction or even elimination of cold chain storage, and continuous production. Several relevant aspects determine the suitability of a particular drying method.

Table 3.15 General comparison of drying methods for vaccines

	Freeze drying	Spray drying	Spray freeze drying	Vacuum foam drying	Supercritical fluid drying
Atomization stress	+	−	−	+	−
Freezing stress	−	+	+	+	+
Thermal stress	+	−	+	±	±
Process speed	−	+	±	−	+
Crystallization risk	+	−	+	−	±
Particle design	−	+	+	−	+
Capacity	±	+	±	±	+
Industrial experience	+	±	−	+	−

+ favorable,− unfavorable

Adapted from Amorij (2007), Amorij et al. (2008), Arpagaus (2020) with permission from Springer Nature

Table 3.15 compares the general concepts of drying vaccines into favorable and unfavorable characteristics. The most important decision factors are process stresses (e.g., atomization, freezing, heat), process speed, crystallization risk, ease of particle design, capacity, recovery, and current (industrial) experience (Amorij 2007; Amorij et al. 2008; Arpagaus 2020).

At first glance, it becomes clear that each drying technology has typical advantages and disadvantages, the relevance and extent of which can be determined by the applied process conditions and the formulation (e.g., excipients used). The favorable factors for spray drying include no freezing stress, process speed, particle design, capacity, costs, and industrial experience. For vaccines, costs and industrial experience are similar to freeze-drying due to the additional powder filling and limited industrial experience with aseptic processing to date.

Furthermore, Table 3.16 compares the main benefits and challenges of each vaccine drying technology in a compromised way using various arguments based on different literature sources, which are briefly presented here.

- Amorij et al. (2008) presented general concepts for drying biopharmaceuticals with a particular focus on the challenges and opportunities for developing stable influenza vaccine powder formulations.
- Maltesen and van de Weert (2008) reported on drying methods for protein pharmaceuticals and emphasized that the right choice of drying technology has a significant impact on the final pharmaceutical product and the overall economics of the process.
- Abdul-Fattah and Truong (2010) provided an overview of alternative drying methods for biopharmaceuticals and highlighted the potential to improve product performance beyond what can be achieved with a conventional freeze-drying process.

Table 3.16 Comparison of different drying methods for biopharmaceuticals and vaccines

Drying method	Benefits	Challenges
Freeze-drying	+ Low-temperature drying + Low moisture content (controlled) + Elegant cake appearance + No powder filling required (dosed in vials) + Aseptic processing + Industrial experience (established for vaccines)	– Long processing time – No particle engineering – Freezing stress and exposure to ice-water interface – Lower capacity (batch) – Equipment costs (vacuum pump) – Energy consumption (process efficiency)
Spray drying	+ Economic process + Fast drying process + High capacity + Particle design (e.g., free-flowing powder, good aerosolization properties, vaccine encapsulation) + Industrial experience with pharmaceuticals (no spray-dried vaccine commercial) + Scalability	– Atomization stress (shear, air–liquid interface) – Thermal stress – Yield (50 to > 70% lab-scale, > 95% industrial scale) – Higher moisture content than freeze-drying – Special equipment required for aseptic processing – Aseptic powder filling
Spray freeze drying	+ Fast freezing + Particle design (e.g., free-flowing powder, good aerosolization properties, vaccine encapsulation) + High yield	– Freezing stress – Long processing time – Industrial experience – Special equipment required for aseptic processing – Equipment costs
Supercritical fluid drying	+ Fast process + Particle engineering + Mild process condition (mild temperature)	– Shear stress during atomization – CO_2 pressure and exposure to organic solvents – Limited yield (material recovery can be < 100%) – Scalability – Less industrial experience – Special equipment required for aseptic processing – Equipment costs
Foam drying (vacuum drying)	+ Avoids stresses associated with freezing + Economic + Low moisture content (controlled)	– Long processing time – No particle engineering

(continued)

Table 3.16 (continued)

Drying method	Benefits	Challenges
	+ No powder filling required (dosed in vials) + Aseptic processing in a freeze dryer	– Larger vials for expanded foam – Rapid boiling/boil over – Freezing stress during foaming – Heterogeneity in product appearance – Equipment costs (vacuum pump)

Summarized from various literature sources (Amorij et al. 2008; Maltesen and van de Weert 2008; Abdul-Fattah and Truong 2010; Walters et al. 2014; Searles and Cherian 2015; Ohtake et al. 2015; Emami et al. 2018; Andersen 2020; Arpagaus 2020; Thorat et al. 2020; Chaurasiya and Zhao 2020; Sharma et al. 2021)

- Walters et al. (2014) reviewed next-generation drying technologies for pharmaceutical applications and discussed the challenges encountered as alternatives to freeze-drying.
- Ohtake et al. (2015) presented strategies for heat-stable vaccines based on formulation approaches and drying processes to stabilize vaccines and other challenges associated with any stabilization technology that must be overcome, including process loads on the vaccine, scalability, and optimal dosage form to ensure compatibility with the delivery device.
- Searles and Cherian (2015) addressed several promising alternative drying methods for vial lyophilization, such as spray drying, bulk tank freeze-drying, crystallization/precipitation and drying, foam drying, vacuum drying, drying on a fiber matrix, etc.
- Emami et al. (2018) reviewed drying technologies for stability and bioavailability of solid dosage forms of biopharmaceuticals and focused on optimized drying processes.
- Andersen (2020) presented new drying processes for biopharmaceuticals, with a particular focus on aseptic spray drying, continuous vial freeze drying, and spray freeze drying.
- Thorat et al. (2020) reviewed the drying options of vaccines and biomolecules in dry powder form and compared various technological challenges to support the current global health scenario due to the COVID-19 pandemic.
- Sharma et al. (2021) reviewed the most innovative drying technologies for biopharmaceuticals that have not gained popularity for the manufacturing parental biopharmaceuticals but offer continuous production of dry particles.
- Chaurasiya and Zhao (2020) reviewed the most frequently used industrial drying and formulation techniques for dry powder for pulmonary delivery.

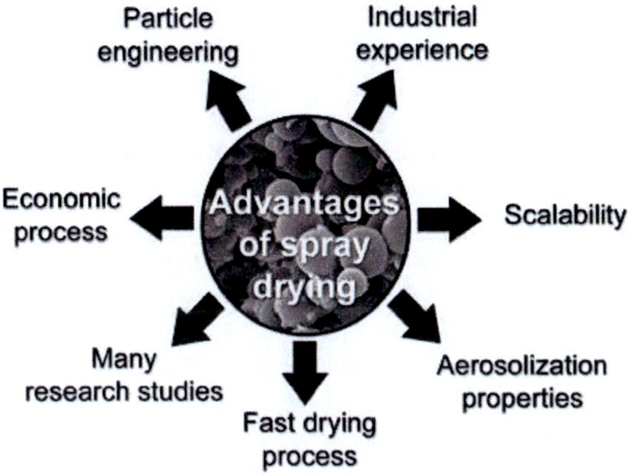

Fig. 3.47 Main advantages of spray drying for vaccines. (1) particle engineering (e.g., free-flowing powder, vaccine encapsulation, particle morphology), (2) industrial experience, (3) scalability, (4) good aerosolization properties (e.g., uniform particle size), (5) fast drying process, (6) many research studies, and (7) economic process

The conclusion is that each drying method has its advantages and limitations that need to be considered when selecting a drying method. While freeze-drying is the number one drying method for vaccines, alternative drying processes under optimal formulation and processing conditions have the potential to improve product performance, for example, particle engineering, economics, rapid drying, uniform particle size, and scalability.

As a final note, Fig. 3.47 reemphasizes the main advantages of spray drying, which has the potential to become a commercially viable process for drying vaccines and, in particular, also to control the powder properties required for different delivery systems, such as particle size for needle-free injections (e.g., epidermal), intranasal, or inhalation applications.

3.9 Excursus: Alternative Drying Technologies for Vaccines

Some alternative drying technologies for vaccines and potential delivery forms for spray-dried particles are presented in this section. In particular, these include dosage forms produced as.

- rapid dissolving oral thin films,
- sugar pills,
- fiber membranes,
- porous sponge inside a syringe,
- powder at room temperature in laminar air, and
- microarray patches.

The functional principles are explained and illustrated with examples in each case, and the main benefits and challenges are described.

3.9.1 Drying of Spray Dried Vaccine Powders in Rapid Dissolving Oral Thin Films

The formulation of spray-dried vaccine particles into a rapidly dissolvable oral thin film is a promising example of an alternative dosage form. Several studies have already been conducted in this area (Mao et al. 2010; Truong-Le et al. 2015; Gala et al. 2017; Nguyen et al. 2017; Bajrovic et al. 2020).

Mao et al. (2010) patented a process for a rapidly dissolvable thin film with bioactive components encapsulated in pH-sensitive polymeric microparticles. Spray-dried powder of rotavirus vaccine was mixed with an aqueous polymer suspension containing sodium alginate, polyvinyl alcohol (PVA), and polyethylene oxide (PEO) and poured onto a flat surface (size 2 cm × 3 cm) coated with polydimethylsiloxane. This suspension was air-dried under laminar flow at 20–30 °C for 5 to 8 h to obtain a rapidly dissolving 100 μm thin film (see Fig. 3.48). The processed thin film strip allowed stabilization of the vaccine at room temperature.

Figure 3.48 shows the basic steps of fabricating a thin film and a stabilized vaccine oral thin film (Truong-Le et al. 2015). The process consists of mixing the vaccine with the excipients and thin-film formers, solvent casting onto a polymer support, convective air drying, and de-lamination of the thin film. As an alternative, the film can be prepared from a dispersion of spray-dried powder (formulated from sucrose or trehalose) and a polymer matrix (Fig. 3.49).

Nguyen et al. (2017) also claimed a method for preparing fast dissolving thin films (within < 2 min under the tongue) with enhanced thermal stability containing bioactive material such as protein, antibody, or vaccine (e.g., rotavirus, typhoid, or influenza vaccine). Drying was performed on a surface (e.g., PET backing liner)

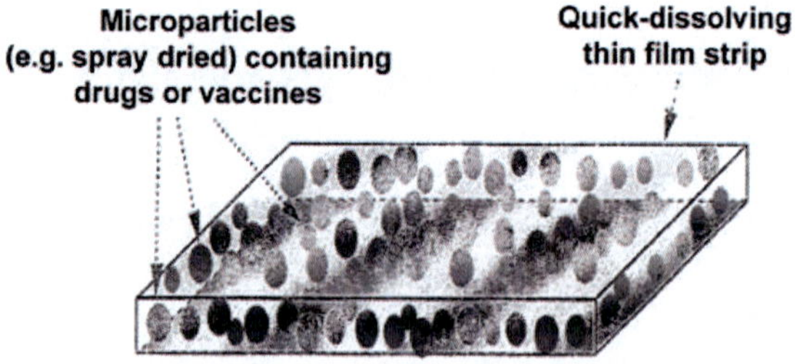

Fig. 3.48 Microparticles (e.g., spray dried) containing drugs or vaccines embedded in quick-dissolving thin film strip made of polymer matrix (Adapted from Mao et al. (2010))

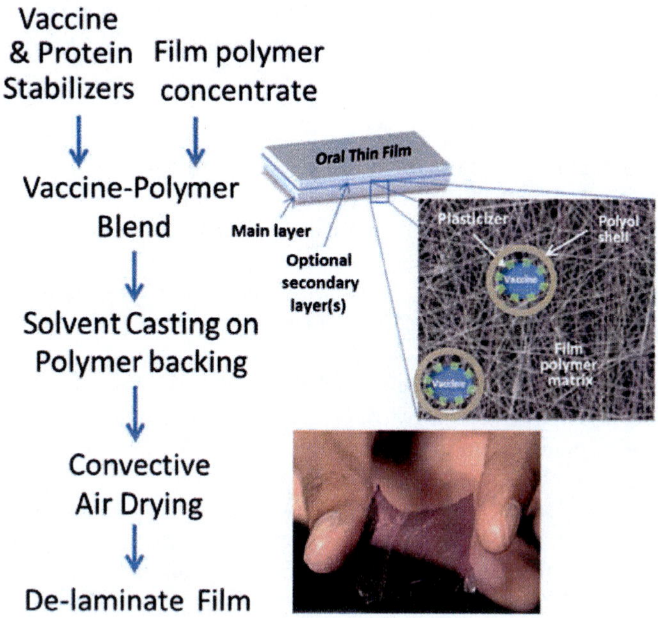

Fig. 3.49 Basic steps in a lab-scale thin-film manufacturing process and the microstructure of the resulting stabilized vaccine film formulation (Adapted from Truong-Le et al. (2015) with permission from Elsevier)

under ambient conditions, with additional heat of warm gas and/or IR light, or under vacuum conditions at low pressure using freeze-drying or vacuum drying. For a rotavirus vaccine mixture, drying into a 50 to 200 μm thin film was performed for 2 h at 60 °C in a convection oven, followed by 24 h of drying at 0.13 mbar at 4 °C.

Gala et al. (2017) reported on an orally disintegrating film (within < 40 s) made of Lycoat® RS720, Neosorb® P60W (sorbitol), and Tween® 80 (polysorbate 80) with incorporated spray-dried microparticles containing live-attenuated measles vaccine. The thin films (2 × 3 cm, 79 μm thickness) were administered to pigs via the buccal route (Sect. 6.4 in Chap. 6 shows further manufacturing process details). All in all, the vaccine formulation induced an efficient immune response.

Recently, Bajrovic et al. (2020) calculated that thin films of vaccines would require much less storage space than packaged vials and ampoules. For example, 350,000 doses of the BCG (Bacille Calmette-Guérin) vaccine would occupy about 600 sheets of thin-film (22 × 28 cm), weighing about 3 kg, whereas the same number of doses in vials would occupy about a soccer field and weigh 2730 kg.

Overall, thin-film drying technology is considered promising because it is relatively easy to implement with current thin-film manufacturing processes. Oral thin-film formulations of vaccines with room temperature stability can increase vaccine coverage by improving patient compliance and facilitating booster vaccinations. In addition, the technology could also enhance the use of current oral vaccines, which include, for example, vaccines against typhoid, cholera, rotavirus, shigellosis, and polio.

3.9.2 Drying of Vaccines by Thin-Film Freezing

Another emerging technology for stabilizing vaccines is thin-film freezing, a combination of spray freezing onto a cryogenic surface and lyophilization. Researchers at the University of Austin have published numerous papers on this drying technology (Engstrom et al. 2008; Overhoff et al. 2009; Zhang et al. 2012a; Thakkar et al. 2017).

Figure 3.50 illustrates the principle, where droplets of a liquid vaccine formulation are sprayed onto a cryogenically cooled rotating drum filled with liquid nitrogen and frozen into thin films (Overhoff et al. 2007, 2009; Li et al. 2015). The surface temperature of the drum can be as low as −100 °C, resulting in ultrafast freezing on the order of 100 ms (Engstrom et al. 2008). These thin films are subsequently freeze-dried to remove the solvent. The particle size and structure of the solid powders are mainly determined by the cooling rate and the solid concentration. Typically, amorphous and highly porous dry powder with a large surface area and low density are achieved.

The thin-film freezing technology is particularly suitable for improving the solubility of poorly water-soluble drugs (Overhoff et al. 2009) and producing dry submicron protein particles for inhalation (Engstrom et al. 2008). Li et al. (2015) reported that vaccines containing aluminum salts could be successfully converted to dry powder using a liquid vaccine suspension containing 2% w/v trehalose without

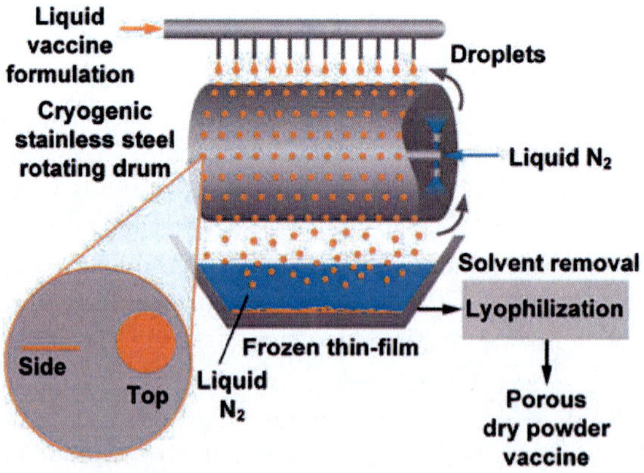

Fig. 3.50 Principle of thin-film freezing: Vaccine droplets fall onto a cryogenically cooled (e.g., by liquid nitrogen) rotating stainless steel drum, where they rapidly freeze into a thin film. Subsequent lyophilization dries the frozen films into porous powders (Adapted from Beteta and Ivanova (2015))

particle aggregation or decrease in immunogenicity of the vaccine after reconstitution. Dried model vaccine powder (ovalbumin-adsorbed aluminum oxyhydroxide) has demonstrated long-term stability up to 6 months at 40 °C and immunogenicity in mice (Thakkar et al. 2017).

The advantages of thin-film freezing over other drying technologies are the ability to produce aerosolized particles of a brittle matrix and nanostructured aggregates suitable for dry powder inhalation, nasal administration, or recirculation to liquid form for injection. In contrast to freeze-drying, particles with a large surface area are produced. In addition, compared to spray freeze drying, protein adsorption at the gas–liquid interfaces is lower, resulting in higher protein stability.

From an industrial perspective, TFF Pharmaceuticals Inc. (USA) has acquired patents for thin-film freezing and inhalation and is now seeking to commercialize the technology. In addition, GreenLight Biosciences Inc. and TFF Pharmaceuticals Inc. have partnered to test the feasibility of a thin film frozen, storage-stable, powdered mRNA-COVID 19 vaccine (TFF 2021). However, what is still lacking is scale-up experience and FDA approvals of freeze-dried thin-film products in the marketplace.

3.9.3 Drying of Vaccines as Low-Cost Sugar Pills

Another method for thermal stabilization of viral vaccines is air-drying into sugar pills. In this context, Leung et al. (2019) reported about the air-drying of DNA (herpes simplex virus type 2) and RNA (influenza A virus) viral vaccines in a mixture of pullulan and trehalose. Pullulan was used as a water-soluble polysaccharide with good film-forming abilities and low oxygen permeability, suitable for coating tablets and granules to extend the shelf life of active ingredients. Both sugars are approved for injection into the human body. Figure 3.51 shows a schematic of the process for pill preparation, including mixing, air-drying, pill formation and storage, and reconstitution.

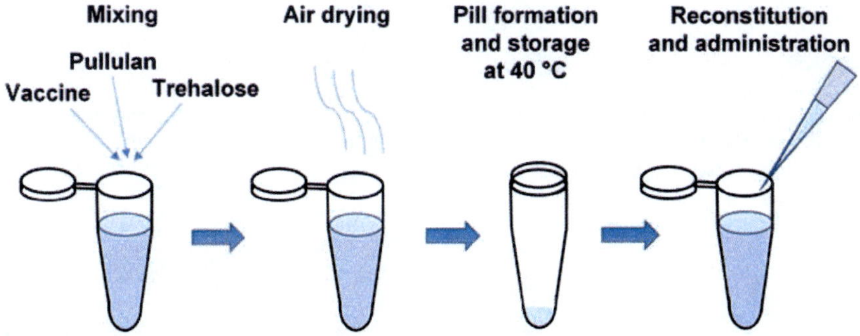

Fig. 3.51 Pill preparation process including mixing a vaccine in pullulan and trehalose, air drying, pill formation and storage, and reconstitution (Adapted from Leung et al. (2019), open access under CC-BY 4.0 license)

The preparation procedure of (Leung et al. 2019) involved mixing 1 μL of a virus-containing solution with 9 μL of a 10 wt% pullulan (200 kDa) solution and 0.5 M trehalose. Samples of 1.7 mL in microcentrifuge tubes were air-dried overnight in a biological safety cabinet. After drying, the tubes were sealed and stored at 40 °C. At certain times, some samples were reconstituted in PBS buffer and used for titration and mouse studies. The results showed that the live-attenuated HSV-2 vaccine maintained efficacy for at least 2 months when stored at 40 °C and inactivated influenza vaccine for at least 3 months.

Overall, the authors have developed a simple drying technology for converting heat-sensitive vaccines into thermostable vaccine pills that can solve the cold chain problem (no refrigeration required) and significantly improve global health by providing access to vaccines for people worldwide. The Canadian research team at McMaster University is now looking at partnerships and more funding to develop the technology further and is also going through the proper regulatory procedures to be approved by the FDA and Health Canada.

3.9.4 Drying of Vaccines on a Fiber Membrane

Drying a vaccine on a fiber matrix, filter, or web of fibers in an expandable hub housing of a hypodermic needle is another approach for ready-to-inject stable vaccines. For example, Fig. 3.52 shows the principle of a vaccine stabilized by sugar glass in a fine mesh of glass fibers of a membrane (Roser 2006). In practical use, the membrane is attached to the hub of a standard syringe drawn up with sterile saline. During the injection into the patient, the saline flows through the membrane in one pass and dissolves the vaccine. The membrane thus provides a large surface-to-volume ratio for rapid reconstitution. In principle, the sugar glass could also contain spray-dried particles with a hydrophobically encapsulated vaccine.

Figure 4.37a shows an SEM image of trehalose sugar glass air-dried on a fiber membrane (Alcock et al. 2010). The membrane is placed in a plastic housing connected to a conventional syringe and needle (Fig. 4.37b). The concept has been

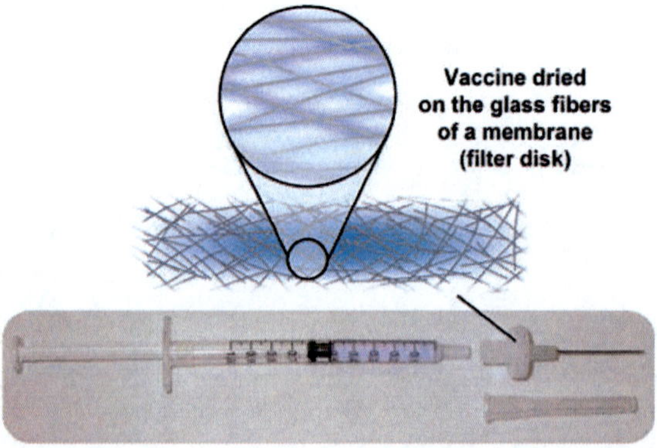

Fig. 3.52 Stabilized vaccine on the glass fibers of a membrane filter disk in the hub of a hypodermic needle (Images from Roser (2006) with permission from Future Medicine Ltd.)

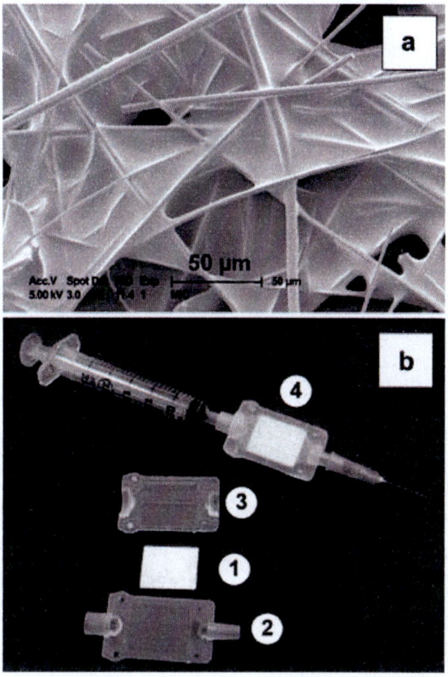

Fig. 3.53 **a** SEM image of trehalose-sucrose glass dried on a glass fiber membrane. The virus vaccine made of 0.5 M trehalose and sucrose was air-dried at 20–25 °C to 2% to 10% RH, **b** HydRIS® concept: (1) fiber membrane with a vaccine, (2) lower and (3) upper parts of the polypropylene housing, and (4) completed device attached to a syringe and needle (Images adapted from Alcock et al. (2010) with permission from The American Association for the Advancement of Science)

commercialized as HydRIS™ (Hypodermic Rehydration Injection System) and is now sold by Nova Laboratories Ltd. (UK) (De Costa et al. 2006; Manufacturing Chemist 2010; De Costa 2015) (Fig. 3.53).

Scientists at Oxford University have stabilized human adenovirus type 5 (AdHu5) and live modified vaccinia virus Ankara (MVA) by formulation with sucrose and trehalose using HydRIS™ (Alcock et al. 2010). Studies with a mouse model observed complete maintenance of titer and immunogenicity after storage at up to 45 °C for up to 6 months. In addition, long-term stability data showed stability of over 12 months at 37 °C (Nova Laboratories 2021).

Dulal et al. (2016) demonstrated the technology for a novel chimpanzee adenovirus-mediated rift valley fever vaccine (ChAdOx1-GnGc) in cattle. The thermostabilized vaccine was stored at 25, 37, or 45 °C for 6 months and elicited comparable neutralizing antibody titers at the same dose of the liquid vaccine stock that must be stored at −80 °C.

More recently, Wang et al. (2018) reported a rabies vaccine with adenovirus vectors stabilized by drying on a fibrous pad. The vaccine induced a strong immune response in mice and was stable at 45 °C storage for 1 month. The results encouraged the authors at Oxford University to conduct a clinical trial with this vaccine candidate in the near future.

Overall, these studies demonstrate that drying vaccines on a fiber matrix is a stabilization technology that enables a wide range of vaccine applications and facilitates storage and distribution at ambient temperatures.

3.9.5 Drying of Vaccines in a Porous Sponge in a Syringe

The UK-based company StablePharma Ltd. has invented a syringe that contains a small compressible porous sponge, in which the vaccine is thermally stabilized in a soluble sugar glass made of trehalose and other excipients (Stablepharma 2013). These pre-dosed syringes based on StablevaX™ technology can be stored at room temperature and comply with the WHO protocol for injection vaccination without additional education from healthcare workers.

Figure 3.54a, b shows the principle of the pre-dosed syringe using a sponge with dried vaccine and the sequence of processing steps. The syringe is filled with sterile water, and the vaccine is released from the sponge. The porous 3D structure of the

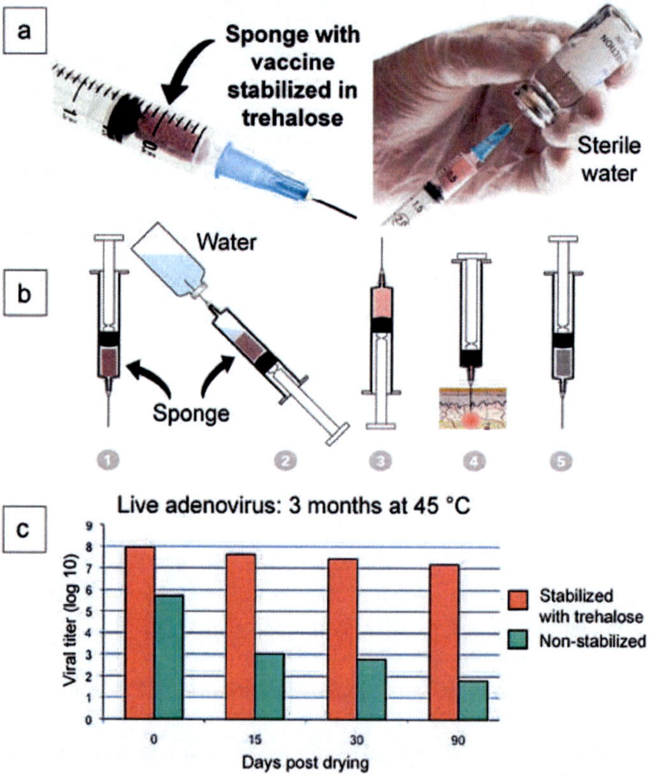

Fig. 3.54 **a** StablevaX™ technology with a compressible sponge inside a medical syringe. The sponge incorporates the stabilized vaccine by air drying in a trehalose solution. The pre-dosed syringe just needs sterile water to release the vaccine. **b** Steps of vaccination using StablevaX™: (1) Removing sterile wrapping, (2) taking up water, (3) blowing out air, (4) injection, and (5) correct vaccine dose injected. **c** Stability data of live adenovirus vector after 3 months at 45 °C stabilized with trehalose compared to non-stabilized vector (Images by courtesy of StablePharma Ltd. UK)

sponge provides a large surface area and releases the vaccine in seconds to be injected into the patient. The compressible sponge allows expressing the correct total dose.

For example, Fig. 3.54c shows stability data of adenovirus used as a living vector to deliver DNA-based vaccines. The stability remains when stored at 45 °C for 3 months with trehalose compared to the non-stabilized virus (Stablepharma 2020a).

The manufacturing process of StablevaX™ is described as inexpensive, and the technology could be used to stabilize various drugs and vaccines. A liquid vaccine is dried inside the sponge by air at room temperature. The stabilizing sugar trehalose dries quickly to thin shells on the large surface of the sponge structure. In 1985, the protective effect of trehalose for proteins and the like was patented (Roser 1985).

In a recent interview at Irish Tech News (2020), the chairman and founder, Dr. Bruce Roser, explained that the required air drying equipment is modest and relatively small. A pilot plant could be installed in a refrigerated truck and produce 1 million doses per week, providing 24 h of operation a day. By parallelization, the production could be increased. Larger facilities might produce 10–100 million doses per week. One syringe contains around 0.5 mL liquid vaccine with about 50–100 μg antigen and 1 mg of adjuvant (e.g., aluminum hydroxide) to stimulate the immune response. In comparison, StablevaX™ enables higher vaccine loading capacities than microarray patches (max. about 5 μL) and the HydRIS™ fiber membrane technology (about 50 μL).

The claimed advantages of spray drying are lower production costs and higher yield. Air drying makes the production process very cost-effective. Although spray drying allows mass production, it requires comparatively more expensive equipment for sterile processing (Roser 2005, 2006, 2009). The particle engineering capabilities of inhalable powder is an obvious advantage of spray drying. Still, the path to adopting new inhalation technology for vaccines involves training health care workers and patients, unlike StablevaX™, which follows the WHO program for injection.

In contrast to freeze-drying and multiple-dose vials, StablevaX™ is already pre-dosed in the syringe when it leaves the factory (ready-to-inject). In addition, there is the advantage of stockpiling the produced pre-dosed syringes until use. The vaccines can be stored on a shelf and are ready for injection as soon as taken out of the box. Overall, StablevaX™ can be introduced as a practical solution for pandemics and vaccine distribution in general.

So far, the stabilization process used in StablevaX™ has been tested with over 13 vaccines, including most types on the market or in development. In addition, over 90 vaccine candidates have identified suitable use with the stabilization technology (Stablepharma 2020b). By August 2020, StablevaX™ had patents granted in 14 international jurisdictions (Stablepharma 2020b).

A pre-approved Tetanus-Diphtheria (Td) vaccine being processed through GMP was chosen as the launch vaccine to validate the StablevaX™ formulation and device. Depending on the FDA approvals, the first human trials will be possible in

2021 (Stablepharma 2020c). Recently, a collaboration with WHO prequalified European vaccine manufacturer BB-NCIPD Ltd. (belongs to the Ministry of Health of Bulgaria) was launched to develop a thermostable version of the Td vaccine.

In addition, a program was initiated on the thermostability of SARS-COV-2 mRNA and lipid nanoparticles (LNP) containing the SARS-COV-2 mRNA to demonstrate that StablevaX™ can stabilize the mRNA and LNP for storage and transport at elevated temperatures.

3.9.6 Drying of Vaccines at Room Temperature in Laminar Air

The Swedish company Ziccum AB is developing new formulations of heat-stable vaccines in dry powder with its patented LaminarPace™ spray-drying technology. Ziccum was founded in 2017 by its original parent company, Inhalation Sciences AB, and the system for producing dry formulations was invented by Dr. Per Gerde of Karolinska Institutet (Gerde 2013).

Figure 3.55a shows a schematic drawing of the original version of the LaminarPace™ spray dryer (Gerde et al. 2010). The system can be defined as a counter-current spray dryer with separate flows. It consists of two main parts, a drying column, and a dehydration column. The drying column has an inner column (about 90 cm long, 5 cm ID, made of steel) perforated and covered with a vapor-permeable membrane (e.g., approximately 100 μm layer of filter membrane). Atomization of the liquid into a spray of fine droplets is created with a vibrating mesh. A swirling airflow is generated inside the inner column to facilitate droplet movement and mass transfer to the membrane. Simulations and modeling has shown that evaporation occurs shortly after the aerosolization step (Soltani et al. 2015). The inner column flow is also used to control the residence time of the particles. In the outer column, a dry air stream circulates counter-currently around the inner column and transports away the water vapor diffused through the membrane (Soltani et al. 2015). The driving force behind the mass transfer is the vapor pressure difference between the inner and the outer column. The dry air is regenerated when passing through the dehydration column, where silica gel absorbs the water, and air with very low humidity is circulated back into the drying column. At the end of the process, micronized dry powder with a monodispersed size distribution and well-defined shape is collected on a filter paper at the bottom of the drying column (Fig. 3.55b). High powder yield (80–90%) is achieved using laminar flow even for particle size of less than 5 μm.

In the LaminarPace™ system, the drying is done at ambient temperature and pressure. This feature offers mild conditions for the vaccine during drying, and it is not exposed to any high shear stress. Vaccines spray-dried with LaminarPace™ are formulated with typical stabilizers, like trehalose, to embed the vaccine into an amorphous matrix. The amorphous matrix and low moisture content result in high glass transition temperature.

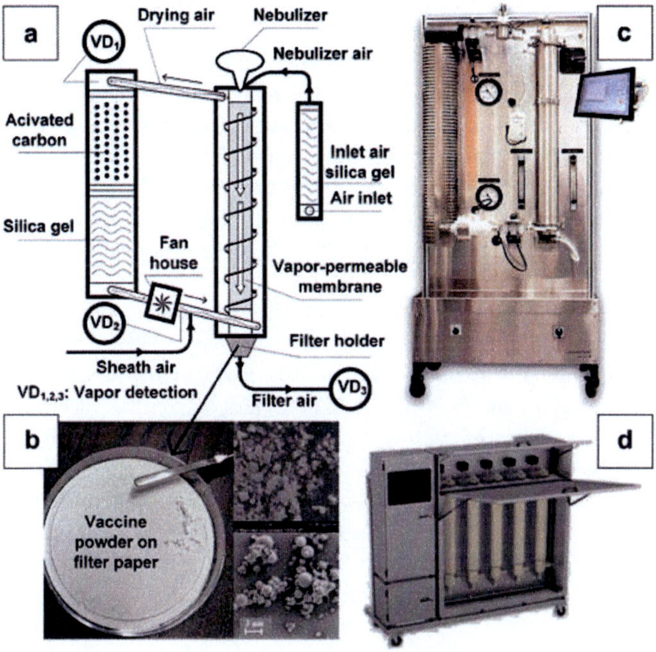

Fig. 3.55 **a** Schematic of the LaminarPace™ spray-dryer principle with drying column with a vapor-permeable membrane, dehydration column, and vapor detectors, **b** Vaccine powder collected on a filter paper, **c** Illustration of a LaminarPace™ spray dryer, **d** LaminarPace™ spray dryer prototype showing how a 5-column operation in a parallel system could look like (Images courtesy of Ziccum AB, Sweden)

Figure 3.55c shows a photograph of the original LaminarPace™ spray dryer. The powder is produced in the drying column to the right, while the drying air is regenerated in the dehydration column to the left, containing activated carbon and silica gel. Organic solvents are used for non-water-soluble compounds, and activated carbon removes the organic vapors.

Ziccum's end goal is to incorporate the system into a full process line, from bulk material to fill and finish, implemented in an industrial GMP environment. The LaminarPace™ development is currently in progress, and one milestone is to have an aseptic version with optimized drying capacity. In addition, some conceptual designs have already started regarding scale-up and scale-out of the system. Figure 3.55d illustrates a 3D model of how a scaled prototype with five drying columns connected in parallel could look like for a higher powder throughput.

3.9.7 Drying of Vaccines as Microarray Patches

Vaccine delivery via the skin into the epidermis or dermis is an attractive approach because of the high density of antigen-presenting cells, such as Langerhans and dermal dendritic cells, which can elicit a strong immune response. Over the past two decades, studies on microneedle array patches (MAPs) have increased significantly (Arya and Prausnitz 2016), demonstrating their potential as an innovative delivery platform for a wide range of vaccines (Marshall et al. 2016; Peyraud et al. 2019; Badizadegan et al. 2020; Nguyen et al. 2020; Prausnitz et al. 2020; Korkmaz et al. 2021).

A MAP consists of many microneedles attached to a patch applied to the skin with finger pressure or an applicator. The microneedles are typically between 100 and 1000 μm long, which is sufficient to penetrate the permeability barrier of the skin (the stratum corneum of about 10–20 μm thickness) and deliver the vaccines to the epidermis or dermis, but short enough not to cause pain. Microfabrication technology has enabled microneedle designs (Kim et al. 2012; Weniger and Papania 2013; Korkmaz et al. 2021). Common MAPs are solid-coated or dissolving microneedles (Gill and Prausnitz 2007; Suh et al. 2014; Prausnitz 2017; Peyraud et al. 2019).

Figure 3.56 shows the structure of the skin and the two MAP principles. Solid-coated MAPs consist of non-biodegradable materials (e.g., silicon) (Henry et al. 1998; Donnelly et al. 2010) that are coated with a liquid vaccine formulation and then dried (Fernando et al. 2018). Upon contact with the skin's interstitial fluid, the dry formulation dissolves and releases the vaccine into the skin. After delivery, the microneedles are removed and discarded (Peyraud et al. 2019). A disadvantage is the biohazardous sharps waste, which can cause safety concerns during disposal. Dissolving MAPs consist of biodegradable polymers or dissolvable sugars that encapsulate the vaccine and dissolve upon contact with the skin's interstitial fluid, releasing the vaccine into the skin and eliminating the risks of biohazardous sharps after usage.

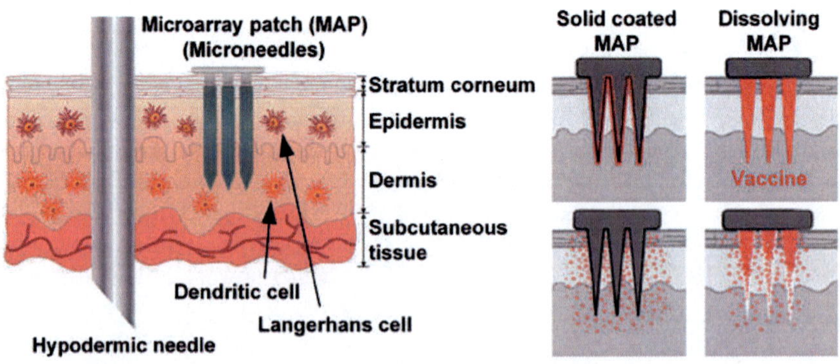

Fig. 3.56 Comparison between conventional intramuscular injection by hypodermic needle and microneedles (i.e., solid coated or dissolving microarray patches, MAP) (Adapted from Suh et al. (2014) and Peyraud et al. (2019) with open access under CC-BY 3.0 and CC-BY-NC-ND license)

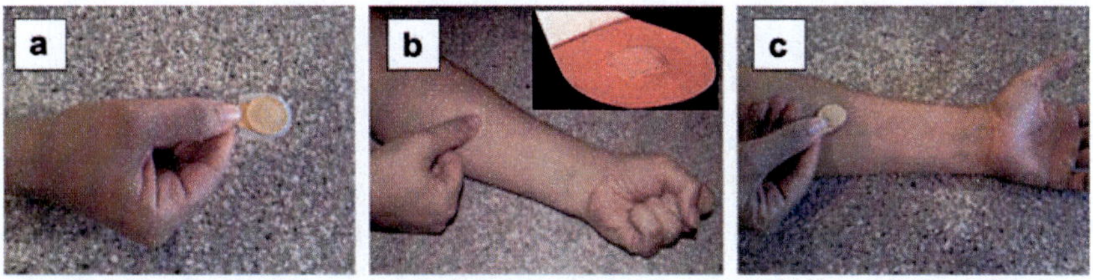

Fig. 3.57 Procedure to apply a microneedle patch to the skin: **a** The protective cap of the patch is removed, **b** The patient forms a fist in the other hand, and the patch is pushed on the forearm until a "click" sound is heard indicating enough force applied, **c** The patch is left on the skin for 20 min, after which the patch is peeled off the skin (Reproduced from Arya et al. (2017) with permission from Elsevier)

Figure 3.57 illustrates the procedure for applying a microneedle patch to the skin. First, the protective cap of the patch is removed. Then, the patch is pressed onto the skin and left on for some minutes to allow the microneedles to dissolve and the vaccine to penetrate the skin. Finally, the patch is peeled off the skin (Arya et al. 2017).

Figure 3.58 shows some close-up images of dissolving microneedles prepared from dried vaccine formulations. Typically, the microneedles are 600–700 μm high, have a 300 μm wide base, and a tip radius of < 10 μm. As a rule of thumb, approximately 0.1–1 μg of vaccine can be administered per microneedle, equating to about 10–100 μg for a patch containing 100 microneedles (Chu et al. 2010; Kim et al. 2012).

Materials often used to design dissolving microneedles include water-soluble polymers that help provide mechanical strength and viscosity during coating, such as

- carboxymethyl cellulose (CMC) (Joyce et al. 2018),
- methylcellulose,
- polyvinyl pyrrolidone (PVP) (Sullivan et al. 2010; Kim et al. 2021),
- polyvinyl alcohol (PVA) (Edens et al. 2015; Joyce et al. 2018),
- hyaluronic acid (Hirobe et al. 2015),
- chondroitin sulfate, dextran, dextrin, or sodium alginate (Prausnitz 2017).

Sugars such as trehalose (Lee et al. 2011), sucrose (Edens et al. 2015; Joyce et al. 2018), and maltose (Miyano et al. 2005) are added to increase microneedle dissolution rate and thermostability (Nguyen et al. 2020). In addition, water-soluble proteins such as gelatin (Vassilieva et al. 2015), albumin, and silk (Prausnitz 2017), and amino acids such as threonine (Edens et al. 2015) or arginine (Mistilis et al. 2015) (Mistilis et al. 2017) are added as vaccine stabilizers.

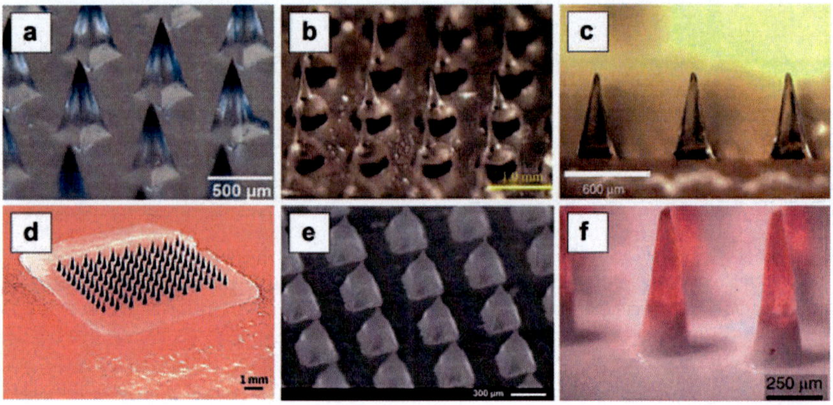

Fig. 3.58 Dissolving microneedles made of water-soluble polymers and sugars with encapsulated vaccines: **a** Microneedles made of sucrose, threonine (amino acid), and CMC and PVA (600 μm, base 300 μm, tip radius < 3 μm) for measles vaccine, shown here with trypan blue dye for better visualization (Reproduced from Edens et al. (2015) with permission from Elsevier), **b** Microneedles (700 μm long) made of sucrose, threonine, CMC, and PVA for measles and rubella vaccine (Reproduced from Joyce et al. (2018) with permission from Oxford University Press), **c** Microneedles (600 μm, base 300 μm) made of CMC for transdermal drug delivery (Reproduced from Lee et al. (2008) with permission from Elsevier), **d** Microneedle patch made of 100 microneedles (650 μm) of gelatin and sucrose with encapsulated influenza vaccine (Reproduced from Rouphael et al. (2017) with permission from Elsevier), **e** Microneedles (600 μm, base 300 μm, tip radius < 10 μm) made of CMC and trehalose with encapsulated human growth hormone (Reproduced from Lee et al. (2011) with permission from John Wiley and Sons), **f** Microneedles (650 μm, tip radius 10 μm) made of PVP with 3 μg of encapsulated inactivated influenza virus vaccine per patch (Reproduced from Sullivan et al. (2010) with permission from Springer Nature)

To develop a thermostable MAP for influenza vaccination, Mistilis et al. (2015) screened 61 different excipient formulations and identified the combination of arginine and calcium heptagluconate (50:50 ratio) as the best stabilizer during storage at room temperature for 6 months. In a subsequent long-term study, trehalose/sucrose and arginine/heptagluconate formulations showed no significant loss of vaccine activity after 24 months of storage at 25 °C (Mistilis et al. 2017). Therefore, as shown in the review paper of Nguyen et al. (2020), mainly trehalose and sucrose are used as stabilizers to improve the thermostability of dissolving MAPs.

Proof of concept for delivery by MAP for seasonal influenza vaccine has been established in Phase I clinical trials of both the dissolving and solid-coated MAP platforms (Hirobe et al. 2015; Rouphael et al. 2017; Fernando et al. 2018; Frew et al. 2020). The results showed similar immune responses compared to intramuscular injection, good safety profiles and relatively low skin reactogenicity, and good acceptance of use (ClinicalTrials.gov, NCT02438423, 100 healthy adults aged 18–49 years, Georgia Institute of Technology, NCT03207763: 33 healthy children aged 6 weeks to 24 months, Emory University).

Currently, live measles-rubella vaccines are being advanced in MAPs toward readiness for clinical trials. The most advanced candidate has demonstrated immunogenicity and protection of MAP vaccination in preclinical studies and thermostability under CTC conditions (Edens et al. 2015; Joyce et al. 2018; Prausnitz et al. 2020). In addition, stability data showed no significant loss of vaccine activity when stored at 40 °C for up to 28 days (Joyce et al. 2018; Prausnitz et al. 2020). Phase I/II human clinical trials of measles-rubella microneedle patches in infants are expected to begin in 2021 (ClincicalTrials.gov, NCT04394689, 225 participants, Micron Biomedical, Inc).

Simultaneously, numerous other vaccine MAPs are in preclinical development, including next-generation vaccines for Rotavirus, Polio, Tetanus Toxoid, Diphtheria, Rabies, Herpes virus, Hepatitis B, Tuberculosis Bacille Calmette Guérin, Human papillomavirus, HIV, Ebola (Poirier et al. 2017; WHO 2018; Peyraud et al. 2019; Nguyen et al. 2020; O'Shea et al. 2021) and Coronavirus (Kim et al. 2020; O'Shea et al. 2021). Overall, there is quite an intense research activity to advance the MAP technology platform for vaccine delivery.

The design and fabrication of a dissolving MAP for vaccine delivery require an interdisciplinary microfabrication approach, vaccine encapsulation, effective delivery into the skin, and safe disposal (Donnelly et al. 2010). In addition, various molding and casting techniques are involved in the manufacturing process (Lee et al. 2008, 2011; Chu et al. 2010; Mistilis et al. 2015, 2017).

Figure 3.59 shows a simplified flowchart of the manufacturing process for dissolvable microneedle patches (Chu et al. 2010; Mistilis et al. 2015). First, a master with a microneedle structure is fabricated using UV lithography and reactive ion etching (Park et al. 2007). The master structure is then used to cast an inverse mold of polydimethylsiloxane (PDMS), whose surface allows higher vaccine activity after drying than stainless steel (Mistilis et al. 2015). The PDMS microneedle mold is then cast with the aqueous liquid vaccine formulation with excipients. This step is followed by drying the vaccine formulation in the mold cavities in the air near room temperature or under centrifugation to remove holes created by water evaporation. These drying times can vary depending on the material and processing conditions. Drying times overnight (Chu et al. 2010; Vassilieva et al. 2015) and in the order of a few hours have been reported (Park et al. 2007; Lee et al. 2008, 2011). Rapid drying is generally desirable to reach higher productivity.

A two-step manufacturing process is often used, in which the mold cavities are first filled with a casting solution containing vaccine to form the microneedles and then with a second casting solution without a vaccine to create the patch carrier. This procedure allows optimization of the microneedle design and localization of the vaccine within the needle tip (Chu et al. 2010). Air drying the backing layer in the second step also requires a few hours.

Finally, the dried solid microneedles are released from the mold by an adhesive layer, assembled into a patch, and placed in suitable packaging (PATH 2020). Residual moisture contents of less than 2 wt% have been reported for microneedles made of fish gelatin and sucrose (Vassilieva et al. 2015). Due to the high moisture sensitivity of MAPs, polymer-based films with low permeability and aluminum

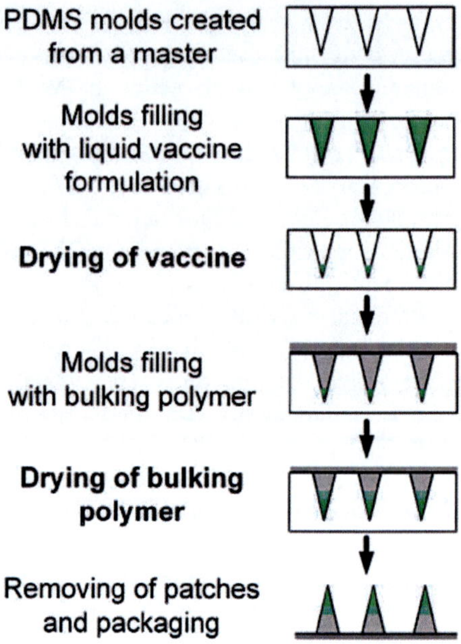

Fig. 3.59 Simplified flow diagram of a typical process for manufacturing dissolvable MAPs by molding and casting techniques (Adapted from Mistilis et al. (2015) with permission from Elsevier)

foils, films, or sheets are recommended for their barrier properties (PATH 2020). Desiccants may be required if the packaging system does not provide sufficient moisture control.

A slightly different process is used to produce polyvinylpyrrolidone (PVP) microneedles at room temperature using photopolymerization of liquid vinylpyrrolidone monomer (Sullivan et al. 2010) (Fig. 3.58f). PVP is chosen as the structural material because it is biocompatible, mechanically stable, and highly soluble in water. For example, an influenza vaccine was incorporated into the polymeric microneedles in a freeze-dried form, avoiding the need for reconstitution before administration.

Recently, Kim et al. (2021) developed a casting method using PVP to encapsulate freeze-dried inactivated H1N1 influenza vaccine into dissolving MAPs. Figure 3.60 shows the principle and an image of the generated microneedles. The lyophilized vaccine powder was suspended in anhydrous organic solvent (i.e., chloroform). PVP use was important because it is soluble in both chloroform for casting and water for dissolution in the skin. In addition, the storage stability of 3 months at 40 °C was achieved. The authors concluded that this two-step fabrication process could simplify and accelerate the development of thermostable dissolvable MAPs, especially since stable lyophilized vaccines already exist. In principle, this process could also be used to incorporate spray-dried vaccine powder into microneedles.

The production of dissolving MAPs involves process steps such as molding, coating, casting, drying, and packaging, which are common in the pharmaceutical

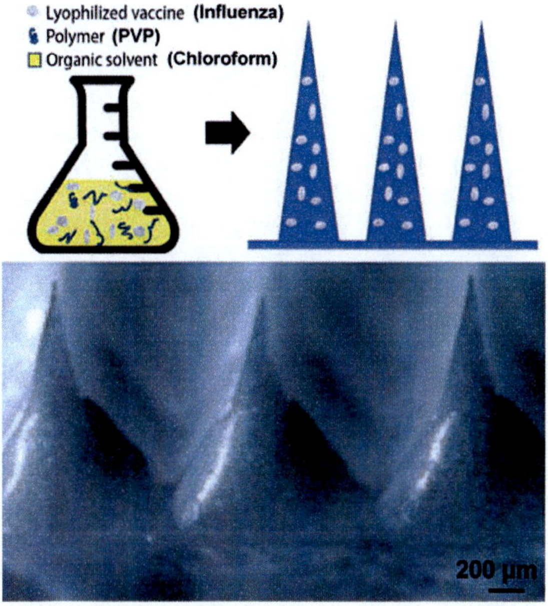

Fig. 3.60 Microneedles made of PVP encapsulating lyophilized influenza vaccine prepared by casting in anhydrous chloroform (Reproduced from Kim et al. (2021) with permission from Springer Nature)

industry and can be mass-produced at low cost (Arya and Prausnitz 2016). For example, an assessment of the potential cost-effectiveness of MAPS in childhood measles vaccination programs revealed that the cost structure could be similar to or even lower than the subcutaneous vaccine delivery by syringes and vials (Adhikari et al. 2016).

As reviewed by Nguyen et al. (2020), several manufacturers are already developing vaccine MAPs. These companies are mainly located in the US (Micron Biomedical, Vaxess, 3 M Corporation), Korea (QuadMedicine, Raphas, JUVIC), and Australia (Vaxxas). While detailed information about their manufacturing methods is not published, 3 M Corporation offers a solid microstructured transdermal system (sMTS) that has undergone FDA approval and is available as a stand-alone device (Hansen et al. 2010; Zhang et al. 2012b; Kommareddy et al. 2013; 3 M 2014). Its proprietary GMP manufacturing and aseptic coating technology have a capacity of up to 10,000 MAPs per day (3 M 2014).

The company Vaxxas, a spin-off of the Australian Institute of Bioengineering & Nanotechnology at the University of Queensland, and Harro Höfliger (Germany) (HARRO Magazine 2019), are working on an affordable and reliable large-scale production of the Nanopatch™. This is a centimeter-sized polymer square with several thousand micro projections of 250 µm in height, coated with vaccines using a patented coating process (Kendall et al. 2018, 2019). The particular challenges lie in the high-precision coating of the needles with a vaccine using a drying gas flow and sealing.

In summary, the main arguments for MAPs are:

- Single-dose delivery without vaccine preparation (no reconstitution of vaccine needed)
- Ease-of-use and administration possible by minimally trained personnel
- Acceptance by patients (Frew et al. 2020)
- Reduced need (elimination) of cold chain storage (thermostable at room temperature for several months, simplifying supply chain)
- Reduction in vaccine dose due to direct response of immune cells from viable skin
- Increased number of doses available in case of vaccine shortage, as less vaccine is needed per MAP compared to injections
- Low cost of vaccination (manufacturing costs are competitive with syringes and vials) (Arya and Prausnitz 2016; Adhikari et al. 2016)

Figure 3.61 clearly shows that MAPs take up much less space than hypodermic needles, syringes, vials of vaccine, vials of diluent, and refrigerators (Arya and Prausnitz 2016). Thus, MAPs do not require continuous refrigeration and enable stockpiling and distributing vaccines, e.g., through the postal system in a pandemic. In addition, dissolving polymer microneedles leave no biohazardous sharps waste.

The remaining critical challenges for MAPs are (Lutton et al. 2015; Marshall et al. 2016; WHO 2018; Peyraud et al. 2019; O'Shea et al. 2021):

- Demonstration of scalability of manufacturing processes to produce millions of MAPs at an acceptable cost,

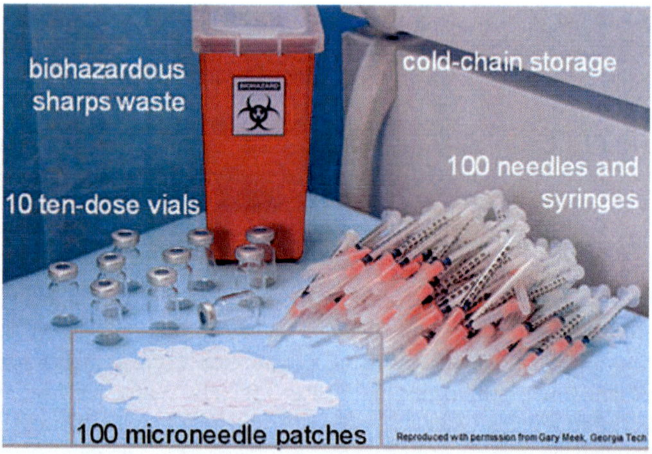

Fig. 3.61 100 microneedle patches (MAPs) for vaccination placed next to 100 needles and syringes, 10 ten-dose vials, a biohazardous sharps waste container, and a refrigerator for cold chain storage (Reproduced from Arya and Prausnitz (2016) with permission from Elsevier)

- Limited vaccine dose loading of coated or dissolving MAPs when the inclusion of adjuvants is also required to elicit an adequate immune response,
- Limited data from human clinical trials,
- Universal acceptance criteria and GMP specifications, and
- Uncertainty of the regulatory pathway.

From a regulatory perspective, dissolvable MAPs are considered a combination product of a drug (vaccine) and a device (potential carriers and/or applicators) (FDA 2018). In contrast, hollow and solid microneedles already have a traditional medical device classification (Marshall et al. 2016). Because MAPs are a new dosage form, product specifications, GMP manufacturing processes, critical quality attributes for clinical use, and regulatory guidance are only recently defined (Lutton et al. 2015).

To address these challenges, WHO has established the Measles-Rubella Microarray Patch (MR-MAP) Working Group, which has developed a target product profile for MR vaccine patches (WHO and UNICEF 2019). The working group explores potential use cases to understand the potential market for MN patches. PATH has established a Center of Excellence for Microarray Patch Technology to accelerate the development of MAPs (PATH 2019). A working group is addressing regulatory issues to identify critical quality attributes, develop test standards for regulatory submissions, and evaluate manufacturing methods for MN patches. In addition, a new initiative has been launched between Gavi, WHO, the Bill & Melinda Gates Foundation, PATH, and UNICEF to prioritize innovative vaccine delivery technologies, such as MAPs. This effort is known as the Vaccination Innovation Prioritization Strategy (VIPS 2020).

In summary, MAPs can change the way vaccines are delivered within immunization programs. As a simple and minimally invasive skin-based delivery method, MAPs do not require medical professionals or a continuous cold chain, making them a promising option, especially for low- and middle-income countries, and as a potential solution for pandemics. Manufacturing MAPs under mild drying conditions close to room temperature minimizes vaccine degradation and increases stability. The possibility of embedding spray-dried powders in dissolving microneedles (similar to lyophilized powders) could further increase the long-term stability of vaccines. The use cases would need to be investigated in more detail.

3.9.8 Summary and Conclusions

In this excursus, alternative drying technologies for vaccines were discussed that enable new delivery forms, such as rapidly soluble oral thin films, sugar pills, fiber membranes, or porous foams as attachments for syringes, powders dried at room temperature in laminar air, and microarray patches that enable vaccine delivery via the skin.

In contrast to freeze-drying and spray-drying, which expose the vaccine to low and high temperatures, drying processes have been developed that can be performed at room temperature and atmospheric pressure, allowing mild drying.

In particular, this excursus shows the technological diversity and possibilities of drying vaccines by stabilizing techniques for encapsulation and preserving vaccines at room temperature. Some drying processes are based on convective air drying, perhaps the simplest of all drying methods. The vaccine solution/suspension is dried by exposure to air near room temperature. On the other hand, longer process times may be expected compared to spray drying, and low final moisture content remains a challenge.

Potential synergies with spray-dried powders exist in the form of pre-prepared suspensions that can be manufactured and embedded into new delivery forms such as fiber membranes, thin films, pills, or microneedle array patches.

These innovative and partially disruptive technologies share the common goal of expanding vaccine distribution via new delivery forms that facilitate administration by enabling immunization by minimally trained healthcare workers.

Compared to syringe injection, the new delivery forms, such as thin films, pills, microarray patches, membranes, and particles in dried form, are very space-saving and enable new routes of administration, such as rapid oral dissolution of thin films, respiratory inhalation in powder form, or as patches via the skin. The common features of these new innovative technologies are in particular:

- extending the shelf life of vaccines at high tropical temperatures,
- allowing safe distribution and storage without loss of efficacy,
- avoiding unnecessary vaccine waste and sharps waste of needles and syringes,
- reducing or even eliminating the need for cold chain storage and transportation, and
- contributing to significantly improving global health through access to vaccines for people around the world.

On the other hand, these innovative technologies often compete for the same resources, and deciding which technology should have priority is a challenge for donors and vaccine manufacturers. Therefore, this judgment must be well weighed.

Challenges include scaling up manufacturing processes and further investment in technology and products. Since the developments mainly originate from university research and have emerged from spin-offs, the technologies often have intellectual properties and patent protection. The research teams are looking for further funding, strategic partnerships and approach investors to develop and scale the innovative technologies.

The next steps are to go through the relevant regulatory processes, conduct human clinical trials that need to be approved by national authorities, and bring the vaccine products to market. Collaboration with existing partners will continue in many ways.

References

3M (2014) 3M solid microstructured transdermal system. Patient-friendly microneedle technology designed to increase your commercial success. https://multimedia.3m.com/mws/media/1004089O/solid-microstructured-transdermal-system-smts-sell-sheet.pdf

Abdul-Fattah AM, Truong VL (2010) Drying process methods for biopharmaceutical products: an overview. In: Jameel H (ed) Formulation and process development strategies for manufacturing biopharmaceuticals. John Wiley & Sons Inc., Hoboken, NJ, USA, pp 705–738

Abdul-Fattah AM, Truong-Le V, Yee L et al (2007a) Drying-Induced variations in physico-chemical properties of amorphous pharmaceuticals and their impact on stability II: stability of a vaccine. Pharm Res 24:715–727. https://doi.org/10.1007/s11095-006-9191-2

Abdul-Fattah AM, Kalonia DS, Pikal MJ (2007b) The challenge of drying method selection for protein pharmaceuticals: product quality implications. J Pharm Sci 96:1886–1916. https://doi.org/10.1002/jps.20842

Abdul-Fattah AM, Truong-Le V, Yee L et al (2007c) Drying-Induced variations in physico-chemical properties of amorphous pharmaceuticals and their impact on stability (I): stability of a monoclonal antibody. J Pharm Sci 96:1983–2008. https://doi.org/10.1002/jps.20859

Abdul-Fattah AM, Lechuga-Ballesteros D, Kalonia DS, Pikal MJ (2008) The impact of drying method and formulation on the physical properties and stability of methionyl human growth hormone in the amorphous solid state. J Pharm Sci 97:163–184. https://doi.org/10.1002/jps.21085

Adali MB, Barresi AA, Boccardo G, Pisano R (2020) Spray freeze-drying as a solution to continuous manufacturing of pharmaceutical products in bulk. Processes 8. https://doi.org/10.3390/PR8060709

Adhikari BB, Goodson JL, Chu SY et al (2016) Assessing the potential cost-effectiveness of microneedle patches in childhood measles vaccination programs: the case for further research and development. Drugs R D 16:327–338. https://doi.org/10.1007/s40268-016-0144-x

Adler M, Lee G (1999) Stability and surface activity of lactate dehydrogenase in spray-dried trehalose. J Pharm Sci 88:199–208. https://doi.org/10.1021/js980321x

Adler M, Unger M, Lee G (2000) Surface composition of spray-dried particles of bovine serum albumin/trehalose/surfactant. Pharm Res 17:863–870. https://doi.org/10.1023/A:1007568511399

Afkhami S, LeClair DA, Haddadi S et al (2017) Spray dried human and chimpanzee adenoviral-vectored vaccines are thermally stable and immunogenic in vivo. Vaccine 35:2916–2924. https://doi.org/10.1016/j.vaccine.2017.04.026

Agarkhedkar S, Kulkarni PS, Winston S et al (2014) Safety and immunogenicity of dry powder measles vaccine administered by inhalation: a randomized controlled Phase I clinical trial. Vaccine 32:6791–6797. https://doi.org/10.1016/j.vaccine.2014.09.071

Ajmera A, Scherließ R (2014) Stabilisation of proteins via mixtures of amino acids during spray drying. Int J Pharm 463:98–107. https://doi.org/10.1016/j.ijpharm.2014.01.002

Alcock R, Cottingham MG, Rollier CS et al (2010) Long-Term thermostabilization of live poxviral and adenoviral vaccine vectors at supraphysiological temperatures in carbohydrate glass. Sci Transl Med 2:19ra12–19ra12. https://doi.org/10.1126/scitranslmed.3000490

Alexeenko A, Topp E (2020) Future directions. In: Ohtake S, Izutsu K-I, Lechuga-Ballesteros D (eds) Drying Technologies for biotechnology and pharmaceutical applications. Wiley, pp 355–372

Alhajj N, O'Reilly NJ, Cathcart H (2021) Designing enhanced spray dried particles for inhalation: a review of the impact of excipients and processing parameters on particle properties. Powder Technol 384:313–331. https://doi.org/10.1016/j.powtec.2021.02.031

Allison SD, Chang B, Randolph TW, Carpenter JF (1999) Hydrogen bonding between sugar and protein is responsible for inhibition of dehydration-induced protein unfolding. Arch Biochem Biophys 365:289–298. https://doi.org/10.1006/abbi.1999.1175

Al-Zoubi N, Gharaibeh S, Aljaberi A, Nikolakakis I (2021) Spray drying for direct compression of pharmaceuticals. Processes 9:267. https://doi.org/10.3390/pr9020267

Amidi M, Pellikaan HC, Hirschberg H et al (2007) Diphtheria toxoid-containing microparticulate powder formulations for pulmonary vaccination: preparation, characterization and evaluation in guinea pigs. Vaccine 25:6818–6829. https://doi.org/10.1016/j.vaccine.2007.05.064

Amorij J-P (2007) The development of stable influenza vaccine powder formulations for new needle-free dosage forms. PhD Thesis, University of Groningen, The Netherlands. https://www.researchgate.net/publication/50843621

Amorij J-P, Saluja V, Petersen AH et al (2007) Pulmonary delivery of an inulin-stabilized influenza subunit vaccine prepared by spray-freeze drying induces systemic, mucosal humoral as well as cell-mediated immune responses in BALB/c mice. Vaccine 25:8707–8717. https://doi.org/10.1016/j.vaccine.2007.10.035

Amorij J-P, Huckriede A, Wilschut J et al (2008) Development of stable influenza vaccine powder formulations: challenges and possibilities. Pharm Res 25:1256–1273. https://doi.org/10.1007/s11095-008-9559-6

Andersen SK (2020) New drying processes for biopharmaceuticals. In: Bright. 5 May 2020. https://www.brighttalk.com/webcast/10519/373808/new-drying-processes-for-biopharmaceuticals

Arduini M (2016) Freeze drying market analysis. In: Aseptic processing symposium, Buffalo, USA, September 14–16, 2016, pp 1–32

Arpagaus C (2007) Part B, Spray drying. In: The laboratory assistent. Büchi Labortechnik AG, Flawil, pp 67–87

Arpagaus C (2012) A novel laboratory-scale spray dryer to produce nanoparticles. Dry Technol 30:1113–1121. https://doi.org/10.1080/07373937.2012.686949

Arpagaus C (2018) Pharmaceutical particle engineering via nano spray drying—process parameters and application examples on the laboratory-scale. Int J Med Nano Res 5. https://doi.org/10.23937/2378-3664.1410026

Arpagaus C (2020) Spray drying of vaccines: a review. In: IDS 2020, 22nd International drying symposium worcester, Massachusetts, USA, June 28-July 1, 2020. pp 1–14

Arpagaus C, Meuri M (2010) Laboratory scale spray drying of inhalable particles: a review. Respir Drug Deliv 469–476

Arpagaus C, John P, Collenberg A, Rütti D (2017) Nanocapsules formation by nano spray drying. In: Jafari SM (ed) Nanoencapsulation technologies for the food and nutraceutical industries. Elsevier, pp 346–401

Arpagaus C, Collenberg A, Rütti D et al (2018) Nano spray drying for encapsulation of pharmaceuticals. Int J Pharm 546:194–214. https://doi.org/10.1016/j.ijpharm.2018.05.037

Arya J, Prausnitz MR (2016) Microneedle patches for vaccination in developing countries. J Control Release 240:135–141. https://doi.org/10.1016/j.jconrel.2015.11.019

Arya J, Henry S, Kalluri H et al (2017) Tolerability, usability and acceptability of dissolving microneedle patch administration in human subjects. Biomaterials 128:1–7. https://doi.org/10.1016/j.biomaterials.2017.02.040

Asgreen C, Knopp MM, Skytte J, Löbmann K (2020) Influence of the polymer glass transition temperature and molecular weight on drug amorphization kinetics using ball milling. Pharmaceutics 12:483. https://doi.org/10.3390/pharmaceutics12060483

AskNature (2017) VitRIS and HydRIS vaccine stabilization technologies. Nova Laboratories Ltd, 6 Feb 2017. In: www.asknature.org. https://asknature.org/idea/vitris-and-hydris-vaccine-stabilization-technologies/

Audouy SAL, van der Schaaf G, Hinrichs WLJ et al (2011) Development of a dried influenza whole inactivated virus vaccine for pulmonary immunization. Vaccine 29:4345–4352. https://doi.org/10.1016/j.vaccine.2011.04.029

Badizadegan K, Goodson JL, Rota PA, Thompson KM (2020) The potential role of using vaccine patches to induce immunity: platform and pathways to innovation and commercialization. Expert Rev Vaccines 19:175–194. https://doi.org/10.1080/14760584.2020.1732215

Bahamondez-Canas TF, Cui Z (2018) Intranasal immunization with dry powder vaccines. Eur J Pharm Biopharm 122:167–175. https://doi.org/10.1016/j.ejpb.2017.11.001

Bajrovic I, Schafer SC, Romanovicz DK, Croyle MA (2020) Novel technology for storage and distribution of live vaccines and other biological medicines at ambient temperature. Sci Adv 6: eaau4819. https://doi.org/10.1126/sciadv.aau4819

Batens M, Massant J, Teodorescu B, Van den Mooter G (2018) Formulating monoclonal antibodies as powders for reconstitution at high concentration using spray drying: models and pitfalls. Eur J Pharm Biopharm 127:407–422. https://doi.org/10.1016/j.ejpb.2018.02.002

Berthelot M (1859) Ueber trehalose und mycose. Ann Der Chemie Und Pharm 109:34–35. https://doi.org/10.1002/jlac.18591090104

Beteta O, Ivanova S (2015) Cool down with liquid nitrogen. CEP Mag www.aiche.org/cep, pp30–35

Bhambhani A, Antochshuk V (2020) Chapter 5: vaccines and microorganisms. In: Ohtake S, Izutsu K-I, Lechuga-Ballesteros D (eds) Drying technologies for biotechnology and pharmaceutical applications. Wiley-VCH Verlag GmbH & Co, KGaA, pp 121–136

Bhandari BR, Datta N, Howes T (1997) Problems associated with spray drying of sugar-rich foods. Dry Technol 15:671–684. https://doi.org/10.1080/07373939708917253

Bosshammer F (2014) Lyo-Motion, Etablierte Technologie vs Innovation in Der Gefriertrocknung. Technopharm 161:156–161

Bowen M, Turok R, Maa Y-F (2013) Spray drying of monoclonal antibodies: investigating powder-based biologic drug substance bulk storage. Dry Technol 31:1441–1450. https://doi.org/10.1080/07373937.2013.796968

Brandau DT, Jones LS, Wiethoff CM et al (2003) Thermal stability of vaccines. J Pharm Sci 92:218–231. https://doi.org/10.1002/jps.10296

Breit J, DuBose D (2013) Spray-Dry manufacture of vaccine formulations. In: Bioprocess Int. https://bioprocessintl.com/manufacturing/monoclonal-antibodies/spray-dry-manufacture-of-vaccine-formulations-347400/

Broadhead J, Rouan SKE, Hau I, Rhodes CT (1994) The effect of process and formulation variables on the properties of spray-dried β-galactosidase. J Pharm Pharmacol 46:458–467. https://doi.org/10.1111/j.2042-7158.1994.tb03828.x

Bronshtein V (2004) Preservation by foam formulation: an alternative to freeze-drying. Pharm Technol 28:86–92

Brown WR (1996) Enteric immunization: promises and challenges. Dig Dis 14:192–200. https://doi.org/10.1159/000171550

Büchi Labortechnik AG (2014) Ultrasonic package, maximize your particle size, ultrasonic nozzle for mini spray dryer B-290, Brochure 115593290 en 1405. https://static1.buchi.com/sites/default/files/downloads/Ultrasonic_Package_Product_Brochure_en_C_0.pdf

Büchi Labortechnik AG (2020) Spray-Drying—basic theory and applications, pp 1–28

Bullich R (2015) Continuous freeze drying. In: Innovation forum in pharmaceutical process professional, PHARMAprocess Forum, 27–28 October 2015, Barcelona, Spain, pp 1–38

Burger JL, Cape SP, Braun CS et al (2008) Stabilizing formulations for inhalable powders of live-attenuated measles virus vaccine. J Aerosol Med Pulm Drug Deliv 21:25–34. https://doi.org/10.1089/jamp.2007.0658

Bürki K, Jeon I, Arpagaus C, Betz G (2011) New insights into respirable protein powder preparation using a nano spray dryer. Int J Pharm 408:248–256. https://doi.org/10.1016/j.ijpharm.2011.02.012

Cape SP, Villa JA, Huang ETS et al (2008) Preparation of active proteins, vaccines and pharmaceuticals as fine powders using supercritical or near-critical fluids. Pharm Res 25:1967–1990. https://doi.org/10.1007/s11095-008-9575-6

Cape SP, McAdams DH, Manion JR et al (2010) Development of stable, immunogenic, and protective measles vaccines for needle-free administration. In: 2010 FIP pharmaceutical sciences world congress and aaps annual meeting and exposition, 14–18 Nov 2010, New Orleans, LA, USA

Carpenter JF, Crowe JH (1989) An infrared spectroscopic study of the interactions of carbohydrates with dried proteins. Biochemistry 28:3916–3922. https://doi.org/10.1021/bi00435a044

Carpenter JF, Prestrelski SJ, Anchordoguy TJ, Arakawa T (1994) Interactions of stabilizers with proteins during freezing and drying. In: Cleland JL, Langer R (eds) Formulation and delivery of proteins and peptides, ACS symposium series, vol 567. American Chemical Society. Washington, DC, pp 134–147

Carpenter JF, Chang BS, Garzon-Rodriguez W, Randolph TW (2002) Rational design of stable lyophilized protein formulations: theory and practice. In: Pharmaceutical biotechnology, pp 109–133

Carrigy NB, Ordoubadi M, Liu Y et al (2019) Amorphous pullulan trehalose microparticle platform for respiratory delivery. Int J Pharm 563:156–168. https://doi.org/10.1016/j.ijpharm.2019.04.004

Carrigy NB, Liang L, Wang H et al (2020) Trileucine and pullulan improve anti-campylobacter bacteriophage stability in engineered spray-dried microparticles. Ann Biomed Eng 48:1169–1180. https://doi.org/10.1007/s10439-019-02435-6

Cattenati C (2019) Business case on low powder dosages, Romaco s.r.l., 20th September 2019

Cavalier G, Fertel C, De Paoli F et al (2021) Cold chain technology brief: vaccines

Cavalier G (2020) The cold chain: a logistical challenge at the heart of the distribution of the Covid-19 vaccine. https://iifiir.org/en/news. https://iifiir.org/en/news/the-cold-chain-a-logistical-challenge-at-the-heart-of-the-distribution-of-the-covid-19-vaccine

Chan MY, Dutill TS, Kramer RM (2017) Lyophilization of adjuvanted vaccines: methods for formulation of a thermostable freeze-dried product. In: Methods in molecular biology, pp 215–226

Chang LL, Pikal MJ (2009) Mechanisms of protein stabilization in the solid state. J Pharm Sci 98:2886–2908. https://doi.org/10.1002/jps.21825

Charlesworth DH, Marshall WR (1960) Evaporation from drops containing dissolved solids. AIChE J 6:9–23. https://doi.org/10.1002/aic.690060104

Chaurasiya B, Zhao Y-Y (2020) Dry powder for pulmonary delivery: a comprehensive review. Pharmaceutics 13:31. https://doi.org/10.3390/pharmaceutics13010031

Chen D, Endres RL, Erickson CA et al (2000a) Epidermal immunization by a needle-free powder delivery technology: immunogenicity of influenza vaccine and protection in mice. Nat Med 6:1187–1190. https://doi.org/10.1038/80538

Chen T, Fowler A, Toner M (2000b) Literature review: supplemented phase diagram of the trehalose-water binary mixture. Cryobiology 40:277–282. https://doi.org/10.1006/cryo.2000.2244

Chen D, Maa Y-F, Haynes JR (2002) Needle-free epidermal powder immunization. Expert Rev Vaccines 1:265–276. https://doi.org/10.1586/14760584.1.3.265

Chen D, Endres R, Maa Y-F et al (2003) Epidermal powder immunization of mice and monkeys with an influenza vaccine. Vaccine 21:2830–2836. https://doi.org/10.1016/S0264-410X(03)00175-0

Chen D, Kapre S, Goel A et al (2010) Thermostable formulations of a hepatitis B vaccine and a meningitis A polysaccharide conjugate vaccine produced by a spray drying method. Vaccine 28:5093–5099. https://doi.org/10.1016/j.vaccine.2010.04.112

Cheow WS, Ng MLL, Kho K, Hadinoto K (2011) Spray-freeze-drying production of thermally sensitive polymeric nanoparticle aggregates for inhaled drug delivery: effect of freeze-drying adjuvants. Int J Pharm 404:289–300. https://doi.org/10.1016/j.ijpharm.2010.11.021

Chu LY, Choi S-O, Prausnitz MR (2010) Fabrication of dissolving polymer microneedles for controlled drug encapsulation and delivery: bubble and pedestal microneedle designs. J Pharm Sci 99:4228–4238. https://doi.org/10.1002/jps.22140

Cicerone MT, Douglas JF (2012) β-Relaxation governs protein stability in sugar-glass matrices. Soft Matter 8:2983. https://doi.org/10.1039/c2sm06979b

Cicerone MT, Tellington A, Trost L, Sokolov A (2006) Plasticized hydrophilic glasses for improved stabilization of biological agents. Patent US7101693B2, pp 1–22

Cicerone MT, Pikal MJ, Qian KK (2015) Stabilization of proteins in solid form. Adv Drug Deliv Rev 93:14–24. https://doi.org/10.1016/j.addr.2015.05.006

Clapp T, Siebert P, Chen D, Jones Braun L (2011) Vaccines with aluminum-containing adjuvants: optimizing vaccine efficacy and thermal stability. J Pharm Sci 100:388–401. https://doi.org/10.1002/jps.22284

Colaço CALS, Roser B (1994) Trehalose-a multifunctional additive for food preservation. Food packaging and preservation. Springer, US, Boston, MA, pp 123–140

Costantino HR (2004) Chapter 5: excipients for use in lyophilized pharmaceutical peptide, protein, and other bioproducts. In: Costantino HR, Pikal MJ (eds) Lyophilization of biopharmaceuticals. AAPS Press, pp 139–228

Crommelin DJA, Anchordoquy TJ, Volkin DB et al (2021) Addressing the cold reality of mRNA vaccine stability. J Pharm Sci 110:997–1001. https://doi.org/10.1016/j.xphs.2020.12.006

Cross RM, Kaplan C, Mcclean D (1957) The heat resistance of dried smallpox vaccine. Lancet 269:446–448. https://doi.org/10.1016/S0140-6736(57)90521-4

Crowe LM, Crowe JH, Rudolph A et al (1985) Preservation of freeze-dried liposomes by trehalose. Arch Biochem Biophys 242:240–247. https://doi.org/10.1016/0003-9861(85)90498-9

Crowe JH, Hoekstra FA, Crowe LM (1992) Anhydrobiosis. Annu Rev Physiol 54:579–599. https://doi.org/10.1146/annurev.ph.54.030192.003051

Crowe LM, Reid DS, Crowe JH (1996) Is trehalose special for preserving dry biomaterials? Biophys J 71:2087–2093. https://doi.org/10.1016/S0006-3495(96)79407-9

Crowe JH, Carpenter JF, Crowe LM (1998a) The role of vitrification in anhydrobiosis. Annu Rev Physiol 60:73–103. https://doi.org/10.1146/annurev.physiol.60.1.73

Crowe JH, Clegg JS, Crowe LM (1998b) Anhydrobiosis: the water replacement hypothesis. In: Reid D (ed) The properties of water in foods ISOPOW 6. Springer, US, Boston, MA, pp 440–455

CureVac (2020a) Clinical trial protocol, COVID-19: a phase 2b/3, randomized, observer-blinded, placebo-controlled, multicenter clinical study evaluating the efficacy and safety of investigational SARS-CoV-2 mRNA Vaccine CVnCoV in Adults 18 Years of Age and Older, 11 Nove, pp 1–152

CureVac (2020b) CureVac's COVID-19 vaccine candidate, CVnCoV, suitable for standard fridge temperature logistics. Press Release, 12 Nov 2020

Dani B, Platz R, Tzannis ST (2007) High concentration formulation feasibility of human immunoglubulin G for subcutaneous administration. J Pharm Sci 96:1504–1517. https://doi.org/10.1002/jps.20508

De Costa S (2014) Advanced delivery devices—new drug delivery & stabilization platforms gaining global interest. Drug Dev. Technol. https://drug-dev.com/advanced-delivery-devices-new-drug-delivery-stabilization-platforms-gaining-global-interest/

De Costa S (2015) Particle engineering for bioavailability enhancement. Pharm. Technol. https://www.pharmtech.com/view/particle-engineering-bioavailability-enhancement

De Costa SD, Roser B, Sen S (2006) Pharmaceutical device for the administration of substances to patients, Patent EP1976483B1. Date of Filing 21(11):2006

De Meyer L, Van Bockstal P-J, Corver J et al (2015) Evaluation of spin freezing versus conventional freezing as part of a continuous pharmaceutical freeze-drying concept for unit doses. Int J Pharm 496:75–85. https://doi.org/10.1016/j.ijpharm.2015.05.025

De Swart RL, LiCalsi C, Quirk AV et al (2007) Measles vaccination of macaques by dry powder inhalation. Vaccine 25:1183–1190. https://doi.org/10.1016/j.vaccine.2006.10.019

Dean HJ, Chen D (2004) Epidermal powder immunization against influenza. Vaccine 23:681–686. https://doi.org/10.1016/j.vaccine.2004.06.041

Debus K (2017) Lyophilization experiences growth spurt in pharma. Pharm Process 32:22–24

DeMarco F, Renzi E (2017) Bulk freeze drying using spray freezing and stirred drying, pp 1–17

Desobry SA, Netto FM, Labuza TP (1997) Comparison of Spray-drying, drum-drying and freeze-drying for β-carotene encapsulation and preservation. J Food Sci 62:1158–1162. https://doi.org/10.1111/j.1365-2621.1997.tb12235.x

Dobry DE, Settell DM, Baumann JM (2015) Chapter 9: spray drying and scale-up. In: Newman A (ed) Pharmaceutical amorphous solid dispersions. John Wiley & Sons Inc., Hoboken, NJ, USA, pp 1–26

Domínguez JM (2011) Drying. In: Comprehensive Biotechnology. Elsevier, pp 727–735

Donnelly RF, Singh TRR, Woolfson AD (2010) Microneedle-based drug delivery systems: Microfabrication, drug delivery, and safety. Drug Deliv 17:187–207. https://doi.org/10.3109/10717541003667798

Drake AC, Lee Y, Burgess EM et al (2018) Effect of water content on the glass transition temperature of mixtures of sugars, polymers, and penetrating cryoprotectants in physiological buffer. PLoS ONE 13:e0190713. https://doi.org/10.1371/journal.pone.0190713

Dulal P, Wright D, Ashfield R et al (2016) Potency of a thermostabilised chimpanzee adenovirus rift valley fever vaccine in cattle. Vaccine 34:2296–2298. https://doi.org/10.1016/j.vaccine.2016.03.061

Eber FJ, Yazdan Panah B, Sewing S et al (2021) Dry powder composition comprising long-chain RNA, Patent US 2021/0030683 A1, Filed: June 18, 2020, pp 1–42

Edens C, Collins ML, Goodson JL et al (2015) A microneedle patch containing measles vaccine is immunogenic in non-human primates. Vaccine 33:4712–4718. https://doi.org/10.1016/j.vaccine.2015.02.074

Emami F, Vatanara A, Park E, Na D (2018) Drying technologies for the stability and bioavailability of biopharmaceuticals. Pharmaceutics 10:131. https://doi.org/10.3390/pharmaceutics10030131

Engstrom JD, Lai ES, Ludher BS et al (2008) Formation of stable submicron protein particles by thin film freezing. Pharm Res 25:1334–1346. https://doi.org/10.1007/s11095-008-9540-4

FDA (2018) Combination product definition combination product types. https://www.fda.gov/combination-products/about-combination-products/combination-product-definition-combination-product-types

Feng AL, Boraey MA, Gwin MA et al (2011) Mechanistic models facilitate efficient development of leucine containing microparticles for pulmonary drug delivery. Int J Pharm 409:156–163. https://doi.org/10.1016/j.ijpharm.2011.02.049

Fernando GJP, Hickling J, Jayashi Flores CM et al (2018) Safety, tolerability, acceptability and immunogenicity of an influenza vaccine delivered to human skin by a novel high-density microprojection array patch (Nanopatch™). Vaccine 36:3779–3788. https://doi.org/10.1016/j.vaccine.2018.05.053

Fourie P, Germishuizen W, Wong Y-L, Edwards D (2008) Spray drying TB vaccines for pulmonary administration. Expert Opin Biol Ther 8:857–863. https://doi.org/10.1517/14712598.8.7.857

Frew PM, Paine MB, Rouphael N et al (2020) Acceptability of an inactivated influenza vaccine delivered by microneedle patch: results from a phase I clinical trial of safety, reactogenicity, and immunogenicity. Vaccine 38:7175–7181. https://doi.org/10.1016/j.vaccine.2020.07.064

Gala RP, Popescu C, Knipp GT et al (2017) Physicochemical and preclinical evaluation of a novel buccal measles vaccine. AAPS PharmSciTech 18:283–292. https://doi.org/10.1208/s12249-016-0566-3

Ganguly A, Demarco FW, Renzi E, Debo D (2019) Bulk freeze drying using spray freezing and agitated drying with heating, pp 1–25

Garmise RJ, Staats HF, Hickey AJ (2007) Novel dry powder preparations of whole inactivated influenza virus for nasal vaccination. AAPS PharmSciTech 8:2–10. https://doi.org/10.1208/pt0804081

Gaspar F, Vicente J, Neves F, Authelin J-R (2014) Chapter 8: spray drying: scale-up and manufacturing. In: Shah N et al. (ed) Amorphous solid dispersions, advances in delivery science and technology. Controlled Release Society, pp 261–302

GEA (2021) Spray dried vaccines, customer case. In: Website, Accessed 28 Jan 2021. https://www.gea.com/en/customer-cases/spray-dried-vaccines.jsp

GEA Niro (2017) Aseptic spray dryers: an effective method for sterile solid drug manufacture, BNA 971/GB 1011 CBC

Geng O (2018) SPX FLOW aseptic spray drying, 13 March 2018. https://file.vogel.com.cn/2018/08/31/5b8905c261000.pdf

Gerde P (2013) System and method for producing dry formulations, Patent US 8347525 B2. Appl Filed Inhalation Sci Sweden AB 24(2008):1–14

Gerde P, Sjöberg C-O, Acevedo F (2010) The laminarpace spray dryer: producing small portions of fine powders at ambient temperatures in high yields. In: Respir. Drug Deliv. 2010, Orlando, Florida, USA. http://www.inhalation.se/wp-content/uploads/2014/08/Form-Respiratory-drug-delivery-20101.pdf

Germishuizen WA, Venter L, Khosa A et al (2010) Establishing a platform for spray drying inhalable vaccines in South Africa. Eur Cells Mater 19:16

Gill HS, Prausnitz MR (2007) Coated microneedles for transdermal delivery. J Control Release 117:227–237. https://doi.org/10.1016/j.jconrel.2006.10.017

Gladd T (2014) Using spray drying for flu vaccines: what challenges do we need to overcome? 14 April 2014. In: www.bioprocessonline.com, https://www.bioprocessonline.com/doc/using-spray-drying-for-flu-vaccines-what-challenges-do-we-need-to-overcome-0001

Gomez M, Archer M, Barona D et al (2021a) Microparticle encapsulation of a tuberculosis subunit vaccine candidate containing a nanoemulsion adjuvant via spray drying. Eur J Pharm Biopharm 163:23–37. https://doi.org/10.1016/j.ejpb.2021.03.007

Gomez M, McCollum J, Wang H et al (2021b) Development of a formulation platform for a spray-dried, inhalable tuberculosis vaccine candidate. Int J Pharm 593:120121. https://doi.org/10.1016/j.ijpharm.2020.120121

Gordon M, Taylor JS (1952) Ideal copolymers and the second-order transitions of synthetic rubbers. i. non-crystalline copolymers. J Appl Chem 493–500. https://doi.org/10.1002/jctb.5010020901

Grasmeijer N, Stankovic M, de Waard H et al (2013) Unraveling protein stabilization mechanisms: vitrification and water replacement in a glass transition temperature controlled system. Biochim Biophys Acta Proteins Proteomics 1834:763–769. https://doi.org/10.1016/j.bbapap.2013.01.020

Grasmeijer N, Tiraboschi V, Woerdenbag HJ et al (2019) Identifying critical process steps to protein stability during spray drying using a vibrating mesh or a two-fluid nozzle. Eur J Pharm Sci 128:152–157. https://doi.org/10.1016/j.ejps.2018.11.027

Greb E (2009) Is spray drying a viable alternative to lyophilization. PharmTech, pp 1–2

Green JL, Angell CA (1989) Phase relations and vitrification in saccharide-water solutions and the trehalose anomaly. J Phys Chem 93:2880–2882. https://doi.org/10.1021/j100345a006

Gribbon EM, Sen SD, Roser BJ, Kampinga J (1996) Stabilisation of vaccines using trehalose (Q-T4) technology. Dev Biol Stand 87:193–199

Hageman MJ (1988) The role of moisture in protein stability. Drug Dev Ind Pharm 14:2047–2070. https://doi.org/10.3109/03639048809152002

Hancock BC, Zografi G (1994) The relationship between the glass transition temperature and the water content of amorphous pharmaceutical solids. Pharm Res J Am Assoc Pharm Sci 11:471–477

Hancock BC, Zografi G (1997) Characteristics and significance of the amorphous state in pharmaceutical systems. J Pharm Sci 86:1–12. https://doi.org/10.1021/js9601896

Hancock BC, Shamblin SL, Zografi G (1995) Molecular mobility of amorphous pharmaceutical solids below their glass transition temperatures. Pharm Res an off J Am Assoc Pharm Sci 12:799–806. https://doi.org/10.1023/A:1016292416526

Handscomb CS, Kraft M, Bayly AE (2009) A new model for the drying of droplets containing suspended solids. Chem Eng Sci 64:628–637. https://doi.org/10.1016/j.ces.2008.04.051

Hansen KJ, Duan D, Determan A et al (2010) Transdermal delivery of vaccines and proteins using a hand-applied, polymeric microstructured array (sMTS-hand and Press&Patch), poster. In: Controlled Release Society, CRS 2010

Hansen LJJ, Daoussi R, Vervaet C et al (2015) Freeze-drying of live virus vaccines: a review. Vaccine 33:5507–5519. https://doi.org/10.1016/j.vaccine.2015.08.085

HARRO Magazine (2019) A micro revolution

Henry S, McAllister DV, Allen MG, Prausnitz MR (1998) Microfabricated microneedles: a novel appraoch to transdermal drug delivery. Proc Control Release Soc 87:30–31. https://doi.org/10.1021/js980042

Hinrichs WL, Prinsen M, Frijlink H (2001) Inulin glasses for the stabilization of therapeutic proteins. Int J Pharm 215:163–174. https://doi.org/10.1016/S0378-5173(00)00677-3

Hirobe S, Azukizawa H, Hanafusa T et al (2015) Clinical study and stability assessment of a novel transcutaneous influenza vaccination using a dissolving microneedle patch. Biomaterials 57:50–58. https://doi.org/10.1016/j.biomaterials.2015.04.007

Hosokawa Micron BV (2019) Batch and continuous drying, process technologies for tomorrow, brochure HMBV.DRY.ENG.19.12.A40

Hosokawa Micron BV (2020) How hosokawa micron's active freeze drying solution simplifies validation for pharmaceutical companies. In: News. https://www.hosokawa-micron-bv.com/news-and-events/publications/how-active-freeze-drying-simplifies-validation-for-pharmaceutical-companies.html

Huang J, Garmise RJ, Crowder TM et al (2004) A novel dry powder influenza vaccine and intranasal delivery technology: induction of systemic and mucosal immune responses in rats. Vaccine 23:794–801. https://doi.org/10.1016/j.vaccine.2004.06.049

Huang J, D'Souza AJ, Alarcon JB et al (2009) Protective immunity in mice achieved with dry powder formulation and alternative delivery of plague F1-V vaccine. Clin Vaccine Immunol 16:719–725. https://doi.org/10.1128/CVI.00447-08

IAC (2018) Vaccines with diluents: how to use them, immunization action coalition, item #P3040 (8/18). https://www.immunize.org/catg.d/p3040.pdf

IMA Life (2019a) LYNFINITY: continuous aseptic spray freeze drying. Process, Technology and Product Characterization (White Paper)

IMA Life (2019b) LYnfinity just a drop away (White Paper)

IMA Life (2020) LYNFINITY: continuous aseptic spray-freeze-drying. https://ima.it/pharma/paper/lynfinity-continuous-aseptic-spray-freeze-drying-process-technology-and-product-characterization/

Irish Tech News (2020) Dr. Bruce Roser the founder and chairman of StablePharma appears on this Episode of the Irish Tech News Podcast (www.irishtechnews.ie), 11 June 2020

Jameel F, Pikal MJ (2010) Design of a formulation for freeze drying. In: Jameel H (ed) Formulation and process development strategies for manufacturing biopharmaceuticals. John Wiley & Sons Inc., pp 459–492

Jangle R, Pisal S (2012) Vacuum foam drying: an alternative to lyophilization for biomolecule preservation. Indian J Pharm Sci 74:91. https://doi.org/10.4103/0250-474X.103837

Jiang G, Joshi SB, Peek LJ et al (2006) Anthrax vaccine powder formulations for nasal mucosal delivery. J Pharm Sci 95:80–96. https://doi.org/10.1002/jps.20484

Jin T, Tsao E (2013) Powder vaccines for pulmonary delivery. In: Flower DR, Perrie Y (eds) Immunomic discovery of adjuvants and candidate subunit vaccines. Springer, New York, New York, NY, pp 291–301

Jin TH, Tsao E, Goudsmit J et al (2010) Stabilizing formulations for inhalable powders of an adenovirus 35-vectored tuberculosis (TB) vaccine (AERAS-402). Vaccine 28:4369–4375. https://doi.org/10.1016/j.vaccine.2010.04.059

Jovanović N, Bouchard A, Hofland GW et al (2004) Stabilization of proteins in dry powder formulations using supercritical fluid technology. Pharm Res 21:1955–1969. https://doi.org/10.1023/B:PHAM.0000048185.09483.e7

Joyce JC, Carroll TD, Collins ML et al (2018) A microneedle patch for measles and rubella vaccination is immunogenic and protective in infant rhesus macaques. J Infect Dis 218:124–132. https://doi.org/10.1093/infdis/jiy139

Kankala RK, Zhang YS, Wang S-B et al (2017) Supercritical Fluid technology: an emphasis on drug delivery and related biomedical applications. Adv Healthc Mater 6:1700433. https://doi.org/10.1002/adhm.201700433

Kanojia G, Willems G-J, Frijlink HW et al (2016) A design of experiment approach to predict product and process parameters for a spray dried influenza vaccine. Int J Pharm 511:1098–1111. https://doi.org/10.1016/j.ijpharm.2016.08.022

Katkov II, Levine F (2004) Prediction of the glass transition temperature of water solutions: comparison of different models. Cryobiology 49:62–82. https://doi.org/10.1016/j.cryobiol.2004.05.004

Kauzmann W (1948) The nature of the glassy state and the behavior of liquids at low temperatures. Chem Rev 43:219–256. https://doi.org/10.1021/cr60135a002

Kemp IC, Wadley R, Hartwig T et al (2013) Experimental study of spray drying and atomization with a two-fluid nozzle to produce inhalable particles. Dry Technol 31:930–941. https://doi.org/10.1080/07373937.2012.710693

Kemp IC, Hartwig T, Herdman R et al (2016) Spray drying with a two-fluid nozzle to produce fine particles: atomization, scale-up, and modeling. Dry Technol 34:1243–1252. https://doi.org/10.1080/07373937.2015.1103748

Kendall MAF, Fernando G, Chen X, Prow T (2018) Coating method. Vaxxas Pty Ltd, Patent US 10022322 B2, Filed. 12 Nov 2015, pp 1–56

Kendall MAF, Meliga F, Depelsenaire S et al (2019) Delivery device. Vaxxas Pty Ltd, Australian Patent AU 2019201051 A1, Date of Filing, 14 Feb 2019, pp 1–255

Kim Y-C (2017) Chapter 30: skin vaccination methods: gene gun, jet injector, tattoo vaccine, and microneedle. Percutaneous penetration enhancers physical methods in penetration enhancement. Springer, Berlin Heidelberg, Berlin, Heidelberg, pp 485–499

Kim YC, Park JH, Prausnitz MR (2012) Microneedles for drug and vaccine delivery. Adv Drug Deliv Rev 64:1547–1568. https://doi.org/10.1016/j.addr.2012.04.005

Kim E, Erdos G, Huang S et al (2020) Microneedle array delivered recombinant coronavirus vaccines: immunogenicity and rapid translational development. EBioMedicine 55:102743. https://doi.org/10.1016/j.ebiom.2020.102743

Kim YC, Lee JW, Esser ES et al (2021) Fabrication of microneedle patches with lyophilized influenza vaccine suspended in organic solvent. Drug Deliv Transl Res 11:692–701. https://doi.org/10.1007/s13346-021-00927-4

Kisich KO, Higgins MP, Park I et al (2011) Dry powder measles vaccine: particle deposition, virus replication, and immune response in cotton rats following inhalation. Vaccine 29:905–912. https://doi.org/10.1016/j.vaccine.2010.10.020

Kommareddy S, Baudner BC, Bonificio A et al (2013) Influenza subunit vaccine coated microneedle patches elicit comparable immune responses to intramuscular injection in guinea pigs. Vaccine 31:3435–3441. https://doi.org/10.1016/j.vaccine.2013.01.050

Kompella UB, Koushik K (2001) Preparation of drug delivery systems using supercritical fluid technology. Crit Rev Ther Drug Carr Syst 18:28. https://doi.org/10.1615/CritRevTherDrugCarrierSyst.v18.i2.20

Korkmaz E, Balmert SC, Sumpter TL et al (2021) Microarray patches enable the development of skin-targeted vaccines against COVID-19. Adv Drug Deliv Rev 171:164–186. https://doi.org/10.1016/j.addr.2021.01.022

Kunda NK, Wafula D, Tram M et al (2016) A stable live bacterial vaccine. Eur J Pharm Biopharm 103:109–117. https://doi.org/10.1016/j.ejpb.2016.03.027

Langford A, Ohtake S (2019) Spray drying of biotherapeutic compounds, Chapter 43. In: Nema S, Ludwig JD (eds) Parenteral medications. Taylor & Francis Incorporated, pp 929–945

Langford A, Balthazor B, Bhatnagar B et al (2018a) Beyond freeze-drying of biologics: vacuum-foam drying and spray freeze-drying. In: Proceedings of 21th international drying symposium, IDS'2018a, València, Spain, 11–14 September 2018. Universitat Politècnica València, Valencia, pp 11–14

Langford A, Bhatnagar B, Walters R et al (2018b) Drying of biopharmaceuticals: recent developments, new technologies and future direction. Japan J Food Eng 19:15–24. https://doi.org/10.11301/jsfe.18514

Langford A, Bhatnagar B, Walters R et al (2018c) Drying technologies for biopharmaceutical applications: recent developments and future direction. Dry Technol 36:677–684. https://doi.org/10.1080/07373937.2017.1355318

Langford A, Luy B, Ohtake S (2018d) Current needs of the pharmaceutical industry: opportunities and challenges for implementing novel drying technologies. In: Proceedings of 21th international drying symposium. Universitat Politècnica València, Valencia, pp 11–14

Langford A, Ohtake S, Lechuga-Ballesteros D, Izutsu K (2020) Chapter 1: introduction. In: Ohtake S, Izutsu K-I, Lechuga-Ballesteros D (eds) Drying technologies for biotechnology and pharmaceutical applications. Wiley-VCH Verlag GmbH & Co, KGaA, pp 1–7

Leathers TD (2003) Biotechnological production and applications of pullulan. Appl Microbiol Biotechnol 62:468–473. https://doi.org/10.1007/s00253-003-1386-4

Lechuga-Ballesteros D, Charan C, Stults CLM et al (2008) Trileucine improves aerosol performance and stability of spray-dried powders for inhalation. J Pharm Sci 97:287–302. https://doi.org/10.1002/jps.21078

LeClair DA, Cranston ED, Xing Z, Thompson MR (2016a) Evaluation of excipients for enhanced thermal stabilization of a human type 5 adenoviral vector through spray drying. Int J Pharm 506:289–301. https://doi.org/10.1016/j.ijpharm.2016.04.067

LeClair DA, Cranston ED, Xing Z, Thompson MR (2016b) Optimization of spray drying conditions for yield, particle size and biological activity of thermally stable viral vectors. Pharm Res 33:2763–2776. https://doi.org/10.1007/s11095-016-2003-4

LeClair DA, Li L, Rahman N et al (2019) Stabilization of HSV-2 viral vaccine candidate by spray drying. Int J Pharm 569:118615. https://doi.org/10.1016/j.ijpharm.2019.118615

Ledet GA, Graves RA, Bostanian LA, Mandal TK (2015) Spray-Drying of biopharmaceuticals. In: Varshney D, Singh M (eds) Lyophilized biologics and vaccines. Springer, New York, New York, NY, pp 273–297

Lee JW, Park J-H, Prausnitz MR (2008) Dissolving microneedles for transdermal drug delivery. Biomaterials 29:2113–2124. https://doi.org/10.1016/j.biomaterials.2007.12.048

Lee JW, Choi S-O, Felner EI, Prausnitz MR (2011) Dissolving microneedle patch for transdermal delivery of human growth hormone. Small 7:531–539. https://doi.org/10.1002/smll.201001091

Leung V, Mapletoft J, Zhang A et al (2019) Thermal stabilization of viral vaccines in low-cost sugar films. Sci Rep 9:7631. https://doi.org/10.1038/s41598-019-44020-w

Li X, Anton N, Arpagaus C et al (2010a) Nanoparticles by spray drying using innovative new technology: the Buchi Nano Spray Dryer B-90. J Control Release 147:304–310

Li X, Anton N, Arpagaus C et al (2010b) Nanoparticles by spray drying using innovative new technology: the Buchi Nano Spray Dryer B-90. J Control Release 147:304–310. https://doi.org/10.1016/j.jconrel.2010.07.113

Li X, Thakkar SG, Ruwona TB et al (2015) A method of lyophilizing vaccines containing aluminum salts into a dry powder without causing particle aggregation or decreasing the immunogenicity following reconstitution. J Control Release 204:38–50. https://doi.org/10.1016/j.jconrel.2015.02.035

LiCalsi C, Maniaci MJ, Christensen T et al (2001) A powder formulation of measles vaccine for aerosol delivery. Vaccine 19:2629–2636. https://doi.org/10.1016/S0264-410X(00)00503-X

Lin W-H, Griffin DE, Rota PA et al (2011) Successful respiratory immunization with dry powder live-attenuated measles virus vaccine in rhesus macaques. Proc Natl Acad Sci 108:2987–2992. https://doi.org/10.1073/pnas.1017334108

Lloyd J (2000) Technologies for vaccine delivery in the 21st century, WHO/V&B/00.35, December 2000

Lovalenti PM, Truong-Le V (2020) Foam drying. In: Ohtake S, Izutsu K-I, Lechuga-Ballesteros D (eds) Drying technologies for biotechnology and pharmaceutical applications. Wiley, pp 257–282

Lovalenti PM, Anderl J, Yee L et al (2016) Stabilization of live attenuated influenza vaccines by freeze drying, spray drying, and foam drying. Pharm Res 33:1144–1160. https://doi.org/10.1007/s11095-016-1860-1

Lutton REM, Moore J, Larrañeta E et al (2015) Microneedle characterisation: the need for universal acceptance criteria and GMP specifications when moving towards commercialisation. Drug Deliv Transl Res 5:313–331. https://doi.org/10.1007/s13346-015-0237-z

Luy B, Stamato H (2020) Spray freeze drying. In: Ohtake S, Izutsu K-I, Lechuga-Ballesteros D (eds) Drying Technologies for Biotechnology and pharmaceutical applications. Wiley, pp 217–237

Luy B, Plitzko M, Struschka M (2012) Process line for the production of freeze-dried particles, pp 1–36

Lv F, Lu Y, Hao Z et al (2016) Preparation and heat resistance study of porcine reproductive and respiratory syndrome virus sugar glass vaccine. Vaccine 34:3746–3750. https://doi.org/10.1016/j.vaccine.2016.06.003

Maa Y-F, Hsu CC (1996) Effect of high shear on proteins. Biotechnol Bioeng 51:458–465. https://doi.org/10.1002/(SICI)1097-0290(19960820)51:4%3c458::AID-BIT9%3e3.0.CO;2-H

Maa Y-F, Hsu CC (1997) Protein denaturation by combined effect of shear and air-liquid interface. Biotechnol Bioeng 54:503–512. https://doi.org/10.1002/(SICI)1097-0290(19970620)54:6%3c503::AID-BIT1%3e3.0.CO;2-N

Maa Y-F, Costantino HR, Nguyen P-A, Hsu CC (1997) The effect of operating and formulation variables on the morphology of spray-dried protein particles. Pharm Dev Technol 2:213–223. https://doi.org/10.3109/10837459709031441

Maa Y-F, Nguyen P-A, Sit K, Hsu CC (1998a) Spray-drying performance of a bench-top spray dryer for protein aerosol powder preparation. Biotechnol Bioeng 60:301–309. https://doi.org/10.1002/(SICI)1097-0290(19981105)60:3%3c301::AID-BIT5%3e3.0.CO;2-L

Maa Y-F, Nguyen P-AT, Hsu SW (1998b) Spray-Drying of air-liquid interface sensitive recombinant human growth hormone. J Pharm Sci 87:152–159. https://doi.org/10.1021/js970308x

Maa YF, Nguyen PA, Sweeney T et al (1999) Protein inhalation powders: spray drying vs spray freeze drying. Pharm Res 16:249–254

Maa Y-F, Zhao L, Payne LG, Chen D (2003) Stabilization of alum-adjuvanted vaccine dry powder formulations: mechanism and application. J Pharm Sci 92:319–332. https://doi.org/10.1002/jps.10294

Maa Y, Ameri M, Shu C et al (2004) Influenza vaccine powder formulation development: spray-freeze-drying and stability evaluation. J Pharm Sci 93:1912–1923. https://doi.org/10.1002/jps.20104

Maa Y-F, Ameri M, Shu C et al (2007) Hepatitis-B surface antigen (HBsAg) powder formulation: process and stability assessment. Curr Drug Deliv 4:57–67. https://doi.org/10.2174/156720107779314758

Maltesen MJ, van de Weert M (2008) Drying methods for protein pharmaceuticals. Drug Discov Today Technol 5:e81–e88. https://doi.org/10.1016/j.ddtec.2008.11.001

Mangal S, Meiser F, Tan G et al (2015) Relationship between surface concentration of l-leucine and bulk powder properties in spray dried formulations. Eur J Pharm Biopharm 94:160–169. https://doi.org/10.1016/j.ejpb.2015.04.035

Manion JR, Cape SP, McAdams DH et al (2012) Inhalable antibiotics manufactured through use of near-critical or supercritical fluids. Aerosol Sci Technol 46:403–410. https://doi.org/10.1080/02786826.2011.634453

Manufacturing Chemist (2010) Nova Laboratories acquires Cambridge Biostability assets, 24 Feb 2010. https://www.manufacturingchemist.com/news/article_page/Nova_Laboratories_acquires_Cambridge_Biostability_assets/42772

Mao H-Q, Yu CK, Truong VL et al (2010) Quick-dissolving oral thin film for targeted delivery of therapeutic, Patent WO 2010/002418 A2. Priority Date 1:1–53

Marshall S, Sahm LJ, Moore AC (2016) The success of microneedle-mediated vaccine delivery into skin. Hum Vaccin Immunother 12:2975–2983. https://doi.org/10.1080/21645515.2016.1171440

Massant J, Fleurime S, Batens M et al (2020) Formulating monoclonal antibodies as powders for reconstitution at high concentration using spray-drying: trehalose/amino acid combinations as reconstitution time reducing and stability improving formulations. Eur J Pharm Biopharm 156:131–142. https://doi.org/10.1016/j.ejpb.2020.08.019

Masters K (1985) Spray drying handbook, 4th edn. George Godwin, London

Masters K (1991) Spray drying handbook, 5th edn. Longman Scientific & Technical, Harlow, USA

Masters K (2002) Spray drying in practice. SprayDryConsult International ApS, Denmark

Maury M, Murphy K, Kumar S et al (2005) Effects of process variables on the powder yield of spray-dried trehalose on a laboratory spray-dryer. Eur J Pharm Biopharm 59:565–573. https://doi.org/10.1016/j.ejpb.2004.10.002

McAdams DH (2011) Production, optimization and characterization of pharmaceutical and excipient powders produced by carbon dioxide-assisted nebulization with a bubble dryer (CAN-BD). In: PhD Thesis, Univ. Color. https://scholar.colorado.edu/downloads/0z708w64z

McAdams D, Chen D, Kristensen D (2012) Spray drying and vaccine stabilization. Expert Rev Vaccines 11:1211–1219. https://doi.org/10.1586/erv.12.101

Meridion (2020) Freeze drying of microspheres by spray freezing and dynamic bulk freeze drying

Mikszta JA, Sullivan VJ, Dean C et al (2005) Protective immunization against inhalational anthrax: a comparison of minimally invasive delivery platforms. J Infect Dis 191:278–288. https://doi.org/10.1086/426865

Miller DP, de Pablo JJ (2000) Calorimetric solution properties of simple saccharides and their significance for the stabilization of biological structure and function. J Phys Chem B 104:8876–8883. https://doi.org/10.1021/jp000807d

Miller DA, Gil M (2012) Chapter 10: spray-drying technology. In: Williams RO, Watts A, Miller D (eds) Formulating poorly water soluble drugs, AAPS advances in the pharmaceutical sciences series, vol 3. Springer. New York, NY, pp 363–442

Miller DP, de Pablo JJ, Corti H (1997) Thermophysical properties of trehalose and its concentrated aqueous solutions. Pharm Res 14:578–590

Mistilis MJ, Bommarius AS, Prausnitz MR (2015) Development of a thermostable microneedle patch for influenza vaccination. J Pharm Sci 104:740–749. https://doi.org/10.1002/jps.24283

Mistilis MJ, Joyce JC, Esser ES et al (2017) Long-term stability of influenza vaccine in a dissolving microneedle patch. Drug Deliv Transl Res 7:195–205. https://doi.org/10.1007/s13346-016-0282-2

Miyano T, Tobinaga Y, Kanno T et al (2005) Sugar micro needles as transdermic drug delivery system. Biomed Microdevices 7:185–188. https://doi.org/10.1007/s10544-005-3024-7

Moderna (2020) Fact sheet for healthcare providers administring vaccine (vaccination providers) emergency use authorization (EUA) of the moderna COVID-19 vaccine to prevent corona virus disease 2019 (COVID-19) in Individuals 18 years of age and older, pp 1–5

Morgan BA, Manser M, Jeyanathan M et al (2020) Effect of shear stresses on adenovirus activity and aggregation during atomization to produce thermally stable vaccines by spray drying. ACS Biomater Sci Eng 6:4304–4313. https://doi.org/10.1021/acsbiomaterials.0c00317

Mumenthaler M, Hsu CC, Pearlman R (1994) Feasibility study on spray-drying protein pharmaceuticals: recombinant human growth hormone and tissue-type plasminogen activator. Pharm Res 11:12–20. https://doi.org/10.1023/a:1018929224005

Murugappan S, Patil HP, Kanojia G et al (2013) Physical and immunogenic stability of spray freeze-dried influenza vaccine powder for pulmonary delivery: comparison of inulin, dextran, or a mixture of dextran and trehalose as protectants. Eur J Pharm Biopharm 85:716–725. https://doi.org/10.1016/j.ejpb.2013.07.018

News10 (2020) Local company looks to freeze dry covid vaccine, posted, 11 Dec 2020, www.news10.com/news/coronavirus. https://www.news10.com/news/coronavirus/local-company-looks-to-freeze-dry-covid-vaccine/

Nguyen V, Lovalenti PM, Truong-Le V et al (2017) Method for preparation of quick dissolving thin films containing bioactive material with enhanced thermal stability, Patent WO 2017/214187 A1, Priority date, 7 June 2016, pp 1–69

Nguyen TT, Oh Y, Kim Y et al (2020) Progress in microneedle array patch (MAP) for vaccine delivery. Hum Vaccines Immunother 17:1–12. https://doi.org/10.1080/21645515.2020.1767997

Nicholls H (2004) Cash injection for thermostable vaccines. Drug Discov Today 9:945. https://doi.org/10.1016/S1359-6446(04)03281-7

Nova Laboratories (2021) Stabilization technologies. In: Website, Accessed 28 Jan 2021. https://www.novalabs.co.uk/stabilisation-technologies

O'Shea J, Prausnitz MR, Rouphael N (2021) Dissolvable microneedle patches to enable increased access to vaccines against SARS-CoV-2 and future pandemic outbreaks. Vaccines 9:320. https://doi.org/10.3390/vaccines9040320

Oakley DE (2004) Spray dryer modeling in theory and practice. Dry Technol 22:1371–1402. https://doi.org/10.1081/DRT-120038734

Ogata S, Nakamura K, Itou K et al (2013) Vacuum freeze-drying apparatus and method for vacuum freeze drying, pp 1–8

Ohtake S, Wang YJ (2011) Trehalose: current use and future applications. J Pharm Sci 100:2020–2053. https://doi.org/10.1002/jps.22458

Ohtake S, Martin RA, Yee L et al (2010) Heat-stable measles vaccine produced by spray drying. Vaccine 28:1275–1284. https://doi.org/10.1016/j.vaccine.2009.11.024

Ohtake S, Martin R, Saxena A et al (2011a) Room temperature stabilization of oral, live attenuated Salmonella enterica serovar Typhi-vectored vaccines. Vaccine 29:2761–2771. https://doi.org/10.1016/j.vaccine.2011.01.093

Ohtake S, Martin RA, Saxena A et al (2011b) Formulation and stabilization of francisella tularensis live vaccine strain. J Pharm Sci 100:3076–3087. https://doi.org/10.1002/jps.22563

Ohtake S, Lechuga-Ballesteros D, Truong-Le V, Patzer EJ (2015) Strategies for heat-stable vaccines. In: Vaccine development and manufacturing. Wiley, pp 287–318

Ohtake S, Izutsu K, Lechuga-Ballesteros D (2020) Drying technologies for biotechnology and pharmaceutical applications. Wiley, VCH

Overhoff KA, Engstrom JD, Chen B et al (2007) Novel ultra-rapid freezing particle engineering process for enhancement of dissolution rates of poorly water-soluble drugs. Eur J Pharm Biopharm 65:57–67. https://doi.org/10.1016/j.ejpb.2006.07.012

Overhoff KA, Johnston KP, Tam J et al (2009) Use of thin film freezing to enable drug delivery: a review. J Drug Deliv Sci Technol 19:89–98. https://doi.org/10.1016/S1773-2247(09)50016-0

Park J-H, Yoon Y-K, Choi S-O et al (2007) Tapered conical polymer microneedles fabricated using an integrated lens technique for transdermal drug delivery. IEEE Trans Biomed Eng 54:903–913. https://doi.org/10.1109/TBME.2006.889173

Passos ML, Mujumdar AS (2005) Mathematical models for improving spray drying processes for foods. Stewart Postharvest Rev 1:1–12. https://doi.org/10.2212/spr.2005.4.6

PATH (2019) The PATH center of excellence for microarray patch technology. https://path.azureedge.net/media/documents/MAPs-fs-Jan2020.pdf

PATH (2020) Microarray patch packaging: an exploration of technical, usability, and general design considerations, MAPs for PrEP Project: dissolving microarray patches (MAPs) for long-acting HIV and pregnancy prevention. https://path.azureedge.net/media/documents/MAP_Packaging_Report_final.pdf

Patil HP, Murugappan S, Ter Veer W et al (2014) Evaluation of monophosphoryl lipid A as adjuvant for pulmonary delivered influenza vaccine. J Control Release 174:51–62. https://doi.org/10.1016/j.jconrel.2013.11.013

Patil HP, Murugappan S, De Vries-Idema J et al (2015) Comparison of adjuvants for a spray freeze-dried whole inactivated virus influenza vaccine for pulmonary administration. Eur J Pharm Biopharm 93:231–241. https://doi.org/10.1016/j.ejpb.2015.04.004

Peyraud N, Zehrung D, Jarrahian C et al (2019) Potential use of microarray patches for vaccine delivery in low- and middle-income countries. Vaccine 37:4427–4434. https://doi.org/10.1016/j.vaccine.2019.03.035

Pfizer-BioNTech (2020) Fact sheet for healthcare providers administring vaccine, emergency use authorization (EUA) of the Pfizer-BioNTech COVID-19 vaccine to prevent coronavirus disease 2019 (COVID-19) in individuals 16 years of age and older, pp 1–6

Pisal S, Wawde G, Salvankar S et al (2006) Vacuum foam drying for preservation of LaSota virus: effect of additives. AAPS PharmSciTech 7:E30–E37. https://doi.org/10.1208/pt070360

Pisano R, Arsiccio A, Capozzi LC, Trout BL (2019) Achieving continuous manufacturing in lyophilization: technologies and approaches. Eur J Pharm Biopharm 142:265–279. https://doi.org/10.1016/j.ejpb.2019.06.027

Plitzko M, Struschka M, Gebhard T, Luy B (2012) A process line for the production of freeze-dried particles, pp 1–29

Poirier D, Renaud F, Dewar V et al (2017) Hepatitis B surface antigen incorporated in dissolvable microneedle array patch is antigenic and thermostable. Biomaterials 145:256–265. https://doi.org/10.1016/j.biomaterials.2017.08.038

PowderPro AB (2013) PowerPro freeze granulation, brochure, pp 1–2

Prausnitz MR (2017) Engineering microneedle patches for vaccination and drug delivery to skin. Annu Rev Chem Biomol Eng 8:177–200. https://doi.org/10.1146/annurev-chembioeng-060816-101514

Prausnitz MR, Goodson JL, Rota PA, Orenstein WA (2020) A microneedle patch for measles and rubella vaccination: a game changer for achieving elimination. Curr Opin Virol 41:68–76. https://doi.org/10.1016/j.coviro.2020.05.005

Price DN, Kunda NK, Ellis R, Muttil P (2020) Design and optimization of a temperature-stable dry powder BCG vaccine. Pharm Res 37:11. https://doi.org/10.1007/s11095-019-2739-8

Research and Markets (2020) Freeze drying equipment—global market trajectory and analytics

Richards A, Krakowka S, Dexter L et al (2002) Trehalose: a review of properties, history of use and human tolerance, and results of multiple safety studies. Food Chem Toxicol 40:871–898. https://doi.org/10.1016/S0278-6915(02)00011-X

Roe KD, Labuza TP (2005) Glass transition and crystallization of amorphous trehalose-sucrose mixtures. Int J Food Prop 8:559–574. https://doi.org/10.1080/10942910500269824

ROMACO (2018) Aseptic powder dosing solutions, brochure 2018/06. https://www.romaco.com/files/Dokumente/delivering-solutions/romaco_aseptic_powder_filling-EN.pdf

Roos Y (1993) Melting and glass transitions of low molecular weight carbohydrates. Carbohydr Res 238:39–48. https://doi.org/10.1016/0008-6215(93)87004-C

Roos Y, Karel M (1991) Water and molecular weight effects on glass transitions in amorphous carbohydrates and carbohydrate solutions. J Food Sci 56:1676–1681. https://doi.org/10.1111/j.1365-2621.1991.tb08669.x

Roser B (1985) UK Patent GB 2187191 B: protection of proteins and the like, pp 1–38

Roser B (1991) Trehalose, a new approach to premium dried foods. Trends Food Sci Technol 2:166–169. https://doi.org/10.1016/0924-2244(91)90671-5

Roser B (2005) Sterile spray drying for stable liquid 21st century pharmaceuticals. In: Innov. Pharm. Technol. http://iptonline.com/articles/public/IPT/19/2005_p50+nonprintable.pdf

Roser B (2006) Stable liquid vaccines and drugs for the 21st century. Future Microbiol 1:21–31. https://doi.org/10.2217/17460913.1.1.21

Roser B (2009) Aseptic spray drying for stable liquid 21st century pharmaceuticals. Cambridge Biostability Ltd, UK. https://freelancecopy.files.wordpress.com/2009/09/cb-aseptic-spray-drying.pdf

Rosini R, Barocchi MA, Rappuoli R (2011) Microbial vaccine design: the reverse vaccinology approach. Development of vaccines. John Wiley & Sons Inc., Hoboken, NJ, USA, pp 1–18

Roth F, Scherließ R (2021) Protein DPI production: discriminating destabilising influences during spray drying. In: Drug Deliv. to Lungs. https://ddl-conference.com/wp-content/uploads/2021/11/03.Roth_.pdf

Rouphael NG, Paine M, Mosley R et al (2017) The safety, immunogenicity, and acceptability of inactivated influenza vaccine delivered by microneedle patch (TIV-MNP 2015): a randomised, partly blinded, placebo-controlled, phase 1 trial. Lancet 390:649–658. https://doi.org/10.1016/S0140-6736(17)30575-5

Saboo S, Tumban E, Peabody J et al (2016) Optimized formulation of a thermostable spray-dried virus-like particle vaccine against human papillomavirus. Mol Pharm 13:1646–1655. https://doi.org/10.1021/acs.molpharmaceut.6b00072

Saleki-Gerhardt A, Zografi G (1994) Non-isothermal and isothermal crystallization of sucrose from the amorphous state. Pharm Res 11:1166–1173

Saluja V, Amorij J-P, Kapteyn JC et al (2010) A comparison between spray drying and spray freeze drying to produce an influenza subunit vaccine powder for inhalation. J Control Release 144:127–133. https://doi.org/10.1016/j.jconrel.2010.02.025

Schaefer J, Lee G (2015) Arrhenius activation energy of damage to catalase during spray-drying. Int J Pharm 489:124–130. https://doi.org/10.1016/j.ijpharm.2015.04.078

Scherließ R, Ajmera A, Dennis M et al (2014) Induction of protective immunity against H1N1 influenza A(H1N1)pdm09 with spray-dried and electron-beam sterilised vaccines in non-human primates. Vaccine 32:2231–2240. https://doi.org/10.1016/j.vaccine.2014.01.077

Schmid K, Arpagaus C, Friess W (2009) Evaluation of a vibrating mesh spray dryer for preparation of submicron particles. Respir Drug Deliv 323–326

Schmid K, Arpagaus C, Friess W (2011) Evaluation of the nano spray dryer B-90 for pharmaceutical applications. Pharm Dev Technol 16:287–294. https://doi.org/10.3109/10837450.2010.485320

Schüle S, Schulz-Fademrecht T, Garidel P et al (2008) Stabilization of IgG1 in spray-dried powders for inhalation. Eur J Pharm Biopharm 69:793–807. https://doi.org/10.1016/j.ejpb.2008.02.010

Schwartzbach H (2011) Achieving aseptic drying with spray drying technologies. Pharm Technol Eur 23:90–92

Searles J, Cherian M (2015) Alternatives to vial lyophilization. Lyophilized biologics and vaccines. Springer, New York, New York, NY, pp 257–271

Sebastião IB, Bhatnagar B, Tchessalov S et al (2019) Bulk dynamic spray freeze-drying part 2: model-based parametric study for spray-freezing process characterization. J Pharm Sci 108:2075–2085. https://doi.org/10.1016/j.xphs.2019.01.011

Sellers SP, Clark GS, Sievers RE, Carpenter JF (2001) Dry powders of stable protein formulations from aqueous solutions prepared using supercritical CO_2-assisted aerosolization. J Pharm Sci 90:785–797. https://doi.org/10.1002/jps.1032

Seville PC, Learoyd TP, Li H-Y et al (2007) Amino acid-modified spray-dried powders with enhanced aerosolisation properties for pulmonary drug delivery. Powder Technol 178:40–50. https://doi.org/10.1016/j.powtec.2007.03.046

Shamblin SL, Tang X, Chang L et al (1999) Characterization of the time scales of molecular motion in pharmaceutically important glasses. J Phys Chem B 103:4113–4121. https://doi.org/10.1021/jp983964+

Sharma A, Khamar D, Cullen S et al (2021) Innovative drying technologies for biopharmaceuticals. Int J Pharm 609:121115. https://doi.org/10.1016/j.ijpharm.2021.121115

Sievers RE (1999) Formation of aqueous small droplet aerosols assisted by supercritical carbon dioxide. Aerosol Sci Technol 30:3–15. https://doi.org/10.1080/713834046

Sievers R (2015) Needle-Free delivery of stable respirable powder vaccine, grand challenges in global health: 2005–2015 retrospective. https://drupal.grandchallenges.org/sites/default/files/files/robertsievers.pdf

Sievers RE, Karst U (1994) Method and apparatus for fine particle formation, Patent EP 1160018 B1, Priority: 08.04.1994 US 224764, pp 1–21

Sievers RE, Sellers SP, Carpenter JF (1999) Supercritical fluid-assisted nebulization and bubble drying, Patent EP 1185248 B1, Priority: 09.06.1999 US 138394 P, pp 1–37

Sievers RE, Huang ETS, Villa JA et al (2001) Low-temperature manufacturing of fine pharmaceutical powders with supercritical fluid aerosolization in a Bubble Dryer®. Pure Appl Chem 73:1299–1303. https://doi.org/10.1351/pac200173081299

Sievers RE, Huang ETS, Villa JA et al (2003) Micronization of water-soluble or alcohol-soluble pharmaceuticals and model compounds with a low-temperature Bubble Dryer®. J Supercrit Fluids 26:9–16. https://doi.org/10.1016/S0896-8446(02)00188-2

Sievers RE, Bennett DJ, Cape SP et al (2007a) Micronization of measles vaccine and siRNA by CAN-BD for aerosol delivery by air expansion of powders with a PuffHaler. RDD Europe 2007, April 17–20. France, Paris, pp 231–234

Sievers RE, Quinn BP, Cape SP et al (2007b) Near-critical fluid micronization of stabilized vaccines, antibiotics and anti-virals. J Supercrit Fluids 42:385–391. https://doi.org/10.1016/j.supflu.2007.03.001

Slade L, Levine H (1995) Water and the glass transition—dependence of the glass transition on composition and chemical structure: Special implications for flour functionality in cookie baking. J Food Eng 24:431–509. https://doi.org/10.1016/0260-8774(95)90766-5

Slade L, Levine H, Reid DS (1991) Beyond water activity: recent advances based on an alternative approach to the assessment of food quality and safety. Crit Rev Food Sci Nutr 30:115–360. https://doi.org/10.1080/10408399109527543

Smith TG, Siirin M, Wu X et al (2015) Rabies vaccine preserved by vaporization is thermostable and immunogenic. Vaccine 33:2203–2206. https://doi.org/10.1016/j.vaccine.2015.03.025

Snyder HE, Lechuga-Ballesteros D (2008) Spray drying: theory and pharmaceutical applications, chapter 7. In: Augsburger LL, Hoag SW (eds) Pharmaceutical dosage forms: tablets, 3 edn, volu 1: unit operations and mechanical properties. Informa Healthcare USA, Inc., New York, NY, pp 227–260

Soltani S, Gerde P, Acevedo F, Rasmuson A (2015) Counter-current spray drying with stream separation: computational modeling of a novel dryer design. Chem Eng Res Des 93:163–173. https://doi.org/10.1016/j.cherd.2014.05.023

Sou T, Morton DAV, Williamson M et al (2015) Spray-Dried influenza antigen with trehalose and leucine produces an aerosolizable powder vaccine formulation that induces strong systemic and mucosal immunity after pulmonary administration. J Aerosol Med Pulm Drug Deliv 28:361–371. https://doi.org/10.1089/jamp.2014.1176

Sou T, Forbes RT, Gray J et al (2016) Designing a multi-component spray-dried formulation platform for pulmonary delivery of biopharmaceuticals: the use of polyol, disaccharide, polysaccharide and synthetic polymer to modify solid-state properties for glassy stabilisation. Powder Technol 287:248–255. https://doi.org/10.1016/j.powtec.2015.10.008

Stablepharma (2013) Improved injections, Patent WO 2013/110956 A1, pp 1–45

Stablepharma (2020a) Illustrative examples of extraordinary stability of vaccines when dried in trehalose, June 2020, pp 1–11

Stablepharma (2020b) Patent pack, August 2020, pp 1–17

Stablepharma (2020c) Introduction to stablepharma

Ståhl K, Claesson M, Lilliehorn P et al (2002) The effect of process variables on the degradation and physical properties of spray dried insulin intended for inhalation. Int J Pharm 233:227–237. https://doi.org/10.1016/S0378-5173(01)00945-0

Struschka M, Plitzko M, Gebhard T, Luy B (2012) Rotary drum for use in a vacuum freeze-dryer, pp 1–28

Suh H, Shin J, Kim Y-C (2014) Microneedle patches for vaccine delivery. Clin Exp Vaccine Res 3:42. https://doi.org/10.7774/cevr.2014.3.1.42

Sullivan VJ, Mikszta JA, Laurent P et al (2006) Noninvasive delivery technologies: respiratory delivery of vaccines. Expert Opin Drug Deliv 3:87–95. https://doi.org/10.1517/17425247.3.1.87

Sullivan SP, Koutsonanos DG, del Pilar MM et al (2010) Dissolving polymer microneedle patches for influenza vaccination. Nat Med 16:915–920. https://doi.org/10.1038/nm.2182

Teekamp N, Tian Y, Visser JC et al (2017) Addition of pullulan to trehalose glasses improves the stability of β-galactosidase at high moisture conditions. Carbohydr Polym 176:374–380. https://doi.org/10.1016/j.carbpol.2017.08.084

Teixeira J (2005) Lyophilization: growing with biotechnology. Genet Eng News 25:52–54

TFF (2021) Feasibility arrangement for shelf-stable powder form of messenger RNA COVID-19 vaccine candidate announced by greenlight biosciences and TFF pharmaceuticals, Press Release, 9 March 2021. https://ir.tffpharma.com/news-releases/news-release-details/feasibility-arrangement-shelf-stable-powder-form-messenger-rna

Thakkar SG, Ruwona TB, Williams RO, Cui Z (2017) The immunogenicity of thin-film freeze-dried, aluminum salt-adjuvanted vaccine when exposed to different temperatures. Hum Vaccin Immunother 13:936–946. https://doi.org/10.1080/21645515.2016.1259042

Thorat BN, Sett A, Mujumdar AS (2020) Drying of vaccines and biomolecules. Dry Technol 1–23. https://doi.org/10.1080/07373937.2020.1825293

Tillequin F (2009) Le Tréhala, une drogue à la croisée des chemins. Rev Hist Pharm (paris) 96:163–172. https://doi.org/10.3406/pharm.2009.22043

Tlaxca JL, Ellis S, Remmele RL (2015) Live attenuated and inactivated viral vaccine formulation and nasal delivery: potential and challenges. Adv Drug Deliv Rev 93:56–78. https://doi.org/10.1016/j.addr.2014.10.002

Tonnis WF, Amorij J-P, Vreeman MA et al (2014) Improved storage stability and immunogenicity of hepatitis B vaccine after spray-freeze drying in presence of sugars. Eur J Pharm Sci 55:36–45. https://doi.org/10.1016/j.ejps.2014.01.005

Truong VL, Abdul-Fattah AM (2010) The impact of formulation and drying processes on the characteristics and performance of biopharmaceutical powders. In: Jameel H (ed) Formulation and process development strategies for manufacturing biopharmaceuticals. John Wiley & Sons Inc., pp 555–585

Truong-Le V (2006) Preservation of bioactive materials by freeze dried foam, Patent US 7135180 B2, Filed: April 10, 2003, pp 1–27

Truong-Le V, Pham B (2007) Preservation of bioactive materials by spray drying, Patent US 7258873 B2, filed, 10 April 2003, pp 1–30

Truong-Le V, Ohtake S, Martin RA et al (2012) Sonic low pressure spray drying, Patent US 8268354 B2, Filed, 6 Nov 2008, pp 1–17

Truong-Le V, Ohtake S, Martin RA et al (2014) Sonic low pressure spray drying, Patent US 8673357 B2, Filed, 14 Aug 2012, pp 1–18

Truong-Le V, Lovalenti PM, Abdul-Fattah AM (2015) Stabilization challenges and formulation strategies associated with oral biologic drug delivery systems. Adv Drug Deliv Rev 93:95–108. https://doi.org/10.1016/j.addr.2015.08.001

ULVAC (2010a) ULVAC solutions for micro-powder dry system. https://www.ulvac.co.jp/eng/data/solution/mpd/mpd.html

ULVAC (2010b) Micro powder dry system for powder process, News, 24 June 2010. In: News. https://www.ulvac.co.jp/en/news/micro_powder_dry_system_for_powder_process_ulvac_has_launched_a_fully_packaged_process_line_integrat/

Upperton (2015) Spray Drying—an alternative to freeze drying? In: Youtube-Video, 23 Feb 2015. https://www.youtube.com/watch?v=uz2Hf_6u-N8

Van der Wel PG (2012) Stirred freeze drying, pp 1–19

Van Oort MM, Sacchetti M (2007) Chapter 11: spray-drying and supercritical fluid particle generation techniques. In: Hickey AJ (ed) Inhalation aerosols: physical and biological basis for therapy, 2nd edn. Informa Healthcare USA Inc., New York, NY, pp 307–346

Vass P, Démuth B, Hirsch E et al (2019) Drying technology strategies for colon-targeted oral delivery of biopharmaceuticals. J Control Release 296:162–178. https://doi.org/10.1016/j.jconrel.2019.01.023

Vassilieva EV, Kalluri H, McAllister D et al (2015) Improved immunogenicity of individual influenza vaccine components delivered with a novel dissolving microneedle patch stable at room temperature. Drug Deliv Transl Res 5:360–371. https://doi.org/10.1007/s13346-015-0228-0

Vehring R (2007) Dry dosage forms produced by vaccum foam drying. In: 2CBI Vaccine Summit, Philadelphia, PA, 2007. https://sites.ualberta.ca/~vehring/PE/Page/Conferences/Vehring_CBI_Vaccine_Summit_2007_Dry_Dosage_Forms_Produced_by_Vacuum_Foam_Drying.pdf

Vehring R (2008) Pharmaceutical particle engineering via spray drying. Pharm Res 25:999–1022. https://doi.org/10.1007/s11095-007-9475-1

Vehring R, Foss WR, Lechuga-Ballesteros D (2007) Particle formation in spray drying. J Aerosol Sci 38:728–746. https://doi.org/10.1016/j.jaerosci.2007.04.005

Vehring R, Snyder H, Lechuga-Ballesteros D (2020) Spray Drying. In: Ohtake S, Izutsu K, Lechuga-Ballesteros D (eds) Drying technologies for biotechnology and pharmaceutical applications. Wiley, pp 179–216

VIPS (2020) The VIPS prioritisation process: methodology and outcomes (vaccine innovation prioritisation strategy), pp 1–24

Vodak DT, Dobry DE, Falk R et al (2012) Enhancing the immune response to intranasal vaccination with norovirus VLPs—size matters. In: Drug delivery to the lungs (DDL 23), Edinburgh, pp 46–49

Walters RH, Bhatnagar B, Tchessalov S et al (2014) Next generation drying technologies for pharmaceutical applications. J Pharm Sci 103:2673–2695. https://doi.org/10.1002/jps.23998

Walton DE (2000) The morphology of spray-dried particles a qualitative vieW. Dry Technol 18:1943–1986. https://doi.org/10.1080/07373930008917822

Walton DE, Mumford CJ (1999) Spray dried products—characterization of particle morphology. Chem Eng Res Des 77:21–38. https://doi.org/10.1205/026387699525846

Wang SH, Kirwan SM, Abraham SN et al (2012) Stable dry powder formulation for nasal delivery of anthrax vaccine. J Pharm Sci 101:31–47. https://doi.org/10.1002/jps.22742

Wang C, Dulal P, Zhou X et al (2018) A simian-adenovirus-vectored rabies vaccine suitable for thermostabilisation and clinical development for low-cost single-dose pre-exposure prophylaxis. PLoS Negl Trop Dis 12:e0006870. https://doi.org/10.1371/journal.pntd.0006870

Wanning S, Süverkrüp R, Lamprecht A (2015) Pharmaceutical spray freeze drying. Int J Pharm 488:136–153. https://doi.org/10.1016/j.ijpharm.2015.04.053

Weisburd S (1988) Death-Defying dehydration. Sci News 133:107. https://doi.org/10.2307/3972291

Weniger BG, Papania MJ (2013) Alternative vaccine delivery methods. Vaccines 1200–1231. https://doi.org/10.1016/B978-1-4557-0090-5.00063-X

WHO (2018) Global vaccine action plan, monitoring, evaluation & accountability, secretariat annual report 2018

WHO (2021) WHO prequalified vaccines. Online database: https://extranet.who.int/gavi/PQ_Web, Accessed on 20 March 2021. https://extranet.who.int/gavi/PQ_Web

WHO, UNICEF (2019) Measles-rubella microarray patch (MR-MAP) target product profile, June 2019. https://www.who.int/immunization/research/ppc-tpp/WHO_MR_MAP_TPP.pdf

Williams ML, Landel RF, Ferry JD (1955) The Temperature dependence of relaxation mechanisms in amorphous polymers and other glass-forming liquids. J Am Chem Soc 77:3701–3707. https://doi.org/10.1021/ja01619a008

Wong Y-L, Sampson S, Germishuizen WA et al (2007) Drying a tuberculosis vaccine without freezing. Proc Natl Acad Sci 104:2591–2595. https://doi.org/10.1073/pnas.0611430104

Wright T (2019) Rising demand for lyophilized products, 01.25.19. Contract Pharma January/Fe

Yoshioka S, Aso Y (2007) Correlations between molecular mobility and chemical stability during storage of amorphous pharmaceuticals. J Pharm Sci 96:960–981. https://doi.org/10.1002/jps.20926

Zhang M, Li H, Lang B et al (2012a) Formulation and delivery of improved amorphous fenofibrate solid dispersions prepared by thin film freezing. Eur J Pharm Biopharm 82:534–544. https://doi.org/10.1016/j.ejpb.2012.06.016

Zhang Y, Brown K, Siebenaler K et al (2012b) Development of lidocaine-coated microneedle product for rapid, safe, and prolonged local analgesic action. Pharm Res 29:170–177. https://doi.org/10.1007/s11095-011-0524-4

Zhou D, Zhang GGZ, Law D et al (2002) Physical stability of amorphous pharmaceuticals: importance of configurational thermodynamic quantities and molecular mobility. J Pharm Sci 91:1863–1872. https://doi.org/10.1002/jps.10169

Zhu C, Shoji Y, McCray S et al (2014) Stabilization of HAC1 influenza vaccine by spray drying: formulation development and process scale-Up. Pharm Res 31:3006–3018. https://doi.org/10.1007/s11095-014-1394-3

Aseptic Spray Drying Technology

4

Abstract

This chapter is dedicated to aseptic spray drying technology. The reader will learn the key elements for the design and components for constructing an aseptic spray dryer and the cleaning and sterilization procedures. Open-loop and closed-loop configurations are schematically illustrated and discussed. The qualification tests to be performed are described. Another focus is sterile powder processing and aseptic powder filling after spray drying, i.e., handling the powder during powder recovery, removal, filling, and packaging. Various apparatuses are presented and explained. Several commercial and laboratory scale aseptic spray drying systems are illustrated with operating maps and figures. Multiple options for working with a Mini Spray Dryer B-290 at low bioburden levels are presented in an excursus. Although similar considerations apply to other biopharmaceutical products, an attempt is made to focus on vaccine-related issues. Finally, the technological maturity of aseptic spray drying for vaccine clinical trials is discussed.

4.1 A Short History of Aseptic Spray Drying

The first description of the design of aseptic spray dryers within an aseptic process can be traced back to 1975 (Masters and Vestergaard 1975; Masters 1976). An aseptic spray drying system has become possible by developing high-temperature-resistant high-efficiency particulate air (HEPA) filters, sterile liquid filters combined with contamination-free atomization, powder discharge systems, and the use of utmost care in the spray dryer design. The design and construction of an aseptic spray dryer require various precautions and extreme care to ensure a high standard of cleanliness and obtain a powder free from contamination (Masters 2002). In this way, the aseptically produced powder can be filled directly into various dosage forms such as vials, pouches, blisters, or other medical devices during packaging.

C. Arpagaus, *Spray Drying of Vaccines*,
https://doi.org/10.1007/978-3-031-24323-3_4

In principle, spray drying is more worthwhile for high powder quantities (Andersen 2016, 2020). Since vaccines are considered high-value products, the focus is on product quality, sterility, and powder recovery. Energy savings are less important in the development of vaccine powders.

An aseptic spray drying system is a must-have criterion for injectable vaccines to maintain sterile conditions during the powder production and obtain a dry powder suitable for clinical use without terminal sterilization (Siew 2016). Oral, pulmonary, or nasal vaccines require controlled processing with low bioburden, and sterility in manufacturing is not a prerequisite. Although terminal sterilization by steam, dry heat, or irradiation of a finished medicinal product, an active substance, an excipient, and a primary container is the preferred technique according to the (EMA 2019) guideline, this is not always possible for sensitive biologics such as vaccines. Therefore, sterile filtration and aseptic processing are accepted for sensitive biological products under validated and controlled conditions.

Terminal sterilization of spray-dried vaccine formulations using heat or irradiation may adversely affect the structure and activity of the products (McAdams et al. 2012). Nevertheless, there are attempts with tailored stabilizing and protective excipient solutions in the vaccine formulation (e.g., amino acid-based) to protect biological materials from irradiation (Scholz and Altenhofen 2011; Tscheliessnig et al. 2012; Ajmera and Scherließ 2014; Scherließ et al. 2014; Altrichter et al. 2019).

When it comes to cost comparison between aseptic spray drying and freeze drying, it is not straightforward as it depends strongly on the scale and is typically customer-specific. Aseptic spray dryers for low molecular weight drugs are available and in regular use all over the world. However, it is still a niche technology that is not yet widely used for marketed biopharmaceutical products. In principle, an aseptic spray dryer is developed together with the customer. Therefore, an exact cost calculation cannot be provided in this book. However, a few indications of investment and operating costs are given in Chap. 3 when comparing spray drying and freeze drying principles.

Overall, it can be assumed that the investment in an aseptic spray dryer system with powder filling is comparable to freeze drying, although the investment costs for large-scale spray drying might be somewhat lower with regards to equipment, facility, and utilities (Greb 2009; Gladd 2014). Powder filling is generally more expensive than liquid filling so that the overall costs balance out again (Andersen 2016, 2020). Operationally, freeze drying is simpler compared to overall powder handling in aseptic spray drying. Besides, the spray dryer may require a second dryer (e.g., vacuum dryer or microwave dryer) to reduce residual moisture to lower levels (Masters 1991; Miller and Gil 2012; Dobry et al. 2015).

In 2009, the Cambridge Biostability plant in Leicester (UK) was the world's first aseptic spray drying plant for the cGMP production of sterile vaccines (Roser 2009). Today, aseptic spray drying is still seen as an emerging technology that is expected to be increasingly commercialized in the biopharmaceutical industry in the coming years (Gladd 2014; Searles and Cherian 2015; Andersen 2016, 2020).

In the last years, there have been several inventions in the field of aseptic spray drying. In addition, several patents have been filed, which prove the feasibility of aseptic spray drying systems and processes for the production of dried pharmaceutical and active biological ingredients (Hill 1984; Chickering et al. 2003; Truong-Le and Pham 2007; Turok and Rasmussen 2010; Fragale and Brueck 2015; Gill et al. 2017; Teo et al. 2021).

4.2 Clean Areas for Aseptic Spray Drying

The conversion of a spray dryer to an aseptic spray dryer requires compliance with aseptic processing guidelines, such as the FDA Guidance for Industry (FDA 2004) or the ISPE Baseline® Guide on Sterile Product Manufacturing Facilities (ISPE 2018). In-depth knowledge of aseptic manufacturing techniques is essential for the successful maintenance of such facilities and equipment.

When building a truly aseptic spray dryer, particular attention has to be paid to quality standards such as ASME Bioprocessing Equipment (BPE) (ASME 2019) and cGMP regulations (McGuire 1998; FDA 2004; EMEA 2006; ICH 2010) which concern personnel, buildings, equipment, production controls, records, and other aspects of the vaccine manufacturing process (Artaud et al. 2019). In addition, filtration, draining, cold spots, steam supply, materials science, CIP/SIP equipment, and zoning are all topics that need particular attention (Mogavero 2014). Table 4.1 shows the zoning definition and classification of clean areas for the production of sterile pharmaceutical products according to WHO guidelines (WHO 2002).

Table 4.1 Comparison of different airborne particulate classification systems for clean areas to manufacture sterile pharmaceutical products and limits for microbiological contamination (average values) according to WHO guidelines (WHO 2002)

WHO (GMP) EEC (GMP)	United States (customary)	ISO/TC (209)	Max. number of particles permitted per m^3 (at rest)		Limits for microbiological contamination		
			0.5 to 5.0 μm	> 5.0 μm	Air sample (CFU per m3)	Settle plates (diameter 90 mm) (CFU per 4 h)	Contact plates (diameter 55 mm) (CFU per plate)
Grade A	Class 100	ISO 5	3500	0	< 3	< 3	< 3
Grade B	Class 1000	ISO 6	35,000	200	10	5	5
Grade C	Class 10,000	ISO 7	350,000	2000	100	25	25
Grade D	Class 100,000	ISO 8	3,500,000	20,000	200	50	50

EEC European Commission, *ISO/TC 209* International organization for standardization/technical committee 209 on cleanrooms and associated controlled environments

The clean areas are classified into four grades:

- Grade A: The local zone for high-risk operations, e.g., aseptic powder filling and making aseptic connections. Typically, a laminar-airflow workstation provides such conditions with a homogeneous airspeed of approximately 0.45 m/s ± 20% (guidance value) at the working position.
- Grade B: In aseptic preparation and filling, the background environment for the grade A zone.
- Grades C and D: Clean areas for carrying out less critical stages in manufacturing sterile products.

Spray drying systems designed for processing active pharmaceutical ingredients (API) under validated cGMP conditions utilize a sealed system with HEPA filters, and the spray dryer unit operates in a cleanroom environment, typically Class 10,000 or 100,000 (ISO 7 or 8, Grade C or D, see Table 4.1).

The FDA (2004) recommends that the area immediately adjacent to an aseptic processing line meet, at a minimum, Class 10,000 (ISO 7, Grade C) standards under dynamic conditions. Monitoring the microbiological cleanliness should be frequent, and methods such as settle plates and volumetric air and surface sampling (e.g., swabs and contact plates) should be used. For further details, the reader is referred to WHO guidelines (WHO 2002). Furthermore, the control sequence must be exact to create a sterile environment. Each manufacturing process requires an appropriate cleanliness level to minimize the risks of particulate or microbiological contamination of the product.

In summary, the requirements for an aseptic spray drying system are as follows:

- Inside the spray dryer system (Grade A)
- Powder handling (Grade A)
- Powder filling (Grade A)
- Environment (Grade C or better).

A typical aseptic powder transfer is shown schematically in Fig. 4.13.

4.3 System Configurations for Aseptic Spray Dryers

In the following, an attempt is made to summarize and visualize the main design points of an aseptic spray dryer, based in particular on further and recommended literature on the subject (Masters and Vestergaard 1975; Masters 2002; Chickering et al. 2003; Turok and Rasmussen 2010; Gladd 2014; Mogavero 2014; Pakowski and Mujumdar 2014; Arpagaus 2020; Vehring et al. 2020).

4.3.1 Concept of an Aseptic Spray Dryer System

Figure 4.1 illustrates the main concept of an aseptic spray drying system (Masters 2002). It features high-temperature resistant HEPA filters for the drying and atomizing gas, an indirect electrical heater for the drying gas, a nozzle atomizer, sterile liquid feed filters combined with contamination-free atomization, a spray drying chamber, a cyclone for separating the gas and the particles, and an outlet bag filter (police filter) separating dust from the exhaust gas, and remote powder discharge and handling systems.

Due to the short residence time in the spray dryer, a small amount of moisture, typically 2–10 wt%, remains in the spray-dried particles collected from the cyclone. Thus, in addition to the spray dryer, a secondary drying step (e.g., in a vacuum dryer or microwave dryer) may optionally be required to reduce the residual moisture content to lower levels and meet product safety and quality requirements (Masters 1991; Miller and Gil 2012; Dobry et al. 2015) (see Chap. 5). Finally, the powder flows into further processing under cleanroom conditions, such as sterile powder filling.

The following sections illustrate the most commonly used aseptic spray drying system configurations in open and closed operations.

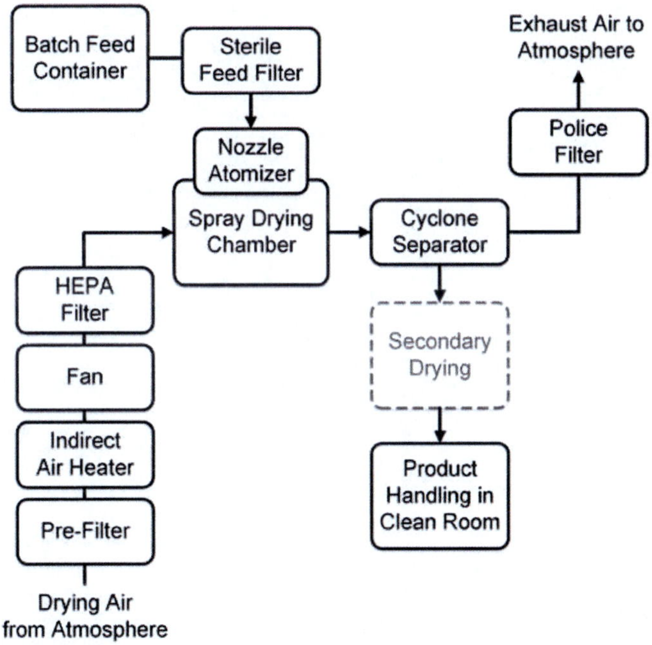

Fig. 4.1 Concept of an aseptic spray drying system operating slightly above atmospheric pressure with an optional secondary drying step (Adapted from Masters (2002) with kind permission from SprayDryConsult, Denmark)

4.3.2 Open-Cycle System

Most aseptic spray dryers are operated in an open-loop for aqueous formulations, such as vaccines and other biological compounds. Basically, an open cycle differs from a closed-loop system in that it handles aqueous feeds and has lower investment costs. Figure 4.2 shows a flow diagram of a typical open-loop aseptic spray drying system, described in various literature sources (Masters and Vestergaard 1975; Masters 1976, 1985; Mujumdar and Alterman 1997; Masters 2002; BETE 2005; Turok and Rasmussen 2010; Pakowski and Mujumdar 2014; Manufacturing Chemist 2016; GEA Niro 2017).

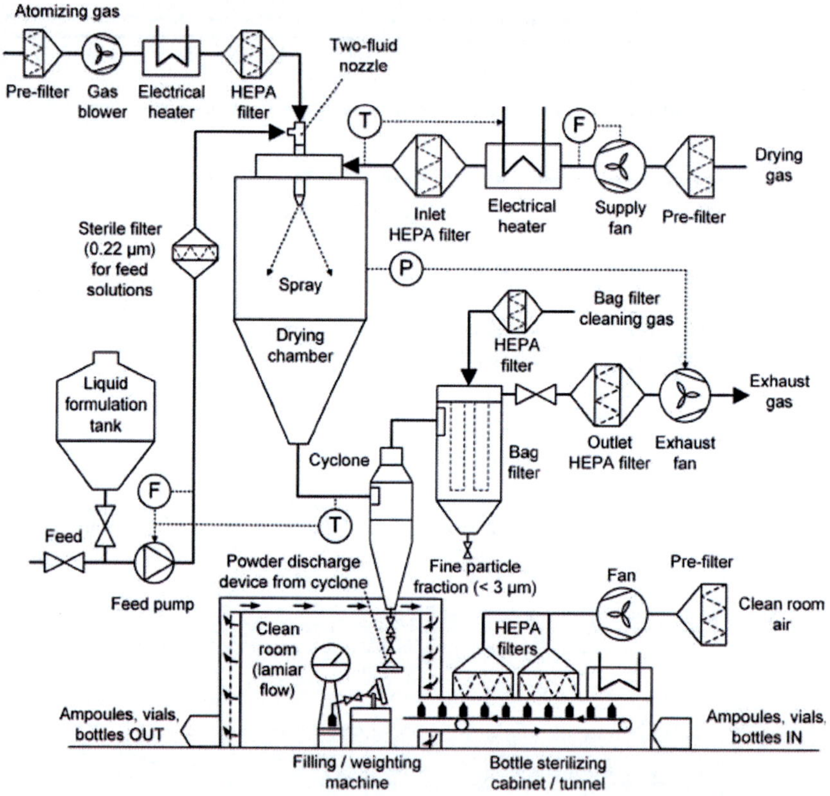

Fig. 4.2 Flow diagram of an aseptic open-cycle spray drying system with two-fluid nozzle (atomizer), feed pump, sterile filter for feed solutions, prefilter for drying and atomization gas (air), supply and exhaust fans, electrical heater, high-efficiency particulate air (HEPA) filters, cyclone separator, powder discharge device from the cyclone (collection vessel), packaging area with clean room, filling/weighing machine, and sterilizing cabinet/tunnel (Based on various literature sources (Masters and Vestergaard 1975; Masters 1976, 1985, 2002; Mujumdar and Alterman 1997; BETE 2005; Turok and Rasmussen 2010; Pakowski and Mujumdar 2014; Manufacturing Chemist 2016; GEA Niro 2017; Arpagaus 2020))

Sterile air is used as drying gas because it is the most cost-effective solution for providing large process gas volumes and satisfactory protection against microbial contamination (Pakowski and Mujumdar 2014). In most cases, however, nitrogen is preferred to eliminate any potential risk of contamination, even though this involves a minor additional cost.

The inlet filters can be combined with pre-filters for catching larger particles and increasing the lifetime of the HEPA filters. The drying gas is prefiltered, heated to the desired inlet temperature (e.g., 100 to 180 °C depending on the vaccine media to be processed), and then passes through a sterilizing HEPA filter before entering the drying chamber via a specially constructed air distributor at the drying chamber ceiling assuring a flow pattern suitable for drying of the media to be processed. The gas is sterile after having passed through the HEPA filter. A temperature sensor monitors the inlet gas temperature before the hot drying gas is distributed in the drying chamber by the gas disperser.

An aqueous-based feed is fed to the atomizer at low pressure through a feed line and solutions through an in-line sterile filter. Suspensions of vaccines (e.g., peptides, bacterial) are made sterile during their preparation under aseptic conditions.

The atomizer is installed at the top of the spray drying chamber so that the sterile feed enters the spray drying chamber through the top (in co-current mode). In the drying chamber, the atomized droplets are dried continuously (e.g., for up to 5 days (De Costa 2014)) to a powder with well-defined properties. The transformation time from the liquid state to a dry powder is in the order of a few seconds, depending on the droplet size and drying chamber size. Thus, the whole spray drying process is carried out within a minute, a distinct advantage compared to other drying technologies for vaccines (see Chap. 3).

Furthermore, droplet evaporation keeps the temperature low initially, which is suitable for heat-sensitive products. The spray-dried particles reach at most the outlet drying gas temperature, which is in the range of about 40–70 °C (40 °C at low capacity) for water applications. Finally, the shear stress to which the biomolecules are subjected by atomization is worth mentioning. Studies show that about 15% of the bioactivity can be lost through atomization alone, necessitating excipients for protection.

The gas exits the cyclone and flows to the bag filter, where the remaining fines of the powder are collected. The dried powder product is discharged from the cyclone collector through a rubber pinch valve placed in a laminar-flow cabinet or cleanroom. The powder container is connected to the pinch valve using a sealing mechanism, assuring a dust-free powder discharge into sterile vessels. At the appropriate time, the dried powder is then transferred to an aseptic powder filler that fills the resulting spray-dried powder in vials or other formats, then capped and sealed (Fragale and Brueck 2015).

Small-sized spray dryers are constructed as stand-alone units, but larger capacity installations comply with the "through the wall" concept, separating the equipment supplying the dryer utilities from the clean product contact areas with packaging (Masters 2002).

4.3.3 Closed-Cycle System

A closed-loop configuration using nitrogen as process gas is often used in the pharmaceutical industry when:

- organic solvents are required for the formulation of poorly water-soluble drugs (nitrogen eliminates explosion risks of flammable organic solvents),
- recovery of the solvent is necessary (flammable, toxic, or valuable organic solvent is used),
- pollution is not permitted (e.g., due to vapor, particulate emissions, or odor), or
- the powder will degrade by oxidation during drying in the air (requires inert gas drying).

A closed-loop configuration is less common for spray-drying of vaccines, as most are aqueous solutions, suspensions, or emulsions that can be processed in an open system. However, it may be useful to dry certain vaccines under inert conditions to protect them from oxidation.

Figure 4.3 shows a schematic of a closed-cycle aseptic spray dryer system based on various literature (Masters and Vestergaard 1975; BETE 2005; Thybo et al. 2008; Pakowski and Mujumdar 2014; Andersen 2016; Manufacturing Chemist 2016; GEA Niro 2017; Geng 2018; GEA Process Engineering A/S 2020a).

Essentially, the same installation elements as for open-cycle spray dryers are used (Pakowski and Mujumdar 2014). An inert gas, typically nitrogen, is used as a drying and processing gas. The humid effluent gas is directed through a condenser to recover most of the solvent. This vapor stream with reduced partial vapor pressure is reheated in a cleanable heater and fed back (recycled) through a HEPA filter into the dryer inlet.

Such a closed system will produce a powder with elevated residual water content than an open-loop system due to the residual vapor present in the recycle stream (Vehring et al. 2020). Nevertheless, closed-cycle spray dryers offer the possibility of maintaining proper sanitary conditions (Pakowski and Mujumdar 2014). A closed-loop system is also more complex in terms of validation procedures (Masters 2002). Besides, their higher costs are justified for some special applications, such as toxic pesticides or explosive dyes (Broadhead et al. 1992; BETE 2005) and pharmaceuticals, e.g., poorly soluble drugs (Ziaee et al. 2019; Poozesh and Bilgili 2019).

For example, the new generation of GEA Niro AsepticSD™ spray dryers using nitrogen in a closed-cycle set-up offers the same wide range of possibilities for capacity, powder engineering, batch size flexibility, etc., as open-cycle spray dryers GEA Niro (2017).

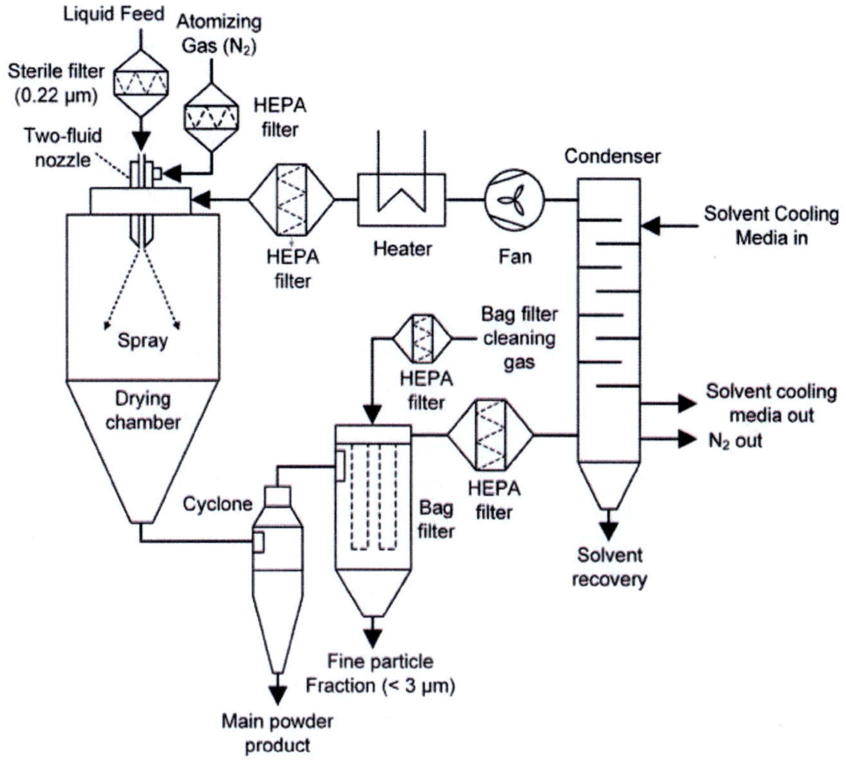

Fig. 4.3 Schematic illustration of an aseptic closed-cycle spray dryer system using nitrogen (Based on literature (Masters and Vestergaard 1975; BETE 2005; Thybo et al. 2008; Pakowski and Mujumdar 2014; Manufacturing Chemist 2016; GEA Niro 2017; Geng 2018; GEA Process Engineering A/S 2020a))

4.4 Components of an Aseptic Spray Dryer

After discussing the actual circuit configurations (open or closed) of an aseptic spray dryer, this section briefly describes the requirements for the main components, namely drying chamber and connections, nozzle, HEPA filters, sterile liquid filters, and cyclone collection.

4.4.1 Drying Chamber and Connections

Another critical element is that all parts should consist of as few parts as possible, e.g., piping with as few connections as possible to avoid leaks and cold spots. The drying chamber equipment must be leak-tight and pressure-resistant (Masters 2002) because the drying process operates under slight positive pressure (about 0.1 bar) (Vehring et al. 2020) to ensure that ambient air cannot enter the process.

All surfaces of the product-contacting materials on the nozzle, inner chamber walls, cyclone, ducts, etc., must be as smooth as possible to prevent clogging and deposits and ensure easy cleaning. In addition, the internal surface finish of materials in contact with the product must meet the standards applicable to equipment used in aseptic pharmaceutical processing (FDA 2004; ISPE 2018).

Good manufacturing practices also include ensuring that all corners are rounded, as crevices can serve as potential areas for material buildup and must be avoided. Connections between ducts and ports should be kept as short as possible to prevent powder accumulation and facilitate cleaning. All product-contact surfaces must be capable of being cleaned (CIP) and sterilized (SIP) in place, the latter with saturated steam (e.g., 2 to 3 bar, 121–135 °C) or hot air, which can heat all product-contact surfaces to 200 °C. Therefore, an aseptic spray dryer is somewhat heavier than a standard spray dryer, as it must withstand the higher pressure required for steam sterilization. High temperatures are needed to achieve a thoroughly sterilized spray dryer in the shortest possible production downtime. Finally, rotating or sliding parts should be avoided in spray dryer systems, as mechanical friction of parts can create small metallic contaminants in the product. The equipment needs to be properly grounded to avoid static charging of the powders and improve the powder yield.

Couplings between sections of a spray dryer should be selected to maintain sterile conditions. Examples include Tri-Clover™ or equivalent flanges and gaskets (Chickering et al. 2003). The Tri-Clover BioPharm product range consists of a comprehensive selection of tubes and fittings with internal surfaces of an average roughness (Ra) < 0.4 to 0.8 μm, either electropolished or mechanically polished. Pipes and fittings from this range comply with dimensional standards DIN 11850, ISO 1127, and ASME BPE.

4.4.2 Nozzles

Several atomizer systems can produce a spray inside an aseptic drying chamber. These atomizers may be classified according to the kind of energy used and include two-fluid nozzles (kinetic energy), pressure nozzles (pressure energy), ultrasonic nozzles (acoustic energy) (Truong-Le et al. 2012, 2014; Büchi Labortechnik AG 2014), and rotary atomizers (centrifugal energy) (Masters 1985, 1991, 2002; Broadhead et al. 1992; Cal and Sollohub 2010; Miller and Gil 2012; Gaspar et al. 2014; Vehring et al. 2020).

Nozzles must produce a fine, homogeneous spray with a spray cone angle that combines with the hot drying gas flow from the gas disperser to maintain clean wall operations. In the pharmaceutical industry, two-fluid nozzles and pressure nozzles are generally best suited for the aseptic processing of vaccines (Mogavero 2014), as rotary atomizers can cause the formation of metal/lubricant particles through mechanical action, which may contaminate the final product. Besides, two-fluid nozzles and pressure nozzles deliver a spray pattern that does not direct the droplet

Table 4.2 Comparison of pressure and two-fluid nozzles

Pressure nozzles	**Two-fluid nozzles**	
Pressure swirl	External mixing	Internal mixing
Feed flow > 5 kg/h (not suitable for lab units)	High atomization gas consumption restricts the production of small particles at larger scales	Efficient atomizationat low gas-to-liquid ratios
Flexible spray angles: 40° to 70°	Narrow spray angles: about 20°	Flexible spray angles: 20° to 60°
Less suitable for high viscous feeds (< 750 cP)	Insensitive to viscosity	More sensitive to viscosity
Weak manipulation of droplet size (for the same throughput)	High degree of droplet size control	
Favors flowability	Fine powder with low densities	
Mean droplet size: 20 to 100 µm Pressure: 15 to > 100 bar	Droplet mean size 5 to 50 µm Air–liquid ratio (kg/kg): 0.5:1 to > 3:1	
Narrow distribution, span 1.4 to 1.8	Wide distribution, span 2.0 to 2.4	
Mainly pilot and industrial scale	Lab, pilot, and industrial scale	

Based on Masters (2002), Miller and Gil (2012), Gaspar et al. (2014)

trajectory towards the drying chamber wall. Table 4.2 summarizes pressure and two-fluid nozzles (Masters 2002; Miller and Gil 2012; Gaspar et al. 2014).

Two-fluid nozzles are the most common atomizers in laboratory and pilot-scale spray dryers due to their simplicity and flexibility using air or nitrogen for atomization (Broadhead et al. 1992). The two major types of two-fluid nozzles on the market are external mixing and internal mixing nozzles. External mixing nozzles are operated with a low-pressure liquid supply through an inner channel, while the atomization gas is fed by an external annular opening around the liquid orifice. On the other hand, internal mixing nozzles take advantage of gas expansion at the nozzle outlet, which is more energy efficient in regards to the gas to liquid ratio and therefore preferred for larger-scale spray dryers, and especially when small particle sizes are required (e.g., < 10 µm) (Miller and Gil 2012).

A significant advantage of two-fluid nozzles is controlling the droplet size since feed and gas flow rates can be adjusted independently (Gaspar et al. 2014). With ultrasonic atomizers, the droplet size can be tuned by the ultrasound frequency and the design of the nozzle tip. As a result, the droplet size distribution is very uniform and ranges from about 20 to 100 µm (Gaspar et al. 2014). The major drawback of these devices is the limited throughput of about 50 mL/min, which restricts their applicability mainly to the laboratory (Büchi Labortechnik AG 2014) and pilot-scale units (Gaspar et al. 2014).

Pressure nozzles are more common at the industrial scale when larger droplet sizes and narrower particle size distributions are required (Miller and Gil 2012). Pressure nozzles achieve atomization by transforming pressure (15 to > 100 bar) into kinetic energy. The pressure nozzles are designed to produce hollow swirling sprays. Attention must be paid to suspensions that may lead to nozzle erosion (Miller and Gil 2012). As the liquid feed needs to be accelerated, pressure nozzles are less suitable for high viscous feeds. The viscosity limit is about 750 cP (Gaspar et al. 2014).

4.4.3 Filtration

Crucial in aseptic spray drying is maintaining sterility during manufacturing (Siew 2016). Such requirements exist in all stages of drying pharmaceutical products (Pakowski and Mujumdar 2014). Potential sources of contamination during spray drying are the liquid feed and the gas used for drying and atomization. Therefore, all gas entering and leaving the sterile drying zone and the liquid formulation must be adequately filtered. In addition, air quality must be Class 100 compliant (less than 100 particles of 0.5 µm per cubic foot, ISO 5) (FDA 2004).

High-efficiency particulate air (HEPA) filters, which can be heat sterilized, are critical components used to trap particles 0.3 µm and larger, i.e., most known bacteria. All gas supplied to an aseptic spray dryer passes through HEPA filters to remove all contaminants. An often used HEPA filter for hot process gases is a Termikfil 2000 (Turok and Rasmussen 2010; Camfil 2020), made of glass fibers and has an efficiency of 99.99% at 0.3 µm and meet FDA requirements. Besides, these HEPA filters are approved for continuous operation at high temperatures up to 350 °C.

Figure 4.4a shows a sterile filter arrangement for inlet process gas with stainless steel housing and filter cartridge and Figure 4.4c some details from an exhaust filter arrangement from an ASEPTICSD™ spray dryer (GEA Niro 2017).

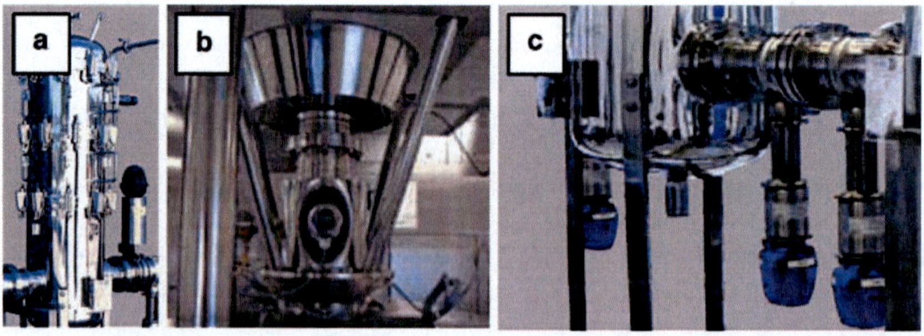

Fig. 4.4 **a** Sterile filter arrangement for inlet process gas with stainless steel housing and filter cartridge, **b** sterile powder discharge from a cyclone in a clean room, **c** detail from the exhaust filter arrangement (Image courtesy by GEA Niro (2017) from an ASEPTICSD™ spray dryer)

Gas filtration can be done in multiple stages to protect HEPA filters from clogging (Searles and Mohan 2010). At least one prefilter is added to remove dust from the air before entering the HEPA filter. For example, the drying air is pre-filtered, heated to the target temperature, passed through a HEPA filter, and then distributed into the drying chamber. The atomizing gas is pre-filtered, compressed, and passed through a HEPA filter before entering the nozzle.

In this context, Turok and Rasmussen (2010) presented an invention where the process gas of a spray drying system is first sterile filtered and then heated in a heating device, which does not release particles. In such a system, the spray drying chamber can be supplied with sterile gas from a filtering system that filters all viruses and bacteria before the gas passes the heater (i.e., cold gas filtration).

In spray dryer systems with cyclones, a bag filter with several filter bags removes the fines not collected in the cyclone. The number of filter bags depends on the drying gas flow rate and the filter bags' material.

HEPA filters and laminar flow units are installed in the packaging area to remove airborne microbes from the air that feeds the packaging area (Pakowski and Mujumdar 2014). All personnel and ancillary equipment must be placed downstream from the sterile product.

The liquid vaccine formulation is pre-sterilized with a sterilizing grade filter before spray drying (Masters 1976; Searles and Mohan 2010). Membrane filters with 0.22 µm pore size are used in the final step of sterile product preparations (Armenante and Kirpekar 1997). The specified pore size of 0.2 or 0.22 µm is not taken literally but is an estimate based on the bubble point performance test of what cannot pass through. Regulatory expectations for sterilizing-grade filters are described in the FDA Aseptic Processing Guidelines (FDA 2004) and the ASTM F838-20 standard test method for determining bacterial retention of membrane filters utilized for liquid filtration (ASTM 2020). The criterion for passing the test is when the filter on exposure to the tiny bacterium *Brevundimonas diminuta* produces a sterile effluent with a minimum concentration of 10^7 colony forming units (CFU) per cm^2 of effective filter area.

Several companies offer a complete range of membrane filters with sterilizing quality for this purpose, such as Merck Millipore or Pall Corporation. Aqueous vaccine formulations are easily filtered through 0.22 µm sterilizing grade filters using polyvinylidene fluoride (PVDF)-based membrane filters. These achieve low protein binding and high throughputs. Suitable are the hydrophilic Durapore® 0.22 µm membranes from Merck Millipore (2020) (Fig. 4.5) or Fluorodyne® II DFL 0.2 µm membranes from PALL (2017).

If sterile filtration of the vaccine formulation is not possible, sterility must be ensured by another means, such as preparing the formulation under aseptic conditions using sterile procedures (Gladd 2014). Typically, the supply line is aseptically connected to the spray nozzle. A common practice is also to start the spray drying run with sterile water to achieve thermal equilibration of the spray drying system before switching to the feed supply solution (Searles and Mohan 2010).

Filter integrity testing is a critical process parameter in sterile manufacturing products and should be maintained to ensure aseptic conditions. Guidance for

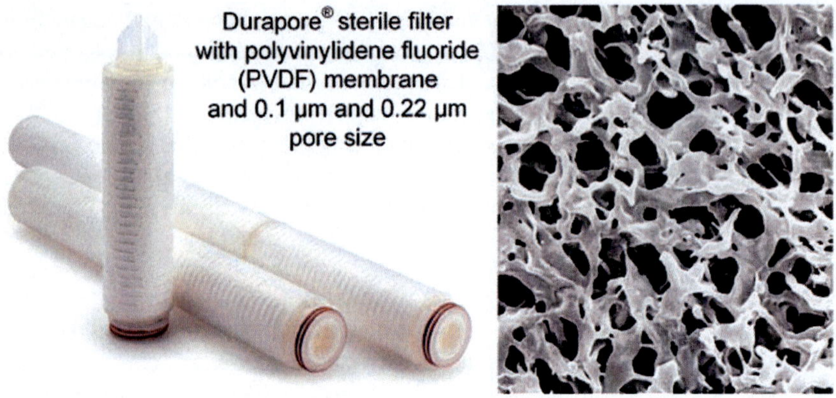

Fig. 4.5 Durapore® sterile filters made of hydrophilic polyvinylidene fluoride (PVDF) membrane (0.1 μm and 0.22 μm pore size) provide low protein binding, high throughput, and broad chemical compatibility (Adapted from Merck Millipore® website and catalog (Merck Millipore 2020))

validation is available from FDA (2004), PDA (2008), ASTM (2020). Various factors, such as particle load, fluctuations in pressure and temperature, and cleaning steps, which may involve high temperatures or strong pH changes, can cause wear to the filter elements (Baumfalk and Finazzo 2007). Regulatory agencies like the FDA and the Parenteral Drug Association (PDA) recommend integrity testing of the filters before processing (pre-use) and routinely post-use to check for filter leaks or damage by an appropriate non-destructive test such as bubble point, diffusive flow, water intrusion, or pressure hold test.

A filter integrity test is typically performed at the beginning and end of each cycle in aseptic spray dryers to ensure that the filters function as intended (Schwartzbach 2011). A challenge can be that relatively large filter banks must be tested for integrity (Searles and Cherian 2015). The integrity testing methods should be agreed on between the manufacturer and user. Filter validation should be conducted using the worst-case conditions, such as maximum filter use time and pressure (FDA 2004). In practice, the flow rate of a gas is determined under specified conditions and compared with the manufacturer's tolerance data. A gas flow higher than that specified indicates a loss of filter integrity.

The water intrusion test is an established test for air filtration elements based on hydrophobic membranes (e.g., PTFE) (Baumfalk and Finazzo 2007). During this test, the penetration of ultrapure water into the hydrophobic membrane is measured under pressure and evaluated by comparing the results to a limit value.

The bubble point and diffusion test are established test methods (Armenante and Kirpekar 1997). The diffusive flow through a wetted hydrophilic membrane is measured against increasing differential pressure. The resulting bubble point is inversely proportional to the membrane's pore size. Thus, a bubble point lower than the specified bubble point indicates integrity failure.

4.4.4 Cyclone Collection

The most common powder collection systems used in aseptic spray dryers are cyclones and bag filters. Cyclones are simple, efficient, easy to clean-in-place, and widely available in the laboratory (Arpagaus and Schwartzbach 2008; Büchi Labortechnik AG 2020b) and commercial scales (Al-Khattawi et al. 2018). However, disadvantages can be the hardware scaling and particle size-dependent separation efficiency, which must be considered during scale-up (Snyder and Lechuga-Ballesteros 2008; Vehring et al. 2020). Figure 4.6a illustrates the principle function of a reverse-flow cyclone.

The particles are separated from the drying gas by inertia into a particle fraction and a gas fraction in the cyclone. The drying gas carrying the dried particles enters the cyclone tangentially, creating a downward vortex that, through centrifugal forces, sweeps the material against the walls. At the bottom of the cyclone, the gas makes a 180-degree turn upward into the cyclone center. Due to their inertia, most particles will continue to move downward and are collected below the cyclone. Only the smallest and lightest particles will reverse direction and travel upward with the gas through the cyclone center (inner vortex). The bag filter then collects the finest particles.

The cyclone separation efficiency depends on the cyclone design, the operating conditions (e.g., gas velocity), and the particle properties of the entering product such as size distribution, bulk density, abrasion resistance, self-bonding properties between particles, stickiness, and powder loading in the gas (Masters 2002).

For example, Fig. 4.6b compares the separation efficiency of a standard and a high-performance cyclone used in a laboratory Mini Spray Dryer B-290 (Brandenberger 2003). In the example, ammonium salt solution was spray-dried at different concentrations to obtain particles between 3.2 and 5.7 μm measured by laser diffraction. The high-performance cyclone has a higher separation efficiency and is particularly suitable for small particle sizes and high-value products in small quantities.

A spray drying plant in the pharmaceutical industry is often equipped with a single cyclone. However, multiple cyclones can be used in parallel or series to increase the separation efficiency and to avoid scale-up issues (Masters 2002; Miller and Gil 2012). Figure 4.6c shows a multi-cyclone layout arranged parallel with a common powder collection hopper and single-point powder discharge for downstream handling (Geng 2018). Conical hoppers are typically fitted to cyclones to reduce the re-entrainment of powder caused by the intense vortex flow within the cyclone base.

Bag filters work by extracting the particles from the gas stream through either a depth filter or a size-exclusion membrane filter (Snyder and Lechuga-Ballesteros 2008; Plummer and Barreto 2012; Galvin 2013). The gas phase flows through the filter material while the powder forms a layer on the filter surface. The filters are typically cylindrical geometries mounted in a parallel arrangement within the containment vessel. The advantage of a bag filter for pharmaceutical spray drying is the high separation efficiency, even for particles < 3 μm. Disadvantages include

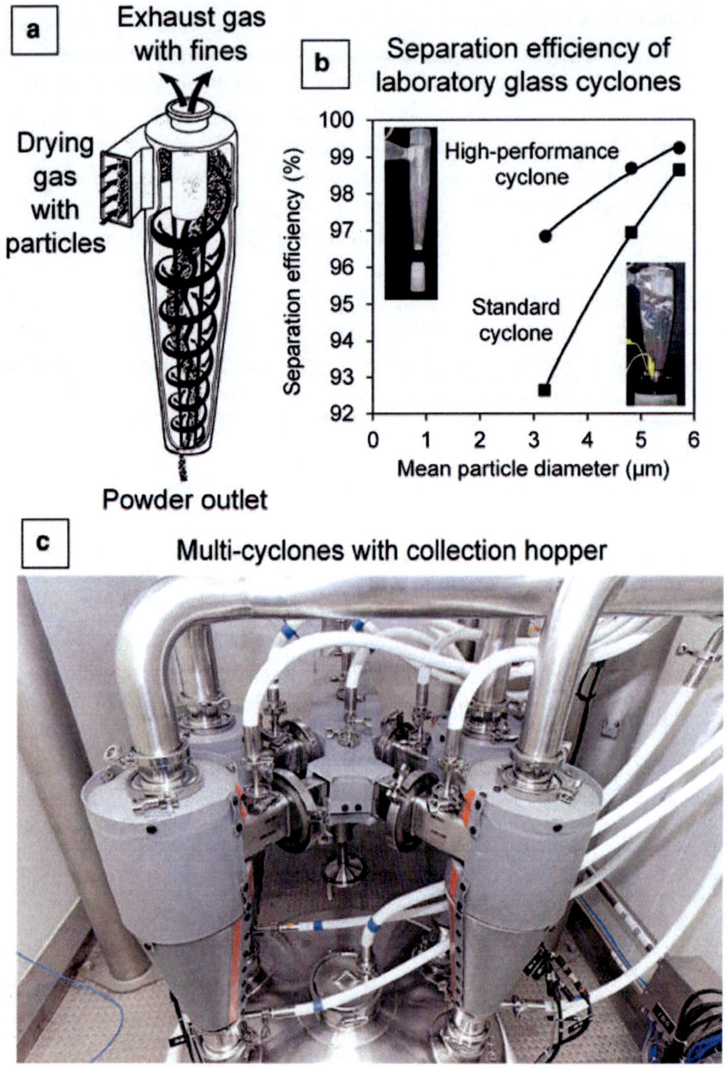

Fig. 4.6 **a** Schematic representation of a reverse flow cyclone showing the gas and particle trajectories. **b** The separation efficiency of the standard and high-performance glass cyclones of the Mini Spray Dryer B-290 for spray-dried ammonium salt $(NH_4)_2SO_4$ at 1%, 5% and 20% solid concentration (drying parameters: Tin: 160 °C, Tout 85 °C (standard cyclone), 72 °C (high-performance cyclone), aspirator: 100%, pump: 35%) (Data and images from Brandenberger (2003) with kind permission from Büchi Labortechnik AG, Switzerland). **c** Multi-cyclones layout with common collection hopper (Image adapted from Geng (2018))

powder compaction due to cake formation, hardware complexity, pulsating pressure fluctuations in the system, possible detachment of the filter media into the product, and cleaning to allow a validated process. In addition, prolonged contact with the filter surface at high temperatures may negatively impact product quality.

The performance of bag filters depends on the product, the powder loading in the gas, the bag fabric material, and the cleaning method. To meet hygienic operational standards, bag filter designs for spray drying application must meet continuous operation requirements with in-place dry cleaning and in-place wet cleaning. A common form of bag filter cleaning is a pulsed air jet (Masters 2002).

The selection of suitable filter material is essential, as the composition must be compatible with the application and robust enough to withstand the temperatures and moisture contents used in spray drying and cleaning. Bag filters with PTFE membrane can outperform other filter constructions in many applications. PTFE repels water, is resistant to aggressive cleaning cycles, and has a very low friction coefficient (easily removed particles) and high-temperature resistance. Figure 4.7a shows a cross-section of microporous PTFE laminate filter membrane (Plummer and Barreto 2012), which keeps particles out of the fabric.

Particles are collected on the surface of the membrane (Fig. 4.7b) and are released during cleaning cycles, even under high-moisture conditions. Examples of such membranes are nanofiber Ultra-Web®, PTFE Torit-Tex®, polyester spunbond, or high-temperature Thermo-Web® (Galvin 2013). Today's bag filters allow continuous operation through pressure monitoring, gas flow regulation, and automatic pulse cleaning (Masters 2002). However, manual tasks are still required to replace spent filters.

Bag-in/bag-out (BIBO) HEPA filter housings are used in a cleanroom, laboratory, and healthcare application to enable filter exchange, ensuring maximum safety for the public and service personnel. The term BIBO refers to special bags and handling procedures that allow an operator to change filters without direct exposure to the highly potent biopharmaceutical powder in the collector. Figure 4.8 illustrates the main components of a BIBO HEPA filter assembly, including the filter

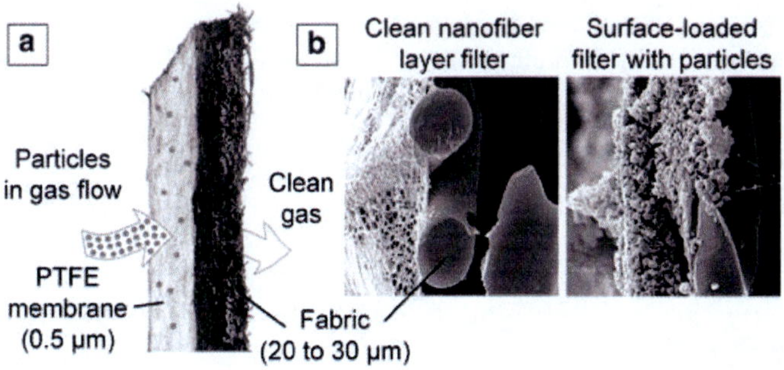

Fig. 4.7 **a** Cross-section of a microporous PTFE laminate filter membrane. **b** Particles collected on the surface of a nanofiber layer filter can be cleaned off while the filter substrate stays clean (left: clean filter membrane, right: surface-loaded membrane with dust particles) (Images adapted from Plummer and Barreto (2012), Galvin (2013), and www.baghouse.com)

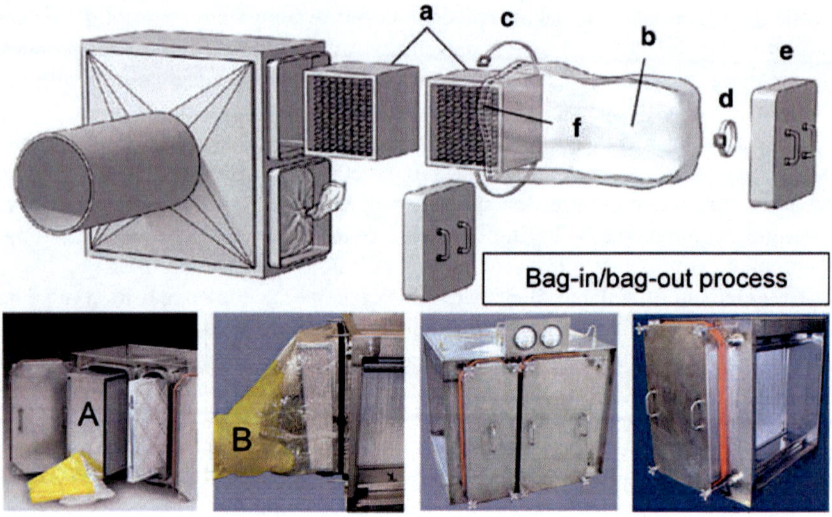

Fig. 4.8 Principle of a Bag-In/Bag-Out (BIBO) filter exchange to minimize worker exposure to potentially harmful dust and reduce atmospheric exposure to pollutants. The upper HEPA filter assembly is exploded out to show that the filter unit (**a**), removal bag (**b**), safety straps (**c**), cinching straps (**d**), and housing cover (**e**) are arranged so that a contaminated filter (**f**) can be removed with the pre-packed bag already in the housing. The lower filter assembly is shown only with the housing cover removed to illustrate how the filter and removal bag is packed into the housing (Images adapted from CDC/NIH (2020)). The photographs show a BIBO HEPA filter housing for critical clean air exhaust applications. Inside the doors of each unit, a ribbed inlet collar provides for a bag attachment. The bag forms a barrier seal between maintenance personnel and the contaminated filters. An initial bag kit is included with each system (Images from EnviTec, USA)

unit, removal bag, safety straps, cinching straps, and housing cover (CDC/NIH 2020). The BIBO process uses bags connected to the collector with integral collars. Inside the doors of each HEPA filter unit, a ribbed inlet collar provides for a bag attachment. The bags are attached to the collars with rubber gaskets and rubberized band clamps to create a dust-tight barrier seal between the BIBO bag and the collector (Galvin 2013). Overall, the bags should only be replaced by contamination experts and have high Occupational Exposure Limits (OEL) ratings. Possible bag filter suppliers are Sefar AG (Switzerland), EnviTec Corporation (USA), or Donaldson Company (USA).

4.5 Cleaning, Sterilization, and Testing of an Aseptic Spray Dryer

The previous sections have mainly covered the concepts and components of an aseptic spray dryer. The following section presents and discusses the processes of cleaning and sterilization.

From a processing point of view, three basic steps must be considered to achieve aseptic conditions in a spray drying process (Schwartzbach 2011; Vehring et al. 2020):

- initial plant cleaning and sterilization of product contact surfaces,
- maintaining sterility during the production of spray-dried powder, and
- aseptic extraction of the spray-dried powder from the drying system and powder filling.

All process steps must be performed in compliance with cGMP. In addition, the spray dryer equipment is validated through Installation Qualification (IQ), Operation Qualification (OQ), and Performance Qualification (PQ). For more detailed information on validating a pharmaceutical process and especially cleaning procedures, the reader is referred to the book of Ostrove (2016).

Before starting production, the spray dryer system must be cleaned, sterilized, and depyrogenated (Geng 2018). Depyrogenation refers to removing pyrogens, which induce fever when entering the bloodstream. As pyrogens such as endotoxins are not removed by steam sterilization, they are washed off from product contact surfaces of the equipment and ductwork with clean water (e.g., water for injection) (Turok and Rasmussen 2010).

For large-scale spray dryers where disassembly of the components is not feasible, appropriate semi-automatic or fully automatic cleaning-in-place (CIP) and sterilization-in-place (SIP) systems are required for the initial plant cleaning and sterilization of product contact surfaces (Vehring et al. 2020). This involves an integrated cleaning system, sterilization in place, and filter sterilization of inlet gas and fluid HEPA filters. Traditionally, the approaches for spray drying systems include steam or dry-heat sterilization (Schwartzbach 2011). After sterilization, the system is cooled down to operating temperature and is ready for spray drying. Overall, the requirements for CIP, SIP, and construction of a pressure-rated drying chamber make the aseptic spray drying system inherently more complex than a standard pharma (GMP) spray dryer (Mogavero 2014).

4.5.1 Cleaning-In-Place (CIP)

A CIP system for cleaning purposes is a well-known operation for bioprocessing equipment (ASME 2019). In an aseptic spray dryer, the initial plant cleaning is done via an integrated cleaning system (ICS) consisting of a series of spraying nozzles (e.g., static spray balls) throughout the drying system wetting all product contact surfaces and ensuring effective wall cleaning results (Roser 2009; Mogavero 2014). The spraying devices are installed at specific locations to ensure maximum coverage. The dewatering systems are designed so that gravity helps to dewater the CIP fluids. In addition, the drainage points are selected so that the washing liquids can drain off quickly enough; otherwise, a dead-zoning problem may occur. A fully automatic CIP sequence includes (Turok and Rasmussen 2010; ASME 2019):

- Rinsing with potable water,
- Washing with caustic detergent,
- Rinsing with potable water,
- Washing with acid detergent, and
- Rinsing with highly purified water (e.g., water for injection) to remove the remaining cleaning solutions.

The endpoints of the rinsing can be determined by analytical measurements such as conductivity sensors. The cleaning variables to be considered are the time of exposure (i.e., contact time) to wash and rinse solutions, the temperature of wash and rinse solutions, the chemical concentration of wash solutions, and fluid hydraulics. The reader can find more details on CIP liquid skid design in the standard of (ASME 2019).

4.5.2 Sterilization-In-Place (SIP)

The two main types of sterilization used are steam and dry heat sterilization. In practice, sterilization using saturated steam is the preferred and proven method for in-process sterilization of aseptic spray dryers (Turok and Rasmussen 2010; Schwartzbach 2011; Mogavero 2014; GEA Niro 2017; SPX FLOW Inc. 2017; Vehring et al. 2020). Saturated steam of at least 2 to 3 bar is used for sterilization, including the feed line, atomization gas line, process gas line, nozzle, spray drying chamber, and sterile HEPA filters (Mogavero 2014). Figure 4.9 shows the relationship between saturated steam pressure and sterilization temperature.

SIP is performed as an evacuation and heating sequence followed by steam sterilization at a minimum of 121 °C for 20 min (Schwartzbach 2011), with some

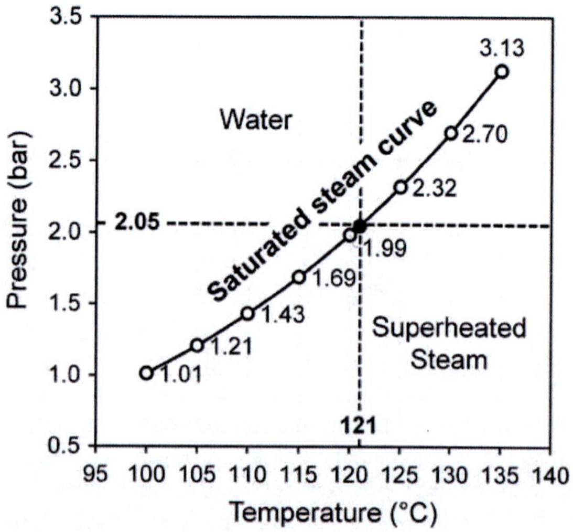

Fig. 4.9 Relationship of saturated steam pressure to sterilization temperature

applications using temperatures as high as 131 °C (Mogavero 2014). SIP operation at 2 bar requires constructing a pressure-rated ASME vessel (American Society of Mechanical Engineers) (Armenante and Kirpekar 1997). Then, it is combined with cGMP and aseptic design requirements and operates under vacuum conditions.

Vacuum cycles are used to remove non-condensables and ensure that the air does not prevent steam from accessing all the internal voids within the plant (Mogavero 2014; Geng 2018). In addition, the SIP system approach requires a system pressure check and appropriate equipment accommodations such as condensate traps and thermal insulation. These demanding specifications are, for example, met by the new generation of GEA Niro® ASEPTICSD™ spray dryers (GEA Niro 2017) or Anhydro™ MicraSpray® Aseptic spray dryers (SPX FLOW Inc. 2017) for processing heat sensitive products, such as peptides, proteins, vaccines, antibiotics, hormones, allergens, and other biopharmaceutical products requiring aseptic processes. More details on commercially available aseptic spray dryer equipment are described in Sect. 4.7.

Sterilization by dry heat is another option, which has the advantage that a spray dryer is already designed for hot gas generation. Spray dryers and ductwork can be sterilized by maintaining the installation at elevated temperatures for specified periods. The U.S. Pharmacopoeia specifies dry heating of all surfaces at 160–170 °C and holding for 2–4 h to ensure sterility (Pakowski and Mujumdar 2014). The challenge with dry heat sterilization is to do depyrogenation simultaneously. In addition, dry gas heat has a low heating capacity, making it difficult to heat the inside surface of the system and ductwork where cold bridges are present (Mogavero 2014).

Besides, a spray dryer has a large mass of stainless steel and requires a lot of energy to bring the entire system adequately to temperature. In addition, any cold area requires a considerable increase in insulation. Therefore, supplementary electric surface heaters, also called blanket heaters, can be installed to cover the surface of the drying chamber, the outlet pipes, and the cyclone (Turok and Rasmussen 2010). Blanket heaters can be specially adapted to the reactor and pipe geometry and installed in specific sites with cold bridges to ensure aseptic conditions. However, validation is challenging to verify that all surfaces were hot enough. Compared with dry heat, steam sterilization exposes the equipment to lower temperatures but elevated pressure for a shorter period. In conclusion, steam sterilization validation is more accessible by monitoring the pressure.

4.5.3 Sterility Testing with Bacteria

The effectiveness of thermal sterilization is tested by placing test bacteria samples in strategic locations in the installation. The guideline for aseptic processing suggests a "death rate" of 6-logs reduction (sterility assurance level, SAL of 10^{-6}) of a suitable heat resistant biological indicator (FDA 2004; Mogavero 2014). A value of 10^{-6} means a probability of a nonsterile unit greater than one in a million after a sterilization process (FDA 2004). *Bacillus subtilis* is typically used for hot air sterilization and spores of *Bacillus stearothermophilus* for steam sterilization testing. *Bacillus stearothermophilus* is one of the most heat-resistant organisms commonly encountered.

Figure 4.10 shows the typical death rates of *Bacillus subtilis* and *Bacillus stearothermophilus* after steam sterilization. The death by moist heat is caused by denaturation and coagulation of essential cell proteins (Armenante and Kirpekar 1997). As can be seen, there is a semi-logarithmic relationship between the fraction of surviving organisms (N/N_0) and sterilization time at different steam temperatures. As can be seen, the kinetics of the sterilization process increases dramatically with temperature. Spores of *Bacillus stearothermophilus* withstand 121 °C steam for up to about 12 min, making this organism ideal for thermal steam sterilization testing. Enough time must be allowed to enable all parts of the spray dryer to reach the sterilization temperature. Then, this temperature is maintained for the required time, typically 20 min at a sterilization temperature of 121 °C (Armenante and Kirpekar 1997). After sterilization, the samples are sown onto culture media in Petri dishes and observed for surviving spores and growth. The absence of any growth indicates that sterilization was successful.

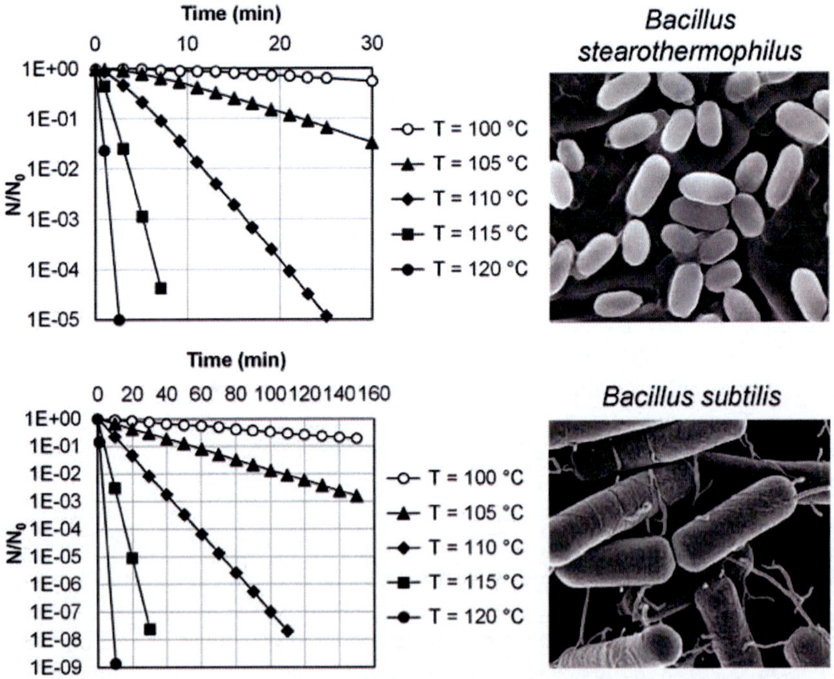

Fig. 4.10 Typical death-rate plots for *Bacillus stearothermophilus* and *Bacillus subtilis* after steam sterilization. N/N_0: fraction of surviving organisms, N_0: initial number of viable organisms per mL, N = total number of viable organisms per mL after some time of sterilization (Data adapted from Armenante and Kirpekar (1997) with permission from Springer Nature)

4.6 Sterile Powder Processing

This section is about the sterile powder processing after spray drying has been completed, i.e., how the powder is handled during powder collection, extraction, filling, and packaging.

4.6.1 Aseptic Handling of the Spray-Dried Powder

Aseptically spray-dried product collection has a drying chamber with single-point powder discharge into a cyclone or bag filter for downstream handling. The powder is captured at the bottom of the cyclone in a collection vessel, sealed within the process stream. The collection vessel includes a valve assembly, a split butterfly valve, or a containment valve that allows for the sterile transfer of the powder from the cyclone of the spray dryer to the collection vessel and then from the collection vessel to an aseptic powder filler. Dried powder handling is conducted under a continuing aseptic environment in laminar flow clean rooms under closely controlled temperature, humidity, and air quality conditions (Masters 2002). Figure 4.11 shows the powder collection of custom-designed PHARMA-SD® type PSD-1 and PSD-4 spray dryers (low bioburden plants, non-aseptic) in a cleanroom at GEA's spray drying facility in Soeborg (Denmark).

The introduction of an aseptic safety sieving step to separate larger or compacted particles after spray drying may be an additional option (Gladd 2014). Overall, the transfer of aseptically spray-dried powder materials out of the spray dryer is one of the most significant potential sources of contamination. For prevention, facility design needs to ensure that the area between a spray dryer and a powder filling line provides Class 100 (ISO 5, Grade A) protection (FDA 2004). In addition, transport and loading procedures should afford the same protection conditions.

Implementing a split butterfly valve provides a safe aseptic transfer of the highly potent powder from the lower cyclone exit to the collector container. In addition, the split butterfly valve enables the spray drying process to continue when charging and discharging the powder, and it protects both people and the local environment.

Figure 4.12 shows a split butterfly valve assembly docking system comprising an active and passive half that can independently seal two different containers (Dunne 2020). Each half contains one-half of a butterfly valve disc and cannot be opened unless docked together. The active unit is secured to the cyclone of the spray dryer and the passive unit to the collection vessel, bag, bottle, or intermediate bulk container. The safety interlock system prevents the active part from being separated from the passive part while the valve assembly is open. During docking, the passive unit is connected to the active unit and sterilized with vaporized hydrogen peroxide (H_2O_2) to decontaminate previously exposed faces of the active and passive units. Next, the valve disc opens to allow the powder transfer from the spray dryer cyclone into the collection vessel.

Fig. 4.11 Powder collection in a cleanroom. Custom-designed PHARMA-SD® type PSD-1 (top) and PSD-4 (bottom) at the GEA Spray Drying Facility located in Soeborg, Denmark (Images by courtesy of GEA Processing A/S, Denmark)

The typical charge bottles are robust 1 to 20-L stainless steel powder containers with optional pressure rating and sight glass viewing ports. In addition, unique disposable bag systems, such as Hicoflex® (GEA 2016) or ChargeBag® (ChargePoint Technology 2019), are other highly contained docking solutions for the aseptic handling and storage and transfer of powders.

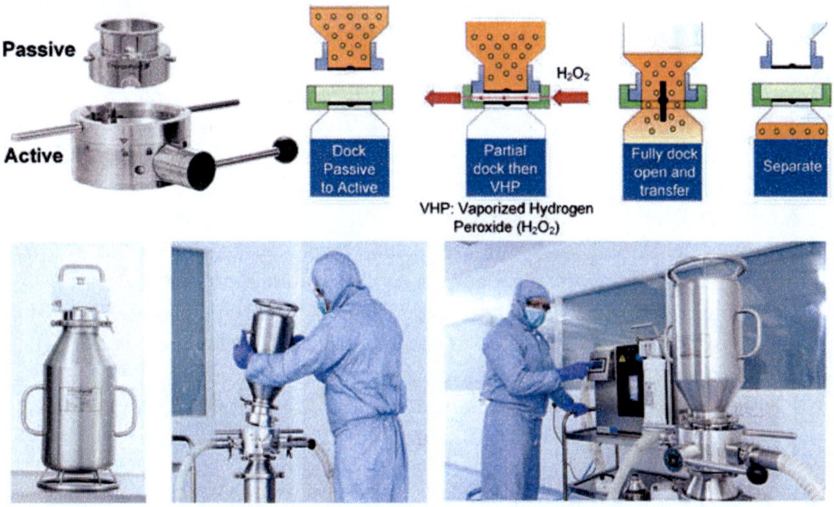

Fig. 4.12 Split butterfly valve (Series 3) with the principle of operation used for both highly potent active pharmaceutical ingredient (API) and aseptic processing (Images by courtesy of ChargePoint Technology Ltd., UK)

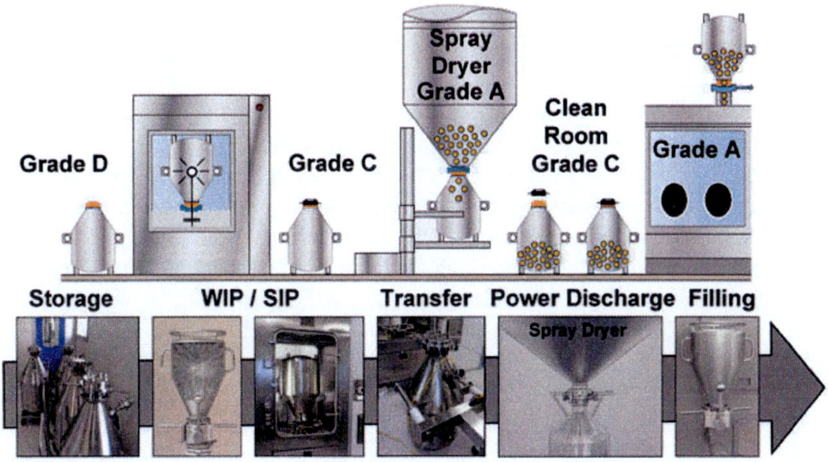

Fig. 4.13 Aseptic powder transfer from the spray dryer discharge point to the powder filling line via charge bottles. Empty bottles are stored in standard Grade D (Class 100,000, ISO 8) conditions, then washed (WIP, washing-in-place), sterilized (SIP, sterilization-in-place), and transferred to the spray dryer clean room in Grade C (Class 10,000, ISO 7) (Images by courtesy of ChargePoint Technology Ltd., UK)

The operation sequence with an aseptic transfer valve is as follows (ChargePoint Technology 2019):

(1) First, the active half is sterilized using a SIP passive unit.
(2) Then, the two halves are locked in place to form a single sealed disk unit. The previously exposed interfaces are sealed together to create a single butterfly valve disc.
(3) Next, the disc faces are exposed to decontamination gas (e.g., vaporized hydrogen peroxide H_2O_2) within a sealed chamber to ensure decontamination of all critical areas, providing a validated 6-log reduction (SAL of 10^{-6}).
(4) Then, the disc opens to allow powder to transfer through the valve. The active and passive-interface are sealed to ensure no material can penetrate the critical area. Once the transfer has taken place, the valve is closed.
(5) Finally, the active and passive units are unlocked and decoupled, exposing the previously closed interfaces and ensuring a clean powder transfer.

A range of split butterfly valves and docking systems are available for the aseptic transfer of solid powders on the market, such as the AseptiSafe® aseptic transfer valves (ChargePoint Technology 2019) and the BUCK® high-containment valves from GEA (2016). In addition, the AseptiSafe® Bio valve offers a unique decontamination in-place approach to achieve a 6-log reduction in microbiological contamination (SAL of 10^{-6}).

4.6.2 Aseptic Powder Transfer to Filling Line

Figure 4.13 illustrates the main process steps in transferring the collection vessel to an aseptic powder filler. First, the empty storage containers (e.g., stainless-steel charge bottles) must be washed and sterilized. For this purpose, tank cleaning technologies such as rotary spray heads and rotary jet heads offer exceptional cleanability and reduce cleaning cycle times during the Washing-In-Place (WIP) procedure. In addition, thorough cleaning and sterilization (SIP) of contact materials ensure sterility (Gladd 2014).

The mechanical interlock of AseptiSafe® Bio valves (ChargePoint Technology 2019) ensures safety before, during, and after powder material transfer. In addition, the critical area is maintained within the valve, which allows the charge bottles to be stored under less strict cleanroom standard Grade D (Class 100,000, ISO 8, see Table 4.1) and eliminates the need for higher quality controlled areas (EU Commission 2020).

A double-door ensures transfer from a lower to a higher classified area (FDA 2004). In addition, airlocks and interlocking doors facilitate better control of air balance throughout the aseptic processing facility. After transferring the charging bottle into the spray dryer cleanroom Grade C (Class 10,000, ISO 7), the bottle is docked to the spray dryer, and the powder product can be discharged into the container.

After that, the powder-charged bottle is transferred to the aseptic powder filling line installed in isolators or RABS (Restricted Area Barrier Systems) (ISPE 2018). With the advancement of technology, a facility can control sterility during production, which offers improved sterility assurance for the product and the efficient protection of the staff.

4.6.3 Aseptic Powder Filling into Dosage Formats

The final and increasingly important step is the aseptic filling of the spray-dried powder into various desired product formats (Siew 2016; Vehring et al. 2020), including vials, bottles, blisters, capsules, pouches, syringes, inhalers and other medical containers. Sometimes the goal is to produce a flowable spray-dried powder to mix and blend the powder aseptically with other substances subsequently.

In particular, for dry powder inhalers, this is a challenge that requires extra care in the design of the powder filling process. For inhalers, most spray-dried powders are packaged in:

- blister packages (e.g., single doses containing a few mg of powder) (LiCalsi et al. 1999; Vehring et al. 2007; Cape et al. 2010; Searles and Mohan 2010; McAdams et al. 2012),
- capsules (Weniger 2011; Weniger and Papania 2013), or
- single-use disposable inhalers (see Chap. 6, Sect. 6.2.4).

This protects the powder from exposure to humidity and sunlight and reduces contamination risk. A fundamental difference between using an aseptic spray drying process and liquid and freeze-dried formulations is the drying and filling step sequence, as shown in Fig. 4.14.

Aseptic spray drying occurs first, followed by aseptic powder filling, whereas with freeze drying, the vaccine solution is first filled into vials with sterile filtration and then dried within the final vials into a powder cake (Schwartzbach 2011; Gladd 2014; Andersen 2016, 2020). The unit dose is locked in when the vial is filled (Luy and Stamato 2020). In comparison, a spray-dried powder allows the formulation to

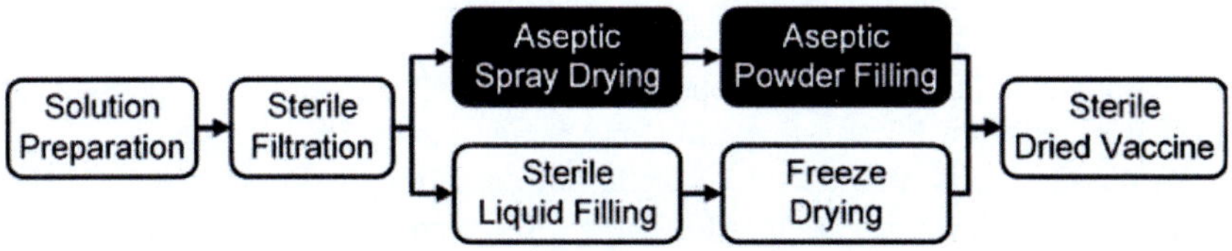

Fig. 4.14 Process steps of aseptic spray drying with aseptic powder filling (top) in contrast to sterile liquid filling and freeze drying (bottom) (Adapted from Andersen (2016, 2020) with kind permission from the author)

be loaded into final containers at a variable amount. However, a secondary drying step might be needed to further reduce the final moisture content. It must first be milled and sieved to convert a lyophilized product into a powder (Siew 2016). On the other hand, sterile water can easily be filled into a vial to reconstitute a lyophilized vaccine.

From an application point of view, aseptic powder filling is an established technology in the pharmaceutical industry and has been in use for many years for capsule and vial filling of antibiotics such as penicillin or cephalosporins, including (Searles and Cherian, 2015):

- cefazolin and ceftriaxone for the treatment of several bacterial infections,
- piperacillin in combination with tazobactam for the treatment of serious hospital-acquired infections, or
- penems, which are penicillin derivatives, inhibit cell wall biosynthesis in the bacterial organism.

Even cytotoxic drugs can be manufactured this way, including cyclophosphamide used in chemotherapy to suppress the immune system, bleomycin used to treat cancer, and mitomycin for treating various disorders associated with the growth and spread of cells (Searles and Cherian, 2015).

Aseptic powder filling technology exists today on a manufacturing scale reproducible for dosages down to 1 mg weight of formulated powder, including the active ingredient and excipients (Searles and Mohan 2010; McAdams et al. 2012). For example, this has been demonstrated using spray-dried inhalable insulin Exubera® (Pfizer, USA) within 1 and 3 mg dosing blisters to treating diabetes.

Wong et al. (2007) reported aseptic powder filling experiments with spray-dried leucine in vials on a commercial modular aseptic compact line (MAC) from IMA Life. The leucine powder had similar flow and aerosolization properties as Bacille Calmette-Guérin (BCG) vaccine to treat tuberculosis. The MAC equipment was designed to fill vials at a maximum rate of 4200–6000 vials per hour. In the experiments, 2 kg of spray-dried powder was placed in the hopper of the filling equipment, and the filling chamber volume and vacuum were adjusted to a filling target of 30 and 45 mg, respectively.

Figure 4.15 illustrates that the average fill weights of the batches produced, which were 29.4 $\pm$ 1.26 mg ($\pm$ 0.4 wt%) and 46.2 $\pm$ 0.97 mg ($\pm$ 0.2 wt%). Thus, overall, the study demonstrated reproducible aseptic powder filling and accurate dosing in vials in relevant masses for commercial application of spray-dried tuberculosis vaccine powder.

As shown in Fig. 4.16, an aseptic powder filling line into vials consists of washing, sterilizing, filling, stoppering, and capping the vials.

There are several providers for aseptic powder filling systems on the market, such as IMA Life (Italy), ROMACO (Italy), Syntegon Technology GmbH (formerly Bosch Packaging Technology GmbH, Germany), NKP Pharma (India), Cazzoli (USA), Harro Höfliger (Germany), Dec Group (Switzerland), or 3P

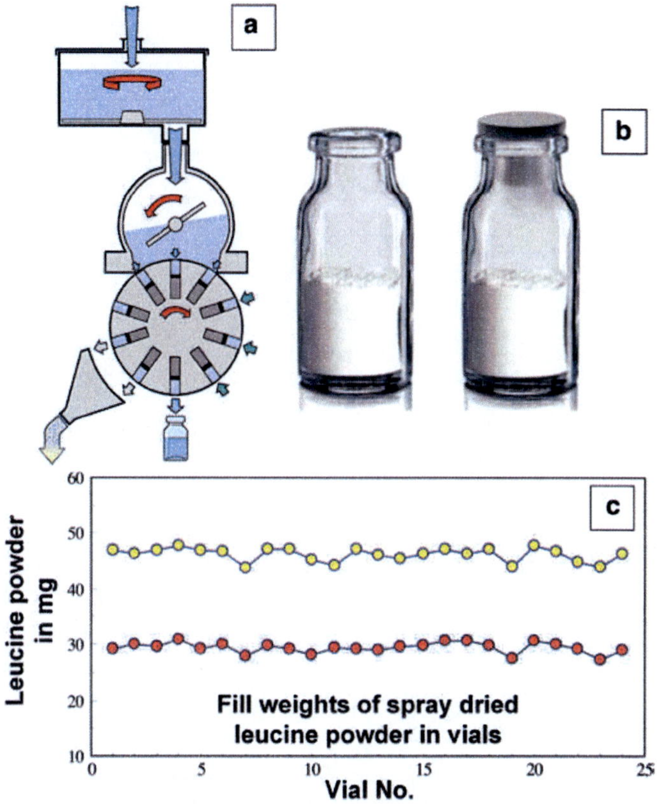

Fig. 4.15 **a** Schematic of dosing disc filling of powder into vials (Adapted from IMA Life (2012)). **b** Powder-filled vials with a rubber stopper (Adapted from IMA Life (2012)). **c** Fill weights of spray-dried leucine powder in 24 vials on a commercial aseptic powder filling line from IMA Life. The leucine powder had similar flow and aerosolization properties as the Bacille Calmette-Guérin (BCG) vaccine to treat tuberculosis by pulmonary administration. Each point represents a single fill weight with target weights of 30 and 45 mg. The achieved average values were 29.4 ± 1.26 mg (± 0.4%) and 46.2 ± 0.97 mg (± 0.2%) (Data adapted from Wong et al. (2007) with permission from Proceedings of the National Academy of Sciences, PNAS, USA)

Innovation (UK). Table 4.3 lists a selection of aseptic powder filling systems for small fill weights.

Figure 4.17 illustrates the aseptic powder filling lines SMART PWD (IMA Life), MicroMaxX (ROMACO), AFG (Syntegon), and MicroVave (Harro Höfliger) installed in closed RABS (Restricted Access Barrier System), providing a class ISO 5 (Grade A) aseptic environment.

The AFG 5000 series from Syntegon offers output rates from 9600 up to 28,800 vials per hour with 100% in-process control (IPC) of dosing weights (Syntegon 2021). The automatic Macofar MicroMaxX series from ROMACO (2018) can fill up to 24,000 vials per hour. For this purpose, the MicroMaxX 24 model is equipped

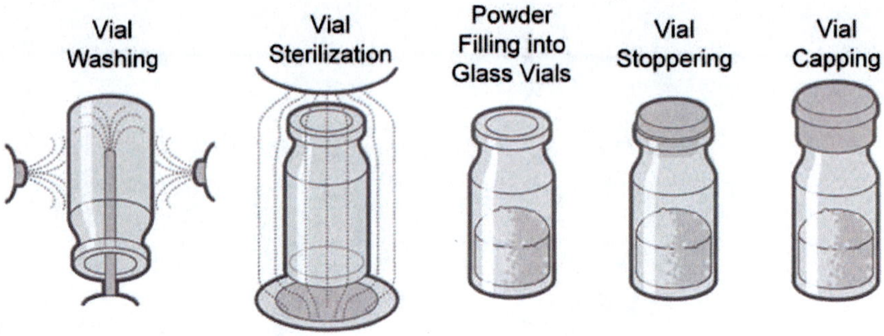

Fig. 4.16 Steps of an aseptic powder filling line into vials with washing, sterilization (depyrogenation), filling, stoppering, and capping (Images adapted from ROMACO (2018))

Table 4.3 A selection of aseptic powder filling systems available on the market for small fill weights

Company (Country)	Aseptic powder filling systems	Min. fill weight (Accuracy, RSD)	Max. throughput (vials per hour)	Remarks	References
ROMACO (Italy)	Macofar MicroMaxX, MicroRobot 50	20 to 30 mg (± 2%)	Up to 24,000	Dual dosing disc, vacuum-pressure technology	ROMACO (2018), Cattenati (2019)
IMA Life (Italy)	SMART PWD, Microfill, MD	50 mg (± 2%)	7200, 9000 to 24,000	Double-dosing disks filling up to 4 vials	IMA Life (2012, 2020)
Syntegon (Germany)	AFG 5000	50 mg (± 7.5%)	9600 to 28,800	100% in-process control of dosing weights	Syntegon (2021)
NKP (India)	NKPF	50 mg (± 1%)	3600 to 14,400	Single to four doses	NKP (2020)
Cazzoli (USA)	PF	50 mg (± 1%)	Up to 18,000	2 to 16 dosing heads	Cazzoli (2019)
Harro Höfliger (Germany)	Microvibe SDM (auger dosing)	1 mg (< 3%) 50 mg (< 3%)	600 to 4800 1200 to 4800	Dosing time < 2 s for ~10 mg Dosing time < 2 s for 500 mg	Harro Höfliger (2021)
Dec Group (Switzerland)	µPTS	< 1 mg (± 2%)	900 to 3600	Single head/cavity can be parallelized	Dec Group (2021a)
3P Innovation (UK)	Fill2Weight	1 mg (± 3%)	1800	Dosing time: around 2 s	3P Innovation (2020)

Fig. 4.17 Aseptic powder filling lines with powder dosing units installed in closed RABS (Restricted Access Barrier System). **a** SMART PWD from IMA Life with a speed of up to 7200 vials per hour, **b** MicroMaxX from ROMACO for up to 24,000 vials per hour, **c** AFG 5000 from Syntegon with output rates up to 28,800 vials per hour, **d** MicroVibes from Harro Höfliger with smallest powder weights dosing from 1 mg based on the fill-to-weight principle (Images by courtesy of the suppliers)

with a dual powder dosing disc which allows multiple dosing into the same vial. Alternatively, the product can be filled in two separate process steps. Besides, the MicroRobot 50 robotic filling and capping machine with patented vacuum-assisted dosing wheels has just been launched, filling hygroscopic and cytotoxic API powders into vials. It enables the filling of up to 3000 vials per hour, even 20–30 mg powder quantities per dose, with a filling accuracy of 2% (RSD, relative standard deviation) (Cattenati 2019).

The SMART PWD from IMA Life (2020) has up to two dosing units and enables individual dosing from 50 mg of different powders into the same vial with an operating speed of 7200 vials per hour. The Microfill and MD series (IMA Life 2012) offer double-dosing disks that fill up to four vials with powder at a time. This provides an increased output and a dosing range of 9000 to 24,000 vials per hour. The Cazzoli PF powder filler series are available with 2 to 16 dosing heads and can reach speeds of up to 300 containers per minute (18,000 vials per hour), with accuracies of ± 1% (RSD) of fill volume (Cazzoli 2019). Powder quantities of 50 mg to 75 g can be filled in a single metered dose. A multi-dosing feature is available for larger fill volumes. The dry powder filling machines NKPF 300 with a rubber stoppering from NKP Pharma provide outputs from about 3600 to 14,400 vials per hour for single and four doses at fill sizes of 50 mg to 6 g (± 1% RSD) (NKP 2020).

Harro Höfliger Verpackungsmaschinen GmbH (Germany) is a leading manufacturer of automatic powder dosing and filling technologies for inhalation products (Höfliger 2021). The product portfolio includes machines for blister packaging of dry powder inhalers and inhalers with unit-dose cartridges or multi-dose discs (Fig. 4.18a, b). Depending on the powder properties and container closure systems, different filling modules are used, like membrane fillers or drum fillers, which enable reliable dosing of free-flowing and cohesive powders even in minimal quantities of less than 1 mg at high speed. Production machines for blister filling have output rates of up to 2400 blister cups per minute. Figure 4.18c shows an overview of key machine features offered by Harro Höfliger, including blister forming, filling and sealing, in-line weight inspection, X-ray inspection, capsules filling, micro-dosing, device assembly, parts handling, automation, barrier technology, and others. Overall, Harro Höfliger has significant experience in packaging tricky powders (HARRO 2020a).

Unlike dry powder pharmaceuticals for parenteral injection, inhalable dry powders do not require aseptic filling but controlled bioburden because the lungs have an immune system and can fight off contamination. Insulin dry powder for inhalation is a good example. For MannKind Corporation (USA), (Harro Höfliger 2015) engineered and constructed the powder filling and assembly lines for the inhaled insulin Afrezza®. The production line is based on the FLP-200 machine platform and enables powder dosing in 400 cartridges per minute with filling weights of 3 to 15 mg per cartridge at a filling tolerance of maximum ± 5 wt% (Harro Höfliger 2015).

Figure 4.18d shows the diabetes therapy system consisting of a dry powder inhaler (called Dreamboat) and powder-filled cartridges for individual application.

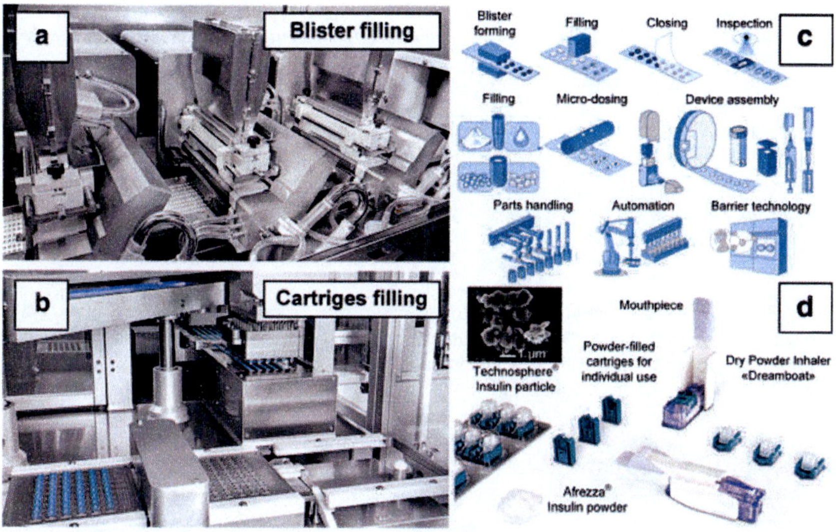

Fig. 4.18 **a** Production machine for commercial blister filling of blister strips for dry powder inhalers. The blister cavities are filled at the same time with several drum dosage units. This method enables precise powder micro-dosing at high output rates of up to 2400 blister cups per minute. **b** FLP-200 production platform for filling cartridges for dry powder inhalers and product handling of filled and sealed cartridges onto larger trays (400 cartridges per minute, 3 to 15 mg fill weights per cartridge, verified by sensor field and electrical capacity). A camera system inspects that powder is present in each cartridge. **c** Illustration of various key machine features of Harro Höfliger's powder dosing and filling systems. **d** The dry powder inhaler is called "Dreamboat," and cartridges are filled with insulin powder Afrezza®. The Technosphere® insulin particles have a mean size of about 2 to 3 µm (Richardson and Boss 2007) (Images by courtesy of Harro Höfliger Verpackungsmaschinen GmbH, Germany)

Afrezza® is available in single-dose cartridges containing 4, 8, and 12 units of insulin powder (MannKind 2020). MannKind uses the Technosphere® delivery platform (Pfützner et al. 2002; Pfützner and Forst 2005; Richardson and Boss 2007; Angelo et al. 2009; Neumiller and Campbell 2010; Goldberg and Wong 2015; Castagna 2018), which is based on an organic molecule called fumaryldiketopiperazine (FDKP) for rapid dissolution in the lungs and elimination from the body without metabolism. Figure 4.18d shows an SEM image of typical Technosphere® insulin particles with an average aerodynamic diameter of about 2–3 µm (Steiner et al. 2002; Richardson and Boss 2007; Angelo et al. 2009).

Another best practice example is the blister filling for the Papillon dry powder inhaler from Hovione (HARRO 2020b). The inhaler consists of two reusable plastic components and an inserted blister. The Papillon can be reused multiple times and for different applications. The design as a double blister configuration makes it possible to inhale two separate powders in parallel and simultaneously. The right powder filling technology plays a decisive role, and Harro Höfliger provides the know-how to scale up on an industrial scale.

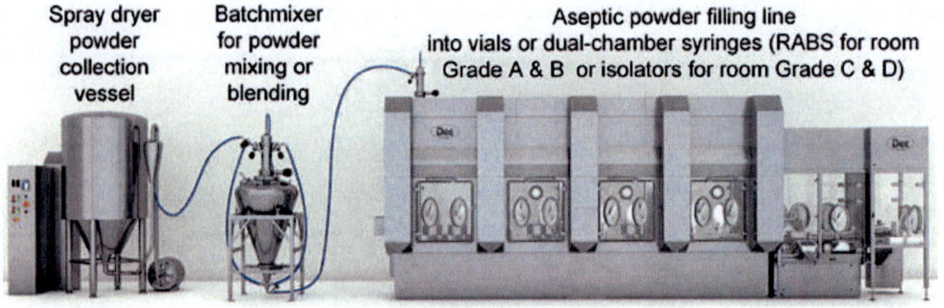

Fig. 4.19 Interlinked, continuous, and closed sterile processing from the spray dryer outlet through homogenization in a batch mixer to the aseptic powder fill/finish process into vials or dual-chamber syringes (RABS for room Grade A & B or isolators for room Grade C & D) (Image by courtesy of Dec Group, Switzerland)

A leading global provider of contained powder handling systems is the Dec Group, headquartered in Lausanne, Switzerland. The PTS Powder Transfer System® (Dec Group 2021b) with aseptic design uses both vacuum and pressure to move powder across multiple floors. Figure 4.19 shows a conceptual diagram for integrating aseptic powder handling from a spray dryer outlet through a PTS Batchmixer® (Dec Group 2021c) (for possible homogenization and/or blending of other powders) to a powder filling line, keeping the whole process closed and ensuring continuous manufacturing.

The patented micro-dosing device units based on PTS technology (μPTS) allow precise and rapid dosing of milligram powder quantities (< 1 mg to 500 mg) within less than one second (Dec Group 2021a). High accuracy of up to 2% (RSD) can be achieved even for powders with different bulk densities, as the powder is filled in a calibrated chamber of adjustable volumes by the action of vacuum and emptied mechanically by a piston. Improved accuracy can be achieved by two-step dosing, where a larger powder quantity is dosed first and a second smaller quantity (typically 10 to 30% of the first dose) is dynamically adjusted and dosed on top. In this way, < 1% (RSD) accuracy can be achieved, as illustrated in Fig. 4.20, showing some test data for a target fill weight of 15 mg powder.

Another enabling technology for accurate powder filling is Fill2Weight™ from 3P innovation Ltd. based in the UK (3P Innovation 2020). 3P innovation supplies flexible, scalable filling solutions and complete production lines for the pharmaceutical and medical industries. Fill2Weight enables fast and highly accurate gravimetric filling in an aseptic environment, suitable for spray-dried and freeze-dried powders. The technology is used for a wide range of products, including injection devices, cartridges, vials, pre-filled syringes, inhalers, capsules, or blisters with 100% weight verification (Strothers and Seaward 2020). The dosages are configurable from 1 mg to 2 g with an accuracy of < 3% (RSD). According to specifications, the dosing cycle time per dose is around 2 s, the tare weight up to 40 g, and the weighing resolution 10 μg (3P innovation 2020). An application list of a series of lyophilized powder fillings with 10 mg fills weight

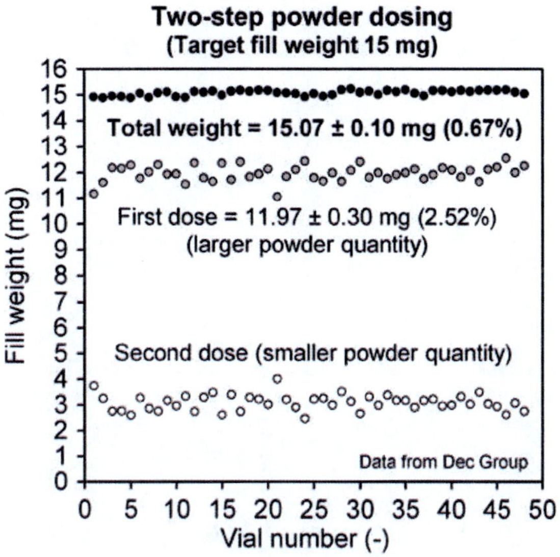

Fig. 4.20 Data of two-step powder dosing with a micro-dosing device based on PTS technology (μPTS). Target powder fill weight: 15 mg. Achieved average total weight: 15.07 ± 0.10 mg (0.67% RSD) (Data by courtesy of Dec Group, Switzerland)

showed an average fill weight accuracy of 0.7–2.2% (RSD) (Bailey and Seaward 2013). As shown in Fig. 4.21, Fill2Weight is a scalable technology by adding additional components in parallel and suitable for use in the laboratory, early phase clinical trials to commercial manufacturing (Bailey and Seaward 2013; Strothers and Seaward 2020). In addition, the powder dispensing heads can be integrated in aseptic enclosures.

In general, it is worth adding that gravimetric dosing systems are based on weighing and therefore require time per sample, compared to volumetric dosing systems (such as drum dosing), which are fast and allow scale-up of larger sample quantities, as shown in the examples for blisters or cartridges.

There is also an increasing market trend towards prefilled dual-chamber disposable syringes as a potential new way of dry powder vaccine delivery. This parenteral dosage form contains a sterile medicinal product, for example, a lyophilized or potential spray-dried vaccine, in one chamber and a reconstitution diluent in a separate chamber (Fig. 4.22). The liquid flows through a bypass in the syringe barrel to the powder during syringe activation and is mixed. This practice shows clear advantages as it increases the stability of the vaccine, provides longer shelf life at room temperature and reduces the need for a cold chain (Dec Group 2020). Figure 4.22 shows the process steps of aseptic powder and liquid filling into a dual-chamber syringe. The filling is performed in enclosures that provide an ISO 5 class (Grade A) environment.

An example of a commercial product using the prefilled dual-chamber syringe is Abilify Maintena® from Otsuka Pharmaceutical Ltd. (Otsuka 2020), an atypical antipsychotic for treating schizophrenia in adults and maintenance monotherapy bipolar I disorder in adults. The API aripiprazole (300 or 400 mg) is filled into the

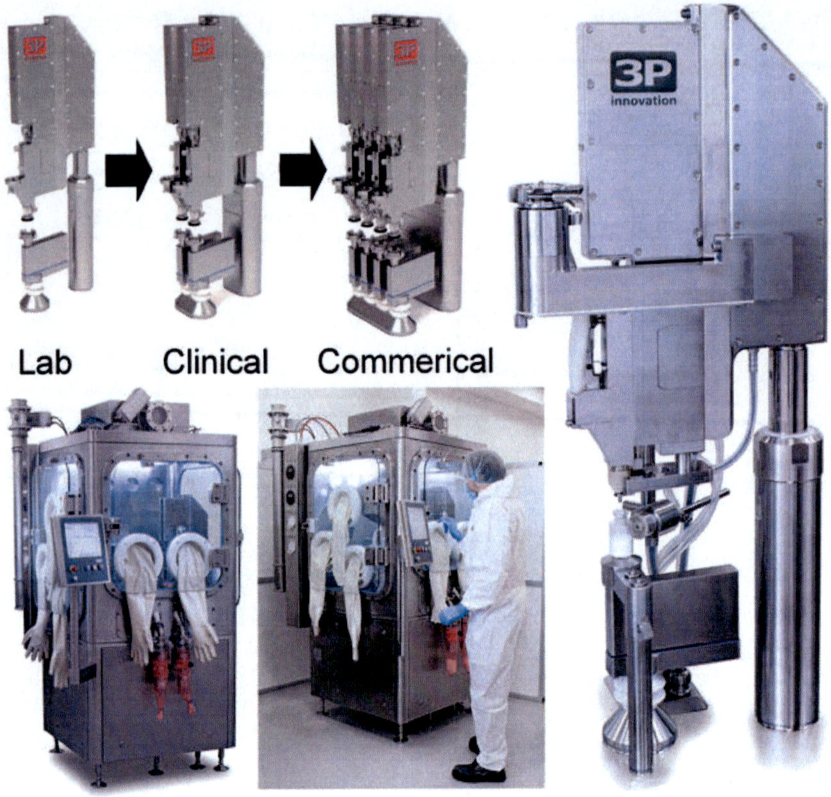

Fig. 4.21 Fill2Weight™ is a high-speed, high-accuracy gravimetric filling technology for aseptic environments suitable for spray-dried and lyophilized powders. The technology is scalable and flexible, measuring, controlling, and recording the weight of every dose. A single Fill2Weight head used for lab and clinical supply can be multi-tracked for commercial applications to provide a scalable solution. The images show twin Fill2Weight heads installed inside 3P's R1000 robotic capsule and container filler, including a negative pressure containment isolator with local HEPA filtration (Images by courtesy of 3P innovation Ltd., UK)

syringe as a lyophilized powder with excipients carboxymethylcellulose sodium, mannitol, H_4NaO_5P, and NaOH along with sterile water for injection. The product can be stored at room temperature below 30 °C.

Potential biopharma companies that could use dual-chamber syringe technology to reconstitute dried vaccine powder with diluents are GSK, Sanofi Pasteur, and Merck. This relates specifically to freeze-dried vaccines listed in a handout from the Immunization Action Coalition (IAC 2018), including freeze-dried ActHib® (Sanofi), Hiberix® (GSK), Imovax® (Sanofi), M-M-R II®, (Merck), Menveo® (GSK), Pentacel® (Sanofi), ProQuad® (Merck), RabAvert® (GSK), Rotarix® (GSK), Shingrix® (GSK), Varivax® (Merck), YF-VAX® (Sanofi), and Zostavax® (Merck). A more general overview of vaccine manufacturers producing lyophilized

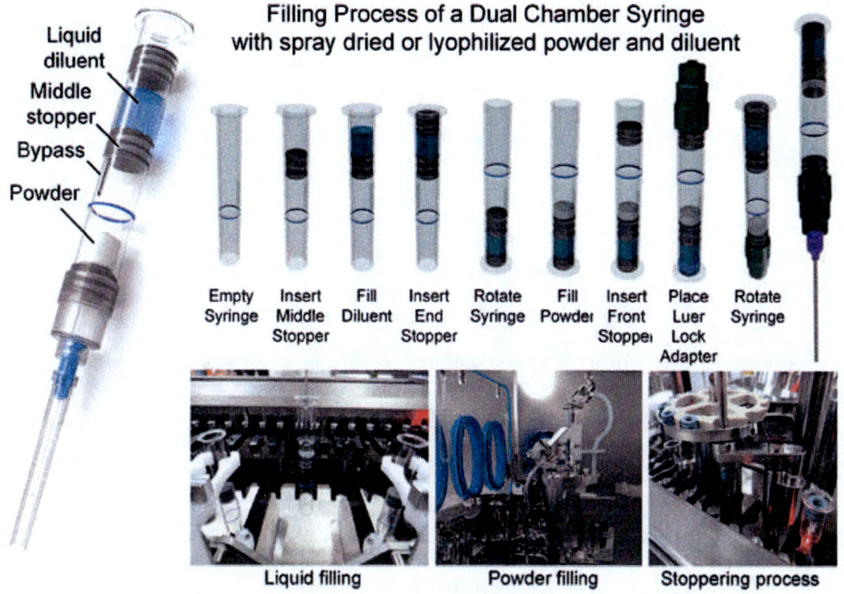

Fig. 4.22 Process steps for pre-filled dual chamber disposable syringes for parental injection of a sterile medicinal product, for example, a potential spray-dried or lyophilized powder vaccine and diluent (Images by courtesy of Dec Group, Switzerland)

vaccines is provided in Chap. 2 (Sect. 2.2). Overall, a prefilled dual-chamber syringe, which allows for complete reconstitution within the syringe, eliminates the transfer step from the vial to the syringe, facilitating vaccine administration.

An alternative to packaging two separately stored components are dual-chamber tubes. The Swiss packaging specialist Neopac develops and produces tubes for healthcare applications. Figure 4.23 shows the so-called Fleximed® Easymix™ tube, a transparent, flexible medical tube made of synthetic high-barrier laminate, enabling safe and straightforward mixtures for parenteral, topical, or oral medicines (Neopac 2020). The two chambers are separated by a frangible seam that opens towards the back chamber under pressure. In this way, a liquid and a powder component or two liquids can be mixed immediately before administration. In addition, a tube head with a Luer lock connector allows the attachment of syringes for parenteral use or a nozzle head suitable for oral or topical administration. Fill volumes are possible from approx. 15 mL down to 1 mL per chamber and up to 240 tubes per minute can be filled on a production machine. The dual-chamber tube is especially suitable for the reconstitution of aseptically prepared APIs, which have to be stored in a dried powder state. Examples of potential applications for Easymix™ are antibiotics for intravenous administration, modern recombinant therapeutics (e.g., for autoimmune diseases), topical wound care applications, oral vaccines, or ophthalmic applications.

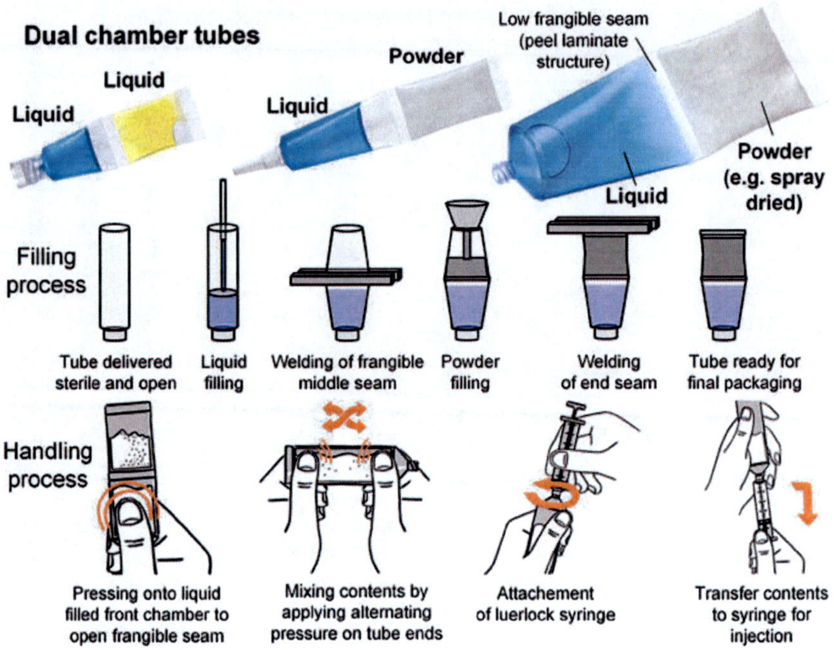

Fig. 4.23 Dual chamber tubes (Fleximed® Easymix®, Neopac) for parental injection of sterile medicinal products, for example, a potential spray-dried or lyophilized powder vaccine and diluent. The process steps for filling and administration are shown. Fill volumes from approximately 15 mL down to 1 mL per chamber are possible with production rates of up to 240 tubes per minute (Images by courtesy of Neopac, Switzerland)

To conclude this section, it can be stated that although there are already numerous aseptic powder filling systems available on the market with high accuracy and speed, applications of filling for spray-dried vaccine powders, including their validation, are still lacking, at least in the literature. It is assumed that such information is rather not published in the public and scientific domain.

4.7 Laboratory-Scale Spray Dryers Used in Vaccine Research

Laboratory-scale spray dryers are particularly useful for feasibility studies at the early stages of development to demonstrate the compatibility of the technology with the biopharmaceutical product. The drying chamber and cyclone are frequently constructed of glass for these small-scale systems, allowing direct observation of the drying and separation processes. The liquid feed is fed via a peristaltic pump into the nozzle atomizer and sprayed into the drying gas.

The main limitation of the laboratory equipment is the limited particle size due to the small dimensions of the drying chambers. The residence time is limited, and therefore the droplets must be small to be completely dried before leaving the drying chamber (Miller and Gil 2012; Gaspar et al. 2014). An exception is ProCepT's 4M8-TriX spray dryer, which operates under laminar flow conditions and whose drying chamber can be extended from 1.4 to 2.1 m (providing 5 s residence time) to dry larger droplets (100 μm and larger with ultrasonic nozzle) (Schaefer and Lee 2015). Thus, these devices can mimic the size and morphology of spray-dried particles on a larger scale (ProCepT 2020, 2021).

According to CFD simulations in a Mini Spray Dryer B-290 (Fig. 4.24), the average residence time of a droplet is about 0.2–0.35 s (Kemp et al. 2016). The time required to evaporate a water droplet is about 0.03 s for a 10 μm droplet and 3 s for a 100 μm droplet. Some recirculation can be seen in the upper part of the drying chamber near the walls, but the fast-moving core flow has the primary effect.

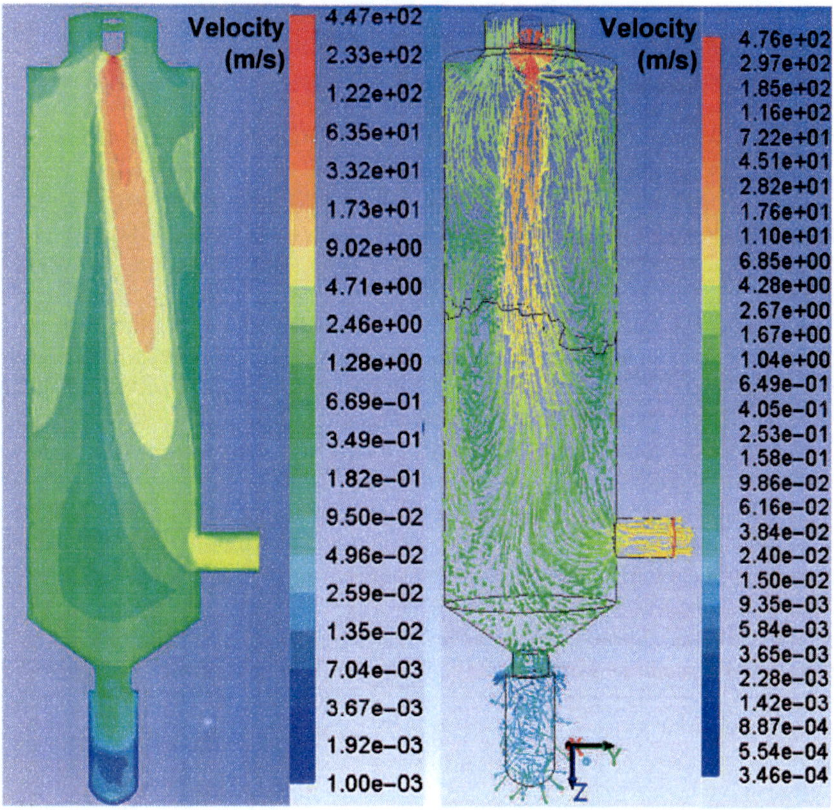

Fig. 4.24 Predicted air-flow pattern (gas velocity in m/s) by CFD (ANSYS FLUENT™) on a mini spray dryer B-290 (diameter 0.165 m, length approx. 0.5 m). Air mass flow rates: 0.004 kg/s (left) and 0.006 kg/s (right) (Images adapted from Kemp et al. (2016) with permission from Taylor & Francis)

The high-velocity gas flow from atomization persists considerably down the drying chamber. These drying times may increase for a spray-dried particle that forms a crust (Büchi Labortechnik AG 2020a).

Thus, most laboratory-scale equipment produces powders with average particle sizes below 10–20 μm, often between 3 and 10 μm (Gaspar et al. 2014). Table 4.4 lists the most common laboratory-scale spray dryers used for vaccine research, shown by photographs in Fig. 4.25.

The laboratory spray dryers are generally equipped with process controls for setting the drying gas flow rate, inlet temperature, atomization gas, and feed flow rates. The maximum drying gas flow rate is in the order of 40 kg/h, which enables the evaporation of about 1 L/h of water. Thus, it is not surprising that lab-scale units are used to produce commercial quantities of several 100 g of low-volume products (Gaspar et al. 2014). Since most lab-scale spray dryers must be stopped to collect the powder products, they cannot run for long periods to produce large development batches. However, continuous product discharge could be achieved, for example, by installing a rotary valve or slide gate. Such adaptations are well-known technical solutions (Gursch et al. 2015). For example, the GEA Niro SDMicro™ can run continuously and produce smaller GMP batches. It features a bag filter consisting of four PTFE bags and continuous pulse-jet cleaning (GEA 2013).

In the pilot scale, the GEA Niro MOBILE MINOR™ is the most commonly used spray dryer (Miller and Gil 2012). It can be operated continuously at a water evaporation rate of up to 3 to 4 kg/h (at 200 °C inlet temperature) (GEA Process Engineering A/S 2020b). These units are similar to production-scale units in terms of configuration and construction materials (e.g., stainless steel, CIP, pressure nozzles).

For lab-scale units, two-fluid nozzles are the standard atomization system. In addition, in some models, ultrasonic nozzles (4M8-TriX Spray Dryer, Mini Spray Dryer B-290) (Büchi Labortechnik AG 2014; ProCepT 2020), piezoelectric mesh vibration (Nano Spray Dryer B-90) (Arpagaus et al. 2009, 2017, 2018; Arpagaus 2018, 2019), or monodisperse nozzles (4M8-TriX Spray Dryer) (Schaefer and Lee 2015) are available.

The design of the drying chamber and powder collection system depends on the scale. Cyclones, together with an outlet filter, are most commonly used. High-performance glass cyclones are available for smaller particles and small powder batches.

The 4M8-TriX Spray Dryer and the Nano Spray Dryer B-90 can process minimal sample quantities in the range of 0.5 to 2 mL with relatively high yield for milligrams of powder (Arpagaus et al. 2017, 2018; Arpagaus 2018, 2019; Ousset et al. 2019). The Nano Spray Dryer B-90 is a laboratory instrument capable of producing particles in the submicron range with an electrostatic particle collector and recovery yields of up to 90%. Extensive use of nano spray drying for pharmaceutical applications has been well reported in the literature (Arpagaus et al. 2017, 2018). In addition, accessories for closed-loop operation with the Mini Spray Dryer B-290 and Nano Spray Dryer B-90 are available, such as the Inert Loop B-295 with nitrogen and the Dehumidifier B-296 for removal of residual moisture.

Table 4.4 Most common laboratory-scale spray dryers used for vaccine research

Model	Manufacturer	Drying gas flow rate (kg/h)	Water evaporation capacity (L/h)	Max. inlet drying gas temperature (°C)	Possible particle size (μm)	Smallest sample size (mL)	Closed-loop possible	Powder collection system
Mini Spray Dryer B-290	Büchi	35	1.0	220	2 to 25 (10 to 60)*	30	Yes	Cyclone + Filter
Mini Spray Dryer B-191	Büchi	35	1.0	220	2 to 25	30	No	Cyclone + Filter
Nano Spray Dryer B-90	Büchi	10	0.2	120	0.2 to 5	2	Yes	Electrostatic
SDMicro	GEA Niro	30	1.0	200	2 to 30	100	No	Cyclone + Bag filter
MicraSpray MS-35	SPX Anhydro	38	1.6	200	2 to 40	100	No	Cyclone + Filter
4M8-TriX Spray Dryer	ProCepT	40	0.9	200	1 to 30 (1 to 100)*	0.5	Yes	Cyclone + Filter

[a]with ultrasonic nozzles

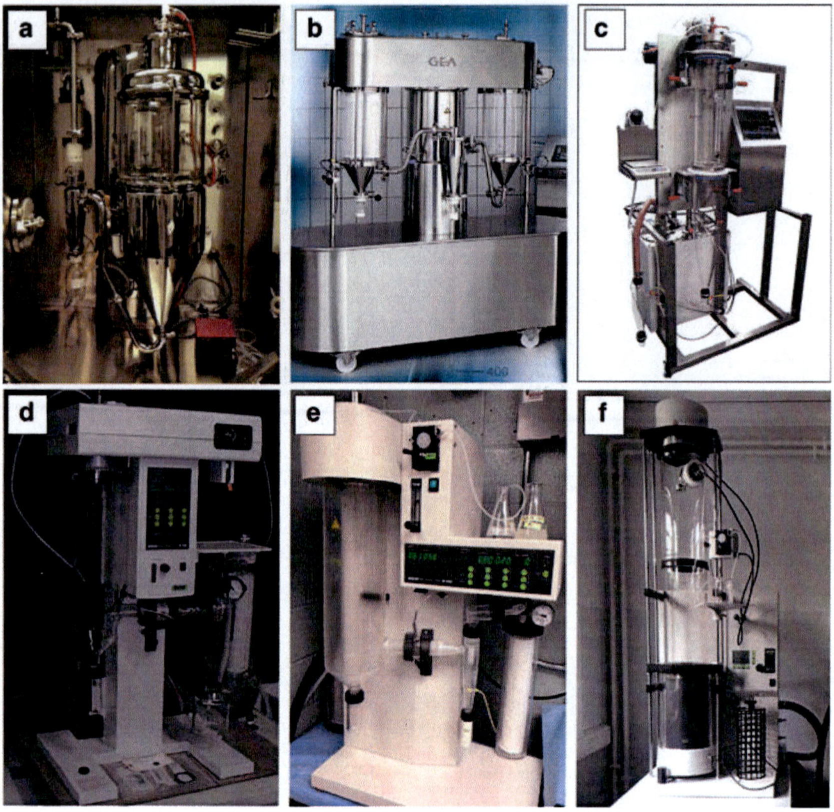

Fig. 4.25 Photographs of laboratory-scale spray dryers. **a** Aseptic MicraSpray MS-35 (SPX Anhydro) (Adapted from Gikanga et al. (2015) with kind permission of Dr. Yuh-Fun Maa from Genentech), **b** SDMicro (Image by courtesy of GEA Niro, Denmark), **c** 4M8-TriX Spray Dryer (Adapted from Tseng (2015), Image by courtesy of ProCepT NV, Belgium), **d** Mini Spray dryer B-191 (Image courtesy by Büchi Labortechnik AG, Switzerland), **e** Mini Spray dryer B-290 with High-Performance cyclone (Image courtesy by Büchi Laboratechnik AG, Switzerland), **f** Nano Spray Dryer B-90 (Image courtesy by Büchi Labortechnik AG, Switzerland)

Many publications show that the Mini Spray Dryer B-290 and its predecessor models B-191 and B-190 (Arpagaus et al. 2010) are the most frequently used laboratory spray dryers for vaccines research. SPX Anhydro and GEA Niro are probably the only companies providing laboratory spray dryers for the cGMP manufacturing of powders. The sources for truly aseptic spray dryers are limited to the MicaSpray MS-35 aseptic dryer (SPX FLOW Inc. 2017) and GEA Niro SDMicro (GEA 2013).

4.8 Excursus: Mini Spray Dryer B-290 Options for Working with Low Bioburden

The Mini Spray Dryer B-290 manufactured by Büchi Labortechnik AG (Switzerland) has been designed and built as a laboratory instrument and for batch production for feasibility studies. If the instrument is used with potentially toxic or hazardous substances, it must be installed inside a closed fume hood or glove box (Büchi Labortechnik AG 2020b). In such cases, the complete processing and system handling must be performed within an isolator to avoid toxication and other hazardous situations to the user and the environment. Furthermore, the operation of the instrument needs to respect standard laboratory safety rules, i.e., as stated in the latest version of the WHO Laboratory Biosafety Manual (WHO 2004), which provides practical guidance on biosafety techniques for use in laboratories at all levels.

Figure 4.26 illustrates some available accessories to work with the Mini Spray Dryer B-290 under low bioburden conditions, which can be proposed to carry out spray drying trials with vaccines.

4.8.1 Sterilization of Components

The glassware, connectors, tubing, nozzle, and mechanical parts of the spray dryer need to be separately sterilized before the spray drying trials in an autoclave or sterilizer. A standard method to sterilize glassware via dry heat in an oven is at least 170 °C for 2 h or longer. Alternatively, the glassware can be autoclaved at 121 °C with moist heat for at least 30 min. The sterilized items must be handled and stored to remain noncontaminated until used for reassembly.

Later on, after reassembly, the laboratory spray dryer can be sterilized in-place (SIP) using clean steam or dry heat. This takes advantage of the fact that a laboratory spray dryer is designed for drying gas inlet temperatures of up to 220 °C providing saturated steam (Büchi Labortechnik AG 2020b). It is common practice to start the spray drying experiment with purified (sterile) water to achieve thermal equilibration of the system at a specific inlet and outlet temperature before switching to the feed solution (Searles and Mohan 2010). Gas sterilization with vaporized hydroperoxide (H_2O_2) or other agents is also possible. The system must have a sterile path throughout processing. A comprehensive validation program is instituted for sterile spray drying of small batches of vaccines to comply with cGMP requirements and ensure reliability (Searles and Mohan 2010).

4.8.2 Pressure Mode

The spray dryer is to be operated in a cycle with slight overpressure. For this purpose, the aspirator is installed at the inlet line of the spray dryer to push the

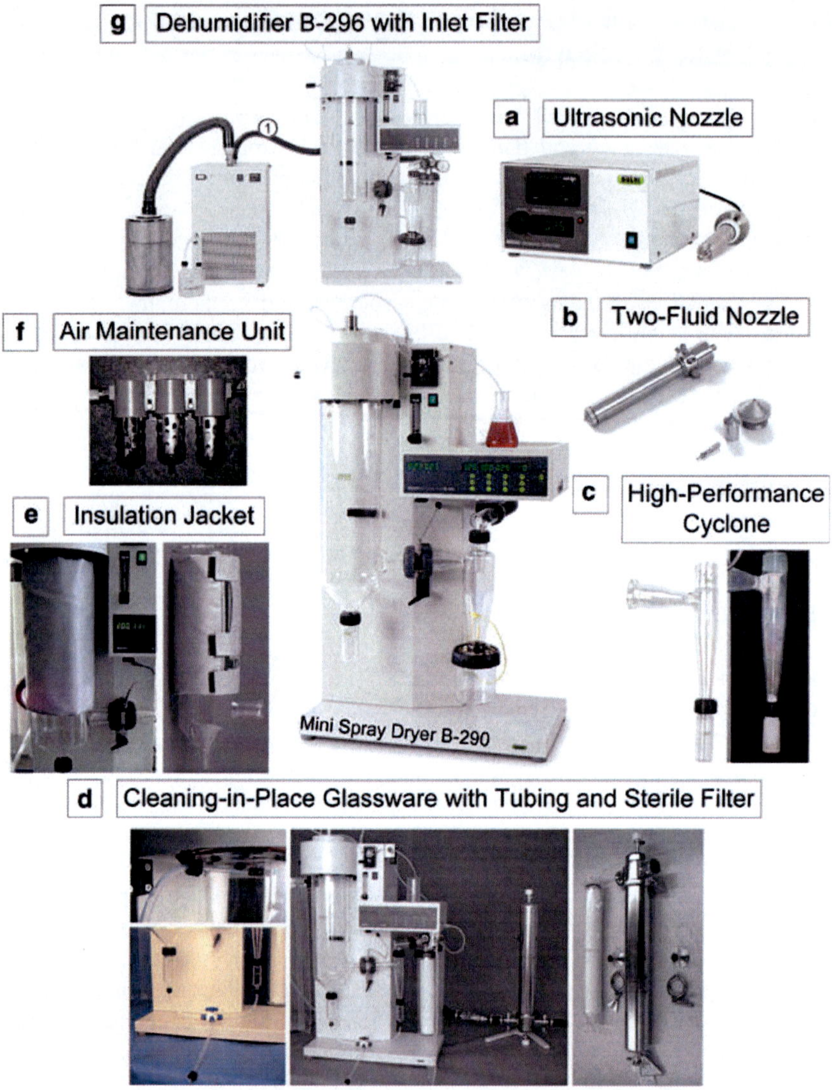

Fig. 4.26 Mini Spray dryer B-290 with accessories to work under low bioburden conditions in the laboratory. **a** Ultrasonic nozzle, **b** two-fluid nozzle with different nozzle tips and caps, **c** High-performance glass cyclone, **d** pre-cleaning glass-set for decontamination and disinfection with sterile filter, **e** insulation for spray drying chamber to reduce heat losses, **f** air maintenance unit for clean spray gas, **g** inlet HEPA filter and Dehumidifier B-296 in open-mode configuration (Images by courtesy of Büchi Labortechnik AG, Switzerland)

drying gas into the system (pressure mode) (Büchi Labortechnik AG 2020b). This creates a slight overpressure in the instrument, preventing contamination from the external environment. The pressure value on the manometer indicates the overpressure.

4.8.3 Inlet HEPA Filter

A HEPA filter needs to be installed at the heater inlet to prevent even the finest particles from the ambient air from entering the spray dryer system. For conditioning, the inlet drying air, and reducing inlet humidity, the installation of a Dehumidifier B-296 is recommended (Fig. 4.26g). The accessory device enables drying under stable and reproducible humidity conditions and cools the air to a dew point of about 3 to 5 °C (Büchi Labortechnik AG 2020b).

4.8.4 Outlet HEPA Filter

A HEPA filter, e.g., 0.2 μm EMFLON filter from PALL (2020), needs to be installed downstream from the outlet filter at the process gas outlet to purify the exhaust gas (Jin et al. 2010; Tumban et al. 2015; Kunda et al. 2016; Price et al. 2020). The sterile filter works as a safety element to hold back the finest product residues from the process gas. It typically consists of a PTFE membrane, which is hydrophobic, chemically inert, and explicitly designed to remove contaminating bacteria and viruses. Figure 4.26d shows a typical set-up with a HEPA filter fixed with a stand. Besides, the HEPA filter can be tested in-situ by a forward flow integrity test or a water intrusion test.

4.8.5 Sterile Compressed Air for Atomization

A compressed air maintenance unit is recommended to decontaminate the pressurized air for atomization (Fig. 4.26f). The air maintenance unit consists of three cartridges (5 μm prefilter, 0.03 μm submicron filter, and 0.01 μm charcoal filter). This way, the pressurized air is of sterilized quality.

4.8.6 Sterile Liquid Filtration

Before entering the atomizing nozzle, the liquid feed solution containing the vaccine and excipients must be pre-filtered with a sterile filter (0.22 μm pore size or smaller). If the feed formulation can not be sterile filtered, the formulation needs to be established under aseptic conditions. Then, the vaccine solution is added aseptically to the excipient solution, and the mixture is ready for spray drying. Frequently used sterile filters in laboratory-scale are, for example, Millidisk® cartridges and Millipak® disposable capsules from EMD Millipore (2015) that are stacked disc filters designed to remove particles and microorganisms from liquids and gases. After spray drying the vaccine solution, the feed tubing is rinsed with purified water and detergent solutions.

4.8.7 Sterilization of Nozzles

The two-fluid nozzle and ultrasonic nozzle are made of stainless steel and need to be cleaned and sterilized before use. The standard two-fluid nozzle has a nozzle tip of 0.7 mm in diameter. Nozzle tips with sizes of 0.5, 1.4, and 2.0 mm are optionally available (Büchi Labortechnik AG 2020b) (Fig. 4.26b). Alternatively, the ultrasonic nozzle can produce larger particles closer to the industrial scale (Fig. 4.26a) (Büchi Labortechnik AG 2014). Particle sizes of 10 to 60 μm are achievable depending on the frequency, feed rate, and solids concentration. Cooling the nozzle and the feed vessel with cooling water is common practice to reduce the thermal load of the feed formulation into the spray dryer.

4.8.8 Powder Collection

The powder collection vessel can be a small glass container or vial and needs to be cleaned and sterilized before use. After spray drying, the collection vessel containing the dry product is closed with a cap inside the biosafety cabinet and transferred to a glove box, where the powder is filled into vials under a nitrogen atmosphere. Some users prefer to use a separate attachment for the cyclone to collect the powder directly in pre-sterilized scintillation vials (Tumban et al. 2015; Saboo et al. 2016; Kunda et al. 2016). Finally, the vial is removed and capped. All further processing of the powder is carried out in a biosafety cabinet adjacent to the spray dryer. The vials are then typically stored at 2 to 8 °C for further analysis (Ingvarsson et al. 2013; Scherließ et al. 2014).

4.8.9 Optional High-Performance Glass Cyclone

On a laboratory scale, the limited recovery of powder yield due to deposits on the drying chamber walls and cyclone is challenging (Searles and Cherian 2015). Therefore, a high-performance cyclone has been developed for small sample volumes (Brandenberger 2003) (Fig. 4.26c). It has a higher separation efficiency than the standard cyclone and is particularly suitable for small particle sizes in the range of 3 to 5 μm, e.g., for dry powder inhalation, thus optimizing the spray drying process for high-quality products in small quantities. For example, the high-performance cyclone has been used in vaccine studies against tuberculosis (Wong et al. 2007), influenza (Scherließ et al. 2014), pneumonia (Kunda et al. 2015), pertussis (Kanojia 2018; Kanojia et al. 2018), or herpes simplex (LeClair et al. 2019).

4.8.10 Cleaning of Glass Parts, Metal Parts, and O-rings

After every experiment, the spray dryer instrument needs to be cleaned with detergent solutions and a brush. First, the spray dryer is dismantled, and the individual glass items are cleaned and washed with soapy water before disinfecting all components, e.g., with 70% v/v isopropanol/water mixtures (Scherließ et al. 2014). Cleaning is made easy by washing the glass parts in a sink. The metal parts to clean are:

- the heater housing entrance to the spray chamber (because of back-flow of gas and particles due to turbulence),
- the metal holder for the spray cylinder,
- the metal connecting piece between spray cylinder and cyclone,
- the metal cover of the cyclone to the product collection vessel,
- the metal entrance duct for the nozzle, and
- the complete nozzle (i.e., tip, cap, needle, and nozzle body).

The sealings and O-rings must also be cleaned appropriately and need to be replaced from time to time. Finally, after spray drying, the general disinfection of the space must be carried out. Multiple sets of glassware and a second two-fluid nozzle are recommended to work more parallel.

4.8.11 Optional Cleaning-In-Place (CIP) Glassware

The decontamination and disinfection glassware is an optional decontamination package for CIP (Fig. 4.26d) (Büchi Labortechnik AG 2020b). The glassware consists of a modified spray cylinder with an opening at the top for connecting a washing tube, a glass separator, and a particle collection container with openings for connecting hoses, as well as tubings and a three-way valve. The washing tube has small spray openings and is inserted at the top into the spray cylinder. By pumping disinfectant solution into the spray chamber, the glassware is washed, and agglomerated particles from the spray cylinder and the cyclone glass wall are bound in liquid form. A disinfectant proven effective against viruses, bacteria, and fungi, is Rely + On™ Virkon™ (LANXESS 2017). The entire spray dryer can be flooded with disinfectant. The glassware is soaked for at least 30 min. to completely decontaminate the spray dryer before the cleaning liquid is drained from the spray dryer by opening the three-way valve.

4.8.12 Optional Insulation for the Spray Drying Cylinder

Heat losses over the spray cylinder wall are more significant on a small scale and can be minimized using a cylinder insulation jacket as an accessory, leading to a higher outlet temperature and higher heating energy efficiency (Fig. 4.26e). Especially when working over 160 °C, this effect takes place. The process conditions are closer to up-scaled instruments where the gradient between the drying gas and cylinder walls is smaller. Thus, fewer differences in drying behavior will occur. Besides, the inner wall of the glass cylinder is warmer, and hence the temperature difference to the gas is smaller. This may help during the sterilization process using clean steam or dry heat. Furthermore, the warmer wall may lead to fewer particle deposits on the glass wall since the relative humidity on the cooler glass wall is higher, resulting in an eventually stickier product.

4.8.13 Optional Brown Glass for Protection of Light-Sensitive Products

The transparent glass parts of the spray dryer can be replaced by brown glass, which is an option for light-sensitive products. Brown glass can prevent photochemical reactions triggered by absorbed UV light, such as photodegradation, i.e., the decay of materials due to exposure to sunlight.

4.8.14 Materials in Contact with the Product

For validation, the materials in contact with the spray-dried product need quality certification. Official and authorized installation, operation, and performance qualification (IQ/OQ/PQ) documentation is available (Büchi Labortechnik AG 2000), which helps validate the spray dryer. The materials in contact with the product are polymers of the feed tubing, O-rings, sealing parts, stainless steel of the heater housing, spray chamber holder, central tube, connector, and nozzle parts borosilicate glass 3.3 of the spray cylinder and cyclone. The tubing, O-rings, and sealings are typical replacement parts.

4.8.15 Installation in a Biosafety Cabinet and Enclosures

For laboratory spray dryers, off-line sterilization of the demountable components in an autoclave or sterilization oven is feasible. The parts can be dismounted, cleaned, sterilized, and then reassembled under sterile conditions inside a Class 100 (ISO 5) isolator system or a biosafety cabinet (Biosafety Level BL-2 or BL-3) (DiBerardinis et al. 2013) capable of maintaining sterility during spray drying (Gladd 2014). However, this requires appropriate laboratory facilities to handle the sterilized components and sterilize the interior of the enclosure (Vehring et al. 2020).

Figure 4.27 shows a photograph of a Mini Spray dryer B-290 installed inside a BioPROTECT® III Jr. biosafety cabinet to minimize the risk of product contamination and prevent laboratory staff exposure to the spray-dried material of low to moderate risk (Arpagaus 2020). The airflow shows how the air is filtered via HEPA filters inside the biosafety hood. Additionally, a glove box insulator can be installed inside the biosafety cabinet for subsequent powder weighing and filling into vials under a nitrogen atmosphere.

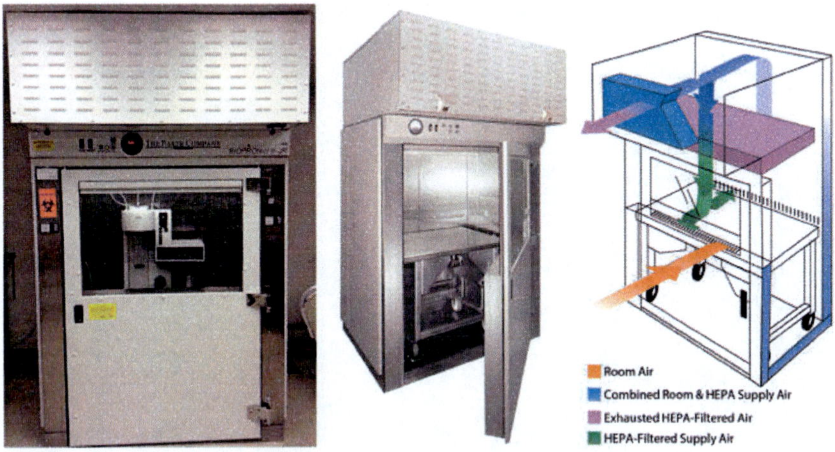

Fig. 4.27 Buchi Mini Spray Dryer B-290 installed in a BioPROTECT® III Jr. containment enclosure (The Baker Company 2010) to protect the operator and the dry powder vaccine from contamination (Images by courtesy of Prof. Pavan Muttil, College of Pharmacy, University of New Mexico, Albuquerque, USA, and kind permission from The Baker Company, USA). With this spray dryer system, among others, research studies on thermostable BCG dry powder vaccine (Price et al. 2020), virus-like particle vaccine against Human Papilloma Virus (HPV) (Tumban et al. 2015; Saboo et al. 2016), and attenuated live bacterial vaccines (Kunda et al. 2016) were performed

This setup was used in research studies for the aseptic preparation on thermostable BCG dry powder vaccine (Price et al. 2020), virus-like particle vaccine against Human Papilloma Virus (HPV) (Tumban et al. 2015; Saboo et al. 2016), and attenuated live bacterial vaccines (Kunda et al. 2016).

Figure 4.28 shows an enclosure from Flow Sciences Inc. (2020) for a Mini Spray Dryer B-290 for API processing. The enclosure effectively contains compounds with Occupational Exposure Limits (OEL) of less than 1 μg/m^3 (Flow Sciences Inc. 2020). In addition, humidity can be controlled by introducing inert gases into the chamber, e.g., nitrogen. Besides, the enclosure provides space for weighing API and dissolving the sample in various solvents, which are then spray-dried.

Polypropylene is used for the frame structure due to its strength and chemical resistance. The size of the enclosure, including the cart with master casters, is 90″ (2.3 m) outer width, approx. 47″ (1.2 m) depth, and 109″ (2.8 m) height. Acrylic glass windows ensure visibility around the enclosure. Glove ports are installed on all four sides to remove and clean parts ergonomically. Bag-in/bag-out dual HEPA fan filter units are mounted on top of the enclosure.

Further features are a tapered sink with a drain to accommodate the first washing of the spray cylinder and other glassware internally, a mini-helic pressure gauge, an integrated velocity alarm, and rubber membrane iris ports for connections and tubing. In addition, a pass-through opening on the side allows the transfer of a 1-L

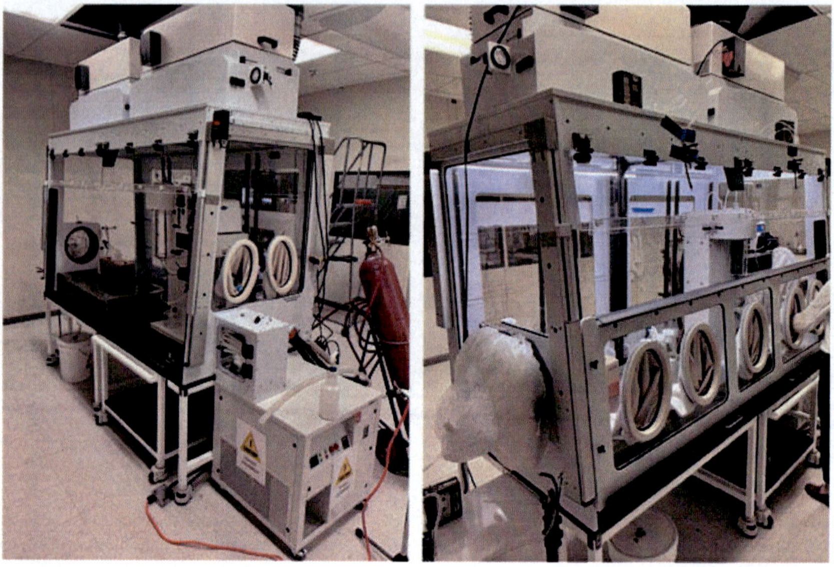

Fig. 4.28 Enclosure for Mini Spray Dryer B-290 for processing active pharmaceutical ingredients (API). The enclosure effectively contains compounds with Occupational Exposure Limits (OEL) of less than 1 μg/m^3. It provides space for weighing API and dissolving the sample in various solvents, which are then spray-dried (Images by courtesy of Flow Sciences Inc., USA)

bottle. An N_2 connection for the spray gas has also been integrated. Finally, a cutout for a cart allows the spray dryer equipment to be rolled in and out to facilitate cleaning and maintenance between runs.

A spray dryer user of such an enclosure would typically go through the following processing steps:

(1) Introducing the prepared feed solution containing the vaccine in a sealed container into the enclosure through the open door pass-through.
(2) Performing the spray drying experiment and collecting the powder in the collection vessel.
(3) Detaching the collection vessel from the cyclone, closing it with a cap, and removing the vessel from the enclosure via the pass-through on the right side.
(4) Dismantling the glassware of the spray dryer and using the mist wand to remove residues.
(5) Removing the glassware out through the continuous liner, wrapping it for autoclaving, and transferring it for sterilization. Replacing the HEPA filters if necessary.

Table 4.5 summarizes a selection of design features for enclosures for laboratory spray dryers. Figure 4.27 shows different variants of enclosures from Flow Sciences Inc. (2020) designed for various API toxicity levels up to complete barrier containment in a Grade A environment (Fig. 4.29).

Table 4.5 Features for designing enclosures for laboratory spray dryers

• Floor-mounted enclosure	• Sink table with drain line for washing glassware
• Enclosure cart with master casters	• Misting wand
• Equipment cart with master casters	• Mating rib
• Polypropylene frame with acrylic panels	• Waste chute
• Removable sliding doors	• Leveling pads
• removable sliding sash for equipment loading	• Iris port for a sink drain line and hose exhaust
• Raisable front doors for equipment loading	• Iris ports for entry/removal of equipment cables
• Oval glove ports and sleeves	• Sanitary fittings (interior and exterior)
• Bag-in/bag-out dual 99.99% HEPA filters	• Quick connect nitrogen fitting
• Minihelic pressure gauges	• Pass-through
• Integrated velocity alarm	• Continuous liner shelf
• LED light switch	• Access door openings

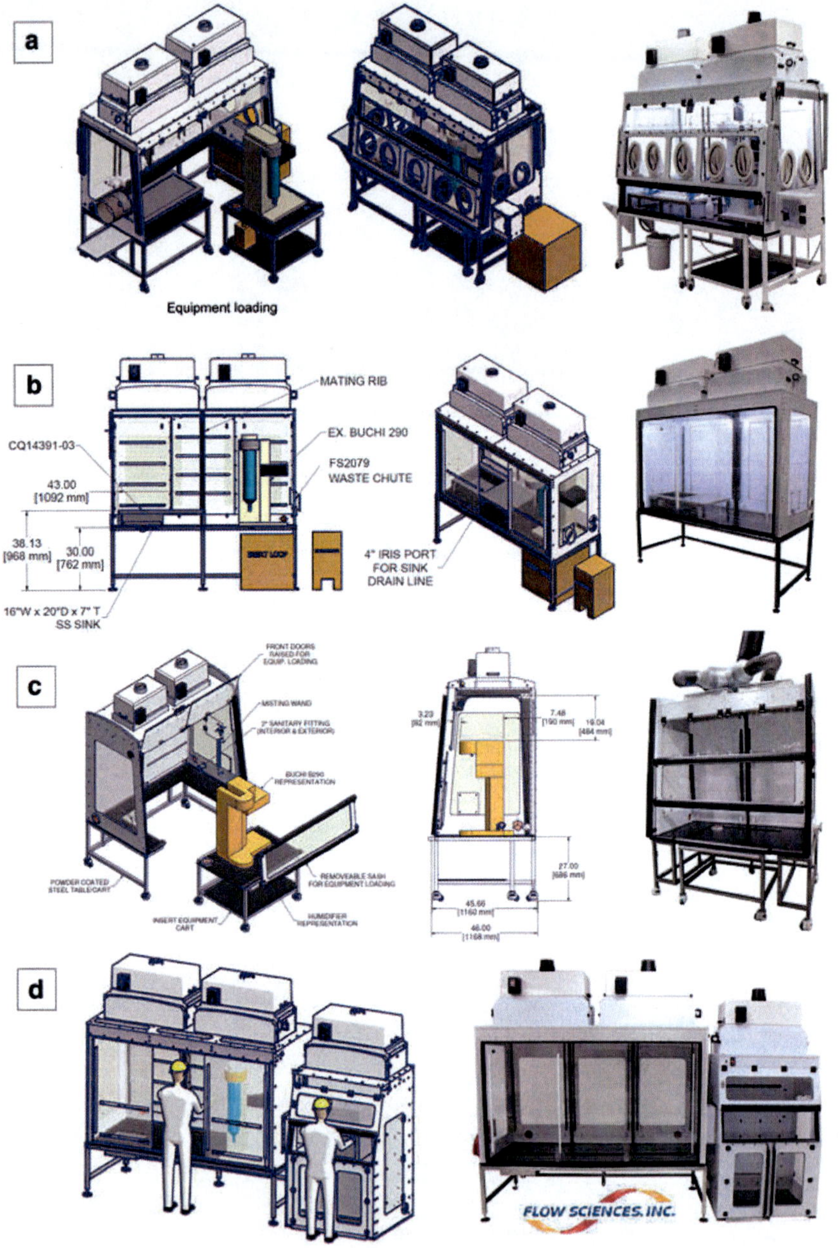
a
Equipment loading
b
MATING RIB
EX. BUCHI 290
FS2079
WASTE CHUTE
CQ14391-03
43.00
[1092 mm]
38.13
[968 mm]
30.00
[762 mm]
16"W x 20"D x 7" T
SS SINK
4" IRIS PORT
FOR SINK
DRAIN LINE
c
FRONT DOORS
RAISED FOR
EQUIP. LOADING
MISTING WAND
2" SANITARY FITTING
(INTERIOR & EXTERIOR)
BUCHI B290
REPRESENTATION
REMOVEABLE SASH
FOR EQUIPMENT LOADING
POWDER COATED
STEEL TABLE/CART
INSERT EQUIPMENT
CART
HUMIDIFIER
REPRESENTATION
3.23
[82 mm]
7.48
[190 mm]
19.04
[484 mm]
27.00
[686 mm]
45.66
[1160 mm]
46.00
[1168 mm]
d
FLOW SCIENCES, INC.

◀ **Fig. 4.29** Enclosures designed by Flow Sciences, Inc. for the Mini Spray Dryer B-290 for API processing. **a** Enclosure made of polypropylene frame with acrylic panels, removable sash with oval glove ports and sleeves, primary filter (bag-in/bag-out), 99.99% HEPA PTFE filters, minihelic pressure gauges, velocity alarm, LED light switch, sink table, thimble connection for house exhaust connections, sanitary fittings, quick connect N_2 fitting, pass-thru to the recovery condenser Inert Loop B-295, continuous liner shelf, and carts for the enclosure and equipment with master casters, **b** containment of polypropylene frame with removable sliding doors (max. one door open), dual bag-in/bag-out 99.99% HEPA filters, integrated alarm, mating rib, waste chute, and iris port for sink drain line, **c** containment of polypropylene frame and acrylic panels, raisable front doors and removable sliding sash for equipment loading, 99.99% HEPA filters, pressure gauges, velocity alarm, sanitary fitting (interior and exterior), misting wand, insert equipment cart, and access door on the side, **d** two enclosures connected together with a pass-thru. The containment on the right has a side-hinged door opening (Images by courtesy of Flow Sciences, Inc., USA)

4.9 Commercial-Scale Spray Dryer Equipment

This section highlights the spray dryer market based on literature research and direct contact with spray dryer manufacturers. Several examples of pilot and commercial scale aseptic spray dryers are presented with operating maps and figures. Finally, an excursus presents two important particle technologies from the UK company Nova Laboratories Ltd. using spray drying and vaccine formulation.

4.9.1 Spray Dryer Equipment Market

According to several market analyses, the global spray drying equipment market was estimated to be approximately US$ 4.3 to 5.1 billion in 2020 and is expected to reach US$ 6.0 to 7.2 billion by 2025, at a CAGR (Compound Annual Growth Rate) of 5.7–7.1%. Figure 4.30 illustrates the market trend.

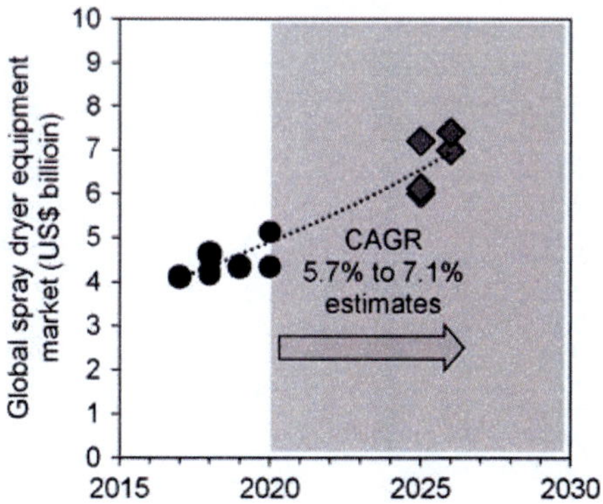

Fig. 4.30 The global spray dryer equipment market size is projected to grow at a CAGR (Compound Annual Growth Rate) of about 5.7–7.1% and reaches about US$ 6.0 to 7.2 billion by 2025 (Data based on various market studies (ReportLinker 2018, 2020, Global View Research 2019, Grand View Research 2019, Reports and Data 2019, Market Data Forecast 2020, Markets and Markets 2020, Million Insights 2020, Verified Market Research 2020)

Food and dairy is the largest application segment for spray drying handling heat-sensitive feedstocks. Growing preference for functional and ready-to-eat foods and focusing on products with improved shelf life drive the growth of the food processing market.

The pharmaceutical spray drying industry is driven by the increasing manufacturing of APIs in the form of tablets, capsules, inhalable drugs, and injectables (ReportLinker 2018) and is expected to grow to well beyond US$ 1 billion by 2025 (Markets and Markets 2020). The impact of the COVID-19 pandemic will probably also affect the growth of the spray drying equipment market for the production of therapeutics and vaccines. However, the mRNA vaccines will most likely be freeze-dried. The ability of spray drying to produce fine, uniform, micron-sized particles and stabilize heat-sensitive vaccines are significant factors that are projected to encourage the usage of spray drying equipment in the pharmaceutical industry (Markets and Markets 2020). Aseptic spray drying is likely to be one of the key focus areas in this sector (ReportLinker 2014).

The spray drying equipment market is highly competitive, with many global and local players. Table 4.6 provides an overview of the leading players in the spray drying equipment market according to various market reports over the past two years.

Table 4.6 Leading spray dryer manufacturers and service providers for pharmaceutical products in Europe, the USA, India, and China

Location of headquarters	Spray dryer manufacturers	Spray drying service providers
Europe	• Buchi Labortechnik AG (Switzerland) • European Spraydry Technologies Ltd (UK) • GEA Group AG (Germany) • G. Larsson Starch Technology AB (Sweden) • Labplant (UK) • ProCepT (Belgium) • SiccaDania A/S (Denmark) • Tetra Pak International S.A. (Switzerland)	• Ajinomoto (Omnichem) (Belgium) • Almac Group (UK) • Coriolis Pharma (Germany) • Evotec (Germany) • Hovione (Portugal) • Idifarma (Spain) • Juniper Pharma Services (Catalent) (UK) • Leukocare AG (Germany) • Micro-Sphere S.A. (Switzerland) • Nova Laboratories (UK) • Siegfried (Switzerland) • Ufag Laboratorien (Switzerland) • Upperton Pharma Solutions (UK) • Vectura (UK) • Xedev (Belgium)

(continued)

Table 4.6 (continued)

Location of headquarters	Spray dryer manufacturers	Spray drying service providers
USA	• Carrier Vibrating Equipment Inc • C. E. Rogers Company • Dedert Corporation • Freund Vector Corporation • SPX FLOW Inc • Swenson Technology Inc • Techni Process Group • Yamato Scientific America Inc	• AMRI Global • Aridis Pharmaceuticals Inc • Avara Pharmaceutical Services Inc • Catalent Pharma Solutions • Formex LLC (Biotech Investment Group) • Formurex Inc • Lonza (Capsugel / Bend Research) • North Star Processing LLC • Norwich Pharma Services • Patheon • Pulse Combustion Systems • Richman Chemical Inc • Spray-Tek Inc • Summit Custom Spray Drying
India	• Acmefil Engineering Systems Pvt. Ltd • Advanced Drying Systems • Hemraj Enterprise • Labultima Process Technologies Pvt. Ltd • New AVM Systech Pvt. Ltd • Raj Process Equipments and Systems Pvt. Ltd • Saka Engineering Systems Pvt. Ltd	No information available
China	• Changzhou Lemar Drying Engineering Co., Ltd • Changzhou Jinqiao Spray Drying and Engineering Co., Ltd • Shandong Tianli Drying Technology & Equipment Co., Ltd • Shandong Shungeng Drying Equipment Co., Ltd • Wuxi Shuangrui Machinery Co., Ltd	No information available

The companies are listed in alphabetical order. Information is based on various market reports (ReportLinker 2018, 2020; Global View Research 2019; Grand View Research 2019; Reports and Data 2019; Market Data Forecast 2020; Markets and Markets 2020; Million Insights 2020; Verified Market Research 2020)

Over 20 manufacturers have developed more than 80 spray dryer models for use in the pharmaceutical industry (ReportLinker 2018). These models offer variation in the scale of operation (i.e., laboratory, pilot, and industrial-scale), evaporation rate, inlet gas temperature, and other technical parameters.

Over 50 companies currently offer spray drying services and contract manufacturing for pharmaceutical products (ReportLinker 2014, 2018). Notable service providers based in Europe and the USA are listed in alphabetical order in Table 4.6. Unfortunately, no information is available from India and China. Many CDMOs (Contract Development and Manufacturing Organization) have installed spray dryers developed by GEA Niro™ (Denmark) in their facilities. The product range of GEA Niro includes several spray dryers for pharmaceutical application from pilot to production scale, the so-called Pharma-SD® (PSD™ types) with nominal drying gas rates ranging from 100 to 5000 kg/h (GEA Process Engineering A/S 2020a). Two examples of PSD-1 and PSD-4 installations with powder collection in a low-bioburden cleanroom are shown in Fig. 4.11. The PSD and Mobile Minor™ models are most commonly used (GEA Process Engineering A/S 2020a).

Besides, laboratory-scale spray dryers developed by Büchi Labortechnik AG (Switzerland) (Mini Spray Dryer B-290™ and Nano Spray Dryer B-90™) are used by many service providers and research institutes. Furthermore, Anhydro™ MicraSpray dryers offered by SPX FLOW, Inc. (US) and 4M8-TriX™ by ProCepT (Belgium) are also gaining popularity.

According to ReportLinker (2018), the pharmaceutical spray drying services market is expected to grow at an annualized rate of around 20% between 2018 and 2028. However, not many spray dryer manufacturers offer aseptic spray dryers or CDMOs providing aseptic spray drying services. Furthermore, with the COVID-19 pandemic and the increasing number of therapeutic drugs and vaccines, this market is expected to develop.

4.9.2 Aseptic Spray Drying Equipment and Service Providers

This section will focus on the most well-known manufacturers and CDMOs in the aseptic spray drying market, including GEA Process Engineering A/S (Denmark), SPX FLOW Inc. (USA), Nova Laboratories Ltd. (UK), Hovione (Portugal), and Upperton Pharma Solutions (UK). However, many other CDMOs on the market, as shown in Table 4.6, have experience with spray drying biopharmaceuticals at low bioburden.

As mentioned before, the market for aseptic spray drying equipment is led by GEA Process Engineering A/S and SPX FLOW Inc., which provide hardware and support for GMP and aseptic installations across a wide range of processing scales (GEA Niro 2017; SPX FLOW Inc. 2017).

The GEA Niro AsepticSD™ spray dryers are designed to produce sterile pharmaceutical products such as vaccines, antibiotics, and other products requiring aseptic conditions (GEA Niro 2017). The processing scales of AsepticSD™ spray dryers in terms of capacity and batch size are based on the PharmaSD™ models, but with particular attention to aseptic design, additional HEPA filters, hot gas sterilization, and so on. Indeed, aseptic spray dryers are, however, customized products. Besides, for small GMP batch processing of 100 to 200 mL samples, the GEA Niro SDMicro™ spray dryer is available (GEA 2013), which can also be set up to run continuously (Scrafford 2020).

SPX FLOW Inc. (USA) offers the Anhydro MicraSpray aseptic series of spray dryers in various scales for pharmaceutical products that require sterile processing. The aseptic spray dryers are based on the same equipment standard as the Anhydro MicraSpray A-Class series, combined with several special features to ensure a truly aseptic process, such as mirror-polished surfaces and comprehensively documented qualification services. The aseptic spray dryers are designed for peptides, proteins, vaccines, and other biopharmaceutical products requiring aseptic processes like live cells, inhalable powders, and antibiotics (SPX FLOW Inc. 2017; Geng 2018). Figure 4.31 shows actual SPX Anhydro MicraSpray MS 1250, MS 150, and MS 35 aseptic dryers and suggests scale-up capability. Table 4.7 summarizes the available Anhydro MicraSpray aseptic dryers series with nominal drying gas rates of 35 to 2500 kg/h. Figure 4.32 (top) shows the water evaporation rates at 70 °C drying gas outlet temperature. For example, the water evaporation capacity of a model MS 1250 is approximately 48 times greater than that of a type MS 35.

Back in 2010, SPX Anhydro supplied an aseptic spray dryer to the Aeras Global TB Vaccine Foundation (2018 acquired by the International AIDS Vaccine Initiative, IAVI) for aseptic processing of Bacille Calmette-Guérin (BCG) tuberculosis vaccine and pulmonary administration (Pharmaceutical Technology 2010). Anhydro enabled this spray dryer to accept biosafety level BL-2 hazard infectious materials (WHO 2004; CDC/NIH 2020). The spray-dried powder was fast and cheap to make, and it was stable at room temperature, unlike most traditional vaccines that require cold storage. It was stated that, in principle, the developed aseptic spray drying technology could make vaccines against any disease, not just tuberculosis (Pharmaceutical Technology 2010).

The size of an aseptic spray dryer is best described by the drying gas flow rate that the plant can handle. Figure 4.32 (bottom) shows the water evaporation rate of the PSD-1 to PSD-6 spray dryers (GEA Process Engineering A/S 2020a). The curves also apply to the AsepticSD™ spray dryers' models, as they are based on the same systems. The lines show the water evaporation rate as a function of the inlet drying gas temperature at a constant outlet drying gas temperature of 90 °C. Therefore, the curves can be used to estimate the powder production rate. For example, at 200 °C inlet and 90 °C outlet temperature, a PSD-4 evaporates about 40 kg of water per hour, which results in roughly 10 kg/h of powder when spray drying a 20 wt% feed solution. Similarly, the water evaporation rate for the Anhydro MicraSpray dryers can be roughly estimated.

Nova Laboratories Ltd. (UK) is one of the leading independent contract manufacturers producing spray-dried vaccines and proteins and has built a global reputation for aseptic spray drying of sterile injectable products for human use (Nova Laboratories 2021).

The aseptic spray drying facility was developed in association with GEA (Roser 2005a, 2009) and is based on a GEA Niro AsepticSD™ size ASD-1, capable of producing batches of 5000 to 10,000 doses (GEA 2021; Nova Laboratories 2021). The plant was installed within a sterile Class 100 isolator (Grade A) (Roser 2009), which was located in a separate Class 100,000 room (Grade D) to fully quarantine

Fig. 4.31 Aseptic spray drying systems from SPX FLOW Inc. (US) showing the scale-up capability. **a** Commerical-scale Anhydro MicraSpray 1250 aseptic dryer, **b** Anhydro MicraSpray 150 aseptic dryer (SPX FLOW Inc. 2017). **c** Anhydro MicraSpray 35 aseptic dryer (Geng 2018), (Images by courtesy of SPX FLOW, Inc., USA)

Table 4.7 Anhydro MicraSpray aseptic dryer series with information on the configurations and main process parameters (Data based on the product datasheets published by SPX FLOW Inc., USA)

MicraSpray dryer model	MS 35	MS 75	MS 150	MS 400	MS 750	MS 1250	MS 2500
Aseptic configurations	Inert gas single pass			Atmospheric, inert gas single pass, closed cycle			
Max. inlet air temperature (Tin) (°C)	200	200	350	325	350	325	325
Max. water evaporation capacity (kg/h) (at outlet temperature 70 °C)	1.6	3	14	37	77	115	463
Nominal drying gas rate (kg/h)	35	75	150	400	750	1250	2500
Drying chamber diameter (m)	0.3	0.6	0.9	1.15	1.6	2.15	2.7
Estimated particle size (μm) (Geng 2018)	5 to 20	10 to 30	20 to 50	40 to 80	60 to 90	80 to 100	100 to 150

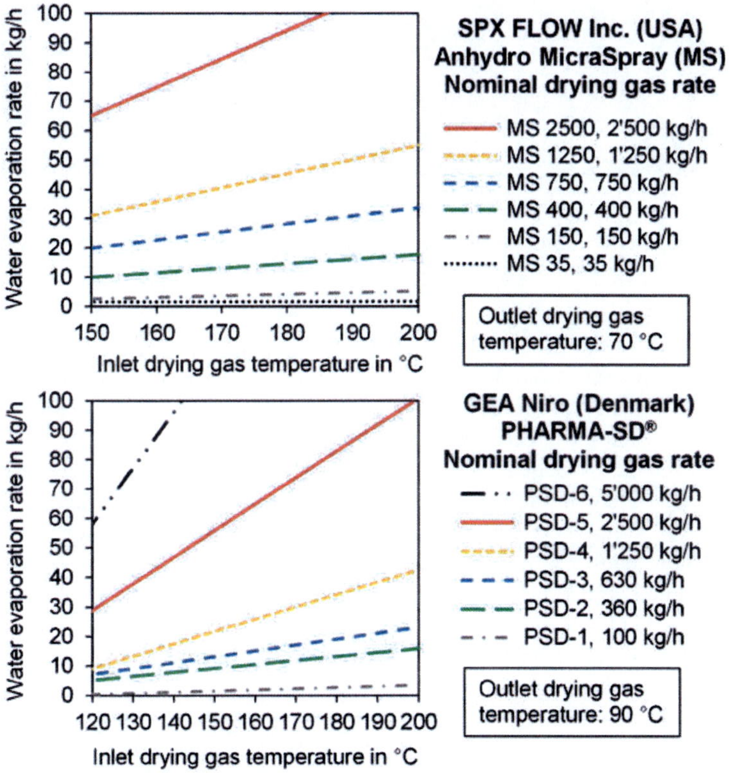

Fig. 4.32 Water evaporation rate of different commercial-scale spray dryers. Top: Anhydro MicraSpray dryers MS 35 to MS 2500 from SPX FLOW Inc. (USA) (Data from available product datasheets from the website: www.spxflow.com/anhydro/products/micraspray-dryers). Bottom: PHARMA-SD™ PSD-1 to PSD-6 spray dryers from GEA Niro (Data from GEA Process Engineering A/S (2020a)

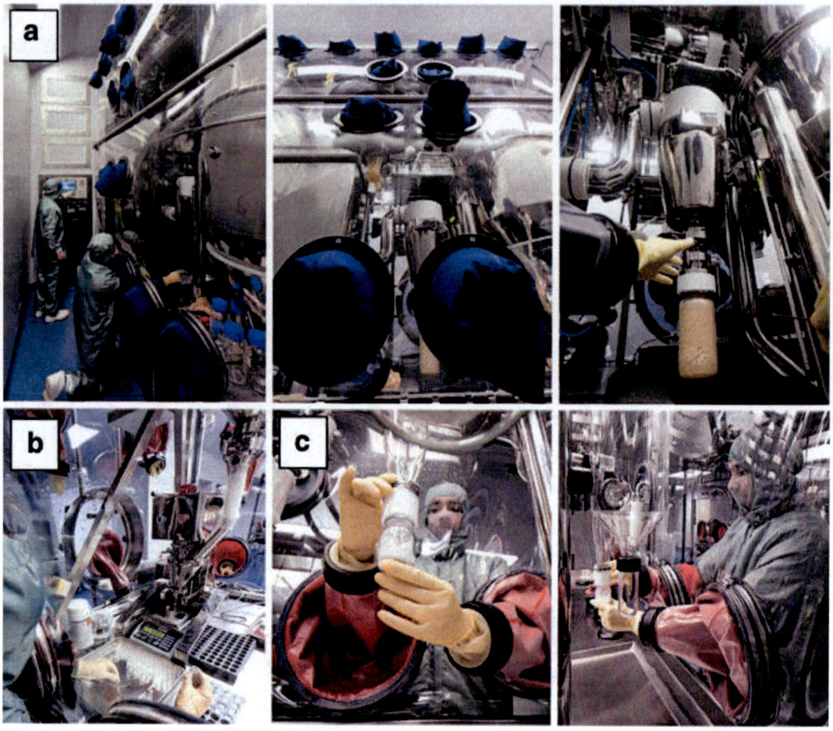

Fig. 4.33 **a** Aseptic spray dryer on commercial-scale with powder collection. **b** Aseptic powder filling within a sterile and low-humidity environment for manual filling into vials. **c** Small-scale spray dryer in enclosure (Images by courtesy of Nova Laboratories Ltd., Leicester, UK)

vaccine manufacture from all other pharmaceutical products in the facility (Roser 2005a). Pharmaceutical grade nitrogen is used as drying and atomization gas.

CIP is obtained using conventional spray ball equipment and standard pharmaceutical cleaning protocols. SIP between runs is achieved by attaining temperatures of 180 °C by superheating the drying nitrogen gas throughout the drying chambers and maintaining the temperatures for over 6 h to ensure sterility (Roser 2005a, 2009). This has required ancillary heating in certain sites and a considerable increase in insulation. In addition, the smaller demountable parts are designed to be autoclaved and remounted under sterile conditions.

Figure 4.33 illustrates an aseptic spray drying system with powder collection and filling into vials within a sterile environment at Nova Laboratories Ltd. The spray drying facility has undergone numerous full-scale media and process validation cycles and is available for contract manufacturing (Nova Laboratories 2021). Pharmaceutical biologics can be aseptically sprayed continuously for up to 5 days to manufacture large kilogram quantities of power (De Costa 2014).

An example of Nova Labortories' experience in scaling up an aseptic spray drying manufacturing process is Raplixa™, developed by ProFibrix BV, a subsidiary of The Medicines Company based in The Netherlands. Raplixa™ is the first aseptically spray-dried fibrin sealant powder to control bleeding in adults during surgery and was approved by the FDA and EMA in 2015. Commercial supplies of Raplixa™ sealing powder are manufactured at the sterile manufacturing facilities of Nova Laboratories Ltd. (Manufacturing Chemist 2015). Raplixa™ is a mixture of thrombin and fibrinogen that are individually spray-dried with trehalose as a stabilizer. The resulting free-flowing microparticles are then aseptically mixed and filled into a vial under sterile conditions (McKeage 2015).

An example of a professional CDMO (Contract Development and Manufacturing Organization) provider is Upperton Pharma Solutions, based in Nottingham in the UK. Founded in 1999, Upperton offers a complete CDMO service with cGMP spray drying from the early development phase through scale-up to clinical trials with all the necessary documentation. Formulation development covers many applications, including peptides, proteins, monoclonal antibodies (mAbs), vitamins, biologics, and vaccines. More and more customers see spray drying as a more cost-effective alternative to freeze drying with the added benefits of speed, scalability, as well as particle engineering capabilities.

As shown in Fig. 4.34, the spray dryer equipment includes numerous established spray dryers like GEA Niro PSD-1, GEA Niro SD Micro, ProCepT 4M8-TriX, and several Büchi Mini Spray Dryer B-290 models with associated powder handling and dosage form development capabilities. The focus is on dry powder formulations for capsules, tablets, and pulmonary and nasal devices. The range of spray dryer systems covers batch sizes from a few milligrams to several kilograms of powder.

Upperton has worked on many projects for biologics and therefore has significant experience in this area. Some of the results are also published, such as a recent study on novel GMP manufacturing processes for obtaining thermostable HIV-1 gp41 virosomes under solid forms for various mucosal vaccination routes (Amacker et al. 2020).

As a leading CDMO with over 60 years of experience in pharmaceutical spray drying, Hovione provides services from drug development to manufacturing intermediates and drug products. In the area of inhalation, Hovione offers a full range of services, including formulation, particle engineering, spray drying, powder handling, and capsule and blister filling for various disposable and reusable dry powder inhalers (Fernandes and Villax 2019; HARRO 2020b; Hovione 2021a, b). The cGMP spray drying facilities include laboratory-, pilot-, and commercial-scale systems at four FDA-inspected manufacturing sites in the USA (New Jersey), China (Macau), Ireland (Cork), and Portugal (Loures), as well as R&D centers in Portugal (Lumiar) and the USA (New Jersey). The production capacity of the Irish spray drying facilities is up to 300 tons/year of high potency APIs. As announced in 2018, Hovione is investing in spray drying technology for pharmaceutical applications over the next years, expanding capacity at its sites in Portugal, Ireland, and New Jersey. Figure 4.35 illustrates a laboratory-scale Büchi Mini Spray Dryer

Fig. 4.34 Spray dryer equipment of a professional CDMO (Contract Development and Manufacturing Organization) providing comprehensive services from proof-of-concept studies in the laboratory through drug development and manufacturing for clinical trials. **a** GEA Niro PharmaSD™ PSD-1 spray dryer, **b** GEA Niro SDMicro spray dryer, **c** Büchi Mini Spray Dryer B-290 Advanced, **d** ProCepT 4M8-TriX spray dryer (Images by courtesy of Upperton Pharma Solutions, UK)

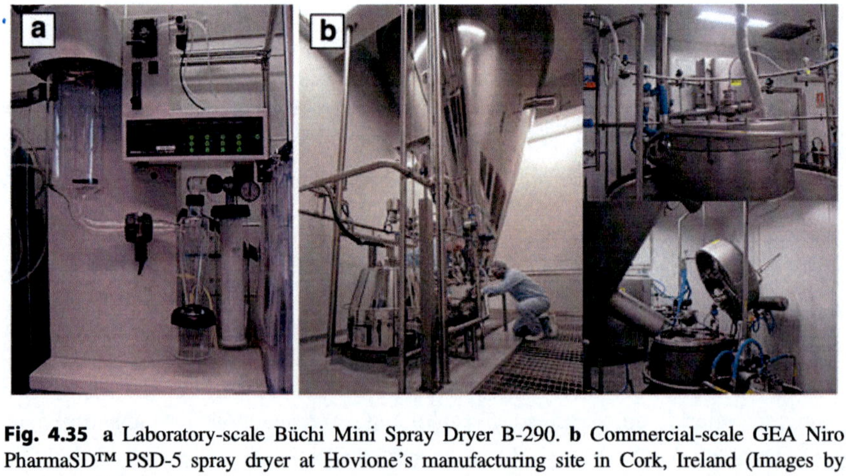

Fig. 4.35 **a** Laboratory-scale Büchi Mini Spray Dryer B-290. **b** Commercial-scale GEA Niro PharmaSD™ PSD-5 spray dryer at Hovione's manufacturing site in Cork, Ireland (Images by courtesy of Hovione)

B-290 Advanced and a commercial scale GEA Niro PharmaSD™ PSD-5 spray dryer at Hovione's manufacturing site in Cork, Ireland.

Many competencies have been built in robust scale-up methodologies based on scientific principles and practical experience. The methodology has been tested in numerous spray-dried formulations at all production scales. The findings on scale-up methods for pharmaceutical spray drying and Quality by Design (QbD) approaches for inhalation formulations have partially been published (Gil et al. 2010; Miller and Gil 2012; Costa et al. 2014; Gaspar et al. 2014). Besides, Hovione is also involved in organizing training, workshops, and seminars on spray drying (ECA Academy 2020, 2021).

4.9.3 Excursus: Particle Engineering Technologies from Nova Laboratories Ltd.

In this excursus, the particle technologies VitRIS™ (Vitrified Readily Injectable Suspension) and HydRIS™ (Hypodermic Rehydration Injection System) (Manufacturing Chemist 2010) from Nova Laboratories Ltd, are presented, which are based on spray drying and used for vaccine formulation. The two technologies were acquired in 2010 from Cambridge Biostability Ltd, a UK developer of temperature-stable liquid vaccines, including intellectual property.

The VitRIS™ platform uses aseptic spray drying to stabilize vaccines and other pharmaceutical products in water-soluble amorphous glass microspheres (Tg well over 70 °C) (Roser 2005a, b) of about 3 to 20 μm diameter (Nova Laboratories 2021). In a second step, the dried powder containing the stabilized vaccine is suspended in an inert anhydrous liquid, such as medically approved high-density fluorocarbons, hydrofluoroethers, or low-density metabolizable oils (Roser 2005b, 2006, 2009; Manufacturing Chemist 2013; De Costa 2015). The result is a homogeneous, thermostable, density-matched, and ready-to-inject suspension that can be stored and transported at ambient temperature without losing its efficacy.

Since the stable liquid formulations are anhydrous, they are inherently bacteriostatic, as bacteria require water to multiply. Thus, the need for preservatives becomes redundant (Das 2004; Roser 2006). The vaccine suspensions can be injected immediately without reconstitution. When injected, the glass microspheres dissolve in body fluids releasing the potent vaccine, and the anhydrous liquids are exhaled in the breath or rapidly metabolized (Roser 2006). The sugar microspheres can also be engineered to dissolve more slowly, thereby releasing vaccines over time (Roser 2005a). This would remove the need for vaccine boosters (BIONITY 2004; Das 2004; AskNature 2017).

Trials have shown that the VitRIS™ technology can be applied successfully to vaccines, insulin, monoclonal antibodies (mAbs), recombinant growth hormones, proteins, enzymes, and nucleic acids (Manufacturing Chemist 2010; De Costa 2014, 2015; AskNature 2017; Nova Laboratories 2021).

In a further development, Nova Laboratories Ltd. has acquired the expertise to engineer hollow microspheres (Roser and Moos 2006), known as aerospheres, by adding blowing agents (e.g., ammonium bicarbonate) during spray drying to create a thin shell wall (De Costa 2015). The aerospheres have a larger surface area and therefore dissolve up to a factor of 10 faster than conventionally spray-dried powder (De Costa 2014).

When drying the sprayed droplets, the blowing agent decomposes into biocompatible gas bubbles that trap inside the dried glassy solid powders. As a result, the degree of inflation of the aerospheres and their density is controllable within tight limits by adjusting the spray drying conditions (e.g., drying temperature, feed rate) and the quantity of blowing agent used (Roser and Moos 2006).

Figure 4.36a shows spray-dried glass microspheres suspended in Crodamol™ GTCC oil (caprylic/capric triglycerides), used as an alternative for mineral or vegetable oil in personal care and pharmaceutical applications. The spray-dried solid microspheres sink in the oil (0, left) without blowing agent, while the hollow microspheres spray-dried with above 0.3 molar blowing agent float on oil. The controlled density-match is achieved at around 0.2 mol preventing the powder from floating or sinking (Roser 2005a, 2006). Figure 4.36b, c illustrates the microspheres by confocal fluorescence microscopy, while solid glass microspheres were produced by spray drying without blowing agent and hollow spheres with a single central gas bubble by the decomposition of the blowing agent (Roser 2006).

A further benefit while in suspension is that the sugar particles prevent the interaction of vaccines before injection. Therefore, multiple or multivalent vaccines can also be produced using the VitRIS™ technology and given in the same shot, making delivery faster and easier (BIONITY 2004; Das 2004; AskNature 2017). In addition, this allows customization of the polyvalent mixture to local conditions by individual countries or even local health authorities (Roser 2006).

In 2004, Panacea Biotec Ltd., one of India's leading biotechnology companies, announced that it would start clinical trials in India with a pentavalent vaccine against the childhood diseases diphtheria, tetanus pertussis, Haemophilus influenzae type b, and hepatitis B (Das 2004). One year later, Panacea Biotec Ltd. launched Easyfive-TT® (diphtheria, pertussis, tetanus, hepatitis B, and Hib) as the world's first fully liquid whole-cell pertussis (wP) based pentavalent vaccine in the Indian market. With WHO prequalification in 2008, more than 100 million doses have been delivered to more than 50 countries worldwide (Panacea Biotec 2019a). According to the package insert (Panacea Biotec 2019b), EasyfiveTT® is a uniform liquid suspension with a shelf-life of 3 years when stored at 5 ± 3 °C.

However, the author of this book lacks information on which formulation technology was used to produce the particle suspension. Presumably, the development of a spray-dried formulation has not yet reached this stage but could be an alternative option.

The other key technology, HydRIS™, enables coating of a mixture of glass-forming sugars with a dispersed vaccine onto a filter paper-like membrane consisting of an inert fiber matrix, which is then air-dried and enclosed in a plastic casing with ports at either end for a needle and syringe (De Costa et al. 2006; De Costa 2015).

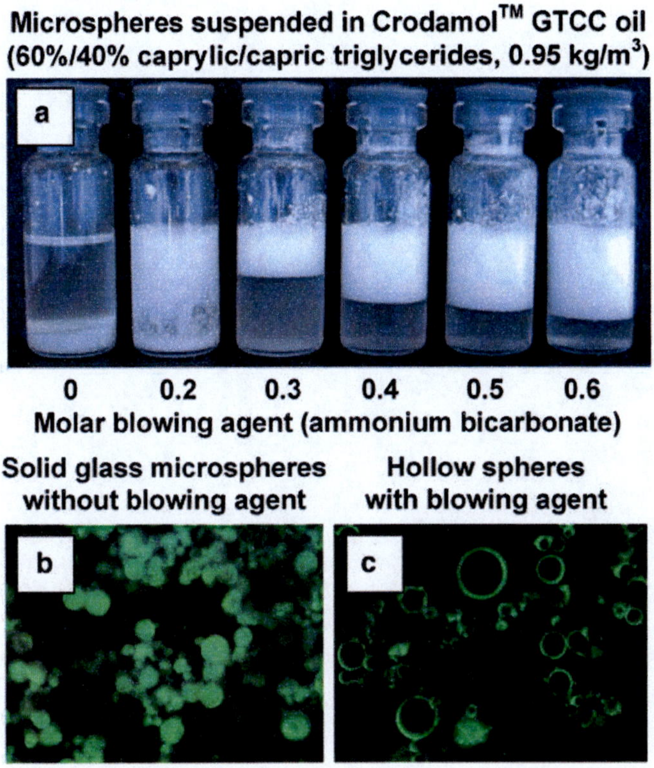

Fig. 4.36 **a** Suspended microspheres in Crodamol GTCC oil (60%/40% caprylic/capric triglycerides, 0.95 kg/m^3). Solid microspheres spray-dried without blowing agent sink in oil (0, left). Hollow microspheres spray-dried with above 0.3 molar blowing agent float on oil, and the volume of particles increase with the agent's molar concentration. The density-match is achieved at around 0.2 M. **b** Confocal fluorescence micrograph of spray-dried solid glass microspheres without blowing agent (about 2 μm to 15 μm in size). **c** Spray-dried hollow microspheres with the blowing agent (about 20 μm) (Images from Roser (2006) with permission from Future Medicine Ltd.)

Figure 4.37a shows an SEM picture of trehalose-sucrose glass dried on a fiber membrane forming a thin layer of web-like structures between and around the individual fibers (Alcock et al. 2010). The air-dried sugar glass may also contain spray-dried particles with an encapsulated vaccine. The membrane thus provides a high surface-to-volume ratio for rapid air drying and a support matrix for the resulting glass and access for buffer during reconstitution.

Figure 4.37b illustrates the concept of the HydRIS™ delivery device (Alcock et al. 2010). After drying of sugar glass containing the vaccine onto the support membrane (1), it is placed between the lower (2) and upper (3) parts of an injection-molded medical-grade polypropylene housing, which are then welded together ultrasonically under sterile conditions. When required for use, the completed HydRIS™ delivery device (4) is attached to a conventional syringe and

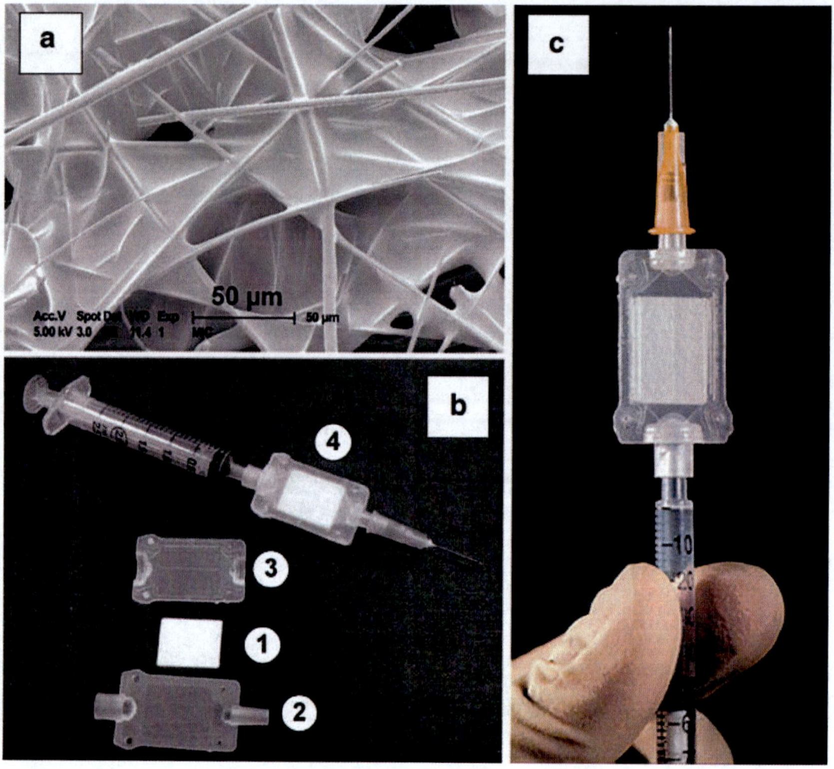

Fig. 4.37 **a** SEM picture of trehalose-sucrose glass dried and stabilized on a glass fiber membrane. Virus vaccine formulated in 0.5 M trehalose and sucrose was pipetted onto the membrane and air-dried at ambient temperatures (20–25 °C, 2% to 10% RH), **b** concept of the HydRIS® delivery device: After drying of sugar glass containing vaccine suspension onto the support fiber membrane (1), it is placed between lower (2) and upper (3) parts of the injection-molded medical-grade polypropylene housing, which are then welded together ultrasonically under sterile conditions. The completed device (4) is attached to a conventional syringe and needle, (Images A and B adapted from Alcock et al. (2010) with permission from The American Association for the Advancement of Science). **c** HydRIS® delivery device ready-to-inject a vaccine (Image reproduced from De Costa (2014) with kind permission from Nova Laboratories Ltd., Leicester, UK)

needle (Fig. 4.37c). The buffer fluid secured in the syringe floods the device, and the dried vaccine-containing sugar glass on the membrane is almost instantly redissolved, and the vaccine flows into the needle for administration.

Scientists at Oxford University have shown that it is possible to dry and stabilize human adenovirus type 5 (AdHu5) and live modified vaccinia virus Ankara (MVA) by formulation with sucrose and trehalose on glass fiber membranes of the HydRIS® delivery device (Alcock et al. 2010). Complete retention of titer and immunogenicity after storage at up to 45 °C for up to 6 months was observed in a mice model. In addition, long-term stability data showed stability of over 12 months at 37 °C (Nova Laboratories 2021).

Overall, the HydRIS™ platform has proven highly successful in stabilizing a broad range of conventional vaccines, labile products, and highly sensitive live viruses, facilitating distribution and storage at room temperature. However, evaluation of whether and to what extent commercialization of VitRIS™ and HydRIS™ technologies for use in human clinical trials has progressed would need to be discussed directly with the manufacturing company.

4.10 Summary

This chapter covered the key aspects of design, construction, components, manufacturers, and CDMOs for aseptic spray dryers. These topics were explicitly discussed based on in-depth information from various literature. In addition, numerous schematics, photographs, and diagrams visualize the respective subjects to create a better understanding.

Historically, the concept of aseptic spray drying has been known for almost 50 years, and the technology is successfully used for pharmaceutical applications, such as peptides, proteins, vaccines, antibiotics, hormones, allergens, and other biopharmaceutical products requiring aseptic processes.

However, it took until about 2010 for the first FDA-approved aseptically spray-dried vaccine to be disseminated. To the best of the author's knowledge, there is no spray-dried vaccine available on the market today. Potential barriers include the complexity of aseptic equipment, regulatory acceptance, and the conservative pharmaceutical industry. What would help further dissemination is certainly better knowledge and technology transfer (KTT) from the research results to commercialization, disclosure of patents, and more qualified engineers to achieve further process improvements.

For sure, the construction and design of aseptic spray dryers require the most advanced and customized equipment designs within the biopharmaceutical spray drying industry. Compared to a pharmaceutical spray dryer, aseptic spray dryers are not standard equipment, but custom-made products include:

- temperature resistant HEPA filters for the process gas with bag-in/bag-out replacement principle,
- sterile filters for the feed liquids based on resistant membranes with 0.22 μm pore size,
- cleaning-in-place (CIP) systems and cleaning procedures with detergents, acids, alkaline solutions, and purified water,
- sterilization-in-place (SIP) systems and pressure-monitored sterilization cycles with saturated steam of 121 °C for about 30 min, and
- polished internal steel surfaces and short pipes.

A major take out of this chapter is that some selected aseptic spray dryer series are available on the market. The drying gas flow rate extends from about 50 to 2500 kg/h, allowing water evaporation of up to 100 kg/h. A market analysis has shown that there are only a few manufacturers of aseptic spray dryers. SPX Anhydro and GEA Niro play a pioneering role and have experience with vaccine drying. There are several service providers and CDMOs, whereas Nova Laboratories is particularly worth mentioning the field of vaccines and Hovione, with a lot of experience with the scale-up of spray drying technology for inhalable drugs. There are other professional suppliers and service providers, but their efforts are less known to the public.

In addition to aseptic spray drying, aseptic powder transfer from the spray dryer to a powder filling line with sterile containers and airlock systems is also critical, as feasibility depends significantly on being able to fill the spray-dried powder into vials, blisters, capsules, syringes, inhalers, or other medical containers.

Several aseptic powder filling systems are available on the market and allow for high throughputs in the milligram range with robotics, helping the breakthrough of aseptic spray drying. In addition, the increasing development of precise powder filling systems opens up the possibility of flexible dosing for a specific batch of vaccine powder into variable containers.

A handful of spray dryer models in the laboratory sector are used for vaccine research studies. Leading manufacturers are Büchi Labortechnik AG, GEA Niro, SPX Anhydro, and ProCepT. There are also possibilities to work with vaccines under low bioburden conditions in the laboratory. Various options for safe operation in biosafety cabinets and enclosures to protect the operator and the environment from the highly potent vaccines were highlighted.

Aseptic spray drying is expected to expand its position in the vaccine industry further and gain market share in the coming years as an alternative to traditional freeze drying. The COVID-19 pandemic is also contributing to the development of new thermostable vaccines. Today, freeze drying is the only method used to produce dry vaccines. However, it is a time-consuming batch process and a technique with high energy consumption. Aseptic spray drying is a viable alternative as a continuous production process for some dry vaccine products, especially when particle engineering capabilities are needed with product quality and for high volume products. The potential of stabilization in dry powder form has not yet been exhausted. Dry powder offers storage advantages and reduces dependence on the cold chain. Particle technology also provides new opportunities for pulmonary and nasal delivery.

Finally, it is highly expected that biopharmaceutical companies, spray dryer manufacturers, powder fillers, service providers, researchers, and other market players will join forces to improve solutions further and invest in aseptic spray dryer technologies.

References

3P Innovation (2020) Fill2Weight: aseptic powder filling. Product brochure. https://www.3pinnovation.com/machines/powder-filling/fill2weight

Ajmera A, Scherließ R (2014) Stabilisation of proteins via mixtures of amino acids during spray drying. Int J Pharm 463:98–107. https://doi.org/10.1016/j.ijpharm.2014.01.002

Alcock R, Cottingham MG, Rollier CS et al (2010) Long-term thermostabilization of live poxviral and adenoviral vaccine vectors at supraphysiological temperatures in carbohydrate glass. Sci Transl Med 2:19ra12-19ra12. https://doi.org/10.1126/scitranslmed.3000490

Al-Khattawi A, Bayly A, Phillips A, Wilson D (2018) The design and scale-up of spray dried particle delivery systems. Expert Opin Drug Deliv 15:47–63. https://doi.org/10.1080/17425247.2017.1321634

Altrichter J, Kemter K, Ströver R, Scholz M (2019) Exploring amino acid-based and stable spray-dried vaccinations. EBR—Eur Biopharm Rev. https://en.calameo.com/books/0061133855dbf0ab162d7

Amacker M, Smardon C, Mason L et al (2020) New GMP manufacturing processes to obtain thermostable HIV-1 gp41 virosomes under solid forms for various mucosal vaccination routes. NPJ Vaccines 5:41. https://doi.org/10.1038/s41541-020-0190-9

Andersen SK (2016) Challenges in spray drying process and formulation development. In: Lyophilisation Europe, 5 July, 2016, London, UK, pp 1–31

Andersen SK (2020) New drying processes for biopharmaceuticals. In: Bright, 5 May, 2020. https://www.brighttalk.com/webcast/10519/373808/new-drying-processes-for-biopharmaceuticals

Angelo R, Rousseau K, Grant M et al (2009) Technosphere® insulin: defining the role of technosphere particles at the cellular level. J Diabetes Sci Technol 3:545–554. https://doi.org/10.1177/193229680900300320

Armenante PM, Kirpekar AC (1997) Chapter 11: Sterilization in the pharmaceutical and biotechnology industry. In: Handbook of downstream processing. Chapman & Hall, pp 261–308

Arpagaus C (2018) Pharmaceutical particle engineering via nano spray drying—process parameters and application examples on the laboratory-scale. Int J Med Nano Res 5. https://doi.org/10.23937/2378-3664.1410026

Arpagaus C (2019) Production of food bioactive-loaded nanoparticles by nano spray drying. Nanoencapsulation Food Ingredients by Spec Equip:151–211. https://doi.org/10.1016/B978-0-12-815671-1.00004-4

Arpagaus C (2020) Spray drying of vaccines: a review. In: IDS 2020, 22nd International drying symposium Worcester, Massachusetts, USA, June 28–July 1, 2020, pp 1–14

Arpagaus C, Schwartzbach H (2008) Scale-up from the Büchi Mini spray dryer B-290 to the Niro mobile minor. best@buchi Information Bulletin, Number 52/2008

Arpagaus C, Friess W, Schmid K (2009) Nano spray drying in laboratory scale. GIT Lab J 11–12:30–31

Arpagaus C, Schafroth N, Meuri M (2010) Laboratory scale spray drying of lactose: a review. best@buchi Information Bulletin, Number 57/2010

Arpagaus C, John P, Collenberg A, Rütti D (2017) Nanocapsules formation by nano spray drying. In: Jafari SM (ed) Nanoencapsulation technologies for the food and nutraceutical industries. Elsevier, pp 346–401

Arpagaus C, Collenberg A, Rütti D et al (2018) Nano spray drying for encapsulation of pharmaceuticals. Int J Pharm 546:194–214. https://doi.org/10.1016/j.ijpharm.2018.05.037

Artaud C, Kara L, Launay O (2019) Vaccine development: from preclinical studies to phase 1/2 clinical trials. In: Methods in molecular biology, pp 165–176

AskNature (2017) VitRIS and HydRIS vaccine stabilization technologies. Nova Laboratories Ltd, 6 Feb, 2017. www.asknature.org. https://asknature.org/idea/vitris-and-hydris-vaccine-stabilization-technologies/

ASME (2019) Bioprocessing equipment, ASME BPE-2019, The American Society of Mechanical Engineers, ISBN 978-0-7918-7225-3, pp 1–380

ASTM (2020) ASTM F838-20, Standard test method for determining bacterial retention of membrane filters utilized for liquid filtration. ASTM International, West Conshohocken, PA, 2020, American Society of Testing Materials, www.astm.org. https://www.astm.org/Standards/F838.htm

Bailey T, Seaward D (2013) 3P's aseptic Fill2Weight powder dispenser: an enabling technology for parenteral reconstitution devices. www.3Pinnovation.com. ONdrugDelivery. http://www.ondrugdelivery.com/publications/45/Issue_45_Hi_Res.pdf

Baumfalk R, Finazzo J (2007) Filter integrity testing in the pharmaceutical process environment. Pharm Technol Eur 19:28–33

BETE (2005) Spray dry manual. BETE Fog Nozzle, Inc

BIONITY (2004) Braking the cold chain: a new technology based on processes two billion years old promises to revolutionise the international vaccine programme, 21.10.2004. Bionity.com, News. https://www.bionity.com/en/news/40384/ © LUMITOS AG

Brandenberger H (2003) Development of a novel high-performacne cyclone to increase the yield in a mini spray dryer. In: Best@Buchi Inf. Bull. Number 27/2003. https://www.buchi.com/sites/default/files/downloads/B-290_Bb_27_EN_01.pdf

Broadhead J, Edmond Rouan SK, Rhodes CT (1992) The spray drying of pharmaceuticals. Drug Dev Ind Pharm 18:1169–1206. https://doi.org/10.3109/03639049209046327

Büchi Labortechnik AG (2000) BUCHI laborator equipment: requirements prior to an IQ/OQ

Büchi Labortechnik AG (2014) Ultrasonic package, maximize your particle size, ultrasonic nozzle for mini spray dryer B-290, Brochure 115593290 en 1405. https://static1.buchi.com/sites/default/files/downloads/Ultrasonic_Package_Product_Brochure_en_C_0.pdf

Büchi Labortechnik AG (2020a) Spray-drying—basic theory and applications, pp 1–28

Büchi Labortechnik AG (2020b) Operation manual mini spray dryer B-290, Document number: 093001/Q en, Publication data: 01.2020. https://www.buchi.com/en/products/spray-drying-and-encapsulation/mini-spray-dryer-b-290

Cal K, Sollohub K (2010) Spray drying technique. i: hardware and process parameters. J Pharm Sci 99:575–586. https://doi.org/10.1002/jps.21886

Camfil (2020) Termikfil 2000, High temperature HEPA filter. Product Datasheet. https://www.camfil.com/product/documents/dam/30482/Product-Documentation-Termikfil-2000.pdf

Cape SP, McAdams DH, Manion JR et al (2010) Development of stable, immunogenic, and protective measles vaccines for needle-free administration. In: 2010 FIP pharmaceutical sciences world congress and AAPS annual meeting and exposition, 14–18 Nov 2010, New Orleans, LA, USA

Castagna M (2018) TECHNOSPHERE®: drug delivery platform. In: United Ther Science Day, 24 Sept, 2018. https://pipeline.unither.com/product/tyvaso-dpi-inhaled-treprostinil/presentations/

Cattenati C (2019) Business case on low powder dosages, Romaco s.r.l., 20th Sept 2019

Cazzoli (2019) PF—powder filler series, Data Sheet. https://www.cozzoli.com/wp-content/uploads/2019/01/PF-Data-Sheet.pdf

CDC/NIH (2020) Biosafety in microbiological and biomedical laboratories, 6th ed, Revised June 2020, pp 1–574

ChargePoint Technology (2019) AseptiSafe contained aseptic transfer valves, Brochure 04/2019, pp 1–12

Chickering DE, Keegan MJ, Randall G et al (2003) Spray drying apparatus and method of use, pp 1–10

Costa E, Neves F, Andrade G, Winters C (2014) Scale-up & QbD approaches for spray-dried inhalation formulations. ONdrugDelivery. https://www.ondrugdelivery.com/scale-qbd-approaches-spray-dried-inhalation-formulations/

Das P (2004) Revolutionary vaccine technology breaks the cold chain. Lancet Infect Dis 4:719. https://doi.org/10.1016/S1473-3099(04)01222-8

De Costa S (2014) Advanced delivery devices—new drug delivery & stabilization platforms gaining global interest. Drug Dev Technol. https://drug-dev.com/advanced-delivery-devices-new-drug-delivery-stabilization-platforms-gaining-global-interest/

De Costa S (2015) Particle engineering for bioavailability enhancement. Pharmaceutical Technology. https://www.pharmtech.com/view/particle-engineering-bioavailability-enhancement

De Costa SD, Roser B, Sen S (2006) Pharmaceutical device for the administration of substances to patients. Patent EP1976483B1, Date of filing: 21.11.2006

Dec Group (2020) Aseptic fill & finish operation of dry powder and liquids into dual chamber syringes, poster. In: 2020 PDA universe pre-filled syringes inject devices, 5–8 Oct, 2020, Las Vegas, Nevada, USA. https://www.pda.org/global-event-calendar/event-detail/2020-pda-universe-of-pre-filled-syringes-and-injection-devices

Dec Group (2021a) Microdosing µPTS. Product Datasheet. https://www.dec-group.net/micropts

Dec Group (2021b) PTS powder transfer system®. Product datasheet. https://www.dec-group.net/pts-powder-transfer-system

Dec Group (2021c) Blending PTS Batchmixer®. Product datasheet. https://www.dec-group.net/pts-batchmixer

DiBerardinis LJ, Baum JS, First MW et al (2013) Guidelines for laboratory design: health, safety, and environmental considerations. John Wiley & Sons

Dobry DE, Settell DM, Baumann JM (2015) Chapter 9: Spray drying and scale-up. In: Newman A (ed) Pharmaceutical amorphous solid dispersions. John Wiley & Sons Inc., Hoboken, NJ, USA, pp 1–26

Dunne C (2020) Aseptic spray dryer application. ChargePoint Technology, www.thechargepoint.com, pp 1–10

ECA Academy (2020) Spray drying: solutions for the pharmaceutical industry, seminar, 12–14 May 2020, Copenhagen, Denmark

ECA Academy (2021) Spray drying solutions for the pharmaceutical industry. In: Live online conference, 2/3 March 2021. https://www.gmp-navigator.com/seminare/gmp-seminar-konferenz/spray-drying

EMA (2019) Guideline on the sterilisation of the medicinal product, active substance, excipient and primary container, 6 March 2019, EMA/CHMP/CVMP/QWP/850374/2015. https://www.ema.europa.eu/en/documents/scientific-guideline/guideline-sterilisation-medicinal-product-active-substance-excipient-primary-container_en.pdf

EMD Millipore (2015) Data sheet Millidisk® cartridges and Millipak® disposable capsules, for the sterile filteration of liquids in pharmaceutical applications, Lit No. DS57770000 V1.0, PS-13-08603, 08/2015, pp 1–8

EMEA (2006) IHC Topic Q7 good manufacturing practice for active pharmaceutical ingredients. Step 5:1–48

EU Commission (2020) EU GMP Annex 1 revision: manufacture of sterile medicinal products (draft). Annex 1 Consult Doc. December 2020. https://www.gmp-compliance.org/guidemgr/. https://www.gmp-compliance.org/guidelines/gmp-guideline/eu-gmp-annex-1-revision-manufacture-of-sterile-medicinal-products-draft

FDA (2004) Guidance for industry, sterile drug products produced by aseptic processing—current good manufacturing practice

Fernandes JV, Villax P (2019) Hovione Technology - Dry powder inhalers: towards, effective, affordable, sustainable respiratory healthcare, 8th November 2019. ONdrugDelivery. https://ondrugdelivery.com/dry-powder-inhalers-towards-effective-affordable-sustainable-respiratory-healthcare/

Flow Sciences Inc. (2020) Buchi B-290 mini-spray dryer enclosure for API processing. https://flowsciences.com/buchi-b-290-mini-spray-dryer-enclosure-for-api-processing

Fragale C, Brueck D (2015) Spray drying vancomycin, Patent US 9023258 B2, Filed: April 28, 2014, pp 1–11

Galvin T (2013) Collecting pharma dust—technology advancements for improved safety and efficiency. Tech Artic Donaldson Torit. www.donaldson.com. https://www.donaldson.com/en-us/industrial-dust-fume-mist/technical-articles/collecting-pharma-dust-technology-advancements/

Gaspar F, Vicente J, Neves F, Authelin J-R (2014) Chapter 8: Spray drying: scale-up and manufacturing. In: Shah N et al (ed) Amorphous solid dispersions, advances in delivery science and technology. Controlled Release Society, pp 261–302

GEA (2013) GEA Niro spray dryer SDMicro, ideal for research and development, pp 1–4

GEA (2016) Containment technology for solid dosage processing, pp 1–20

GEA (2021) Spray dried vaccines, customer case. https://www.gea.com/en/customer-cases/spray-dried-vaccines.jsp. Accessed 28 Jan 2021

GEA Niro (2017) Aseptic spray dryers: an effective method for sterile solid drug manufacture, BNA 971/GB 1011 CBC

GEA Process Engineering A/S (2020a) GEA pharmaceutical spray drying, brochure BTC 0012-EN 052020, pp 1–16

GEA Process Engineering A/S (2020b) GEA spray drying: small-scale solutions for R&D and production, brochure BTC0100-EN 0520

Geng O (2018) SPX FLOW aseptic spray drying, March 13, 2018. https://file.vogel.com.cn/2018/08/31/5b8905c261000.pdf

Gikanga B, Turok R, Hui A et al (2015) Manufacturing of high-concentration monoclonal antibody formulations via spray drying–the road to manufacturing scale. PDA J Pharm Sci Technol 69:59–73. https://doi.org/10.5731/pdajpst.2015.01003

Gil M, Vicente J, Gaspar F (2010) Scale-up methodology for pharmaceutical spray drying. Chim Oggi/Chem Today 28:18–22

Gill D, Saigal N, Kale S et al (2017) Thermostable spray dried rotavirus vaccine formulation and process thereof, Patent US 2017/0173145 A1, PCT filed: July 17, 2015, pp 1–17

Gladd T (2014) Using spray drying for flu vaccines: what challenges do we need to overcome? April 14, 2014. www.bioprocessonline.com. https://www.bioprocessonline.com/doc/using-spray-drying-for-flu-vaccines-what-challenges-do-we-need-to-overcome-0001

Global View Research (2019) Spray drying equipment market size, share, & trends analysis report by cycle (open, closed), by product (rotary, nozzle atomizer), by stage (single-, two-stage), by flow, by application, and segment forecasts, 2019–2025, Published: June, 2019, pp 1–101

Goldberg T, Wong E (2015) Afrezza (Insulin Human) inhalation powder. P&T 40(11):735–741. https://www.ncbi.nlm.nih.gov/pmc/articles/PMC4634344/

Grand View Research (2019) Spray drying equipment market size, share, and trends analysis report by cycle (open, closed), by product (rotary, nozzle atomizer), by stage (single, two-stage), by flow, by application, and segment forecasts, 2019–2025, Published: June, 2019, pp 1–101

Greb E (2009) Is spray drying a viable alternative to lyophilization. PharmTech 1–2

Gursch J, Hohl R, Armenante ME et al (2015) Continuous drying of small particles for pharmaceutical applications—an evaluation of selected lab-scale systems. Org Process Res Dev 19:2055–2066. https://doi.org/10.1021/acs.oprd.5b00309

HARRO (2020a) Packaging tricky powders: unfeasible does not exist! HARRO Mag Ed 9. https://www.harro-magazine.com/en/09-2020/packaging-tricky-powders

HARRO (2020b) Papillon: as light as the flap of a butter-fly's wings. HARRO Mag Ed 11. https://www.harro-magazine.com/en/11-2020/papillon-as-light-as-the-flap-of-a-butterflys-wings/

Harro Höfliger (2015) Inhaling insulin instead of injecting. HARRO Mag Ed 1, Oct 2015. https://www.harro-magazine.com/en/wp-content/uploads/sites/19/2019/09/Harro_01_EN.pdf

Harro Höfliger (2021) Dosing & filling: sophisticated processes, individual solutions, machines. https://www.hoefliger.com/en/machines-technology/dosing-filling

Hill SA (1984) Pharmaceutical composition, pp 1–5

Hovione (2021a) TwinCaps—World's largest selling disposable DPI. http://www.hovionetechnology.com/en/portfolio/product-family/twincaps/

Hovione (2021b) Large dose DPI-disposable DPI for high dose drug delivery (TwinMax, 8Shot). http://www.hovionetechnology.com/en/portfolio/product-family/large-dose-dpi/

IAC (2018) Vaccines with diluents: how to use them, immunization action coalition, Item #P3040 (8/18). https://www.immunize.org/catg.d/p3040.pdf

ICH (2010) ICH qualty guideline Q7: good manufacturing practice guide for active pharmaceutical ingredients. https://www.ich.org/page/quality-guidelines

IMA Life (2012) Microdosing series: powder microdosers for aseptic environments, Brochure February 2012-991ZS021101. 1–12

IMA Life (2020) SMART PWD powder filling & stoppering machine. https://ima.it/pharma/machine/smart-pwd/

Ingvarsson PT, Schmidt ST, Christensen D et al (2013) Designing CAF-adjuvanted dry powder vaccines: spray drying preserves the adjuvant activity of CAF01. J Control Release 167:256–264. https://doi.org/10.1016/j.jconrel.2013.01.031

ISPE (2018) ISPE baseline guide: sterile product manufacturing facilities

Jin TH, Tsao E, Goudsmit J et al (2010) Stabilizing formulations for inhalable powders of an adenovirus 35-vectored tuberculosis (TB) vaccine (AERAS-402). Vaccine 28:4369–4375. https://doi.org/10.1016/j.vaccine.2010.04.059

Kanojia G (2018) Drying made easy: spray drying a promising technology for the production of stable vaccine and therapeutic protein formulations, PhD thesis. University of Groningen

Kanojia G, Raeven RHM, van der Maas L et al (2018) Development of a thermostable spray dried outer membrane vesicle pertussis vaccine for pulmonary immunization. J Control Release 286:167–178. https://doi.org/10.1016/j.jconrel.2018.07.035

Kemp IC, Hartwig T, Herdman R et al (2016) Spray drying with a two-fluid nozzle to produce fine particles: atomization, scale-up, and modeling. Dry Technol 34:1243–1252. https://doi.org/10.1080/07373937.2015.1103748

Kunda NK, Alfagih IM, Miyaji EN et al (2015) Pulmonary dry powder vaccine of pneumococcal antigen loaded nanoparticles. Int J Pharm 495:903–912. https://doi.org/10.1016/j.ijpharm.2015.09.034

Kunda NK, Wafula D, Tram M et al (2016) A stable live bacterial vaccine. Eur J Pharm Biopharm 103:109–117. https://doi.org/10.1016/j.ejpb.2016.03.027

LANXESS (2017) Rely+ON and Virkon, Broad spectrum virucidal disinfectant, Brochure UK/06.17/G, pp 1–4

LeClair DA, Li L, Rahman N et al (2019) Stabilization of HSV-2 viral vaccine candidate by spray drying. Int J Pharm 569:118615. https://doi.org/10.1016/j.ijpharm.2019.118615

LiCalsi C, Christensen T, Bennett JV et al (1999) Dry powder inhalation as a potential delivery method for vaccines. Vaccine 17:1796–1803. https://doi.org/10.1016/S0264-410X(98)00438-1

Luy B, Stamato H (2020) Spray freeze drying. In: Ohtake S, Izutsu K-I, Lechuga-Ballesteros D (eds) Drying technologies for biotechnology and pharmaceutical applications. Wiley, pp 217–237

MannKind (2020) Afrezza (human insulin inhalation powder): getting started with Afrezza. www.afrezza.com. https://afrezza.com/wp-content/uploads/2020/12/US-AFR-1341-Afrezza-Patient-Getting-Started-Guide-digital-December-2020.pdf

Manufacturing Chemist (2010) Nova Laboratories acquires Cambridge biostability assets, 24 Feb 2010. https://www.manufacturingchemist.com/news/article_page/Nova_Laboratories_acquires_Cambridge_Biostability_assets/42772

Manufacturing Chemist (2013) Meeting modern development needs by spray drying, 27 March 2013. https://www.manufacturingchemist.com/news/article_page/Meeting_modern_development_needs_by_spray_drying/87136

Manufacturing Chemist (2015) Nova Laboratories manufactures first aseptically spray dried biologic, 13 May 2015. https://www.manufacturingchemist.com/news/article_page/Nova_Laboratories_manufactures_first_aseptically_spray_dried_biologic/108345

Manufacturing Chemist (2016) A green light for cGMP aseptic spray drying, 21 Jan 2016. https://www.manufacturingchemist.com/news/article_page/A_green_light_for_cGMP_aseptic_spray_drying/114844

Market Data Forecast (2020) Spray drying equipment market by type (rotary atomizer, nozzle atomizer, fluidized, closed loop, and centrifugal), application (food, pharmaceutical, and chemical), drying stage, cycle type, flow type by region—global Industry analysis, size, share, gro, pp 1–175

Markets and Markets (2020) Spray drying equipment market, global forecast to 2025, Published on October 2020, pp 1–192

Masters K (1976) Chapter 16: Applications in the pharmaceutical-biochemical industry. In: Spray drying: an introduction to principles, operational practice and applications, 2nd ed. John Wiley & Sons, New York

Masters K (1985) Spray drying handbook, 4th edn. George Godwin, London

Masters K (1991) Spray drying handbook, 5th ed. Longman Scientific & Technical, Harlow, USA

Masters K (2002) Spray drying in practice. SprayDryConsult International ApS, Denmark

Masters K, Vestergaard I (1975) Aseptic and closed cycle spray drying in pharmaceuticals/bio-chemicals manufacture. Process Biochem 10:21–29

McAdams D, Chen D, Kristensen D (2012) Spray drying and vaccine stabilization. Expert Rev Vaccines 11:1211–1219. https://doi.org/10.1586/erv.12.101

McGuire MT (1998) WHO good manufacturing practices for pharmaceutical products: main principles. who-GMP-Annex 2, pp 170–170

McKeage K (2015) Raplixa™: a review in improving surgical haemostasis. Clin Drug Investig 35:519–524. https://doi.org/10.1007/s40261-015-0307-5

Merck Millipore (2020) Durapore® membrane filter, 0.22 µm, Durapore family guide, portfolio of sterile membrane filters, brochure, 09/2020. https://www.emdmillipore.com/US/en/product/Durapore-Membrane-Filter-0.22m,MM_NF-GVWP04700?bd=1

Miller DA, Gil M (2012) Chapter 10: Spray-drying technology. In: Williams RO, Watts A, Miller D (eds) Formulating poorly water soluble drugs, AAPS advances in the pharmaceutical sciences series, vol 3. Springer. New York, NY, pp 363–442

Million Insights (2020) Spray drying equipment market analysis report by cycle, product, stage, flow, application and segment—forecast from 2019 to 2025, Published: August 2020, pp 1–101

Mogavero M (2014) Insight into the aseptic spray drying process, 17 Dec 2014. www.powderbulksolids.com, pp 4–7

Mujumdar AS, Alterman DS (1997) Chapter 10: Drying in the pharmaceutical and biotechnology fields. Handbook of downstream processing. Chapman & Hall, London, UK, pp 235–260

Neopac (2020) Easymix™ Tube product presentation: mixing made easy! www.neopac.ch, pp 1–15

Neumiller JJ, Campbell RK (2010) Technosphere® insulin. BioDrugs 24:165–172. https://doi.org/10.2165/11536700-000000000-00000

NKP (2020) Product catalogue 2019–2020, N.K.P. Pharma Pvt. Ltd. https://www.nkppharma.com

Nova Laboratories (2021) Stabilization technologies. https://www.novalabs.co.uk/stabilisation-technologies. Accessed 28 Jan 2021

Ostrove SA (2016) How to validate a pharmaceutical process. Elsevier

Otsuka (2020) ABILIFY MAINTENA (aripiprazole): full prescribing information, Otsuka Pharmaceutical Co., Ltd., Japan, www.abilifymaintena.com, https://www.otsuka-us.com/sites/g/files/qhldwo3021/files/media/static/Abilify-M-PI.pdf

Ousset A, Bassand C, Chavez P-F et al (2019) Development of a small-scale spray-drying approach for amorphous solid dispersions (ASDs) screening in early drug development. Pharm Dev Technol 24:560–574. https://doi.org/10.1080/10837450.2018.1534861

Pakowski Z, Mujumdar A (2014) Drying of pharmaceutical products. In: Mujumdar AS (ed) Handbook of industrial drying, 4th ed. CRC Press

PALL (2017) Fluorodyne II filters, brochure 11/2017. https://shop.pall.com/us/en/biotech/filtration/sterile-liquid-filters-0-2-micron-/zidgri78lwi

PALL (2020) Emflon® PFR filters, Datasheet USD 1477c

Panacea Biotec (2019a) Panacea Biotec receives USD 24.32 million award of pentavalent vaccine from U.N. Agencies, News Update: 19 Dec 2019. https://www.panaceabiotec.com/panacea-biotec-receives-usd-24.32-million-award-of-pentavalent-vaccine-from-u.n.-agencies

Panacea Biotec (2019b) Package insert of easyfive-TT (fully liquid pentavalent vaccine): Diphtheria, Tetanus, Pertussis (Whole Cell), Hepatitis B (rDNA) and Haemophilus influenzae Type b Conjugate Vaccine (Adsorbed) IP, Last updated on June 2019, pp 1–5

PDA (2008) Sterilizing filtration of liquids. Technical report no. 26 (revised 2008). 01 Jan 2008, 62(5 Suppl TR26):2-60, PMID: 19642423. PDA J Pharm Sci Technol. https://journal.pda.org/

Pfützner A, Forst T (2005) Pulmonary insulin delivery by means of the Technosphere™ drug carrier mechanism. Expert Opin Drug Deliv 2:1097–1106. https://doi.org/10.1517/17425247.2.6.1097

Pfützner A, Mann AE, Steiner SS (2002) Technosphere™/Insulin—a new approach for effective delivery of human insulin via the pulmonary route. Diabetes Technol Ther 4:589–594. https://doi.org/10.1089/152091502320798204

Pharmaceutical Technology (2010) Anhydro designs aseptic spray dryer for tuberculosis vaccines, Press Release, 29 April 2010. https://www.pharmaceutical-technology.com/contractors/water_treatment/anhydro1/pressreleases/pressaseptic-spray-dryer-for-tuberculosis-vaccines/

Plummer J, Barreto V (2012) Solving baghouse performance problems in spray-drying systems, April 2012. Powder Bulk Eng. www.powderbulk.com, pp 1–3

Poozesh S, Bilgili E (2019) Scale-up of pharmaceutical spray drying using scale-up rules: a review. Int J Pharm 562:271–292. https://doi.org/10.1016/j.ijpharm.2019.03.047

Price DN, Kunda NK, Ellis R, Muttil P (2020) Design and optimization of a temperature-stable dry powder BCG vaccine. Pharm Res 37:11. https://doi.org/10.1007/s11095-019-2739-8

ProCepT (2020) Spray dryer 4M8-TriX brochure. www.procept.be, pp 1–4

ProCepT (2021) Spray drying of monoclonal antibodies for oral applications. Xedev Application Note, www.xedev.com, p 1

ReportLinker (2014) Pharmaceutical spray drying market, 2014–2024, Published: July 17, 2014. www.reportlinker.com

ReportLinker (2018) Pharmaceutical spray drying market, 2nd ed., 2018–2028, Published: April 2018, pp 1–210

ReportLinker (2020) Spray drying equipment market by product type, application, cycle type, drying stage, flow type, region—global forecast to 2025, October 2020, pp 1–188

Reports and Data (2019) Spray drying equipment market analysis, by type (nozzle, rotary atomizer, fluidized, centrifugal) by stage type (single-stage, two-stage, multi-stage) by flow type by cycle (open, closed) by application forecasts to 2026, published: October, 2019, pp 1–150

Richardson PC, Boss AH (2007) Technosphere® insulin technology. Diabetes Technol Ther 9:65–72. https://doi.org/10.1089/dia.2007.0212

ROMACO (2018) Aseptic powder dosing solutions, Brochure 2018/06. https://www.romaco.com/files/Dokumente/delivering-solutions/romaco_aseptic_powder_filling-EN.pdf

Roser B (2005a) Sterile spray drying for stable liquid 21st century pharmaceuticals. Innov Pharm Technol. http://iptonline.com/articles/public/IPT192005_p50+nonprintable.pdf

Roser B (2005b) Single-shot stable liquid vaccines. Innov Pharm Technol. http://www.iptonline.com/articles/public/IPT_16_2005_p58_62x.pdf

Roser B (2006) Stable liquid vaccines and drugs for the 21st century. Future Microbiol 1:21–31. https://doi.org/10.2217/17460913.1.1.21

Roser B (2009) Aseptic spray drying for stable liquid 21st century. Pharmaceuticals, Cambridge Biostability Ltd, UK. https://freelancecopy.files.wordpress.com/2009/09/cb-aseptic-spray-drying.pdf

Roser B, Moos D (2006) Improvements in the stabilization of bilogical materials. Patent EP 1928422 B1, Date of filing: 31.08.2006, pp 1–9

Saboo S, Tumban E, Peabody J et al (2016) Optimized formulation of a thermostable spray-dried virus-like particle vaccine against human papillomavirus. Mol Pharm 13:1646–1655. https://doi.org/10.1021/acs.molpharmaceut.6b00072

Schaefer J, Lee G (2015) Making large, flowable particles of protein or disaccharide in a mini-scale spray dryer. Pharm Dev Technol 21:1–9. https://doi.org/10.3109/10837450.2015.1063649

Scherließ R, Ajmera A, Dennis M et al (2014) Induction of protective immunity against H1N1 influenza A(H1N1)pdm09 with spray-dried and electron-beam sterilised vaccines in non-human primates. Vaccine 32:2231–2240. https://doi.org/10.1016/j.vaccine.2014.01.077

Scholz M, Altenhofen W (2011) Stabilising biologics for vaccine production. Manufacturing Chemist. https://manufacturingchemist.com/news/article_page/Stabilising_biologics_for_vaccine_production/63953

Schwartzbach H (2011) Achieving aseptic drying with spray drying technologies. Pharm Technol Eur 23:90–92

Scrafford J (2020) An integrated approach to spray drying: from scale-up to manufacturing. Catalent, Pharm Technol Spec Spons Sect 1–6

Searles J, Cherian M (2015) Alternatives to vial lyophilization. Lyophilized biologics and vaccines. Springer, New York, New York, NY, pp 257–271

Searles J, Mohan G (2010) Spray drying of biopharmaceuticals and vaccines. Formulation and process development strategies for manufacturing biopharmaceuticals. John Wiley & Sons Inc., Hoboken, NJ, USA, pp 739–761

Siew A (2016) Exploring the use of aseptic spray drying in the manufacture of biopharmaceutical injectables. Pharm Technol 40:24–27

Snyder HE, Lechuga-Ballesteros D (2008) Spray drying: theory and pharmaceutical applications. In: Augsburger LL, Hoag SW (eds) Pharmaceutical dosage forms: tablets, vol 1: Unit operations and mechanical properties, Ch 7, 3rd ed. Informa Healthcare USA, Inc., New York, NY, pp 227–260

SPX Flow Inc (2017) Anhydro spray drying for the pharmaceutical industry, brochure ANH-116-GB. Issued 12(2017):1–8

Steiner S, Pfützner A, Wilson BR et al (2002) Technosphere™/insulin—proof of concept study with a new insulin formulation for pulmonary delivery. Exp Clin Endocrinol Diabetes 110:17–21. https://doi.org/10.1055/s-2002-19989

Strothers S, Seaward D (2020) How to reduce development and clinical manufacturing timescales with robotic filling innovation. White Paper, 3P innovation Ltd. https://www.3pinnovation.com/media-events/publications

Syntegon (2021) AFG 5000—filling and sealing machines for pharmaceutical powder and antibiotics. https://www.syntegon.com/products/afg-5000-filling-and-sealing-machines-pharmaceutical-powder-and-antibiotics?t=1644&tg=1587&vpa=&kp=1559&prod=1508&ind

Teo C, Sun X, Wang X (2021) Method and System for producing high purity vancomycin hydrochloride. Patent US 2021/0023011 A1, Filed: Oct 12, 2020, pp 1–23

The Baker Company (2010) BioPROtect® III class II style walk-in clean air and containment enclosures. Datasheet, 134-043-10, 11/10, pp 1–6

Thybo P, Hovgaard L, Lindeløv JS et al (2008) Scaling up the spray drying process from pilot to production scale using an atomized droplet size criterion. Pharm Res 25:1610–1620. https://doi.org/10.1007/s11095-008-9565-8

Truong-Le V, Pham B (2007) Preservation of bioactive materials by spray drying. Patent US 7258873 B2, filed: April 10, 2003, pp 1–30

Truong-Le V, Ohtake S, Martin RA et al (2012) Sonic low pressure spray drying. Patent US 8268354 B2, Filed: Nov 6, 2008, pp 1–17

Truong-Le V, Ohtake S, Martin RA et al (2014) Sonic low pressure spray drying. Patent US 8673357 B2, Filed: Aug 14, 2012, pp 1–18

Tscheliessnig R, Zörnig M, Herzig EM et al (2012) Nano-coating protects biofunctional materials. Mater Today 15:394–404. https://doi.org/10.1016/S1369-7021(12)70166-9

Tseng YC (2015) Solving solubility issues with amorphous solid dispersions. Eur Pharm Rev 20:28–32

Tumban E, Muttil P, Escobar CAA et al (2015) Preclinical refinements of a broadly protective VLP-based HPV vaccine targeting the minor capsid protein, L2. Vaccine 33:3346–3353. https://doi.org/10.1016/j.vaccine.2015.05.016

Turok RH, Rasmussen SS (2010) Process gas filtration, pp 1–7

Vehring R, Foss WR, Lechuga-Ballesteros D (2007) Particle formation in spray drying. J Aerosol Sci 38:728–746. https://doi.org/10.1016/j.jaerosci.2007.04.005

Vehring R, Snyder H, Lechuga-Ballesteros D (2020) Spray drying. In: Ohtake S, Izutsu K, Lechuga-Ballesteros D (eds) Drying technologies for biotechnology and pharmaceutical applications. Wiley, pp 179–216

Verified Market Research (2020) Global spray drying equipment market size by dryer type, by application, by flow type, by geographic scope and forecast. Published: October 2020, pp 1–108

Weniger BG (2011) Research on new vaccine delivery methods: focus on influenza. In: Second WHO consultation on the global action plan for influenza vaccines (GAP-II), 12–14 July 2011, Geneva, Switzerland. https://www.who.int/influenza_vaccines_plan/resources/weniger.pdf

Weniger BG, Papania MJ (2013) Alternative vaccine delivery methods. Vaccines:1200–1231. https://doi.org/10.1016/B978-1-4557-0090-5.00063-X

WHO (2002) WHO expert committee on specifications for pharmaceutical preparations, Annex 6: good manufacturing practices for sterile products. WHO Tech Rep Ser No 902:76–93

WHO (2004) Laboratory biosafety manual. In: SpringerReference, 3rd ed. World Health Organization 2004, Geneva, pp 1–178

Wong Y-L, Sampson S, Germishuizen WA et al (2007) Drying a tuberculosis vaccine without freezing. Proc Natl Acad Sci 104:2591–2595. https://doi.org/10.1073/pnas.0611430104

Ziaee A, Albadarin AB, Padrela L et al (2019) Spray drying of pharmaceuticals and biopharmaceuticals: critical parameters and experimental process optimization approaches. Eur J Pharm Sci 127:300–318. https://doi.org/10.1016/j.ejps.2018.10.026

Design of Experiment Studies and Scale-Up

5

Abstract

This chapter provides an overview of design of experiment (DOE) studies of spray-dried vaccines and approaches for scaling the spray-drying process from laboratory to pilot and commercial scale. The chapter aims to present a clearer picture of the field and highlight the factors influencing particle design and vaccine activity. Key aspects of scale-up and the methodology are described along with application examples where scale-up for vaccine powders has been performed. As a result, readers can take away the concept of quality-by-design (QbD) for developing a quality product with predefined goals and understanding the critical process and product parameters based on risk analysis considerations.

5.1 Quality by Design (QbD) Approach

Regulatory authorities such as the ICH (International Conference on Harmonization) and the U.S. FDA (Food and Drug Administration) have called for the pharmaceutical industry to gain a comprehensive understanding of its manufacturing processes and an accurate assessment of their robustness and reliability. As a result, emphasis has been placed on the quality-by-design (QbD) concept, defined as a "systematic approach to development that begins with predefined objectives and emphasizes product and process understanding and process control, based on sound science and quality risk management" (ICH 2009). The major points of QbD are (Yu et al. 2014; Andersen 2018):

- Quality should be built into the product rather than tested afterward.
- Understanding the product and the process by which it is developed and manufactured.

C. Arpagaus, *Spray Drying of Vaccines*,
https://doi.org/10.1007/978-3-031-24323-3_5

- Knowledge of the risks associated with manufacturing the product and how best to mitigate those risks.

A QbD approach includes the following three key elements:

- Quality target product profile (QTPP),
- Critical quality attributes (CQA), and
- Critical process parameters (CPP).

In addition, a robust control strategy can be developed based on previous product and process experience (ICH 2009).

Applying the QbD approach to spray drying is critical due to the multidimensional nature of the process throughout the life cycle, from development to validation (Gaspar et al. 2014; Andersen 2018; Scrafford 2020; Baumann et al. 2021).

The first step in developing a spray-dried vaccine powder begins with defining the QTPP to ensure the safety and efficacy of the vaccine dry powder. In an aerosol formulation, the most critical QTPP is to deliver the optimal vaccine dose to the right lung segment to achieve the desired immunization.

Then, the critical CQAs are identified so that product characteristics can be studied and fully controlled. For example, for an inhalable vaccine powder, the CQAs include particle size, yield, and moisture content, impacting product quality (Ingvarsson et al. 2013b; Andersen 2018).

The next step involves identifying the functional relationships between spray drying process CPPs and CQAs using risk assessment and design of experiments (DOE) studies, then determining an optimal operational design space (Scrafford 2020). CPPs are the key variables that should be monitored to detect deviations in standardized production processes. The essential CPPs for vaccine particle size are feed viscosity, atomization gas flow, drying gas flow rate, feed flow rate, and outlet temperature, which need to be systematically tested.

DOE is recognized as the primary tool for identifying critical and non-critical parameters that can affect product quality and allows quantification of interactions between critical parameters on quality responses.

Overall, the QbD framework for the spray drying process will improve the vaccine developer's understanding of the interplay between process parameters and product quality. Furthermore, this will increase the chances of a successful commercial manufacturing process outcome (Scrafford 2020). Examples of the application of the QbD approach to the development of an inhalable dry powder through spray drying processes include:

- Solid lipid microparticles as drug carriers (Ignjatović et al. 2021),
- Cancer drug erlotinib hydrochloride (Soni et al. 2020),
- Emulsions of antioxidant drug resveratrol (Benjasirimongkol et al. 2019),
- siRNA-loaded lipidoid polymer hybrid nanoparticles (Dormenval et al. 2019)
- Nanosuspensions of antipsychotic drug risperidone (Nair et al. 2019)
- Nanosuspensions of anti-inflammatory drug indomethacin (Kumar et al. 2014)

- Antibiotic drug ciprofloxacin (Karimi et al. 2016)
- General dry powder inhalation formulations (Costa et al. 2014)
- Liposomal adjuvant for vaccines (Ingvarsson et al. 2013b)
- Mannitol and trehalose (Baldinger et al. 2012)
- Insulin (Maltesen et al. 2008).

Figure 5.1 shows an Ishikawa (fishbone) diagram based on various literature sources (ICH 2009; Baldinger et al. 2012; Ingvarsson et al. 2013b; Karimi et al. 2016; Benjasirimongkol et al. 2019; Dormenval et al. 2019; Soni et al. 2020; Ignjatović et al. 2021), with possible factors influencing the outcome of a spray-dried dry powder vaccine.

The factors are divided into several groups, including:

- Equipment design,
- Process parameters,
- Product characteristics,
- Analytical methods,
- Preparation method,
- Formulation parameters,

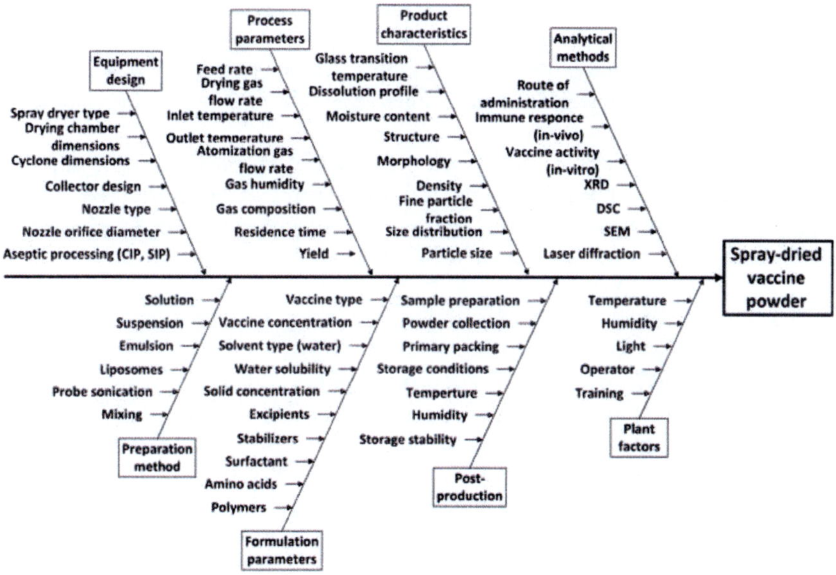

Fig. 5.1 Ishikawa (fishbone) diagram to visualize the factors affecting a spray-dried vaccine powder quality. Researchers can use the chart for risk analysis and evaluation of the critical process parameters (CPPs) impact on the critical quality attributes (CQAs). The diagram was compiled based on various literature sources (ICH 2009; Baldinger et al. 2012; Ingvarsson et al. 2013b; Karimi et al. 2016; Benjasirimongkol et al. 2019; Dormenval et al. 2019; Soni et al. 2020; Ignjatović et al. 2021)

- Post-production, and
- Plant factors.

The Ishikawa diagram can serve as a basis for a risk analysis that allows a cross-functional team of experts to collaboratively identify potential sources of variability that may impact desired product quality. For example, certain factors can be kept constant, such as the designed spray drying equipment, water as a solvent, or the humidity of the drying gas, while others have such as process and formulation parameters have to be optimized depending on the vaccine to meet the targeted product profile.

5.2 Design of Experiments (DOE) Studies of Spray-Dried Vaccines

Design of Experiments (DOE) is the fastest and most cost-effective approach to systematically study the effects of multiple factors and identify critical and non-critical parameters and their respective interactions for a production process (Kanojia et al. 2017; Kanojia 2018). Thus, implementing the DOE approach to spray drying vaccines allows the determination of optimal process conditions, thereby reducing costs and material consumption. Besides, Quality by Design (QbD) plays a vital role in process optimization and scale-up. Therefore, the primary goal is process and product understanding, leading to a design space suitable for scale-up that guarantees the right quality according to the critical quality attributes (Arpagaus 2020).

As discussed in Chap. 3, spray drying consists of numerous process and formulation parameters tuned for optimal results. Several studies have used a DOE approach to investigate and optimize the operating parameters of vaccine spray drying, including

- Tuberculosis (Price et al. 2020),
- Polio virus (Kanojia et al. 2018),
- Influenza virus (Lovalenti et al. 2016; Kanojia et al. 2016),
- Rotavirus (Gill et al. 2017),
- Human type 5 adenoviral vector (LeClair et al. 2016),
- Herpes simplex virus (LeClair et al. 2019),
- Human papillomavirus (Saboo et al. 2016; Peabody et al. 2017),
- Pneumococcus (Anish et al. 2014), and
- Bovine albumin as model protein (Chablani et al. 2012).

Table 5.1 gives an overview of DOE studies on spray-dried vaccines and provides valuable information on the varied input parameters and analyzed output parameters, as well as the number of experiments used and the spray dryer models. The industry is striving to perform DOE on laboratory-scale spray dryers because of

Table 5.1 Overview of design of experiment (DOE) studies of spray-dried vaccines with information on the varied input and analyzed output parameters (■) as well as the number of experiments and spray-dryer models

Varied input parameters						Analyzed response parameters										Applications and references	
Feed flow rate	Inlet temperature	Drying gas flow rate	Atomization gas flow rate	Solid concentration	Formulation composition	Particle size	Water/moisture content	Vaccine activity (in-vitro)	Immune response (in-vivo)	Thermal storage stability	Powder yield	Particle morphology	Glass transition temperature	Number of experimental runs	Spray dryer model (laboratory scale)	Vaccine type and route of administration (I: inhalation, O: oral, N: nasal, IM: intramuscular)	References
					■	■	■	■	■	■	■		■	25	B-290	Tuberculosis (I)	Price et al. (2020)
■	■	■	■	■	■	■	■		■		■	■	■	19	B-290	Tuberculosis (I)	Ingvarsson et al. (2013a)
■	■				■	■	■			■				30	B-290 +B-296	Polio virus (O)	Kanojia et al. (2018)
■	■	■	■		■	■	■			■	■	■		23	B-290 +B-296	Influenza virus (I)	Kanojia et al. (2016)
					■		■		■	■			■	30	B-190	Influenza virus (N)	Lovalenti et al. (2016)
■	■	■	■		■	■	■	■		■				66	B-290	Rotavirus (O)	Gill et al. (2017)
■	■	■	■	■		■	■	■			■	■		28	B-290	Adenoviral vector (I)	LeClair et al. (2016)
	■		■	■	■		■	■		■			■	17	B-290	Herpes simplex virus (I)	LeClair et al. (2019)
■	■		■		■	■	■	■	■	■	■			19	B-290	Human Papillomavirus (IM, O)	Saboo et al. (2016), Peabody et al. (2017)
■	■	■	■			■	■		■		■		■	20	LU-227	Pneumococcus (IM)	Anish et al. (2014)
					■	■					■	■		16	B-191	Bovine albumin (O)	Chablani et al. (2012)

their flexibility and because high-value vaccines are only available in small quantities at the early stages of development. The more representative the lab-scale design, the better the parameters, models, and conclusions can be transferred to scale-up.

For spray-dried vaccines, the primary response parameters include:

- particle size,
- water/moisture content,
- vaccine activity (in-vitro),
- immune response (in-vivo), and
- storage stability.

Related or secondary parameters are:

- powder yield,
- particle morphology, and
- glass transition temperature.

The particle size is decisive for the dosage form and application. For inhaled use in dry powder inhalers, the target aerodynamic particle size ranges from 1 to 5 μm (Byron 1986; Patton et al. 2004; Kanojia et al. 2017). Larger particles are also suitable for other dosage forms, such as nasal administration or via the skin or oral. The maximum particle size is determined by the size of the spray dryer, available drying length, and process conditions, respectively (see Chap. 3, Sect. 3.4). For vaccines, of course, the aim is to maintain as high activity as possible after the spray drying process. High powder yield lowers the final product cost, the low moisture content is beneficial for long-term storage stability and preventing microbial growth, and optimized particle morphology and size ensure high and uniform loading of the vaccine.

A key element in ensuring vaccine activity after spray drying and for long-term storage is a formulation with the addition of suitable excipients. Table 5.2 summarizes commonly used excipients added to the formulation of spray-dried vaccines and their stabilization mechanism. The information is based on various literature (Maa et al. 1998; Adler and Lee 1999; Truong-Le and Pham 2007; Vehring 2008; Hasija et al. 2013; Ajmera and Scherließ 2014; Saboo et al. 2016; Lovalenti et al. 2016; Kunda et al. 2016; Kanojia et al. 2017; Ziaee et al. 2019; Arpagaus 2020; Vehring et al. 2020; Gomez et al. 2021).

The range of excipients goes from simple small buffer molecules to more complex and sophisticated polymers. Each component plays an essential role in the feed formulation.

- Inorganic salts are typically added to vaccine liquid formulation to adjust tonicity and thus prevent pain upon injection.
- Buffering agents are added to maintain solution pH and stabilize antigens adsorbed to aluminum salts.

Table 5.2 Common used excipients in spray-dried vaccines

Category	Examples of excipients	Mechanism of action
Inorganic salts	NaCl, KCl, $MgCl_2$	Adjust tonicity, specific stabilization
Buffering agents	Phosphate buffer (KPO_4), succinate buffer, sodium borate	Maintain solution pH, modulation of antigen binding to aluminum salt adjuvants
Surfactants	Polysorbate 20 and 80, Triton X-100, Pluronic F68	Prevent protein aggregation at air–liquid interfaces, increase solubility, surface coverage on droplets
Amino acids	L-glutamate, L-arginine, L-histidine, L-leucine, Trileucine, glycine, methionine	Specific interaction with proteins, charge shielding of protein surfaces, stabilizing the conformations of proteins, reducing aggregation, increasing dispersibility of dry powder, antioxidant, buffering agent
Sugars and polyols	Sucrose, trehalose, inulin, mannitol, dextran, maltodextrin, inositol, sorbitol, pullulan	Glassy state stabilization, water replacement, preferential hydration, low hygroscopicity, interfacial tension in lipid bilayers
Proteins	Human and bovine serum albumin, gelatin	Inhibition of protein aggregation by competing for surface coverage on the droplets and reducing the heterogeneity of protein distribution inside the powder particles
Antioxidants	Ethylenediaminetetraacetic acid (EDTA), citric acid, ascorbic acid, metabisulphate	Chelating agents, oxygen scavenger
Polymers	Eudragit, cellulose acetate phthalate, hydroxypropyl methylcellulose acetate succinate (HPMCAS), poly-lactic-co-glycolic acid (PLGA), polyvinylpyrrolidone (PVP)	Enteric coating polymers, encapsulating polymers

Based on various literature (Maa et al. 1998; Adler and Lee 1999; Truong-Le and Pham 2007; Seville et al. 2007; Vehring 2008; Hasija et al. 2013; Ajmera and Scherließ 2014; Saboo et al. 2016; Lovalenti et al. 2016; Kunda et al. 2016; Kanojia et al. 2017; Ziaee et al. 2019; Arpagaus 2020; Vehring et al. 2020; Gomez et al. 2021)

- Nonionic surfactants prevent aggregation due to mechanical stresses during droplet atomization, compete for surface coverage of proteins on liquid droplets, and prevent adsorption to different surfaces.
- Amino acids are well-known stabilizers for protein-based pharmaceuticals and vaccines and act by diverse mechanisms, including preferential hydration, prevention of aggregation, charge shielding of protein surfaces, and direct binding to protein molecules.

- Sugars and polyols give structure and protection to the powder particles, stabilize the preferential hydration, and increase the activation energy for protein unfolding.
- Human and bovine serum albumin and animal-derived gelatin are other frequent stabilizers of vaccines competing for surface coverage and reducing the heterogeneity of protein distribution in the powder particles.
- Several antioxidants are available as pharmaceutical excipients to prevent oxidation.
- Encapsulating polymers protect antigens in the enteric environment and control release for delivery via the oral route.

In this context, a review paper by Foged (2011) also pointed out that for the future, particulate vaccine delivery systems such as virosomes, emulsions (water-in-oil, oil-in-water), liposomes, and others should be used to better tailor vaccine formulations to specific human diseases. Ultimately, the development of thermostable vaccines is needed to increase vaccine coverage worldwide, especially in low-income countries, which can be achieved through drying techniques such as freeze drying and spray drying.

The following sections discuss the respective studies and highlight the most significant findings. Recently, Price et al. (2020) developed a temperature-stable BCG dry powder vaccine against Mycobacterium tuberculosis by optimizing the composition of several excipients through factorial design using Design Expert® Software (Stat-Ease, Inc., USA). The powders were characterized by particle size, powder yield, residual moisture, hygroscopicity, glass transition temperature, and aerosol behavior. The factorial design generated 25 experimental runs at different excipient ratios. An optimal excipient mixture of L-leucine, mannitol, polyvinylpyrrolidone (PVP), bovine serum albumin (BSA), and trehalose (at weight ratios 75.2/10.0/5.0/4.8/5.0) provided storage stability of 12 months at 25 °C (60% RH). In addition, the aerodynamic size of the powders was in the respirable range, indicating principal suitability for pulmonary administration. L-leucine was added to increase the dispersibility of the dry powder and yield, PVP to increase Tg, mannitol provided low hygroscopicity compared to the other excipients, and BSA prevented recrystallization of the powder. Overall, the authors concluded that the DOE studies proved helpful in strategically finding the most effective formulations. The vaccine retained its immunogenicity when healthy mice were immunized intradermally with reconstituted BCG powders stored for 2 years.

Ingvarsson et al. (2013a) formulated liposomes as delivery vehicles for vaccines by spray drying to dry powder and applied as a delivery system for tuberculosis vaccine. Based on a full factorial design (SAS JMP®8 software) with 19 experiments, the excipients lactose, trehalose, and mannitol were analyzed for optimal stabilization. The ability to form a glassy matrix around the liposomes was crucial and was best met by trehalose due to the high Tg. A high drying rate was beneficial to avoid lipid accumulation on the surface of the microparticles. Finally, immunization studies in mice with encapsulated tuberculosis vaccine maintained adjuvant activity in vivo. Overall, the spray-dried powder liposome formulations were seen as promising delivery systems for future designs of stable vaccine formulations.

Further excipient screening with various sugars, plasticizers, proteins, amino acids, surfactants, chelators, metal ions, and buffers was performed by Lovalenti et al. (2016) to develop stable formulations of live-attenuated influenza vaccines of type A H1N1 and B strains. To get an overview of the most promising components, 30 formulations with 3 to 10 ingredients were spray dried in a Mini Spray Dryer B-190 using a partial factorial DOE. Each formulation contained at least one sugar and one buffer. Pluronic F68 protected the vaccine from destabilizing stresses during the spray drying process, including reducing mechanical stresses during atomization and preventing aggregation at the droplet surface during drying due to the surface activity of H1N1 A and B strains. Spray drying transformed all liquid formulations into fine powders with residual moisture contents between 1 and 5 wt %. In general, a moderate process-induced reduction in viral titer was observed after spray drying, with a mean process loss of 0.9 ± 0.5 log $TCID_{50}$/mL, which tended to be higher than for foam-dried formulations. The authors concluded that the appropriate choice of excipients plays a significant role in achieving optimum storage stability while minimizing process losses for a given drying process. However, the best formulation for process loss reduction is not necessarily the best formulation for storage stability.

Kanojia et al. (2018) designed an inactivated Sabin polio vaccine that maximized antigen recovery after spray drying and storage at 40 °C for 7 days. Using the DOE approach (MODDE® 12.0 software, Umetrics, Sartorius), a full factorial experimental design with 19 experimental runs was performed for excipients screening, followed by a fractional factorial design with 11 runs to identify the most promising monovalent serotype formulations. The choice of a fractional rather than a full-factorial experimental design was made for efficiency reasons, as fewer experimental runs would provide the same amount of information, and less vaccine would be needed. The results showed that the trehalose/monosodium glutamate and maltodextrin/arginine combinations were most suitable for stabilizing serotypes 1 and 2, respectively. For serotype 3, the best formulation contained a cell culture medium (M199 media, mixture of stabilizing excipients), glutathione, and maltodextrin.

In an earlier study, Kanojia et al. (2016) used a DOE approach (MODDE® 10.0 software) to predict an optimal design space for spray-dried trehalose-based whole inactivated influenza virus vaccine with inhalable particle size (1 to 5 μm), powder yield > 70%, outlet temperature < 100 °C, residual moisture content between 1 and 3%, and vaccine antigenicity. In addition, the interaction of inlet air drying temperature, feed flow rate, and atomization pressure was evaluated based on a full factorial design with 23 experimental runs. Figure 5.2 shows the predicted design space with optimal spray drying parameters. In the green area, all design space criteria were met (sweet spot). Moreover, the optimized influenza vaccine powder was thermostable (in-vitro) when stored at 60 °C for 3 months.

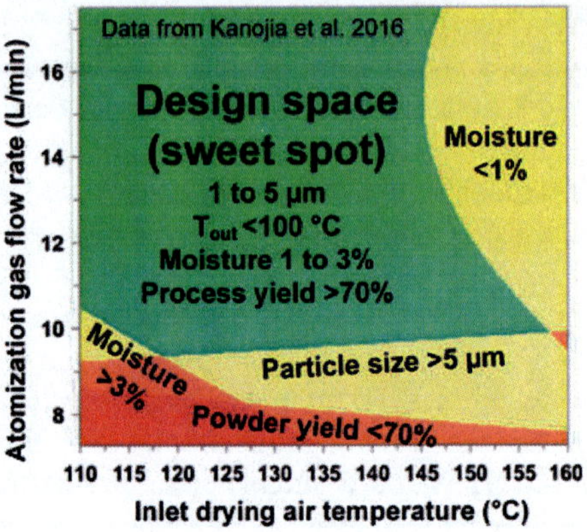

Fig. 5.2 Predicted design space for whole inactivated influenza virus (WIV) vaccine stabilized with trehalose in a Mini Spray Dryer B-290 (closed-loop configuration with N_2 drying gas). Feed flow rate (1.0 mL/min) and feed concentration (100 mg/mL) were fixed. The design space criteria were 1 to 5 µm particle size, < 100 °C outlet temperature, 1 to 3 wt% residual moisture content, and > 70% process yield. In the green area, all criteria are met (sweet spot) (Graph adapted from Kanojia et al. (2016), open access under CC-BY 4.0 license)

Gill et al. (2017) presented an invention focused on improved thermostabilization of spray-dried rotavirus vaccine for oral administration. Using DOE, the formulation was optimized with excipients like trehalose, maltodextrin, NaCl, glutamic acid, histidine, $CaCl_2$, and HEPES buffer. Potency (loss of potency < 0.5 log) was maintained for 4 weeks at 45 °C, 24 weeks at 37 °C, and for 24 months at 2 to 8 °C and met the WHO criterion for maximum heat stability (Vaccine Vial Monitor VVM 30 meaning 30 days at 37 °C) (WHO/UNICEF 2014).

LeClair et al. (2016) encapsulated recombinant human type 5 adenovirus vector in mannitol and dextran (67/33 weight ratio, Tg of 130 °C) by spray drying and optimized the parameters for 1 to 5 µm aerodynamic particle size, maximum powder yield, and minimal viral AdHu5LacZ titer loss (tissue culture). A Box-Wilson central composite design was used with 28 experiments (Design Expert 7.0 software). Figure 5.3 shows the effects of the process parameters on the total titer loss, which varied from less than 0.5 log to 4.0 log loss compared to the original viral vector stock (control). The authors concluded that a wide range of spray drying parameters could result in a stable powder containing adenovirus vector, with less than 1 log loss considered good but more than 2 log loss considered poor. The titer loss increased predictably at high inlet temperatures and atomization gas flow rates. The decrease in stability was attributed to shear forces during atomization and increased protein interactions at the air–water interface of the droplets and the thermal load. Overall, the trends and interactions presented and discussed here can serve as a guide for future studies.

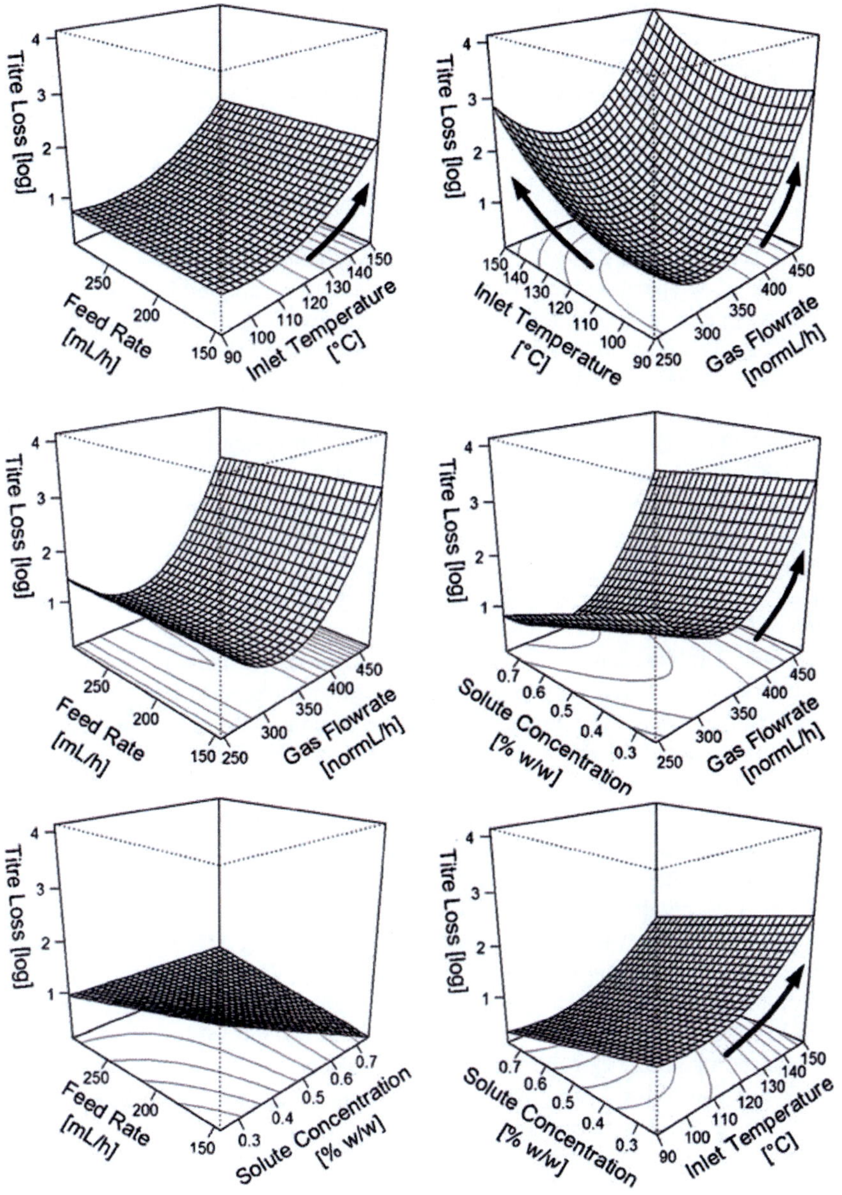

Fig. 5.3 Surface plots of titer loss (log) of recombinant human type 5 adenovirus vector (AdHu5LacZ) after spray drying in mannitol and dextran (weight ratio 67:33) at different process parameter combinations. The aspiration rate was set constant at 100% (35 m^3/h). Titer loss increased with inlet temperatures and gas flow rate (Data adapted from LeClair et al. (2016) with permission from Springer Nature)

A more recent DOE study by LeClair et al. (2019) with 17 experiments and a Box-Wilson Central Composite Design (Design-Expert 7.0 software) demonstrated the thermostability of herpes simplex virus type 2 (HSV-2) vaccine by spray drying in suitable excipients. Vaccine activity loss was measured by in-vitro plaque-forming assay with Vero cell line and accelerated storage tests for 10 days at 45 °C. Excipient screening showed that the use of trehalose and sucrose resulted in the lowest overall activity losses. The trehalose-based dry powder vaccine provided the most suitable thermal stability. The optimized process conditions using a Mini Spray Dryer B-290 (with high-performance cyclone) were 105 °C inlet temperature, 18 wt% solids concentration, and 360 L/h gas flow rate. Increasing the solids concentration reduced the process loss. Besides, a technology transfer study with a tenfold larger batch size was carried out in another facility to verify the predictive model and the robustness of the optimized formulation and process.

Saboo et al. (2016) used a DOE approach to optimize adjuvant composition and spray drying parameters for thermostable virus-like particles (VLP) against human papillomavirus (HPV) (Fig. 5.4). A half-factorial design (2^{5-1}) with 19 experiments was performed (Design-Expert 8.0 software) to maximize yield, minimize moisture content, and optimize particle size. Variations were made in inlet temperature, liquid feed, and gas flow rate. The optimized formulation consisted of a combination of three sugars (i.e., mannitol, trehalose, and dextran) and the amino acid L-leucine. The addition of L-leucine served to protect against moisture uptake. Figure 5.4 shows that the spray-dried VLPs were immunogenic in a mouse model even after 14 months of storage at 37 °C. After intramuscular administration of the reconstituted powders, the IgG titers were comparable to the results with fresh liquid VLP vaccine stored at 4 °C.

In a follow-up study by Peabody et al. (2017), inositol was added to the formulation consisting of mannitol, trehalose, dextran, and L-leucine due to a stabilizing effect during the spray drying process (Jin et al. 2010; Tumban et al. 2015). Even after 34 months of storage at room temperature, two (5 μg) doses of reconstituted spray-dried VLPs elicited high antibody titers similar to those of mice immunized with fresh liquid VLPs. More importantly, the mice showed significant protection against vaginal infection with HPV pseudovirus 16 (representative of HPV16) 4 months after immunization. Thus, the longevity of protection in mice and the thermostability of the vaccine powder suggested spray-dried VLPs to be suitable in Third World countries with less advanced refrigeration facilities, where most HPV-associated cancers occur.

Anish et al. (2014) presented a study on the formulation of spray-dried biodegradable polylactide (PLA, 45 kDa) microparticles encapsulating pneumococcal surface protein A (PspA). The PLA particles were prepared by a double emulsion solvent evaporation method, where the internal aqueous phase consisted of PspA and sucrose, the organic phase of PLA in dichloromethane, and the external aqueous phase of PVA and sucrose (Fig. 5.5). The addition of surfactants improved the colloidal stability and reduced particle aggregation. The presence of PVA in the outer aqueous phase contributed to particle size regulation and

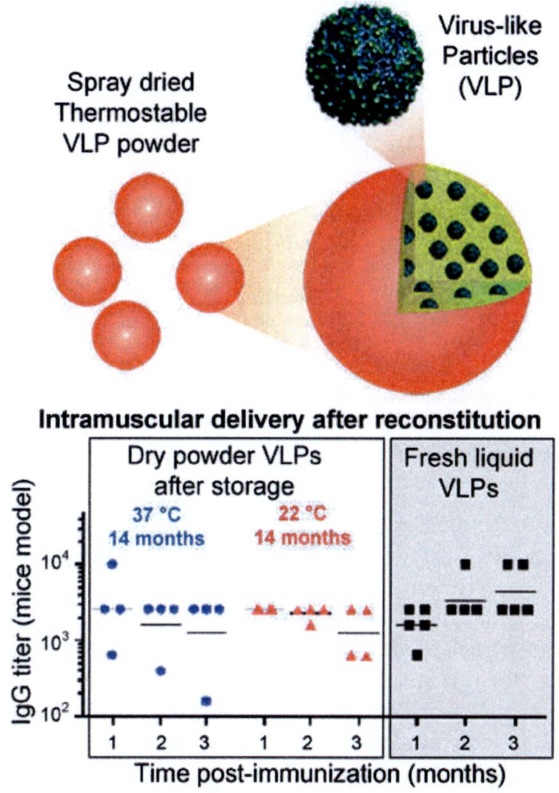

Fig. 5.4 Illustration of virus-like particle (VLP)-based vaccine against human papillomavirus (HPV) infection and immunogenicity data of intramuscularly immunized mice. The reconstituted spray-dried VLPs exhibited similar IgG titers to freshly prepared liquid VLPs (powder reconstituted in phosphate-buffered saline, PBS). Mice were given a single dose of 5 μg VLPs. The spray-dried VLP powders were stored at 37 and 22 °C for 14 months. Sera were collected 1, 2, and 3 months after immunization, and end-point dilution ELISA determined anti-L2 IgG titers. Each dot refers to an individual immunized mouse; lines denote geometric mean IgG titers (Data adapted from Saboo et al. (2016) with permission from the American Chemical Society)

influenced the burst release of protein from the particles. The emulsions were spray-dried in an LU-227 spray dryer (Labultima, India) based on a half-factorial design (2^{5-1}) with 20 experiments, optimizing inlet temperature and feed rate, drying air flow rate, and atomization pressure. The vaccine powders elicited robust IgG responses in intramuscularly immunized mice and rats. Vaccine loading in the microparticles correlated with antibody titer, indicating maintenance of protein integrity during spray drying.

Chablani et al. (2012) used spray drying to develop vaccine microparticles for oral administration that protect the antigen protein from acidic conditions in the stomach. The authors tested bovine albumin as a model vaccine protein. The

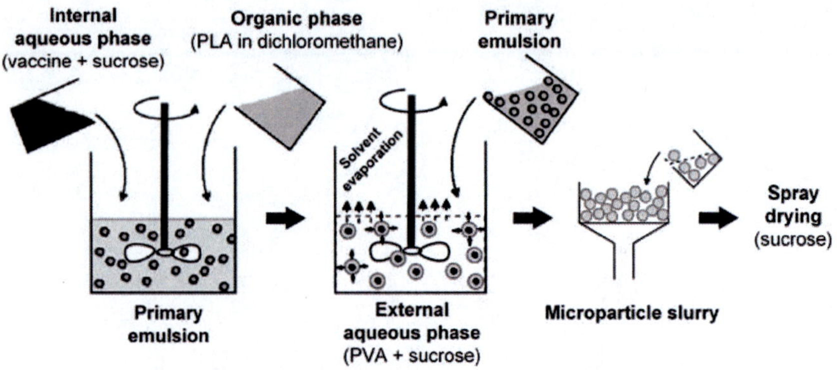

Fig. 5.5 Illustration of the process steps to prepare polymeric microparticles by double emulsions solvent evaporation method and spray drying the microparticle slurry (Adapted from Anish et al. (2014) with permission from Elsevier)

microparticles were formulated with enteric polymers such as Eudragit® FS 30 D and HPMCAS and BSA based on DOE. Studies on M-cell uptake proved the entry of the particles into the Peyer's patches of the small intestine and elicited a desired immune response in mice.

Moreover, the particles offered prolonged release for 8 h. The use of such vaccine particles for oral administration has been suggested to be suitable for numerous diseases such as pneumonia, influenza, HPV, HIV, and cancer therapies. Oral administration is an attractive route of immunization due to ease of administration, low manufacturing cost, and patient compliance.

In summary, the spray drying studies reviewed in this section demonstrate that DOE is a suitable methodology for process improvements and building predictive models for critical product characteristics. However, there is a gap in the literature examining scale-up and technology transfer to larger powder volumes of spray-dried vaccines. Therefore, the next section is dedicated to scale-up studies that shed light on technology transfer.

5.3 Analytical Methods to Characterize Spray-Dried Vaccines

Table 5.3 summarizes the most common analytical methods used to characterize spray-dried vaccines. For rapid progress in formulation development and identification of leading vaccine candidates, it is essential to perform analysis of each spray-dried batch to ensure it meets product specifications. The characterization of spray-dried dry vaccine formulations involves several aspects of performance, including solid-state characterization of the powder, in vitro solution tests, and potency tests.

Table 5.3 Commonly used analytical methods for the characterization of spray-dried vaccines

Analytical method	Purpose	References
Solid-state characterization		
Differential scanning calorimetry (DSC)	Glass transition temperature, crystalline structure	Abdul-Fattah et al. (2007), Jin et al. (2010), McAdams (2011), LeClair et al. (2016a, 2019), Kanojia et al. (2016), Afkhami et al. (2017), Jones et al. (2017), Kanojia (2018)
X-ray diffraction (XRD)	Amorphous versus crystalline structure, morphology	McAdams (2011), LeClair et al. (2016a), Kanojia et al. (2016, 2018a)
Scanning electron microscopy (SEM)	Particle size and shape	Sou et al. (2015), Kanojia et al. (2016, 2018a), Afkhami et al. (2017), Jones et al. (2017), Kanojia (2018), Gomez et al. (2021c)
Laser diffraction	Particle size distribution	Garcia-Contreras et al. (2008), Jin et al. (2010), Kanojia et al. (2016), Afkhami et al. (2017), Kanojia (2018)
Anderson cascade impactor	Aerosolization performance, fine particle fraction	Wong et al. (2007), Burger et al. (2008), Garcia-Contreras et al. (2008), Sou et al. (2015)
X-ray spectroscopy (XPS)	Surface composition	Abdul-Fattah et al. (2007a), Mangal et al. (2015), Li et al. (2016), LeClair et al. (2016a)
BET (Brunauer, Emmett, and Teller) surface area analyzer	Specific surface area	Abdul-Fattah et al. (2007, 2007b, 2008), Truong and Abdul-Fattah (2010), Langford et al. (2018c)
Karl Fisher titration	Moisture content	Sou et al. (2015), LeClair et al. (2016a, 2019), Kanojia et al. (2016), Jones et al. (2017), Kanojia (2018), Gomez et al. (2021c)
Humidified incubator	Hygroscopicity, moisture uptake	Price et al. (2020)
In vitro solution tests		
Stirring and shaking	Reconstitution time	Dani et al. (2007), Bowen et al. (2013), Batens et al. (2018), Massant et al. (2020)
Dynamic light scattering (DLS)	Particle size after reconstitution, polydispersity index	Ingvarsson et al. (2013a), Scherließ et al. (2014), Kanojia et al. (2016), Kanojia (2018), Gomez et al. (2021c)
UV spectrometry	Turbidity after reconstitution	Jiang et al. (2006), Breit and DuBose (2013)
Sodium dodecyl sulfate–polyacrylamide gel electrophoresis (SDS-PAGE)	Protein mass, antigen integrity, and stability	Baras et al. (2000a, b), Maa et al. (2003), Scherließ et al. (2014), Kunda et al. (2015a, b), Jones et al. (2017), Gomez et al. (2021c)

(continued)

Table 5.3 (continued)

Analytical method	Purpose	References
Size exclusion chromatography (SEC)	Protein aggregation	Maury et al. (2005b), Jiang et al. (2006), Abdul-Fattah et al. (2007b), Schüle et al. (2008), Scherließ et al. (2014), Jones et al. (2017)
Circular dichroism (CD)	Antigen integrity and structure	Jiang et al. (2006), Kunda et al. (2015a, b)
Potency assays associated with vaccines		
Enzyme-linked immunosorbent assay (ELISA)	Antibody titers of animal sera (in-vitro), immunogenicity	Baras et al. (2000b), Chen et al. (2010), Año et al. (2011), Agarkhedkar et al. (2014), Kanojia et al. (2018b), Garcea et al. (2020)
Tissue culture infectious dose (TCID50)	Antibody response	Ohtake et al. (2010), Jin et al. (2010), Lovalenti et al. (2016), LeClair et al. (2016a, b, 2019)
Cell culture infectious dose (CCID50)	Antibody response	WHO (2006b), Cape et al. (2010), McAdams (2011), Kumru et al. (2014)
Plaque assay	Antibody response	Burger et al. (2008), Cape et al. (2010), Agarkhedkar et al. (2014), Verch et al. (2018), LeClair et al. (2019)
Fluorescent focus assay (FFA)	Antibody response of multivalent vaccines	Wainwright (2006), Truong-Le and Pham (2007), Naik et al. (2017)
Haemagglutination inhibition assay (HAI)	Hemagglutinin antigen content	Scherließ et al. (2014), Sou et al. (2015), Kanojia (2018), Amorij et al. (2007)

Solid-state characterization includes differential scanning calorimetry (DSC) to determine thermal properties (e.g., glass transition temperature and melting point) (Abdul-Fattah et al. 2007; Jin et al. 2010; McAdams 2011; LeClair et al. 2016a, 2019; Kanojia et al. 2016; Afkhami et al. 2017; Jones et al. 2017; Kanojia 2018) and X-ray diffraction (XRD) for morphological structure analysis and detection of the amorphous or crystalline state of the powders (McAdams 2011; LeClair et al. 2016a; Kanojia et al. 2016, 2018a).

Scanning electron microscopy (SEM) is routinely utilized to analyze the spray-dried powder's surface morphology and size distribution (Sou et al. 2015; Kanojia et al. 2016, 2018a; Afkhami et al. 2017; Jones et al. 2017; Kanojia 2018; Gomez et al. 2021c). In addition, laser diffraction is performed to analyze the particle size distribution (Garcia-Contreras et al. 2008; Jin et al. 2010; Kanojia et al. 2016; Afkhami et al. 2017; Kanojia 2018). For pulmonary applications, an Anderson Cascade Impactor simulating the human respiratory tracts is used to determine the aerosolization performance of the powder (Wong et al. 2007; Burger et al. 2008; Garcia-Contreras et al. 2008; Sou et al. 2015). The fine particulate fraction (FPF) is calculated as a percentage of the emitted dose (ED).

The surface properties of dried vaccine powders also play a role in storage stability. The most commonly used technique to measure surface composition is X-ray photoelectron spectroscopy (XPS) (Abdul-Fattah et al. 2007a; Mangal et al. 2015; Li et al. 2016; LeClair et al. 2016a). The powder's specific surface area (SSA) is determined by the BET (Brunauer, Emmett, and Teller) surface area analyzer. (Abdul-Fattah et al. 2007; Abdul-Fattah et al. 2007b, 2008; Truong and Abdul-Fattah 2010; Langford et al. 2018c). The residual moisture content (RMC) of the spray-dried vaccine powders is usually determined by colorimetric Karl Fisher titration (Sou et al. 2015; LeClair et al. 2016a, 2019; Kanojia et al. 2016; Jones et al. 2017; Kanojia 2018; Gomez et al. 2021c) or thermogravimetric analysis (TGA) measuring the weight loss in a drying oven (Ingvarsson et al. 2013b). In addition, the moisture uptake from the atmosphere is recorded by placing the powder in a humidified incubator and measuring the weight gain over time (Price et al. 2020). Combining these analyzing studies provides a deeper understanding of the vaccine formulation in the dry state and offers clues for the next-step processing conditions, stability, and performance (Breit and DuBose 2013).

The reconstitution time is determined by stirring and shaking until complete powder dissolution (Dani et al. 2007; Bowen et al. 2013; Batens et al. 2018; Massant et al. 2020). Dynamic light scattering (DLS) is used to measure the particle size (e.g., of liposomes, nanoemulsions) after reconstitution and polydispersity index (Ingvarsson et al. 2013a; Scherließ et al. 2014; Kanojia et al. 2016; Kanojia 2018; Gomez et al. 2021c) and light absorption with a UV spectrophotometer to measure the turbidity of the reconstituted solution compared to the pure solution (Jiang et al. 2006). Turbidity assays during resuspension-rate analysis can indicate loss of vaccine integrity during spray drying or resuspension issues induced by excipients or antigen concentration (Breit and DuBose 2013). Sodium dodecyl sulfate–polyacrylamide gel electrophoresis (SDS-PAGE) (Baras et al. 2000a, b; Maa et al. 2003; Scherließ et al. 2014; Kunda et al. 2015b, a; Jones et al. 2017; Gomez et al. 2021c) and size-exclusion chromatography (SEC) (Maury et al. 2005b; Jiang et al. 2006; Abdul-Fattah et al. 2007b; Schüle et al. 2008; Scherließ et al. 2014; Jones et al. 2017) are methods that allow protein separation by mass and size. In addition, circular dichroism (CD) assays can provide information on the macromolecular structure after spray drying and resuspension (Jiang et al. 2006; Kunda et al. 2015a, b).

Vaccine potency and stability testing are the cornerstones of vaccine formulation development (Kumru et al. 2014). Historically, vaccines were released using in vivo potency assays requiring the immunization of dozens of animals (Verch et al. 2018). During early development, vaccine potency is routinely assessed by measuring the total antibody titer generated against the vaccine in immunized mice. However, efforts are underway to replace animal-based potency tests with in vitro potency assays like enzyme-linked immunosorbent assays (ELISA) (Baras et al. 2000b; Chen et al. 2010; Año et al. 2011; Agarkhedkar et al. 2014; Verch et al. 2018; Kanojia et al. 2018b; Garcea et al. 2020) and antigen-specific assays (Breit and DuBose 2013) to quantify antibodies present in the animal sera. As a result, there is increasing reliance on in vitro potency assays, i.e., in vitro cell assays and/or antibody binding assays, as well as purely chemical or physicochemical methods (Ranheim et al. 2015; Verch et al. 2018). The choice of potency assay depends mainly on the vaccine type. For example, live attenuated vaccines often require cell-based assays to measure infectivity, i.e., tissue culture infectious dose (TCID50) (Ohtake et al. 2010; Jin et al. 2010; Lovalenti et al. 2016; LeClair et al. 2016a, b, 2019), cell culture infectious dose (CCID50) (WHO 2006b; Cape et al. 2010; McAdams 2011; Kumru et al. 2014), or plaque-forming units (PFU) (Burger et al. 2008; Cape et al. 2010; Agarkhedkar et al. 2014; Verch et al. 2018; LeClair et al. 2019). For recombinant (subunit) vaccines, potency can be reliably monitored with immunoassays (Verch et al. 2018). For example, the hemagglutination inhibition assay (HAI) quantifies influenza HA antigen content in the formulation (Amorij et al. 2007; Scherließ et al. 2014; Zhu et al. 2014; Sou et al. 2015; Kanojia 2018). In addition, fluorescent focus forming assays (FFA) are used for multivalent vaccines (Wainwright 2006; Truong-Le and Pham 2007; Naik et al. 2017). For more detailed explanations of the respective analytical techniques and vaccine potency tests, the reader is referred to further literature in this field (Ohtake et al. 2010; Miller and Gil 2012; Breit and DuBose 2013; Agarkhedkar et al. 2014; Zhu et al. 2014; Ranheim et al. 2015; Searles and Cherian 2015; Lovalenti et al. 2016; LeClair et al. 2016b, 2019; Kanojia 2018; Verch et al. 2018; Arpagaus et al. 2018; Kanojia et al. 2018b; Price et al. 2020).

Accelerated stability studies with forced degradation are performed in the early stages of vaccine development to identify potential stabilizers and optimal conditions for long-term storage (Hasija et al. 2013). In this regard, the quality guideline Q1A(R2) of the International Conference on Harmonization (ICH 2003) suggests the following accelerated temperature and relative humidity (RH) conditions:

- 25 °C/60% RH or 30 °C/65% RH for at least 12 months for long-term studies,
- 30 °C/65% RH for at least 6 months for intermediate studies, and
- 40 °C/75% RH for at least 6 months for accelerated studies.

Consideration of effects in open and closed containers is necessary for vaccine powder stability studies.

5.4 Scale-Up of Spray-Dried Vaccines

Spray drying of vaccine formulations in dry powder form has proven successful in numerous applications, as described in the previous section based on various published literature. However, a major challenge in product development is transferring spray-dried vaccines to a larger scale. This task requires skilled engineering and practical experience.

A simple possibility to increase the throughput is the parallel arrangement of several identical spray dryers, as used in the ProCepT SD3 concept. Even though this is a relatively expensive solution for scale-up, it is suitable for smaller quantities.

Figure 5.6 shows the typical scales of spray dryers used in the various phases of development for a spray drying process. Formulation development and identifying critical process parameters for scale-up usually start with experimental work at the laboratory scale. This is necessary to evaluate the optimum product and process parameters, such as inlet and outlet temperatures and feed flow rate. In proof-of-concept studies at laboratory scale, powders are produced in the milligram to gram range (e.g., 50 mg to several 100 g), while pilot spray dryers can produce several kilograms of powder per batch. In further developing the spray drying process, the scale-up leads to commercialization and clinical manufacturing through pilot studies. On a commercial scale, spray dryers can produce several hundreds of kilograms of powder per batch. Also shown are the typical powder production rates, although the different scales can overlap qualitatively in the various product development phases.

There is extensive literature available on the scale-up of spray dryers. Table 5.4 summarizes the most important literature of recent years structured by type of publication, i.e., books, book chapters, review papers, research papers, and articles.

Masters (1985, 1994, 2002) particularly emphasized the importance of close collaboration between the end-user and the spray dryer supplier to prevent misunderstandings in scale-up in practice. Skilled design engineers and practical experience are required to scale up spray dryers.

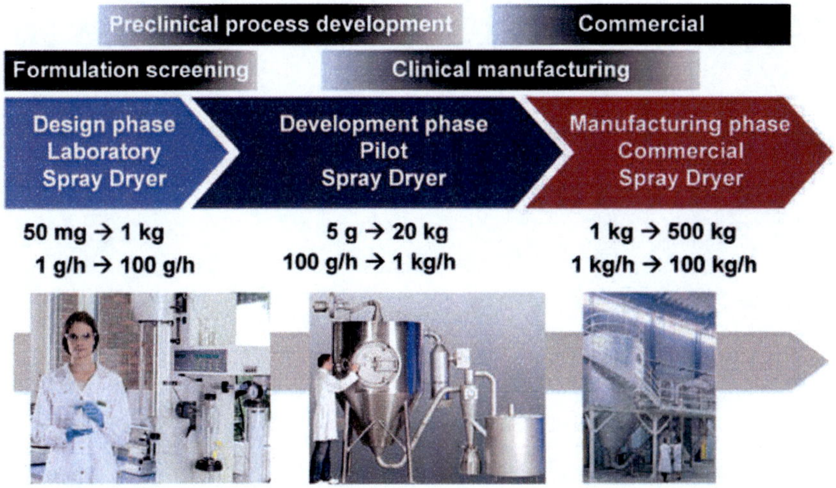

Fig. 5.6 Spray drying scale-up with equipment scales, batch sizes, production rates, and phases across scales (Adapted from Baumann et al. (2013) and Dobry et al. (2015) with images from Poozesh and Bilgili (2019) with permission from Elsevier)

Table 5.4 Relevant literature on the scale-up of spray dryers structured by type of publication and number of pages

Title of the publication	Type of publication	No. of pages	References
Spray drying handbook	Books	697	Masters (1985)
Spray drying in practice		465	Masters (2002)
Spray drying technology	Book chapters	80	Miller and Gil (2012)
Spray drying scale-up and manufacturing		42	Gaspar et al. (2014)
Spray drying and scale-up		26	Dobry et al. (2015)
Drying of pharmaceutical products		24	Pakowski and Mujumdar (2014)
Spray drying of biotherapeutic compounds		17	Langford and Ohtake (2019)
Scale-up of pharmaceutical spray drying using scale-up rules: a review	Review papers	22	Poozesh and Bilgili (2019)
The design and scale-up of spray-dried particle delivery systems		17	Al-Khattawi et al. (2018)
Spray drying formulation of amorphous solid dispersions		24	Singh and Van den Mooter (2016)
Scale-up of spray dryers	Research papers	23	Masters (1994)
Dryer scale-up methodology for the process industries		12	Genskow (1994)
Scaling up the spray drying process from pilot to production scale using an atomized droplet size criterion		11	Thybo et al. (2008)
A model-based methodology for spray drying process development		10	Dobry et al. (2009)
Spray drying with a two-fluid nozzle to produce fine particles: atomization, scale-up, and modeling		10	Kemp et al. (2016)
Technological considerations related to the up-scaling of protein microencapsulation by spray drying		5	Johansen et al. (2000)
Spray drying of vaccines: a review	Articles	14	Arpagaus (2020)
Spray-dried dispersions: efficient scale-up strategy for spray-dried amorphous dispersions		9	DuBose et al. (2013)
Scale-up from the Büchi Mini Spray Dryer B-290 to the Niro Mobile Minor		8	Arpagaus and Schwartzbach (2008)
An integrated approach to spray drying: from scale-up to manufacturing		6	Scrafford (2020)

(continued)

Table 5.4 (continued)

Title of the publication	Type of publication	No. of pages	References
Scale-up & QBD approaches for spray-dried inhalation formulations		6	Costa et al. (2014)
Scale-up of spray dried amorphous solid dispersions		5	Santos et al. (2013)
Scale-up methodology for pharmaceutical spray drying		5	Gil et al. (2010)

For example, Thybo et al. (2008) demonstrated the scale-up from the pilot (Mobile Minor, PSD-1) to production scale (PSD-4, approx. 18-fold scale-up in feed rate) for formulations of acetaminophen (also known as paracetamol, reducing fever and pain) and PVP-K30 by matching droplet size distribution across scales and concluded that some production scale experiments are still required to obtain similar product characteristics to pilot scale.

Genskow (1994) discussed a general approach to process the scale-up of various dryers as unit operations. Pakowski and Mujumdar (2014) emphasized the hygienic requirements of a typical spray dryer for an aseptic operation. Miller and Gil (2012) provided an in-depth review of spray drying technology and discussed its application in the formulation of poorly water-soluble drugs. Gil et al. (2010) outlined a scale-up methodology for pharmaceutical spray drying, and Costa et al. (2014) showed how to apply the scale-up and QbD approach for spray-dried inhalation formulations. Gaspar et al. (2014) demonstrated the scale-up by using a QbD methodology to produce amorphous solid dispersions. The development of predictive models provided a reliable design space fundamental for a better understanding of the scale-up process. Singh and Van den Mooter (2016) discussed scale-up challenges, QbD, and downstream processing of spray-dried amorphous solid dispersions.

Dobry et al. (2009, 2015) and DuBose et al. (2013) presented a model-based methodology for manufacturing spray-dried amorphous dispersions. Science-based approaches were proposed, such as modeling and experimental strategies. Similarly, Scrafford (2020) outlined an integrated approach to spray drying from scale-up to manufacturing to increase the bioavailability of poorly soluble APIs. Langford and Ohtake (2019) summarized some spray dryers for applying biotherapeutic compounds at different scales from R&D to commercial systems and those offering aseptic processing capabilities.

A recent review by Poozesh and Bilgili (2019) examined the scale-up rules for pharmaceutical spray drying with non-dimensional parameters and multi-scale analysis. Al-Khattawi et al. (2018) reviewed the design and scale-up of spray-dried particle delivery systems and summarized the current industrial approaches for upscaling pharmaceutical spray drying. Kemp et al. (2016) looked into the atomization, scale-up, and modeling of spray drying with a two-fluid nozzle to produce fine particles and better understand the drying times to the droplet in spray dryers.

Last but not least, Arpagaus and Schwartzbach (2008) presented a simple procedure on how to perform a scale-up from a laboratory-scale Mini Spray Dryer B-290 to a pilot-scale Niro Mobile Minor™. An example will illustrate this procedure in Sect. 5.4. Arpagaus (2020) summarized some scale-up studies of spray-dried vaccines.

The scale-up method used by many authors is typically divided into stages. The first stage is about atomizing the droplets with particle formation during drying to obtain a similar particle size and structure during scale-up. The second stage elaborates a robust thermodynamic operating space that can be easily scaled up. Such a scale-up method is relatively straightforward to apply and has been tested in numerous spray-dried formulations across all production scales (Dobry et al. 2009, 2015; Gil et al. 2010; Baumann et al. 2013; Breit and DuBose 2013; DuBose et al. 2013; Costa et al. 2014; Gaspar et al. 2014).

Atomization is typically considered first because it plays an essential role in controlling droplet size and, along with solids concentration (see Eq. 3.1 in Sect. 3.4.4), controls the final particle size of the dried powder. This step aims to study similar particles production in different scales of spray dryers. Matching of droplet size distributions from nozzles of different sizes is typically evaluated by droplet size measurement using laser diffraction.

For a given product, empirical correlations between atomization parameters (i.e., feed composition, viscosity, nozzle geometry, atomization pressure, spray gas, and feed flow rates) and droplet size can be developed to achieve the desired particle size over a variety of scales (Masters 1985, 2002; Arpagaus 2007; Gaspar et al. 2014; Poozesh and Bilgili 2019). The goal is typically to keep the target particle size constant across scales.

As an example, Fig. 5.7 shows for an aqueous HPMCAS solution that the target droplet size can be kept relatively constant when the liquid feed rate of the corresponding spray dryers are increased, and the applied atomization gas pressures for the nozzles are adjusted from about 15 bar at the pilot-scale (GEA Niro PharmaSD™ PSD-1) to more than 100 bar at production scale (PSD-4).

The second scale-up step considers thermodynamics. A common practice is to characterize a robust thermodynamic design space and revalidate it when scaling up the spray drying process. The design space is used to select a suitable drying gas outlet temperature (T_{OUT}) and a feed rate ($\dot{M}_{FEED}$). The design space can be determined by simple mass and energy balance around the drying chamber as a control volume (Dobry et al. 2009, 2015). The energy required for evaporation is compared to the energy lost by the drying gas into the system (Eqs. 5.1 and 5.2, neglecting heat losses).

$$\dot{M}_{FEED} \cdot (1 - x_{SOLIDS}) \cdot \Delta H_{EVAP} = \dot{M}_{GAS} \cdot c_P \cdot (T_{IN} - T_{OUT}) \tag{5.1}$$

where ΔH_{EVAP} is the heat of solvent evaporation, x_{SOLIDS} the solids mass fraction in the feed, and c_P the heat capacity of the drying gas. Solving for T_{OUT} results

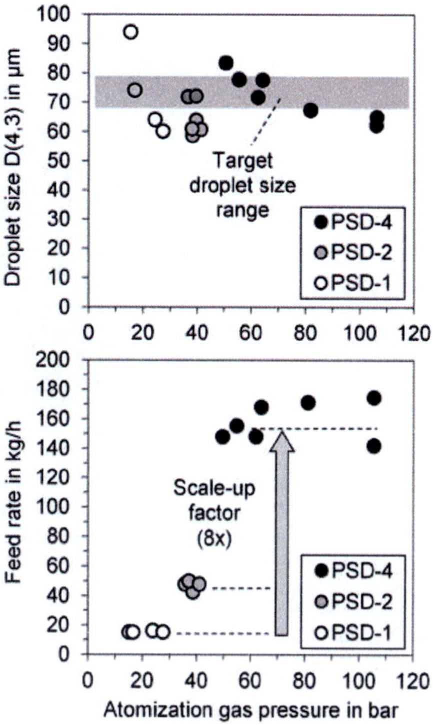

Fig. 5.7 Droplet size (top) and feed rate (bottom) of an aqueous hydroxypropyl methylcellulose acetate succinate (HPMCAS) solution as a function of atomization pressure in pressure-swirl nozzles at increasing scales for GEA Niro PharmaSDTM PSD-1, PSD-2, and PSD-4 spray dryers (Data adapted from DuBose et al. (2013) with permission from Drug Development & Delivery)

$$T_{OUT} = T_{IN} - \frac{\dot{M}_{FEED}}{\dot{M}_{GAS}} \cdot \frac{(1 - x_{SOLIDS}) \cdot \Delta H_{EVAP}}{c_P} \quad (5.2)$$

From this simplified equation, T_{OUT} and, with further equations, Y_{OUT}, the vapor concentration at the dryer outlet can be calculated as a function of T_{IN}, $\dot{M}_{GAS}$ and $\dot{M}_{FEED}$. Since $\dot{M}_{GAS}$ is typically fixed at a particular drying scale, the specific ratio of $\dot{M}_{FEED}/\dot{M}_{GAS}$ can be used to compare different scales of spray dryers. As a remark, heat losses should usually be considered, especially when scaling up from laboratory scale (typically glass) to pilot scale when stainless steel is used.

An illustrative example of a thermodynamic design space for acetone is shown in Fig. 5.8 (Dobry et al. 2009, 2015). The specific drying ratio $\dot{M}_{FEED}/\dot{M}_{GAS}$ is plotted as a function of the drying gas inlet temperature T_{IN} of the spray dryer, resulting in contours of constant dryer outlet temperature T_{OUT} and vapor concentration at the outlet Y_{OUT}.

An operating design space can be defined by applying process constraints (Dobry et al. 2009, 2015). The overlaid constraints in Fig. 5.8 include

- a process efficiency constraint specifying a minimal specific drying ratio $\dot{M}_{FEED}/\dot{M}_{GAS}$ to fulfill the process throughput targets in the operating space,

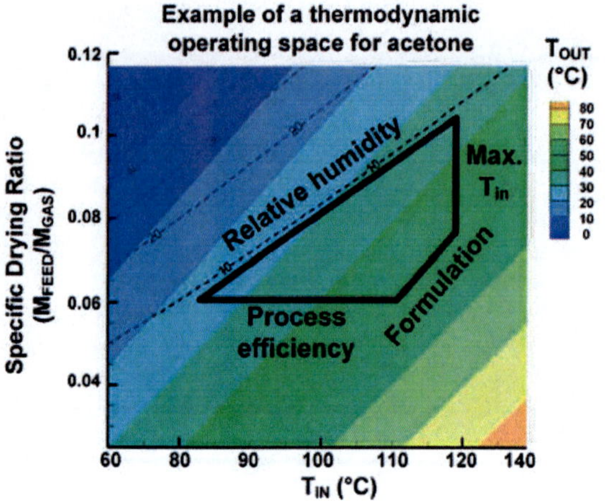

Fig. 5.8 Example of a thermodynamic operating space of a spray drying process with acetone as a base for the production of amorphous solid dispersions. The colored contours represent the drying gas outlet temperature (T_{OUT}) and the dashed lines constant relative humidities (Y_{OUT}) in the outlet drying gas. Constraints are shown for the feed rate for process efficiency, the relative humidity for complete drying, the maximal outlet temperature for thermal stability of the formulation, and the maximal inlet temperature reflecting the chemical stability of the formulation (Image adapted from Dobry et al. (2009, 2015) with permission from Springer Nature and John Wiley and Sons)

- a formulation constraint along with a T_{OUT} contour line describing the maximum acceptable thermal stability of the formulation to prevent degradation or inactivation,
- a maximal T_{IN} constraint to minimize process risks such as degradation or discoloration and to prevent deposits of particles at the gas disperser inlet into the spray dryer, and
- a relative humidity constraint (e.g., at 10% line) preventing insufficient drying, yield problems, and excessive moisture content in the final powder.

For aqueous vaccine formulations, an operating space could be derived similarly. An approximate feed rate conversion factor from acetone to water based on the different heat of evaporation would be around 0.22 (Arpagaus and Schwartzbach 2008).

An important scale-up criterion is to keep the vapor concentration in the gas phase at the outlet constant across scales, as this correlates with the residual moisture content in the powder product (Arpagaus and Schwartzbach 2008; Dobry et al. 2009, 2015; DuBose et al. 2013; Costa et al. 2014; Gaspar et al. 2014; Al-Khattawi et al. 2018). For this purpose, the outlet drying gas temperature is controlled by adjusting the inlet temperature, feed and drying gas flow rates. The residual moisture in the powder is of particular importance as it strongly affects its glass transition temperature, stability of an encapsulated vaccine, and microbiology.

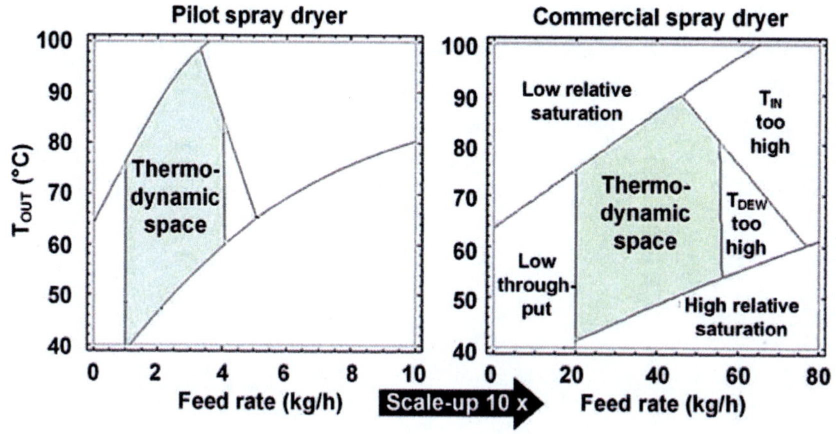

Fig. 5.9 Thermodynamic operating space at pilot and commercial scales (Adapted from Gaspar et al. (2014) with permission from Springer Nature)

Another example of a thermodynamic operating space is shown in Fig. 5.9 (Gaspar et al. 2014). The diagram on the left shows the thermodynamic operating space in a pilot spray dryer, and the chart on the right shows a commercial-scale spray dryer with an approximately 10 times larger feed rate. Various imposed constraints limit the operating space, and these relationships can be seen at different scales. Constraints include low throughput (e.g., the process is inefficient), too low and too high relative saturation of the outlet gas (e.g., low and high residual moisture content in the powder), inlet temperature too high (e.g., equipment limitations), feed rate too high leading to too high dew points (T_{DEW}) (e.g., condensation expected), product requirements (e.g., T_{OUT} limited by product degradation, stability, or sticking behavior), and other process constraints (e.g., drying gas flow rate limited by the gas disperser or flow requirements). A thermodynamic analysis is a powerful tool that often replaces experiments at different scales.

Published studies on process scale-up of spray-dried vaccines are generally scarce (Arpagaus 2020). Table 5.5 summarizes some studies by vaccine type, spray dryer model, and scale-up factor. The main findings are highlighted (Johansen et al. 2000; Chen et al. 2010; Bowen et al. 2013; Zhu et al. 2014; Gikanga et al. 2015; Tlaxca et al. 2015).

Zhu et al. (2014) performed a scale-up from a laboratory Mini Spray Dryer B-290 to a custom-built pilot-scale spray dryer BLD-1 (Bend Research) to formulate recombinant hemagglutinin influenza antigen HAC1. During scale-up, the drying temperatures were kept constant at T_{IN} 100 °C and T_{OUT} between 58 and 62 °C. After excipient screening, a lead formulation consisting of 7.5 wt% trehalose and 2.5 wt% hydrolyzed gelatin was selected and transferred to the pilot spray dryer. The typical batch size at the laboratory scale was 10 mL (about 0.5 g dried powder) for the first screening and 65 mL for a thermostability study. Process yields ranged from approximately 70 to 88%, and the mean particle sizes were

Table 5.5 Scale-up studies of spray-dried vaccines

Vaccine (substance)	Spray dryer models for scale-up	Scale-up factor (powder production)	T_{IN}/T_{OUT} (yield)	References
Influenza antigen HAC1	B-290 → BLD-1 (36 mL/h, 0.5 g) → (240 mL/h, 50 g)	100× (powder batch size) 6.7× (feed rate)	100 °C/58 to 62 °C (> 90%)	Zhu et al. (2014)
Respiratory syncytial virus (RSV)	B-190 → MS75 (1 mL/min) → (10 mL/min)	10× (feed rate)	n.a./50 to 60 °C (> 80%)	Tlaxca et al. (2015)
Tuberculosis	B-290 → MS35 (15 g/h) → (60 g/h) powder	4× (powder production)	65 to 125 °C/34 to 50 °C (70 to 85%)	Jin et al. (2010), Jin and Tsao (2013)
Hepatitis B Meningitis A	Mobile minor	1 L batch (5000 to 10,000 doses)	90 to 95 °C/60 to 71 °C (> 90%)	Chen et al. (2010)
Tetanus toxoid	B-191 → B-191 (same equipment) (1 g) → (10 g)	10× (powder batch size)	45 °C/n.a. (ethyl formate) (n.a.)	Johansen et al. (2000)
Monoclonal antibodies (mAbs)	MS35 → MS150 (15 mL/min) → (50 mL/min)	3.3× (feed rate)	180 to 190 °C/80 to 90 °C (> 95%)	Bowen et al. (2013), Gikanga et al. (2015)

Adapted from Arpagaus (2020)

between 3.4 and 7.2 µm. The batch size on the pilot spray dryer was about 100 times larger with 500 mL containing 50 g solids. The study reported process yields of 90% and higher. The feed rate increased from about 36 to 240 mL/h from laboratory to pilot scale, corresponding to a scale-up factor of about 6.7. All powder batches maintained an amorphous state (Tg of 120 °C) with a water content of 4.0 ± 0.5 wt% and were physically stable for at least 6 months of storage at elevated temperatures. SRID (single radial immunodiffusion) assay tests showed little or no loss of potency during a 6-month stability study. In mice, the spray-dried powders induced equivalent immune responses to the HAC1 vaccine control after the powders were stored at temperatures up to 50 °C for 3 months. Overall, the pilot-scale scale-up process was reproducible, did not alter the immunogenicity of the HAC1 antigen, and the vaccine formulation was thermostable.

Tlaxca et al. (2015) spray-dried a live-attenuated intranasal vaccine (MEDI-559) to prevent lower respiratory tract disease caused by a respiratory syncytial virus (RSV). A Mini Spray Dryer B-190 and a pilot-scale Anhydro MicraSpray 75 were used. The feed rate was approximately 1 mL/min, and process yields were typically 50% at the laboratory scale. Feed rates ranged from 7 to 11 mL/min at the pilot scale, with powder yields exceeding 80%. Residual moisture contents of 3 to 4 wt% were achieved at outlet temperatures of 50 and 60 °C. The formulation consisted of trehalose for stabilization, phosphate buffer to maintain pH and ionic strength, and a KCl salt for tonicity. Other excipients such as mannitol or dextran were investigated for bulk matrix formation and phospholipids and surfactants to stabilize the viral lipid membrane. The initial viral titer loss of the powder was approximately 1 $\log_{10}$ FFU/mL (fluorescent forming units) when stored at 25 °C and stabilized after 2 months. The authors emphasized that the formulation composition and residual moisture content affected the stability and required more detailed studies.

Jin and Tsao (2013) reported that a MicraSpray MS-35 was set up in Aeras (today IAVI) facility to manufacture powdered tuberculosis vaccines. The spray dryer was custom designed by Anhydro to meet the FDA requirement of the aseptic concept to produce human vaccines. The aseptic spray dryer produced 180–250 g of powder within 2–4 h (Table 5.6). Compared to a Mini Spray Dryer B-290, the MicraSpray MS-35 cGMP spray dryer increased the recovery yield by 20% (i.e., from 50 to 65% to 70 to 85%) and the powder production by a factor of four (i.e., from 15 to 60 g/h) (Jin and Tsao 2013).

Comparing the cost-effectiveness between the MicraSpray MS-35 cGMP spray dryer and a traditional lyophilizer (shelf area of 40 ft^2 or 3.7 m^2) showed a shorter processing cycle and larger evaporative capability (Table 5.7). Other cost factors like labor, materials, consumables (e.g., process gases, electricity, cooling water), and equipment were comparable (Jin and Tsao 2013).

Table 5.6 Performance of the SPX Anhydro MicraSpray MS-35 cGMP spray dryer when producing vaccine powders

Vaccine	Spray drying period per batch (h)	Yield of powder collection at the end of each run (g)	Dosages per batch (10 mg/dosage)
Bacteria/virus (live)	2 to 4	180	18,000
Protein/peptide	2 to 4	250	25,000

Data from Jin and Tsao (2013) with permission from Springer Nature

Table 5.7 Comparison of cost-effectiveness between an MS-35 cGMP spray dryer and a traditional lyophilizer with a shelf area of 40 ft^2 or 3.7 m^2

Equipment	Process time (h)	Labor cost	Materials cost	Equipment cost	Consumables (gas, electricity, cooling water)	Output dosages[a] (250 g solid)
Spray dryer MS-35	2 to 4	Comparable	Comparable	Comparable	$ 300	10,000
Lyophilizer (Shelf area 3.7 m^2)	2 to 72	Comparable	Comparable	Comparable	$ 300	10,000

[a]25 mg per dose or vial, 0.5 mL filling for lyophilizer vial, 25 mg for spray-dried powder (Data from Jin and Tsao (2013) with permission from Springer Nature)

In comparison, the freeze dryer has a higher yield than the MS-35 spray dryer, whose recovery was reported to be 70–85%. For the application of spray drying in vaccine development, Jin and Tsao (2013) pointed out essential aspects to be thoroughly investigated:

- An optimal formulation needs to be found for each vaccine, closely related to good stability, low hygroscopicity, and fast powder filling of the final product.
- A low inlet temperature is required to avoid denaturation or inactivation of proteins or live active ingredients during the drying process, which affects the yield of the final products due to reduced feed rate.
- Aseptic design of spray dryer for vaccine production.
- Requirement of precise powder filling.
- Effective and economical dry powder inhalers to enable mass vaccination.

These challenges still exist today, although progress has been made in all areas over the last years.

Chen et al. (2010) developed thermostable formulations of a hepatitis B subunit vaccine and a meningitis A polysaccharide conjugate vaccine by spray drying in a Mobile Minor™. The batch size per experiment was approximately 1 L, sufficient to produce 5000–10,000 vaccine doses. Trehalose was chosen as the stabilizer and sodium hydrogen phosphate (NaH_2PO_4) as an excipient. The process parameters selected were a feed rate of 0.52 kg/h, a solid concentration of 11 wt%, and an inlet temperature of 90–95 °C with an outlet temperature of 60–71 °C. Spray drying produced free-flowing fine white powders with particle sizes in the desired range of 5 to 10 μm and residual moisture contents ranging from 2.2 to 3.4 wt%. The process yield was above 90% on the pilot scale. The hepatitis B vaccine showed good storage stability (based on bioassays) for at least 2 years at 37 °C, while the meningitis A vaccine formulations showed stability up to 60 °C for 2 weeks. The particle size of the vaccines also did not change after storage. Furthermore, antibody responses in Balb/c mice induced by a single immunization with reconstituted spray-dried formulation after 15 months of storage at 37 °C were not significantly different from those induced by the standard liquid vaccine on day 0 or after 15 months of storage at 4 °C.

In a previous study, Johansen et al. (2000) prepared clinical samples of a tetanus vaccine for single administration based on biodegradable poly(lactide-co-glycolide) PLGA polymer microspheres. Protein solutions in water and PLGA in ethyl formate or dichloromethane were emulsified by ultrasound before spray drying on a Mini Spray Dryer B-191. The inlet temperature of the spray dryer was set at 45 °C. Finally, a research-scale batch (1 g of product) and a development-scale batch (10 g) were compared. Thus, approximately 300–400 human doses of the microsphere-based tetanus vaccine could be produced. The vaccine loading efficiency ranged from 32 to 36%. The results showed that upscaling the vaccine's encapsulation by spray drying was feasible. The process and product quality were mainly affected by the type of solvent, polymer concentration, additives, the ratio of aqueous to the organic phase, and product removal.

For monoclonal antibodies (mAbs), Bowen et al. (2013) and Gikanga et al. (2015) demonstrated the feasibility of scale-up from a MicraSpray MS-35 spray dryer to a pilot-scale MS-150 system. Based on the drying air throughput (38 to 154 kg/h), the capacity could be increased by a factor of 4. The feed rate was increased from 15 mL/min to 50 mL/min in comparison, corresponding to a scale-up factor of about 3.3. Suitable trehalose-based formulations resulted in acceptable stability and a good powder yield of over 95%. A low sugar concentration favored the yield of the spray drying process and allowed a lower viscosity of the solution during reconstitution. The typical reconstitution times of the spray-dried mAb powders were less than 2 min. The properties of the powders, including residual water content, particle size, morphology, and storage stability, were found to be similar at both scales. Moisture contents of about 5 wt% were achieved at 180 to 190 °C inlet and 80 to 90 °C outlet temperatures. Despite the high inlet temperature, the physical stability of the spray-dried mAbs was comparable to freeze-dried samples (Bowen et al. 2013). The activity of the mAb formulations after spray drying was comparable to that of the respective liquid formulations before spray drying. These studies showed that spray drying is an option for the production of highly concentrated mAb powders and allows bulk storage.

5.5 Excursus: Scale-Up from Mini Spray Dryer B-290 to Mobile Minor

The following example shows a step-by-step procedure for upscaling an aqueous formulation from a laboratory Mini Spray Dryer B-290 to a Mobile Minor on a pilot scale. The calculations are based on the work of Arpagaus and Schwartzbach (2008). Figure 5.10 illustrates the stepwise scale-up procedure.

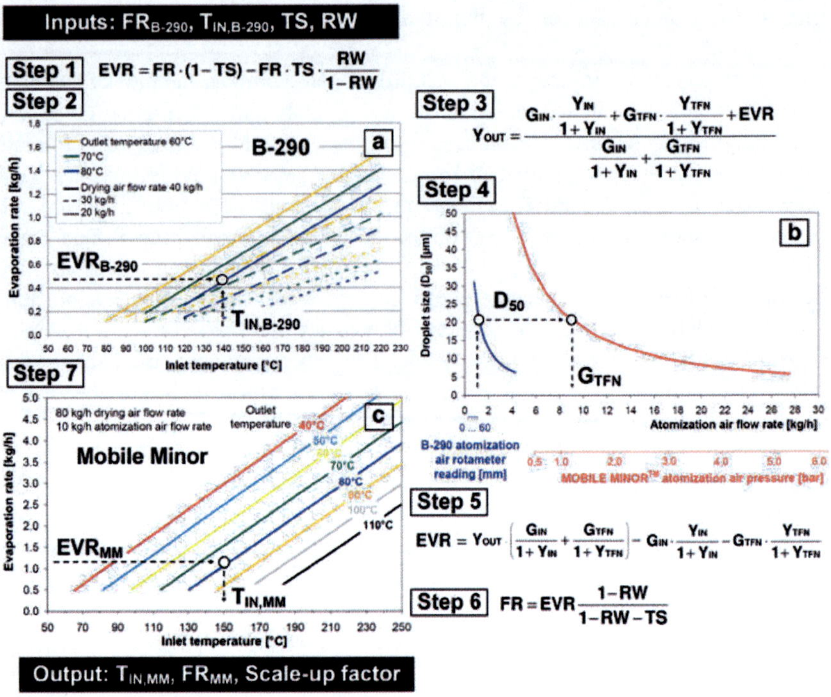

Fig. 5.10 Scale-up procedure from a Mini Spray Dryer B-290 to a Mobile Minor (Data adapted from Arpagaus and Schwartzbach (2008))

The starting point is when good process conditions have been found with the Mini Spray Dryer B-290. Then, the task is to produce a powder with similar residual moisture content and particle size on the Mobile Minor but with a higher production rate.

The outlet air temperature (T_{OUT}), the droplet size (D_{50}), and the outlet vapor concentration (Y_{OUT}) are the critical parameters for scale-up. T_{OUT} is a measured parameter in every spray dryer and is kept constant during scale-up. The moisture content in the final powder (*RW*) depends on the humidity of the air at the outlet of the drying chamber. The water evaporation rate (*EVR*) is calculated from a simple mass balance (Eq. 5.3) with the feed rate (*FR*), the total solids in the feed (*TS*), and *RW* (Step 1).

$$EVR = FR \cdot (1 - TS) - FR \cdot TS \cdot \frac{RW}{1 - RW} \tag{5.3}$$

The *FR* can be accurately determined by weighing the feed container before and after an experimental run and dividing the weight by the time. *RW* can be determined by a moisture analyzer.

The calculation example assumes a feed rate of 0.6 kg/h, a total solids content of 0.25 kg solids per kg feed, and 0.05 kg residual water content per kg final powder. These parameters result in a water evaporation rate of 0.44 kg/h. The drying air flow rate in the Mini Spray Dryer B-290 is adjusted by the aspirator setting, an indicative reading. However, this is not sufficiently accurate for the scale-up calculations as the flow rate may change with the pressure drop across the spray dryer, e.g., due to powder build-up in the outlet filter. Instead, the accurate inlet and outlet air temperature readings are used to estimate the drying air flow rate (in kg/h) from the evaporative capacity diagram (design space) of the Mini Spray Dryer B-290 (Fig. 5.10a) (Step 2). An evaporation rate of 0.44 kg/h at 140 °C inlet and 80 °C outlet temperature corresponds to a drying air flow rate ($G_{IN,B-290}$) of about 40 kg/h (including water vapor). Now, the outlet drying air vapor concentration (Y_{OUT}) can be calculated from a simple mass balance across the spray dryer (Step 3) (Eq. 5.4).

$$Y_{OUT} = \frac{G_{IN} \cdot \frac{Y_{IN}}{1+Y_{IN}} + G_{TFN} \cdot \frac{Y_{TFN}}{1+Y_{TFN}} + EVR}{\frac{G_{IN}}{1+G_{IN}} + \frac{G_{TFN}}{1+G_{TFN}}} \tag{5.4}$$

If the drying gas is ambient air, the water concentration may vary from day to day, ranging from about 2 to 20 g water per kg dry air. Local weather stations provide humidity data. A dehumidifier (e.g., B-296) can be used to lower the relative humidity of the inlet air, which improves the drying performance, especially for hygroscopic materials. The compressed air used for the two-fluid nozzle is typically dry. In the calculation example, 10 g of water vapor per kg of dry air is used for the inlet drying gas and 2 g for the atomizing air. The typical atomizing air flow rate in the Mini Spray Dryer B-290 is around 0.5 kg/h (including water vapor), resulting in about 20 μm sized droplets. Using this information, Eq. 5.4 calculates an Y_{OUT} of 0.021 kg of vapor per kg of dry air. This value is kept constant during scale-up.

The next step is to find spray conditions to achieve the same particle size. A simple approach is to operate the nozzle of the Mobile Minor at conditions providing similar droplet sizes as in the B-290 nozzle. For this purpose, Fig. 5.10b shows the mean droplet size for water (D_{50}) obtained with both two-fluid nozzles as a function of the atomization air flow rate (G_{TFN}). The higher the atomization air flow rate, the smaller the mean droplet size. The effects of liquid feed rate are minor within the relevant flow range and can be disregarded in this approximate method. The curves can be used to estimate the required G_{TFN} (or supplied atomization air pressure) to achieve a similar droplet size in the Mobile Minor as is the Mini Spray Dryer B-290. If larger particles are required, the Mobile Minor offers further options like a larger drying chamber with longer drying times or other atomization systems. In the example, the required atomization air flow rate for 20 μm droplets is about 0.5 kg/h in the B-290 and about 9 kg/h in the Mobile Minor (Step 4). The atomization gas pressures were comparable at around 1.3 bar.

In most spray dryers, the drying air flow rate is kept within a range to maintain correct loading and velocity in the air disperser, powder transporting ducts, cyclone, and filters. Spray dryers are typically optimized for a specific drying air flow rate. In a Mobile Minor (*MM*), a typical drying air flow rate used is about 80 kg/h ($G_{IN,MM}$). With this information, the *EVR* in the Mobile Minor can be calculated according to Eq. (5.5) (Step 5). The example results in a water evaporation rate (EVR_{MM}) of 1.04 kg water per hour. With the given *RW* and *TS* values, a feed rate (FR_{MM}) of 1.41 kg/h is calculated for the Mobile Minor with Eq. 5.6 (Step 6).

$$EVR = Y_{OUT} \cdot \left(\frac{G_{IN}}{1 + Y_{IN}} + \frac{G_{TFN}}{1 + Y_{TFN}} \right) - G_{IN} \cdot \frac{Y_{IN}}{1 + Y_{IN}} - G_{TFN} \cdot \frac{Y_{TFN}}{1 + Y_{TFN}} \tag{5.5}$$

$$FR = EVR \cdot \frac{1 - RW}{1 - RW - TS} \tag{5.6}$$

Finally, the drying air inlet temperature of the Mobile Minor ($T_{IN,MM}$) can be estimated from Fig. 5.10c to be about 150 °C (Step 7). The graph shows the design space at a drying airflow rate of 80 kg/h and an atomizing air flow rate of 10 kg/h. The final results of the scale-up example are summarized in Table 5.8. The scale-up factor achieved is 2.35 based on the feed flow rate.

Table 5.8 Summary of the results of the scale-up calculation example

	Parameters	Mini spray dryer B-290	Mobile minor
Fixed	Total solids content in the feed (*TS*)	0.25 kg solids per kg feed	
	Residual moisture in the final powder (*RW*)	0.05 kg water per kg powder	
	Vapor concentration of inlet drying air (Y_{IN})	0.01 kg vapor per kg dry air	
	Atomization air vapor concentration (Y_{TFN})	0.002 kg vapor per kg dry air	
Similar	Mean droplet size (D_{50})	20 µm	
	Outlet drying air temperature (T_{OUT})	80 °C	
	Vapor concentration of outlet drying air (Y_{OUT})	0.021 kg vapor per kg dry air	
Scaled	Atomization air flow rate (G_{TFN})	0.5 kg/h	9 kg/h
	Inlet drying air temperature (T_{IN})	140 °C	150 °C
	Water evaporation rate (*EVR*)	0.44 kg water per h	1.04 kg water per h
	Feed flow rate (*FR*)	0.60 kg feed per h	1.41 kg feed per h
	Scale-up factor	2.35	

Adapted from Arpagaus and Schwartzbach (2008)

In practice, it is recommended to first bring the spray dryer to uniform temperature conditions by atomizing pure water. The estimated inlet temperature from Fig. 5.10c can be adjusted to match the evaporation rate during this start-up period. Then, the pump is switched to the actual liquid feed when steady-state conditions are reached. Adjustment of the feed rate is probably necessary since the feed now contains solids.

5.6 Summary and Conclusions

In summary, this review of published DOE studies on spray-dried vaccines shows that experimental design is readily used in vaccine research to understand better the interaction between spray drying parameters such as the temperature of the inlet drying gas, liquid feed rate or atomization gas flow rate, and formulation parameters on dry powder properties such as particle size, vaccine activity, immune response, storage stability, moisture content, and product yield.

About a dozen publications had been published on DOE studies of spray-dried vaccines and could be summarized as examples. Spray drying of vaccines with good yield and retained activity is possible, as demonstrated for various applications, including:

- tuberculosis (Price et al. 2020),
- polio virus (Kanojia et al. 2018),
- influenza virus (Lovalenti et al. 2016; Kanojia et al. 2016),
- Rotavirus (Gill et al. 2017),
- human type 5 adenoviral vector (LeClair et al. 2016),
- herpes simplex virus (LeClair et al. 2019),
- human papillomavirus (Saboo et al. 2016; Peabody et al. 2017),
- pneumococcus (Anish et al. 2014),
- heat-sensitive bovine albumin as model protein (Chablani et al. 2012), and monoclonal antibodies (Bowen et al. 2013; Gikanga et al. 2015).

Most studies have been carried out on a laboratory scale to demonstrate the feasibility. The bench-top design and small dimensions of laboratory spray dryers offer quickly drying of small powder amounts. About 20–30 experimental runs already provide a sufficient basis for decision-making to identify critical process and formulation parameters and derive optimization potentials and conclusions.

The DOE approach helps save time in optimizing spray drying conditions and reduce material costs of vaccines in the early stages of development. Some DOE studies created predictive models for spray drying parameters and design space to better understand the interaction between process parameters and formulation.

In particular, DOE is used for screening excipients to stabilize vaccine powders, as vaccines are exposed to process stresses such as heat, shear, and desiccation during spray drying, which can degrade the vaccine. The optimal selection and composition of excipients is a playground for formulation specialists, and there is a great deal of expertise in this area. Combining the right excipient mixture and delivery vehicles such as emulsions, suspensions, liposomes, or solutions are diverse. Typical vaccine formulations include a combination of a polyol (sugar), a polymer, a surfactant, and an amino acid, each with its specific stability properties.

In addition, DOE studies enable the identification of key drying parameters to maximize storage stability at room temperature, maintain immunization activity with minimal loss of potency, achieve high powder yield to increase productivity, and achieve suitable particle size for vaccine loading and potency.

The most common analytical methods used to characterize spray-dried vaccines involves solid-state characterization of the powder (e.g., by DSC, XRD, SEM, Laser diffraction, Anderson cascade impactor, XPS, BET surface area analyzer, Karl Fisher titration, Humidified incubator), in vitro solution tests (e.g., by stirring and shaking, DLS, UV spectrometry, SDS-PAGE, SEC, CD), and potency assays (e.g., by ELISA, TCID50, CCID50, plaque assay, FFA, HAI).

The case studies of spray-dried powder vaccine formulations generally demonstrate improved thermal storage stability compared to liquid formulations. Even after months of powder storage at room temperature, reconstituted spray-dried vaccines induced similarly high antibody titers in animal models as immunization with the fresh liquid vaccine. However, it appears that studies in humans have not been reported in the literature so far. Based on the available literature, there is no approved spray-dried vaccine on the market yet. Pointing in this direction, Agarkhedkar et al. (2014) reported a Phase I clinical trial in which people were immunized with dry powder measles vaccine by inhalation. Despite these promising data, it is premature to claim that spray-drying technology is an established process for vaccine production.

Furthermore, spray drying enables particle engineering, i.e., designing particles with the desired functionality. Accordingly, spray-dried vaccine formulations can be an alternative to conventional lyophilized vaccines. Additional routes of administration such as pulmonary delivery or oral administration with enteric polymer coatings for protection are approached.

In the future, the DoE approach is expected to increase as both regulatory agencies and the industry place higher demands on vaccine quality management, which could also accelerate the approval of spray-dried vaccines. In addition, DOE studies are part of the continued and increased focus on QbD for companies looking to demonstrate product and process understanding.

The second part of this section highlighted the topic of the scale-up of spray-dried vaccines. Numerous pieces of literature have already been published on spray dryer scale-up, focusing on pharmaceutical applications and scale-up methodologies. Some scale-up studies also showed the technology transfer of spray-dried vaccine powders from laboratory to pilot scale up to 100 g powder quantities per batch (around 1 L of feed). Scale-up factors of 10 have been reported based on the feed rate.

It is recognized that working out the design space of a spray drying process from mass and energy balances is essential to understand and support the scale-up of the spray drying process. In most cases, the biggest challenge in scale-up is related to particle size and residual moisture content. Robust scale-up methods are vital for controlling critical quality attributes and improving powder properties during vaccine development. Several authors used a scale-up method in two steps, a thermodynamic step and an atomization and particle formation step. This method is relatively easy to apply and test in numerous spray-dried formulations. In most cases, an attempt is made to derive the thermodynamic design space from laboratory scale data and keep the relative vapor concentration of the outlet gas constant. The droplet size is adjusted by optimizing the feed rate and atomizing air of the nozzle.

A calculation example outlined and discussed here for the scale-up from a Mini Spray Dryer B-290 to a Mobile Minor can be used as a guide for future studies, although the exact answers will vary for different formulations and spray dryer setups. However, an operator can already understand the scale-up situation by measuring the outlet temperature, the moisture content in the final powder, and simple calculation formulas based on mass and energy balances. The case studies also show that the product yield at the pilot scale reaches 90% and higher, which should be considered when comparing with other drying technologies.

A look into the future indicates that spray drying of vaccines is an evolving area of research. Spray-dried vaccine formulations could be an alternative to conventional lyophilized vaccines. Compared to freeze drying, spray drying is fast and cost-effective for drying biological materials. However, some challenges still need to be overcome before spray-dried vaccines can be marketed for commercial use. First, the scale-up of spray-dried vaccines is challenging but feasible, as shown by several studies. Still, there is a gap in the literature to explore scale-up. Second and equally important is developing optimal formulations for each vaccine to retain activity during the spray-drying process and ensure good shelf-life. A third challenge is the precise aseptic filling of small powder quantities in capsules or inhaler devices, although progress has been made in all areas in recent years.

A final point to consider is that a commercial spray-dried vaccine product requires industrial-scale spray drying equipment and facilities qualified to manufacture products in an aseptic process. Recent advances in aseptic spray drying technology are described in Chap. 4. It is expected that the increasing availability of aseptic spray dryers and service providers will generate greater industry interest in expanding the spray drying process for vaccines and other biological products.

References

Adler M, Lee G (1999) Stability and surface activity of lactate dehydrogenase in spray-dried trehalose. J Pharm Sci 88:199–208. https://doi.org/10.1021/js980321x

Agarkhedkar S, Kulkarni PS, Winston S et al (2014) Safety and immunogenicity of dry powder measles vaccine administered by inhalation: a randomized controlled Phase I clinical trial. Vaccine 32:6791–6797. https://doi.org/10.1016/j.vaccine.2014.09.071

Ajmera A, Scherließ R (2014) Stabilisation of proteins via mixtures of amino acids during spray drying. Int J Pharm 463:98–107. https://doi.org/10.1016/j.ijpharm.2014.01.002

Al-Khattawi A, Bayly A, Phillips A, Wilson D (2018) The design and scale-up of spray dried particle delivery systems. Expert Opin Drug Deliv 15:47–63. https://doi.org/10.1080/17425247.2017.1321634

Andersen SK (2018) Integration of quality-by-design into qualification and validation of spray drying processes. In: ECA training course, 10–12 April 2018, Lisbon, Portugal, pp 1–18

Anish C, Upadhyay AK, Sehgal D, Panda AK (2014) Influences of process and formulation parameters on powder flow properties and immunogenicity of spray dried polymer particles entrapping recombinant pneumococcal surface protein A. Int J Pharm 466:198–210. https://doi.org/10.1016/j.ijpharm.2014.03.025

Arpagaus C (2007) Part B, spray drying. In: The laboratory assistent. Büchi Labortechnik AG, Flawil, pp 67–87

Arpagaus C (2020) Spray drying of vaccines: a review. In: IDS 2020, 22nd international drying symposium Worcester, Massachusetts, USA, June 28-July 1, 2020, pp 1–14

Arpagaus C, Schwartzbach H (2008) Scale-up from the büchi mini spray dryer B-290 to the Niro MOBILE MINOR, best@buchi Information Bulletin, Number 52/2008

Baldinger A, Clerdent L, Rantanen J et al (2012) Quality by design approach in the optimization of the spray-drying process. Pharm Dev Technol 17:389–397. https://doi.org/10.3109/10837450.2010.550623

Baumann J, Dobry D, Ray R (2013) Amorphous dispersion formulation development: phase-appropriate integrated approaches to optimizing performance, manufacturability, stability & dosage form. Drug Dev Deliv. https://drug-dev.com/formulation-development-amorphous-dispersion-formulation-development-phase-appropriate-integrated-approaches-to-optimizing-performance-manufacturability-stability-dosage-form/

Baumann JM, Adam MS, Wood JD (2021) Engineering advances in spray drying for pharmaceuticals. Annu Rev Chem Biomol Eng 12:annurev-chembioeng-091720-034106. https://doi.org/10.1146/annurev-chembioeng-091720-034106

Benjasirimongkol P, Piriyaprasarth S, Moribe K, Sriamornsak P (2019) Use of risk assessment and Plackett-burman design for developing resveratrol spray-dried emulsions: a quality-by-design approach. AAPS PharmSciTech 20:14. https://doi.org/10.1208/s12249-018-1220-z

Bowen M, Turok R, Maa Y-F (2013) Spray drying of monoclonal antibodies: investigating powder-based biologic drug substance bulk storage. Dry Technol 31:1441–1450. https://doi.org/10.1080/07373937.2013.796968

Breit J, DuBose D (2013) Spray-dry manufacture of vaccine formulations. Bioprocess Int. https://bioprocessintl.com/manufacturing/monoclonal-antibodies/spray-dry-manufacture-of-vaccine-formulations-347400/

Byron PR (1986) Prediction of drug residence times in regions of the human respiratory tract following aerosol inhalation. J Pharm Sci 75:433–438. https://doi.org/10.1002/jps.2600750502

Chablani L, Tawde SA, D'Souza MJ (2012) Spray-dried microparticles: a potential vehicle for oral delivery of vaccines. J Microencapsul 29:388–397. https://doi.org/10.3109/02652048.2011.651503

Chen D, Kapre S, Goel A et al (2010) Thermostable formulations of a hepatitis B vaccine and a meningitis A polysaccharide conjugate vaccine produced by a spray drying method. Vaccine 28:5093–5099. https://doi.org/10.1016/j.vaccine.2010.04.112

Costa E, Neves F, Andrade G, Winters C (2014) Scale-up & QbD approaches for spray-dried inhalation formulations. ONdrugDelivery. https://www.ondrugdelivery.com/scale-qbd-approaches-spray-dried-inhalation-formulations/

Dobry DE, Settell DM, Baumann JM et al (2009) A model-based methodology for spray-drying process development. J Pharm Innov 4:133–142. https://doi.org/10.1007/s12247-009-9064-4

Dobry DE, Settell DM, Baumann JM (2015) Chapter 9: Spray drying and scale-up. In: Newman A (ed) Pharmaceutical amorphous solid dispersions. John Wiley & Sons Inc., Hoboken, NJ, USA, pp 1–26

Dormenval C, Lokras A, Cano-Garcia G et al (2019) Identification of factors of importance for spray drying of small interfering RNA-loaded lipidoid-polymer hybrid nanoparticles for inhalation. Pharm Res 36:142. https://doi.org/10.1007/s11095-019-2663-y

DuBose D, Settell D, Baumann J (2013) Spray-dried dispersions: efficient scale-up strategy for spray-dried amorphous dispersions. Drug Dev Technol 13:54–62

Foged C (2011) Subunit vaccines of the future: the need for safe, customized and optimized particulate delivery systems. Ther Deliv 2:1057–1077. https://doi.org/10.4155/tde.11.68

Gaspar F, Vicente J, Neves F, Authelin J-R (2014) Chapter 8: Spray drying: scale-up and manufacturing. In: Shah N et al (ed) Amorphous solid dispersions, Advances in delivery science and technology. Controlled Release Society, pp 261–302

Genskow LR (1994) Dryer scale-up methodology for the process industries. Dry Technol 12:47–58. https://doi.org/10.1080/07373939408959949

Gikanga B, Turok R, Hui A et al (2015) Manufacturing of high-concentration monoclonal antibody formulations via spray drying–the road to manufacturing scale. PDA J Pharm Sci Technol 69:59–73. https://doi.org/10.5731/pdajpst.2015.01003

Gil M, Vicente J, Gaspar F (2010) Scale-up methodology for pharmaceutical spray drying. Chim Oggi/chemistry Today 28:18–22

Gill D, Saigal N, Kale S, et al (2017) Thermostable spray dried rotavirus vaccine formulation and process thereof, Patent US 2017/0173145 A1, PCT filed: July 17, 2015. 1–17

Gomez M, McCollum J, Wang H et al (2021) Development of a formulation platform for a spray-dried, inhalable tuberculosis vaccine candidate. Int J Pharm 593:120121. https://doi.org/10.1016/j.ijpharm.2020.120121

Hasija M, Li L, Rahman N, Ausar SF (2013) Forced degradation studies: an essential tool for the formulation development of vaccines. Vaccine Dev Ther 11. https://doi.org/10.2147/VDT.S41998

ICH (2009) ICH Quality Guideline Q8(R2) Pharmaceutical development. https://www.ich.org/page/quality-guidelines

Ignjatović J, Đuriš J, Cvijić S et al (2021) Development of solid lipid microparticles by melt-emulsification/spray-drying processes as carriers for pulmonary drug delivery. Eur J Pharm Sci 156:105588. https://doi.org/10.1016/j.ejps.2020.105588

Ingvarsson PT, Schmidt ST, Christensen D et al (2013a) Designing CAF-adjuvanted dry powder vaccines: spray drying preserves the adjuvant activity of CAF01. J Control Release 167:256–264. https://doi.org/10.1016/j.jconrel.2013.01.031

Ingvarsson PT, Yang M, Mulvad H et al (2013b) Engineering of an inhalable DDA/TDB liposomal adjuvant: a quality-by-design approach towards optimization of the spray drying process. Pharm Res 30:2772–2784. https://doi.org/10.1007/s11095-013-1096-2

Jin T, Tsao E (2013) Powder vaccines for pulmonary delivery. In: Flower DR, Perrie Y (eds) Immunomic discovery of adjuvants and candidate subunit vaccines. Springer, New York, New York, NY, pp 291–301

Jin TH, Tsao E, Goudsmit J et al (2010) Stabilizing formulations for inhalable powders of an adenovirus 35-vectored tuberculosis (TB) vaccine (AERAS-402). Vaccine 28:4369–4375. https://doi.org/10.1016/j.vaccine.2010.04.059

Johansen P, Merkle HP, Gander B (2000) Technological considerations related to the up-scaling of protein microencapsulation by spray-drying. Eur J Pharm Biopharm 50:413–417. https://doi.org/10.1016/S0939-6411(00)00123-5

Kanojia G (2018) Drying made easy: spray drying a promising technology for the production of stable vaccine and therapeutic protein formulations. PhD thesis. University of Groningen

Kanojia G, Willems G-J, Frijlink HW et al (2016) A design of experiment approach to predict product and process parameters for a spray dried influenza vaccine. Int J Pharm 511:1098–1111. https://doi.org/10.1016/j.ijpharm.2016.08.022

Kanojia G, ten Have R, Soema PC et al (2017) Developments in the formulation and delivery of spray dried vaccines. Hum Vaccin Immunother 13:2364–2378. https://doi.org/10.1080/21645515.2017.1356952

Kanojia G, ten Have R, Brugmans D et al (2018) The effect of formulation on spray dried Sabin inactivated polio vaccine. Eur J Pharm Biopharm 129:21–29. https://doi.org/10.1016/j.ejpb.2018.05.021

Karimi K, Pallagi E, Szabó-Révész P et al (2016) Development of a microparticle-based dry powder inhalation formulation of ciprofloxacin hydrochloride applying the quality by design approach. Drug Des Devel Ther 10:3331–3343. https://doi.org/10.2147/DDDT.S116443

Kemp IC, Hartwig T, Herdman R et al (2016) Spray drying with a two-fluid nozzle to produce fine particles: atomization, scale-up, and modeling. Dry Technol 34:1243–1252. https://doi.org/10.1080/07373937.2015.1103748

Kumar S, Gokhale R, Burgess DJ (2014) Quality by design approach to spray drying processing of crystalline nanosuspensions. Int J Pharm 464:234–242. https://doi.org/10.1016/j.ijpharm.2013.12.039

Kunda NK, Wafula D, Tram M et al (2016) A stable live bacterial vaccine. Eur J Pharm Biopharm 103:109–117. https://doi.org/10.1016/j.ejpb.2016.03.027

Langford A, Ohtake S (2019) Spray drying of biotherapeutic compounds. In: Nema S, Ludwig JD (eds) Parenteral medications, Chap. 43. Taylor & Francis Incorporated, pp 929–945

LeClair DA, Cranston ED, Xing Z, Thompson MR (2016) Optimization of spray drying conditions for yield, particle size and biological activity of thermally stable viral vectors. Pharm Res 33:2763–2776. https://doi.org/10.1007/s11095-016-2003-4

LeClair DA, Li L, Rahman N et al (2019) Stabilization of HSV-2 viral vaccine candidate by spray drying. Int J Pharm 569:118615. https://doi.org/10.1016/j.ijpharm.2019.118615

Lovalenti PM, Anderl J, Yee L et al (2016) Stabilization of live attenuated influenza vaccines by freeze drying, spray drying, and foam drying. Pharm Res 33:1144–1160. https://doi.org/10.1007/s11095-016-1860-1

Maa Y-F, Nguyen P-AT, Hsu SW (1998) Spray-drying of air-liquid interface sensitive recombinant human growth hormone. J Pharm Sci 87:152–159. https://doi.org/10.1021/js970308x

Maltesen MJ, Bjerregaard S, Hovgaard L et al (2008) Quality by design—spray drying of insulin intended for inhalation. Eur J Pharm Biopharm 70:828–838. https://doi.org/10.1016/j.ejpb.2008.07.015

Masters K (1985) Spray drying handbook, 4th edn. George Godwin, London

Masters K (1994) Scale-up of spray dryers. Dry Technol 12:235–257. https://doi.org/10.1080/07373939408959955

Masters K (2002) Spray drying in practice. SprayDryConsult International ApS, Denmark

Miller DA, Gil M (2012) Chapter 10: Spray-drying technology. In: Williams RO, Watts A, Miller D (eds) Formulating poorly water soluble drugs, AAPS Advances in the pharmaceutical sciences series, vol 3. Springer. New York, NY, pp 363–442

Nair A, Khunt D, Misra M (2019) Application of quality by design for optimization of spray drying process used in drying of Risperidone nanosuspension. Powder Technol 342:156–165. https://doi.org/10.1016/j.powtec.2018.09.096

Pakowski Z, Mujumdar A (2014) Drying of pharmaceutical products. In: Mujumdar AS (ed) Handbook of industrial drying, 4th ed. CRC Press

Patton JS, Bukar JG, Eldon MA (2004) Clinical pharmacokinetics and pharmacodynamics of inhaled insulin. Clin Pharmacokinet 43:781–801. https://doi.org/10.2165/00003088-200443120-00002

Peabody J, Muttil P, Chackerian B, Tumban E (2017) Characterization of a spray-dried candidate HPV L2-VLP vaccine stored for multiple years at room temperature. Papillomavirus Res 3:116–120. https://doi.org/10.1016/j.pvr.2017.03.004

Poozesh S, Bilgili E (2019) Scale-up of pharmaceutical spray drying using scale-up rules: a review. Int J Pharm 562:271–292. https://doi.org/10.1016/j.ijpharm.2019.03.047

Price DN, Kunda NK, Ellis R, Muttil P (2020) Design and optimization of a temperature-stable dry powder BCG vaccine. Pharm Res 37:11. https://doi.org/10.1007/s11095-019-2739-8

Saboo S, Tumban E, Peabody J et al (2016) Optimized formulation of a thermostable spray-dried virus-like particle vaccine against human Papillomavirus. Mol Pharm 13:1646–1655. https://doi.org/10.1021/acs.molpharmaceut.6b00072

Santos JL, Cordeiro P, Temtem M (2013) Scale-up of spray dried amorphous solid dispersions. Eur Ind Pharm:4–8

Scrafford J (2020) An integrated approach to spray drying: from scale-up to manufacturing. Catalent, Pharm Technol Spec Spons Sect 1–6

Seville PC, Learoyd TP, Li H-Y et al (2007) Amino acid-modified spray-dried powders with enhanced aerosolisation properties for pulmonary drug delivery. Powder Technol 178:40–50. https://doi.org/10.1016/j.powtec.2007.03.046

Singh A, Van den Mooter G (2016) Spray drying formulation of amorphous solid dispersions. Adv Drug Deliv Rev 100:27–50. https://doi.org/10.1016/j.addr.2015.12.010

Soni G, Yadav KS, Gupta MK (2020) QbD based approach for formulation development of spray dried microparticles of erlotinib hydrochloride for sustained release. J Drug Deliv Sci Technol 57:101684. https://doi.org/10.1016/j.jddst.2020.101684

Thybo P, Hovgaard L, Lindeløv JS et al (2008) Scaling up the spray drying process from pilot to production scale using an atomized droplet size criterion. Pharm Res 25:1610–1620. https://doi.org/10.1007/s11095-008-9565-8

Tlaxca JL, Ellis S, Remmele RL (2015) Live attenuated and inactivated viral vaccine formulation and nasal delivery: potential and challenges. Adv Drug Deliv Rev 93:56–78. https://doi.org/10.1016/j.addr.2014.10.002

Truong-Le V, Pham B (2007) Preservation of bioactive materials by spray drying, Patent US 7258873 B2, filed: April 10, 2003. 1–30

Tumban E, Muttil P, Escobar CAA et al (2015) Preclinical refinements of a broadly protective VLP-based HPV vaccine targeting the minor capsid protein, L2. Vaccine 33:3346–3353. https://doi.org/10.1016/j.vaccine.2015.05.016

Vehring R (2008) Pharmaceutical particle engineering via spray drying. Pharm Res 25:999–1022. https://doi.org/10.1007/s11095-007-9475-1

Vehring R, Snyder H, Lechuga-Ballesteros D (2020) Spray drying. In: Ohtake S, Izutsu K, Lechuga-Ballesteros D (eds) Drying technologies for biotechnology and pharmaceutical applications. Wiley, pp 179–216

WHO/UNICEF (2014) Vaccine vial monitor (VVM) assignments for different WHO-prequalified vaccines and their proper handling, July 2014. In: Infomation Bulletin. https://www.who.int/immunization/programmes_systems/service_delivery/EN_Information_Bulletin_VVM_assignments.pdf

Yu LX, Amidon G, Khan MA et al (2014) Understanding pharmaceutical quality by design. AAPS J 16:771–783. https://doi.org/10.1208/s12248-014-9598-3

Zhu C, Shoji Y, McCray S et al (2014) Stabilization of HAC1 influenza vaccine by spray drying: formulation development and process scale-up. Pharm Res 31:3006–3018. https://doi.org/10.1007/s11095-014-1394-3

Ziaee A, Albadarin AB, Padrela L et al (2019) Spray drying of pharmaceuticals and biopharmaceuticals: critical parameters and experimental process optimization approaches. Eur J Pharm Sci 127:300–318. https://doi.org/10.1016/j.ejps.2018.10.026

Applications of Spray-Dried Vaccines 6

Abstract

This chapter overviews numerous application studies on spray-dried vaccines, from laboratory-scale research to clinical trials. The main focus is on vaccines against influenza, measles, hepatitis B (HepB), tuberculosis (TB), human papillomavirus (HPV), cholera, anthrax, typhoid, diphtheria, tetanus, pertussis, rotavirus, poliovirus, vesicular stomatitis virus (VSV), herpes simplex virus (HSV), norovirus, human immunodeficiency virus (HIV), Newcastle disease virus, tularemia, pneumococcal disease, meningococcal disease, and mRNA vaccines. This high number of applications illustrates the wide variety of spray-dried vaccines. The development status of the respective application is presented and summarized by vaccine type and route of administration. The focus is on the spray drying process parameters, formulation strategies, particle design properties, storage stability results, and efficacy. To begin with, an excursus is given on commercialized spray-dried pharmaceutical products that have already been approved, which may pave the way for spray-dried vaccines. In particular, a brief overview of spray-dried inhalable drugs and single-use dry powder inhalers for pulmonary delivery is provided. In addition, highly recommended review articles and books on (bio)pharmaceutical spray drying are presented.

6.1 Excursus: Commercialized Spray-Dried Pharmaceutical Products

Table 6.1 lists some marketed pharmaceutical products processed by spray drying technology. The data has been summarized from various literature sources (Guntur and Dhand 2007; Konstan et al. 2011; Kawabata et al. 2011; Lam et al. 2013; Brough and Williams 2013; Huang and Dai 2014; Jermain et al. 2018; Anderson

C. Arpagaus, *Spray Drying of Vaccines*,
https://doi.org/10.1007/978-3-031-24323-3_6

et al. 2018; Arpagaus 2018; Ziaee et al. 2019; Vehring et al. 2020; Baumann et al. 2021; Bhujbal et al. 2021; Sharma et al. 2021).

As can be seen in Table 6.1, the first commercial spray-dried product was Prograf®, a tacrolimus-containing immunosuppressive drug that was FDA approved in 1994. The homogeneous amorphous spray dried dispersions are filled directly into capsules or processed by blending and tableting to tablets. Over the past decade, spray-dried drugs have become a mature technology platform (Baumann et al. 2021). The main spray-dried drugs that have successfully reached the market are:

- oral solid dispersions, e.g., HIV medicine (etravirine), treatments for cystic fibrosis (ivacaftor/lumacaftor/ivacaftor), Hepatitis C (telaprevir), and immunosuppressant as prophylaxis for organ transplants rejection (tacrolimus/everolimus),
- inhalable dry powders to treat, e.g., diabetes (insulin), cystic fibrosis and respiratory tract infections (tobramycin, mannitol), Parkinson's disease (levodopa), and
- biological products, such as fibrin sealant formulation comprising fibrinogen and thrombin, control bleeding during surgery (Raplixa™).

For oral delivery, cellulosic polymers (e.g., hypromellose acetate succinate HPMCAS, hypromellose HPMC) and synthetic polymers (e.g., PVP/VA copolymer, PVP, or PEG) are the first choices as carriers in the preparation of stable formulations because they are resistant to water absorption and have high T_g (Huang and Dai 2014). These properties correlate directly with the low mobility of dispersed drug molecules in such glassy polymers and result in superior physical stability, long shelf-life (2–4 years at controlled room temperature), and easier shipping and storage (Friesen et al. 2008). Thus, concerning spray drying of vaccine formulations, these pharmaceutical products have a particular exemplary character and can thus pave the way for possible spray-dried vaccine approvals.

Raplixa™ sealant powder, developed by ProFibrix BV, is an excellent example of a biological product using aseptic spray drying technology (Siew 2016). It was approved by the FDA in 2015 and was the first aseptically spray-dried fibrin seal powder used to control adult bleeding during surgery (McKeage 2015). The powder is applied to the sampling site directly from the vial or via a low-pressure spray device (RaplixaSpray™). The fibrin sealant then dissolves in the blood and begins to clot blood. Raplixa™ is a blend of human plasma-derived fibrinogen and thrombin, spray-dried individually with a protective coating of trehalose to prevent the conversion of fibrinogen to fibrin via reacting with thrombin (McKeage 2015). The resulting free-flowing microparticles are then aseptically mixed and filled under sterile conditions into a vial (McKeage 2015). One gram of ready-to-use Raplixa™ powder contains 79 mg fibrinogen and 699 IU thrombin. This mixture eliminates the need to combine fibrinogen and thrombin before use, and the product can be stored at room temperature (Manufacturing Chemist 2016). Commercial supplies of Raplixa™ sealing powder are produced at the sterile production facilities of Nova Laboratories (UK). The ability to aseptically spray dry the biologic product avoids the need for terminal sterilization, offering a quality and cost advantage.

Table 6.1 Examples of commercialized spray-dried pharmaceutical products

Trade name	Drug (weight)	Application	Polymer, excipient[a]	Drug dose[a] (administration route)	Company	Shelf life[a]	Year of approval
Prograf	Tacrolimus (804 g/mol)	Immune suppressant (liver and kidney transplantation)	HPMC, croscarmellose, lactose, magnesium stearate	0.5, 1 or, 5 mg per capsules in blisters (oral use)	Astellas Pharma	3 years	1994 (FDA) 2017 (EMA)
Micardis	Telmisartan (515 g/mol)	Hypertension treatment	PVP, sodium hydroxide, meglumine, sorbitol, magnesium sterate	40 or 80 mg per tablet (oral use)	Boehringer Ingelheim	4 years	1998 (FDA/EMA)
Crestor	Rosuvastatin (482 g/mol)	High cholesterol treatment	HPMC, lactose	5 mg per tablet (oral use)	Astra Zeneca	3 years	2002 (FDA) 2004 (EMA)
Exubera	Insulin human (5805 g/mol)	Diabetes Mellitus	Sodium citrate, mannitol, glycine, sodium hydroxide	1–3 mg per capsule **(inhalable powder)**	Nektar/Pfizer	2 years	2006 (EMA/FDA)
Intelence	Etravirine (435 g/mol)	HIV infections	HPMC	100 or 200 mg per tablet (oral use)	Janssen	2 years	2008 (EMA/FDA)
Modigraf	Tacrolimus (804 g/mol)	Immune suppressant (organ transplants)	Lactose, HPMC, croscarmellose sodium	0.2 or 1 mg granules for oral suspension	Astellas Pharma	3 years	2009 (EMA)
Zortress/ Certican	Everolimus (985 g/mol)	Immune suppressant (organ transplants)	HPMC, lactose	0.75 mg per tablet (oral use)	Novartis	3 years	2010 (FDA) 2015 (EMA)
Aridol (Osmohale) Bronchitol	Mannitol (182 g/mol)	Cystic fibrosis, asthma	None	40 mg per capsule **(inhalable powder)**	Pharmaxis	3 years	2010 (FDA) 2012 (EMA)
Incivek (US) Incivo (EU)	Telaprevir (680 g/mol)	Hepatitis C	HPMCAS, SLS	375 mg per tablet (oral use)	Vertex	2 years	2011 (EMA/FDA)
Kalydeco	Ivacaftor (392 g/mol)	Cystic fibrosis	HPMCAS, SLS	150 mg per tablet (oral use)	Vertex	4 years	2012 (EMA/FDA)

(continued)

Table 6.1 (continued)

Trade name	Drug (weight)	Application	Polymer, excipient[a]	Drug dose[a] (administration route)	Company	Shelf life[a]	Year of approval
TOBI Podhaler	Tobramycin (468 g/mol)	Cystic fibrosis, respiratory tract infections	DSPC, calcium chloride, sulfuric acid (for pH adjustment)	28 mg per capsule **(inhalable powder)**	Novartis	3 years	2011 (EMA) 2013 (FDA)
Harvoni	Ledipasvir (889 g/mol)/ Sofosbuvir (529 g/mol)	Hepatitic C	Copovidone, HPMC, amino-methacrylate copolymer, colloidal silica, croscarmellose sodium, lactose, magnesium stearate, cellulose	90/400 mg per tablet, 33.75/150 mg or 45/200 mg coated granules in sachet (oral use)	Gilead Sciences	6 years	2014 (FDA/EMA)
Orkambi	Lumacaftor (452 g/mol)/ Ivacaftor (392 g/mol)	Cystic fibrosis	HPMCAS, PVP, SLS	200/125 mg per tablet, 150/188 mg per packet (oral use)	Vertex	3 years	2015 (FDA/EMA)
Raplixa	Fibrinogen (340 kDa)/ Thrombin (364 g/mol)	Bleeding control during surgery	Trehalose, calcium chloride, albumin, sodium chloride, sodium citrate, L-arginine-hydrochloride	79 mg fibrinogen and 726 IU thrombin per gram powder (topical use)	ProFibrix BV, Nova Laboratories	3 years	2015 (FDA/EMA)
Zepatier	Elbasvir (882 g/mol)/ Grazoprevir (777 g/mol)	Hepatitis C	Copovidone, lactose, sodium, cellulose	50/100 mg per tablet (oral use)	Merck	3 years	2016 (FDA/EMA)
Epclusa	Sofosbuvir (529 g/mol)/ Velpatasvir (883 g/mol)	Hepatitic C	Copovidone, cellulose, croscarmellose sodium, magnesium stearate	400/100 mg or 200/50 mg per tablet, 200/50 mg or 150/37.5 mg oral pellets in packet	Gilead Sciences	4 years	2016 (FDA) 2021 (EMA)
Pifeltro	Doravirine (426 g/mol)	Antiviral/ HIV infections	HPMCAS, croscarmellose sodium, lactose, magnesium stearate, cellulose, silica	100 mg per tablet (oral use)	Merck Sharp & Dohme	2.5 years	2018 (FDA, EMA)

(continued)

Table 6.1 (continued)

Trade name	Drug (weight)	Application	Polymer, excipient[a]	Drug dose[a] (administration route)	Company	Shelf life[a]	Year of approval
Symdeko	Tezacaftor/ Ivacaftor and Ivacaftor	Cystic fibrosis	HPMCAS, HPMC, SLS, cellulose, croscarmellose sodium, magnesium stearate	100/150 mg per tablet, 150 mg per tablet (oral use)	Vertex	3 years	2018 (FDA, EMA)
Erleada	Apalutamide (447 g/mol)	Prostate cancer	HPMCAS, croscarmellose sodium, magnesium stearate, cellulose, silica	60 mg per tablet (oral use)	Janssen-Cilag	3 years	2018 (FDA) 2019 (EMA)
Inbrija	Levodopa (197 g/mol)	Parkinson disease	Colfosceril palmitate (DPPC), sodium chloride	42 mg per capsule **(inhalable powder)**	Acorda Therapeutics/ Catalant	3 years	2018 (FDA) 2019 (EMA)
XTANDI	Enzalutamide (464 g/mol)	Prostate cancer	HPMCAS, croscarmellose sodium, cellulose, colloidal silica, magnesium stearate	40 or 80 mg per tablet (oral use)	Astellas/Pfizer	3 years	2018 (EMA) 2019 (FDA)
Trikafta (US) Kaftrio (EU)	Elexacaftor (598 g/mol)/ Tezacaftor (521 g/mol)/ Ivacaftor (392 g/mol)	Cystic fibrosis	HPMC, HPMCAS, SLS, croscarmellose sodium, cellulose, magnesium stearate	100/50/75 mg per tablet (oral use)	Vertex	2 years	2019 (FDA) 2020 (EMA)

Data summarized from various literature sources (Guntur and Dhand 2007; Konstan et al. 2011; Kawabata et al. 2011; Lam et al. 2013; Brough and Williams 2013; Huang and Dai 2014; Jermain et al. 2018; Anderson et al. 2018; Arpagaus 2018; Ziaee et al. 2019; Vehring et al. 2020; Baumann et al. 2021; Bhujbal et al. 2021; Sharma et al. 2021)

[a]Information on shelf life, drug dose, and polymer/excipients are based on product labels. Abbreviations: *HPMC* hydroxypropyl methylcellulose, *HPMCAS* hydroxypropyl methylcellulose acetate succinate, *DSPC* 1,2-distearoyl-sn-glycero-3-phosphocholine, *DPPC* 1,2-dipalmitoyl-sn-glycero-3-phosphatidylcholine (surfactant), *PVP* polyvinylpyrrolidone, *SLS* sodium lauryl sulfate (surfactant), *FDA* U.S. Food and Drug Administration (www.accessdata.fda.gov/daf), *EMA* European Medicines Agency (www.ema.europa.eu), *MHRA* UK's Medicines and Healthcare products Regulatory Agency.

As described in Chap. 4, aseptic spray drying technology is available to produce a sterile pharmaceutical powder that could be directly injected into the systemic circulation in humans or animals (Broadhead et al. 1992; Masters 2002). For example, streptomycin is an antibiotic used to treat many bacterial infections (Masters 1976). Therefore, it is on the WHO Model List of Essential Medicines (WHO 2019a) and classified as critically important for human medicine (WHO 2018a). In addition, numerous other pharmaceutical/biochemical substances are suitable and commercialized for aseptic spray-drying technology, including (Masters and Vestergaard 1975; Masters 1976, 1985):

- Aminosalicylic acid (e.g., as an antibiotic to treat tuberculosis)
- Bacitracin (e.g., as an antibiotic for topical treatment of skin and eye infections)
- Blood plasma (e.g., globulins, fibrinogen, thrombin)
- Blood serum (e.g., albumin, red cells)
- Casein hydrolysate (e.g., for cultivating various bacteria)
- Chloramphenicol succinate salts (e.g., as antibiotics to treat bacterial infections)
- Culture media (for multiple applications)
- Dextran (e.g., as antithrombotic to reduce blood viscosity)
- Enzymes (biocatalysts in various applications)
- Iron dextran (e.g., for the correction of anemia with kidney disease)
- Lysine (e.g., as essential amino acid in dietary supplements)
- Oleandomycin (e.g., as an antibiotic to inhibit upper respiratory tract infections)
- Penicillin (effective against many bacterial infections)
- Serum hydrolysate
- Streptomycin and salts (antibiotics to treat bacterial infections)
- Tetracyclines (antibiotics to treat bacterial infections)
- Vitamins (e.g., ascorbic acid)
- Yeasts (e.g., food yeasts or brewer's yeast).

Overall, this whole range of application examples shows that there are already numerous spray-dried (bio)pharmaceuticals on the market that have successfully passed through the clinical phases with humans. The list of spray-dried applications ranges from small molecules to high molecular weight biological materials, such as therapeutic peptides, proteins, enzymes, DNA/RNA, antibodies, blood products (e.g., plasma, serum), hormones, viruses, live cells, bacteria, vaccines, and others (Masters 2002; Celik and Wendel 2005; Greb 2009; Siew 2016; Geng 2018).

These FDA- and EMA-approved products provide a clear green light for companies considering aseptic spray drying as a manufacturing option, making this technology a viable alternative to established drying methods such as lyophilization. As regulatory acceptance of aseptic spray drying increases, which will be discussed in more detail in Chap. 4, other applications such as vaccines in spray-dried powder form are also expected to increase in the coming years. Although oral drug formulations as tablets and capsules still dominate (Al-Zoubi et al. 2021), there is a rising number of spray-dried inhalable drugs like Exubera®, Aridol/Bronchitol®, TOBI Podhaler®, or Inbrija®. The next section highlights other examples of spray-dried inhalable drugs.

6.2 Excursus: Examples of Spray-Dried Inhalable Drugs

The pulmonary route offers excellent potential for drug delivery due to its rich blood supply, large surface area (about 80–120 m^2), increased permeability, and immune tolerance of the lung. In particular, studies have shown that particle size control is fundamental to the efficiency of delivering any aerosol to the deep lung. For example, Usmani et al. (2005) showed that total lung deposition was more significant with monodisperse 1.5 μm albuterol aerosol (56%) than with 3 μm (50%) and 6 μm aerosol (46%) (Fig. 6.1). In addition, peripheral lung deposition was significantly greater with 1.5 μm (25%) than 3 μm (17%) and 6 μm (10%) aerosol (Laube and Dolovich 2014).

Figure 6.2a, b illustrate the relationship between aerodynamic particle diameter and fractional lung deposition (Byron 1986; Patton et al. 2004; Laube et al. 2011). Aerosol deposition to the deep lung is critically dependent on particle size.

Most particles larger than 10 μm are deposited in the oropharynx region and upper airways (Yu et al. 1979; Patton et al. 2004). Particles of about 4–5 μm deposit primarily in the bronchial/conducting airways whereas smaller particles remain in the airstream and are carried into the peripheral airways and the alveolar region (Laube et al. 2011). The optimal particle size for delivery to the alveoli is around 1–3 μm in aerodynamic diameter. In the periphery of the lung, airflow velocity is reduced to zero. Therefore, particles in this region have a longer residence time and are primarily deposited by gravitational sedimentation (particles

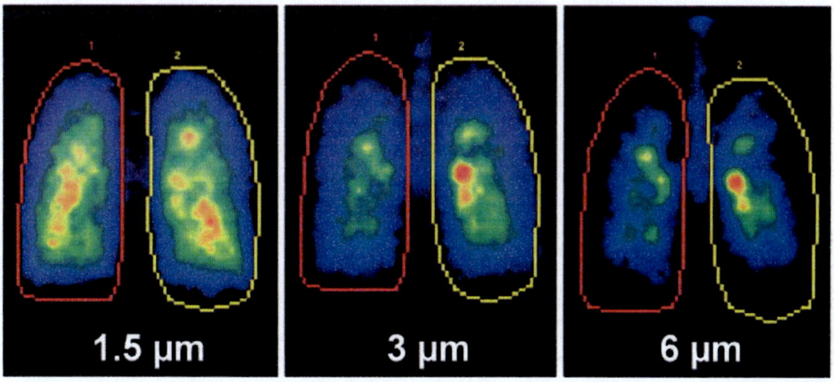

Fig. 6.1 Posterior thorax images of aerosol deposition in the lungs using technetium-99 m-labeled albuterol particles of 1.5, 3, and 6 μm mass median aerodynamic diameter. Red areas indicate regions of highest radioactivity and black areas of least radioactivity (Figure adapted from Laube and Dolovich (2014) with permission from Elsevier)

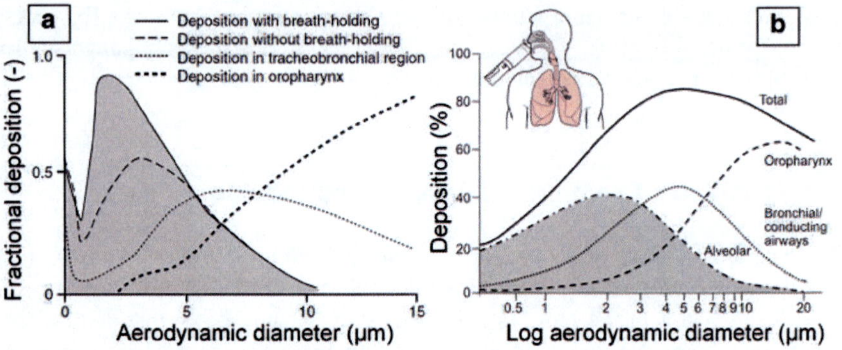

Fig. 6.2 **a** Simulated fractional deposition to the lung as a function of aerodynamic diameter with a slow inspiratory flow rate of 22.5 L/min with and without breath-holding (Figure reproduced from Patton et al. (2004) with permission from Springer Nature), **b** The International Commission on Radiological Protection model showing the relationship between aerodynamic diameter and lung deposition (Figure reproduced from Laube et al. (2011) with permission from the European Respiratory Society)

down to 0.5 µm) and diffusion (Brownian motion below 0.5 µm) (Labiris and Dolovich 2003). In addition, holding the breath during inhalation increases the residence time and allows additional deposition of aerosol particles that would otherwise be exhaled (Byron 1986).

In 2006, Exubera® (Pfizer/Nektar Therapeutics) was the first spray-dried inhalable insulin approved by the FDA and EMA to treat adult patients with Type 1 or Type 2 diabetes (White et al. 2005; Guntur and Dhand 2007). The powder was manufactured on a commercial scale spray dryer with an inlet temperature of 181 °C and outlet temperature of 87 °C (Sadrzadeh et al. 2010) and packaged in blister packs containing 1 or 3 mg of insulin. Unit doses were delivered via a mechanical inhaler and were equivalent to 3 or 8 units of subcutaneously administered insulin (Fig. 6.3b, c).

The powder contained 60% recombinant human insulin with sodium citrate, mannitol, glycine, and sodium hydroxide (Stevenson and Bennett 2014). The sodium citrate buffered the bulk aqueous solution at a pH of 7.2–7.4 before spray drying (Sadrzadeh et al. 2010; Ledet et al. 2015) and provided consistent T_g of about 80 °C after drying with a residual water content of approximately 2 wt% (White et al. 2005; Sadrzadeh et al. 2010). Similarly, glycine increased the aqueous solution's buffering capacity and pH control before spray drying. Mannitol served as a lyoprotectant to stabilize insulin upon water evaporation during spray drying (Stevenson and Bennett 2014). Exubera® was declared stable for more than 2 years at room temperature (25 °C) (Weers et al. 2007; Sadrzadeh et al. 2010).

As shown in Fig. 6.3a, the particle morphology was uniform and raisin-like even after storage at elevated humidity (75% RH, 25 °C) for 36 h (Stevenson and Bennett 2014). In addition, the mean aerodynamic mass diameter was about 3 µm, which is an optimal size for enhanced alveolar deposition (Yu et al. 1979; Byron 1986; Patton et al. 2004; Guntur and Dhand 2007; Laube et al. 2011).

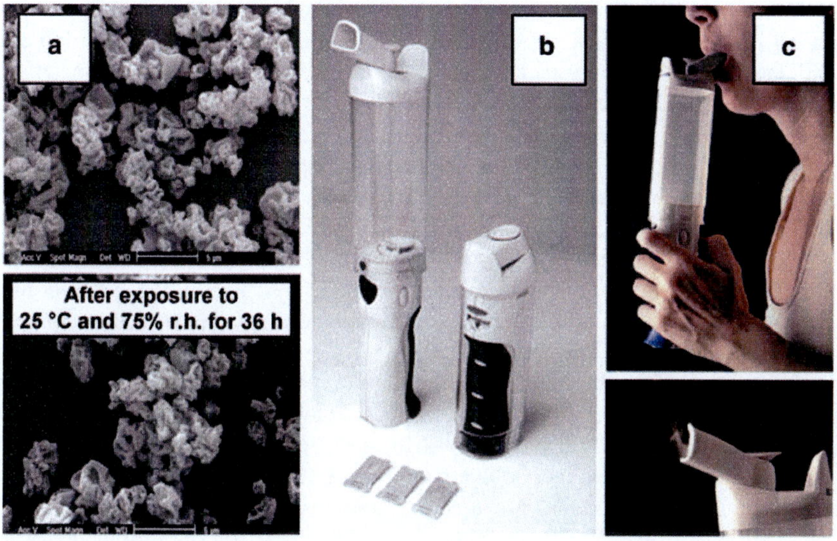

Fig. 6.3 **a** SEM pictures of spray-dried insulin powder before (top) and after (bottom) unprotected exposure to 25 °C and 75% relative humidity for 36 h (Stevenson and Bennett 2014). **b** Exubera pulmonary insulin delivery system base, transJector, and chamber (with mouthpiece) extended and collapsed, and three unit-dose blisters (Stevenson and Bennett 2014). **c** During inhalation (Mack 2007; Kling 2008). All pictures reproduced with permission from Springer Nature

However, despite the promise of a new delivery system for dry powder insulin, Exubera® did not find a profitable niche in the insulin market and was voluntarily withdrawn by Pfizer in October 2007 (Mack 2007). The decision was not based on safety or efficacy but the lack of sales of just $12 million worldwide in the first 9 months in 2007 (Mack 2007; Bailey and Barnett 2007). The withdrawal was estimated to cost Pfizer US$ 2.8 billion (Bailey and Barnett 2007) and resulted in the loss of hundreds of jobs (Stevenson and Bennett 2014).

The failure of Exubera® was attributed to several factors, including the high cost and the large size of the inhaler device, the mg-range dosing, which may have confused patients who had been receiving conventional insulin therapy, and a warning from the FDA regarding an increased incidence of lung cancer cases observed in ongoing clinical studies, especially among former smokers (Heinemann 2008; Kling 2008; Neumiller and Campbell 2010). In this context, Pfizer's data showed 6 lung cancer cases in 4740 Exubera®-treated patients versus 1 in the 4292 control patients (Kling 2008). In addition, in a follow-up study of patients who had previously participated in controlled clinical trials, an increased risk of lung cancer-related mortality associated with the use of Exubera could not be excluded (Gatto et al. 2019). However, if real, the absolute increased risk of lung cancer-related mortality was explained as small (0.48 cases per 1000 person-years).

From a patient's point of view, Exubera® also had a big plus when it came to storing the insulin blisters. The insulin in this formulation was stable for quite a while at room temperature and did not need to be stored in a refrigerator, which reduced the handling effort in everyday life. However, it appeared that such positive aspects were overshadowed by negative ones (Heinemann 2008).

In the end, Exubera® highlighted both technical and non-technical challenges of inhalable dry powder medicine and eventually led to the discontinuation of other similar dry powder insulin programs by Novo Nordisk and Eli Lilly (Kling 2008). On the other hand, these developments also paved the way for further research on pulmonary spray-dried drugs and other large molecules. In addition, new formulation strategies are being further explored with pure insulin without excipients (Balducci et al. 2014; Quarta et al. 2020).

The only company that continued the development of its inhaled dry powder insulin was MannKind Corporation. The product Afrezza® (Technospheres® insulin) was approved by the FDA in 2014 and is currently the only inhaled dry powder insulin on the market (Goldberg and Wong 2015). Afrezza® employs a new inhaler, smaller and easier to use than the previously available Exubera® device (Richardson and Boss 2007; Neumiller and Campbell 2010). The formulation was based on Technosphere® drug carrier technology (Pfützner et al. 2002; Pfützner and Forst 2005; Angelo et al. 2009), in which insulin was encapsulated in small precipitated particles of fumaryl diketopiperazine (FDKP), a proprietary excipient. The well-characterized small organic molecule self-assembles in a mild acid environment into particles with a mean diameter of about 2–3 μm and large surface area (Steiner et al. 2002; Richardson and Boss 2007; Angelo et al. 2009). The precipitated microparticles were then freeze-dried and suitable for pulmonary delivery to the systemic circulation. Once inhaled, the particles dissolved in the pH-neutral environment of the deep lung and rapidly released insulin into the systemic blood circulation (Pfützner and Forst 2005).

In 2010 and 2012, Pharmaxis commercialized spray-dried inhalable mannitol powders Aridol/Osmohale® and Bronchitol® for diagnosis of asthma by detecting active airway inflammation through measuring bronchial hyper-responsiveness (Anderson et al. 2018). Powder with a median particle size of around 3 μm was superior to larger sizes. Mannitol was chosen because of its crystallinity after spray drying attributed to its low T_g of about 11 °C and physical stability when exposed to elevated humidity during storage. Thus, it rehydrates the airway/lung surface and promotes a productive cough. This way, it helps increase mucus clearance and improves lung function and the quality of life of people living with cystic fibrosis (Hurt and Bilton 2012).

In 2013, the FDA approved Novartis TOBI® Podhaler™ with tobramycin inhalation powder to treat cystic fibrosis patients with Pseudomonas aeruginosa bacteria in the lungs (FDA 2013). The powder was manufactured using PulmoSphere™ technology, an emulsion-based spray drying process that produces spherical hollow-porous particles for pulmonary drug delivery (Geller et al. 2011; Weers and Tarara 2014). Figure 6.4 illustrates the particle morphology and the inhaler device. The dried particles had a geometric mean diameter of about

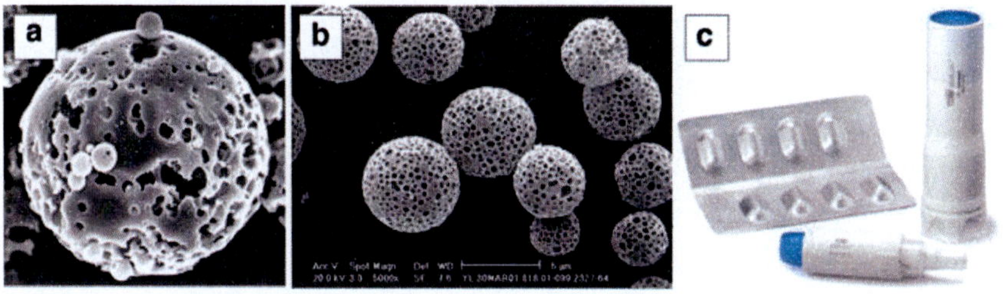

Fig. 6.4 Tobramycin inhalation powder. **a, b** SEM pictures of hollow and porous PulmoSphere™ particles, **c** Novartis TOBI® Podhaler™ for dry powder inhalation with day's supply of Tobramycin inhalation powder with 2 separate blister packs of 4 × 28 mg capsules (Image a reproduced from Konstan et al. (2011) with permission from John Wiley and Sons, Images b and c reproduced from Lam et al. (2013), open access under CC BY-NC 3.0 license, ©Novartis AG)

1–2.7 μm and a mean mass diameter < 4 μm, which was ideal for delivering the drug to the lower respiratory part (Konstan et al. 2011; Lam et al. 2013). First, the active substance tobramycin was stabilized with distearyl phosphatidylcholine (DSPC), calcium chloride, and sulfuric acid. Next, the drug was packaged in a hypromellose capsule containing 28 mg of the active ingredient. Finally, the capsules were stored individually in aluminum blister packs for moisture protection.

In 2018 and 2019, levodopa inhalation powder (Inbrija®) from Acorda Therapeutics has been approved for the intermittent treatment of "off episodes" in adults with Parkinson's disease. The patients were treated with regular carbidopa/levodopa medicine in the USA and levodopa/dopa-decarboxylase inhibitor in the EU (Paik 2020). Levodopa inhalation powder was administered via the breath-actuated Inbrija® inhaler. Each capsule contained 42 mg levodopa as a spray-dried powder, 1,2-dipalmitoyl-sn-glycero-3-phosphocholine (DPPC, a pulmonary surfactant), and sodium chloride. The dry powder particles (5–10 μm in diameter) were homogeneous, low density, and highly porous for optimal aerosolization and lung deposition (Stocchi et al. 2018).

Just recently, Catalent acquired the GMP facility of Acorda Therapeutics in Chelsea, USA (Catalent 2021). The site operates best-in-class spray drying capacity, including GEA NIRO® PSD-1, PSD-4, and PSD-7 spray dryers, with the latter unit being the largest equipment of its kind in North America.

In addition, several large pharmaceutical companies such as AstraZeneca, Vertex, Pfizer, GSK/Novartis, Janssen, and Nova Laboratories have already invested in setting up their own spray drying facilities for the commercial production of their spray-dried APIs.

Various other spray-dried inhaled biotherapeutics have already been tested in phase 2 clinical trials (Vehring et al. 2020), such as leuprolide for the treatment of endometriosis, calcitonin used for osteoporosis, human growth hormone, and Pitrakinra or VR942 antibodies for asthma.

6.3 Excursus: Single-Use Dry Powder Inhalers for Pulmonary Delivery

Expectations for pulmonary vaccine delivery by dry powder inhalers (DPI) are currently high (Sou et al. 2011; Zhou et al. 2014; de Boer et al. 2017). Compared with vaccine injections, the main advantages of a dry powder form are (Tonnis et al. 2012, 2013):

- improved stability in the dry state,
- absence of reconstitution with sterile water, and
- non-invasive administration without needles.

The DPIs on the market are primarily designed for the treatment of pulmonary bronchospasm and inflammation such as asthma and chronic obstructive pulmonary disease (COPD), infections such as cystic fibrosis, or diabetes and Parkinson's disease, and are thus reusable inhalers for the delivery of multiple doses (e.g., for daily treatment) (Ibrahim et al. 2015; Berkenfeld et al. 2015; Young 2018; Kukut Hatipoglu et al. 2018; Levy et al. 2019; Munir et al. 2021) (see application examples with spray-dried products in Table 6.1).

Dry powder inhalers are used with multiple capsules to deliver a high dose, with each capsule emptied in a single inhalation (Sibum et al. 2018). For example, the TOBI® Podhaler™ by Novartis AG that was approved by the FDA on March 22, 2013 (FDA 2013) is available in monthly packs of 4 weekly packs. A weekly pack contains 56 hard capsules containing 28 mg tobramycin powder for inhalation and 1 inhaler. Each inhaler may only be used for 7 days. A smaller version of a capsule inhaler is the Turbospin® for the product Colobreathe® (Schuster et al. 2013). A dose comprises a single capsule containing approximately 145 mg of colistimethate sodium (without excipients), of which around 125 mg is emitted (Young 2018; Sibum et al. 2018). The device requires minimal cleaning and maintenance and is disposable at the end of each 28-day pack (Schuster et al. 2013). An issue with capsule inhalers is that they need to be loaded, prepared, and then inhaled.

However, vaccination usually involves single administration. Therefore, single-use disposable DPIs delivering high-doses are of particular interest, which are used once and then discarded (De Boer et al. 2006; Friebel and Steckel 2010; Zhou et al. 2014; De Boer and Hagedoorn 2015; Berkenfeld et al. 2015; Foged 2016; Sibum et al. 2018; Heida et al. 2021; Sawatdee and Srichana 2021; Son et al. 2021).

A single-use disposable DPI must be pre-loaded, simple, compact, cost-effective, highly reliable, and effective to enable mass vaccination (De Boer and Hagedoorn 2015). If the administration of the vaccine powder via the pulmonary route fails, protection will be insufficient. Therefore, an effective monitoring system of the delivered dose is necessary. In addition, a low price per vaccine dose is essential to be competitive with alternative devices and dosage forms. Furthermore, hygienic design and patient-friendly functionality are crucial to minimizing the need for trained healthcare professionals.

As pointed out by Tonnis et al. (2012, 2013), a combined approach of DPI design and dry powder vaccine formulation is essential. In particular, the older age groups (i.e., adolescents, adults, elderly) may be considered appropriate target groups for pulmonary vaccination by DPIs (De Boer and Hagedoorn 2015).

Several single-use disposable DPIs are available for high-dose pulmonary administration in the milligram range. Figure 6.5 shows some examples, including the TwinCaps®, the TwinMax®, the 8Shot®, the Orbital®, the Twincer®, the Cyclops®, the DoseOne®, the ICOone®, PillHaler®, SOLO®, and the Perlamed®-BLISTair.

The list does not claim to be exhaustive, as there are undoubtedly other single-use disposable DPIs that may be of interest for pulmonary vaccination, but little or no information has been published on their performance. For more detailed overviews of DPIs for single-use, the interested reader is referred to Friebel and Steckel (2010), Tonnis et al. (2013), De Boer and Hagedoorn (2015), Berkenfeld et al. (2015), Sibum et al. (2018) and Sawatdee and Srichana 2021).

The TwinCaps® device by Hovione (Fig. 6.5a) is the world's first market-approved and largest-selling single-use disposable DPI (Hovione 2021a). It has been on the market in Japan since 2010 for the treatment and prophylaxis of

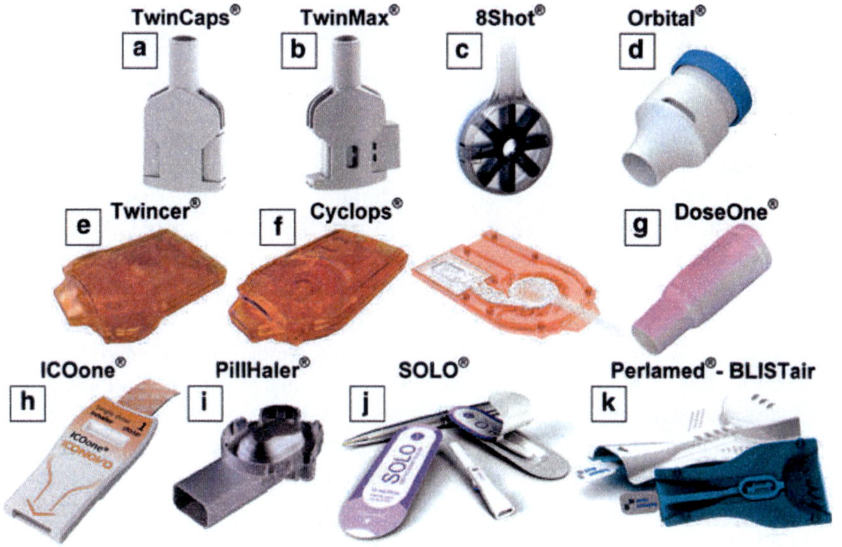

Fig. 6.5 An overview of eight single-use disposable DPIs capable of high dosing. **a** TwinCaps®, **b** TwinMax® and **c** 8Shot® by Hovione (Portugal) available for licensing and supply, **d** Orbital® by Pharmaxis (Australia), **e** Twincer® and **f** Cyclops® developed at the University of Groningen (The Netherlands) and produced by PureIMS, **g** DoseOne® by DoseOne LLC (USA), **h** ICOone® by Iconovo (Sweden), **i** PillHaler® by HollyCon Italy S.r.l., **j** SOLO® inhaler by Manta Devices LLC (USA), and **k** PERLAMED™-BLISTair by Perlen Packaging AG (Switzerland) (Images courtesy of all manufacturers, Twincer® and Cyclops® adapted from Sibum et al. (2018), open access under CC-BY 4.0 license)

influenza A and B infection with Daiichi-Sankyo's antiviral Inavir® (laninamivir octanoate). Millions of patients are treated with this innovative inhaler technology every year. TwinCaps® features two independent pre-filled doses up to 2 × 30 mg, and it is designed for applications such as anti-infectives, vaccines, analgesics, military and rescue treatments. TwinCaps® consists of only two plastic components manufactured by injection molding, making it highly affordable. The device contains two separate, pre-filled dose compartments. The patient pushes the dosing compartment sideways to align the first dose with the inhaler's mouthpiece to start inhalation. Then, the dose tray is simply slid into the other direction to inhale the second dose. Since the delivery of Inavir® using TwinCaps® has been approved for children under 10 years old (Sawatdee and Srichana 2021), this device has good aerosolization performance at lower flow rates.

Hovione's TwinMax® and 8Shot® DPI (Fig. 6.5b, c) enable single or repeated delivery of high doses of antibiotics, antivirals, vaccines, peptides, and other medications to the lungs (Fernandes and Villax 2019). The TwinMax® features two pre-filled compartments compatible with doses up to 2 × 50 mg. 8Shot® is the world's first 8 unit-dose, factory pre-filled, disposable DPI that enables drug delivery in high doses of up to 400 mg (8 × 50 mg) (Hovione 2021b). The circular configuration allows multiple drugs to be delivered independently. The device is fully injection molded from two plastic components.

The Orbital® disposable DPI by Pharmaxis Ltd. (Australia) is designed to deliver high doses via multiple breaths with a single actuation step (Fig. 6.5d) (Young et al. 2014, 2015; Young 2018). The inhaled medicine is released from a spinning "puck" (aerosol sample chamber) that contains up to 400 mg spray-dried powder. This approach eliminates the need for repeated actuation and loading cycles of multiple capsules, and the emitted dose can be tuned so that the inhaled formulation mass per inhalation is tolerable (Young et al. 2014). In addition, precision-machined outlet orifices control the release rate (Zhu et al. 2015; Young 2018). The DPI has been initially developed for Pharmaxis' cystic fibrosis drug Bronchitol® (mannitol). Around 200 mg of spray-dried ciprofloxacin/mannitol and azithromycin/mannitol powder can be delivered to patients via approximately 10 inhalation maneuvers (Zhu et al. 2016). In August 2021, Pharmaxis and Aptar Pharma entered into an agreement under which Aptar Pharma obtained exclusive option to develop and promote Pharmaxis' proprietary inhaler Orbital® worldwide (Pharmaxis 2021).

The Twincer® and the Cyclops® (Fig. 6.5e, f) are single-use disposable DPI devices developed at the University of Groningen (De Boer et al. 2006; Hoppentocht et al. 2015). The DPIs were designed as low-cost, single-use inhalers, making them attractive for vaccines that need to be administered only once. The powder dose is stored in a blister to protect it from moisture and prevent possible negative effects on inhaler efficiency. In addition, the Twincer® has been shown to effectively deagglomerate 25 mg of powder using air classifier technology (De Boer et al. 2006). In addition, in single dose pilot studies with the Twincer®, inhalation of 83.3% colistin sulfomethate (X50 = 1.6 μm) and 16.7% lactose (150–200 μm) was well tolerated by healthy volunteers (Westerman et al. 2007a) and cystic fibrosis patients (Westerman et al. 2007b).

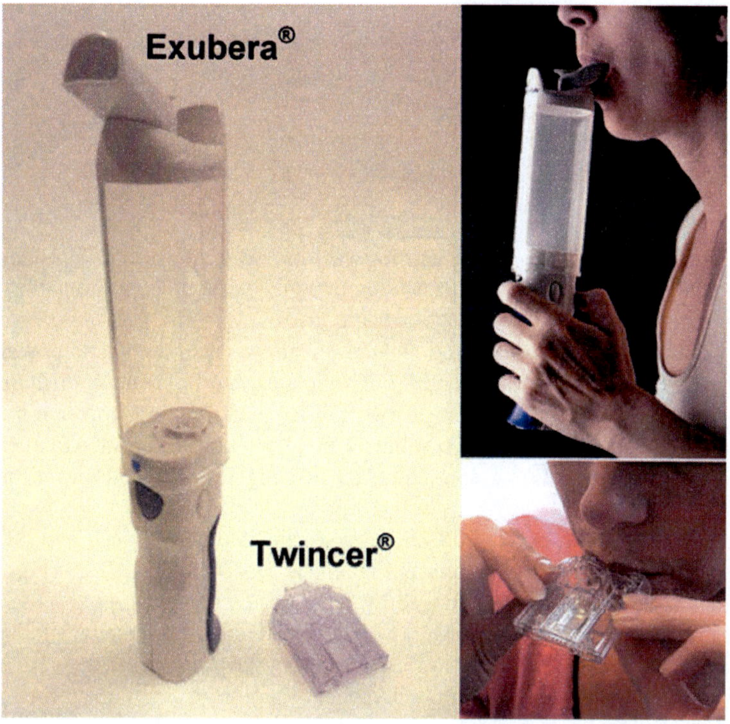

Fig. 6.6 The difference in complexity and size between the Exubera® and Twincer® inhaler (Images reproduced from De Boer and Hagedoorn (2015) and Mack (2007) with permission from Taylor & Francis and Springer Nature, *Photo credit* Indes B.V., NL)

Figure 6.6 illustrates the difference in complexity and size between the Exubera® inhaler and the disposable Twincer® dry powder inhaler for a single dose (Mack 2007; De Boer and Hagedoorn 2015). The small pocket-sized Twincer® inhaler represents the patient's convenience in comparison.

The Cyclops® is a modified version of the Twincer® inhaler and extends the capacity up to 50 mg of pure, spray-dried tobramycin in a single inhalation procedure (Hoppentocht et al. 2015). The device comprises three plate-shaped parts stacked and clicked together and a cyclonic-based blister system for the drug powder (Fig. 6.5f). Recently, Sibum et al. (2019) assessed the aerosol performance of spray-dried isoniazid (anti-TB drug) and found that the use of the Cyclops® versus a Twincer® improved the emitted dose of formulations containing 3%, 5%, and 3% tri-leucine by approximately 43%, 25%, and 10%, respectively.

The Dutch start-up company PureIMS (Pure Inhalation Medication Systems) produces DPIs based on the Cyclops™ platform, and several products are in the pipeline, including (PureIMS 2021):

- Levodopa Cyclops™ as rescue therapy for OFF episodes in Parkinson's disease,
- Epinephrine Cyclops™ for impending (food) allergy-induced anaphylaxis,
- Colistin Cyclops™ for the treatment of cystic fibrosis,
- Tobramycin Cyclops™ to treat pulmonary pseudomonas aeruginosa infections,
- Amikacin Cyclops™ for the treatment of tuberculosis, and
- Hydroxychloroquine for intervention in SARS-CoV-2 and possible future coronavirus outbreaks.

PureIMS has a GMP manufacturing facility licensed for the production of Cyclops™. Harro Höfliger (Germany) supported the scale-up and automation of the powder filling process (HARRO Magazine 2020). An Omnidose powder dosing machine with vacuum drum filling technology allows semi-automatic powder filling of single doses into the Cyclops™ bottom plate. Measurements by (Sibum et al. 2020) with spray-dried (Mini Spray Dryer B-290) isoniazid and amikacin powders resulted in nominal doses of 150 mg and 57 mg, respectively. This led to maximum fine particle doses (FPD) of approximately 107 mg for isoniazid and 47 mg for amikacin.

The DoseOne® disposable DPI has been designed and developed by GFE LLC (US) and then spun off into its current company DoseOne LLC (US) (Fig. 6.5g). The basic technology is fully covered by issued patents (Ganem et al. 2013). The disposable inhaler uses well-known, low-risk capsule-based storage technology and allows a wide range of doses up to 100 mg. The DPI consists of a simple three-piece assembly of injection-molded plastic parts with a manufacturing cost of approximately $0.16 at high volumes (DoseOne LLC 2011). The emitted dose (93%) and fine particle fraction (56%) with spray-dried formulations indicate efficient dose delivery that can compete with other single-dose DPI devices. DoseOne® is a dry powder delivery system for vaccines, biologics, and single-dose rescue or prevention therapies. Moreover, Dose One LLC is exploring various business opportunities, including license agreements and intellectual property sales, and is ready for use in Phase I clinical trials.

The Swedish company Iconovo has developed ICOone® (Fig. 6.5h), a single-dose, disposable DPI consisting of only one injection molded part (Iconovo 2021). The patented DPI is simple and cost-effective for delivering vaccines and biomolecules to the lungs (Lastow and Arvidsson 2016). ICOone® can be designed with two cavities for combination products. A high dose of up to 50 mg per unit can be administered with a 60–70% respirable fraction. The powder is sealed in an aluminum foil with integrated desiccant. All the user has to do is pull a tab and inhale. As announced in July 2021, Iconovo has signed an agreement with the start-up company Immune System Regulation (ISR), which has developed a dry powder vaccine against Covid-19 based on spike proteins and adjuvant using spray drying technology. Clinical trials are currently underway with the ICOone® disposable inhaler.

Furthermore, the PillHaler® is a patented and registered disposable inhaler by HollyCon Italy S.r.l. to administer a single dose of dry powder (Fig. 6.5i) (HollyCon Italy 2021). The device has a pipe shape and consists of only two pieces, an inhalation body, and a mini blister cartridge. By simply pressing the top of the mini blister with a finger, the powder dose is released and can be inhaled. The device is easy to use while ensuring proper dosing of the product. It is constructed of recyclable plastic material with low environmental impact. A grid favors the formation of air turbulence and uniform dispersion of the particles (Wu and Wu 2018).

The SOLO® inhaler developed by Manta Devices LLC (USA) is another single-dose disposable DPI that combines ergonomic simplicity, high-performance dispersion, and dose stability into a cost-effective single-use solution (Manta Devices LLC 2017a). This DPI integrates Manta's patented blister opening and metering dose chamber technologies (Jones and Miller 2012, 2017, 2018). The inhaler consists of a standard foil blister and two injection molded parts. Removing the foil packaging from the device automatically opens the dispersion mechanism for inhalation. The highly efficient dispersion mechanism delivers large or small doses (e.g., 15 mg) for various applications, including asthma and COPD, vaccines, antibiotics, and rescue medications. In addition, Manta's also offers the Torus™ Large Dose DPI, which can dispense larger payloads of spray-dried or blended drug formulations up to 40 mg (particles < 5 μm) (Manta Devices LLC 2017b). The device can be prefilled for single-use applications such as vaccines or fast-acting therapies, or it can be reloaded by the patient for multi-dose applications such as antibiotics. The drug dose is stored in a toroidal, injection-molded plastic chamber incorporating Manta's dose metering and dispersion technologies. Experiments with pyrazinoic acid and excipients (50 mg nominal mass) resulted in an emitted dose of 92% and a fine particle dose and fraction (below 4.46 μm) of 68% and 62%, respectively (Hickey et al. 2019).

Finally, PELAMED™-BLISTair is the world's first disposable thermoformed inhaler developed by Perlen Packaging AG (Switzerland) (Fig. 6.5f) (Perlen Packaging 2021). It is a patented single-use DPI made of a plastic high-barrier primary packaging film against oxygen and water (Beller 2018, 2021). The device consists of a bottom part, an aluminum lid that seals the pocket containing the powder, and an upper part. The bottom part is made of Ultrapotect 360 high barrier film 360, which, in conjunction with the aluminum lid, provides an extremely high barrier to protect the powder. The upper part is made of PVC film that can be easily thermoformed into the desired shapes. The dosing range of the active powder product is 5–100 mm^3. This also depends on which of the five available sizes of BLISTair is used. The primary packaging material (PERLALUX® Ultra Protect 360 made of PVC, PE, and PVDC) is simultaneously the inhalation device through its double funnel shape (overall dimension is 70 × 47 × 10 mm). Figure 6.7 shows the basic structure of the Perlamed™-BLISTair with typical flow patterns. Especially in the early development stage, BLISTair offers great conceptual advantages for flexible adaptation of the inhaler design to the powder requirements.

Handling of the device is exceptionally intuitive. There are no moving parts. All it takes is to pull out the aluminum tab to open the powder chamber and inhale the drug. After inhalation, the Perlamed™-BLISTair is simply discarded. Perlamed™-BLISTair is also available as a dual-chamber option to deliver two different powders into the same device.

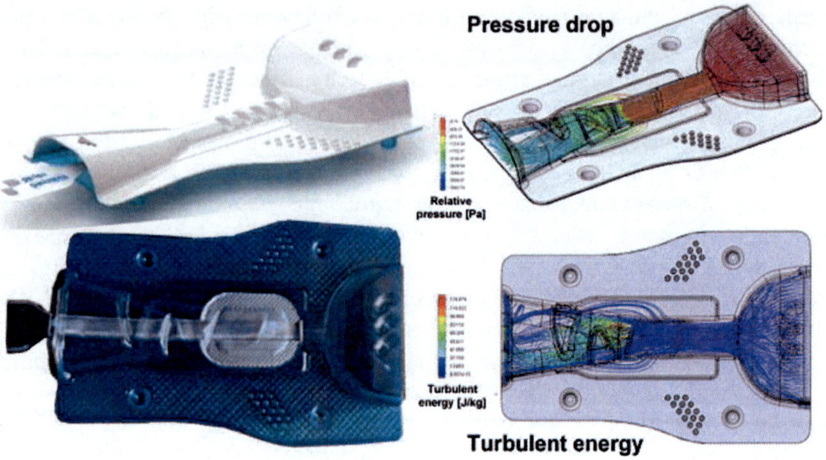

Fig. 6.7 PERLAMED™-BLISTair made of high barrier packaging film PERLALUX® (Ultra Protect 360) with CFD simulation of turbulent energy and pressure drop (Images courtesy of Perlen Packaging AG, Switzerland)

For example, test results of budesonide powder as a model drug for the inhalation market with crystalline lactose as carrier and 1% magnesium stearate as additive have shown that optimization of inhaler shape and geometry allows a significant increase in emitted dose and fine particle fraction.

Overall, it is a platform for the pulmonary delivery of a wide range of drug substances in powder form, including applications like (Perlen Packaging 2021):

- scheduled single-use: vaccination, flu prevention,
- emergencies: allergic reactions, cardiovascular events, tocolysis,
- occasional use: migraine attacks, break-through-pain, anti-emetic, and
- all kinds of short-term therapy.

To sum up, this brief review shows that several single-use disposable DPIs for pulmonary delivery of dry powder vaccines are available. It is essential to coordinate the dry powder vaccine requirements with the inhaler manufacturer at an early stage to prepare for upscaling and developing an optimal inhaler geometry. Some DPI devices are ready for licensing to vaccine manufacturers and for use in clinical trials. The simple production by injection molding and thermoformed high-barrier films make single-dose dry powder inhalers cost-competitive. In addition, the usually spray-dried powder form allows better stability during storage at room temperature compared to liquid formulations and thus independence from the cold chain. The main challenge with pulmonary delivery of dry powder vaccines is a lack of human clinical trials.

On the other hand, pulmonary delivery of dry powder products is available for respiratory diseases (e.g., asthma, COPD, cystic fibrosis), diabetes, or Parkinson's disease. The example of TwinCaps® for the treatment of influenza infections in Japan is a success story for using single-use disposable DPIs. A further boost could come from new developments in dry powder vaccines against Covid-19.

6.4 Excursus: Review Articles and Books on (Bio)pharmaceutical Spray Drying

For many years, spray drying has played a significant role in the pharmaceutical industry in drying excipients, active pharmaceutical ingredients (APIs), and finished drug formulations. Powder, capsules, and tablets are processed in dried form thanks to spray-drying particle engineering technology. Over time, numerous comprehensive state-of-the-art reviews have been published on (bio)pharmaceutical spray drying. Table 6.2 gives an overview of papers and books that interested readers are highly recommended reading (Masters 1976, 1985, 2002; Broadhead et al. 1992; Celik and Wendel 2005; Ré 2006; Ameri and Maa 2006; Snyder and Lechuga-Ballesteros 2008; Vehring 2008; Chen and Kristensen 2009; Sollohub and Cal 2010; Cal and Sollohub 2010; Searles and Mohan 2010; Bowey and Neufeld 2010; Schwartzbach 2011; Yang 2011; McAdams et al. 2012; Ledet et al. 2015; Lowinger et al. 2015; Singh and Van den Mooter 2016; Kanojia et al. 2017; Arpagaus et al. 2018; Arpagaus 2018, 2020; Langford and Ohtake 2019; Putranto and Chen 2019; Ziaee et al. 2019; Ferreira et al. 2020; Salama 2020; Vehring et al. 2020; Pinto et al. 2021; Preston and Randolph 2021; Baumann et al. 2021; Öztürk and Arpagaus 2021; Munir et al. 2021; Samborska et al. 2022; Gomez and Vehring 2022; Ghaemmaghamian et al. 2022).

A literature review by Al-Khattawi et al. (2018) of 5853 research articles mentioning spray drying and drug delivery showed that the number of reports increased 11-fold from 2010 to 2016 compared with the period from 1990 to 1999. The main drivers of this remarkable increase were:

- improvement of solubility (bioavailability) of poorly water-soluble APIs for oral drug delivery,
- formulation of pulmonary drug delivery systems, and
- stabilization of biopharmaceuticals.

Thus, spray drying is an enabling technology to produce amorphous solid dispersions, drug delivery systems for controlled drug release, and fine powders for inhalation.

Table 6.2 Recommended review papers and books on (bio)pharmaceutical spray drying sorted by year of publication

Title	Pages	References
Applications in the Pharmaceutical-Biochemical Industry	20	Masters (1976)
Spray Drying Handbook	697	Masters (1985)
The Spray Drying of Pharmaceuticals	38	Broadhead et al. (1992)
Spray Drying in Practice	465	Masters (2002)
Spray Drying and Pharmaceutical Applications	29	Celik and Wendel (2005)
Formulating Drug Delivery Systems by Spray Drying	14	Ré (2006)
Spray Drying of Biopharmaceuticals: Stability and Process Considerations	7	Ameri and Maa (2006)
Spray Drying: Theory and Pharmaceutical Applications	34	Snyder and Lechuga-Ballesteros (2008)
Pharmaceutical Particle Engineering via Spray Drying	24	Vehring (2008)
Opportunities and Challenges of Developing Thermostable Vaccines	11	Chen and Kristensen (2009)
Spray Drying Technique. I: Hardware and Process Parameters	12	Cal and Sollohub (2010)
Spray Drying Technique: II. Current Applications in Pharmaceutical Technology	11	Sollohub and Cal (2010)
Spray Drying of Biopharmaceuticals and Vaccines	23	Searles and Mohan (2010)
Systemic and Mucosal Delivery of Drugs within Polymeric Microparticles Produced by Spray Drying	19	Bowey and Neufeld (2010)
Achieving Aseptic Drying with Spray Drying Technologies	4	Schwartzbach (2011)
Spray Drying Pharmaceuticals	6	Yang (2011)
Spray Drying and Vaccine Stabilization	9	McAdams et al. (2012)
Practical Considerations for Spray Dried Formulation and Process Development	53	Lowinger et al. (2015)
Spray Drying of Biopharmaceuticals	25	Ledet et al. (2015)
Spray drying formulation of amorphous solid dispersions	24	Singh and Van den Mooter (2016)
Developments in the formulation and delivery of spray dried vaccines	15	Kanojia et al. (2017)
Nano Spray Drying for Encapsulation of Pharmaceuticals	21	Arpagaus et al. (2018)
Pharmaceutical Particle Engineering via Nano Spray Drying—Process Parameters and Application Examples on the Laboratory-Scale	15	Arpagaus (2018)
The design and scale-up of spray dried particle delivery systems	17	Al-Khattawi et al. (2018)
Spray Drying of Biotherapeutic Compounds	17	Langford and Ohtake (2019)

(continued)

Table 6.2 (continued)

Title	Pages	References
Drying in Biotechnology	14	Putranto and Chen (2019)
Spray Drying of Pharmaceuticals and Biopharmaceuticals: Critical Parameters and Experimental Process Optimization Approaches	19	Ziaee et al. (2019)
Spray Drying for the Formulation of Oral Drug Delivery Systems	32	Ferreira et al. (2020)
Spray Drying	38	Vehring et al. (2020)
Spray Drying of Vaccines: A Review	14	Arpagaus (2020)
Spray Drying as an Advantageous Strategy for Enhancing Pharmaceuticals Bioavailability	12	Salama (2020)
Engineering Advances in Spray Drying for Pharmaceuticals	24	Baumann et al. (2021)
Spray Drying: Inhalable Powders for Pulmonary Gene Therapy	30	Munir et al. (2021)
Progress in Spray Drying of Protein Pharmaceuticals: Literature Analysis of Trends in Formulation and Process Attributes	33	Pinto et al. (2021)
Stability of Lyophilized and Spray Dried Vaccine Formulations	12	Preston and Randolph (2021)
Nano Spray-Dried Drugs for Oral Administration: A Review	30	Öztürk and Arpagaus (2021)
Innovations in spray drying process for food and pharma industries	23	Samborska et al. (2022)
Spray Drying and Particle Engineering in Dosage Form Design for Global Vaccines	18	Gomez and Vehring (2022)
Stabilizing vaccines via drying: Quality by design considerations	47	Ghaemmaghamian et al. (2022)

6.5 Statistics of Publications on Spray-Dried Vaccines

Over the past 25 years, numerous research studies have been published on spray drying of vaccines. Figure 6.8a shows the development of the number of publications identified in the online databases Web of Science and PubMed using the search term ALL FIELDS [spray drying AND vaccine]. The search yielded 270 hits in Web of Science and 189 hits in PubMed (accessed on 8 August, 2022). Over time, there has been a steady increase in publications, indicating a lively activity in R&D, with an average of 15–20 additional publications per year. The patent situation is not considered here.

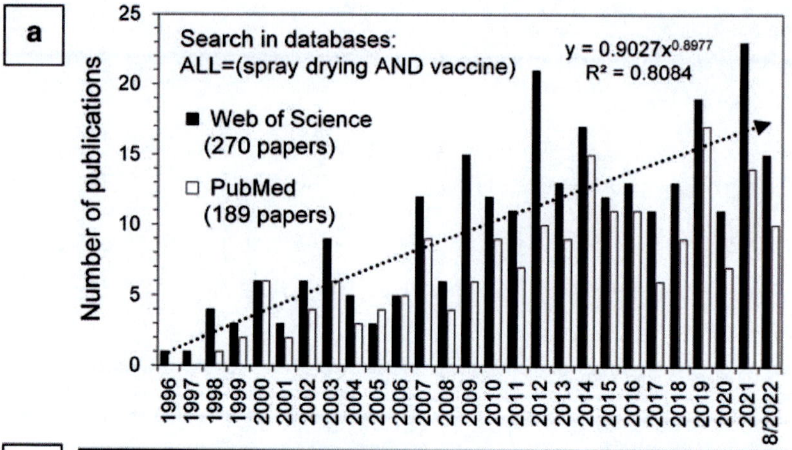

b

No.	Publication Titles (based on Web of Science, accessed 8/8/2022)	Record Count	% of 270
1	INTERNATIONAL JOURNAL OF PHARMACEUTICS	30	11.1%
2	VACCINE	26	9.6%
3	JOURNAL OF CONTROLLED RELEASE	19	7.0%
4	PHARMACEUTICAL RESEARCH	17	6.3%
5	EUROPEAN JOURNAL OF PHARMACEUTICS AND BIOPHARMACEUTICS	16	5.9%
6	JOURNAL OF PHARMACEUTICAL SCIENCES	11	4.1%
7	JOURNAL OF MICROENCAPSULATION	10	3.7%
8	JOURNAL OF DRUG TARGETING	6	2.2%
9	AAPS JOURNAL	4	1.5%
10	AAPS PHARMSCITECH	4	1.5%
11	ADVANCED DRUG DELIVERY REVIEWS	4	1.5%
12	EUROPEAN JOURNAL OF PHARMACEUTICAL SCIENCES	4	1.5%
13	EXPERT REVIEW OF VACCINES	4	1.5%
14	JOURNAL OF AEROSOL MEDICINE AND PULMONARY DRUG DELIVERY	4	1.5%
15	VACCINES	4	1.5%
16	ACS NANO	3	1.1%
17	DRYING TECHNOLOGY	3	1.1%
18	NPJ VACCINES	3	1.1%
19	PHARMACEUTICS	3	1.1%

Fig. 6.8 **a** Number of publications identified in the Web of Science and PubMed online databases using the search term ALL FIELDS [spray drying AND vaccine] (accessed on 8 August, 2022). Web of Science provided 270 hits, PubMed 189 hits. **b** Top journal titles and number of publications related to spray drying of vaccines. The first 19 journals titles cover about 65% of all publications. **c** Analysis of search results from Web of Science by organization and country Influenza virus vaccines

In terms of journal titles, Fig. 6.8b shows that the 19 most used journals account for about 65% of all publications on the topic. The leading journals are:

- International Journal of Pharmaceutics (Elsevier, 30)
- Vaccine (Publisher: Elsevier, Paper Count: 26),
- Journal of Controlled Release (Elsevier, 19)
- Pharmaceutical Research (Springer, 17),
- European Journal of Pharmaceutics and Biopharmaceutics (Elsevier, 16),
- Journal of Pharmaceutical Sciences (Elsevier, 11),
- Journal of Microencapsulation (Taylor & Francis, 10), and
- Journal of Drug Targeting (Taylor & Francis, 6).

Moreover, Fig. 6.8c shows the number of publications by organization and country. Most research activity is in the USA, the Netherlands, Canada, UK, Denmark, Switzerland, Belgium, Australia, and France. The list of organizations is headed by the University of Groningen (NLD), Mercer University (USA), and ETH Zurich (SUI), followed by Ghent University (BEL), University of New Mexico (USA), McMaster University (CAN), University of North Carolina (USA), University of Alberta (CAN), and University of London (UK).

For comparison, searches in the Web of Science on [freeze drying AND vaccine], [lyophilization AND vaccine], [spray freeze drying AND vaccine], and [foam drying AND vaccine] yielded 558, 293, 65, and 16 hits, respectively. Notable efforts in R&D of spray drying vaccines have been conducted by several organizations and research partners, including Active-Dry (USA), Aridis Pharmaceuticals (USA), BD Technologies (USA), Cambridge Biostability Limited (UK), Nova Laboratories (UK), MEND (USA), and PATH (USA) (Kristensen 2008).

Finally, numerous review articles and book chapters on vaccine stabilization by spray drying have been published in recent years (see Table 6.2) and are worth reading, including Chen and Kristensen (2009), Searles and Mohan (2010), Kristensen (2012), McAdams et al. (2012), Kanojia et al. (2017), Kanojia (2018), Putranto and Chen (2019), Langford and Ohtake (2019), Ziaee et al. (2019), Arpagaus (2020), Vehring et al. (2020), Pinto et al. (2021), Preston and Randolph (2021), Gomez and Vehring (2022), Ghaemmaghamian et al. (2022).

6.6 Influenza Virus Vaccines

6.6.1 A Short Introduction to Seasonal Influenza

Seasonal influenza is among the most severe respiratory diseases in humans, resulting in high morbidity and mortality rates. In addition, it causes an enormous economic burden in hospitalization costs, lost productivity due to missed workdays, and lives lost (Molinari et al. 2007). Influenza occurs worldwide with an estimated annual attack rate of 5–10% in adults and 20–30% in children (WHO 2012a). These annual epidemics are estimated to result in about 3–5 million cases of severe illness and approximately 290,000–650,000 respiratory deaths (Iuliano et al. 2018; WHO 2018b). Therefore, vaccination is considered the most effective means of mitigating influenza virus infection.

As described in the WHO (2017a) report on preferred product characteristics for next-generation influenza vaccines, advances are foreseen in the next decade, including greater availability of needle-free administration for vaccine delivery in low-income countries and improved thermostability to allow vaccines to be stored at ambient temperatures and obviate the need for a refrigerated cold chain. In addition, today's influenza vaccines are formulated as liquids (see Chap. 2, Sect. 2.2).

Dried influenza vaccines have the inherent advantage of being more stable than liquid vaccines, enabling stockpiling and significantly improving pandemic preparedness (Tomar et al. 2016). However, because the influenza virus is seasonal and changes its characteristics, it is essential to produce and distribute the vaccine as quickly as possible, similar to the new COVID-19 vaccines. Today, influenza vaccine manufacturers produce a vaccine each year based on circulating strains. However, research is also underway to develop a universal influenza vaccine that is not strain-specific. For next-generation vaccines based on stabilized antigens, spray drying would be attractive for stockpiling. In addition, researchers should perform a cost–benefit analysis of manufacturing procedures.

This section provides an update on research studies on spray-dried influenza vaccines, covering its challenges and possibilities, including recent developments and achievements in this field. Then, the discussion will focus on specific studies of spray-dried influenza vaccines, stabilization, and spray drying methods.

6.6.2 Influenza Vaccine Types

There are four types of seasonal influenza viruses (A, B, C, and D), where influenza A and B viruses circulate and cause disease epidemics. Influenza A viruses are further classified into subtypes according to the combination of hemagglutinin (HA) and neuraminidase (NA), the glycoproteins on the surface of the virus. The viruses contain negative-stranded segmented RNA encapsulated by the nucleoprotein to form a ribonucleotide-nucleoprotein complex. These complexes are surrounded by a shell of matrix protein enveloped by a lipid bilayer. HA is the major antigenic surface structure responsible for attachment of the virus to sialic acid-containing receptors on the host cell surface and fusion of the viral and endosomal membranes.

Figure 6.9 schematically shows the four main types of influenza vaccines: whole inactivated virus vaccine (live-attenuated), subunit vaccine, split virus vaccine, and virosomal vaccine (Amorij et al. 2008).

Most influenza vaccines are supplied as split virus vaccines, produced from the chemically disrupted influenza virus, or as subunit vaccines containing predominantly purified HA and NA. In contrast, whole inactivated virus vaccines contain

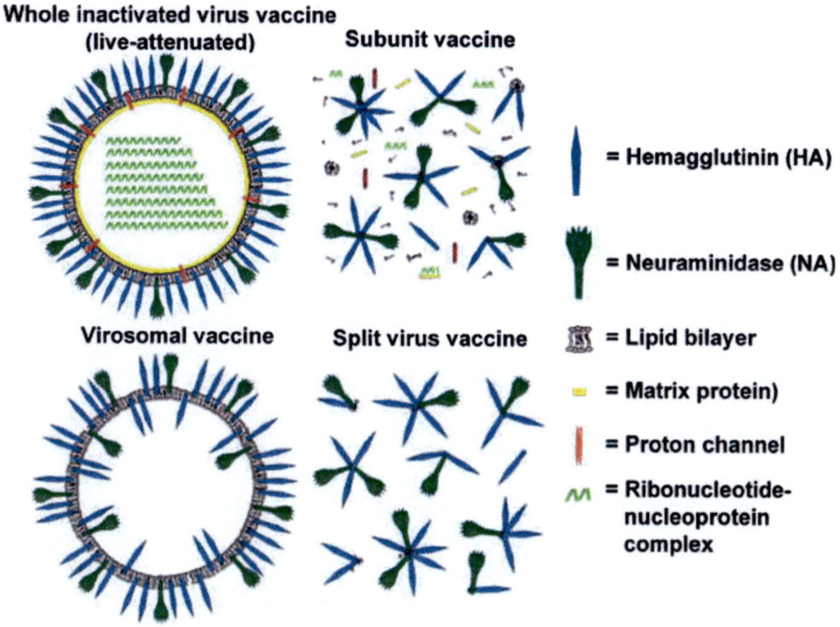

Fig. 6.9 Schematic drawings of the influenza vaccine types: Whole inactivated (live-attenuated) virus vaccine, subunit vaccine, split virus vaccine, and virosomal vaccine (Adapted from Amorij et al. (2008) with permission from Springer Nature)

inactivated virus particles that retain the receptor binding and membrane fusion activity of the native virus. This type of vaccine is approved for intranasal vaccination in many countries. Virosomes are virus-like particles consisting of reconstituted influenza virus envelopes that do not contain the genetic material of the native virus. Inflexal® V from Crucell (formerly Berna Biotech Ltd.) is an approved inactivated influenza vaccine formulated with virosomes (Mischler and Metcalfe 2002; Herzog et al. 2009). Mymetics (Switzerland) has virosomal vaccines in development.

6.6.3 Research on Spray-Dried Influenza Vaccines

The development of spray-dried influenza vaccines has been studied by numerous researchers over the past 20 years, mainly from the United States (Bot et al. 2000, 2001; Smith et al. 2003; Abdul-Fattah et al. 2007a, b; Shastri et al. 2012; Zhu et al. 2014; Lovalenti et al. 2016), The Netherlands (Amorij 2007; Amorij et al. 2008; Saluja et al. 2010; Kanojia et al. 2016), Germany (Ajmera and Scherließ 2014; Scherließ et al. 2014), and Australia (Sou et al. 2015).

The main reasons for the research with the influenza vaccines were (1) to improve stability in dry powder form and extend shelf-life as an alternative to traditional lyophilized vaccines, (2) particle engineering for needle-free delivery forms by pulmonary, nasal, skin, as well as oral administration, and (3) investigation of enhancing vaccine efficacy.

Table 6.3 summarizes published research studies of spray-dried influenza vaccines from 2000 to today with information on the vaccine type, excipients, stabilizers, buffers, spray dryer model, process parameters, particle size, morphology, powder yield, residual moisture content, route of administration, animal model, thermostability (shelf-life tested), references and research partners.

As shown in Table 6.3, the vaccine types studied by spray drying were mainly subunit protein vaccines, followed by whole inactivated, live-attenuated, and split virus vaccines. It is noticeable that the laboratory studies were conducted predominantly with a Mini Spray Dryer model B-190, B-191, and B-290. The researchers tried to operate mild drying conditions to protect the thermally sensitive vaccine products and maintain an efficient drying process at the same time. The average inlet and outlet temperatures were about 102 ± 36 °C and 60 ± 18 °C, respectively.

The produced particles were in a size range of about 1–10 μm with typically spherical and shriveled (raisin-like) morphology, as shown in the SEM images in Fig. 6.10. After spray drying, the residual moisture content of the powder was about 1–5 wt%, which depended on the formulation and drying conditions.

The selection of excipients for vaccine stabilization is of crucial importance. Here, the formulation specialists have a playground ahead on optimizing stability. Glass-forming sugars, such as trehalose (Scherließ et al. 2014; Zhu et al. 2014; Sou et al. 2015; Kanojia et al. 2016), sucrose (Abdul-Fattah et al. 2007a, b; Lovalenti et al. 2016), inulin (Amorij 2007; Amorij et al. 2008; Saluja et al. 2010), and amino acid-based formulations (Ajmera and Scherließ 2014; Scherließ et al. 2014; Lovalenti et al. 2016) have been used as leading stabilizing excipients in spray drying of influenza vaccine. The carbohydrates are especially suitable to protect the antigenic properties by rapid vitrification. For more details on formulation strategies, the reader is referred to Chap. 3, where lessons learned from other drying technologies are discussed. The selected solid concentrations with optimized buffers were mainly 5–15 wt%. The addition of surfactants helped stabilize droplets' surface properties during spray drying (Abdul-Fattah et al. 2007a, b; Lovalenti et al. 2016). Surfactants tend to populate the air–water interface blocking viral particle migration and protecting viral proteins from surface denaturation, which is particularly important for the high specific surface area of spray-dried formulations.

Most reported studies show extended shelf life of spray-dried influenza vaccines on the order of several months when stored at 20–60 °C (Zhu et al. 2014; Sou et al. 2015; Kanojia et al. 2016). Furthermore, since influenza viruses enter the body via the respiratory tract, spray-dried powder vaccine formulations offer a promising platform to deliver the vaccine via the pulmonary and intranasal routes. This feature is crucial compared to freeze drying, which is more established for vaccine drying but less suitable for pulmonary powder delivery. In addition, several animal studies have shown systemically induced immunity comparable to parenteral immunization (Amorij et al. 2007; Saluja et al. 2010; Sou et al. 2015).

Table 6.3 Research studies on spray-dried influenza vaccines with information of vaccine type, excipients, stabilizers, spray dryer model and process parameters, particle size, morphology, powder yield, residual moisture content, route of administration and animal model, thermostability (shelf-life tested to reach 1 log loss in titer), and references and research partners

Influenza vaccine type	Excipients, stabilizers, buffers	Spray dryer model, process parameters	Particle size, morphology, powder yield residual moisture	Route of administration, animal model	Thermo-stability (shelf-life tested)	Reference, research partners
Whole inactivated virus	Trehalose (50–150 mg/mL, 5–15 wt%), hemagglutinin (HA): trehalose (ratio 1:400 w/w)	B-290 with high-performance cyclone and Dehumidifier B-296 (closed-loop with N_2 drying gas), T_{in}/T_{out}: 110–160 °C/ 48–91 °C, F: 1–4.5 mL/min (5–15%), G: Aspirator setting 100%, A: 7.3–17.5 L/min (5 bar)	2.4–8.5 µm, spherical and smooth surface, 42–82% yield, 1.2–4.9 wt% RMC	No animal test	3 months (60 °C) (bioassay)	Kanojia et al. (2016), Intravacc, University of Groningen, Leiden University
Live-attenuated virus	Trehalose, sucrose, Pluronic F-68, sorbitol, histidine, $ZnCl_2$	B-190 with two-fluid nozzle and ultrasonic vibrations, T_{in}: ~60 °C, T_{out} nozzle: ~45 °C, F: 1 mL/min, A: up to 1.65 bar	> 50% yield, 1–4 wt% RMC	Ferrets, nasal instillation (0.1 mL per nare)	1 week (37 °C) 4 weeks (25 °C) 18 months (4 °C)	Lovalenti et al. (2016), Aridis Pharmaceuticals, SRI International
Inactivated virus	Trehalose (9 g), L-leucine (1 g), HA (10–100 µg) 5 wt% solids concentration (10 g in 200 mL)	B-190 with 0.5 mm two-fluid nozzle, T_{in}/T_{out}: n.a./70 °C, F: 6.67 mL/min, G: aspirator setting: 20, A: 800 L/h	49.3% FPF (< 5 µm) 3.2–4.6 wt% RMC	Rats, pulmonary by liquid instillation, subcutaneous	2 months (40 °C)	Sou et al. (2015), Monash University, Gribbles Veterinary Pathology
Subunit protein (HAC1)	Trehalose (7.5%), hydrolyzed gelatin (2.5%), HAC1 protein (360 µg/mL) (recombinant monomeric hemagglutinin protein)	B-290 with 0.7 mm two-fluid nozzle, T_{in}/T_{out}: 100 °C/58–62 °C, F: ~0.6 mL/min (2% pump), A: 35 mm setting (0.3 bar N2), Batch size: 10 and 65 mL, BLD-1 (pilot-scale custom-built with two-fluid nozzle), T_{in}/T_{out}: ~100 °C/58 °C, F: 4 mL/min, G: 30 kg/h N2, A: 0.6–2.1 bar, batch size: 500 mL	B-290: 3.4–7.2 µm, shriveled, raisin-like, ~ 75–85% yield (~0.5 g powder), ~4% (w/v) RMC BLD-1: 0.5–30 µm, ≥ 90% yield (~50 g powder), 4.0 ± 0.5% RMC, T_g of 120 °C	Mice, intramuscular	6 months (50 °C)	Zhu et al. (2014), PATH, Fraunhofer USA Center for Molecular Biotechnology, Bend Research Inc

(continued)

Table 6.3 (continued)

Influenza vaccine type	Excipients, stabilizers, buffers	Spray dryer model, process parameters	Particle size, morphology, powder yield residual moisture	Route of administration, animal model	Thermo-stability (shelf-life tested)	Reference, research partners
Split virus Pandemrix protein (HA)	Trehalose (160 mg/mL), SPS formulation (80 mg/mL) containing different amino acids (e.g. arginine, glycine), glycyrrhizinic acid (2 mg/mL), chitosan-HCl (2 mg/mL)	B-290 with two-fluid nozzle and high-performance cyclone, T_{in}/T_{out}: 120 °C/50–55 °C, F: 5–7.5 ml/min, G: 35 m^3/h air (100%), A: 470 L/h	12.8 µm, 7–8 wt% RMC	Macaques, reconstituted, intramuscular	3 months (25 °C) (60% RH)	Ajmera and Scherließ (2014, Scherließ et al. (2014), Kiel University, Public Health England, Leukocar AG
Whole inactivated virus	Eudragit S 100 (1% w/v), trehalose, lectin (0.0025% w/v)	B-191, Tin: 110 °C F: 10 mL/min, A: 5.5 bar air	1–6 µm, 72–78% powder yield	Mice, oral	6 months (4 °C)	Shastri et al. (2012), Mercer Uni, Emory Uni, Kyung Hee Uni, Georgia State University
Subunit protein (HA)	Inulin (HA/inulin 1:200 w/w), phosphate buffer saline (PBS), 5.5% w/v solid concentration	B-190 with 0.5 mm two-fluid nozzle, T_{in}/T_{out}: 130 °C/70 °C, F: 5 mL/min, G: aspirator 20, A: 800 L/h air, heating: 8.5, batch size: 80 mL	1–6 µm, 37% FPF (< 5 µm)	Mice, pulmonary	> 3 years (20 °C) (10% RH)	Saluja et al. (2010), University of Groningen, Solvay Biologicals
Subunit protein (HA)	Inulin (4 kD, T_g of 156 °C) (55 mg/mL), HA (275 µg/ml), inulin:HA (200:1), 5.5% w/v inulin solution, PBS buffer	B-190 with 0.5 mm two-fluid nozzle, process parameters not reported	3 µm (average), spherical and smooth surface	Balb/c mice, reconstituted, intramuscular	No data reported	Amorij (2007), Amorij et al. (2008), University of Groningen
Live-attenuated virus	Sucrose (10% w/v), surfactant (Pluronic F68) (0.2% w/v), vaccine:sucrose (1:100 w/w)	B-190 with custom-made high-pressure nozzle, T_{in}/T_{out}: 40 °C/30 °C, F: 2 mL/min, A: ~90 bar N2, post-drying: 30 min in collector tube at 34–35 °C	4–5 wt% RMC after spray drying, 1–2 wt% RMC after additional vacuum drying at 35 °C for 1 h	No animal test	20 weeks (25 °C) 1 week (37 °C)	Abdul-Fattah et al. (2007a, b), University of Connecticut, MedImmune Vaccines, AMGEN
Whole inactivated virus & subunit protein	Lipid-based emulsion DPPC/HES = ~8:1 w/w (~20, ~10 mg/mL), DPPC: dipalmitoyl-phosphatidylcholine, HES: hydroxyethlstarch	B-191, T_{in}/T_{out}: 85 °C/61 °C, F: 2.5 ml/min, G: 100% aspirator, A: 2800 L/h	1–5 µm (highly porous particles), density: ≤ 0.2 g/cm3	Balb/c mice, Sprague Dawley rats, pulmonary	No data reported	Bot et al. (2000, 2001), Smith et al. (2003), ManKind, Alliance Pharmaceutical, Inhale Therapeutics Systems

Mini Spray Dryer models B-190, B-191 and B-290 (Büchi Labortechnik AG), *HA* Hemagglutinin, T_{in}/T_{out} Inlet/outlet drying gas temperature, *F* Feed flow rate, *G* Drying gas flow rate, *A* Atomization gas flow rate, *RH* relative humidity, *RMC* residual moisture content

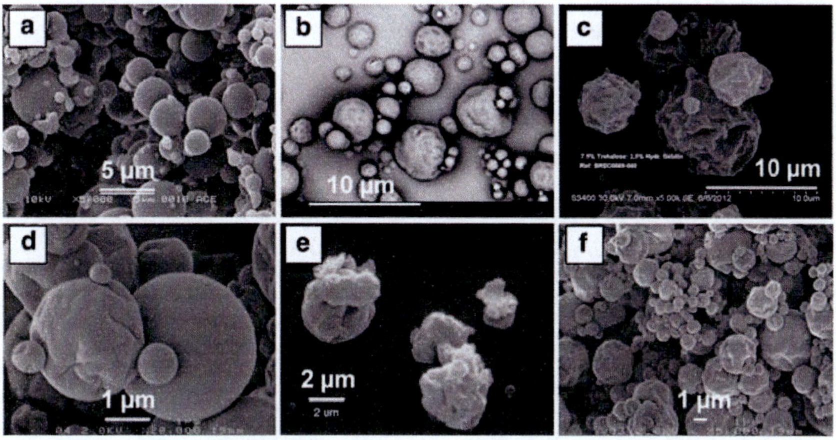

Fig. 6.10 **a** Spray-dried whole inactivated influenza virus stabilized with trehalose (Adapted from Kanojia et al. (2016), open access under CC-BY 4.0 license). **b** Spray-dried trehalose/leucine formulations containing 10 µg HA/mg antigen (Adapted from Sou et al. (2015), © Mary Ann Liebert, Inc.). **c** Recombinant hemagglutinin (HA) influenza vaccine formulated with 7.5% trehalose and 2.5% hydrolyzed gelatin after storage at 50 °C for 2 months (Adapted from Zhu et al. (2014) with permission from Springer Nature). **d** Spray-dried influenza subunit vaccine (average particle size is around 3 µm) stabilized with inulin (Adapted from Amorij et al. (2008) with permission from Springer Nature). **e** Spray-dried microparticles loaded with whole UV-inactivated influenza virus using biocompatible lung surfactant phospholipid (1,2-dipalmitoylphosphatidylcholine, DPPC) as the major excipient (Adapted from Smith et al. (2003) with permission from Elsevier). **f** Spray-dried subunit vaccine stabilized with inulin in PBS buffer (1:200 w/w HA:inulin) (Adapted from Saluja et al. (2010) with permission from Elsevier)

6.6.4 Discussion of Spray-Dried Influenza Vaccine Results

Some studies listed in Table 6.3 are discussed in more detail in the following. Kanojia et al. (2016) demonstrated in a Design of Experiment (DoE) approach the effects of the main process parameters on the product characteristics of a spray-dried influenza vaccine, as shown in Fig. 6.11.

Based on a parameter study with 23 experimental runs on a Mini Spray Dryer B-290 in a closed-loop configuration, the authors predicted a design space (sweet spot) in which trehalose-based influenza vaccine powder can be produced with inhalable particle size (1–5 µm), powder yield above 70%, and residual moisture between 1 and 3 wt%. The parameter study covered inlet drying gas temperature between 110 and 160 °C, atomization gas flow rate between 7.3 and 17.5 L/min, and feed flow rate between 1 and 4.5 mL/min. The trehalose concentration was varied between 100 and 150 mg/mL.

The spray-dried powders were spherical with a smooth surface, as shown in Fig. 6.10a. Besides, the produced influenza vaccine powder retained its antigenicity (no significant loss of HA titer) and was thermostable on storage at 60 °C for 3 months.

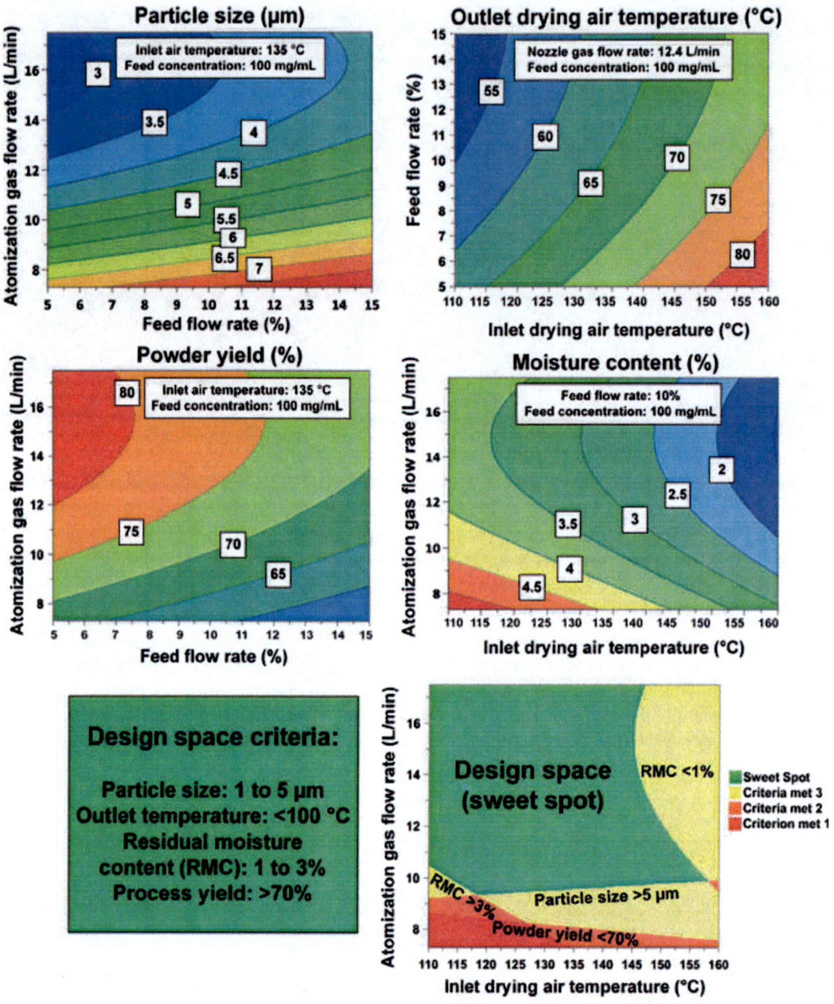

Fig. 6.11 Example of a predicted design space for influenza vaccine stabilized with trehalose in a Mini Spray Dryer B-290 (closed-loop configuration with N_2 drying gas). Fixed process parameters: feed flow rate: 1.0 mL/min, feed concentration: 100 mg/mL. The design space criteria were set to: particle size: 1–5 µm, outlet temperature: < 100 °C, residual moisture content: 1–3 wt %, process yield: > 70%. In the green area, all criteria are met (sweet spot) (Adapted from Kanojia et al. (2016), open access under CC-BY 4.0 license)

Figure 6.12 shows the improved long-term stability of a live-attenuated B/harbin influenza vaccine stabilized by spray drying for 14 months at 25 °C compared to its liquid counterpart based on a Fluorescent Focus Assay (FFA assay) (Truong-Le and Pham 2007; Ohtake et al. 2015). The vaccine formulation contained 5% sucrose, 2% trehalose, 2% arginine, methionine, 0.2% Pluronic F68, KPO_4 buffer, and EDTA (ethylenediaminetetraacetic acid).

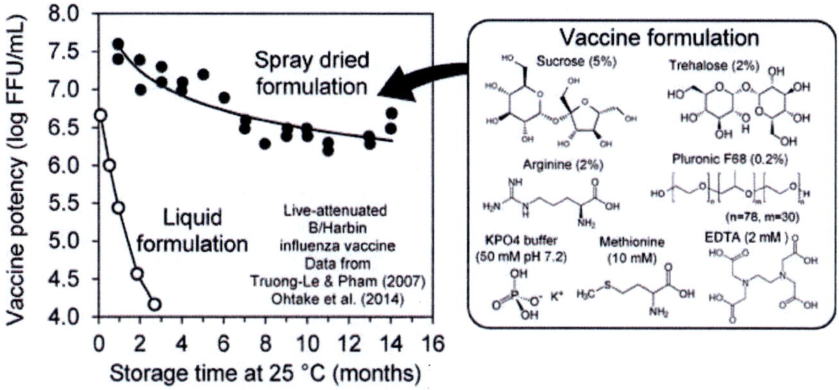

Fig. 6.12 Spray-dried formulation of live-attenuated B/Harbin influenza virus vaccine shows improved long-term stability at 25 °C (Slope = −0.087 log loss per month) over its liquid counterpart. A Fluorescent Focus Assay (FFA assay) was used as biological activity assay. One Fluorescent Focus Unit per mL (FFU/mL) is approximately equal to one Tissue Culture Infectious Dose per mL ($TCID_{50}$/mL). The vaccine was formulated with 5% sucrose, 2% trehalose, 2% arginine, 0.2% Pluronic F68, 50 mM pH 7.2 KPO_4 buffer, 10 mM methionine, and 2 mM EDTA (ethylenediaminetetraacetic acid) (Data adapted from Truong-Le and Pham (2007) and Ohtake et al. (2015) with permission from John Wiley and Sons)

Each component played an essential role in the feed formulation. Sucrose and trehalose acted as polyols and gave structure and protection to the powder particles. The amino acids arginine and methionine were used to scavenge oxidizing and deamidating agents and stabilize the conformations of proteins. Pluronic F68, a polypropylene glycol surfactant, was added to increase solubility and KPO_4 buffer to regulate pH. Finally, EDTA was included to reduce aggregation of formulation ingredients and/or scavenge metal ions to initiate destructive free-radical chemistries (Truong-Le and Pham 2007). This example demonstrates that under optimal formulation and processing conditions, spray drying can stabilize live-attenuated virus vaccines for one year or longer at 25 °C.

In another study, Sou et al. (2015) prepared a spray-dried influenza vaccine with trehalose as stabilizing excipient. In addition, the amino acid L-leucine was included in the formulation to enhance the dispersibility of the resulting powder (Seville et al. 2007). In-vitro HA inhibition assays verified that the dry influenza vaccines were stable for 2 months when stored at 40 °C. After reconstitution of the powder in saline, the vaccine was injected into the lungs of rats by liquid instillation, which elicited strong mucosal and systemic immunity comparable to the subcutaneously administered liquid vaccine. Histological examination of lung tissue showed no evidence of pulmonary inflammation.

Ajmera and Scherließ (2014) screened different amino acids (i.e., arginine, glycine, histidine, leucine, lysine, methionine, phenylalanine, and tyrosine) and their combinations for stabilizing proteins during spray drying. The combination of arginine and glycine showed the most promising results for stabilizing catalase

(250 kDa molecular weight). The authors transferred this stabilizing concept to other protein structures, such as model vaccine antigen lysozyme (14 kDa) and influenza haemagglutinin (HA) (230 kDa) from the commercially available inactivated influenza split vaccine Pandemrix® (GSK). The antigenic component of Pandemrix (containing 15 μg/mL of HA) was spray-dried in combination with arginine and glycine (1:1) in a mass ratio of 2:1 amino acid to protein together with trehalose (160 mg/mL) as a bulking excipient. The powder showed a better HA titer when stored for 1 month at 25 °C (60% RH) than the original liquid vaccine. The mean particle size of 3.3–4.8 μm enabled rapid redispersion before injection and would be suitable for inhalation.

In a subsequent study, Scherließ et al. (2014) spray-dried influenza split virus antigen using trehalose (160 mg/mL) and a proprietary aqueous stabilization and protection solution (SPS®) developed by Leukocare AG (Germany) consisting of a tailored amino acid-based formulation (Tscheliessnig et al. 2012; Ajmera and Scherließ 2014; Altrichter et al. 2019). In addition, the authors added small amounts of glycyrrhizinic acid and chitosan-HCl to enhance the immune response in-vivo due to stabilizing and adjuvant effects (Scherließ et al. 2014). For example, Fig. 6.13 shows the principle stabilization mechanism of the protective layer on the surface of an IgG antibody after drying (Scholz and Altenhofen 2011).

After spray drying the vaccine on a Mini Spray Dryer B-290 with a high-performance cyclone at 120 °C inlet and 50–55 °C outlet temperature, the powder samples were aseptically filled in glass vials and terminally sterilized by β-irradiation with 25 kGy (electron beam) (Scherließ et al. 2014). The vaccine

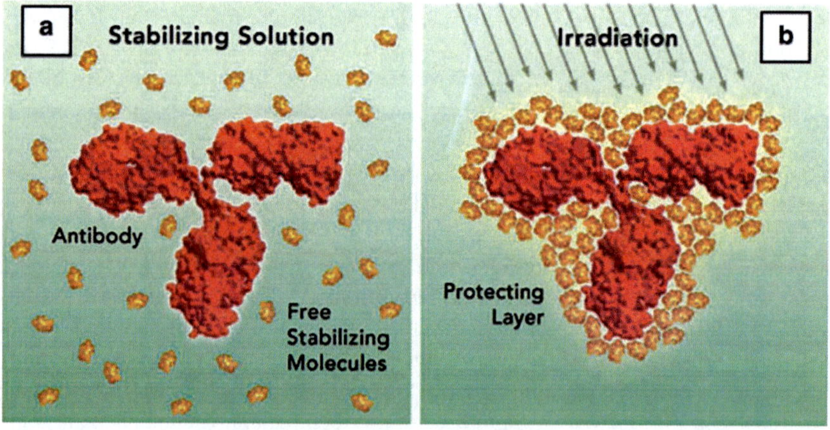

Fig. 6.13 Schematic representation of the stabilization mechanisms by the aqueous stabilization and protection solution (SPS®) from Leukocare AG (Germany), successfully applied to stabilize vaccines. For example, an IgG antibody is embedded in the stabilizing solution (**a**). After drying (**b**), the small molecules form a protective layer around the antibody, resulting in extended shelf life and protection from irradiation during terminal sterilization. Furthermore, the protective layer is reversible by rinsing (Images adapted from Scholz and Altenhofen (2011) with permission from HPCi Media Ltd., UK)

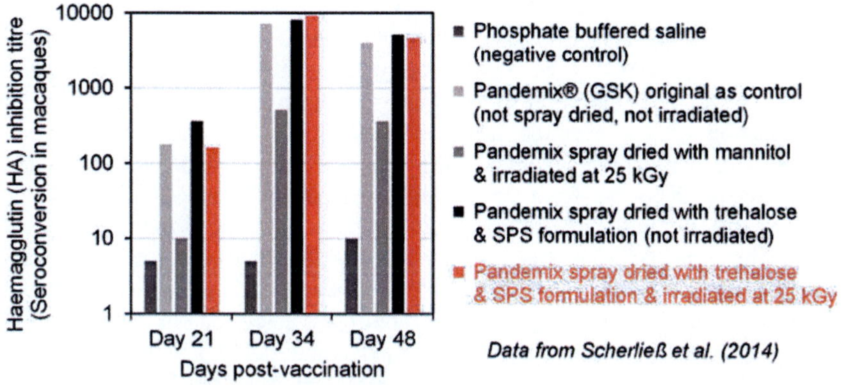

Fig. 6.14 Immune responses (HA inhibition assay titers) in non-human primate model (*Macaca fascicularis*) to influenza vaccine (Pandemrix®, GSK) after 21, 34, and 48 days. Influenza powder vaccine spray-dried with trehalose and SPS solution (tailored amino acid-based formulation) shows highly efficient immune response even after β-irradiation terminal sterilization (electron beam, 25 kGy sterilization dose) (Data adapted from Scherließ et al. (2014) with permission from Elsevier)

powder remained stable at 25 °C (60% RH) for at least 3 months. A preclinical seroconversion study in non-human primates (*Macaca fascicularis*) showed that the SPS/trehalose-stabilized and β-irradiated powder vaccines (redispersed and administered intramuscularly) fully retained the immune response of the animals compared to non-irradiated original Pandemrix® (Fig. 6.14). The SPS-solution inhibited sterilization-mediated protein denaturation and provided a viable strategy to produce highly stable and effective influenza vaccine powders by terminal sterilization.

Enabling terminal sterilization relieves the constraints of aseptic spray drying, which is a challenge for most CDMOs, although some companies, such as Nova Labs in the UK, offer aseptic spray drying (see Chap. 3, Sect. 3.9, for more details).

Some years earlier, Amorij et al. (2008) prepared a spray-dried influenza subunit vaccine powder using the oligosaccharide inulin (4 kD, T_g of 156 °C) as a stabilizer. A 5.5% w/v inulin solution containing 275 μg/ml subunit vaccine (inulin:HA, 200:1 ratio) in PBS solution was spray-dried using a Mini Spray Dryer B-190 with a 0.5 mm two-fluid nozzle (Amorij 2007). The spray-dried powders were spherical and smooth with an average particle size of 3 μm (Fig. 6.10d). In-vivo studies in Balb/c mice immunized intramuscularly with 5 μg of subunit antigen from reconstituted spray-dried powder had no adverse effect on the immunogenicity of the subunit antigen. In addition, serum IgG titers and serum HI titers in Balb/c mice were comparable to the unprocessed original vaccine (Amorij et al. 2008).

Saluja et al. (2010) continued these studies and developed an inhalable influenza subunit vaccine dry powder consisting of the surface glycoprotein HA and stabilized by inulin. The dispersion from the dry powder inhaler Twincer® (De Boer et al. 2006) resulted in a fine particle fraction (aerodynamic particle size < 5 μm) of

37% (specific surface area of 1.87 m^2/g), as shown in Fig. 6.10f. The spray-dried powders were administered to mice via the pulmonary route using a Dry Powder Insufflator™ (Model DP-4 M, Penn-Century Inc., USA). Overall, the immunogenicity of the vaccine powder formulations was comparable to the intramuscular immunization of a conventional liquid formulation. Pulmonary administration resulted in higher IgG titers in mice. Long-term tests showed complete stabilization of the vaccine in powder form for at least 3 years at 20 °C.

In an earlier study, Smith et al. (2003) developed spray-dried lipid-based microparticles loaded with whole inactivated virus or split-subunit influenza vaccine for pulmonary delivery. A fluorocarbon-in-water emulsion (approximately 0.2 µm, prepared by high-pressure homogenization) was stabilized with a biocompatible phospholipid (1,2-dipalmitoylphosphatidylcholine, DPPC) present in normal lung. Hydroxyethyl starch was added to the aqueous solution. Lactose, calcium chloride, and the biocompatible surfactant tyloxapol were used to improve the release of the antigens from the microparticles and the immune profile (Smith et al. 2003). The authors used a Mini Spray Dryer B-191 at 85 °C inlet and 61 °C outlet temperature (Bot et al. 2000, 2001) to generate highly porous particles of 1–5 µm size and 50–300 nm pore diameter ($\leq$ 0.2 g/cm^3 density). Rats were immunized by intratracheal dry powder insufflation. The microparticles were delivered directly into the lung tree via the trachea using an insufflator (Penn-Century Inc., USA) loaded with 1 mg of powder and pushed into the lungs with 2 mL of air. The data highlight a simple strategy to convert a conventional parenterally administered influenza virus subunit vaccine to pulmonary administration using biocompatible and immunologically inert excipients.

Oral administration of spray-dried influenza vaccine has been investigated by Shastri et al. (2012) as an alternative to parenteral injection because it is easy to administer, has higher compliance, allows mass vaccination, and reduces the cost of vaccine production. The authors encapsulated inactivated influenza virus by spray drying into enteric-coated microparticles of Eudragit S100 and trehalose. The powders of 1–6 µm size were prepared at 110 °C drying air inlet temperature. The enteric polymer coating protected the incorporated vaccine from the acidic gastric conditions and released it only in the intestine. Tuning the composition of Eudragit and trehalose could adjust the antigen release profile. In a mouse model, oral administration of the vaccine-induced antigen-specific immunoglobulins protected against viral challenge compared to naive control mice.

More recently, Lovalenti et al. (2016) compared spray drying, freeze drying, and foam drying processes to produce a stable solid dosage form of live-attenuated influenza vaccines. In a DoE analysis, 30 excipient combinations were evaluated using a Mini Spray Dryer B-190. The excipient screening was based on previous experience, including different sugars, plasticizers, proteins, amino acids, surfactants, chelators, metal ions, and buffers. The researchers used a two-fluid nozzle with ultrasonic vibration to reduce atomization stresses and minimize process loss (Truong-Le et al. 2014). The process parameters were set to maintain a nozzle outlet temperature of 45 °C. Inlet temperatures were typically around 60 °C. Figure 6.15 shows representative images of foam-dried, freeze-dried, and spray-dried vaccine

Fig. 6.15 **a** Typical appearance of foam-dried, freeze-dried, and spray-dried vaccine formulations (Image adapted from Lovalenti et al. (2016) with permission from Springer Nature). **b** Morphologies of sucrose-rich dried formulations observed by SEM (Images adapted from Abdul-Fattah et al. (2008) with permission from Elsevier)

samples (a) and morphologies (b). The drying method significantly impacted the specific surface area (Abdul-Fattah et al. 2007a, 2008). Freeze-dried powders were typically composed of thin plates, foam-dried powders of larger plates, and spray-dried powders of spherical particles with the largest surface area (Table 6.4, B).

Besides, Fig. 6.16 compares the storage stability profiles at 37 °C based on a $TCID_{50}$ potency test (50% infectious dose in tissue culture) (Lovalenti et al. 2016). The spray-dried and freeze-dried formulations lost more than 1 log of potency after one week at 37 °C, while the titer of the foam-dried formulation decreased slowly (see values in Table 6.4, A) and eventually lost the same amount after about 20 weeks. The authors attributed the higher stability to the low specific surface area provided by foam drying (Abdul-Fattah et al. 2007a, b; Truong and Abdul-Fattah 2010; Langford et al. 2018) and a reduced air–solid interface exposure for the vaccine during drying (Truong and Abdul-Fattah 2010; Lovalenti and Truong-Le 2020). The researchers also noted that samples with non-ionic surfactant Pluronic F68 had less process loss than formulations without, suggesting that the surfactant may have a shielding effect against destabilizing during spray drying. The immunogenicity of samples from all three drying methods was comparable to the liquid control when administered intranasally (after reconstitution) to ferrets.

Abdul-Fattah et al. (2007a, b) investigated the stability of a genetically engineered live-attenuated bivalent vaccine candidate against human parainfluenza virus

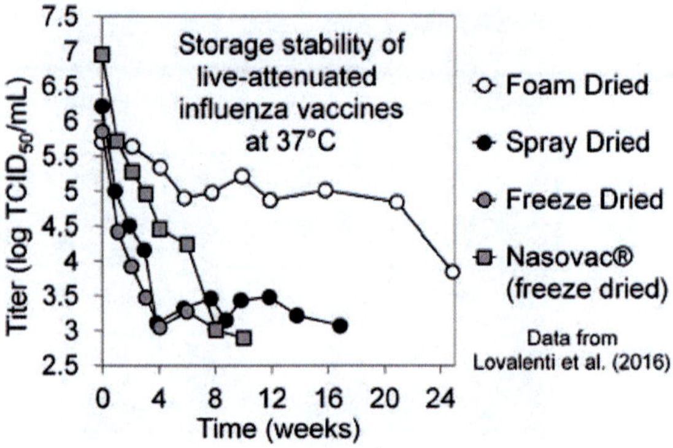

Fig. 6.16 Stability profiles at 37 °C of foam-dried, freeze-dried, and spray-dried formulations of live-attenuated influenza vaccine (spray-dried with sucrose, trehalose, and surfactant Pluronic F-68). Nasovac™ is a commercial freeze-dried product marketed by the Serum Institute of India (Data adapted from Lovalenti et al. (2016) with permission from Springer Nature)

Table 6.4 Comparison of process-associated loss and storage stability for dry vaccine formulations prepared by foam drying, spray drying, and freeze drying

(A) Live-attenuated influenza vaccine (H1N1) Lovalenti et al. (2016)	Process loss (log $TCID_{50}$/mL)	Rate of titer loss at 37 °C (log $TCID_{50}$/mL/week)	
Foam drying	0.4 ± 0.2	0.055 ± 0.011	
Spray drying	0.3 ± 0.2	0.72 ± 0.09	
Freeze drying	0.5 ± 0.2	0.66 ± 0.12	
Nasovac™ (freeze-dried)	n.a	0.58 ± 0.09	
(B) Live-attenuated virus vaccine candidate for human parainfluenza virus (Medi 534) Abdul-Fattah et al. (2007a, b)	Process loss (log $TCID_{50}$/mL)	Rate of potency loss at 37 °C (log $TCID_{50}$/mL/week$^{0.5}$)	Specific surface area (m^2/g)
Foam drying	0.8	2.31 ± 0.19	0.13 ± 0.02
Foam drying (with surfactant Pluronic F-68)	0.9	1.20 ± 0.19	0.10 ± 0.01
Spray drying (with Pluronic F-68)	0.4	2.25 ± 0.19	1.26 ± 0.03
Spray drying	0.8	2.76 ± 0.19	2.80 ± 0.07
Freeze drying	1.4	2.65 ± 0.30	1.42 ± 0.03

(A) Live-attenuated influenza vaccine (H1N1), (B) Genetically engineered live-attenuated bivalent vaccine candidate against human parainfluenza virus (Medi 534)
Data adapted from Lovalenti et al. (2016) and Abdul-Fattah et al. (2007a, b) with the permission from Springer Nature and Elsevier

(Medi 534) using spray drying, freeze drying, and foam drying with sucrose-based formulations with and without surfactant Pluronic F68. A custom-made high-pressure nozzle was used for spray drying, providing approximately 4–20 µm droplets (Truong-Le and Scherer 2008). The atomizing gas was N_2 at a pressure of about 90 bar, which allowed to spray dry at a low inlet temperature of 40 and 30 °C outlet temperature. To further reduce the moisture content from 4 to 5 wt% down to 1 to 2 wt%, the powders were subjected to additional vacuum drying at 35 °C for 1 h. Stability studies at 25 and 37 °C with bioassays showed that the vaccine was most stable in the foam-dried preparation with surfactant. The order of stability was: foam-dried (with surfactant) > spray-dried (with surfactant), foam-dried (without surfactant) > spray-dried (without surfactant). The improvement in stability with foam drying was associated with very low specific surface area (about 0.1 m^2/g see Table 6.4), low vaccine surface coverage, and high enthalpy recovery (i.e., low molecular mobility). On the other hand, the specific surface area was highest, about 2.8 m^2/g, in the spray-dried formulation without surfactant.

These findings were confirmed by other studies of Truong and Abdul-Fattah (2010) with the live B/Harbin influenza virus vaccine, where an optimized formulation was dried by freeze drying, spray drying, and foam drying. The best-case stability of vaccine storage at 37 °C was 0.4 months for spray-dried powder and 4.2 months for foam-dried vaccine, respectively. This correlated with the specific surface area, i.e., spray-dried powder from 2.6 to 3.1 m^2/g and foam-dried powder was 0.06–0.09 m^2/g. In conclusion, different drying methods applied to the same formulation resulted in different solid-state structures and properties.

Last but not least, to bring a thermostable spray-dried vaccine to market, scale-up validation is necessary to ensure that the product maintains the intended properties when produced in large quantities. As scale-up is not a linear process, it can be challenging due to differences in batch size, run time, particle size, and temperature gradient. However, a successful example of scale-up by about 100-fold was reported by Zhu et al. (2014), where a subunit influenza hemagglutinin vaccine with trehalose and gelatin hydrolysate was developed on a Mini Spray Dryer B-290 with approximately 10–65 mL batch size (0.5 g powder samples per batch) and transferred to a pilot-scale BLD-1 (custom dryer, Bend Research, Inc. USA) with a batch size of 500 mL (50 g samples with a powder yield of > 90%).

Most particles were between 5 and 10 µm in size and exhibited a shriveled, raisin-like morphology (see Fig. 6.10c). The powders showed stable physical properties and retained their in vitro potency when stored for 6 months at temperatures ranging from −20 °C to + 50 °C. In mice, the formulations elicited immunogenic responses (HAI antibody titers) equivalent to those of a frozen batch of the bulk vaccine used as stability control.

6.6.5 Conclusions on Spray-Dried Influenza Vaccines

Overall, the results of spray-dried influenza vaccines show improved thermostability in dry powder form in the order of several months at room temperature, making spray drying an attractive technology for stockpiling and rapid deployment in public emergencies, including pandemics, as well as for routine vaccination in regions with insufficient cold chain. In addition, the engineering of particles in powder form opens up alternative delivery routes without needles through pulmonary, nasal, skin, and oral administration. Finally, spray drying technology can also be applied to other types of vaccines, as numerous other examples in this book demonstrate.

6.7 Measles Vaccines

6.7.1 A Short Introduction to Measles Vaccines

Measles is caused by a virus from the paramyxovirus family and is usually transmitted through direct contact and air (WHO 2017b). The virus infects the respiratory tract and then spreads throughout the human body. The measles virus is a single-stranded, enveloped, negative-sense RNA virus with about 120–250 nm diameter (Strebel et al. 2018).

As noted by Rota et al. (2016), measles is a technically and biologically eradicable disease and could join smallpox and soon polio on the list of eradicated human viral diseases. Routine measles vaccination for children has been in place for nearly 60 years. It is safe, effective, and inexpensive. In fact, it costs about one U.S. dollar to vaccinate a child against measles (WHO 2019b).

In 2019, about 85% of children worldwide received the first dose of measles-containing vaccine (MCV1) by their first birthday, and 71% received a second dose (MCV2) (WHO/UNICEF 2020) (see global coverage in Chap. 1). From 2000 to 2018, measles vaccination prevented an estimated 23.2 million deaths worldwide, with measles deaths decreasing from an estimated 536,000 in 2000 to 142,000 in 2018, mostly among children under the age of five (WHO 2019b).

Several live-attenuated measles vaccines are available either as monovalent or in combination with rubella, mumps, or varicella vaccines. When the combined measles-rubella (MR) vaccine, measles-mumps-rubella (MMR) vaccine, or measles-mumps-rubella-varicella (MMRV) vaccine is used, the immune response to each vaccine antigen is mainly unchanged (WHO 2017b).

Current formulations of measles vaccines are freeze-dried and require cold chain storage at 2–8 °C until use. As summarized in the WHO reports of Galazka (1989) and Galazka et al. (1999) on the thermostability of vaccines, measles vaccines in the lyophilized form are stable at subzero temperatures and are not damaged by freezing and refreezing. At 2–8 °C minimum potency is retained for more than 2 years, at room temperature (20–25 °C) for at least 1 month, and at 37 °C for at least 1 week (i.e., virus titer loss $< 1.0 \log_{10}$ infectious units), which is a WHO requirement (WHO 1994; Strebel et al. 2018). This information is based specifically on results from Allison et al. (1981) and Mann et al. (1983), who tested the stability of more than 21 freeze-dried live-attenuated measles vaccines from nine manufacturers at different storage temperatures.

After reconstitution with the sterile diluent supplied by the manufacturer, the measles vaccines should be administered immediately by needle and syringe by trained medical staff. If the reconstituted vaccine is not used within 6 h, it must be discarded due to the risk of bacterial contamination and loss of potency with exposure to light and increased temperature (Strebel et al. 2018). Notably, the reconstituted measles vaccine loses about 50% of its potency after 1 h at 20 °C, and almost all potency when stored at 37 °C for 1 h (WHO 2017b).

Attenuated live measles virus vaccines have an impressive record of safety, efficacy, and ability to induce lifelong immunity to measles infection (Zuniga et al. 2007). A growing pipeline of attenuated measles virus is also a vector against other viral infections (Gerke et al. 2019). A particular focus is currently on the Chikungunya vaccine as the first measles-vectored vaccine that has demonstrated the technology's safety, immunogenicity, and functionality in humans (Phase II clinical trials are ongoing). In addition, there are many preclinical candidates for various pathogens using the measles vector platform technology (Gerke et al. 2019). It may be particularly relevant for future vaccine development considering that viral (adenovirus) vectors are used for some current mRNA-based vaccines for COVID-19.

Despite tremendous success in global measles control, the virus still circulates, causing infections, severe illness, and deaths. Although the measles vaccine is highly safe and effective, coverage could be improved with a vaccine that is easier to administer and transport (Ohtake et al. 2010; Lin et al. 2011; Agarkhedkar et al. 2014). In particular, delivery by needle injection prevents sustained high coverage in many low-income countries. In this context, the existing barriers are mainly (Lin et al. 2011):

- shortage of trained medical staff to sterilely reconstitute and safely inject the vaccine,
- limited cold chain infrastructure in low-income countries that causes waste of unused freeze-dried vaccine in multidose vials (i.e., 5–10 doses are commonly used), which after reconstitution lose potency within a short time, and
- risks for transmission of bloodborne diseases associated with the use and disposal of contaminated needles and syringes.

Significant advances in development are skin vaccination through microarray patches, enabling house-to-house vaccination, and allowing administration without medically trained personnel (WHO 2017b). This solution would greatly benefit countries with limited human resources, and it would increase the likelihood of success in achieving regional elimination goals. According to WHO and UNICEF (2019), measles-rubella microarray patches are considered a potentially disruptive technology that can change the existing market. The potential role of using vaccine microarray patches as a platform and pathways for innovation and commercialization were recently reviewed by Badizadegan et al. (2020).

Pre-clinical data demonstrated that microarray patches loaded with measles vaccine (Edens et al. 2015) and measles-rubella vaccine (Joyce et al. 2018)

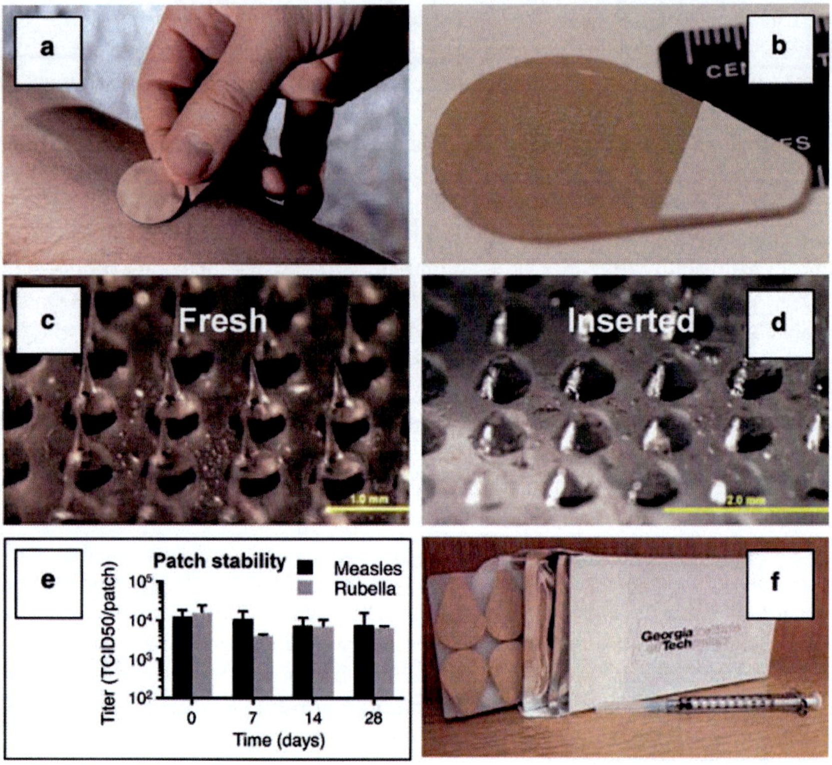

Fig. 6.17 a, b Microarray patch (3.5 × 2.2 cm in size) is applied to the skin. **b** Microneedle array of 1 cm2 size consisting of 100 units of 700 μm long microneedles enclosing freeze-dried measles and rubella vaccine in water-soluble excipients (including sucrose). **c**, **d** Microneedles before and after application and dissolution in the skin. Once inserted, the needles dissolve entirely, leaving only a blunt base structure. **e** Stability of measles and rubella vaccine in microneedle patches stored at 40 °C for up to 28 days without significant loss of vaccine activity. Data show mean titer ± standard deviation (n = 4 replicates). Abbreviation: $TCID_{50}$, 50% tissue culture infective dose. **f** Box of 50 microneedle patches next to a 1-mL syringe. (Images adapted from Joyce et al. (2018) and Prausnitz et al. (2020) of the Georgia Institute of Technology in Atlanta USA with permission from Oxford University Press and Elsevier, open access under CC BY 4.0 license)

produced protective titers of neutralizing antibodies in infant rhesus macaques and afforded complete protection from measles virus challenge.

Figure 6.17a, b visualizes the method of administration of dissolvable microarray patches on the skin and (c, d) a close-up of the microneedles before and after application. The microarray patches contain microscopic (500–800 μm), water-soluble microneedles made from a dried formulation of the vaccine, which is pressed onto the skin and left for 10–15 min to allow the microneedles to dissolve.

The main benefits of the microarray patch compared to conventional subcutaneous injection methods are (WHO and UNICEF 2019; Peyraud et al. 2019; Prausnitz et al. 2020):

- fewer trained medical personnel are needed,
- thermostability allowing transport and storage in a controlled temperature chain (CTC) (i.e., vaccines capable of tolerating at least 40 °C for a minimum of 3 days before use),
- no reconstitution is required,
- minimal storage and disposal due to single-dose, single-use, small volume, and no liquids,
- no sharp waste, and
- painless vaccination.

As shown in Fig. 6.17e, stability data of measles-rubella vaccines in microarray patches revealed no significant loss of vaccine activity when stored at 40 °C for up to 28 days (Joyce et al. 2018; Prausnitz et al. 2020). Besides, microarray patch vaccines could be cost-competitive to traditional injected vaccines, as evaluated by Adhikari et al. (2016), especially when the analysis accounts for cost-savings associated with minimally trained personnel, thermostability, reduced vaccine waste, elimination of sharps waste, and no reconstitution (Fig. 6.17f).

Human clinical trials with measles-rubella microneedle patches in adults, vaccinated infants, and naïve infants are anticipated to begin in 2021 to assess the safety and immunogenicity (ClincicalTrials.gov, Identifier NCT04394689, Micron Biomedical, Inc).

In comparison, microneedle patches have already been successfully tested in clinical trials (Hirobe et al. 2015; Rouphael et al. 2017; Fernando et al. 2018). The results showed similar immune responses to intramuscular injection, good safety profiles, and low skin reactogenicity. Thus, the main challenges are to scale up manufacturing processes to enable the production of millions of microarray patches at an acceptable cost of goods (Fernando et al. 2018) and to gain more experience from human clinical trials (Prausnitz 2017).

For many years, the respiratory route has been proposed as a promising alternative way for the targeted delivery of measles vaccine, either as a liquid or dry powder aerosol (Rota et al. 2016). Studies with wet mist aerosol delivery were conducted as early as the 1960s in Japan, the USA, and the former Union of Soviet Socialist Republics (USSR) (Sabin 1983), later on especially in Mexico (Bennett et al. 2002; Valdespino-Gómez et al. 2006), South Africa (Dilraj et al. 2000), and India (Low et al. 2015).

The feasibility of aerosol vaccination in humans was proposed particularly for children and the senior population, low-income countries, and disaster scenarios (Roth et al. 2003). Successful trials of aerosolized liquid measles vaccine in large chambers in the USSR (Terskikh et al. 1971; Danilov 1973) also raised the possibility of vaccinating groups of school children within closed classrooms with parental consent (Cutts et al. 1997).

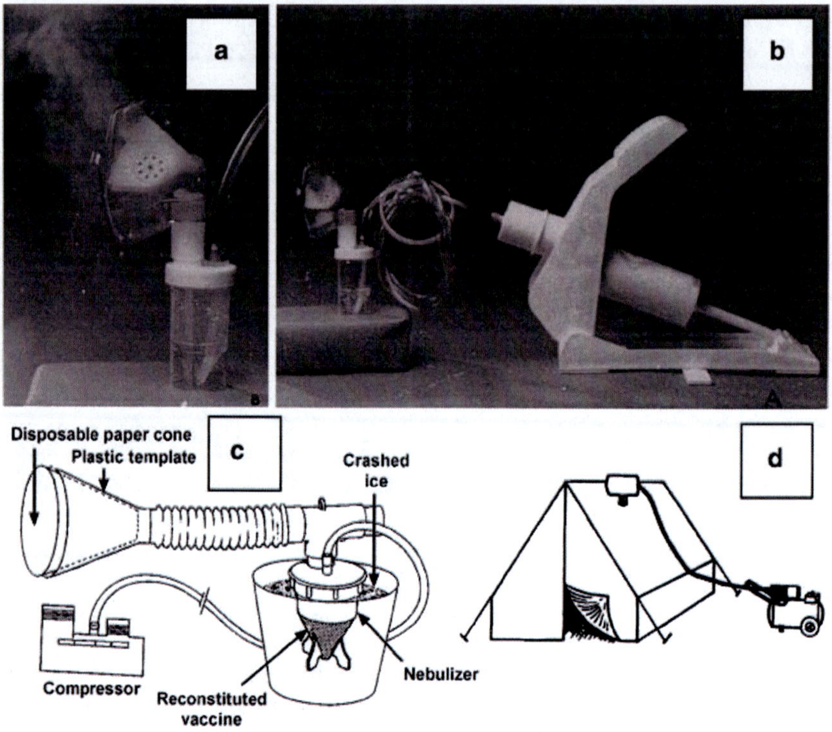

Fig. 6.18 **a** Nebulizer with aerosol emerging from a face mask, **b** Plastic foot pump that could also be used by hand, nebulizer, and face mask (Images adapted from Sabin (1991) with permission from Springer Nature), **c** Aerosol generator for immunization of reconstituted lyophilized vaccine (Classic Mexican Device) (Adapted from Bennett et al. (2002) and Valdespino-Gómez et al. (2006) with permission from WHO, open access license CC BY 3.0 IGO, and Springer Nature), **d** Exposure tent for mass and rapid immunization (Adapted from Roth et al. (2003))

Figure 6.18a, b shows a personal aerosol generator operated by manual pumping (Sabin 1991), (c) the classic Mexican device consisting of a plastic nebulizer, ice-cooled container to keep the vaccine stable during nebulization, electric compressor, and disposable paper mask (Bennett et al. 2002; Valdespino-Gómez et al. 2006), and (d) an exposure tent for mass and rapid immunizations (Roth et al. 2003).

The method of using a nebulizer with reconstituted live-attenuated Edmonston Zagreb measles vaccine has been used in extensive field trials with about 4 million Mexican children, with good results, high public acceptance, low cost, and fewer side effects compared to the usual subcutaneous immunization (Roth et al. 2003). Financial advantages have been demonstrated based on experience with measles vaccination in Brazil (Sabin 1983, 1991). A liquid nebulizer and disposable masks made of thin plastic or waxed paper were described as inexpensive and could be used repeatedly in subsequent years for annual aerosol mass vaccination of infants and children (Sabin 1991).

Several reviews compared published studies of aerosolized measles vaccine trials in children with subcutaneous administration (Cutts et al. 1997; Hiremath and Omer 2005; Valdespino-Gómez et al. 2006; Low et al. 2008, 2015). Hiremath and Omer (2005) concluded that in infants 9 months of age and older, aerosolized measles vaccine was at least as effective or better than via subcutaneous administration. In infants younger than 9 months, seroresponses after subcutaneous delivery exceeded the aerosols of the same vaccine (Cutts et al. 1997). Similarly, Valdespino-Gómez et al. (2006) and Wong-Chew et al. (2006) described that the proportion of 9-month-old infants who developed cellular and/or humoral immunity to measles was lower in the aerosol vaccine group. The differences in seroconversion rates were attributable to the lower aerosol dose of the infants compared with the subcutaneous injection. A systematic review of the literature (Low et al. 2008) concluded that seroconversion rates with aerosolized and subcutaneous vaccines appeared equivalent at around 1 year of age. On the other hand, in a clinical trial in India with infants of 9.0–11.9 months of age, the aerosolized vaccine against measles was immunogenic but statistically inferior to the subcutaneous vaccine concerning the rate of seropositivity (Low et al. 2015). In the aerosol group, 85.4% (662 of 775 children) were seropositive compared with 94.6% (743 of 785 children) in the subcutaneous group, a difference of 9.2%.

Overall, the researchers concluded that more research would be required on standardization of aerosol dosage, administration equipment (i.e., nebulizer devices), efficacy, and safety of respiratory administration of measles vaccine.

Another delivery method for vaccines is inhalation as a dry powder (LiCalsi et al. 1999). In this context, LiCalsi et al. (2001) micronized lyophilized live-attenuated measles vaccine by jet milling to produce particle sizes suitable for pulmonary delivery (1–5 μm aerodynamic diameter). The researchers demonstrated the feasibility of delivering measles vaccines as a powder aerosol for immunization.

6.7.2 Research on Spray-Dried Measles Vaccines

Several studies investigated spray drying as a formulation technology to develop an inhalable dry powder measles vaccine with the aim of higher shelf-life at a higher temperature (De Swart et al. 2007; Ohtake et al. 2010; Kisich et al. 2011; Lin et al. 2011; Agarkhedkar et al. 2014; Gala et al. 2017; Kanojia et al. 2017; Putranto and Chen 2019).

Table 6.5 summarizes the major studies on spray dried live-attenuated measles vaccines.

De Swart et al. (2007) prepared a dry powder measles vaccine of Edmonston-Zagreb strain (Berna Biotech, Switzerland) by jet milling of lyophilized vaccine (LiCalsi et al. 2001) and spray drying. The bulk liquid vaccine was dialyzed to remove salts and then mixed with trehalose for stabilization. However, the authors did not disclose the spray drying process parameters. The mean particle size

Table 6.5 Studies on spray-dried live-attenuated measles virus vaccines described in the literature

Spray drying conditions	Excipients	Particle size, residual moisture	Animal immunization study	Delivery route	Shelf life, stability	References
Spray drying parameters not reported, jet-milled lyophilized vaccine particles blended with lactose carrier	Trehalose	Around 4 µm	Macaques, immunity lower than by injection and nebulization	Pulmonary (Insufflator)	Not reported	LiCalsi et al. (2001), De Swart et al. (2007)
B-190 with ultrasonic nozzle, T_{in}/T_{out}: 60 °C/40 °C, feed rate 0.5 mL/min, atomizing N_2 (1 bar)	Trehalose, sucrose, human serum albumin, L-arginine	5–25 µm distribution, approx. 3% residual moisture	No	Not reported	8 weeks (37 °C), bioassay	Ohtake et al. (2010), Truong-Le et al. 2012, 2014)
CAN-BD (CO_2 assisted nebulization with bubble dryer), 50 °C, feed rate 0.3 mL/min, N_2 gas 30 L/min (83 bar)	Myo-inositol, sorbitol, L-arginine	Fine particle fraction: FPF < 5.8 µm (46%), FF < 3.3 µm (20%), 0.7% residual moisture	Cotton rats, Rhesus macaques, Phase I clinical study with 60 men, immunity comparable to injection	Pulmonary (Puffhaler, Solovent)	> 2 years (2–8 °C), 1 week (37 °C)	Kisich et al. (2011), Lin et al. (2011), Agarkhedkar et al. (2014)
B-191, 0.5 mm nozzle (cooled to −5 °C), T_{in}/T_{out}: 110 °C/74 °C, 1% solution, feed rate 20 mL/h, aspirator 95%, spray air 650 L/h	Bovine serum albumin crosslinked with glutaraldehyde	1–5 µm, irregular shape, 85–95% process yield	Pigs, no comparison of immunity and to other delivery routes	Buccal (particles in oral dissolving film)	Not reported	Bejugam et al. (2008), Uddin et al. (2009), Ubale et al. (2013), Gala et al. (2017)

of the resulting powder was about 4 µm, a suitable size for pulmonary administration. Macaques were vaccinated using a dry powder insufflator (Penn Century, USA). The study demonstrated the proof of principle for dry powder inhalation as a method for measles vaccination, although the delivery route was less immunizing than vaccination by injection or nebulized aerosol. This result was mainly attributed to the fact that most of the powder dose did not reach the deep lung.

Later on, Ohtake et al. (2010) investigated the spray drying of a live-attenuated Edmonston-Zagreb measles vaccine (provided by the Serum Institute of India) in a Mini Spray Dryer B-190 using a custom-made SONICAIR™ nozzle (IVEK 2012) providing fine droplets (< 7 µm mean size) at low pressure (1 bar of nitrogen) and ultrasonic energy (Truong-Le et al. 2012, 2014). Spray drying was performed at a drying gas temperature of 60 °C/40 °C inlet/outlet to minimize heat stress on the vaccine. The best-performing formulation included (Ohtake et al. 2010) (Fig. 6.19):

- Trehalose-sucrose (17% w/v) to stabilize against dehydration stress through hydrogen bonding and vitrification (Crowe et al. 1985, 1992, 1998a, b; Koster et al. 2000),
- L-arginine (4% w/v) to reduce protein-to-protein interactions and thus aggregation,
- Glycerol (1.25% w/v) as a plasticizer to increase the interactions between the glassy matrix of sugars and the virus, and
- Human serum albumin (4% w/v) for stabilization through a combination of increased T_g, increased viscosity, and surfactant-like action.

Moreover, potassium phosphate buffer KPO_4 (50 mM) was used in conjunction with divalent cations ($CaCl_2$ and $ZnCl_2$, 1–4 mM) to reduce process loss at pH 6.0, which helped maintain the virus structure during processing.

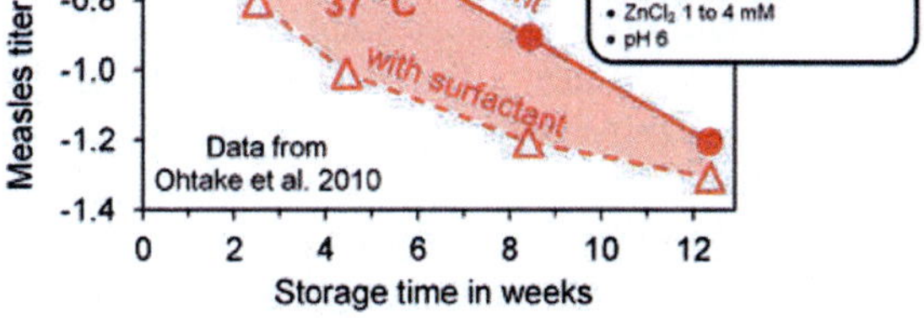

Fig. 6.19 Storage stability of spray-dried measles vaccine stored at 4 and 37 °C with and without surfactant (Pluronic F68) in the formulation based on tissue culture infectious dose ($TCID_{50}$) assay (Data adapted from Ohtake et al. (2010) with permission from Elsevier)

Figure 6.19 shows the storage stability of spray-dried measles vaccine stored at 4 and 37 °C with and without surfactant (Pluronic F68) in the formulation. Based on tissue culture infectious dose assay ($TCID_{50}$), the optimized spray-dried formulation achieved stability of 8 weeks storage at 37 °C (about 0.9 log $TCID_{50}$ titer loss), exceeding the WHO high stability criterion for shelf life at 37 °C for 30 days (WHO/UNICEF 2014). Furthermore, when the surfactant was removed from the formulations containing human serum albumin (HSA), storage stability improved, indicating that for the formulation types tested, the surfactant Pluronic F68 is not compatible with HSA and that HSA is an effective stabilizer for measles vaccine (Ohtake et al. 2010).

Compared to other drying technologies like freeze drying and foam drying, spray drying resulted in the least virus titer loss and the lowest process-associated loss at 37 °C of storage. Overall, these findings were a meaningful improvement in stability over some commercial freeze-dried measles vaccine products, but only an incremental improvement over commercially available freeze-dried Rimevax®, which exhibited 1 log loss in 5 weeks (Ohtake et al. 2010).

In response to the Grand Challenges in Global Health Initiative No. 3 (Varmus 2003), an inhalable needle-free live-attenuated measles virus vaccine was developed for use in low-income countries (Burger et al. 2008). A significant grant of US $ 20 million was awarded over 5 years from 2005 to 2010 for research efforts by an international collaboration including the Serum Institute of India, the U.S. Centers for Disease Control and Prevention, the University of Colorado, Active-Dry LLC, and private companies (Lay 2011; Sievers 2015).

For this purpose, the researchers applied the supercritical CAN-BD process (CO_2-assisted nebulization with a bubble dryer, presented in Chap. 3, Sect. 3.7 in more detail) to convert a liquid measles-clarified virus pool (MCVP-3) supplied by the Serum Institute of India into a stable dry powder suitable for inhalation (Burger et al. 2008; Cape et al. 2010). The liquid MCVP-3 consisted of a live measles vaccine (Edmonston-Zagreb strain), myo-inositol plus hydrolyzed gelatin, amino acids, and buffer (Fig. 6.20).

The optimal formulation in terms of storage stability contained around 50 g/L of myo-inositol as the primary excipient with low hygroscopicity and imparted good aerosolization properties to the powder (McAdams 2011). A total of 120 mL of the vaccine was processed at a 0.3 mL/min flow rate. The pressure of the CO_2 nebulization liquid was 83 bar, and the flow rate of the N_2 drying was 30 L/min. Primary drying was performed at 50 °C for 6 h. The collected powder was post-dried for 30 min (secondary drying) to a residual moisture content of 0.7 wt% (Lin et al. 2011). The CAN-BD process yielded a free-flowing powder (Burger et al. 2008; Cape et al. 2010) with a significant percentage of the powder in the range of 1–5 μm, suitable for pulmonary administration. Figure 6.21a shows the particle shape of the spray-dried measles vaccine powder.

Kisich et al. (2011) demonstrated in cotton rats that the developed dry powder measles vaccine readily entered the lungs of the animals and elicited an immune response comparable to the injectable vaccine. A PuffHaler™ developed and

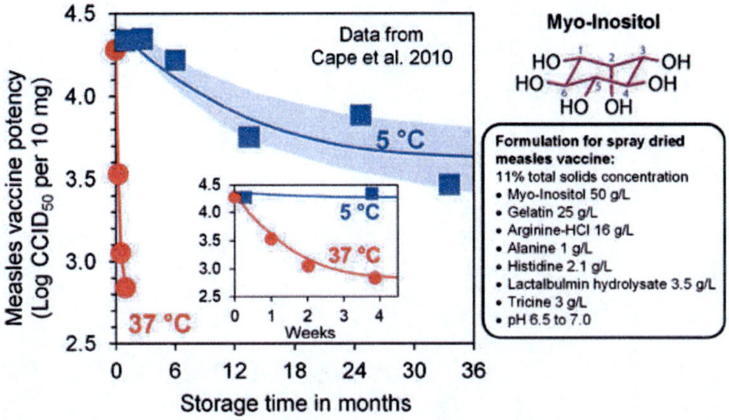

Fig. 6.20 Potency of spray-dried measles vaccine based on cell culture infective dose assay ($CCID_{50}$) during storage at 5 and 37 °C in Active-Dry PuffHaler™ blisters. The formulation contained myo-inositol as the main excipient (50 g/L, 46 wt%) (Data adapted from Cape et al. (2010), University of Colorado, Boulder, CO, USA)

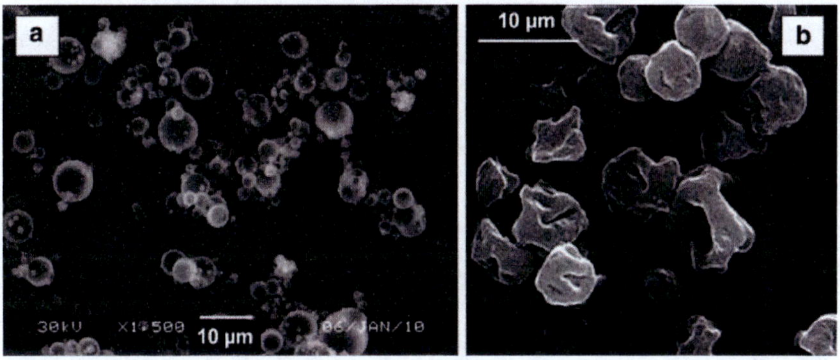

Fig. 6.21 SEM images of spray-dried live-attenuated measles virus vaccine powder. **a** Measles vaccine placebo formulated with myo-inositol (11% total dissolved solids, 50 g/L myo-inositol, 25 g/L gelatin, 16 g/L arginine-HCl, 1 g/L alanine, 2.1 g/L histidine, 3.5 g/L lactalbumin hydrolysate, 3 g/L tricine, pH 6.5–7.0) (Image adapted from Cape et al. (2010), University of Colorado, Boulder, CO, USA). **b** Measles vaccine encapsulated in bovine serum albumin crosslinked with glutaraldehyde (Image adapted from Gala et al. (2017) with permission from Springer Nature)

manufactured by Aktiv-Dry LLC (Fig. 6.22a) was adapted for use with rats by inserting the animals' noses directly into the reservoir. The researchers placed a mask on the animals to inhale the dispersed powder.

In a subsequent study, Lin et al. (2011) administered the dry powder measles vaccine by mask and nasal interface to rhesus macaques (4 groups of 3 monkeys).

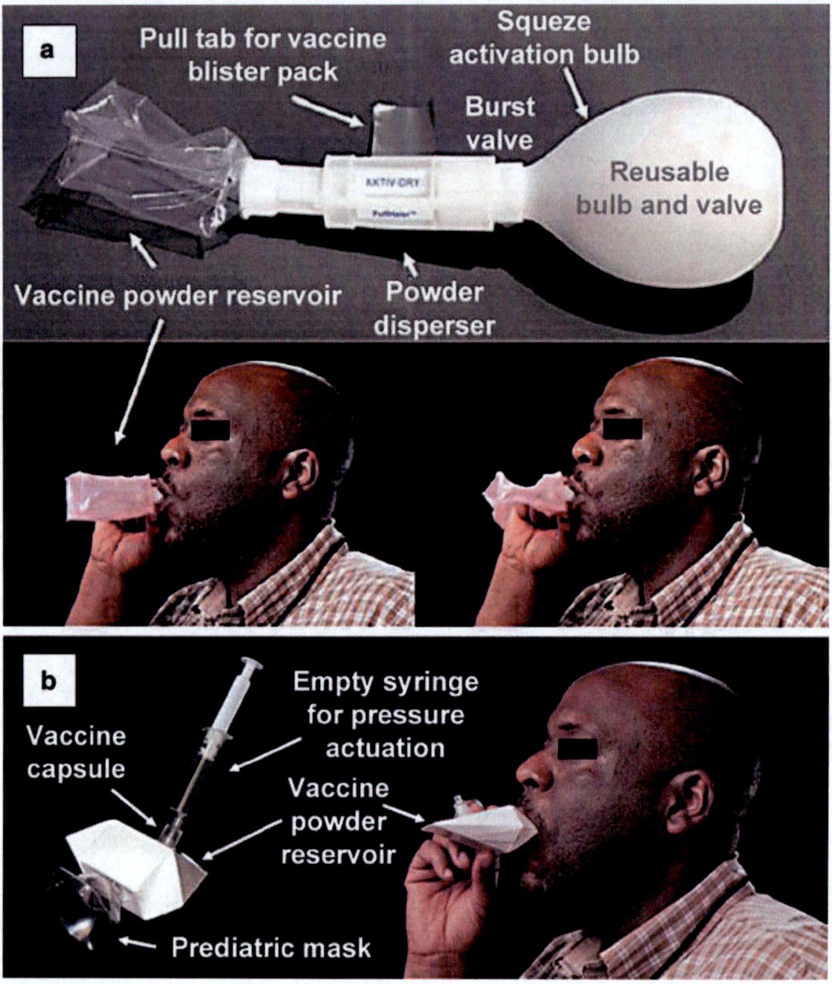

Fig. 6.22 Dry powder inhaler devices developed to deliver measles vaccine in animal studies and a clinical Phase I trial (Images adapted from Agarkhedkar et al. (2014) with permission from Elsevier). **a** The PuffHaler™ (Active Dry LLC, Boulder, CO, USA) is a dry powder inhaler designed to aerosolize powders into a disposable anti-static plastic bag reservoir. From there, the powder is inhaled either directly through a mouthpiece with adults and adolescents or via a disposable mask placed over the nose and mouth with infants and young children (not shown) (Cape et al. 2010; Weniger and Papania 2013). When the reusable squeeze activation bulk is compressed to 2 psi, the burst valve pops open, air flows through the powder in an aluminum foil blister into the disperser, and the aerosol cloud fills the folded plastic bag. The vaccine dose is administered by breathing for about 30 s. **b** Solovent™ (Becton, Dickinson and Company, Franklin Lakes, NJ, USA) is a dry powder inhaler designed to aerosolize powders into a cardstock spacer from which the powder is either directly inhaled from the spacer with adults and adolescents or via a mask placed over the nose and mouth with infants and young children (Weniger and Papania 2013). By depressing the syringe plunger, the air from the syringe ruptures the membrane of the vaccine capsule and releases a plume of dry powder into the disposable cardboard spacer

The PuffHaler™ and Solovent™ inhaler devices (Fig. 6.22b) were connected to silicone-based masks that covered the mouth and nose for free breathing and provided a seal to reduce losses.

The inhaled vaccine powder induced robust responses with no adverse effects that protected the macaques from challenge with wild-type measles virus over one year later, similar to vaccination by injection. In addition, respiratory administration of the live-attenuated measles vaccine was effective, and the vaccine powder maintained its potency for more than 2 years at 2–8 °C storage.

Figure 6.20 shows the potency of the myo-inositol based formulation during storage at 5 and 37 °C in Aktive-Dry PuffHaler™ blisters (Cape et al. 2010). The stability reached less than 1 log loss upon storage for almost 3 years at 2–8 °C or < 1 log loss for 1 week at 37 °C.

These promising results with the dry powder measles vaccine were further evaluated for safety and immunogenicity in a clinical Phase I trial (CTRI/2012/02/002447, Clinical Trials Registry India) (CTRI 2012) in 60 healthy Indian men aged 18–45 years (Agarkhedkar et al. 2014). The volunteers received a single powder dose by either the Puffhaler™ (10 mg) or Solovent™ (10 mg). In addition, a control group received a licensed measles vaccine via the standard subcutaneous route (0.5 mL) (CTRI 2012).

No serious adverse events were observed during 180-day patient monitoring. The inhaled powder vaccine was well tolerated in all subjects and was found to be safe. Moreover, the immunogenicity profiles (i.e., measles IgG antibodies and measles-specific neutralizing antibody titers) via Puffhaler™ and Solovent™ devices were comparable to those of the licensed lyophilized measles vaccine administered subcutaneously after reconstitution (Agarkhedkar et al. 2014).

Thus, the clinical Phase I study in humans is the first reported clinical trial of a spray-dried powder vaccine for inhalation. Additional more extensive clinical trials, particularly in younger subjects without pre-existing measles antibodies, the primary target population for measles vaccination, are needed to further clarify the safety and immunogenicity of the inhaled dry powder measles formulation (Agarkhedkar et al. 2014). However, further development with additional clinical trials is in limbo (Sievers 2015). Thus, further investment would be needed to advance the vaccine towards the market. In summary, the authors could draw several learnings from these studies:

- The CAN-BD process could produce clinical-grade batches of inhalable dry powder vaccines with little effort.
- The stability of the measles vaccine in dry powder form was maintained for almost 3 years at 2–8 °C, similar to that of a lyophilized vaccine. Higher storage temperatures up to about 6 months at 25 °C or 7 days at 37 °C were possible with suitable stabilizers such as myo-inositol.
- The simplest way to deliver spray-dried powder vaccines was by inhalation, i.e., without a needle and purified water for reconstitution. A major advantage is that it could avoid injection.

- The initial Phase I trials in India showed that the inhaled powder vaccine had comparable immunogenicity to a parenterally administered licensed vaccine.

The apparent challenges observed during this project were, in particular, the scalability of the supercritical CAN-BD process to larger powder volumes and the more complex inhalation delivery process compared to the traditional injection method. The production rate of the specially designed CAN-BD plant was equivalent to a lyophilization plant with a capacity of about 100 mL of liquid vaccine (11 wt% solids concentration). Besides, vaccine administration needs to be simplified for the patient, especially infants and children, which would be possible with improved powder inhaler design and training. However, the inhaled form of administration may overcome the potential risks of the injection form.

More recently, Gala et al. (2017) prepared spray-dried microparticles of bovine serum albumin and glutaraldehyde containing live-attenuated measles vaccine (Edmonston Zagreb strain, lyophilized M-VAC™ from the Serum Institute of India) and incorporated them into an oral disintegrating film (ODF) made of Lycoat® RS 720, Neosorb® P60W (Sorbitol), and Tween® 80 (Polysorbate 80). The researchers administered the films to pigs via the buccal route. Figure 6.23 shows the production method of the ODF loaded with measles vaccine microparticles.

Briefly, a 1% solution of sterile bovine serum albumin (BSA) in sterile water was prepared and kept overnight for cross-linking with glutaraldehyde. Next, live-attenuated measles virus (5 wt%) was added to the solution and spray-dried in a Mini Spray Dryer B-191 through a 0.5 mm nozzle (cooled to −5 °C) at a feed rate of 20 mL/h. The inlet temperature of the drying was set to about 110 °C, which resulted in an outlet temperature of around 74 °C (Bejugam et al. 2008; Uddin et al. 2009; Ubale et al. 2013). The process yield was about 85–95%, and the microparticles were between 1 and 5 μm in size with an irregularly shaped surface (Fig. 6.21b).

For the preparation of the ODF, the film-forming polymer Lycoat® RS 720 was mixed in a water solution with Neosorb® P60W and Tween® 80 under mechanical agitation. Measles vaccine microparticles were then dispersed in the suspension, and the ODF formulation was cast into a film of 3 cm width and 79 μm thickness. Finally, film strips of 2 cm length were administered to the pigs for buccal immunization.

Serum samples were isolated and analyzed for specific IgG antibody titers for the measles virus by enzyme-linked immunosorbent assay (ELISA). The data showed a significant increase in serum antibody titer at 2 weeks after the first dose. After a second dose at 2 weeks, a further rise in antibody titer was observed at weeks 4 and 6, associated with the induction of innate immune response. Unfortunately, no comparison of immunity with other routes of administration was performed.

Overall, the study demonstrated that the ODF formulation could induce an efficient immune response translated for global clinical applications. The mucosal surface of the buccal cavity may be ideal for vaccination because it is easily accessible and provides antigen presentation. Other benefits of the ODF formulation include:

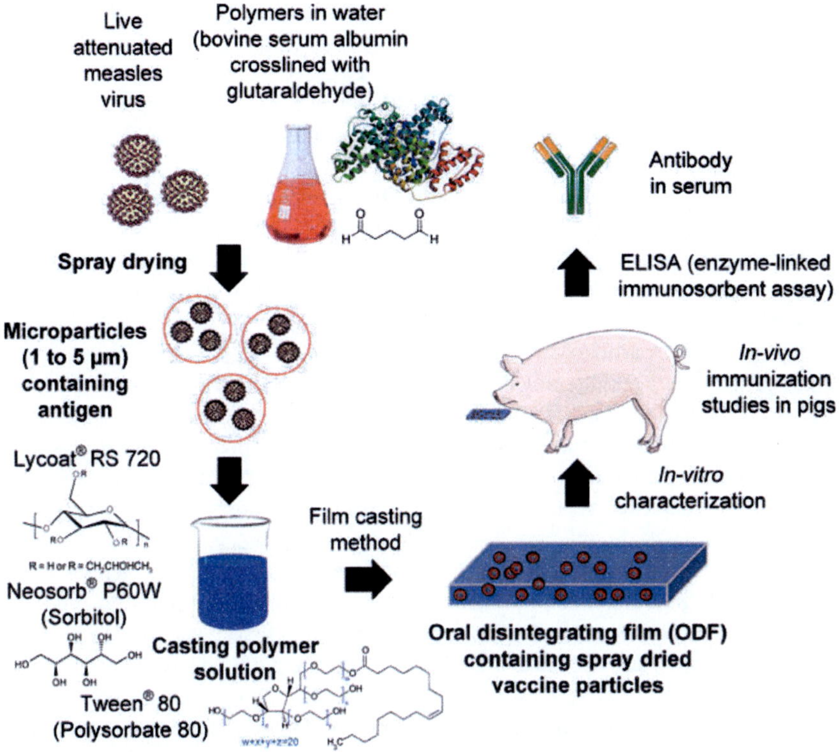

Fig. 6.23 The production method of the orally disintegrating film (ODF) loaded with measles vaccine containing bovine serum albumin microparticles (Images adapted from Gala et al. (2017) with permission from Springer Nature)

- non-invasive vaccination method,
- can be self-administered by individuals (all ages from infants to adults),
- hydrophilic polymers dissolve quickly on the tongue in the mouth,
- ODF formulation is easily adaptable to large-scale production, resulting in affordable costs, and
- cold chain storage may be reduced, enabling rapid and large population immunization.

6.7.3 Conclusions on Spray-Dried Measles Vaccine

A brief review of the literature has shown that the freeze-dried measles vaccine used today is extremely safe and effective. However, easier to administer and transport vaccines could improve global coverage. In addition, microarray patches have made significant advancements in recent years and could trigger a disruption in vaccine delivery in the marketplace. Therefore, increased clinical trials are expected in the coming years.

For many years, the inhalation route has also been proposed as a promising alternative for the targeted delivery of measles vaccine in young children, either as a liquid aerosol or dry powder. Studies of immunization via the aerosol route have been comparable to subcutaneous injection.

The studies presented on spray drying live-attenuated virus measles vaccines demonstrate the particle engineering capability and successful immunization by inhalation in rats, macaques, and humans and dispersion into orally disintegrating films for buccal administration in pigs. The presented data may provide guidelines for further developing spray-dried measles vaccines.

The produced particle sizes are in the range of about 1–5 μm aerodynamic diameter with low residual moisture contents, which allows direct administration of the powder vaccine via the respiratory tract. To minimize process loss during spray drying, gentle atomization, e.g., ultrasonic nozzles at low pressure, nozzle cooling, and low drying temperatures, is crucial. In addition, an essential requirement is to increase the stability of the measles vaccine by optimizing the formulation composition.

The most promising formulations consist of a mixture of sugars, amino acids, proteins, and other stabilizers. Trehalose, sucrose, sorbitol, and myo-inositol have shown to be particularly suitable for forming stabilizing glassy matrix. Amino acids such as L-arginine have been attributed protein stabilizing properties in both liquid and solid states, and glycerol as a plasticizer increases the interactions between the glassy sugar matrix and the virus. Large molecule components such as human serum albumin can slow the migration of live-attenuated virus particles to the air–liquid interface during droplet drying in the spray dryer and enhance stability by combining the increased glass transition temperature and increased viscosity, and surfactant-like action.

Spray-dried measles vaccine powders show stability at 37 °C for at least 1 week, meeting the WHO criterion for storage stability. The loss of potency is comparable to freeze-dried measles vaccines. Extended shelf life at elevated temperatures can overcome cold chain requirements during transport and administration.

A Phase I clinical trial of the spray-dried dry powder measles vaccine administered by inhalation in humans is a major highlight. The clinical study in healthy Indian men showed an immune response comparable to a subcutaneously administered licensed measles vaccine (reconstituted lyophilized powder). In addition, the inhaled spray-dried powder vaccines were well tolerated in all subjects.

A major challenge identified was the scale-up of the supercritical CAN-BD process (CO_2-assisted nebulization with a bubble dryer), which required a secondary drying step. In addition, the reported studies had been limited to a batch size of approximately 100 mL of liquid vaccine.

As a future view, additional clinical trials are needed for further development, particularly in younger subjects who are the primary target population for measles vaccination. A more compact design of dry powder inhalers and user-friendliness is also essential for successful population acceptance. Further investment would be needed to bring a spray-dried measles vaccine to the market. Alternative routes of vaccine administration such as buccal delivery using orally disintegrating films and biocompatible spray-dried microparticles are also feasible and further explored.

6.8 Hepatitis B Vaccines

6.8.1 A Short Introduction to Hepatitis B

Hepatitis B is caused by the hepatitis B virus (HBV) and leads to severe liver disease and liver cancer (Fig. 6.24). Chronic infection with HBV affects approximately 257 million people worldwide, making Hepatitis B a major global health problem. In 2015, there were an estimated 887,000 deaths (WHO 2020a). HBV is transmitted from person to person through contact of mucous membranes, infected blood, or other specific body fluids (e.g., saliva, semen, and vaginal fluid) (WHO 2017c), most commonly from mother to child during labor and delivery (WHO 2020a).

Vaccination against hepatitis B is the most effective prevention strategy (Zhao et al. 2020a). It is estimated that 85–90% of HBV-associated deaths are preventable through vaccination (WHO 2015a). Hepatitis B vaccine (HepB) is available as a monovalent formulation for birth dose (HepB-BD) or vulnerable adults in combination with other vaccines, including DTP, Hib, and IPV. In addition, a combined HepB and HepA vaccine is also available (WHO 2017c).

Conventional HepB is administered as an intramuscular injection as part of the mass immunization program. WHO recommends that all infants receive the HepB-BD within 24 h after birth, followed by 2 or 3 doses at least 4 weeks apart to complete the series. In 2019, coverage with 3 doses reached 85% globally, compared with about 30% in 2000 (WHO 2020a; WHO/UNICEF 2020) (see Fig. 1.9 in Chap. 1. However, coverage of HepB-BD is 43% and geographically uneven, with coverage of only 6% in the WHO Africa region (WHO 2020a). Barriers are primarily attributed to vaccine storage and transport and high rates of home births (Petit et al. 2017).

The current HepB vaccine is a liquid suspension consisting of purified hepatitis B virus surface protein (HBsAg) (Fig. 6.24) adsorbed to aluminum hydroxide or aluminum phosphate adjuvant (WHO 2006). One dose contains about 5–40 μg of recombinant HBsAg, which is flexible for use in different subpopulations (Zhao et al. 2020a).

In general, WHO recommends transporting and storing HepB in the cold chain at 2–8 °C from manufacture to administration to maintain efficacy. Freezing must be avoided as it would lead to dissociation of the antigen from the adjuvant, resulting in loss of potency (Brandau et al. 2003; Kumru et al. 2014; WHO 2017c).

By adding amino acid histidine and phosphate buffer at pH 5.2 to the liquid HepB vaccine formulation, efforts have improved stability for up to 12 months at 37 °C (Braun et al. 2009). Another study reported 9 weeks stability at 55 °C followed by at least 6 months at 45 °C (Jezek et al. 2009).

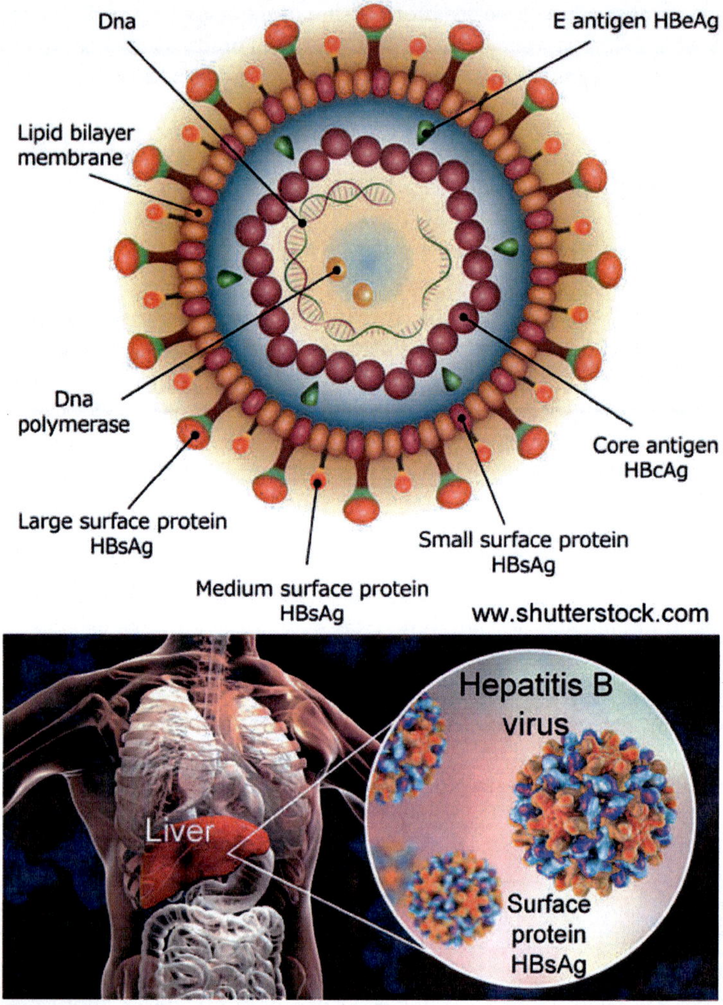

Fig. 6.24 The structure of the hepatitis B virus (Images adapted from www.shutterstock.com, Item IDs: 173727872 and 1103225849)

A review by the WHO of published data based on in vivo and in vitro testing of monovalent HepB vaccines concluded thermostability for up to 4 years at 2–8 °C, several months at 25 °C, several weeks at 40 °C, and 1 week at 45 °C (WHO 2016a). Thus, the results support the WHO recommendation to store HepB in CTC conditions.

Today, seven hepatitis B vaccines are prequalified by the WHO, but none have yet been approved for CTC use (Petit et al. 2017). However, some manufacturers are working toward licensure (WHO 2017d, 2018c; Kahn et al. 2017). In addition, a modeling study by Scott et al. (2018) showed that a CTC strategy for HepB-BD vaccine yields health benefits, improves vaccination coverage, and is cost-effective in various global settings.

More recently, Seaman et al. (2020) and Cui (2020) pointed out that CTC and single-dose compact prefilled auto-disable devices (CPAD) are considered cost-effective solutions to overcome coverage barriers. Figure 6.25 shows an example of a CPAD system with a DTwP-HepB-Hib combination vaccine. CPADs may allow minimally trained health care workers to inject vaccines accurately, mainly when a large proportion of births occur in the home setting. In addition, the auto-disable feature prevents reuse and transmission of bloodborne diseases between patients. CPAD devices are intended for single-use, stored at ambient conditions, and minimize waste.

Another way to improve HBsAg stability is to formulate the vaccine as a dry powder. Dry forms of hepatitis B vaccine can reduce the risk of chemical and

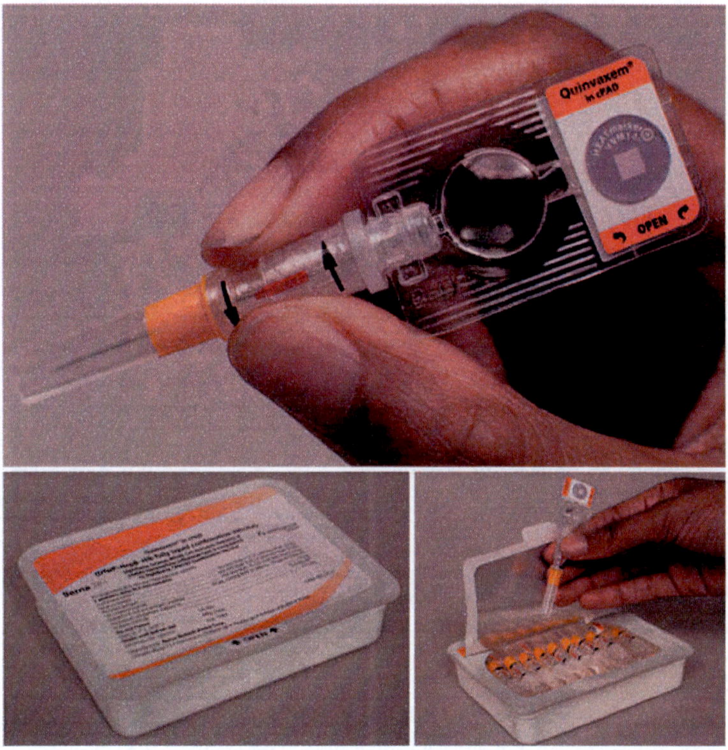

Fig. 6.25 Single dose compact prefilled autodisable device (CPAD) of DTwP-HepB-Hib combination vaccine, Trade Name: Quinvaxem in cPAD, Shelf life: 24 months at 2–8 °C, Manufacturer: Berna Biotech Korea Corporation, a Crucell Company (Image source: www.who.int/immunization_standards/vaccine_quality/pq_283_dtphepbhib_1dose_uniject_Crucell_Korea)

physical damage due to hydrolysis, aggregation, denaturation, degradation, or pH fluctuations of the liquid (McAdams et al. 2012). Examples of HBsAg powder drying include (see Chap. 3):

- spray freeze-drying in an amorphous matrix of inulin or dextran/trehalose for pulmonary delivery (stable for at least 3 months at RT) (Tonnis et al. 2014a),
- spray freeze-drying of 30–60 µm powder with trehalose/mannitol/dextran formulation (3:3:4 w/w) suitable for epidermal powder immunization (EPI) (stable for 16 weeks at 40 °C) (Maa et al. 2007), or
- supercritical drying using the CAN-BD process and trehalose as a stabilizer (Sievers et al. 2007; Cape et al. 2008).

6.8.2 Research on Spray-Dried Hepatitis B Vaccines

The following section focuses on published research studies on spray-dried hepatitis B vaccines (Maa et al. 2003, 2007; Chen et al. 2010; Muttil et al. 2010a; Bhowmik et al. 2012; Tomar et al. 2019). Table 6.6 lists the most important works with information on formulation, spray drying parameters, and storage stability.

Maa et al. (2003) reported the spray drying of aluminum hydroxide-adjuvanted HBsAg in a Mini Spray Dryer B-191 at 130 °C inlet and 70 °C outlet temperature and a liquid feed rate of 10 ml/min. Different excipient compositions of mannitol, trehalose, and dextran were tested, whereby trehalose and dextran were used due to the high T_g (Crowe et al. 1996, 2001) and mannitol because of its capability to serve as bulking agents to modify the flowability and maintain the structural integrity (Maa et al. 2004). As a result, spherical powders from 10 to 14 µm (median particle size) could be produced.

In a follow-up study, Maa et al. (2007) developed an HBsAg dry powder vaccine from trehalose/mannitol/dextran for epidermal powder immunization (EPI). For this purpose, the researchers applied a custom-built spray dryer with a 152 cm long glass chamber and an ultrasonic nebulizer (120 kHz, Sono-Tek, USA). The spray drying conditions were 130–140 °C inlet and 80–85 °C outlet temperature with a 3.5 mL/min liquid feed rate. The dry powder was collected on a filter paper. Large particles of 28–80 µm with a moisture content of 3.5 wt% were obtained. The HBsAg powder lost about 50% of its original potency during storage in a bioassay.

Chen et al. (2010) from PATH spray-dried HBsAg vaccine in a pilot-scale GEA Niro Mobile Minor at a batch size of approx. 1 L per experiment, sufficient to produce 5000–10,000 vaccine doses. Trehalose, mannitol, and sodium hydrogen phosphate (NaH_2PO_4) were used as excipients, while NaH_2PO_4 was added particularly to improve stability (Jezek et al. 2009). Nitrogen was used as drying gas at 90–95 °C inlet and 61–65 °C outlet temperature at a feed rate of 520 g/h. Free-flowing powders of 5–10 µm size were produced (Fig. 6.27a) (Kristensen 2008) with yields between 77 and 93% and water content ranging from 2.4 to 4.1%.

Table 6.6 Spray-dried hepatitis B surface antigen (HBsAg) vaccines described in the literature

Spray dryer process parameters	Excipients, composition	Particle size, residual moisture	Administration route	Shelf life, thermostability	References
B-191, T_{in}/T_{out}: 130/70 °C, F: 10 mL/min, A: 500 L/min air (5.5 bar)	Mannitol, dextran, trehalose	10–14 μm, spherical	Epidermal powder immunization (EPI)	Not reported	Maa et al. (2003)
Custom-made, 15.2 cm diameter, 152 cm height, 120 kHz ultrasonic atomizer, T_{in}/T_{out}: 130–140/80–85 °C, F: 3.5 mL/min	Mannitol, dextran, trehalose	28–80 μm, 50 μm medium, moisture 3.5 wt%	Epidermal powder immunization (EPI)	50% potency (bioassay)	Maa et al. (2007)
Mobile Minor T_{in}/T_{out}: 90–95/61–65 °C, F: 520 g/h, A: 6 kg/h (1.9 bar), 11.2 wt% (1 L/run)	Trehalose, mannitol, NaH_2PO_4	5–10 μm, 2.4–4.1 wt%, yield 77–90%	Mice (intra-peritoneal injection)	24 months (37 °C) (bioassay), 15 months (37 °C) (immunogenicity)	Chen et al. (2010)
GEA Niro atomizer T_{in}/T_{out}: 80/33 °C F: 30 mL/min, A:1.6 L/min D: 98 kg/h	PLGA/PEG nanoparticles in L-leucine (10:90 weight ratio)	FPF 50%, hollow/low-density particles, moisture < 1%	Guinea pigs (pulmonary by insufflation) (IgA response higher than intramuscular)	Not reported	Muttil et al. (2010a)
B-191, process parameters not reported (proprietary)	Chitosan, BSA, HPMCAS, eudragit FS30D, trehalose, Tween-20	0.8 and 1.5 μm, yield 67%	Mice (oral)	Not reported	Bhowmik et al. (2012)
B-290 T_{in}/T_{out}: 100/58 °C F: 5 mL/min, A: 500 L/h, D: 100% aspirator setting	Inulin (4 kDa) (1:250 weigth ratio HBsAg: inulin), 5 wt%	90% < 6 μm, median 2.2–2.7 μm)	Mice (pulmonary by insufflation) (high IgG and IgA titers)	Not reported	Tomar et al. (2019)

Mobile Minor: Pilot-scale spray dryer from GEA Niro, B-290: Mini Spray Dryer B-290 from Büchi Labortechnik AG, *BSA* Bovine serum albumin, *HPMCAS* hydroxypropylmethylcellulose acetate succinate, *PLGA* Poly(lactic-co-glycolic) acid, *PEG* Polyethylene glycol, NaH_2PO_4 Sodium dihydrogen phosphate, *FPF* Fine particle fraction < 5 μm

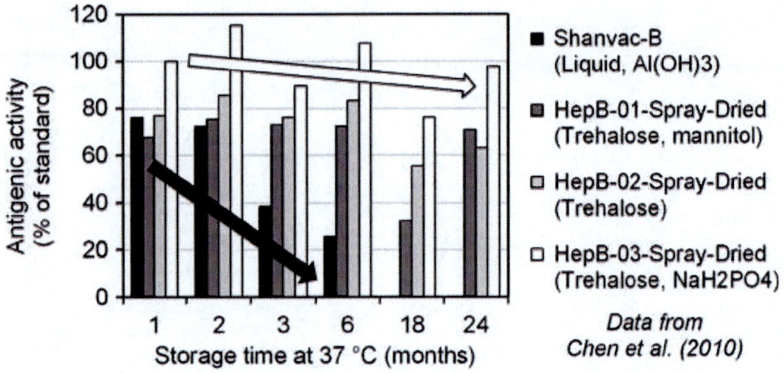

Fig. 6.26 Antigenic activity of spray-dried hepatitis B vaccine over storage time at 37 °C in comparison to original liquid Shanvac-B vaccine. The values are relative to the Shanvac-B vaccine standard stored at 4 °C. The composition of the spray-dried formulations in mg/dose of reconstituted vaccine were HepB-01 (12.4 mg vaccine, 50 mg trehalose, 50 mg mannitol, HepB-02 (12.4 mg vaccine, 50 mg trehalose), and HepB-03 (12.4 mg vaccine, 85 mg trehalose, 15 mg NaH_2PO_4) (Data adapted from Chen et al. (2010) with permission from Elsevier)

The particle size of the vaccine powders remained unchanged after storage at 37 °C. As shown in Fig. 6.26, antigenic activity tests (measured in-vitro by Enzyme-linked Immunosorbent Assay, ELISA) indicated that the spray-dried HepB vaccine formulations were stable for up to 24 months at 37 °C (Kristensen 2008; Chen et al. 2010), this in contrast to the commercially available liquid formulation (Shanvac-B, black bars), which lost about 50% of its activity within 3 months and all activity after 18 months. In addition, the reconstituted spray-dried vaccines exhibited similar antibody responses after injection into mice before and after storage at 37 °C for 15 months.

Muttil et al. (2010a) encapsulated rHBsAg in PLGA/PEG nanoparticles by a double emulsion method and spray-dried the aqueous suspensions with a solution of L-leucine to form a powder for deep lung delivery. The porous nanoparticle aggregates consisted of approximately 140–180 nm PLGA nanoparticles with PEG shell (Pulliam et al. 2007; Sung et al. 2007) (Fig. 6.27b). The inlet and outlet temperatures of the spray dryer (Niro atomizer) were maintained at 80 and 33 °C at a feed rate of 30 ml/min. Dry powder vaccine was administered to guinea pigs via a DP-4 insufflator (Penn Century, Inc.). The vaccination induced significant mucosal immune responses (i.e., high titers of IgG in serum and IgA in bronchioalveolar lavage fluid) in the lungs 24 weeks after the first immunization, similar to and even higher than those after parenteral administration.

More recently, Tomar et al. (2019) investigated pulmonary immunization with fluorescent dye-labeled hepatitis B vaccine powders (HBsAg: inulin 1:250 w/w) in mice using an insufflator and a self-developed aerosol generator (Tonnis et al. 2014b). Powders were prepared using a Mini Spray Dryer B-290 at 110 °C inlet and 58 °C outlet temperature. 90% of the powder was smaller than 6 μm (mean size 2.2–2.7 μm), and SEM images showed intact spherical particles (Fig. 6.27c).

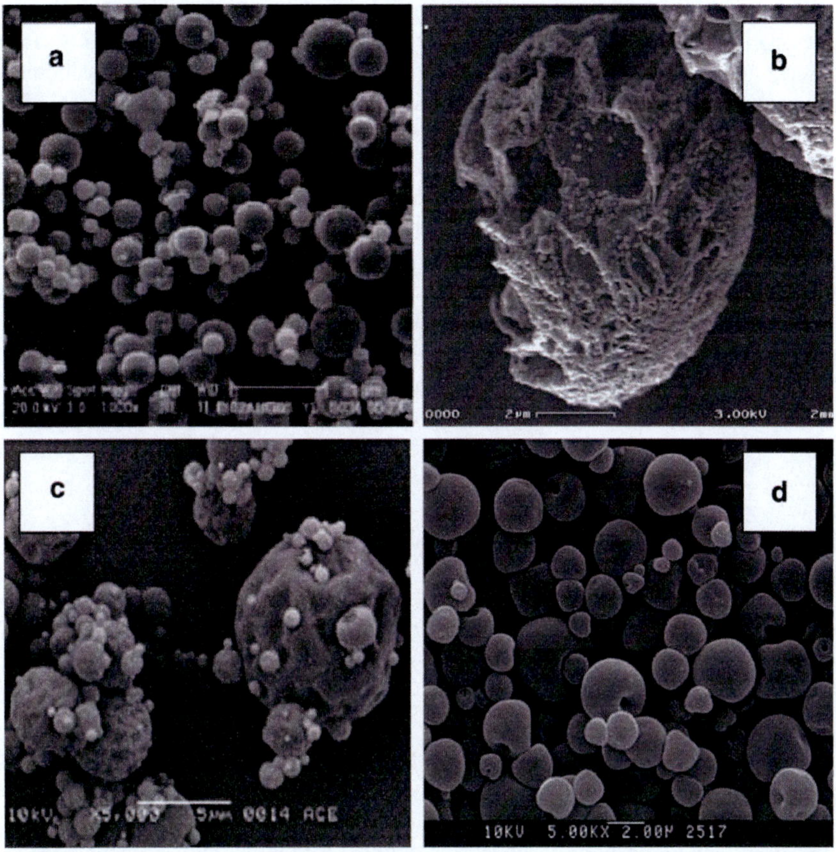

Fig. 6.27 SEM pictures of spray-dried Hepatitis B surface antigen (HBsAg) vaccine powder. **a** Spray-dried powder (Image adapted from Kristensen (2008), PATH, USA), **b** Porous nanoparticle-aggregate particle (PNAP) composed of 170 nm PLGA nanoparticles (25 wt%) and L-leucine (75 wt%) (Image adapted from Pulliam et al. (2007), ©Taylor & Francis), **c** Spray-dried HBsAg vaccine powders in inulin (1:250 weight ratio) (Image adapted from Tomar et al. (2019), open access under CC BY-NC-ND 4.0 license), **d** Spray-dried particles for oral delivery containing plasmid DNA vaccine encoding hepatitis-B surface antigen (Bhowmik et al. 2012)

The researchers used an imaging system to verify powder deposition in the lungs. When the vaccine powder was deposited in the trachea/central airways, no immune responses were induced, whereas the same powder elicited significant immune responses when deposited in the deep lung. These results led to the hypothesis that for non-airborne transmitted infections, such as hepatitis B, deposition of the vaccine in the deep lung appears to be essential to induce the desired immune response. In principle, these findings could be applied to other non-airborne diseases such as cholera or typhoid fever.

Bhowmik et al. (2012) described a novel approach for oral delivery of microparticles containing plasmid DNA (pDNA) encoding the hepatitis B surface antigen gene. The HBsAg-pDNA vaccine was entrapped in biodegradable albumin-based chitosan particles using a Mini Spray Dryer B-191 (proprietary process). The formulation involved preparing a uniform suspension of BSA/HPMCAS/Eudragit FS30D-(1:1:1 ratio) polymers, water-soluble chitosan, trehalose, and Tween-20. In addition, an M-cell-specific ligand called Aleuria Aurantia Lectin (AAL) was incorporated to specifically target the microparticles to Peyer's patch M-cells in the small intestine for enhanced uptake. SEM analysis of the spray-dried particles showed a smooth surface and confirmed a size range between 0.8 and 1.5 μm (Fig. 6.27d). The entrapment efficiency was determined to be 36%, and the yield was approximately 67%. Comparison of serum IgG titers in orally and subcutaneously vaccinated mice showed no significant difference. The immune response was detectable up to 35 weeks and showed an increased response after one priming and four booster vaccinations.

6.8.3 Conclusions on Spray-Dried Hepatitis B Vaccines

Overall, the reviewed laboratory and pilot-scale research studies demonstrated that spray drying is viable for producing HBsAg powder with improved long-term stability while maintaining immunogenicity. Optimized process and formulation parameters have resulted in a spray-dried HBsAg powder stable at 37 °C for more than 24 months.

Pulmonary immunization may also have advantages in non-airborne diseases such as hepatitis B because the lungs provide a large surface area for induction of an immune response and have abundant antigen-presenting cells. Using insufflation, animal studies of mice and guinea pigs showed that pulmonary immunization is feasible with dry powder formulations of hepatitis B antigen. Particle design for deep lung deposition is critical and requires an aerodynamic particle size distribution of 1–5 μm. Consistency of the delivered dose is an additional optimization factor.

Oral delivery of spray-dried vaccine with specific formulation and spray-drying parameters are also possible, as demonstrated in a study of HBsHV-pDNA-encapsulated vaccine in biodegradable albumin-chitosan particles.

However, there are no published clinical trials or experimental data with spray-dried hepatitis B vaccines.

This is mainly because current liquid hepatitis B vaccines can withstand a temperature of 37 °C for at least 1 month, which is suitable for use in a CTC. Thus, pressure to develop a dry vaccine is limited as vaccine manufacturers are reluctant to adopt other production methods. Nevertheless, thermostable dry HBsAg vaccines promise to simplify distribution logistics and increase vaccination coverage.

6.9 Tuberculosis Vaccines

6.9.1 A Short Introduction to Tuberculosis

Until the coronavirus (COVID-19) pandemic, tuberculosis (TB) was the leading cause of death from a single infectious agent (WHO 2021a). Worldwide, about 10 million people become ill with TB each year (WHO 2020b). Although it is a preventable and curable disease, an estimated 1.5 million people die every year (WHO 2021a). TB is caused by the bacillus *Mycobacterium tuberculosis*, which spreads when sick people expel the bacteria into the air in the form of droplets, such as through coughing. The disease mainly affects the lungs (pulmonary TB) but can also affect other sites (extrapulmonary TB) (Fig. 6.28). The treatment usually includes combination therapy with several drugs such as rifampicin, pyrazinamide, isoniazid, and ethambutol over a long dosing period of 6–9 months (Chae et al. 2021).

Bacille Calmette-Guérin (BCG) vaccine, developed in 1921 (Calmette 1931; Zwerling et al. 2011), is the only licensed vaccine against *M. tuberculosis*. BCG is a live-attenuated strain derived from an isolate of *Mycobacterium bovis*. The BCG vaccine is lyophilized and administered by intradermal injection and given to over 130 million newborns every year (Weiss 2020) as part of standard childhood immunization programs in more than 200 countries worldwide (Zwerling et al. 2011).

Storage stability is 2 years in the cold chain at 2–8 °C and several months at 25 °C. However, at 37 °C, it loses more than 20% of its potency after about 1 month (Chen and Kristensen 2009) (see Chap. 2, Tables 2.2 and 2.3). Currently, there are only four manufacturers of BCG vaccines that are WHO prequalified (Price et al. 2016; Angelidou et al. 2020):

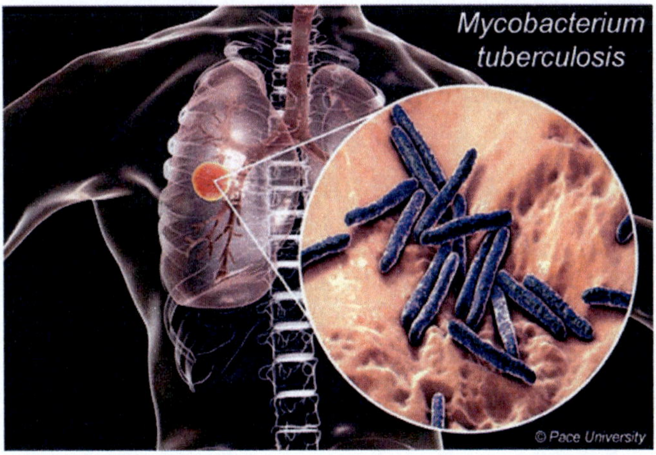

Fig. 6.28 Tuberculosis (TB) is caused by harmful *Mycobacterium tuberculosis*. The bacteria often infect the lungs resulting in severe illness (Image adapted from www.shutterstock.com, Item ID: 1064066351)

- Japan BCG Laboratory (Japan) (prequalified on 01/01/1987)
- BB-NCIPD Ltd. (Bul Bio-National Center of Infectious and Parasitic Diseases Ltd., Bulgaria, distributed by InterVax, Canada) (01/02/1991),
- Serum Institute of India Pvt. Ltd. (India) (29/05/2003), and
- Statens Serum Institut (Denmark) (27/09/1994).

Because of the widespread nature of TB, any improvement in vaccine efficacy over the current BCG vaccines would have a significant impact. Therefore, the development of new TB vaccines remains a global health priority (Summary Group 2015). Moreover, the COVID-19 pandemic has again spotlighted the century-old BCG vaccine (Branswell 2020; Weiss 2020; Aziz et al. 2020; Li et al. 2021). A number of preprint studies (posted online before peer-review) published on the Health Sciences Preprint Server (medRxiv.com) suggest that the BCG vaccine provides non-specific immune protection against other respiratory infections and possibly against the new coronavirus. Country-specific correlations between BCG vaccination and protection against COVID-19 point to this. The WHO seems to agree but has published a scientific report concluding that there is currently no evidence (WHO 2020c).

At the same time, alternative routes of administration for TB vaccines compared to conventional injection are being investigated. One such approach is administering the TB vaccine as a dry powder directly into the lungs by mucosal immunization mimicking the natural transmission route for *M. tuberculosis* (Gomez et al. 2022). In this context, research is being conducted to adjust the particle size optimally to influence the immune response. In addition, several drying processes have been developed to formulate stable dry powder TB vaccine with low residual moisture, including spray drying, spray freeze drying, and supercritical fluid drying. The following section will present the leading research works on spray-dried TB vaccines. For information on spray-dried anti-TB drugs (therapeutics) such as inhalable isoniazid, rifampicin, rifapentine, pyrazinamide, ethambutol, streptomycin, and other powders, the reader is referred to the reviews of Wong et al. (2016), Das and Stewart (2016), and Momin et al. (2018).

6.9.2 Research on Spray-Dried Tuberculosis Vaccines

Several research groups from Canada, the USA, Australia, Denmark, and The Netherlands are researching spray-dried TB vaccines administered by pulmonary delivery. Table 6.7 summarizes the most promising results over the past 15 years, including studies on:

- live-attenuated BCG vaccine (*M. smegmatis* and *M. bovis*) (Wong et al. 2007; Garcia-Contreras et al. 2008; Fourie et al. 2008; Morello et al. 2009; Price et al. 2020),
- adenoviral-vectored TB vaccines (Jin et al. 2010; LeClair et al. 2016a, b; Afkhami et al. 2017), and

Table 6.7 Overview of experimental investigations on spray-dried tuberculosis vaccines described in the literature

Vaccine type	Excipients, composition	Spray dryer process parameters	Administration route (particle size)	In-vitro, In-vivo	Shelf life, thermostability	References (country)
Bacillus Calmette-Guérin, live bacteria (*M. smegmatis*, *M. bovis*)	L-leucine (95:5 leucine:BCG)	B-290 with HP cyclone, T_{in}/T_{out}: 100–125/40 °C, F: 7 mL/min, A: 0.6 L/min, D: 35 m^3/h	Pulmonary (2–3 μm)	Guinea pigs	12 months (4 °C, 20% RH) 9 months (25 °C, 20% RH) 4 months (25 °C, 60% RH), (bacterial viability)	Wong et al. (2007), Garcia-Contreras et al. (2008), Fourie et al. (2008), Morello et al. (2009) (USA)
Bacillus Calmette-Guérin, live bacteria	L-leucine, mannitol, PVP, BSA, trehalose, 1% (w/v)	B-290 in biosafety cabinet, T_{in}/T_{out}: 150/50 °C, F: 4 mL/min (10%), D: 742 L/h (100%)	Pulmonary (2.0–5.3 μm, 0.6–1.5 wt % RMC)	Mice	1 year (25 °C) (viability assay) 2 years (25 °C) (immunogenicity)	Price et al. (2020) (USA)
Ad35-vectored TB AERAS-402, live adenovirus vector	Mannitol, cyclodextrin, trehalose, dextran, L-leucine, sucrose, inositol, histidine, PVP, Tween 80	B-290 with PTFE filter, T_{in}/T_{out}: 65–125/34–50 °C, F: 4.5 mL/min, A: 439–538 L/h (N2), D: 35 m^3/h (N2)	Pulmonary (1 and 5 μm, 1.45 wt% RMC)	In-vitro	45 months (4 °C) 12 months (25 °C) 5 weeks (37 °C) (0.12 log $TCID_{50}$ titer loss, bioassay)	Jin et al. (2010), Jin and Tsao (2013) (USA, The Netherlands)
AdHu5 (Recombinant type 5 adenovirus vector)	Mannitol, dextran (67:33 wt%)	B-290, T_{in}/T_{out}: 90–120/48–65 °C, F: 2.4–3.6 mL/min, A: 7.3–11.2 L/min	Pulmonary (reconstituted, < 10 μm)	Mice	3 months (20 °C, < 10% RH)	LeClair et al. (2016a, b), Afkhami et al. (2017) (Canada)

(continued)

Table 6.7 (continued)

Vaccine type	Excipients, composition	Spray dryer process parameters	Administration route (particle size)	In-vitro, In-vivo	Shelf life, thermostability	References (country)
Recombinant antigen rAg85B (subunit protein)	PLGA, muramyl dipeptide or TDB	B-191, T_{in}/T_{out}: 65/41–43 °C, F: 4.5 mL/min, A: 600 L/h (N2, 3 bar), D: 50% aspirator	Pulmonary (3.4–4.3 µm median size)	In-vitro and guinea pigs	Not reported	Lu et al. (2007, 2010) (USA)
Subunit fusion protein H56 and adjuvant CAF01	Liposomal dispersions (DDA-TDB), trehalose	B-290, T_{in}/T_{out}: n.a./95 °C, F: 1.5 mL/min, A: 615 L/h air, D: 85% aspirator	Pulmonary (reconstituted)	Mice	Not reported	Ingvarsson et al. (2013a, b), Thakur et al. (2018) (Denmark)
Subunit proteins Culp 1–6 and MPT83 conjugated to novel adjuvant Lipokel	Mannitol	B-290, T_{in}/T_{out}: 50/ < 35 °C, F: 1 mL/min, A: 700–800 L/h, D: 38 m^3/h (100%, suction mode)	Pulmonary	Mice	Not reported	Tyne et al. (2013) (Australia)
Recombinant fusion protein ID93 and GLA in a squalene nanoemulsion as adjuvant system	Trehalose, trileucine, Tris buffer, L-leucine, pullulan	Custom-designed spray dryer, twin-fluid nozzle, T_{in}/T_{out}: 65–70/36–49 °C, F: 0.6 mL/min, D: 200–800 L/min air	Pulmonary (4–5 µm MMAD, 1.8–2.6 wt% RMC)	In vitro (bioassays) and mice	26 months (40 °C) (chemical and physical stability) 45% antigen retained after 12 months (50 °C)	Gomez et al. (2021b, a, c, 2022), Gomez and Vehring (2022) (Canada, USA)

BSA bovine serum albumin, *DDA* dimethyl-dioctadecyl ammonium bromide, *GLA* glucopyranosyl lipid A, *MMAD* mass medium aerodynamic diameter, *PLGA* Poly (lactic-co-glycolic acid), *PVP* polyvinylpyrrolidone, *TDB* trehalose dibehenate, T_{in}/T_{out} inlet/outlet drying gas temperature, *F* feed flow rate, *G* drying gas flow rate, *A* atomization gas flow rate, *RMC* residual moisture content

- subunit proteins secreted by *M. tuberculosis* with adjuvants (Lu et al. 2007, 2010; Tyne et al. 2013; Thakur et al. 2018; Gomez et al. 2021a, b, c, 2022; Gomez and Vehring 2022).

In 2007, Wong et al. (2007) presented a study on spray-dried live-attenuated *M. smegmatis* and *M. bovis* bacteria with L-leucine as an excipient. A dry, high-yield, free-flowing, vial-fillable powder was produced using a Mini Spray Dryer B-290 with a high-performance cyclone. The inlet temperature was varied between 100 and 125 °C to maintain a constant outlet temperature of 40 °C with a feed rate of 7 mL/min. The study demonstrated improved stability of 4 months at 25 °C compared to the standard lyophilized formulation. Figure 6.30a shows a microscopic view of the spray-dried powder of *M. smegmatis* formulated with L-leucine. L-leucine was selected to improve the dispersibility of the dry powder and the spray drying yield. It was also demonstrated that the spray-dried powder could be reproducibly filled into vials under sterile conditions, i.e., 30 and 45 mg powder fill weights (see Fig. 4.15 in Chap. 4).

In a follow-up study, Garcia-Contreras et al. (2008) administered the spray-dried BCG vaccine powder (*M. bovis*) as an aerosol to guinea pigs that were subsequently challenged with *M. tuberculosis*. The immunization significantly reduced bacterial burden and pulmonary pathology compared with untreated and control animals (immunized with the standard parenteral BCG vaccine). In addition, the use of L-leucine demonstrated a high emitted dose from an inhaler designed for newborns in low-income settings.

Figure 6.29 shows the procedure applied for direct administration of the spray-dried vaccine powders into the lungs of mice (Morello et al. 2009). This procedure was adapted for other dry powder vaccine candidates and animals, e.g., rats or guinea pigs. Mice were first anesthetized and secured on a rodent work stand. Then, 250–300 μg of spray-dried BCG powder was loaded into a polyethylene tube and inserted into the insufflator. The researchers used puffs of air to force the powder through the cannula into the trachea of the mouse. Detection of the *M. bovis* BCG vaccine in the lungs was obtained by histopathological examination.

The studies by Wong et al. (2007) and Garcia-Contreras et al. (2008) demonstrated the feasibility of spray-dried BCG vaccine as an inhalable powder against TB. Spray drying and filling the powder into vials/capsules were declared relatively inexpensive and pharmaceutically scalable. In addition, spray drying produced a powder with good dispersion quality suitable for targeted delivery to the deep lung. A remaining challenge included dry powder delivery to the infant's lungs (e.g., inhaler design) (Fourie et al. 2008).

More recently, Price et al. (2020) optimized the storage stability of spray-dried BCG dry powder using a factorial design with five stabilizing excipients. The optimal excipient mixture contained L-leucine, mannitol, PVP, bovine serum albumin (BSA), and trehalose in a weight ratio of 75.2/10.0/5.0/4.8/5.0. Each of these excipients was selected to protect the bacteria during spray-drying and long-term storage:

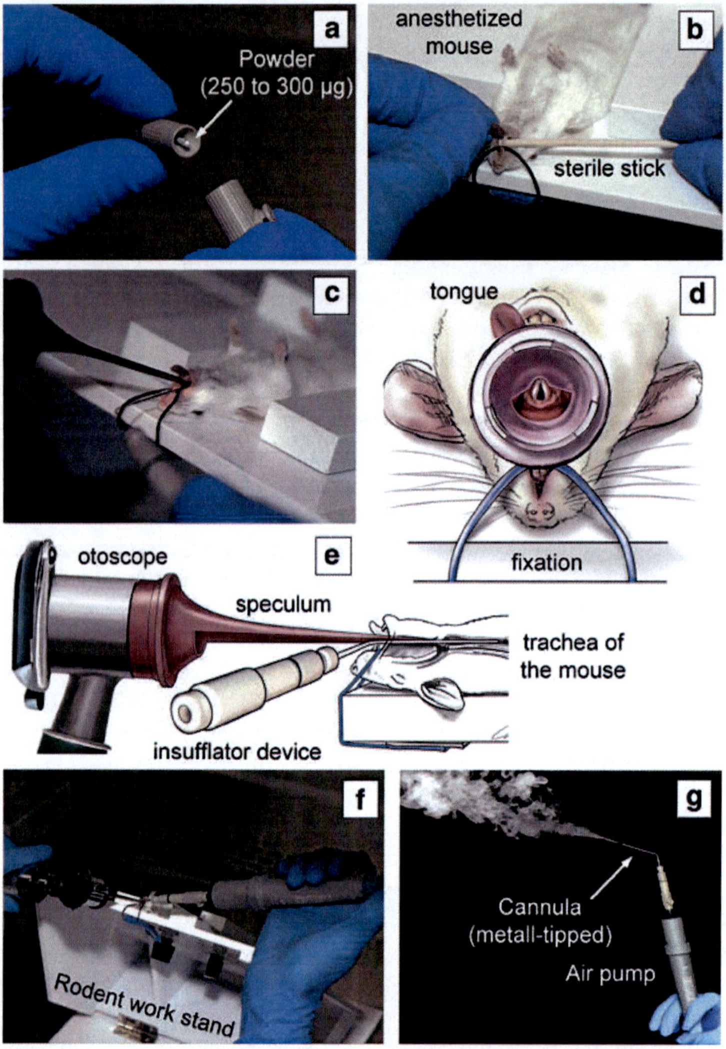

Fig. 6.29 Procedure for pulmonary dry powder insufflation in mice for use in vaccine studies. **a** A polyethylene tubing is loaded with spray-dried powder and placed in the insufflator device (DP-4 M insufflator model, Penn-Century, Inc., USA). **b** An anesthetized mouse is secured to the rodent work stand, and the tongue is rolled gently from the mouth with a sterile stick and grasped between the thumb and forefinger. **c** Insertion of the mouse speculum. **d** Visualization of the tracheal opening obtained with the otoscope. **e** The cannula of the insufflator device is inserted into the tracheal space through the cut-away portion of the speculum and guided down the length of the trachea proximal to the carina. **f** An air pump (AP-1 model, Penn-Century, Inc., USA) is attached to the insufflator device's distal end, and about 4–5 air puffs of 250 mL are delivered to dispense the powder. **g** Dispensing the powder ex vivo for illustration. Finally, the mouse is removed and recovered in a cage warmed to 37 °C (Images adapted from Morello et al. (2009) with permission from Elsevier)

- L-leucine increased the dispersibility of the dry powder and improved the spray drying yield,
- Mannitol had low hygroscopicity,
- Trehalose enabled stabilization in sugar matrix with high glass transition temperature,
- PVP had a high glass transition temperature, and
- BSA prevented the recrystallization of the powder.

The researchers conducted experiments on a Mini Spray Dryer B-290 in a biosafety cabinet (see Chap. 4, Fig. 4.27) at around T_{in}/T_{out} 150/50 °C and feed rate 4 mL/min. The spray-dried BCG powders were stored in a previously developed protective packaging containing a desiccant sachet, an inert N_2 atmosphere, and oxygen scavengers to capture reactive oxygen species generated during storage (Kunda et al. 2016). After 1 year of storage, the spray-dried BCG vaccine showed no loss of viability at 25 °C (up to 60% RH) and about 2-log loss of viability at 40 °C (75% RH). The aerodynamic size of the powder was in the respirable range (Fig. 6.30c). Finally, the vaccine retained its immunogenicity when healthy mice were immunized intradermally with reconstituted spray-dried BCG vaccine powders stored at 25 °C for 2 years. The researchers observed no significant difference in induced cytokine response in mice. There was also no loss of immunogenicity after 4 freeze–thaw cycles (viability test).

Spray-dried vectored vaccines are another area of increasing research. Adenoviral (Ad) vectors are among the most widely utilized vectors for vaccine delivery. Jin et al. (2010) developed a spray-dried recombinant Ad35-vectored TB vaccine (AERAS-402) with mannitol-based stabilizers (i.e., mannitol-cyclodextrin-trehalose-dextran). The researchers used a Mini Spray Dryer B-290 installed in a biosafety cabinet to run the experiments at 65–125 °C inlet and 34–50 °C outlet temperatures. The spray drying process resulted in free-flowing, non-hygroscopic particles with narrow size distribution and median volume diameter of 3.2–3.5 μm, suitable aerosolization properties for pulmonary delivery. The residual moisture content was around 1.5 wt%, and the resistance to water absorption was high. The T_g was > 97 °C providing viral stability for 12 months at 4 and 25 °C without significant change in $TCID_{50}$ titer (bioassay). After storage at 37 °C for 5 weeks, the loss of virus activity was only 0.12 log (Jin et al. 2010; Jin and Tsao 2013).

Proteins secreted by *M. tuberculosis* are a category of subunit antigens that have received significant attention as potential TB vaccine candidates (Gomez et al. 2000). This is because such antigens are among the first molecules interacting with the host immune system during infection. In addition, since most TB patients acquire their primary infection via the pulmonary route, lung macrophages, which are the initial host cells for *M. tuberculosis*, represent a logical site for vaccine delivery. Therefore, several secreted antigens like 85A (Ag85A), 85B (Ag85B), Cutinase-like proteins (Culp 1 and 6), and MPT83, have been investigated as potential adjuvants for spray-dried inhalable dry powder TB vaccines (Lu et al. 2007, 2010; Tyne et al. 2013; Afkhami et al. 2017).

Afkhami et al. (2017) spray-dried human adenoviral (AdHu5)- and chimpanzee adenoviral (AdCh68)-vectored TB vaccines with antigen Ag85A (AdHu5Ag85A and AdCh68Ag85A) in a mannitol/dextran matrix (67/33 wt%). The researchers added dextran to increase the T_g for enhanced thermal stabilization (LeClair et al. 2016a, b). The spray drying was conducted in a Mini Spray Dryer B-290 at T_{in}/T_{out} of 120/65 °C and 3.6 mL/min feed rate. The particles were generally smaller than 10 μm (mean diameters from 4.9 to 5.3 μm) and exhibited high sphericity (Fig. 6.30d). The measured T_g for the mannitol/dextran particles ranged from 134 to 139 °C. After 90 days of storage at 20 °C under low humidity conditions, only minor changes in particle size, morphology, and T_g were observed. Infectivity losses in-vitro were estimated to be low (average loss 0.4 and 0.3 log, respectively) due to the stabilizing effect of mannitol and dextran. In vivo efficacy of reconstituted spray-dried vaccine powders in mice was maintained after 90 days of storage at room temperature, whereas liquid vaccine samples lost immunogenicity during similar storage. Overall, this study showed the feasibility of spray drying AdHu5Ag85A and AdCh68Ag85A vaccines into powders that exhibited long-term thermostability and vaccination efficacy. The authors envisioned future studies investigating aerosol properties and vaccine potency as an inhalable dry powder.

Lu et al. (2007) investigated the use of spray-dried PLGA microparticles as carriers for recombinant antigen 85B (rAg85B) to macrophages to use them for pulmonary delivery of TB vaccine. The authors prepared the powders using a Mini-Spray Dryer B-191 at 65 °C inlet and 41–43 °C outlet temperature. The formulation preparation before spray drying included a solution of PLGA in methylene chloride and sonication with an aqueous phase of rAg85B in sodium phosphate buffer, muramyl dipeptide, or trehalose dibehenate. The spray-dried

Fig. 6.30 Particle morphologies of spray-dried tuberculosis vaccines. **a** *M. smegmatis* spray-dried with 95 wt% L-leucine. The picture reveals live GFP (green fluorescent protein)-labeled bacteria (Reproduced from (Wong et al. 2007) with Copyright (2007) National Academy of Sciences, U.S. A.). **b** *M. smegmatis* with 95 wt% L-leucine. The rod-shaped *M. smegmatis* bacteria appear associated with spherical L-leucine particles (Reproduced from Garcia-Contreras et al. (2008) with Copyright (2008) National Academy of Sciences, U.S.A.). **c** Dead BCG vaccine (heat-inactivated bacteria) encapsulated in an excipient mixture containing L-leucine, mannitol, PVP, BSA, and trehalose (Adapted from Price et al. (2020) with permission from Springer Nature). **d** AdHu5Ag85A encapsulated in mannitol/dextran after 3 months of storage at ambient temperature and low relative humidity (Adapted from Afkhami et al. (2017) with permission from Springer Nature). **e** PLGA particles (75:25 lactide to glycolide ratio) containing muramyl dipeptide and recombinant *M. tuberculosis* antigen 85B (rAg85B) (Adapted from Lu et al. (2007) with permission from Springer Nature). **f** Spray-dried particles containing antigen ID93, adjuvant system GLA-SE, and trehalose after 3 months of storage at 40 °C (Adapted from Gomez et al. (2021a) with permission from Elsevier). **g** Spray-dried particles with encapsulated antigen ID93 and adjuvant system GLA-SE in trehalose after 26 months of storage at 40 °C. The cracked particle indicates a higher concentration of encapsulated nanoemulsion droplets near the surface wall (Adapted from Gomez et al. (2021a) with permission from Elsevier). **h** Inhalable vaccine candidate with highly folded particle surface containing the antigen ID93 and the adjuvant system GLA-SE, spray-dried with trehalose and 3 wt% trileucine (Adapted from Gomez et al. (2021b) with permission from Elsevier) ▶

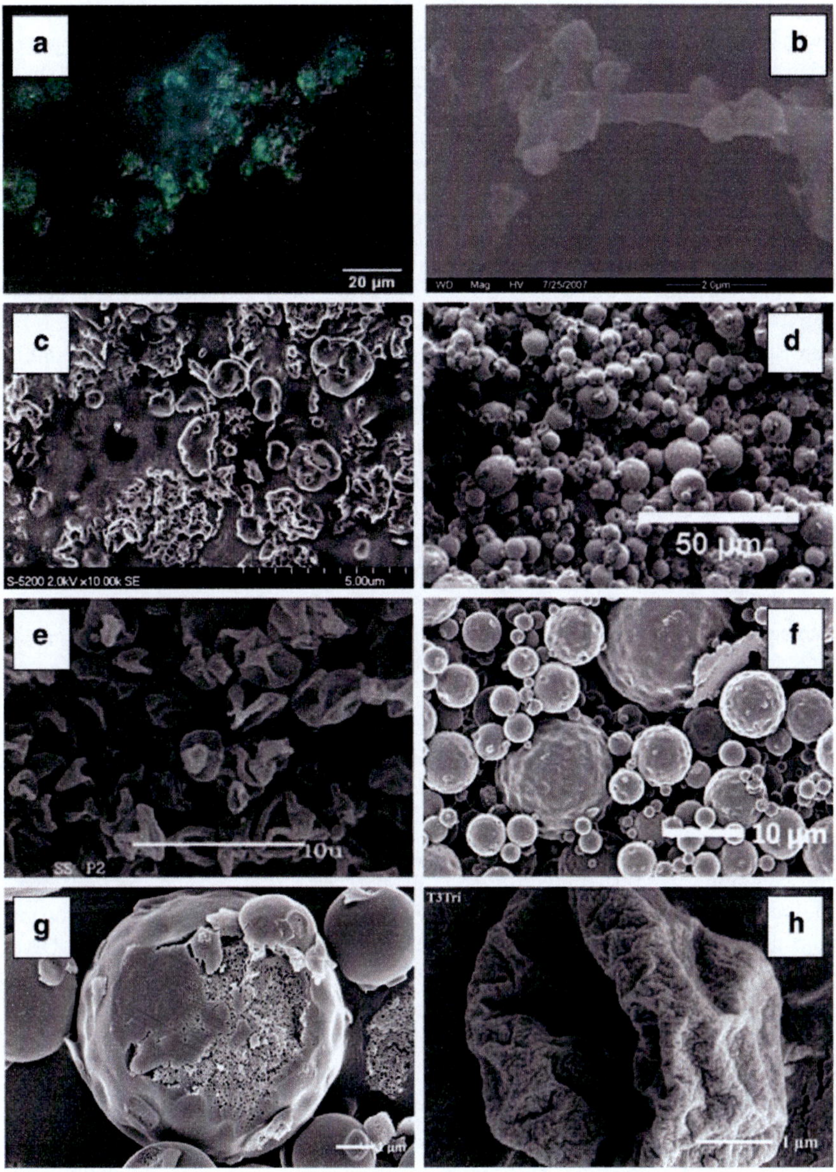

PLGA particles had a raisin-like structure with an average size of 3.4–4.3 μm (Fig. 6.30e). An in-vitro T-cell recognition assay demonstrated that PLGA-rAg85B microparticles induced two orders of magnitude greater antigen-specific T-hybridoma cell response in macrophages than soluble rAg85B. Moreover, the researchers observed a sustained release over 6 days. The results showed that PLGA microparticles could serve as effective carriers for antigen delivery.

In a follow-up study, Lu et al. (2010) investigated the insufflation of the spray-dried PLGA-rAg85B microparticles directly into the lungs of guinea pigs using a DP-3 insufflator model (Penn Century Inc., USA). Multiple doses showed that pulmonary immunization enhanced the protection against TB infection compared to primary vaccination with subcutaneously administered BCG. Thus, the authors concluded that an aerosol boost would be a promising strategy to improve and prolong protective immunity based on BCG-initiated immunity.

Tyne et al. (2013) examined the *M. tuberculosis* antigens Culp 1 and 6 and MPT83 conjugated with a novel adjuvant Lipokel (Lipotek Pty Ltd), a TLR2 ligand (Toll-like receptor 2) that recognizes foreign substances and transmits appropriate signals to cells of the immune system. The conjugated protein-lipocel complexes were spray-dried with mannitol as a dry powder vaccine for pulmonary administration. The researchers chose mannitol as an excipient because of its approval to treat cystic fibrosis by dry powder inhalation (Hurt and Bilton 2012) and run experiments on a Mini Spray Dryer B-290 with a feed rate of 1 mL/min and an inlet temperature of 50 °C. The vaccine powders were delivered directly into the lungs of mice by intratracheal insufflation (DP-4 M insufflator model with manual air pump). Pulmonary administration with either Culp1-6 lipocel or MPT83 lipocel powder vaccines produced protective immune responses in the lungs after an aerosol *M. tuberculosis* challenge (i.e., a significant decrease in bacterial burden compared with non-vaccinated mice). Hence, these successful studies formed the basis for further investigation of TLR2-targeted secreted proteins from *M. tuberculosis* as adjuvants.

Thakur et al. (2018) spray-dried adjuvant CAF01 with multistage subunit vaccine protein H56, a fusion protein of *M. tuberculosis* antigens Ag85B, ESAT-6, and Rv2660. CAF01 comprises dimethyl-dioctadecyl ammonium (DDA) bromide and trehalose dibehenate (TDB) and is known to transport antigens to and activate dendritic cells effectively. The formulation involved preparing a liposome dispersion containing CAF01, trehalose, and adsorbed H56 protein onto the liposomes. The authors prepared a dry vaccine powder by spray drying the liposome nanosuspension (with a size range from 225 to 685 nm) using a Mini Spray Dryer B-290 at a feed rate of 1.5 ml/min and outlet air temperature 95 °C with a powder yield ranging from 60 to 70% (Ingvarsson et al. 2013a, b). Immunization studies in mice demonstrated that the spray-dried and reconstituted H56/CAF01 vaccine elicited cell-mediated and humoral immune responses equivalent to those stimulated by the non-spray-dried vaccine. The authors declared these findings a basis for further studies for pulmonary administration in preclinical TB models.

More recently, Gomez et al. (2021a) reported about spray drying a subunit TB vaccine candidate containing the recombinant fusion protein ID93 and a nanoemulsion adjuvant system. ID93 consists of four *M. tuberculosis* antigens Rv3619, Rv1813, Rv3620, and Rv2608. The adjuvant system was formulated as a nanoemulsion (approximately 100 nm in diameter) of glucopyranosyl lipid A (GLA), a synthetic Toll-like receptor 4 agonist, and squalene.

Trehalose was used as a stabilizer due to its relatively high T_g and ability to stabilize spray-dried proteins and lipid-based dispersions. The authors used a custom-built laboratory-scale spray dryer (Ivey and Vehring 2016; Ivey et al. 2016) with a twin-fluid atomizer (Hoe et al. 2014) and a relatively low drying gas temperature of 65 °C to minimize temperature-dependent degradation. The outlet temperature was around 36 °C, and the feed rate was 0.6 mL/min.

Figure 6.30f, g shows SEM images of the spray-dried vaccine powder. Spray drying resulted in microparticles with geometric diameters ranging from 1 to 20 μm. The interior of the particles appeared foamy within the trehalose matrix, with numerous voids left by the encapsulated nanoemulsion droplets that accumulated near the surface of the drying droplet, as suggested by calculations of the Péclet number from particle formation modeling (Vehring et al. 2007, 2020; Vehring 2008; Boraey and Vehring 2014).

Furthermore, Gomez et al. (2021a) developed a method for packing the powder samples to prevent moisture uptake during storage. Figure 6.31 shows a simplified schematic diagram of the protective packaging. The spray-dried powder was aliquoted into low bind snap cap tubes and placed in an aluminum bag along with a desiccant pouch (e.g., silica gel) at 7% RH to prevent powder moisture uptake. These bags were then placed in a second aluminum bag with a desiccant pouch at 0% RH to minimize the moisture difference scale between the inner package and the outer environment. The aluminum bags were each double heat sealed. Finally, the packaged samples were stored in a humidity-regulated glovebox at 22–24 °C (0% RH). The spray-dried powder measured moisture content of around 2.6 wt% and did not change significantly after 26 months of storage. In addition, the powder maintained its physical stability and the chemical stability of the vaccine components after 26 months of storage at 40 °C (Fig. 6.30g).

In further studies, the authors investigated the addition of the polysaccharide pullulan, the amino acid L-leucine, and the tripeptide trileucine as dispersibility

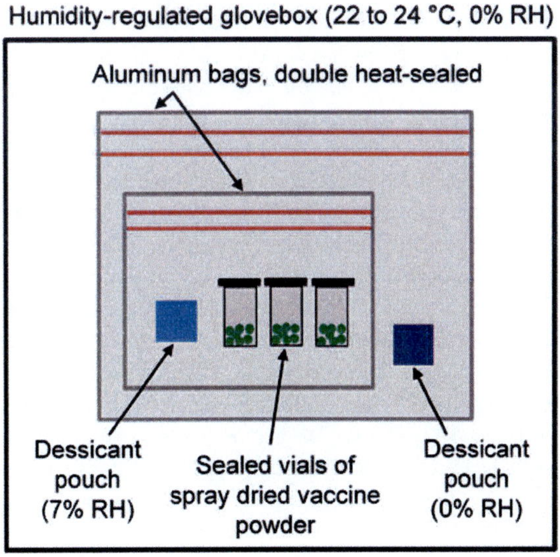

Fig. 6.31 Simplified schematic for packaging spray-dried TB vaccine powder to ensure moisture control during long-term storage (Adapted from Gomez et al. (2021a) with permission from Elsevier)

enhancers to make the spray-dried ID93 + GLA-SE powder suitable for pulmonary delivery (Gomez et al. 2021b, c; Gomez and Vehring 2022).

The results showed that adding L-leucine improved the aerosol properties but increased the aggregation of the emulsion droplets upon reconstitution. The emulsion droplets increased from about 90 to 225 nm. The pullulan's addition preserved the size of emulsion droplets, but the antigen could not be detected after reconstitution. The trileucine-containing formulations (3 and 6 wt%) successfully stabilized the adjuvant system in an inhalable dry powder. The most promising aerosol performance was achieved, which was assessed with a commercial DPI (Seebri Breezhaler, Novartis, approx. 40 mg of powder per capsule) in combination with a human mouth-throat model. The emitted dose (98%) and lung dose (38%) were maintained for all samples after one year of dry storage at temperatures up to 40 °C (Gomez et al. 2021c). The interior and exterior particle morphology was preserved. Approximately 45% of the antigen was still present in the trehalose-trileucine formulation after one year of storage at 25, 40, and 50 °C (in protective packaging to prevent moisture uptake) (Gomez and Vehring 2022). In contrast, the antigen could not be detected in the trileucine-free formulation after 7 months of storage at 40 °C or in the liquid version after 1 month of storage at 37 °C (Gomez et al. 2021c).

Figure 6.30h shows an example of the highly folded particles attributed to forming a trileucine shell during the particle formation process (Gomez et al. 2021b). As predicted by a particle formation model, trileucine accumulated on the particle surface due to the relatively short time to saturation and formed a shell. This property may lead to better moisture protection. In addition, a rougher surface morphology enhanced powder dispersibility and thus improved aerosol performance (Gomez et al. 2021b).

In a recent study, Gomez et al. (2022) investigated the immunogenicity and efficacy of the spray-dried TB vaccine candidate (ID93 + GLA-SE) in a mouse model (in vivo). The vaccine powder was administered as an inhaled aerosol either via the lungs (pulmonary, MMAD of 1.7 ± 0.2 μm) or in the nose (nasal, MMAD of 4.4 ± 0.4 μm) and compared with the reconstituted liquid vaccine administered intranasally or intramuscularly. The smaller particles were designed to promote greater airway penetration. For the inhalation experiments, a custom-built aerosol delivery system was used, which consisted of a dust generator to aerosolize the spray-dried powder vaccine and a nose-only inhalation device to restrain the mice and deliver the aerosolized vaccine. The results showed that both intranasal administration of the reconstitued liquid vaccine and pulmonary administration of the dry powder vaccine resulted in Mtb control in infected mice. Measurements of antigen-specific antibodies in the serum of mice after immunization showed that intramuscular administration with reconstituted ID93 + GLA-SE vaccine had the highest antibody titers (Gomez et al. 2022). These results motivate further research with the spray-dried TB vaccine and investigation of immunogenicity and efficacy in large animal models with greater relevance to humans.

6.9.3 Conclusions and Outlook on Spray-Dried Tuberculosis Vaccines

This section reviewed the advances that have been made in the field of spray-dried TB vaccines in recent years. Spray drying technology offers particle engineering with aerodynamic particle sizes between 1 and 5 μm suitable for inhalation. This characteristic may prove beneficial as the initial infection with *M. tuberculosis* occurs in the deep lung, followed by forming granulomas near the bronchial surfaces and alveoli.

Aerosol administration of a dry powder-based TB vaccine directly into the deep lung could provide an immunologic advantage by inducing mucosal and systemic immune responses. In addition, extended shelf life at ambient conditions in dry powder form becomes essential in the latter stages of the supply chain when cold chain failure is most likely.

In conclusion, this review demonstrated the feasibility of producing a stable dry powder formulation of TB vaccines by spray drying. Several powdered TB vaccines were developed, including:

- reformulated BCG vaccines,
- adenoviral vectorized TB vaccines, and
- secreted proteins from *M. tuberculosis* as antigens with adjuvants such as encapsulation in nanoemulsions, adsorption on liposomes, or encapsulation in PLGA microparticles as carriers of the antigen.

The formulation provides a large playground for formulation specialists and plays a central role in vaccine stabilization, particle morphology, and aerosol properties. An ideal vaccine dry powder consists of the right combination of antigens, adjuvants, optimized excipients, and appropriate packaging to ensure quality and immunogenicity. In addition, each excipient has its specific properties, such as:

- trehalose is added as a bulking agent providing an amorphous sugar matrix with high T_g and stabilizing ability,
- trileucine and L-leucine are surface-active and improve dispersibility and aerosol properties.
- mannitol is used as a base with low hygroscopicity,
- BSA prevents crystallization of the powder,
- pullulan maintains emulsion droplet size and increases T_g in combination with trehalose, and
- PVP and dextran are also added to increase the T_g.

Various studies show improved storage stability over 2 years at 25 °C with preserved immunogenicity comparable to a lyophilized vaccine. The physical nature of the dry powders and the chemical stability of the vaccine components are maintained during storage.

Crucial is the final packaging with low moisture content. Options include individually sealed bags with desiccant or storage of the powder in vials or protective packaging with desiccant, reactive oxygen species scavenger, and inert gas.

Mice and guinea pigs are the preferred animal models for administering the dry powder formulation by the insufflation method, in which the powders are sprayed directly into the respiratory tract.

The most commonly used laboratory spray dryer is a Mini Spray Dryer B-290, equipped with a high-performance cyclone and installed in a biosafety cabinet. The process parameters, such as inlet drying gas temperature, feed rate, and drying gas flow rate, have been set so that the outlet temperatures are between 35 and 50 °C to prevent potential process degradation.

Besides the appropriate formulation and adjusting the optimal inhalable particle size, challenges also include the scale-up of the spray dryer process. Chapter 4 provides some insights into commercial spray dryers for aseptic processing.

As an outlook, further research work is needed to characterize the storage stability of inhalable spray-dried TB vaccines. In addition, preclinical studies need to be further developed to compare the immunogenicity and efficacy of pulmonary vaccine delivery with other routes of administration and larger animal models. The path for future human clinical trials is not yet clearly defined. In any case, the potential is there. In this regard, the selection and characterization of the dry powder inhaler for a dry powder vaccine are also critical because the vaccine and the delivery device are licensed as a single combination product (vaccine and delivery device) rather than as separate, independent products.

6.10 Human Papillomavirus Vaccines

6.10.1 A Short Introduction to Current HPV Vaccines

Human papillomavirus (HPV) is a non-enveloped virus with a double-stranded DNA genome enclosed in an icosahedral capsid, which is composed of two proteins, the major capsid protein (L1) and the minor capsid protein (L2) (Fig. 6.32). The L1 protein forms pentamers, and 72 copies assemble to form a capsid. The L2 protein is suggested to be present as canyons at the vertices of the pentamers. About 12–72 copies of L2 proteins are present on a virion (Yadav et al. 2019).

HPV is the most common sexually transmitted infectious agent worldwide and is responsible for developing several cancers of the genital area, throat, and cervix (WHO 2017e). To date, more than 220 different HPV genotypes are known, of which a limited number of about 19 HPV types are associated with cancer (Zhai and Tumban 2016; Zhai et al. 2019; Yadav et al. 2019). For example, HPV16 and HPV18 are responsible for about 71% of cervical cancer (WHO 2017e). In addition, other strains of HPV, such as HPV6 and HPV11, cause a variety of diseases, including dysplasia, genital warts, and recurrent respiratory papillomatosis, a condition in which tumors grow in the airways leading from the nose and mouth to

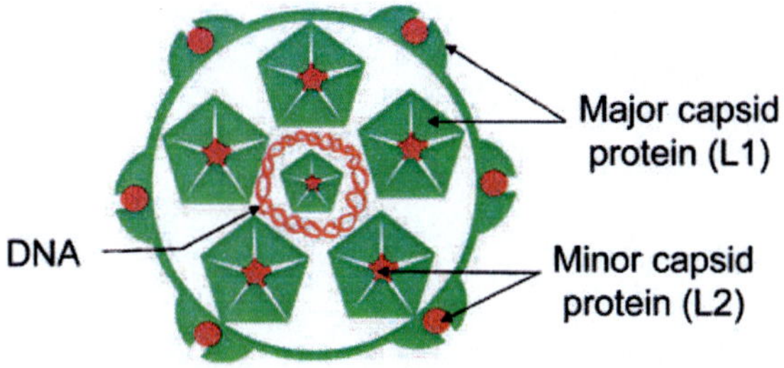

Fig. 6.32 Schematic of a human papillomavirus (HPV) with a double-stranded DNA genome surrounded by the capsid, which is composed of two proteins: the major (L1) and minor (L2) capsid protein. The L1 protein forms pentamers, and the L2 protein is inserted on vertices of the pentamers (Schematic adapted from Yadav et al. (2019), open access under CC-BY 4.0 license)

the lungs (WHO 2020d). However, many other types of HPV do not cause problems, and infections clear up on their own within a few months.

Overall, HPV is very common that almost all sexually active men and women get infected with at some point in their lives (WHO 2020d). In the US, nearly 80 million people (about one in four) are currently infected with HPV (CDC 2020). About 14 million people, including teenagers, become infected with HPV each year. An estimated 35,000 people in the US develop cancer caused by HPV infection each year (CDC 2020). Worldwide, there are about 690,000 new cancer cases per year attributable to HPV: 8.6% in women and 0.8% in men (de Martel et al. 2017, 2020). WHO recommends vaccination of girls aged 9–14 years before becoming sexually active.

Current HPV vaccines consist of virus-like particles (VLP) derived from different combinations of the major capsid protein L1 of HPV (Schiller and Müller 2015). To date, there are four WHO prequalified prophylactic vaccines: Cervarix®, Gardasil® (discontinued in the US), Gardasil 9®, and, most recently, Cecolin® (Table 6.8). The pharmaceutical forms are ready-to-use liquid suspensions for intramuscular injection. Because VLPs are very similar to native HPV particles but do not contain viral DNA, they are not infectious, but the immune system responds accordingly. Therefore, the VLP approach demonstrates applicability to multivalent vaccines (Kols and Sherris 2000), and numerous vaccines based on VLPs are in clinical trials (Donaldson et al. 2018).

As shown in Fig. 6.33, all four HPV vaccines protect against HPV16 and HPV18, which cause about 70% of cervical cancers and other HPV-associated cancer types. Gardasil also protects against HPV6 and HPV11, which cause around 90% of genital warts. In addition, Gardasil-9 additionally protects against five high-risk HPV types (31, 33, 45, 52, and 58), which account for an additional 20% of cervical cancer cases, as well as vulvar, vaginal, anal, and penile disease.

Major barriers to the widespread availability of current HPV vaccines include the logistical and financial burden of administering multiple doses, patient compliance, and the dependence upon continuous refrigeration at 2–8 °C during transport and storage to prevent loss in vaccine efficacy (Zhai and Tumban 2016;

Table 6.8 Storage stability of WHO prequalified Human Papillomavirus (HPV) vaccines

Brand name	Manufacturer	WHO prequalified	HPV types	HPV types	Shelf life, thermostability
Cecolin®	Xiamen Innovax Biotech Co. Ltd	14/10/2021	HPV (Bivalent)	16, 18	2 years (2–8 °C)
Cervarix®	GlaxoSmithKline Biologicals SA	08/07/2009	HPV (Bivalent)	16, 18	5 years (2–8 °C) 3 days (8–25 °C) 1 day (25–37 °C)
Gardasi®	Merck Vaccines	20/05/2009	HPV (Quadrivalent)	6, 11, 16, 18	3 years (2–8 °C) 3 days (8–42 °C)
Gardasil 9®	Merck Vaccines	09/02/2018	HPV (Nonavalent)	6, 11, 16, 18, 31, 33, 45, 52, 58	3 years (2–8 °C) 3 days (8–42 °C) 3 days (0–2 °C)

Data source WHO (2021b)

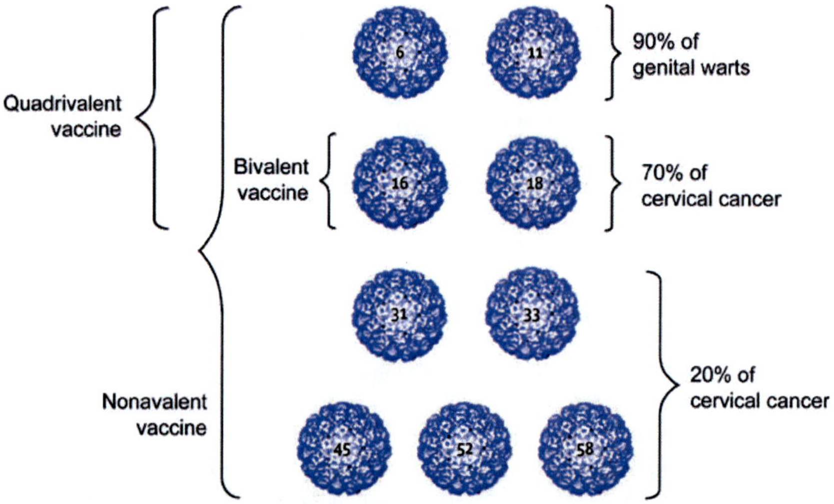

Fig. 6.33 HPV vaccines made up of bivalent, quadrivalent, and nonavalent VLPs derived from HPV types with the proportion of disease attributed to each group (Schematic adapted from Schiller and Müller (2015) with permission from Elsevier)

Kunda et al. 2019; Garcea et al. 2020). Therefore, after being removed from the refrigerator, the vaccines should be administered as soon as possible (WHO 2017e).

The shelf life is 2, 3, or 5 years depending on the vaccine product (Table 6.8). However, for the bivalent vaccine Cervarix, stability for up to 3 days at temperatures between 8 and 25 °C has been demonstrated when stored outside the refrigerator, or up to 1 day between 25 and 37 °C. The quadrivalent vaccine is stable for 3 days at 8–42 °C. For the nonavalent vaccine, stability data indicated up to 3 days when stored at 8–42 °C or 0–2 °C. Consequently, in June 2016, Gardasil (quadrivalent) was approved for CTC use at storage temperatures up to 42 °C for 3 days (WHO 2017d, 2018c). For Gardasil 9, cumulative multiple exposures between 8 and 25 °C are permitted as long as the total duration does not exceed 72 h (Merck 2021). However, these are not recommendations for storage.

Unfortunately, refrigeration and temperature monitoring facilities are less advanced in low-income countries with the highest burden of HPV infection and the need for vaccines (Peabody et al. 2017; de Martel et al. 2017, 2020; Yadav et al. 2019). In addition, distribution in many settings would be simplified if a cold chain was not needed (Schiller and Müller 2015). Researchers generally agree that the next generation of HPV vaccines should be thermostable beyond ambient temperatures and affordable in different regions of the world (Schiller and Müller 2015; Zhai and Tumban 2016; Schellenbacher et al. 2017, 2021). Moreover, current HPV vaccines cost US$ 239 per dose in the private sector in the US (CDC 2021), and two (9–14 years old) or three (15–26 years old) doses are recommended, which is unaffordable for some families.

Another challenge is to achieve broader cross-protection against many HPV types that are not included in the current commercial vaccines. One of the most promising approaches is immunization with the minor HPV capsid protein L2 (Schiller and Müller 2015; Schellenbacher et al. 2017; Yadav et al. 2019). The L2 protein is conserved among different HPV types, suggesting that vaccines targeting L2 produce broadly protective antibodies against various HPV types (Yadav et al. 2019; Huber et al. 2021). Therefore, researchers have focused on immunization with L2 peptides presented multivalently on the surface of VLPs (Yadav et al. 2019). Future studies will specifically assess the longevity of protection at different doses and evaluate the thermostability of the vaccine candidates.

Several researchers have demonstrated to make HPV vaccines more thermostable by converting them into dry powder by lyophilization (in the presence of mannitol or trehalose) (Nieto et al. 2012; Hassett et al. 2015a) or spray freeze drying (using mannitol, trehalose, dextran, and L-leucine) (Zhai et al. 2019). The following section reviews the published studies on spray-dried HPV vaccine formulations (Tumban et al. 2015; Saboo et al. 2016; Peabody et al. 2017; Kunda et al. 2019; Garcea et al. 2020; Rossi et al. 2021). Table 6.9 summarizes the most relevant findings on formulation, spray drying parameters, powder properties, and storage stability.

6.10.2 Research on Spray-Dried HPV Vaccines

In a preclinical study, Tumban et al. (2015) developed a vaccine candidate consisting of bacteriophage VLPs targeting the HPV16 minor capsid protein L2. The VLPs were encapsulated in dry powder by spray drying with a mixture of mannitol,

Table 6.9 Overview of experimental investigations on spray-dried human papilloma virus (HVP) vaccines published in the literature

Vaccine type	Excipients, composition	Spray drying process parameters	Powder results	Administration route, animal model	Shelf life (storage temperature)	References
PfTrx-HPV-L2 antigen	Mannitol (98.5%, 97.83% w/w), GLA (0.17% w/w), PfTrx-HPV-L2 (1.33%, 2.0% w/w), water/ethanol solvent (70:30 v/v ratio), 0.6% w/v solids content	Mini Spray Dryer B-290, 0.7 mm two-fluid nozzle, T_{in}/T_{out}: 125 °C/n.a., F: 3.5 mL/min, A: 601 L/h G: 35 m3/h	1.2–5.3 μm (Span 1.3), MMAD 2.5 μm, mainly spherical, 65–79% yield 81–87% EF, 60–67% RF	Intratracheal with insufflator (mice)	5 month (RT)	Rossi et al. (2021)
Model HPV16 L1 capsomeres	HPV16 L1 capsomers (0.5 mg/mL), histidine HCl (54 mM), trehalose (15% w/v), hydroxyethyl starch (2.5% w/v), NaCl (40 mM), Tween 80 (0.02 mM)	B-290, HP cyclone, parameters not available, 2nd step: atomic layer deposition of alumina in fluidized bed reactor	2–4 μm size, spherical, $\geq$ 80% yield, 5 wt% moisture content	Intramuscular (reconstituted), mice	1 month (50 °C)	Garcea et al. (2020)
Gardasil® 9 (L1 VLPs)	L-Leucine (50 wt%) mannitol (35 wt%), trehalose (5 wt%) *myo*-inositol (5 wt%), dextran (2.5 wt%), PVP (2.5 wt%), Gardasi® 9 liquid suspension (1 wt%), 1% w/v solids content	B-290, T_{in}/T_{out}: 145 ± 10 °C/47 ± 3 °C F: 10% (4 mL/min), A (N2): 742 L/h, G: 100% aspirator	91.5% yield, 70% VLP loading efficiency	Intramuscular (reconstituted in PBS), mice	3 months (40 °C/60% RH)	Kunda et al. (2019)
Bacteriophage L2 VLPs	Mannitol:dextran (76.8:2.1 w/w, 37:1), L-leucine:trehalose (15.8:5.3 w/w, 3:1), 2.5% w/w VLPs, 3% w/v solids content	B-290, T_{in}/T_{out}: 155 °C/55 °C F: 3.6 mL/min A (N2): 450 L/h G (N2): 100% (35 m3/h)	2.7–5.0 μm size (1.0–1.7 span), 1.0–2.5 wt% moisture content, 39–76% yield	Oral (gelatin capsules), rats, i.m. (reconstituted in PBS), mice	14 months (37 °C)	Saboo et al. (2016)

(continued)

Table 6.9 (continued)

Vaccine type	Excipients, composition	Spray drying process parameters	Powder results	Administration route, animal model	Shelf life (storage temperature)	References
Bacteriophage L2 VLPs	Mannitol (85.4 wt%), trehalose (1.71 wt %), dextran (0.85 wt%), L-leucine (7.85 wt%), inositol (4.27 wt%), 1.2 mg/mL VLPs, 2.05% w/v solids content	B-290, PTFE filter T_{in}/T_{out}: 140 °C/47 ± 2 °C F: 10% A (N2): 742 L/h, G (N2): 100%	2–4 µm size, wrinkled particles	i.m. (reconstituted in PBS), mice	34 months (RT) 7.5 months (RT) > 1 month (37 °C)	Tumban et al. (2015, Peabody et al. (2017)

A atomization gas flow rate, *EF* emitted fraction, *F* feed flow rate, *G* drying gas flow rate, *GLA* glucopyranosyl lipid A, *i.m.* intramuscular, *PBS* phosphate-buffered saline, *PfTrx* Pyrococcus furious thioredoxin, *RF* respirable fraction, *RH* relative humidity, *RMC* residual moisture content, *RT* room temperature, T_{in}/T_{out} inlet/outlet drying gas temperature, *VLP* virus-like particle

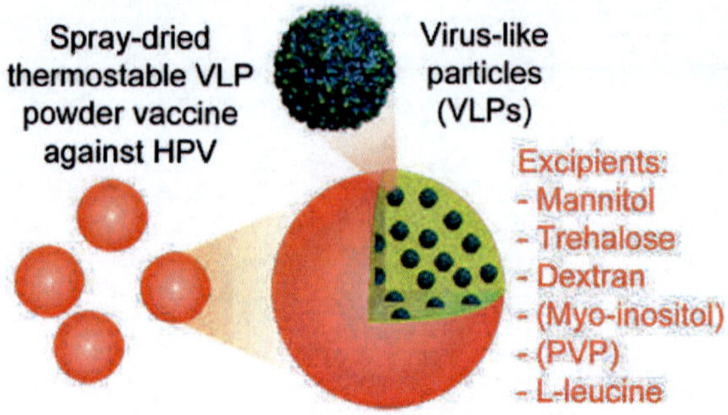

Fig. 6.34 Illustration of virus-like particle (VLP)-based vaccine against HPV infection for oral or intramuscular delivery (Adapted from Saboo et al. (2016) with permission from the American Chemical Society)

trehalose, dextran, L-leucine, and inositol. Figure 6.34 illustrates the principle of encapsulation (Saboo et al. 2016).

- Mannitol resists moisture absorption during storage and prevents powder clumping,
- trehalose provides overall stability of the powder,
- dextran has a high glass transition temperature (T_g) and prevents the crystallization process during long-term storage,
- L-Leucine is an amino acid that gives the dry powder improved flow properties, increases the product yield and the aerosolization of the powder, and
- inositol stabilizes the VLPs during the spray drying process (Jin et al. 2010; Tumban et al. 2015).

A Mini Spray Dryer B-290 was used for the tests and housed in a BioPROtect III Jr.® BSC (Baker Corporation) to protect the dry powder vaccine and operator from contamination. The spray drying parameters were 140 °C/47 ± 2 °C inlet/outlet temperature, 100% (N2) aspiration rate, 742 L/h (N2) spray gas flow rate, and 10% feed rate.

The spray-dried L2 bacteriophage VLP-containing powder showed a distinct wrinkled particle morphology with a particle size of about 2–4 μm (Fig. 6.35a). In addition, spray drying did not affect the integrity of the VLPs after reconstitution in phosphate-buffered saline (PBS). Moreover, spray-dried VLPs vaccines were stable at room temperature for at least 7 months and more than 1 month at 37 °C. Immunization with reconstituted VLPs elicited high titers and antibodies against various HPV types (cross-protecting). The spray-dried vaccine stored at room temperature for 7.5 months protected mice from genital infection with HPV16 pseudoviruses, similar to mice immunized with commercial Gardasil (stored at 4 °C).

In a follow-up study, Saboo et al. (2016) optimized the excipient ratio and spray drying parameters using a Design of Experiments (DoE) approach (Chap. 5). The optimal conditions to maximize yield, minimize moisture content, and optimize

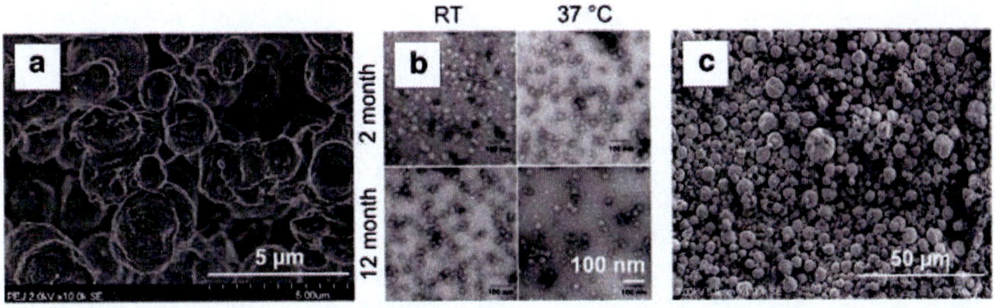

Fig. 6.35 a SEM picture of the spray-dried L2 VLP-based HPV vaccine using stabilizing sugars mannitol, trehalose, dextran, inositol, and amino acid L-leucine (Adapted from Tumban et al. (2015) with permission from Elsevier). **b** TEM images of reconstituted HPV VLP vaccine powder at 2 and 12 months after storage at room temperature (RT) and 37 °C (Adapted from Saboo et al. (2016) with permission from the American Chemical Society). **c** Spray-dried and alumina layer coated particles with narrow particle size distribution and no significant change in single-particle morphology (Adapted from Garcea et al. (2020), open access under CC-BY 4.0 license)

particle size was an L-leucine:trehalose ratio of 3:1 (w/w), mannitol:dextran ratio of 37:1 (w/w), an inlet/outlet temperature of 155 °C and 55 °C, respectively, a gas flow rate of 450 l/h, and a liquid feed rate of 3.6 ml/min. The morphology of the spray-dried VLPs remained the same after 12 months and 2 months of storage at room temperature and 37 °C, as seen under TEM (Fig. 6.35b). After 14 months of storage at 37 °C, intramuscular immunization of mice with a single dose of the reconstituted spray-dried VLP vaccine powder resulted in comparable anti-L2 IgG antibody titers as freshly prepared liquid L2 VLPs (not spray-dried).

Furthermore, tests with oral immunization of VLPs in enteric-coated gelatin capsules elicited lower antibody titers in rats than intramuscularly administered VLPs. This result was attributed to non-uniform VLPs uptake from the small intestine once released from the capsules (Saboo et al. 2016).

A further preclinical study examined the immunogenicity and long-term stability of spray-dried L2 VLPs (Peabody et al. 2017). Two (5 µg) doses of reconstituted spray-dried VLPs stored at room temperature for 34 months elicited high antibody titers in mice, similar to those obtained with fresh VLP suspensions. More importantly, the mice showed significant protection against vaginal infection with HPV pseudovirus 16 (representative of HPV16) 4 months after immunization. The longevity of protection in mice and the thermostability of the vaccine powder suggest that spray-dried VLPs are suitable in countries with less advanced refrigeration facilities.

Kunda et al. (2019) used spray drying to formulate a thermostable HPV vaccine from commercial Gardasil 9. The immunogenicity and protective efficacy of the vaccine were evaluated in mice after 3 months of storage at 4 °C, 25 °C/60% RH, and 40 °C/75% RH. The formulation contained Gardasil®9 vaccine (alum, 1 wt%), L-leucine (50 wt%), mannitol (35 wt%), trehalose (5 wt%), myo-inositol (5 wt%), dextran (2.5 wt%), and PVP (2.5 wt%) at a total solids concentration of 1% w/v. A solution of the excipients without Gardasil 9 was spray-dried as a control. A dry powder yield of 91.5 wt% was obtained with a VLP loading of about 70% efficiency (7 μg per mg dry powder). For storage, the spray-dried powders were loaded into hydroxypropyl methylcellulose (HPMC) capsules and placed into amber-colored serum bottles (Kunda et al. 2016).

As shown in Fig. 6.36, mice immunized twice (10 μg each) with the stored Gardasil 9 dry powder (reconstituted in PBS) and liquid vaccine elicited high-titer IgG antibody responses compared to the PBS control under all storage conditions. Thus, the prospect was raised that future studies should replicate these preclinical results in clinical trials. This could allow approval for a controlled temperature chain (CTC) for at least 3 months and potentially impact HPV vaccination in LMICs (low- and middle-income countries).

Garcea et al. (2020) combined spray drying, freeze drying, and atomic layer deposition (ALD) to produce highly thermostable powders of model HPV16 L1 capsomere antigen. Mixtures of 0.5 mg/mL HPV16-L1 capsomers were formulated in 54 mM histidine HCl along with 15 w/v% endotoxin-free trehalose and 2.5 w/v % hydroxyethyl starch, 40 mM NaCl, and 0.02 mM Tween 80. Some formulations also contained 0.5 mg/mL aluminum from Alhydrogel® (aluminum hydroxide adjuvant) with a final pH of 6.0. Hydroxyethyl starch increased T_g and formed a skin that caused dimples on the particles as drying progressed. The spray-dried microparticles were spherical, with most particles ranging from 1 to 5 μm (Fig. 6.35c). The particles were collected in a high-performance cyclone, and the yield was calculated to be ≥ 80%. The water content was measured to be

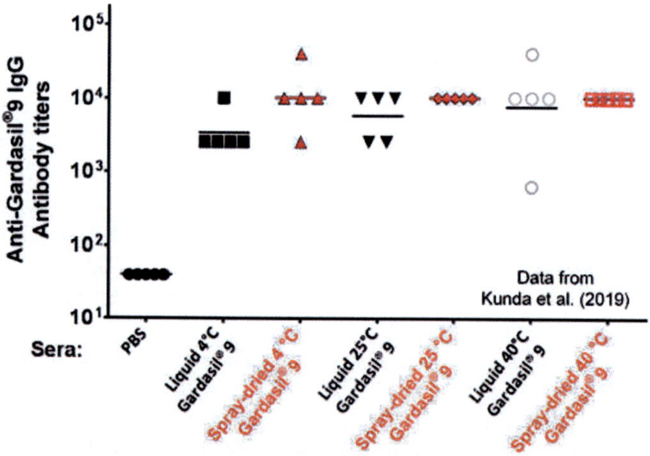

Fig. 6.36 Anti-Gardasil® 9 antibody titers in mice after immunization with Gardasil® 9 stored for 3 months at 4 °C, 25 °C/60% RH, and 40 °C/75% RH. Mice were immunized twice with either liquid (not-spray-dried) Gardasil® 9, spray-dried Gardasil® 9, or PBS (control) at 3-week intervals. Sera were collected 3 weeks after the last immunization, and anti-Gardasil® 9 IgG titers were determined by ELISA (Data adapted from Kunda et al. (2019) with permission from Taylor & Francis)

approximately 5 wt%. The particles were then additionally dried in a lyophilizer at 80 mbar for 16 h at 40 °C to achieve a water content of < 1 wt%.

In a second step, the particles were coated in a custom-made low-pressure fluidized bed reactor with up to 500 molecular layers of alumina (Al_2O_3) by ALD (King et al. 2007, 2012; Liang et al. 2009; Weimer 2019). Alumina (Al_2O_3) was deposited on the surface of the microparticles by alternating pulses of TMA (trimethylaluminum, $Al(CH_3)_3$) and water vapor under argon at 70 °C and 2.7–4 bar, with each cycle, a 2.3 Å thick layer. The deposited alumina nanocoatings serve as an adjuvant whose dissolution can be tuned to deliver a time-delayed booster dose of the antigen at defined times after administration. The HPV16-L1 vaccine microparticles were incubated at 50 °C for 1 month before testing in mice. The study showed that single doses of the ALD-coated vaccine formulations elicited a prime-boost immune response to the L1 antigen with antibody titers that were equivalent or superior to those seen with a standard alum-adsorbed two-dose immunization of the L1 protein. In addition, storage at 50 °C for 1 month did not affect antibody response to the vaccine.

More recently, Rossi et al. (2021) developed a spray-dried HPV dry powder vaccine based on minor capsid protein L2 for inhalation using a multicomponent excipient and spray drying approach (Fig. 6.37). The vaccine particles contained a three-fold repeated peptide epitope derived from HPV16 minor capsid protein L2 presented on Pyrococcus furiosus thioredoxin (PfTrx) as antigen (designated as

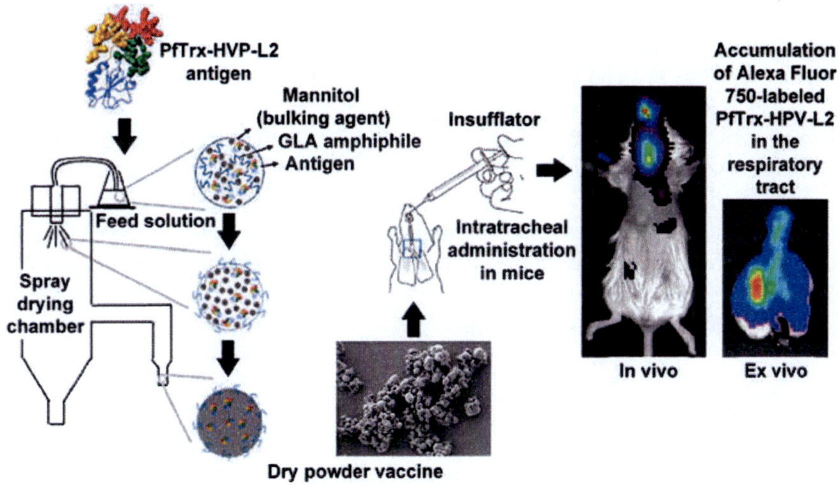

Fig. 6.37 Multi-component excipient and spray-drying approach for particle engineering of a hightly respirable human papillomavirus (PfTrx-HVP-L2) dry powder vaccine with glucopyranosyl lipid A (GLA) as amphiphilic lubricant and immune-adjuvant. The antigen is embedded in GLA-coated particles containing mannitol. In vivo and ex vivo analyses by fluorescence molecular tomography with reactive dye solution (Alexa Fluor 750) showed a fairly uniform antigen distribution in the respiratory tract with a marked accumulation in the lungs (Figures adapted from Rossi et al. (2021) with permission from Elsevier)

PfTrx-HPV-L2 antigen). The Pyrococcus furiosus is a highly thermostable and protease-resistant scaffold. An amphiphilic immune-adjuvant (glucopyranosyl lipid A; GLA) was applied to cover the surface of the spray-dried particles. Mannitol was chosen as the bulking agent and a water–ethanol solution (ratio 70:30 v/v) was used to dissolve the amphiphilic component of the final formulation. The protein solution in 25 mM potassium phosphate buffer (pH 7.4) was added to the water portion. The total solids concentration was maintained at 0.6% (w/v). The PfTrx-HPV-L2 powder vaccine was prepared using a Mini Spray Dryer B-290. The spray-drying parameters were set at 125 °C inlet temperature, 3.5 mL/min feed rate, 601 L/h spray rate, and 35 m^3/h drying air flow rate, resulting in yields ranging from 58 to 79%.

The resulting particles were mainly spherical with a narrow size distribution from about 1.2 to 5.3 μm (span value 1.3). The aerodynamic properties of the powder exhibited a MMAD of around 2.5 μm. The emitted fraction and respirable fraction determined by aerosolization performance analysis were 81–87% and 60–67%, respectively. The dry powder vaccine was always handled under non-refrigerated conditions and did not change its aerodynamic properties and retained fully intact antigen after five months of storage at room temperature.

For intratracheal administration, BALB/c mice were lightly anesthetized and the dry powder vaccine was administered using a modified Penn-Century Dry Powder Insufflator™ (model DP-4 M). The average amount of vaccine powder delivered from the insufflator was 2.43 mg, corresponding to 48.7 μg of antigen. Blood samples were collected to evaluate immune responses in the ELISA and neutralization assays. A short-term immunization study highlighted the ability of the GLA-adjuvanted vaccine powder to induce a systemic anti-L2 immune response comparable to (or even better than) that of the subcutaneously administered liquid-form vaccine. The reasons for this apparent superiority remain to be elucidated. In addition, as shown in Fig. 6.37, in vivo and ex vivo analyses by fluorescence molecular tomography revealed a fairly uniform antigen distribution in the respiratory tract with a marked accumulation in the lungs.

Finally, the authors concluded that the data represent a step forward toward the development of needle-free and easy-to-handle next-generation vaccines. Overall, the data confirmed the favorable aerodynamic properties and suitability of the HPV vaccine powder for pulmonary delivery and that the particle engineering technology could be transferred to other antigens.

6.10.3 Conclusions and Future View on Spray-Dried HPV Vaccines

This review shows that research groups from the United States and Italy are actively developing new spray-dried HPV vaccines, particularly from the University of New Mexico School of Medicine and College of Pharmacy (Albuquerque) (Tumban et al. 2015; Saboo et al. 2016; Peabody et al. 2017; Kunda et al. 2019), the University of Colorado (Boulder) (King et al. 2007, 2012; Liang et al. 2009; Hassett et al. 2015a; Weimer 2019; Garcea et al. 2020; Preston and Randolph 2021), and the University of Parma (Rossi et al. 2021).

Current liquid HPV vaccines are based on L1-VLPs as a platform, with a trend toward L2 protein that provides cross-protection against HPV. A combination of sugars such as trehalose, dextran, mannitol, inositol, and the amino acid L-leucine is used as excipients for spray drying. Under optimal formulation and spray-drying conditions (140–155 °C inlet and 45–55 °C outlet temperature, laboratory scale with the Mini Spray Dryer B-290), the long-term thermostability at room temperature and 37 °C could be improved to more than 1 year. In addition, reconstituted spray-dried VLPs powders resulted in antibody titers comparable to current liquid HPV vaccines in mouse models.

Thus, spray-dried HPV vaccines can be reconstituted in a buffer immediately before intramuscular immunization and are particularly beneficial for remote regions where most cervical cancer patients live and where maintaining the cold chain is a logistical problem for vaccine delivery.

In the future, it would be interesting to determine whether spray-dried HPV-VLP vaccines are compatible with needle-free vaccination methods, such as via pulmonary mucosal immunization. Furthermore, administration of a powder vaccine without reconstitution could be an additional improvement.

Finally, next-generation HPV vaccines should ideally provide cross-protection against multiple HPV types, allow immunogenicity after two or one administration, take advantage of low-cost manufacturing processes, be thermostable for long-term storage, and be suitable for regions with less developed cold chains.

6.11 Cholera Vaccines

6.11.1 A Short Introduction to Current Cholera Vaccines

Cholera is an acute rapidly-dehydrating diarrhoeal disease caused by toxigenic serogroups of the bacterium *Vibrio cholerae* (*V. cholerae*) that, if untreated, can lead to death within hours (WHO 2021c). The bacterium is a curved rod with a single flagellum and is classified into several serogroups that differ in the O side chain of the lipopolysaccharide, whereas the variants O1 and O139 cause epidemic outbreaks (Shin et al. 2011; Pastor et al. 2013b; WHO 2017f).

The cholera disease is transmitted through direct fecal–oral contamination or ingestion of contaminated water or food and is closely associated with poverty, poor sanitation, and lack of safe drinking water (WHO 2017f). The actual global burden of cholera is mainly unknown. However, Ali et al. (2015) estimated that there are about 1.3–4.0 million cases and 21,000–143,000 deaths each year, with the majority in sub-Saharan Africa (60%) and Southeast Asia (29%). India, Ethiopia, Nigeria, Haiti, the Democratic Republic of Congo, Tanzania, Kenya, and Bangladesh have the highest cholera risk. Thus, cholera remains a significant global health challenge (Pastor et al. 2013b).

Table 6.10 WHO prequalified Oral Cholera Vaccines (OCV) with shelf life information

Brand name	Manufacturer	WHO prequalified	Presentation (oral suspension)	Shelf life (storage temperature)
Dukora®	Valneva Sweden AB	25/10/2001	Glass vial and buffer sachet	3 years (2–8 °C) 1 month (37 °C)
Shancho®	Sanofi Healthcare India Private Limited	29/09/2011	Glass vial	30 months (2–8 °C) 14 days (42 °C)
Euvichol® Euvichol-Plus®	EuBiologics Co., Ltd. (South Korea)	23/12/2015 11/08/2017	Glass vial Plastic tube	2 years (2–8 °C)

Data source WHO (2017f, 2021b)

Current strategies to control and prevent cholera are a multidisciplinary approach, including safe drinking water, improved sanitation and hygiene, and mass vaccination campaigns (Shin et al. 2011; Pastor et al. 2013b; Ivers 2017; WHO 2021c).

Today, there are three WHO pre-qualified oral cholera vaccines (OCV) available: Dukoral®, Shanchol®, and Euvichol®/Euvichol-Plus® (Table 6.10). Two doses protect against cholera for at least 3 years, while one dose provides short-term protection (WHO 2021c).

Dukoral® was the first OCV to be licensed globally and contains inactivated whole cells of *V. cholera* serogroup O1 strains with purified recombinant B subunit of cholera toxin. The OCV is administered with a buffer solution of sodium hydrogen carbonate to protect the B subunit from gastric pH. The vaccine has a shelf-life of 3 years at 2–8 °C and remains stable for 1 month at 37 °C (WHO 2017f).

Shanchol® and Euvichol® are essentially the same vaccines based on inactivated whole cells from serogroups O1 and O139 without the B subunit. The omission of the B subunit decreases the cost and complexity of production and removes buffer (Pastor et al. 2013b).

Euvichol-Plus® was developed by EuBiologics Co., Ltd with the help of the International Vaccine Institute (IVI). It is a low-cost OCV, priced at about US$ 1.20 per dose, and is presented in plastic tubes, which reduces the volume by nearly 30% and weight by over 50%, allowing easier transport and distribution of the vaccine and waste management (Odevall et al. 2018). Compared with conventional glass vials, plastic packaging is more straightforward to open and administer. In addition, the shelf life is 2 years at 2–8 °C.

Shanchol® has a shelf life of 30 months at 2–8 °C (WHO 2021b) and maintains safety and immunogenicity when stored for 14 days at elevated temperatures up to 42 °C (Saha et al. 2016; WHO 2017f). Therefore, Shanchol® was prequalified by the WHO for use in a CTC (WHO 2018c), allowing a single excursion not exceeding 40 °C for a minimum of 3 days (VIPS 2020).

A stockpile of OCV (WHO 2013a) ensures rapid access to OCV in outbreak and health emergencies managed by the International Coordinating Group (ICG) and for use in endemic areas managed by the OCV working group of the Global Task Force on Cholera Control (GTFCC) (WHO 2017f).

6.11.2 Research on Spray-Dried Oral Cholera Vaccines

Although safe and effective cholera vaccines are available on the market, the burden of transport and storage remains an issue (Pastor et al. 2014). Thus, research is still needed to develop an OCV that is easy to dose without water and buffer and able to withstand ambient storage temperatures, i.e., would not require a cold chain (Año et al. 2011; Pastor et al. 2013b, 2014). Among the possible strategies, microencapsulation represents a promising approach to improve the uptake and transport of orally administered vaccines.

Microparticles have been widely used in the last decade for the controlled release of drugs and bioactive agents, including the formulation of vaccines with an enteric polymer that protects against degradation by the acidic pH of the stomach (Madani et al. 2020) and elicits a long-lasting immune response. (Gonzalez-Cruz and Gill 2021) reported that the top five diseases studied with particle-based oral vaccines are hepatitis B, cholera, norovirus, diphtheria, and *E. coli* infections.

Polymeric microparticles are gaining considerable interest as vaccine delivery systems (Pastor et al. 2013b). Examples include encapsulation in methacrylic copolymers Eudragit® L30 D-55 and Eudragit® FS 30D (Lin et al. 2003; Año et al. 2011; Tawde et al. 2012, 2016; Pastor et al. 2013a). These copolymers are inert and available in different presentations, including aqueous dispersions for enteric coatings, and provide microparticles with gastro-resistant properties. Eudragit® L30 D-55 dissolves above pH 5.5 and Eudragit® FS 30D at a pH higher than 6.5. Besides, they are of relatively low cost compared with other polymers used for the preparation of microparticles such as PLGA (Poly(lactic-co-glycolic acid) and its derivatives, making them attractive candidates for alternative encapsulation processes (Año et al. 2011).

Several microencapsulation methods are used for the elaboration of polymeric microparticles, including air suspension methods, coacervation (aqueous phase separation), solvent evaporation from water/oil/water emulsion-based systems, or spray drying (Ré 1998; Johansen et al. 2000; Jyothi et al. 2010; Augustin and Sanguansri 2017). Among them, spray drying represents a widely used method because it is a single-step continuous process that can be easily scaled up and that can achieve high vaccine loadings.

As a direct application of this process in cholera vaccines fabrication, Año et al. (2011) investigated the feasibility of spray drying for the microencapsulation of heat-inactivated *V. cholera* strain (El Tor Biotype, Ogawa Serotype) using Eudragit® L30D-55 and FS30D. The vaccine was mixed at 1:10 bacteria-to-polymer ratio and spray-dried with a Mini Spray Dryer B-191 at a feed rate of 5 L/min, 600 L/h air atomization rate, 60–100 °C inlet, and 40–64 °C outlet air temperature

(Table 6.11). The yield was generally 70–80%, with a particle size between 2.7 and 3.4 µm (Fig. 6.38). As a result, uniformly and spherical microparticles were formed, which have a suitable size to allow uptake by M-cells (microfold cells) and thus might favor the antigen targeting to Peyer's patches in the intestine following oral administration. In all cases, 50–55 µg of *V. cholerae* was entrapped per mg of polymer, showing encapsulation efficiencies of around 98.5%.

In vitro release was demonstrated for microparticles prepared with both copolymer types. Under gastric conditions (pH 1.2), less than 5% of the bacteria were released (Fig. 6.39). However, in a neutral medium with a pH of 6.8 (PBS), which is similar to intestinal fluids, the microparticles disintegrated and Eudragit® L30D-55-released 86% after 24 h while Eudragit® FS30D released less than 30%. Eudragit® FS30D, which is soluble only at higher pH, explains the lower release of *V. cholerae* from microparticles at pH 6.8. In addition, a lower release was observed when spray drying at 100 °C inlet temperature, suggesting denser particles.

Rats inoculated with spray-dried microparticles exhibited vibriocidal antibody titers (ELISA inhibition test). Microparticles made of Eudragit® FS30D showed lower immunogenicity, which could be explained by a slower release rate and lower antigenicity maintained after the microencapsulation process.

In addition, microparticles prepared with Eudragit®L 30 D-55 at the highest inlet temperature (100 °C) showed only 88% antigenicity compared to 98% at 60 and 80 °C, which was mainly due to the increased processing temperature. Overall, spray drying was successful for microencapsulation of inactivated *V. cholerae* to obtain OCV with good bacterial loading and gastric resistance suitable for children under 4 years of age (Año et al. 2011).

Based on these results, Pastor et al. (2013a) performed further spray drying studies and focused on the formulation of Eudragit® L30D-55 microparticles combined with mucoadhesive alginate and Carbopol®, thus prolonging residence time in the gut and providing a concentration gradient at the membrane. Microparticles ranging from 7 to 9 µm with high encapsulation efficiencies over 98.5% were obtained. Rats immunized with the spray-dried microparticles showed similar antibody responses to liquid *V. cholera* suspension. Especially Eudragit® L30D-55 alginate microparticles at a 60 or 30 mg dose proved immunogenic. Moreover, after following ICH guidelines on stability testing (ICH 1995, 2003), the microparticles were stable for 12 months at 25 ± 2 °C (60 ± 5% relative humidity) and 6 months at 40 ± 2 °C (75 ± 5% RH), determined by the maintenance of particle size, powder morphology (Fig. 6.39), and gastro-resistance.

In a follow-up study, Pastor et al. (2014) encapsulated heat-inactivated *V. cholerae* in Aquacoat® CPD (cellulose acetate phthalate) as a core polymer using spray drying technique. The addition of alginate was investigated. The result was a fine, white powder with a production yield between 48 and 61%. SEM images showed a smooth surface and a regular shape of the microparticles. The 8.2–8.6% loading corresponded to 89–95% encapsulation efficiency. Subsequent immunization experiments in rats showed that 30 mg Aquacoat® CPD alginate microparticles were particularly capable of eliciting the desired immune response.

Table 6.11 Overview of experimental studies on microencapsulation of heat-inactivated *V. cholera* strains by spray drying

Spray drying process parameters	Excipients composition	Particle size, yield, encapsulation efficiency	Route of administration	Shelf life (thermostability)	References
B-191, 0.7 mm nozzle, T_{in}: 60, 80, 100 °C, T_{out}: 40–64 °C, F: 5 L/min, A (air): 600 L/h	Eudragit® L30 D-55, Eudragit® FS 30D, 1:10 bacteria:polymer ratio	2.7–3.4 μm, 70–80% production yield, 98.5% encapsulation efficiency	Oral, rats	Not reported	Año et al. (2011)
B-191, 0.7 mm nozzle, T_{in}: 80 °C, T_{out}: 60 °C, F: 5 L/min, A (air): 600 L/h	7.5% (w/v) Eudragit® L30 D-55, 0.1% (w/v) alginic acid or Carbopol® (cellulose acetate phthalate), 1:10 bacteria:polymer ratio, 1.5% (w/w) triethylcitrate as plasticizer	7–9 μm particles, mucoadhesive, 9% loading, > 98.5% encapsulation efficiency	Oral, rats	6 months (40 °C 75% RH), 12 months (25 °C, 60% RH)	Pastor et al. (2013a)
B-191, 0.7 mm nozzle, T_{in}: 60 and 80 °C, T_{out}: n.a., F: 5 L/min, A (air): 600 L/h	7.5% (w/v) Aquacoat CPD® (cellulose acetate phthalate), 0.1% (v/w) alginate, 1:10 bacteria:polymer ratio, 10% (w/w) triethylcitrate as plasticizer	6–7 μm particles, smooth surface, mucoadhesive, 48–61% production yield, 8.2–8.6% loading, 89–95% encapsulation efficiency	Oral, rats	Not reported	Pastor et al. (2014)

T_{in}/T_{out} inlet/outlet drying gas temperature, *F* feed flow rate, *A* atomization gas flow rate, *RH* relative humidity, B-191: Mini Spray Dryer B-191, Büchi Labortechnik AG, Switzerland

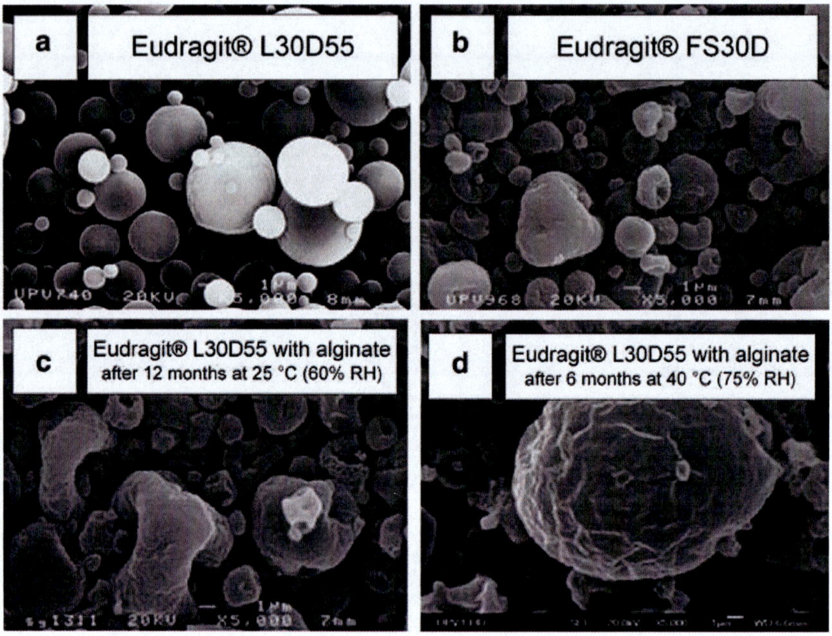

Fig. 6.38 SEM photographs of spray-dried microparticles containing inactivated *V. cholerae* vaccine. **a** Eudragit® L30D55 and **b** Eudragit® FS30D at 60 °C inlet drying air temperature, **c** Eudragit® L30D55 with alginate after 12 months of storage at 25 °C (60% RH), **d** Eudragit® L30D55 with alginate after 6 months of storage at 40 °C (75% RH) (Pictures adapted from Año et al. (2011) and Pastor et al. (2013a) with permission from Elsevier)

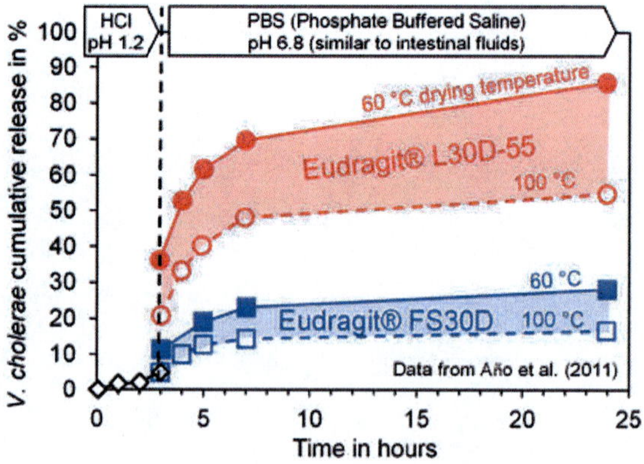

Fig. 6.39 Vibrio cholerae cumulative release profiles from spray-dried Eudragit® L30D-55 and Eudragit® FS30D microparticles (Data adapted from Año et al. (2011) with permission from Elsevier)

6.11.3 Conclusions on Spray-Dried Cholera Vaccines

Although safe and efficient liquid oral cholera vaccines (OCVs) are on the market today, research is still needed to solve transport and storage problems at ambient temperature. Particle-based encapsulation technologies such as spray drying could provide an alternative.

Rat experiments showed that inactivated *V. cholera* bacteria could be successfully stabilized in enteric polymers by spray drying and administered orally without reconstitution. For this purpose, cholera bacteria are microencapsulated with different cellulose and acrylic derivatives (e.g., Eudragit® L30 D-55, Eudragit® FS 30D, Aquacoat® CPD, Carbopol®) as carrier materials, which allow pH-controlled release with variable bacterial loading and high encapsulation efficiency above 98%. In addition, alginate is particularly suitable as a mucoadhesive additive prolonging the residence time in the gut.

The spray-dried microparticles exhibited storage stability of up to 6 months at 40 °C (75% RH) and 12 months at 25 °C (60% RH). In addition, the microparticles sizes are suitable for transport through Peyer's patches in the intestine following oral administration. Immunization experiments on rats with the spray-dried microparticles elicited the desired immune response.

Overall, spray drying could produce OCV powder enabling cold chain-free and adequate transportation and storage. As a powder, the vaccine is also suitable for administration to children. At the same time, tableting or granulation is possible, making the vaccine easier to handle, store and transport. Thus, these studies provide further hope for the fight against cholera.

6.12 Anthrax Vaccines

6.12.1 A Short Introduction to Anthrax

Anthrax is primarily a zoonotic disease of herbivores caused by the spore-forming bacterium *Bacillus anthracis*. The bacterium is rod-shaped and approximately 4 μm by 1 μm in size (WHO 2008). *B. anthracis* produces anthrax toxin, a three-component exotoxin consisting of a cell-binding protein, the protective antigen (PA), and two enzyme components, edema factor (EF) and lethal factor (LF) (Jiang et al. 2006).

The disease is most common in wild and domestic animals, e.g., cattle, sheep, goats, camels, antelope, and other herbivores. In its most common form, cutaneous anthrax (over 95% of cases) (Ashford et al. 2001) creates dark lesions on the skin, from which it derives its name, after the Greek word for coal (Inglesby et al. 2002).

Anthrax spores can survive for long periods in the environment, are easily transmitted, and are associated with high morbidity and mortality rates if untreated with antibiotics. Of the three routes of infection (i.e., cutaneous, gastrointestinal, and inhalation), inhalation anthrax is the deadliest (Fig. 6.40), with an estimated

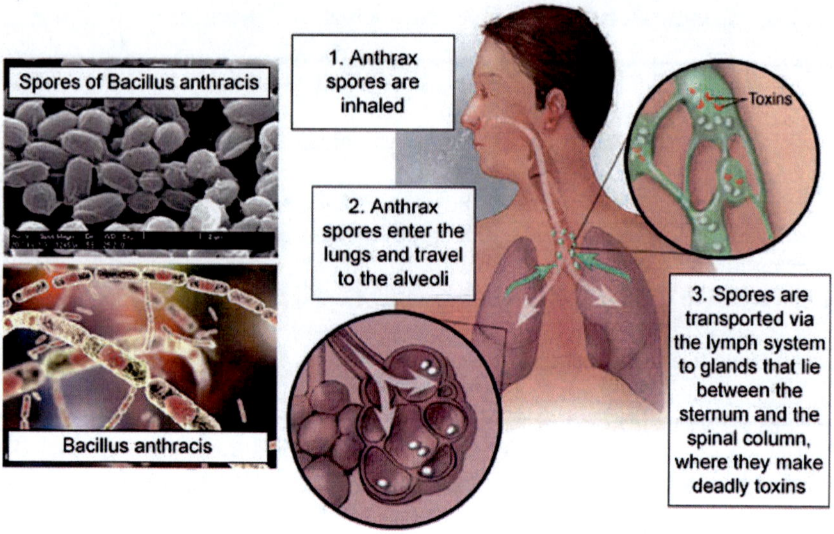

Fig. 6.40 Illustration of spore-forming *Bacillus anthracis* that cause inhalation anthrax, the deadliest form of anthrax infection (Pictures adapted from Mayo Clinic USA, iStock ID 610691058, and Wikimedia Commons, the free media repository)

mortality rate of 75% or more if not treated (Ashford et al. 2001). For gastrointestinal disease, the mortality rate is assumed to be between 25 and 75% if untreated (Brachman et al. 2008).

Humans may acquire the disease directly or indirectly from infected animals or through occupational contact with infected or contaminated animal products. However, rapid and timely antibiotic therapy usually results in a fast recovery of persons or animals infected with anthrax (WHO 2008).

In 1958, the annual worldwide incidence of human cases of anthrax was estimated to be between 20,000 and 100,000 cases (Glassman 1958). In the 1980s and 1990s, the total number worldwide decreased to an estimated 2000 cases per year (Brachman et al. 2008). Today, anthrax is an uncommon disease in humans (Splino et al. 2005), and most people will never be exposed to it.

However, *Bacillus anthracis* has always been high on the list of potential agents for biological warfare and bioterrorism (Ashford et al. 2001; Klietmann and Ruoff 2001). For example, in 2001, letters intentionally containing powdered anthrax spores were distributed through the U.S. Postal Service. As a result, there were 22 cases of anthrax infection, of which 5 resulted in deaths and 17 in illness (Inglesby et al. 2002; Jernigan et al. 2002).

In bioterror scenarios, anthrax spores could get into food or water or be released into the air from a truck, building, or airplane. It would only take a small amount of anthrax to infect many people (CDC 2016). The WHO estimated that 50 kg of *B. anthracis* released upwind of an urban population center of 500,000 could result in 95,000 deaths and 125,000 hospitalizations (WHO 1970). This risk emphasizes the need for research and stockpiling of stable and effective anthrax vaccines for humans in case of a public health emergency, especially in a biodefense situation involving *B. anthracis* (Wang et al. 2012).

Vaccines against anthrax in livestock are available in most countries, and the natural disease is easily controlled and surveilled (WHO 2008). For humans, there are four anthrax vaccines available (WHO 2008); the US Anthrax Vaccine Adsorbed (AVA) vaccine (BioThrax®), the UK anthrax vaccine precipitate (AVP) vaccine, and the live spore vaccines from China and Russia (Table 6.12).

However, the anthrax vaccines are recommended only to specific groups of adults at potential risk of anthrax, including military personnel and laboratory workers who process samples and some people who handle infected animals, such as veterinarians (WHO 2008).

BioThrax® from Emergent BioDefense Corporation (formerly known as BioPort Corporation) is the only FDA-approved anthrax vaccine in the US (Emergent BioSolutions 2015). The vaccine is a purified cell-free filtrate administered in five doses, with an annual booster dose.

Porton Biopharma Ltd. is the sole manufacturer of the anthrax vaccine licensed in the UK (Porton Biopharma Ltd 2015). The vaccine is a sterile filtrate of an alum-precipitated anthrax antigen for injection. The other ingredients are aluminum potassium sulfate, sodium chloride, and purified water. The preservative is thiomersal. The vaccines should be stored in a refrigerator but not frozen because repeated freezing/thawing would reduce the inocula (WHO 2008). The shelf life is

Table 6.12 Commercial human anthrax vaccines

Manufacturer (country)	Vaccine description	Doses, administration route	Storage conditions (shelf-life)
Emergent BioDefense Corporation (formerly BioPort Corporation) (USA)	BioThrax®, suspension of aluminum hydroxide adsorbed filtrate of strain V770 culture	3 doses of 0.5 mL at 0, 1, and 6 months, followed by 2 additional doses at 6 and 12 months, after that annual booster	2–8 °C (4 years)
Porton Biopharma Limited (UK)	Anthrax Vaccine (Alum precipitated sterile filtrate) suspension	4 doses in 6 months, followed by a booster every year, intramuscular injection	2–8 °C (5 years)
Lanzhou Institute of Biological Products (China)	Live spore suspension of strain A16R in 50% glycerol and distilled water	1 dose by scarification of 50 mL in two spots on the skin, followed by a booster after 6 or 12 months, after that annual booster	n.a
Research Institute of Microbiology (Russia)	Lyophilized live spore suspension of strain STI-1	2 doses followed by a single annual booster, scarification of 0.05 mL into the skin, or (in urgent situations) injection of 0.5 mL	n.a

WHO (2008), Emergent BioSolutions (2015), Porton Biopharma Ltd (2015)

between 4 and 5 years when stored at 2–8 °C (Emergent BioSolutions 2015; Porton Biopharma Ltd 2015).

Next-generation anthrax vaccine candidates aim to enhance thermostability and reduce the need for a cold chain during shipping and storage. In this direction, Altimmune, Inc. is currently developing a vaccine, SparVax-L, derived in *E. coli* in a highly stable lyophilized form that can be stored at room temperature (Kondakova et al. 2019). In addition, Emergent BioSolutions Inc. is developing a lyophilized dry formulation of NuThrax™, also known as AV7909, which consists of BioThrax® (anthrax vaccine in adsorbed form) in combination with the immunostimulatory oligodeoxynucleotide CPG 7909 as an adjuvant (Emergent BioSolutions 2018). Phase 1 clinical studies are ongoing investigating the safety, tolerability, and immunogenicity of lyophilized AV7909 vaccine (ClinicalTrials.gov, Identifier NCT04660201).

In addition, the main focus of next-generation anthrax vaccines is on purified recombinant protective antigen (rPA) of the *Bacillus anthracis* as the antigenic component (WHO 2008; Wang et al. 2012; Jones et al. 2017). rPA-based vaccines are well tolerated and effective but generally require a cold chain for storage and distribution (Huang et al. 2007). Therefore, increased thermal stability for long-term storage is a development goal and would allow stockpiling vaccines for various pandemic and biodefense situations (Jiang et al. 2006; Jones et al. 2017). Furthermore, cold chain storage in low-income countries is sometimes challenging to access and unreliable, as calibrated refrigerators are often not available or monitored.

6.12.2 Research on Spray-Dried Anthrax Vaccines

Several studies have demonstrated that a stable dry powder formulation for rPA anthrax vaccine can be prepared by freeze drying (Mikszta et al. 2005; Wimer-Mackin et al. 2006; Klas et al. 2008; Hassett et al. 2015b), spray freeze drying (Mikszta et al. 2005; Jiang et al. 2006; Huang et al. 2007; Wang et al. 2012), and spray drying (Jones et al. 2017).

Unlike a freeze-dried cake, spray drying technologies offer particle design thanks to the high volume-to-surface area ratio of the generated droplets. As a result, the speed of immobilization of a vaccine protein in particulate form is on the order of milliseconds, as opposed to hours as with freeze drying. In addition, the flexibility in excipient type and delivery route make spray drying technologies unique and interesting compared to other solid-state immobilization technologies. Table 6.13 provides an overview of the studies on spray-dried and spray freeze-dried anthrax vaccines.

Mikszta et al. (2005) investigated spray freeze drying of anthrax vaccine powders for cutaneous administration via microneedles and nasal delivery in mice and rabbits. Formulations containing rPA were prepared using trehalose as a glass stabilizer and CpG oligonucleotide adjuvant in an aqueous solution of rPA and CpG in a 1:1 ratio. Chitosan was also added as a biocompatible mucoadhesive polymer

Table 6.13 Overview of preclinical studies with spray-dried and spray-freeze-dried anthrax vaccines

Vaccines type	Excipients composition	Spray (freeze) drying process parameters	Particle size	Animal model (route of administration, dose)	Shelf life (storage conditions)	References (country)
Recombinant protein antigen (rPA) vaccine	Trehalose, trehalose/chitosan, trehalose/CpG, oligonucleotide adjuvant	Spray freeze drying, spray into liquid N_2, freeze-drying 72 h	Approx. 70 µm	Rabbits (Intranasal, 10 and 50 µg rPA per dose)	29 days (40 °C)	Mikszta et al. (2005), Jiang et al. (2006), Huang et al. (2007) (USA)
rPA vaccine	10% (w/v) solute concentration, 2 µg rPA, 8 µg C48/80 per mg trehalose, 1–5 g powder batches	Spray freeze drying, two-fluid nozzle (0.7 mm from B-191), F: 10 mL/min A: 500 L/min N2 (3 bar), spray into liquid N2 (−80 °C), freeze drying 72 h	8–50 µm (D50 of 25 µm, span 1.72), hollow porous structure	Rabbits (intranasal, 30 µg rPA per dose)	2.5 years (at room temperature and stored in desiccant packaging to prevent moisture uptake)	Wang et al. (2012) (USA)
rPA vaccine candidate (pp-dPA83)	Trehalose (~74 wt%), hydrolyzed gelatin (~24 wt%), PBS buffer salts, adjuvant Alhydrogel, Tween 80	Custom build laboratory-scale spray dryer (BLD-35, Bend Research), two-fluid atomizer, custom cyclone, T_{in}: 100–120 °C, T_{out}: 58 °C, F: 4 mL/min, A (air): 8 psi	87–93% yield, 3.8–4.7 wt% moisture content, smooth particles, 1–10 µm	Mice (Intramuscular, 5 µg rPA per dose)	3 months (40 °C, 75% RH)	Jones et al. (2017) (USA)

rPA recombinant protein antigen, T_{in}/T_{out} inlet/outlet drying gas temperature, *F* feed flow rate, *A* atomization gas flow rate, *RH* relative humidity and adjuvant that stimulates immune cells. The solutions were sprayed into liquid nitrogen (Maa et al. 1999), followed by evaporation and freeze-drying under a vacuum for 72 h. Intranasal administration of the powder vaccine to rabbits using a special syringe device with a powder-filled capsule (50 µg per dose) sealed with rupturable film provided complete vaccine protection (100% survival), whereas a liquid formulation provided only partial protection (67% survival).

In a follow-up study, Jiang et al. (2006) showed that the spray freeze-dried powder was highly porous with an average size of about 70 µm (Fig. 6.41a) that was quickly aerosolized and suitable for deposition in the nasal cavity. Often porous particles have a mean aerodynamic mass diameter smaller than their geometric particle size and are more suitable for aerosolization than non-porous particles. Furthermore, the dry powder maintained protein integrity under accelerated storage temperatures of 25 and 40 °C for up to 29 days, whereas the liquid formulation exhibited rapid degradation.

Based on these findings, Huang et al. (2007) demonstrated that intranasally delivered rPA powders containing chitosan and trehalose elicited PA-specific IgG

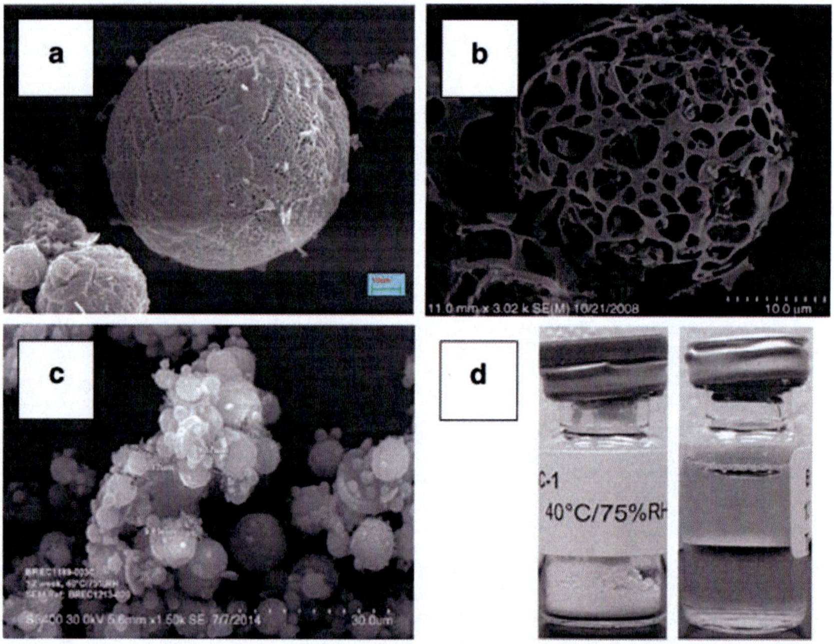

Fig. 6.41 **a** SEM picture of spray freeze-dried rPA vaccine powder (Jiang et al. 2006). **b** Hollow spray freeze-dried rPA powder containing 2 µg rPA and 8 µg C48/80 adjuvant per mg of trehalose (Wang et al. 2012). **c** Spray-dried pp-dPA83 vaccine powder containing ~74% trehalose, ~24% hydrolyzed gelatin, PBS buffer salts, 0.2% antigen, and 0.2% Tween 80 after storage for 3 months at 40 °C (75% RH) (Jones et al. 2017). **d** Spray-dried and reconstituted pp-dPA83 vaccine sample after 3 months of storage at 40 °C (Jones et al. 2017) (Pictures adapted with permission from Elsevier)

and toxin neutralizing antibody (TNA) titers in serum of rabbits equivalent to those of a liquid rPA formulation (10 mg rPA) administered by MicroSprayers™ devices (PennCentury, USA). In addition, the protection against lethal aerosol spores was at least as high as the liquid vaccine administered by injection. Thus, the dry powder vaccine (Jiang et al. 2006) and the noninvasive intranasal delivery method were seen to significantly improve the logistics of vaccine administration, storage, and distribution for biodefense vaccination. Furthermore, the administration of the anthrax vaccine via the nasal cavity produces a durable immune response and is also a convenient route of administration (Bahamondez-Canas and Cui 2018).

Wang et al. (2012) spray freeze-dried rPA vaccine with a novel cell-activator adjuvant C48/80 and trehalose for nasal application. The vaccine formulation with a total solid concentration of 10% (w/v) (2 μg rPA with 8 μg C48/80 per mg trehalose) was sprayed into a stainless-steel tank containing liquid N_2 (−80 °C) at a feed rate of 10 mL/min and an N_2 gas flow rate of 500 L/min (3 bar back pressure) using a two-liquid nozzle (0.7 mm diameter from Mini Spray Dryer B-191). After evaporating the liquid N_2 in a freezer at −80 °C, the frozen particles were transferred to a pre-cooled vessel and freeze-dried for 72 h. The typical batch size was 1–5 g of powder, and the particle size ranged from 8 to 50 μm (D50 of 25 μm, span 1.72) and was suitable for administration into the nasal cavity of rabbits. The addition of protein to the formulation resulted in particles with a hollow porous structure (Fig. 6.41b).

Nasal powder vaccination of rabbits with a UnitDose device (Aptargroup, Inc.) showed that the dried formulations remained as effective as the freshly prepared liquid formulation even after 2.5 years of storage at room temperature. This extraordinary storage stability was achieved by storing the powder with a desiccant and packaging it in a nasal delivery device to prevent moisture absorption.

Jones et al. (2017) developed spray-dried plant-based anthrax vaccine formulations of deglycosylated PA83 purified from Nicotiana benthamiana using a custom-built laboratory-scale spray dryer (BLD-35) (Bend Research, Oregon, now owned by Lonza) at 100–120 °C inlet and 58 °C outlet temperature, 4 mL/min feed rate, and 30 kg/h drying gas flow rate (nitrogen). The main excipients were trehalose (~74 wt%) and hydrolyzed gelatin (~24 wt%). All powder formulations were amorphous (T_g of around 125 °C) and had a low moisture content of 3.8–4.7 wt% and smooth morphology (Fig. 6.41c, d). The powder protein vaccine remained stable for 3 months under accelerated storage conditions of 40 °C (75% RH) and induced high systemic and mucosal immune activity in mice. The toxin-neutralizing antibody response was comparable to reconstituted liquid control achieved with intramuscular administration.

This feasibility study showed that the developed powder formulation could be reconstituted with water and injected with a syringe. In addition, rapid reconstitution resulting in a clear solution (without particles) was a priority, and these criteria were met with this formulation. According to the authors, high powder yields of 87–93% were achieved, making spray drying an attractive technology for stockpiling biologicals and eliminating the need for a conventional cold chain.

6.12.3 Conclusions on Spray-Dried Anthrax Vaccines

Anthrax remains a serious threat as a bioterrorism agent. The ultimate goal is an efficacious, safe, and thermostable human vaccine that can be administered in a few doses over a short time. Next-generation anthrax vaccines are being developed based on recombinant protective antigen (rPA) from *Bacillus anthracis*. Freeze drying, spray freeze drying, and spray drying of vaccine formulations effectively stabilize vaccines and extend shelf life, which is desirable for stockpiling during pandemics or bioterrorism threats. Among drying technologies, freeze-dried vaccines are the most advanced with ongoing clinical trials, followed by spray freeze drying with several research studies and spray drying.

Several preclinical studies in rabbits have shown that immunization with spray freeze-dried rPA protein powder vaccines via the nasal mucosa is feasible. In addition, the particle sizes produced, ranging from 25 to 70 μm, are suitable for intranasal administration or rapid reconstitution for injection due to the high surface area of the particles. In addition, spray drying results have shown that the process is suitable for powder vaccine development and has potential for other vaccines. The formulations described mainly use trehalose as a glass-forming excipient (high T_g of around 120 °C, safe in lungs, non-reducing sugar, no polymorph issues) and adjuvants like chitosan to enhance the immune response.

Overall, dry formulations are a promising way to improve the stability of anthrax vaccines. Under dry storage conditions, shelf lives of up to 2.5 years at room temperature could be achieved. In addition, the combination of improved thermal stability, long-term storage, powder form, and a non-invasive delivery platform enables the feasibility of mass anthrax vaccination.

6.13 Typhoid Vaccines

6.13.1 A Short Introduction to Typhoid

Typhoid, also known as typhoid fever, is an acute intestinal infection caused by the bacterium *Salmonella enterica* serovar Typhi, commonly referred to as *S. Typhi*. The disease is transmitted by contaminated food or water and close contact with an infected person (WHO 2018d). After ingestion, *S. Typhi bacteria* can overcome the gastric acid barrier, colonize the intestinal tract, cross the mucosa, and eventually reach the lymphatic system and bloodstream, causing the illness (Levine and Simon 2018).

Symptoms of infection include persistent high fever, accompanied by weakness, stomach pain, headache, diarrhea or constipation, cough, loss of appetite, and some rash on the body (CDC 2019).

In the era before the introduction of antibiotics, a mortality rate of approximately 10–20% was reported (Levine and Simon 2018), whereas current estimates range from 1 to 4% among those who receive appropriate therapy (WHO 2017g, 2018d).

At present, typhoid fever can be treated effectively with antibiotics, but there are growing rates of antibiotic resistance (Wain et al. 2015; Gibani et al. 2018).

Therefore, typhoid remains a major public health problem in many regions. Worldwide an estimated 11–20 million people get sick from typhoid, and between 128,000 and 161,000 people die from it every year (WHO 2017g, 2018d). High endemicity is observed in South and Southeast Asia, the Middle East, South America, Northeast Africa, sub-Saharan Africa, and some Pacific Islands (Levine and Simon 2018). Typhoid cases are rare in the US and Europe, whereby travelers bring most cases.

Control of typhoid fever relies on clinical information, provision of clean water and improved sanitation, diagnosis through blood testing, understanding the epidemiology of the disease, and targeted vaccination of at-risk populations (WHO 2003, 2018d).

Thus, vaccination is an important tool that helps prevent typhoid cases, limit the additional spread, and reduce the use of antibiotics (Emergent Travel Health Inc. 2020). Currently, there are 3 types of typhoid vaccines licensed (Table 6.14):

Table 6.14 Characteristics of currently available typhoid vaccines

Vaccine information	Live attenuated Ty21a	Unconjugated Vi polysaccharide	Typhoid conjugate
Tradename (Manufacturer)	Vivotif® (Emergent BioSolutions)	Typhim V® (Sanofi Pasteur) Typherix® (GSK)	Typhibev® (Biological E.) Typbar-TCV® (Bharat Biotech)
Administration route	Oral capsules	Intramuscular/ subcutaneous	Intramuscular
Presentation	Enteric-coated (gastro-resistant) gelatin capsules filled with lyophilized live attenuated Ty21a strains	Liquid	Liquid
Age for vaccination	$\geq$ 6 years	$\geq$ 2 years	$\geq$ 6 months
Number of doses	3–4 doses	1 dose	1 dose with boosters every 2–3 years
Interval between doses	About 38 h	–	1–2 months
Efficacy	50–80%	50–80%	79–85%
Duration of efficacy	7 years	Up to 3 years	4 years
Shelf life (storage conditions)	1.5 years (2–8 °C) 14 days (25 °C)	3 years (2–8 °C) 2 years (22 °C) 6 months (37 °C)	2 years (2–8 °C)

Levine and Simon (2018), WHO (2018d, 2021b), Coalition Against Typhoid (2021)

- live attenuated Ty21a vaccine,
- unconjugated Vi polysaccharide, and
- typhoid conjugate vaccine.

Ty21a is an orally administered vaccine based on a live attenuated Ty2 strain of *S. Typhi*, in which several genes have been attenuated by chemically-induced mutagenesis. The resulting vaccine strain, termed Ty21a, lacks the Vi antigen (WHO 2018d). It is marketed as Vivotif® (Typhoid Vaccine Live Oral Ty21a) by Emergent BioSolutions (previously PaxVax Berna GmbH) as an enteric-coated (gastro-resistant) gelatin capsule, each containing one dose of lyophilized vaccine. Vivotif® is supplied in a single foil blister containing 4 doses of vaccine in a single package (Emergent Travel Health Inc. 2020). Ty21a vaccine requires storage at 2–8 °C providing a shelf life of at least 1.5 years and retains potency for approximately 14 days at 25 °C (WHO 2018d).

The unconjugated Vi polysaccharide vaccine is a sterile solution containing purified Vi capsular polysaccharide from the Ty2 *S. Typhi* strain (WHO 2018d). It is marketed as Typhim Vi® by Aventis Pasteur SA for intramuscular injection. The manufacturer-recommended storage temperature is 2–8 °C with a shelf life of 3 years. The vaccine is stable for 6 months at 37 °C, and for 2 years at 22 °C (WHO 2017g, 2018d).

The typhoid conjugate vaccines consist of Vi polysaccharide antigen linked to tetanus toxoid protein (WHO 2018d). Typhibev® (Biological E.) and Typbar-TCV® (Bharat Biotech) are licensed for intramuscular administration of a single dose (WHO 2017g). This type of vaccine can be given to children from the age of 6 months. The shelf life is 2 years at 2–8 °C (WHO 2018d, 2021b). Additional Vi conjugates and new live oral vaccines to prevent typhoid are under development and clinical testing (WHO 2017g; Levine and Simon 2018).

6.13.2 Research on Spray-Dried Typhoid Vaccines

Research studies on spray drying of typhoid vaccines are very scarce (Bejugam et al. 2008; Uddin et al. 2009; Ohtake et al. 2011a). Table 6.15 summarizes the main spray drying parameters and results from laboratory tests.

In a proof-of-concept study, Uddin et al. (2009) encapsulated Typhim Vi® polysaccharide antigen (0.1 and 1 μg/mg) into biodegradable albumin-chitosan microparticles by spray drying. For this purpose, aqueous solutions of 4.5% w/v BSA and 0.5% w/v chitosan were spray-dried using a Mini Spray Dryer B-191 at 110 °C inlet and 73 °C outlet temperature (Bejugam et al. 2008). The production yield was 63–69%, and the mean particle size obtained was approximately 4.2 μm, suitable for incorporating into Peyer's patches and M-cells uptake (Eldridge et al. 1990, 1991) to maximize systemic and mucosal immune responses. In addition, the albumin-chitosan matrix was chosen for its bioadhesive properties and positive zeta potential, which enhances intestinal absorption of the vaccine microparticles. The

Table 6.15 Studies of spray-dried typhoid studies in laboratory scale

Vaccine type	Excipients, composition	Spray drying process conditions	Particle size, morphology, residual moisture	Route of administration	Thermostability	References (country)
Typh	4.5 w/v% bovine serum albumin, 0.5 w/v% chitosan	B-191, T_{in}/T_{out} 110 °C/73 °C, F: 5%, A: 700 NL/h (air), G: 95% aspirator	63–69% process yield, around 4.2 μm particles, crosslinked with 1% glutaraldehyde for 6 h at room temperature	Rats, oral, powder encapsulated in mini-hard gelatin capsules (enteric-coated with Eudragit L 100–55®)	n.a	Bejugam et al. (2008), Uddin et al. (2009) (USA)
Ty21a	7% w/v sucrose	B-190, Sonicair nozzle, T_{in}/T_{out}: n.a./45 °C, F: 0.75 mL/min, A: 15 psi (N2)	Particle size n.a 3.14% residual moisture content	Mice, injection of reconstituted dry powder vaccine	4 weeks (25 °C)	Ohtake et al. (2011a) (USA)

T_{in}/T_{out} inlet/outlet drying gas temperature, *F* feed flow rate, *A* atomization gas flow rate, *G* drying gas flow rate, *n.a.* not available

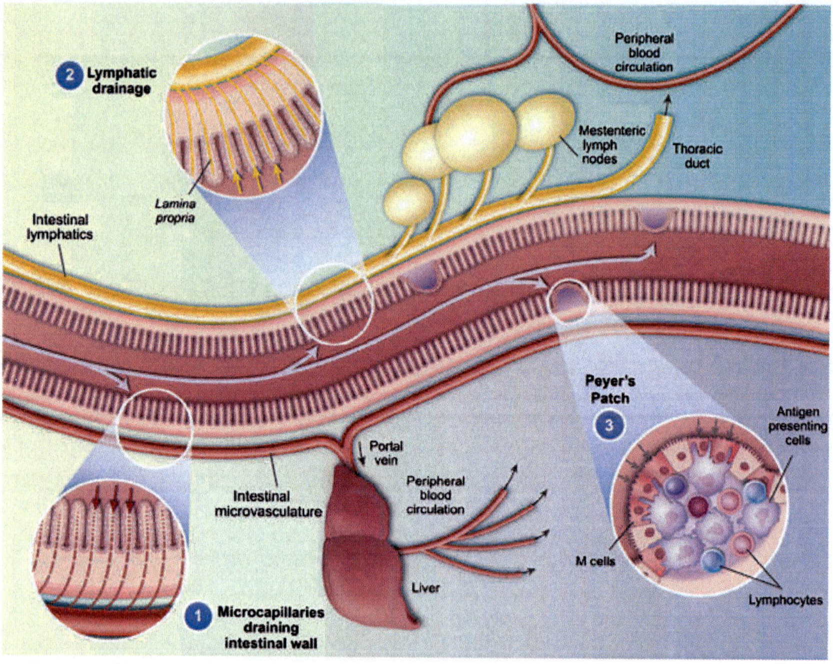

Fig. 6.42 Routes of antigens administered by oral vaccination through the intestine, accessing different parts of the body's immune system. (1) Via microcapillaries into the local and peripheral blood circulation; (2) via the lamina propria and intestinal lymphatics into the mesenteric lymph nodes; (3) direct uptake through M cells into the Peyer's patches (Adapted from New (2019) with permission from John Wiley and Sons)

powder was packaged in enteric-coated mini-hard gelatin capsules and administered orally to rats.

Figure 6.42 shows the general routes of antigens administered by oral vaccination through the intestine leading to different parts of the body's immune system, that is (New 2019):

(1) via microcapillaries into the local and peripheral blood circulation,
(2) via the lamina propria and intestinal lymphatics into the mesenteric lymph nodes, and
(3) direct uptake by M-cells into the Peyer's patches.

The in vivo studies showed that the typhoid vaccine-loaded microparticles were released from the capsules in the small intestine and taken up by M-cells, which induced a multifold increase of antigen-specific IgG and IgA levels 8 weeks after oral immunization (Uddin et al. 2009).

In another study, Ohtake et al. (2011a) investigated foam-drying, freeze-drying, and spray-drying to produce a heat-stable oral live attenuated vaccine against *Salmonella Typhi* (Ty21a). A Mini Spray Dryer B-190 equipped with a

Table 6.16 Comparison of process-associated loss and titer loss after 1 week of storage at 25 °C for spray-dried, freeze-dried, and foam freeze-dried Ty21a vaccine formulations

Drying process	Excipients	Process loss (log_{10} CFU)	Titer loss after 1 week of storage at 25 °C (log_{10} CFU)
Spray drying	7% w/v sucrose	0.7 ± 0.1	1.7 ± 0.1
Freeze drying	28% w/v sucrose, 1% w/v gelatin, 25 mM KPO_4, pH 7	0.8 ± 0.1	0.5 ± 0.0
Foam drying	25% w/v sucrose, 0.5% w/v methionine, 2.4 wt% DMSO, 25 mM KPO4, pH 8	0.3 ± 0.2	0.0 (no loss after 4 weeks at 25 °C)

custom-made Sonicair nozzle (IVEK) was used, atomizing the feed solution at low pressure using sonic energy. The spray drying conditions were 0.75 mL/min feed rate and 45 °C outlet temperature (3.14% residual moisture content) (Table 6.15). Sucrose was used as the main excipient. The drying process-associated losses for the spray-dried, freeze-dried, and foam-dried formulations were 0.8 log_{10}, 0.7 log_{10}, and 0.3 log_{10} CFU, respectively (Table 6.16).

The serum titer losses of mice immunized with dry vaccines stored at 25 °C for 1 week were 0.5 log_{10} and 1.7 log_{10} CFU for the freeze-dried and spray-dried formulations. In contrast, the optimized foam-dried formulation showed no titer loss after 4 weeks of storage at 25 °C and titer stability of about 12 weeks at 37 °C (around 1.0 log_{10} CFU decrease in titer). In addition, foam-dried Ty21a elicited a similar immunogenic response and protection in mice as Vivotif®, the commercial Ty21a vaccine.

The higher titer loss of the spray-dried formulation was explained by the temperature stress and shear stress (i.e., atomization of liquid feed into droplets), leading to some degree of inactivation. On the other hand, the improved storage stability and reduced process loss of the foam-dried vaccine were mainly due to the lower specific surface area and the optimal composition of excipients to produce thermostable formulations.

6.13.3 Conclusions on Spray-Dried Typhoid Vaccines

In conclusion, vaccination is an important preventive measure against typhoid fever. Currently available typhoid vaccines are highly effective but must be stored at 2–8 °C and require a cold chain. Freeze-drying is used for the manufacturing of live attenuated Ty21a vaccine. However, little research has been performed on other drying technologies to extend shelf life. Foam drying offers comparatively higher stability due to the lower specific surface area. Only a few studies have been conducted on spray drying. The produced microparticles of around 5 μm are particularly suitable for oral administration and absorption via M-cells in the Peyer's patches of the intestine. Oral delivery of vaccine powder shows great promise as capsules take up less space than syringes and can be transported in blister packs. In addition, no special medical knowledge is required for administration, and the vaccine can even be taken at home for later self-administration. Without needles, the risk of contamination is eliminated, as is the need for strict safety precautions and controlled waste disposal. Optimizations to powder vaccine formulations are possible to increase shelf life.

6.14 Diphtheria, Tetanus, Pertussis, and Hib Vaccines

6.14.1 A Short Introduction to Diphtheria, Tetanus, Pertussis, and Hib

Vaccine-preventable diseases also include diphtheria, tetanus, pertussis, and Haemophilus influenzae type b (often simply called Hib) (WHO 2020e). According to the WHO monitoring system, there were approximately 145,000 reported cases of pertussis, 23,000 cases of diphtheria, and 15,000 cases of tetanus worldwide in 2019 (WHO 2020e). For Haemophilus influenzae type b (Hib), estimates indicate 340,000 severe infections and 29,800 deaths in 2015 (Wahl et al. 2018; Slack et al. 2020).

- Diphtheria is a highly contagious upper respiratory tract disease caused by the gram-positive *Corynebacterium*, which is transmitted from person to person by droplet infection during coughing and sneezing (Fig. 6.43a). It usually affects the throat, causing pain and swelling that can lead to choking. The bacteria also produce toxins that can damage the heart, kidneys, and nerves (WHO 2017h).
- Tetanus, often called lockjaw, is a life-threatening infectious disease caused by *Clostridium tetani*'s bacterium (Fig. 6.43b). The resistant spores of the bacterium are found almost everywhere, even in road dust or garden soil. If they penetrate wounds, they can multiply and form a toxin that attacks and damages the muscle-controlling nerve cells, causing the typical paralysis and muscle spasms. However, there is no direct infection from person to person (WHO 2017i).
- Pertussis, also known as whooping cough, is a highly contagious respiratory infection caused by the bacterium *Bordetella pertussis* (WHO 2015b). As such, pertussis is highly infectious with an R0 in the range of 15–17, meaning that one primary case can cause up to 17 secondary cases (Heininger 2012). Transmission of the disease occurs via respiratory through droplets produced by coughing or sneezing (Mattoo and Cherry 2005). The disease can occur at any age but is most dangerous in infants and young children. Antibiotics are used to treat the infection, often clarithromycin or azithromycin (Heininger 2012).
- *Haemophilus influenzae* is a gram-negative coccobacillus commonly found in the nasopharynx of many people (WHO 2013b). Infection from person to person

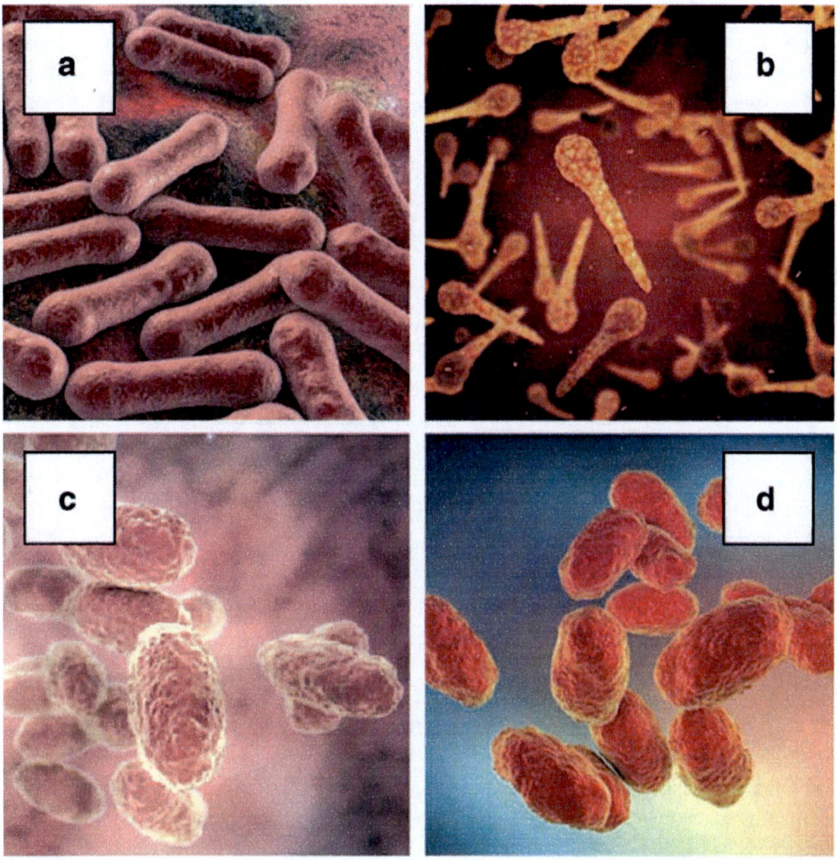

Fig. 6.43 a *Corynebacterium diphtheriae*, b *Clostridium tetani*, c *Bordetella pertussis*, d *Haemophilus influenzae* (Images from www.shutterstock.com, #1889038198, #570888940, #589837346, #595947080)

occurs through droplet infection, e.g., when sneezing, talking, and coughing. Both healthy and ill persons can transmit *Haemophilus influenzae*. Seven different serotypes are known (serotypes a–f). While unencapsulated bacillus strains are relatively harmless, strains with a polysaccharide capsule or shell cause more severe disease. The polysaccharide is the most crucial virulence factor. Of the various types of capsules, type b is responsible for more than 90% of systemic infections. Hib primarily causes pneumonia and meningitis in young children.

Vaccination is the most effective public health intervention for reducing infections (WHO 2013b, 2015b, 2017i, 2017h). The diphtheria-tetanus-pertussis vaccine is routinely administered in infancy and often combined with Hib and HepB.

According to the list of WHO prequalified vaccines, there are 33 DTP vaccines with combinations of HepB and Hib (WHO 2021a). Of these, some dosage forms consist of freeze-dried Hib vaccine that is reconstituted with liquid DTPHepB as a diluent before use. Manufacturers are the Serum Institute of India Pvt. Ltd. (India) and Biological E. Limited (India).

In 2019, approx. 85% of the global target population (about 116 million infants) had received the recommended 3 doses of the DTP3 vaccine. Despite this, 19.7 million infants had still not received their DTP3 dose, indicating a lack of access to immunization (WHO/UNICEF 2020). The goal is to achieve 90% vaccination coverage in the coming years. Global coverage with 3 doses of the Hib vaccine (Hib3) is currently estimated at around 70% (WHO/UNICEF 2020).

Despite the remarkable increase in vaccination coverage rates over the past 10 years, several challenges require the development of improved vaccines. First, vaccines are administered in liquid form by injection via syringes, leading to potential contamination and reuse of needles in parts of the world with mass vaccination. Second, current DTP vaccines must be stored at 2–8 °C, which requires an uninterrupted cold chain (WHO 2006, 2021b; PATH 2012; FDA 2022).

If vaccines were temperature stable, this would be a major advantage for global vaccine distribution, storage, and use. WHO is therefore seeking approval of priority CTC vaccines, including tetanus toxoid-containing vaccines (WHO 2017d).

6.14.2 Research on Spray-Dried Diphtheria, Tetanus, Pertussis, and Hib Vaccines

Dry and encapsulated vaccines in the presence of stabilizing excipients are a promising way to increase the thermostability of vaccines. In this area, research is underway further to develop a technology platform for dry powder vaccine delivery.

For example, Maa et al. (2003) used spray freeze-drying to produce powdered diphtheria and tetanus vaccines encapsulated in trehalose/glycine/dextran suitable for epidermal administration. Another example is supercritical spray drying an aqueous solution of diphtheria toxoid in dextran or chitosan into a supercritical mixture of CO_2 and ethanol for pulmonary delivery of dry powder vaccines (Amidi et al. 2007). In addition, there are a few studies on spray-dried diphtheria, tetanus, pertussis, and Hib vaccines. Table 6.17 summarizes the main experimental studies.

Some attempts have been made to spray-dry biodegradable PLGA microparticles as vaccine delivery systems (Gander et al. 1995a, b, 1996; Johansen et al. 1999a, b, 2000; Boehm et al. 2002; Muttil et al. 2010b). The results demonstrated the technological feasibility of microencapsulating vaccines in biodegradable polymers by spray drying and also showed that this was possible on a larger scale. Aqueous solutions of tetanus and diphtheria toxoid vaccine with trehalose and bovine serum albumin were dispersed in an organic phase of ethyl formate or dichloromethane containing PLGA and spray-dried in a Mini Spray Dryer B-190 or B-191 (Gander et al. 1995a, b, 1996; Johansen et al. 1999a, b, 2000). The inlet drying temperature

Table 6.17 Experimental studies on spray dried diphtheria, tetanus, pertussis, and Hib vaccines

Vaccine type	Excipients	Spray drying process parameters	Particle size, morphology, residual moisture, yield	Animal model, administration route	Thermo-stability	References
Pertussis (OMV from B. pertussis B1917)	Trehalose (10% w/v), 20 mL samples	B-290 with HP cyclone and B-296, T_{in}/T_{out} 135 °C/n.a., F: 10% (3.4 mL/min), A: 12.4 L/min (N2), G: 22 m3/h	5.1 µm (geometric mean), spherical, amorphous, RMC 2.3%	Mice, reconstituted powder pulmonary	4 weeks (40 °C and 65 °C)	Kanojia et al. (2018a)
Diphtheria (CRM-197 antigen)	PLGA nanoparticles (200 ± 50 nm), L-leucine in water (PLGA nanoparticles/L-leucine ratio 25:75)	GEA Niro Atomizer T_{in}/T_{out}: 95 ° C/38 °C, F: 1.8 L/h, G: 100 kg/h (air)	7 µm (volumetric mean), hollow, low-density, FPF 50% (< 5.8 µm), RMC < 1%, 87 ± 5% loading efficiency	Guinea pigs, powder insufflation	n.a	Muttil et al. (2010b)
Hib, DT, TT, PT (tetravalent)	Trehalose/BSA in water, PLA or PLGA (RG502H) in ethyl formate or DCM	B-190, T_{in}: 38–39 °C, F: 2 mL/min	0.5–5 µm, 10% (Hib), 30% (PT), and > 80% (DT, PT, Hib) loading efficiencies	Guinea pigs, reconstituted powder injected subcutaneously	n.a	Boehm et al. (2002)
Tetanus toxoid	Trehalose/BSA in water, PLGA (RG502H) in ethyl formate or DCM	B-191, T_{in}: 45 °C, F: 2.2 mL/min, A: 550 L/min (air)	3.9–6.1 µm, 5–10 g powder batches, 32% loading efficiency, 56–69% yield	n.a	n.a	Johansen et al. (2000)
Diphtheria and tetanus toxoid	Trehalose, PLGA (RG502 and RG752) or PLA (R202 and R206) in ethyl formate or DCM	B-190, T_{in}/T_{out}: 37–45 °C/33–40 °C (solvent depending), F: 2–5 mL/min, A: 400–500 L/h	1–10 µm, (3.4 µm (mean), 32–36% loading efficiency, 50–70% yield	Guinea pigs, reconstituted powder injected subcutaneously	n.a	Gander et al. (1995a, b, 1996), Johansen et al. (1999a, b)

BSA Bovine serum albumin, *DCM* Dichloromethane, *DT* Diphtheria, *FPF* Fine particle fraction, *Hib* Haemophilus influenzae type b, *MMAD* Mass median aerodynamic diameter, *n.a.* not available, *OMV* Outer membrane vesicles, *RMC* Residual moisture content, *TT* Tetanus toxoid, *PT* Pertussis, *PLA* Poly(lactic acid), *PLGA* Poly (lactic-co-glycolic acid) was set at 38–45 °C (depending on the solvent). The efficiency of vaccine loading ranged from 32 to 36%. Immunized guinea pigs showed specific and long-lasting protective antibody responses over 40 weeks, comparable to alum-adjuvanted toxoid.

Boehm et al. (2002) examined the feasibility of spray-drying tetravalent vaccines against Hib, DT, TT, and PT in biodegradable PLA and PLGA. The combination of antigens in spray-dried microparticles improved loading efficiency. Moreover, after a single administration of the microspheres in guinea pigs, high and durable serum titers of antibodies against all four antigens were detected. Multivalent vaccines have the advantage of improving vaccination coverage by reducing the number of vaccinations required for immunization.

Some years later, Muttil et al. (2010b) demonstrated pulmonary immunization of guinea pigs with spray-dried powder vaccine against diphtheria. For this purpose, diphtheria CRM-197 antigen was first encapsulated in PLGA nanoparticles (200 ± 50 nm) by emulsification methods and then spray-dried in aqueous suspensions containing dissolved L-leucine in a Niro atomizer (USA). The inlet and outlet temperatures were 95 and 38 °C, the feed rate was 1.8 l/h, and the airflow rate was 100 kg/h. Dry porous particles of about 7.0 μm with fine particle fractions of about 50% (< 5.8 μm) were obtained. Finally, the dry powder formulations were administered to anesthetized guinea pigs by insufflation (DP-4 Insufflator, Penn Century, USA) (approximately 10 mg powder with 20 or 40 μg vaccine). The immune response elicited in the lungs was determined by serum IgG and BAL-IgA and was significantly higher than when the same vaccine was administered intramuscularly. It was concluded that the generated porous microparticles penetrated the deep lung, collapsed, and released the incorporated nanoparticles with a high surface-to-volume ratio in the alveolar region. Therefore, the lung may be a suitable delivery route for novel dry powder vaccines, offering clinical and pharmaceutical advantages by eliminating the need for needles in a mass vaccination program.

More recently, Kanojia et al. (2018a) developed a spray-dried pertussis vaccine formulation for pulmonary immunization based on outer membrane vesicles (OMVs from *B. pertussis* B1917) and trehalose as a stabilizer. OMVs are non-replicative proteins naturally produced by the outer membrane of *Bordetella pertussis* and consist of a broad spectrum of immunogenic antigens. Samples of 20 mL (89 μg/mL protein concentration) were spray-dried using a Mini Spray Dryer B-290 with high-performance cyclone and Dehumidifier B-296 in a closed-loop configuration using nitrogen as the drying gas. Spherical particles of 5.1 μm average size were produced with amorphous structure as shown by SEM and XRD analyses (Fig. 6.44). The spray drying process did not affect the structural integrity and particle size of omvPV.

The powder was stable at elevated temperatures (4 weeks at 40 and 65 °C) as confirmed by analyses using bioassay, particle size, electrophoresis, and fluorescence spectroscopy (Kanojia et al. 2018a). Mice were immunized twice with a 4-week interval (day 0 and 28) by pulmonary administration of the reconstituted spray-dried powder or liquid vaccine. Pulmonary delivery was performed using a MicroSprayer aerosolizer supplied with a high-pressure syringe (IA-1C and

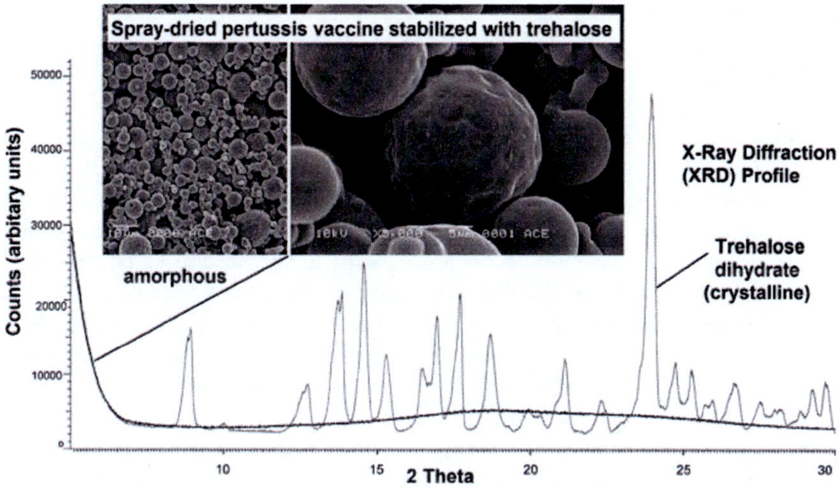

Fig. 6.44 SEM images and X-ray diffraction (XRD) profile of spray-dried OMV pertussis vaccine stabilized with trehalose. The spray-dried vaccine is amorphous (no peaks) compared to crystalline trehalose dihydrate (Adapted from Kanojia et al. (2018a) with permission from Elsevier)

FMJ-250, Penn-Century, USA), as described by Bivas-Benita et al. (2005) and Raeven et al. (2018). Using this method, the vaccine was sprayed into the lungs via the endotracheal route, and the generated droplet size was around 8 μm (MMAD), which is suitable for deposition in the large airways (Bivas-Benita et al. 2005). The reconstituted spray-dried omvPV vaccine showed comparable immune response and protection against challenge with live B. pertussis as the liquid vaccine also administered pulmonary. Overall, the storage stability was improved by spray drying while maintaining effective immunogenic properties.

6.14.3 Conclusions on Spray-Dried DTP/Hib Vaccines

The reviewed studies show that spray drying is a promising technology for powder vaccine production. Microencapsulation in PLGA, trehalose, and L-leucine proved to be a good choice for stabilizing diphtheria, tetanus, pertussis, and Hib vaccines. The powder form enables improved and extended storage stability at higher temperatures than the liquid formulation, greatly benefiting global distribution outside the cold chain and the use of vaccines. A significant advantage of spray drying is the ability of particle design (e.g., encapsulated nanoparticles with vaccine) to develop dry powder vaccines suitable for pulmonary delivery, as animal studies in guinea pigs and mice have shown. Direct inhalation of a spray-dried powder may even bypass the logistics of reconstitution. However, to the author's knowledge, no spray-dried DTPHepBHib formulation has yet reached clinical trials.

6.15 Rotavirus Vaccines

6.15.1 A Short Introduction to Rotavirus Vaccines

Rotavirus is a member of the reovirus family that causes watery diarrhea, vomiting, and severe dehydration in young children (WHO 2018e). In 2016, rotavirus infections were responsible for more than 258 million episodes of diarrhea among children younger than 5 years (Crawford et al. 2017; Troeger et al. 2018). Vaccination is considered the most promising solution to reduce disease burden. Estimates show that the number of rotavirus deaths declined from 528,000 to 128,500 from 2000 to 2016 due to the introduction of rotavirus vaccines (Tate et al. 2016; Troeger et al. 2018; Burke et al. 2019). Today, global rotavirus vaccine coverage is estimated at 46%, and 114 countries have introduced the rotavirus vaccine in their program (WHO 2021d).

Available vaccines against rotavirus are live attenuated rotavirus strains of human or animal origin that replicate in the human intestine to elicit an immune response (WHO 2021e). Table 6.18 summarizes the orally administered vaccines prequalified by the WHO for global use (WHO 2021b). The list includes RotaTeq® (a pentavalent human-bovine reassortant vaccine), Rotarix® (a human monovalent G1P[8] vaccine), Rotavac®, ROTASIIL®, ROTASIIL®Thermo, ROTASIIL-Liquid®, and pentavalent ROTAVAC 5D®. Each vaccine has a unique

Table 6.18 Characteristics of currently prequalified live oral rotavirus vaccines (WHO 2021b)

Commercial name	Manufacturer	Prequalified	Pharmaceutical form	Presentation	Shelf life (storage temperature)
RotaTeq®	Merck	07/10/2008	Liquid	Plastic tube	24 months at 2–8 °C
Rotarix®	GSK	12/03/2009	Oral suspension	Plastic tube, applicator	24 months at 2–8 °C
Rotavac®	Bharat Biotech	05/01/2018	Frozen liquid	Vial	60 months at −20 °C 6 months at 2–8 °C
ROTASIIL®	Serum Institute of India	21/09/2018	Lyophilized	Two vial set	30 months at 2–8 °C
ROTASIIL®Thermo	Serum Institute of India	28/01/2020	Lyophilized	Two vial set	30 months at 25 °C
ROTASIIL-Liquid®	Serum Institute of India	18/02/2021	Liquid	Plastic tube	24 months at 2–8 °C
ROTAVAC 5D®	Bharat Biotech	18/06/2021	Liquid	Vial	24 months at 2–8 °C

presentation as a preformed plastic tube, glass vial, or two vial set (Zehrung 2017, 2018). In addition, there are several other rotavirus vaccine candidates nationally licensed or in the development pipeline (Burke et al. 2019; Kirkwood et al. 2019).

Regarding shelf life, the ready-to-use liquid formulations require a storage and distribution temperature of 2–8 °C to ensure stability and potency (Table 6.18). In contrast, the lyophilized ROTASIIL®Thermo developed at the Serum Institute of India Pvt Ltd offers a shelf life of 30 months at 25 °C. In addition, short-term excursions of up to 37 °C during transport (< 1 week) do not affect the vaccine. Still, the WHO recommends the vaccine be stored below 25 °C, and countries should ensure adequate conditioned facilities with temperatures not exceeding 25 °C (WHO 2021b).

Naik et al. (2017) showed that lyophilized ROTASIIL® is stable for 36 months at 25 °C and 18 months at 40 °C. Sucrose and glycine, each dissolved at 5% (w/v) in Eagle's MEM (Minimum Essential Medium) (Eagle 1959) with Hank's salts (Hanks and Wallace 1949), was found to be optimal excipient composition to stabilize the vaccine during freeze-drying and subsequent storage.

In another study, Madan et al. (2018) developed a heat-stable lyophilized rotavirus vaccine based on the serotypes constituting RotaTeq® and Generally Regarded As Safe (GRAS) excipients, including buffer (HEPES), protectant (sucrose), bulking agent (PVP), salt (NaCl), amino acids (L-arginine, glycine), activating agent ($CaCl_2.2H_2O$ and $ZnCl_2$), and dispersant (Tween 20). As a result, the heat stability profile could be significantly improved to over 36 months at 5 °C, 20 months at 37 °C, and 7 months at 45 °C, which was comparable to lyophilized ROTASIIL® (Naik et al. 2017).

Although thermostable lyophilized rotavirus vaccines are available, there is still research activity focusing on the continued development of alternative formulations, especially the development of heat-stable vaccines that could be transported and stored outside cold chain for an extended period to improve access and coverage in low and middle-income countries (Levine 2011).

6.15.2 Research on Spray-Dried Rotavirus Vaccines

Only a few studies Gill (2014, 2018), Ohtake et al. (2015) and patents (Gill et al. 2017, 2020) have been published on spray-dried rotavirus vaccines. Figure 6.45 shows that spray-dried formulations of live rotavirus vaccines are more stable at room temperature (25 °C) than their liquid counterparts under optimal formulation and processing conditions (Ohtake et al. 2015). Further development involves incorporating stabilized spray-dried rotavirus vaccine powder into a rapidly dissolving oral thin film (Aridis Pharmaceuticals, LLC). The thin film is mucoadhesive, dissolves in seconds, and is designed to deliver the vaccine to the oral cavity efficiently. It is expected that the ease of administration and rapid film dissolution will improve patient compliance, which in turn would facilitate mass vaccinations.

Gill et al. (2017, 2020) from the Hilleman Laboratories (headquartered in New Delhi, India) presented an invention for improved thermostabilization of rotavirus

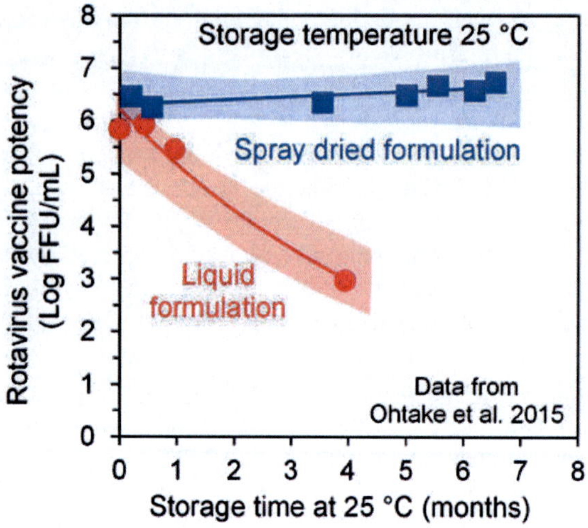

Fig. 6.45 Vaccine potency of spray-dried live rotavirus vaccine formulation compared to liquid formulation showing improved room temperature stability at 25 °C (Data adapted from Aridis Pharmaceuticals LLC, USA, published in Ohtake et al. (2015) with permission from John Wiley and Sons)

Table 6.19 Optimized process parameters that support powder yields > 60% and residual moisture contents < 2%

Process parameters	Optimal values
Inlet temperature	70–75 °C
Outlet temperature	43–48 °C
Drying gas flow (aspirator)	30 Nm^3/h (70–75%)
Feed rate	0.4 mL/min
Spray gas flow	30 mm (Nitrogen or Argon)
Atomization pressure	5 bar
Two-fluid nozzle diameter	1 mm
Feed solids concentration	21 g/100 mL (21% w/v)
Spray dryer model	Büchi Mini Spray Dryer B-290 advanced with HP cyclone and Dehumidifier B-296
Powder yield	> 60%
Residual moisture content	< 2%
Particle size	1–24 µm (D10 to D90), ~ 11 µm (D50)

vaccine by spray drying. Table 6.19 shows the optimized process parameters using a Büchi Mini Spray Dryer B-290 with HP cyclone and Dehumidifier B-296. Inlet and outlet temperatures were between 70 and 75 °C and 43 and 48 °C, respectively, particle size between 1 and 24 µm, powder yield > 60%, and residual moisture content < 2% (Gill et al. 2020). Using DOE, the formulation was optimized with excipients such as trehalose, maltodextrin, NaCl, glutamic acid, histidine, CaCl2, and HEPES buffer. Vaccine potency (loss < 0.5 log) was maintained at 45 °C for 4 weeks, 37 °C for 24 weeks, and 2–8 °C for 24 months and met the WHO criterion for maximum heat stability (Vaccine Vial Monitor VVM 30 meaning 30 days at 37 °C) (WHO/UNICEF 2014).

Along with improved thermostability of the spray-dried rotavirus vaccine, the choice of primary delivery packaging is a key component of optimization. An upcoming technology is integrated reconstitution by physically separating the dry powder (e.g., spray-dried or milled lyophilized powder) and diluent in two chambers (Gill 2014, 2018; Zehrung 2018). Figure 6.46 shows single-dose containers in a two-component prepack format. The two-component pack (a) enables a packaging density of more than 150,000 doses per cubic meter (Gill 2014) and thus saves volume in the cold chain. Figure 6.46b illustrates the design of a dual-chamber pouch with frangible sealing technology for oral administration of vaccine powder or granules and diluent. Figure 6.46c shows the Vaxipak® technology, a low-cost dual-chamber blister device for integrated reconstitution and needle injection (Gill 2018).

According to Gill (2018), Hilleman Laboratories has pilot-scale machines for the dual-chamber device (VaxiPak®) with an output of up to 300 blisters per hour, resulting in an annual production capacity of about 2.5 million single doses. Preclinical development with heat-stable rotavirus vaccine is completed, and Phase I/II clinical trials with adults and children were conducted in Bangladesh from June 2016 to April 2017. No serious adverse safety events were reported, and infant immunogenicity results exhibited similarity to RotaTeq®. Hilleman Laboratories is open to partnerships and collaborations.

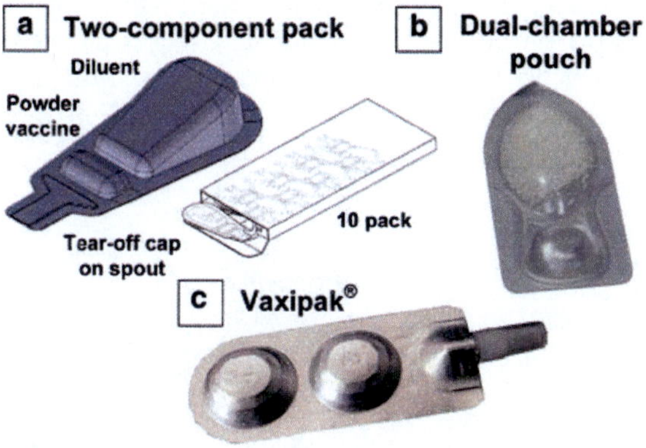

Fig. 6.46 **a** Spray-dried or milled lyophilized powder and diluent in a two-component pack and a 10 pack with a dimension of 130 × 45 × 8 mm enabling a packaging density of > 150,000 doses per cubic meter, **b** Dual-chamber pouch with a frangible seal for oral delivery and integrated reconstitution of vaccine powder or granules and diluent, **c** Dual-chamber blister device VaxiPak® for integrated reconstitution and needle injection (Images a and b adapted from Gill (2014) and c from Gill (2018), *photo credits* Hilleman Laboratories, New Delhi, India)

6.15.3 Conclusions on Spray-Dried Rotavirus Vaccines

The few studies on spray-dried rotavirus vaccines suggest that high heat stability is achieved under optimal formulation and processing conditions, allowing storage outside the cold chain for several months. In addition, the development of single-dose dual-chamber containers with an integrated reconstitution function for powdered vaccines and diluent makes administration easier, safer, more affordable, and reduces packaging size.

6.16 Poliovirus Vaccines

Poliomyelitis (often just called polio) is an acute communicable disease caused by any of the three poliovirus serotypes (type 1, 2, or 3) that largely affects children under 5 years of age (WHO 2016b). Poliovirus is a member of the Picornaviridae family and genus Enterovirus. It is a small virus (30 nm diameter) with no surrounding membrane envelope (Fig. 6.47). Its genome is a 7.5 kb single positive-sense strand of RNA, surrounded and protected by the protein capsid (van den Pol 2013).

The poliovirus attacks the nervous system and can cause paralysis or muscle weakness, most commonly in the legs (WHO 2019c). The virus is transmitted from person to person mainly through the fecal–oral route or, less commonly, through contaminated water or food and multiplies in the intestine (WHO 2019c).

The development of effective vaccines to prevent polio was one of the major medical breakthroughs of the twentieth century. The Global Polio Eradication Initiative (WHO 2021f) uses two types of vaccines to stop polio transmission, i.e., inactivated polio vaccine (IPV) and oral polio vaccine (OPV). OPV is administered as 2 drops (about 0.1 mL) directly into the mouth. IPV is produced from selected wild poliovirus strains like Mahoney or Brunhilde (type 1), MEF-1 (type 2), Saukett (type 3), or Sabin strains (WHO 2016b). In 2020, 83% of infants worldwide received three doses of polio vaccine (WHO 2021d).

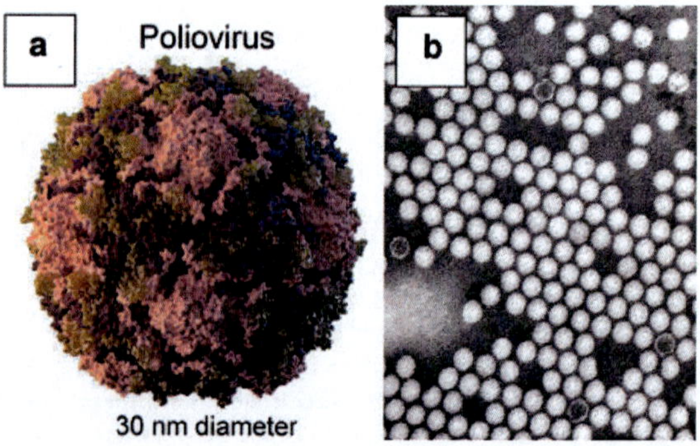

Fig. 6.47 As an outlook 3D presentation of the poliovirus (*Image Source* CDC Public Health Image Library (PHIL), ID#: 22498, *Credit* Meredith Boyter Newlove, M.S.)

Globally, the number of cases caused by wild poliovirus has declined by more than 99% since 1988, from an estimated 350,000 in more than 125 endemic countries to only 33 reported cases in 2018 (WHO 2019c). In 2015 and 2019, wild poliovirus types 2 and 3 were declared as eradicated globally (WHO 2021f). However, wild poliovirus type 1 continued to be detected in parts of Afghanistan and Pakistan, the last endemic reservoir in the world (WHO 2021f).

A drawback of both OPV and IPV, which are marketed as liquid vaccines, is the requirement of a cold chain for their transport and storage (Kanojia et al. 2018b). OPV is highly heat-sensitive and must be kept frozen for long-term storage between −15 and −25 °C (Galazka et al. 1999; WHO 2006; Chen and Kristensen 2009) or, after thawing, at temperatures between +2 and +8 °C for a maximum of 6 months (WHO 2016b). Furthermore, OPV may lose 4–13% of their activity per day at 25 °C, 11–21% per day at 31 °C, and 26–34% per day at 37 °C (Galazka et al. 1999; Hill et al. 2016).

On the other hand, IPV is available either as a stand-alone product or in combination with one or more other vaccine antigens, including DTP, hepatitis B, or Hib (WHO 2016b). When combined with an adjuvanted vaccine, IPV administration is by intramuscular injection. Currently, commercial IPV is formulated as an aqueous solution and is stable for over a year at 4 °C. However, it loses potency in only weeks at 25 or 37 °C (Chen and Kristensen 2009).

Hence, a thermostable IPV formulation is highly desirable to eliminate or at least reduce cold chain requirements during storage and transportation and for use in emergency vaccination and post-eradication stockpiling.

A strategy for stabilizing vaccines is drying in the presence of stabilizing excipients. Therefore, various studies have focused on screening excipients that preserve the vaccine potency. Among others, Kraan et al. (2014) and Qi et al. (2018) studied lyophilized IPV vaccine formulations.

Kraan et al. (2014) converted trivalent IPV into a dry product using freeze-drying (freezing: 2 h at −50 °C, primary drying: 3 h at −45 °C, secondary drying: 24 h at 25 °C, 0.01 mbar). The optimized formulation consisted of 10% (w/v) sorbitol, 8.5% (w/v) MgCl2, and 8.5% (w/v) monosodium glutamate and achieved thermal stability up to 24 weeks at 25 and 37 °C and up to one month at 45 °C.

Qi et al. (2018) showed that the addition of moderate concentrations of urea (e.g., 0.4 M) improved the process and storage stability of freeze-dried IPV serotype 3 vaccines in the presence of sucrose (freezing: 2 h at −45 °C, primary drying: 32 h at −35 °C, secondary drying: 3 h at 30 °C, 0.04 mbar). Urea appeared to improve the preservation of vaccine efficacy during freeze-drying and rehydration by inhibiting the agglomeration of viron particles and preventing conformational changes of capsid proteins that translate into reduced thermal stability after processing.

Studies on spray-drying polio vaccine are scarce, although spray-drying is an attractive alternative to freeze-drying because spray-dried powders open new routes of administration for vaccines, particularly via the lungs and mucous membranes. However, the formulation compositions must be carefully selected to maintain efficacy during and after spray-drying.

The only published study on spray-dried polio vaccine is this one from Kanojia et al. (2018b) in The Netherlands. Kanojia et al. (2018b) developed a spray-dried Sabin inactivated polio vaccine (sIPV) to minimize loss of native conformation (D-antigen) during drying and subsequent storage for 7 days at 40 °C. Table 6.20 summarizes the main experimental results.

For the experimental work a Mini Spray Dryer B-290 with HP cyclone and Dehumidifier B-296 in closed-loop configuration was used. The inlet and outlet drying gas temperatures were kept constant at 110 °C and between 48 to 53 °C, respectively.

Different excipient combinations were extensively screened using DOE studies and factorial experimental designs, including the following compounds:

- non-reducing disaccharide (i.e., trehalose),
- polysaccharide (i.e., maltodextrin),
- divalent metal ion (i.e., $MgCl_2$ magnesium chloride),
- amino acids (i.e., monosodium glutamate, L-arginine, L-glutathione, and glycine),
- a sodium salt of an organic acid (i.e., sodium citrate),
- cell culture medium (i.e., M199), and
- surfactant (i.e., Pluronic®).

The authors recommended to spray-dry the different sIPV serotypes separately and then mix them to obtain a trivalent vaccine powder. Figure 6.48 shows the effects of various excipients on the D-antigen recovery (measure of native conformation) for sIPV serotypes 1, 2, and 3 after spray drying and storage at 40 °C for 1 week.

For serotype 1, trehalose (20% w/v) and monosodium glutamate (MSG) (8% w/v) resulted in 98% D-antigen recovery, which was fully maintained during storage at 40 °C for 1 week. The combination of maltodextrin (20% w/v) and arginine (8% w/v) resulted in 97% D-antigen recovery after spray drying in serotype 2. The positive contribution of arginine was attributed to the reduction of protein–protein interactions, which reduced aggregation. A slight decrease of 18% in D-antigen recovery was seen during 1 week of storage at 40 °C. Serotype 3 exhibited the most prominent loss in D-antigen recovery (56%), which was speculated to depend on the intrinsic thermal instability of sIPV serotype 3 (formalin-inactivated virus). The best formulation contained cell culture medium (M199 medium), glutathione, and maltodextrin. In contrast to freeze-drying, magnesium chloride ($MgCl_2$) was not a beneficial excipient candidate to stabilize sIPV during spray-drying and storage (Kraan et al. 2014).

Overall, the study of Kanojia et al. (2018b) demonstrated the feasibility and limitations of spray drying sIPV in a tailored formulation for each sIPV serotype 1, 2, or 3. As an alternative to freeze-drying, the spray-dried powder formulation offers opportunities for vaccine administration via alternative routes such as intranasal or sublingual, as well as via the lungs and mucous membranes. In addition, a thermostable dry powder vaccine enables stockpiling and distribution in developing countries without a cold chain.

Table 6.20 Process conditions and main results of spray-dried polio vaccine

Vaccine type	Excipients, composition	Spray drying process parameters	Particle size, residual moisture	Administration route	Shelf life, thermostability	References
Poliomyelitis (serotypes 1, 2, and 3)	Trehalose, monosodium glutamate, maltodextrin, arginine cell culture medium M199, Pluronic	Mini Spray Dryer B-290 with HP cyclone, 0.7 mm two-fluid nozzle, Dehumidifier B-296, T_{in}/T_{out}: 110 °C/48–53 °C, F: 3.1 mL/min, A: 12.4 L/min (5 bar, 40 mm) G: 22 m3/h (N2, 100%)	5–7 μm median size, 1–7% RMC	Not investigated	1 week (40 °C)	Kanojia et al. (2018b)

Data based on Kanojia et al. (2018b), open access under CC-BY license
A Spray gas flow rate, *G* Drying gas flow rate, *F* Feed rate, *RMC* Residual moisture content

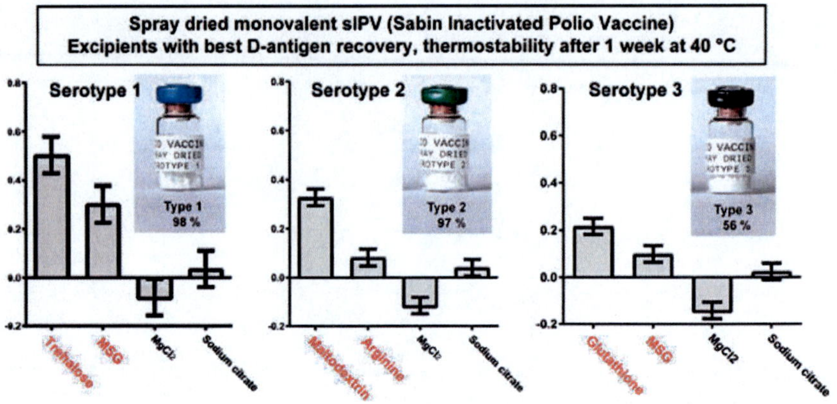

Fig. 6.48 Coefficient plots (D-antigen recovery after spray drying) showing the main excipient effects contributing to the best-fitted models for Sabin Inactivated Polio Vaccine (sIPV) serotypes 1, 2, and 3 (MSG: monosodium glutamate) (Data adapted from Kanojia et al. (2018b), open access under CC-BY license)

6.17 Vesicular Stomatitis Virus Vaccines

Vesicular stomatitis virus (VSV) is a bulled-shaped, enveloped, and single-stranded RNA virus that belongs to the Rhabdoviridae family, the same family as rabies viruses (Fig. 6.49a). VSV transmission occurs through insect bites and can result in painful lesions in horses, cattle, swine, and camelids with symptoms similar to foot and mouth disease (Green 1993; Rodríguez 2002; Peck et al. 2020). VSV infections also occur in humans, although much less frequently than in animals and with much less severe disease symptoms, usually a mild flu-like illness (Timoney 2016; Peck et al. 2020).

For vaccine development, the VSV has proved to be a favorite viral vector platform for infectious diseases (Zemp et al. 2018; Scher and Schnell 2020), including pulmonary tuberculosis (Roediger et al. 2008; Zhang et al. 2017), HIV (Rose et al. 2001), and most successfully Ebola (Geisbert and Feldmann 2011; Wong et al. 2018; Wells et al. 2019), paving the way for future applications. The major benefits of the VSV-vectored platform include (Rose and Clarke 2015; Clarke et al. 2016; Fathi et al. 2019):

- rapid production of high titers and its propagation in almost all mammalian cells,
- lack of reassortment and corresponding potential to undergo a genetic shift in vivo,
- inability of the vector's viral RNA to integrate into the host genome,
- ease of genetic modification with the potential to incorporate one or more antigens,

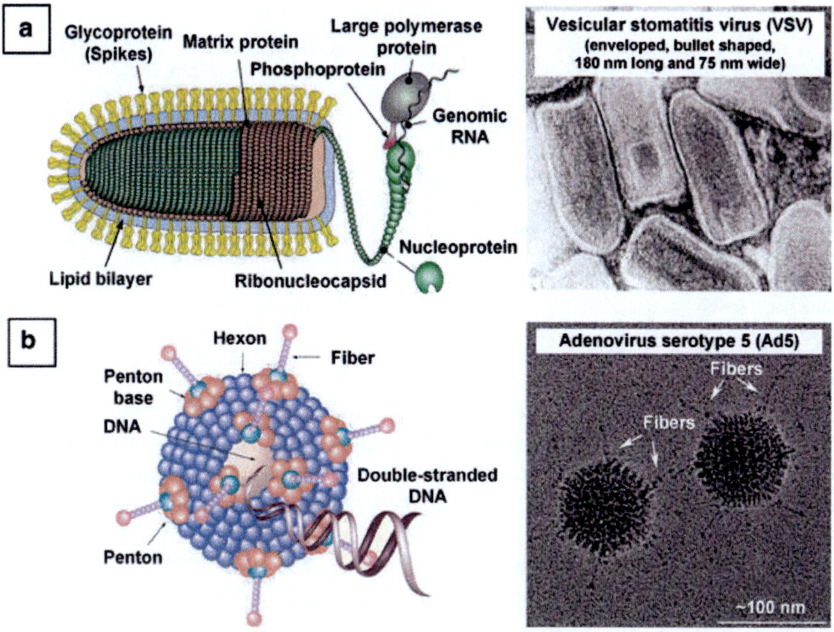

Fig. 6.49 **a** Structure and electron micrograph of vesicular stomatitis virus (VSV). The bullet-shaped, 180 nm long, and 75 nm wide viron capsid is enveloped with a lipid bilayer and glycoprotein (Images adapted from Wikimedia Commons, open access under CC BY-SA 4.0 license in the public domain). **b** Structure and cryo-electron micrograph of adenovirus serotype 5 (Ad5). White arrows highlight the fibers protruding from the vertices of the non-enveloped viron capsid, scale bar approximately 100 nm (Adapted from Goosney and Nemerow (2003) and Rein et al. (2006) with permission from Elsevier and Future Medicine)

- low seroprevalence in the human population,
- mild pathogenicity in humans, and
- induction of humoral and cellular immune responses.

However, VSV-vectored vaccines are typically stored in aqueous media and require refrigeration to maintain their long-term efficacy (Toniolo et al. 2020). Therefore, research is ongoing in developing thermally stable and immunologically active vaccines to mitigate the need for cold-chain storage of liquid vaccines.

In the field of spray drying, Toniolo et al. (2019, 2020) investigated the differences in thermal stabilization of spray-dried viral vectors classified as enveloped (influenza and VSV) and non-enveloped (human adenovirus serotype 5, AdHu5) ((Fig. 6.49b). A Mini Spray Dryer B-290 with HP cyclone was operated at 110 °C inlet temperature using 4 wt% solutions containing trehalose, dextran, mannitol, and blends as excipients (Table 6.21). The viral activity and RNA viability were assessed by accelerated stability studies at 37 °C for up to 30 days.

Table 6.21 Process conditions and main results of spray-dried VSV-vectored vaccine

Vaccine type	Excipients, composition	Spray drying process parameters	Particle size, residual moisture	Administration route	Shelf life, thermostability	References
Vesicular stomatitis virus (VSV) vector	Trehalose, dextran, mannitol, blends, 4 wt% feed solutions	B-290 with HP cyclone T_{in}/T_{out}: 110 °C/n.a A: 439 L/h F: 234 mL/h	n.a	Intramuscular injection (mouse model)	30 days at 37 °C (bioassay), 15 days at 25 °C (bioassay and immunogenictiy)	Toniolo et al. (2019, 2020)

Based on Toniolo et al. (2019, 2020), open access under CC-BY license
A Spray gas flow rate, *F* Feed rate, *n.a.* not available

The results showed that trehalose and trehalose/dextran (3:1) were the best excipients for maintaining viral activity for spray-dried enveloped viral vectors such as VSV and influenza. In contrast, mannitol/dextran mixtures maintained the most activity for the non-enveloped human adenovirus vector AdHu5. Furthermore, the same mannitol/dextran mixture also showed excellent stabilization for chimpanzee adenovirus serotype 68 (AdCh68), as previously reported by the same authors (Afkhami et al. 2017). The authors concluded that it is essential to consider the structure of the viral envelope when selecting excipients for spray drying. The differences in thermal stabilization are due to the vitrifying properties of the sugar excipients and the chemical interactions with the outer surfaces of the viral vectors. In the presence of a lipid membrane in an enveloped virus, trehalose resulted in greater protection due to stronger hydrogen bonds. In contrast, in non-enveloped adenovirus (Fig. 6.49b), mannitol was more effective in replacing the water molecules surrounding the capsid due to its smaller size (Toniolo et al. 2019).

Moreover, after reconstitution of spray-dried VSV with sterile PBS solution, (Toniolo et al. 2020) showed that trehalose conferred higher stability to VSV both in vitro and in vivo. Furthermore, after storage at 37 °C for 15 days, the VSV-vectored vaccine retained its in vivo immunogenicity and potency, whereas a liquid control completely lost its immune-stimulating ability. Overall, the results demonstrated that spray drying with properly tested excipients could stabilize viral vectors such as VSV to be stored long-term at elevated temperatures without depending on the cold chain.

6.18 Herpes Simplex Virus Vaccines

Herpes is an ancient Greek term meaning to creep or crawl, representing a family of viruses (Whitley and Roizman 2001). The most common virus within this family is the herpes simplex virus (HSV), an alpha herpes virus with a double-stranded DNA genome that exists as two biologically distinct serotypes, HSV-1 and HSV-2 (Krishnan and Stuart 2021) (Fig. 6.50). The main components surrounding the

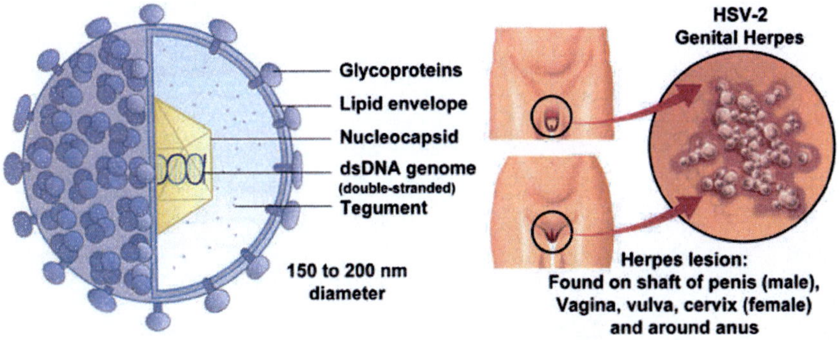

Fig. 6.50 Structure of the herpes simplex virus with a double-stranded DNA genome, icosahedral capsid, an amorphous proteins layer (tegument), and lipid bilayer envelope containing viral glycoproteins (Image left: adapted from Krishnan and Stuart (2021), open access under CC-BY 4.0 license, Image right: adapted from www.mymed.com)

double-stranded DNA core include an icosapentahedral capsid (100–110 nm) anchored by an amorphous layer of proteins called tegument providing structure. The viral capsid is surrounded by a lipid bilayer envelope of polyamines and glycoproteins.

Genital herpes is one of the most prevalent sexually transmitted infections caused by HSV-2, which causes pain, itching, and sores in the genital area. As of 2016, the global health burden of HSV-2 infection was estimated to be 13.2% of the world's population aged 15–49 years (420 million people) and 66.6% (3,752 million people) from HSV-1 (James et al. 2020). After initial infection, the virus remains dormant in the body and can be reactivated several times a year.

An effective HSV-2 vaccine has become more important in recent years due to the increased risk of human immunodeficiency virus (HIV) infection following HSV-2 acquisition (Looker et al. 2015). Although several candidate HSV-2 vaccines had progressed to phase 3 trials (tested in humans), there are no licensed HSV-2 vaccines available, mainly due to low efficacy (WHO 2019d). However, there are various potential vaccine candidates in the pre-clinical and clinical phases of development (Krishnan and Stuart 2021), including whole-virus types (killed, live-attenuated, or genetically disabled), monovalent or multivalent protein subunits with adjuvants, DNA, messenger RNA (mRNA) or live-virus vectors, and peptides and nanoparticles (WHO 2019d).

Although HSV is known to lose activity at elevated temperatures of 22–37 °C rapidly (Plummer and Lewis 1965), a thermostable HSV-2 vaccine is still an unmet medical need and thus would also reduce the burden of HSV-2 and HIV globally (LeClair et al. 2019).

In this research field, Leung et al. (2019) stabilized HSV-2 in a mixture of pullulan and trehalose by air-drying in oral sugar pills and showed that the efficacy was retained for at least 2 months of storage at 40 °C. Furthermore, pullulan served as a water-soluble polysaccharide with good film-forming abilities and low oxygen

Table 6.22 Process conditions and main results of spray-dried herpes simplex virus (HSV-2) vaccine candidate

Vaccine type	Excipients, composition	Spray drying process parameters	Particle size, residual moisture	Administration route	Shelf life, thermostability	References
Herpes simplex virus (HSV-2), live-attenuated	Trehalose and sucrose, 18 wt % and 15 wt% solid concentration	B-290 with HP cyclone, T_{in}/T_{out}: 105 °C/n.a A: 360–450 L/h, F: n.a., D: n.a	n.a 3.6 wt%	In vitro (viral plaque assay)	10 days at 45 °C (trehalose) (< 1.5 log activity loss)	LeClair et al. (2019)

Based on LeClair et al. (2019)
A Spray gas flow rate, *F* Feed rate, *D* Drying gas rate, *n.a.* not available

permeability, suitable for coating tablets and granules to extend the shelf life. Overall, a low-cost and straightforward drying method was developed for converting heat-sensitive vaccines into thermostable vaccine pills.

LeClair et al. (2019) demonstrated the feasibility of spray drying attenuated HSV-2 viral vaccine candidates through a DOE study using a Mini Spray Dryer B-290 and HP cyclone. The thermostability and activity of the HSV-2 viral vaccine were evaluated using an in vitro plaque-forming bioassay with the Vero cell line and accelerated storage conditions at 45 °C for 10 days. Table 6.22 list the process conditions and main results.

Figure 6.51a shows the combined activity losses for formulations containing trehalose, sucrose, lactose, mannitol/dextran, and lactose/trehalose, taking into account pre-spray losses (during formulation mixing manipulation), process losses (during spray drying), and storage losses. Trehalose was the most robust and effective stabilizing excipient in the studies exhibiting the lowest overall HSV-2 activity losses of < 1.5 $\log_{10}$ pfu/mL, and only minor activity losses after storage and during the spray drying process, i.e., < 0.5 $\log_{10}$ pfu/mL.

Sucrose also proved to be an effective stabilizer under some spray-drying conditions but much worse under others, as shown in Fig. 6.51b. The differences in activity losses were attributed to the specific molecular interactions between trehalose/sucrose and HSV-2, the T_g of the respective matrix, and the equilibrium condition to which the particles were spray-dried.

As an outlook, the authors concluded investigating the stability of spray-dried HSV-2 vaccines under different storage conditions (e.g., 5 °C, controlled room temperature) to better define the stability profiles and understand any degradation processes that may occur during storage. Overall, the improved storage stability can solve the cold chain problem (no refrigeration required) and significantly improve global health by providing access to vaccines for people worldwide.

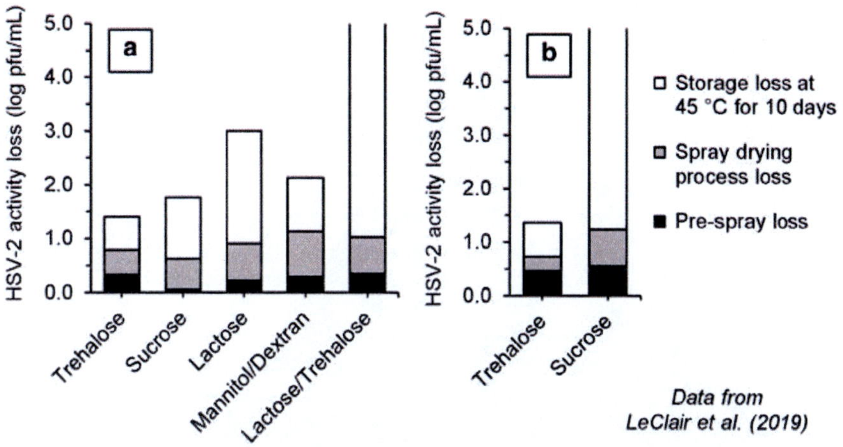

Fig. 6.51 **a** HSV-2 activity losses ($\log_{10}$ pfu/mL) for selected excipient formulations. The activity losses were combined by pre-spray loss due to dilution and mixing, process loss due to spray drying, and storage loss due to storage at 45 °C for 10 days. *denotes viral activity was below the detection limit of > 5 $\log_{10}$ pfu/mL. **b** Combined HSV-2 activity losses ($\log_{10}$ pfu/mL) for trehalose and sucrose formulation spray-dried at 120 °C inlet temperature, 450 L/h spray gas flow rate, and 15% (w/v) solids concentration (Data from LeClair et al. (2019) with permission from Elsevier)

6.19 Norovirus Vaccines

6.19.1 A Short Introduction to Norovirus and Its Disease Burden

Noroviruses are a separate genus in the family Caliciviridae, whose name derives from the calyx (Greek for cup) (Green 2013) and have a great diversity of genogroups (GI to GX), genotypes, and subtypes (Patel et al. 2009; Glass et al. 2009; Chhabra et al. 2019). Most human infections are caused by GI and GII (Baehner et al. 2016), while GIII and GV have been identified only in animals (Cates et al. 2020). The virions are approximately 27–40 nm in diameter (Green 2013), and the outer structure is an unenveloped icosahedral capsid enclosing a 7.5 kb linear, single-stranded positive-sense RNA genome (Cates et al. 2020).

Noroviruses spread in various ways, including person-to-person contact (88%), contaminated food (10%) and water (1.5%), and contaminated surfaces (Green 2013). Less than 100 virus particles can already cause an infection (Patel et al. 2009) and develop severe vomiting and watery diarrhea, also known as acute gastroenteritis (Glass et al. 2009; Tan 2021). Norovirus is ubiquitous and is associated with approximately 18% of all diarrheal cases worldwide. Global estimates predict 684 million annual illnesses and 212,000 deaths caused by norovirus infections (Lopman et al. 2016; Mattison et al. 2018; Cates et al. 2020).

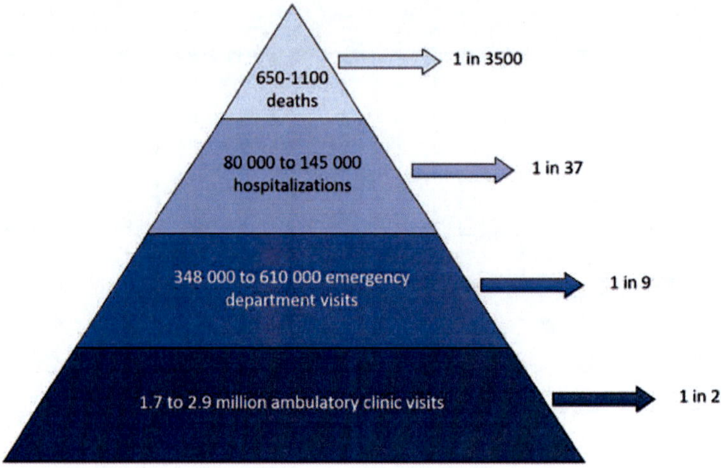

Fig. 6.52 Estimated annual disease burden of norovirus in the United States. The pyramid presents ranges of total annual deaths, hospitalizations, emergency department visits, and ambulatory clinic encounters. The corresponding estimated lifetime risks are noted with arrows (Adapted from Burke et al. (2021), public access)

In the United States, an average of 20 million people become infected with norovirus each year (Hall et al. 2013), resulting in approximately 80,000–145,000 hospitalizations and 650–1100 deaths, primarily among the elderly (Burke et al. 2021) (Fig. 6.52). In addition, estimates suggest 2.3 million outpatient hospitalizations per year and associated healthcare costs of $430 to $740 million (Burke et al. 2021). This means that 1 in 2 Americans will need to visit an ambulance, 1 in 9 will need to visit the emergency room, 1 in 37 will be hospitalized, and 1 in 3,500 will die at some point in life from norovirus.

Thus, noroviruses represent a major global public health threat. Globally, norovirus causes approximately $4.2 billion health system costs and $60.3 billion societal costs annually (Bartsch et al. 2016; Lopman et al. 2016).

6.19.2 Norovirus Vaccines

The development of norovirus vaccines is desired to protect vulnerable populations (e.g., immunocompromised, juvenile, elderly) and high-risk groups, including health care workers, military personnel, and (e.g., cruise ship) travelers experiencing crowding conditions (Lopman et al. 2016; Ludwig-Begall et al. 2021). Several authors have comprehensively reviewed the status of norovirus vaccines in recent years (Lopman 2015; Melhem 2016; Baehner et al. 2016; Lucero et al. 2018; Mattison et al. 2018; Hallowell et al. 2019; Cates et al. 2020; Esposito and Principi 2020; Tan 2021). In addition, the WHO has recognized the development of a norovirus vaccine as a high priority in their 2016 Product Development for Vaccines Advisory Committee (Giersing et al. 2019).

However, so far, there is no norovirus vaccine available on the market (Hallowell et al. 2019; Cates et al. 2020; Ludwig-Begall et al. 2021). Furthermore, the development of a broadly (multivalent) effective norovirus vaccine remains challenging due to the wide genetic and antigenic diversity of noroviruses with multiple circulating variants of various genotypes, the lack of a standard cell culture-based neutralization assay, an efficient animal model, and reliable immune correlates of norovirus protection for vaccine evaluation further hindering the developmental process (Hallowell et al. 2019; Esposito and Principi 2020; Tan 2021). On the other hand, several norovirus vaccine candidates have reached clinical trials, as shown in Table 6.23 (Tan 2021). Main vaccine design strategies include recombinant virus-like particles (VLPs) formed by the main capsid protein VP1 mimicking major antigenic features of the norovirus, subviral particles made from the protruding domain of VP1 (P domain particles), and viral vectors with a norovirus capsid gene insert produced by bioengineering technologies (Lucero et al. 2018; Mattison et al. 2018).

At least four vaccines are currently being tested in clinical trials (Cates et al. 2020), including bivalent GI.1/GII.4 VLP-based intramuscular vaccines from Takeda Pharmaceutical Company Limited (Takeda 2016) and the National Serum & Vaccine Institute China, an oral GI.1/GII.4 vaccine from Vaxart Inc. (Vaxart 2021) on an adenovirus basis, and a quadrivalent VLP vaccine produced by Institut Pasteur in Shanghai and Anhui Zhifei Longcom Biopharmaceutical Co. Ltd. in China. Further details on the trial results are discussed in (Tan 2021).

Among the vaccine candidates, norovirus VLPs are the most common ones, with proven safety, immunogenicity, and protective efficacy, supporting a high success likelihood of a useful norovirus vaccine (Tan 2021). In addition, VLPs are readily manufactured in a reproducible, consistent, and cost-effective manner at a large scale (Richardson et al. 2013).

Several studies explored a dry powder formulation of norovirus VLP vaccine for nasal delivery to increase residence time on the mucosal surface, thereby improving antigen uptake and immune response (Vodak et al. 2012; Springer et al. 2016; Ball et al. 2017; Bahamondez-Canas and Cui 2018). Compared to liquid formulations, a dry powder also offers greater stability (Xu et al. 2021), easier administration, and better patient compliance (Bahamondez-Canas and Cui 2018), making it an attractive alternative to intramuscular injection (Hickey and Garmise 2009; Velasquez et al. 2011; Tlaxca et al. 2015; Vehring et al. 2020).

Springer et al. (2016) investigated a dry powder formulation (mucoadhesive GelVac™) with an in-situ gelling high molecular weight polysaccharide (GelSite™) extracted from Aloe vera for nasal delivery of a bivalent vaccine formulation containing both GI.1 and GII.4 norovirus VLPs. The developed vaccine powder was immunogenic without adjuvants in guinea pigs providing a greater surface area of nasal mucosa for immune induction than mouse models. The GelVac™ vaccine powders were prepared using a freeze-drying and milling method. In a follow-up study, (Ball et al. 2017) evaluated safety, immunogenicity, and possible

Table 6.23 Norovirus vaccine candidates in clinical phase

Company	Vaccine type	Genotype	Vaccine candidate	Adjuvant	Administration route	Clinical phase	ClinicalTrials.gov identifier
Takeda Pharmaceuticals Company Limited (Japan)	VLP	GI.1/GII.4	TAK-214	Chitosan/MPLA, aluminum salt	Intranasal, intramuscular	Phase 2b	NCT03039790
Vaxart, Inc. (USA)	Adenovirus	GI.1/GII.4	VXA-NVV-104	Adenovirus expressing double-stranded RNAs	Oral pill	Phase 1b	NCT04854746
National Serum & Vaccine Institute (China)	VLP	GI.1/GII.4	Hansenula-polymorpha	Aluminum salt	Intramuscular	Phase 1	NCT04188691
Institut Pasteur of Shanghai, Anhui Zhifei Longcom Biopharmaceutical Co. Ltd. (China)	VLP	GI.1/GII.3/GII.4/GII.17	Longkoma	Aluminum salt	Intramuscular	Phase 2a	NCT04563533

Adapted from Tan (2021), open access under CC BY 4.0 license antigenic interference. The dry powder vaccine was found to be safe and well-tolerated and elicited a dose-dependent immune response.

6.19.3 Research on Spray-Dried Norovirus Vaccines

To the best of the author's knowledge, only two studies on spray-dried norovirus vaccines have been published (Velasquez et al. 2011; Vodak et al. 2012). Table 6.24 shows the spray drying process parameters and main results.

Velasquez et al. (2011) prepared GelVac™ dry powder formulations with stabilized VLPs in GelSite® polymer using a Mini Spray Dryer B 290 in a Class 1000 clean room with controlled temperature and humidity. The GelSite® polymer content was 0.25% w/w, and the dry powders had a smooth, spherical appearance with an average particle size of about 20 µm (Fig. 6.53a). Compared with the liquid formulations, the spray-dried GelVac™ powder exhibited better norovirus-specific antibody responses in serum and on the mucosal surface of guinea pigs and induced robust systemic and mucosal antibody titers that were equal to or higher than liquid formulations. The in-situ gelation of the GelVac™ dry powder delayed mucociliary clearance at the nasal mucosal surface, prolonging VLP antigen exposure at immune effector sites.

Vodak et al. (2012) developed a spray-dried norovirus vaccine at a pilot scale using a pressurized vortex atomizer (Niro Mobile Minor™ with extension to increase residence time in the dryer) (Bend Research 2013). The processing conditions were 175 °C/70 °C inlet/outlet temperature, 2.4 kg/h feed rate, and 114 kg/h drying gas flow rate (Table 6.24). The formulation comprised mucoadhesive chitosan glutamate (ChiSys®, Archimedes Development Ltd.) (Watts et al.

Table 6.24 Spray drying parameters and formulation of norovirus VLP-vaccine powder for intranasal delivery

Spray drying process parameters	Formulation	Particle size and morphology	Administration route, animal model	References
Lab-scale Mini Spray Dryer B-290 with two-fluid nozzle, installed in a Class 1000 clean room	GelVac™ dry powder with stabilized VLPs (10 or 25 µg) in 0.25% w/w GelSite® polymer	20–30 µm, smooth, spherical	Intranasal, guinea pigs	Velasquez et al. (2011)
Pilot-scale spray dryer with pressure swirl nozzle atomizer, T_{in}/T_{out}: 175 °C/70 °C, F: 40 g/min, P: 150 psi nozzle pressure, D: 1900 g/min	Chitosan glutamate (ChiSys®), adjuvant monophosphoryl lipid A (MPL®), 150 µg per dose of each GI.1 and GII.4 VLPs	Dv10: 15 µm, Dv50: 35 µm, Dv90: 73 µm, Span 1.7, spherical	Intranasal, rabbits	Vodak et al. (2012)

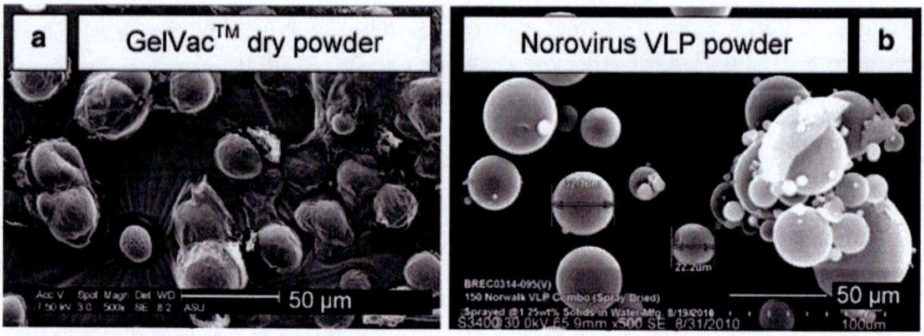

Fig. 6.53 **a** Spray-dried GelVac™ powders formulated with Norwalk virus-like particles (VLP) and mucoadhesive GelSite® polymers for intranasal delivery (Adapted from Velasquez et al. (2011) with permission from Elsevier). **b** Spray-dried norovirus VLP powder with spherical morphology (Adapted from Vodak et al. (2012) presented at DDL2012 conference, Drug Delivery to the Lungs, The Aerosol Society)

2014), adjuvant monophosphoryl lipid A (MPL®, GSK), and 150 µg per dose of each GI.1 and GII.4 VLPs (LigoCyte Pharmaceuticals, Inc., USA). The challenge was to prepare particles small enough to optimize the intranasal immune response but large enough (> 10 µm in diameter) to avoid lung deposition.

Figure 6.53b shows an SEM image of the spray-dried norovirus vaccine particles, which exhibited a spherical, hard, and hollow-shelled morphology. The spray-dried dry powder was loaded into single-dose intranasal delivery devices (Unidose, Aptar Pharma, USA) equipped with small animal adapters and administered to rabbits. The results showed similar VLP-specific IgG titers for all rabbit treatment groups compared to lyophilized/milled vaccine powder (Bend Research 2013). Validation studies of radiolabeling evaluated in-vivo powder deposition. Data analysis confirmed that more than 74 ± 10% of the dose was deposited in the nose, and 1.9 ± 2.0% entered the lungs through the airway.

Based on these data, spray drying proved to be a robust and scalable process for the production of norovirus vaccine powder that provided high product yields (> 80%) while maintaining the integrity of the VLP in the powder and meeting the requirements for narrow particle size distribution (Span < 2).

6.20 Human Immunodeficiency Virus Vaccines

The human immunodeficiency virus (HIV) attacks the cells of the immune system, weakening its ability to fight infections and diseases. If left untreated, HIV can lead to acquired immunodeficiency syndrome (AIDS), the late stage of infection that occurs when the immune system is severely damaged. HIV is transmitted mainly through sexual contacts with the genital and gastrointestinal tracts as the main entry points (Shattock and Moore 2003). Figure 6.54 shows the state of the HIV/AIDS epidemic in 2021, according to global HIV statistics (UNAIDS 2022).

Fig. 6.54 State of the HIV/AIDS epidemic in 2021 relative to 2010 (*Data source* UNAIDS (2022))

In 2021, about 38.4 million people were living with HIV, up from 30.8 million in 2010. Of these, approximately 28.7 million have access to antiretroviral therapy. About 5.9 million people did not know they were living with HIV, and an estimated 1.5 million were newly infected with HIV (about 4100 per day). In 2021, about 0.65 million people died of AIDS-related illnesses, making HIV still one of the leading causes of death worldwide.

Despite numerous trials over many years, there is still no licensed vaccine against HIV (Barré-Sinoussi et al. 2013; Lema et al. 2014; Hsu and O'Connell 2017; Trovato et al. 2018; Ng'uni et al. 2020; Mu et al. 2021; Mymetics 2022). Only one (RV144) out of 6 HIV-1 vaccine trials showed efficacy. Results of ongoing HIV vaccine clinical trials are pending.

The development of thermostabilized liquid or solid dosage forms of HIV vaccines is part of the solution. In this context, an European project called MACIVIVA for "Manufacturing process for Cold-chain Independent Virosome-based Vaccines" was conducted from 2015 to 2018, with the project coordinator Mymetics BV (The Netherlands), and the consortium partners, Mymetris SA (Switzerland), Bachem AG (Switzerland), Upperton Ltd. (UK), Catalent UK Swindon Zydis Ltd (UK), and Chimera Biotec GmbH (Germany) (CORDIS 2018). The project was funded by the EU Horizon 2020 research and innovation program and the Swiss State Secretariat for Education, Research, and Innovation (SERI) (Grant Agreement No. 646122).

Recently, Amacker et al. (2020) presented the main project results of novel GMP manufacturing processes for producing thermostable solid dosage forms of virosome-based vaccines against human HIV-1 for mucosal vaccination. The virosomes served as carriers for the recombinant HIV-1 gp41 protein and adjuvant 3M-052, a TLR7/8 agonist, both anchored to the virosomes surface (Fig. 6.55).

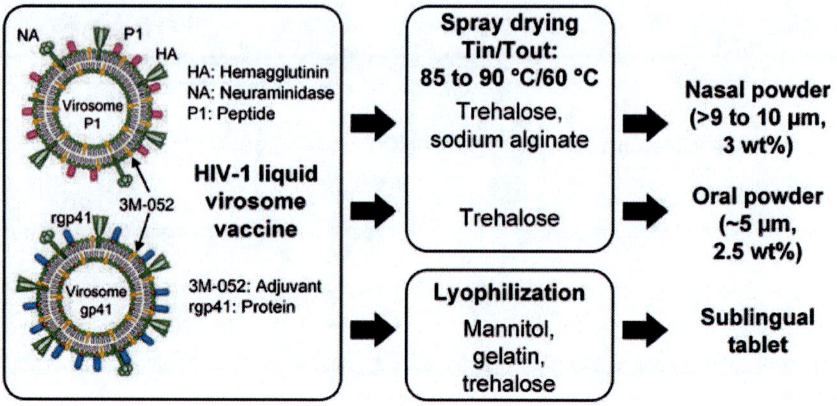

Fig. 6.55 HIV-1 virosomal vaccine MYM-V202 consisting of adjuvanted virosome-P1 (MYM-V111) and virosome-rgp41 (MYM- V112) prepared from isolated viral membrane lipids of inactivated influenza A/H1N1 virus, hemagglutinin (HA), neuraminidase (NA), synthetic lipids with adjuvant 3M-052, and HIV-1 antigen P1 or rgp41 (Adapted from Amacker et al. (2020), open access under a CC BY 4.0 license)

Three solid dosage forms of HIV-1 virosomes were prepared by:

- spray drying for nasal administration,
- spray drying and filling into enteric-coated capsules for oral delivery, and
- lyophilization with mannitol, fish gelatin, and trehalose into rapidly dissolving sublingual tablets.

The spray drying work was performed at the project partner Upperton Ltd. in Nottingham (UK), a professional CDMO (Contract Development and Manufacturing Organization) provider with cGMP spray drying services from the early development phase through scale-up to clinical trials.

Figure 6.56 shows the range of available spray dryers, including GEA Niro PSD-1, GEA Niro SD Micro, ProCepT 4M8-TriX, and several Büchi Mini Spray Dryer B-290 models covering batch sizes from a few milligrams to several kilograms of powder.

The nasal powder formulation was prepared by adding trehalose and sodium alginate (mucoadhesive excipient) to the liquid virosomes formulation at 77 wt% and 8 wt%, respectively. The oral powder formulation was obtained by adding trehalose at 87 wt%. During spray drying, the inlet and outlet temperatures were maintained at 85–90 °C and 60 °C to ensure gentle drying. Spray drying conditions, and formulations were optimized to produce particles with an average volume diameter of > 9–10 µm for nasal administration (since particles < 5 µm are inhaled) and approx. 5 µm for oral powder.

Fig. 6.56 Spray dryer equipment at Upperton Ltd. (Nottingham, UK) for comprehensive services from proof-of-concept studies in the laboratory through drug development and manufacturing for clinical trials. **a** GEA Niro PharmaSD™ PSD-1 spray dryer, **b** GEA Niro SDMicro spray dryer, **c** Büchi Mini Spray Dryer B-290 advanced, **d** ProCepT 4M8-TriX spray dryer (Images courtesy of Upperton Pharma Solutions, UK)

The finished powders were blanketed with nitrogen gas, double bagged in foil pouches for light protection, and stored at 2–8 °C. For stability testing, powder samples were stored at 5 °C, 25 °C/60% RH (relative humidity), and 40 °C/75% RH. A commercially available Nasal Unidose Device from Aptar Pharma was used as a disposable single-dose device (emitted dose > 98%). The residual moisture content of the sublingual tablets was approx. 4 wt%, that of the nasal powder about 3 wt%, and that of the oral powder about 2.5 wt%.

The results demonstrated that the new thermostable spray-dried virosomal vaccine formulations maintained their virosome structure and particle size even after 3 months of storage at 40 °C/75% RH. The immunogenicity of the vaccine was also preserved. Both the nasal and oral formulations were successfully administered to rats. The immunogenicity of the nasal solid HIV-1 vaccine form was comparable to subcutaneous injection of the reference liquid vaccine.

Thus, these "ready-to-use and all-in-one" thermostable, needle-free virosomal HIV-1 mucosal vaccines offer the advantage of simplified logistics with a lower dependence on the cold chain during shipments and distribution.

In addition, Mymetics SA reported the company has intensively started to develop a spray-dried virosome-based vaccine against SARS-CoV-2 (COVID-19), with the goal of a broadly protecting vaccine that is thermostable and can be administered intranasally (Mymetics 2022).

6.21 Newcastle Disease Virus Vaccines

6.21.1 A Short Introduction to Newcastle Disease and Vaccines

Newcastle disease (ND) is a severe and economically important poultry disease that poses an ongoing threat worldwide (Van Boven et al. 2008; Capua and Alexander 2009; Ganar et al. 2014). According to the World Livestock Disease Atlas (2006–2009), ND was the fourth most poultry disease behind highly pathogenic avian influenza, infectious avian bronchitis, and low pathogenic influenza (World Bank; TAFS Forum 2011). When an outbreak of ND occurs, the specific country must report it to the World Organization for Animal Health (OIE), and trading partners can suspend poultry or poultry product imports from that country (Miller and Koch 2020). From 2013 to 2016, 71 of the 181 OIE member countries reported ND in domestic poultry.

The causative agent is the Newcastle disease virus (NDV), synonymous with avian paramyxovirus serotype 1 (APMV-1), which belongs to the Paramyxoviridae family and infects all kinds of birds and poultry species (Holman et al. 2009; Ganar et al. 2014; Dimitrov et al. 2017; Miller and Koch 2020). Based on pathogenicity studies in chickens, strains of Newcastle disease virus have been broadly classified into four pathotypes, including (Ganar et al. 2014; Miller and Koch 2020):

- avirulent (causes no disease),
- lentogenic (low virulence, low mortalities, loss of egg production),
- mesogenic (moderate virulence, mortalities up to 50%, loss of egg production),
- velogenic (high virulence, severe disease with high mortalities).

Figure 6.57 shows a schematic structure and an electron tomographic picture of the NDV (Mast and Demeestere 2009; Miller and Koch 2020). NDV is a pleomorphic enveloped virus of around 200–300 nm diameter and is characterized by a peripheral fringe of glycoprotein spikes. The NDV genome is a negative sense, single-stranded non-segmented RNA, which encodes six major proteins, namely nucleocapsid protein (N), phosphoprotein (P), matrix protein (M), fusion protein (F), hemagglutinin-neuraminidase protein (HN) and the RNA dependent RNA polymerase protein (L) (Rani et al. 2014).

Figure 6.58 shows the major clinical and pathological disease features and gross lesions following infection with velogenic Newcastle disease virus, including (Miller and Koch 2020):

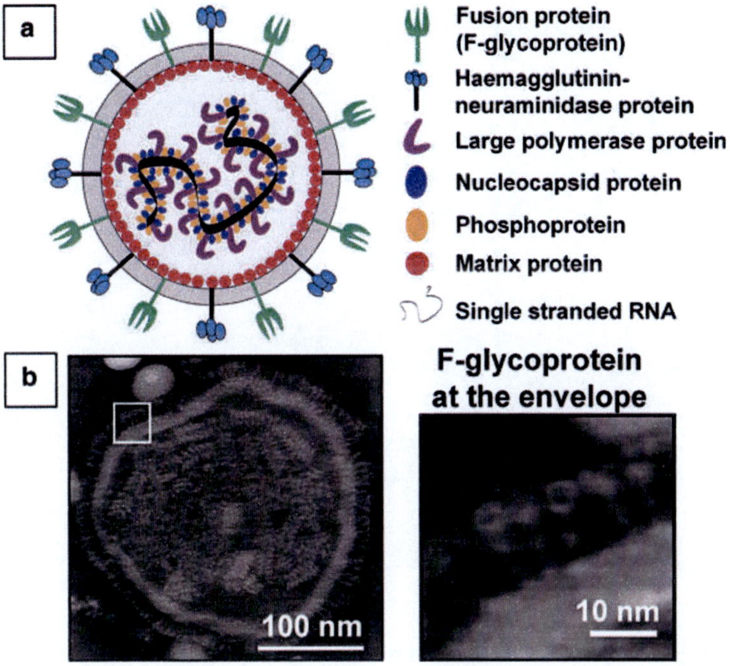

Fig. 6.57 **a** Schematic diagram of the Newcastle disease virus structure (Adapted from Ganar et al. (2014) with permission from Elsevier), **b** Negative staining and electron tomographic analysis of Newcastle disease virus showing F-glycoprotein protruding from the envelope (Adapted from Mast and Demeestere (2009) and Miller and Koch (2020) with permission from BioMedCentral Ltd (open access CC BY 2.0 license) and John Wiley and Sons)

- (A) depressed and lethargic chicken with ruffled feathers,
- (B) crop fluid visible from the oral cavity,
- (C) hemorrhage of the lymphoid tissue of the lower eyelid,
- (D) malformed eggs,
- (E) necrosis and hemorrhage of the cecal tonsils,
- (F) necrosis and hemorrhage of the lymph nodes in the small intestine (Peyer's patches), and
- (G) enlarged, mottled spleen with punctate necrosis, and
- (H) hemorrhage and necrosis of proventriculus.

Vaccination of poultry against Newcastle disease is the only effective way of controlling the disease (Alders and Spradbrow 2001; Alders et al. 2002; Alders 2014; Boumart et al. 2016). Vaccination increases resistance to infection and reduces the amount of NDV shedding, ultimately decreasing its transmission to the poultry population (Miller and Koch 2020). However, analyses have shown that herd immunity can only be achieved if a high proportion of birds (> 85%) have a high antibody titer after vaccination (Van Boven et al. 2008).

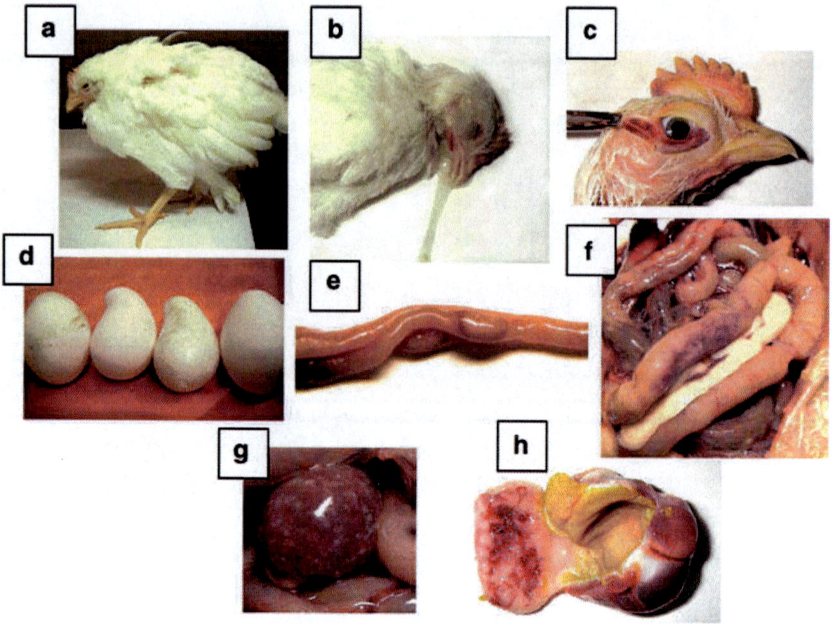

Fig. 6.58 Signs of clinical disease and gross lesions upon infection with velogenic Newcastle disease virus. **a** Birds appear depressed and lethargic with ruffled feathers. **b** Crop fluid visible from the oral cavity. **c** Hemorrhage of the lymphoid tissue of the lower eyelid. **d** Misshapen eggs produced by commercial layers previously vaccinated with four live B1 vaccines. Eggs were produced 30 days after challenge with vNDV as the only clinical sign of ND. **e** Necrosis and hemorrhage of the cecal tonsils. **f** Necrosis and hemorrhage of small intestinal lymphoid patches (Peyer's patches). **g** Enlarged, mottled spleen with pinpoint areas of necrosis. **h** Hemorrhage and necrosis of proventriculus (Photographs reproduced from Miller and Koch (2020) with permission from John Wiley and Sons)

However, vaccination should be used with good management and biosecurity practices (Dimitrov et al. 2017). Inactivated and live NDV vaccines formulated with NDV strains of low virulence, such as B1, LaSota, and more recently Ulster and VG/GA, are the most commonly administered vaccines in the poultry industry (Miller and Koch 2020). In addition, because NDV is nonpathogenic in primates (Bukreyev et al. 2005; Miller and Koch 2020), they are attractive vaccine vectors for other animal and human pathogens (DiNapoli et al. 2007; Kim and Samal 2016).

The most typical vaccine administration techniques in commercial poultry include (Cargill 1999; Miller and Koch 2020):

- drinking water administration on the farm,
- spray administration/nebulization of day-old birds (Fig. 6.59) and birds in a housed environment,
- eye drop administration,

Fig. 6.59 Automated spray vaccination of day-old chicks (primary vaccination) at a hatchery against Infectious Bronchitis and Newcastle Disease. Capacity 900 chick boxes per hour or 90,000 chicks per hour (Photo from Royal Pas Reform 2020)

- transfixion and scarification (cutaneous route in the wing web or foot),
- injection (e.g., intramuscular, subcutaneous) on the farm, or
- in-ovo administration (at 17–19 days of embryonation).

Spray and aerosol vaccination are routine practices in poultry farming to prevent and control the disease (Cargill 1999). However, with live vaccines in drinking water, sprays, or aerosols, mass vaccination is less labor-intensive than administering inactivated vaccines to individual chickens. Therefore, spray vaccination is mainly reserved for priming hatched chicks, which can be performed automated at the hatchery (Fig. 6.59), while aerosol vaccination is performed on 2- and 4-week-old chickens as part of secondary vaccination (Huyge 2012).

However, proper particle size is crucial to ensure an adequate immune response. Broad droplet size distribution generated by nebulizers reduces the efficiency of vaccinations (Corbanie et al. 2007). If the particles are too small, vaccine reactions in the form of respiratory disease may occur because the virus is deposited deep in the lungs. On the other hand, if the particles are too large, the immune response may not be optimal because the virus drops out of the air before the birds become infected (Miller and Koch 2020).

In addition, the lack of vaccines that have adequate thermostability is often a major obstacle to global vaccination efforts, particularly in hot climates with a lack of sufficient cold chains in remote areas and also in moderate climates on hot days to support the transport and storage of effective NDV vaccines (Alders and Spradbrow 2001; Rexroad et al. 2002; Adwar and Lukešová 2008; Boumart et al. 2016; Osman et al. 2021).

As reviewed by Dimitrov et al. (2017) and published on the PoultryMed.com website, current live (freeze-dried), inactivated (water-in-oil emulsions), and vectored NDV vaccines require refrigerated storage at 2–8 °C. Key recommendations to support technically sound and sustainable ND control programs in village chickens include using temperature indicators in the vaccine containers (Alders et al. 2010).

However, very few data are available on the temperature stability of NDV under laboratory or field conditions (Rani et al. 2014; Boumart et al. 2016; Osman et al. 2021). Rani et al. (2014) indicated that the NDV is non-infective above 42 °C and unstable above 72 °C. Boumart et al. (2016) showed that freeze-dried I-2 and LaSota vaccines had the longest theoretical shelf life at 37 °C compared to B1, C30, and VG-GA strains.

In their freeze-dried form, thermotolerant strains like NDV 4-h and I-2 retain activity for up to two months at temperatures between 9 and 29 °C and for two weeks between 30 and 37 °C (Alders et al. 2002). The LaSota vaccine (Thermostable ND, Hester, India) is declared stable at 37 °C for 10 days (Hester 2017). (Osman et al. 2021) tested several commercially available freeze-dried NDV vaccines, claiming increased thermostability. Exposure to elevated temperatures of 51 and 61 °C showed that all dissolved NDV vaccines, including I-2 vaccine strains, were inactivated within 2–4 h. Overall, there is a lack of specific definition and criteria for thermostability (i.e., time, temperature, acceptable losses) of NDV vaccines in the literature and regulatory requirements.

To tackle the issue of thermostability, Lal et al. (2014) developed a consumer-friendly and cost-effective NDV vaccine tablet for smaller poultry herds (50 doses), which is more heat stable than the usual lyophilized product. The vaccine is based on the LaSota strain and is stabilized by dextran, PVP, sucrose, gelatin, and dry milk. The tablets are freeze-dried directly into polymer blisters. As a result, the vaccine remained stable at 4 °C for up to one year with minimal titer loss. Furthermore, it could be stored at 37 °C for up to two weeks. The tablets can be dissolved in tap water in less than 10 s and administered within 2 h by mass administration. This innovation has the potential to make the tablet NDV vaccine available at the village level in remote areas where cold chains are inadequate or non-existent (Alders 2014).

6.21.2 Research on Spray-Dried Newcastle Disease Vaccines

Another approach to increase the cost-effectiveness of mass spray and aerosol application of live attenuated NDV vaccines in poultry is reformulating existing vaccines as readily dispersible spray-dried powders (Corbanie 2007; Corbanie et al. 2007, 2008; Huyge 2012; Huyge et al. 2012; Landman et al. 2015, 2017). The formulation of spray-dried powders enables a defined narrow particle size distribution for more efficient targeting of the upper or lower airways of the poultry (Landman et al. 2015), and it circumvents inactivation during reconstitution and evaporation of droplets (Corbanie 2007). Moreover, spray drying is a faster dehydration process than freeze-drying and is possibly cheaper.

Studies by Corbanie (2007) and Corbanie et al. (2007, 2008) demonstrated that it is possible to produce a spray-dried powder formulation of a live-attenuated NDV vaccine intended for respiratory mass vaccination of broiler chickens as an alternative to liquid spray and aerosol vaccination. The authors used an experimental spray dryer with a 30 μm piezo-actuated nozzle (50 kHz, GEA Niro A/S) and a commercial Mobile Minor D-2000 pilot spray dryer with a 1-mm two-liquid nozzle (GEA Niro A/S) to produce coarse and fine powder factions. Batches of 400 and 1000 mL were spray-dried containing mannitol, trehalose, trehalose-PVP, trehalose-BSA, and trehalose-PVP-BSA to stabilize an attenuated Clone 30 strain of the NDV. In addition, the process parameters were optimized to obtain an outlet temperature of 70 °C and around 70% powder yield. As a result, the spray dryers produced powders with a median size of 28 μm (experimental dryer) and 7 μm (Mobile Minor), suitable for targeting the upper and lower airways of 4-week-old broilers, respectively (Corbanie 2007; Corbanie et al. 2007) (Fig. 6.60a). Trehalose-PVP-BSA (3:1:1) resulted in the highest retention of viral activity (tested by hemagglutination cell assay) when stored for 10 months at 25 °C (Corbanie et al. 2007). In addition, the trehalose-BSA formulation was the most stable, with an average degradation rate of only 0.2 $\log_{10}$ per month. Vaccination of 4-week-old broiler chickens was performed in isolators in groups of 20 chickens.

Overall, the spray-dried powder formulations of the lentogenic live ND vaccine induced high hemagglutination inhibition antibody titers comparable to those of liquid aerosols, suggesting that very good serum antibody responses can also be obtained when respiratory vaccinating with dry powders (Corbanie et al. 2008). However, future research was recommended to optimize the particle size

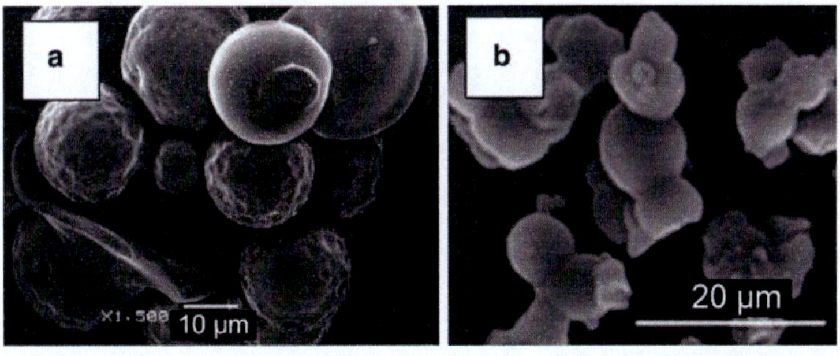

Fig. 6.60 **a** Agglomerate of trehalose particles formed after 1-year storage at 25 °C in glass vials with silica-containing stopper (Adapted from Corbanie et al. (2007) with permission from Elsevier). **b** Spray-dried NDV vaccine powder formulated with mannitol-inositol-BSA (73.3:13.3:13.3 w/w) (Adapted from Huyge et al. (2012) with permission from Elsevier)

distribution for secondary vaccination at 2 weeks of age with particles smaller than 5 μm. In addition, the possibility of dose reduction with powder vaccines should be investigated (Corbanie et al. 2007).

In follow-up studies, Huyge (2012) and Huyge et al. (2012) investigated the suitability of different formulations of NDV vaccine powders by spray drying on a laboratory scale. Mannitol was combined with trehalose or inositol, PVP and/or BSA as stabilizers. Mannitol minimized hygroscopicity, and inositol was added as a substitute for trehalose exhibiting lower hygroscopicity. Aqueous feed solutions containing 5 and 10% w/w solids were spray-dried in a Mini Spray Dryer B-290 at a 7.5 mL/min feed rate with an inlet/outlet temperature of 150 °C/70–75 °C. Powder samples were stored in sealed glass vials at 6 °C (conditioned room) or 25 °C (incubator) packed in airtight aluminum bags. Individual experiments showed that a formulation of mannitol-inositol-BSA (73.3:13.3:13.3 w/w) gave the most suitable results:

- lowest vaccine titer loss after spray drying (1.6–2.0 $\log_{10}$ EID_{50}, 50% egg infectious dose) and after storage for 6 months at 6 °C (conditioned room) (max. 0.8 $\log_{10}$ EID_{50}),
- low moisture content (1.1–1.4 w%),
- low hygroscopicity (1.9–2.1 w% water uptake at 60% relative humidity), and
- good dry dispersibility properties and spherical particles ((Fig. 6.60b).

This powder combination was considered safe, although inhalation experiments should be performed to confirm this (Huyge et al. 2012).

Furthermore, Landman et al. (2015) administered the NDV vaccine strain LZ58 to 4-day-old broiler hens as a spray-dried powder or as a liquid spray. The powder vaccine was based on mannitol-BSA formulations and had a narrow particle size distribution. A coarse (37 μm, span 0.8) powder vaccine was produced using a 4M8-TriX spray dryer (Procept, Belgium) equipped with a 25 kHz ultrasonic nozzle. LZ58 vaccine virus suspensions of 15% w/w solids (60% w/w mannitol and 40% w/w BSA) and 100 mL batch size were spray-dried at an inlet temperature of 100 °C and a feed rate of 2 mL/min, resulting in an outlet temperature of approximately 65 °C. A fine powder vaccine (6.5 μm, span 1.7) was prepared using a Mobile Minor D-2000 (GEA Niro A/S) from 1500 mL batches of a 1% w/w solution containing 80% mannitol and 20% BSA, followed by spray drying at 120 °C/65 °C inlet and outlet temperatures and 12 mL/min flow rate. The broiler chickens were vaccinated in groups of 40 in isolators at 4 days of age and were intratracheally inoculated with *Escherichia coli* (strain 506) at 11 days of age. The immune response was determined by assaying the birds' sera for NDV-HI antibodies 2 weeks after vaccination. In conclusion, the coarse and fine powder vaccines did not differ significantly from their liquid counterparts regarding vaccination response and antibody response (Table 6.25).

Table 6.25 Research studies published on spray drying live-attenuated Newcastle Disease Vaccine (NDV) for mass vaccination of poultry (Summarized from literature Corbanie 2007; Corbanie et al. 2007, 2008; Huyge et al. 2012; Landman et al. 2015)

Vaccine type	Excipients, composition	Spray drying process parameters	Particle size, residual moisture	Administration route	Shelf life, thermostability	References
Clone 30 strain	400 ml batch size, 3% w/w trehalose, 1% w/w PVP, and 1% w/w BSA in water, 5% w/w mannitol in water 1000 mL batch size	Experimental spray dryer, piezo-actuated nozzle: 50 kHz, T_{in}/T_{out}: 90 °C/70 °C, Feed rate: 30 mL/h, Drying gas rate: 25 kg/h Mobile Minor (Niro GEA), 1 mm two-fluid nozzle, atomizing air pressure: 2 bar, T_{in}/T_{out}: 160 °C/70 °C, Feed rate: 25 mL/min, Drying gas rate: 80 kg/h,	Coarse: 28 µm (span 0.7), Fine: 7 µm, 69.2–71.7% yield	Respiratory (broiler chickens in isolators)	10 months at 25 °C (degradation of 0.2 log_{10} per month) (haemagglutination assay)	Corbanie (2007), Corbanie et al. (2007, 2008)
LZ58 strain	100 mL aqueous feed solutions (5%, 10% w/w), mannitol, trehalose, inositol, PVP, BSA, mannitol-inositol-BSA (73.3:13.3:13.3 w/w)	Mini Spray Dryer B-290 (Büchi), 0.7 mm two-fluid nozzle, T_{in}/T_{out}: 150 °C/ 70–75 °C, Feed rate: 7.5 mL/min	2–16 µm, 1.1–1.4% moisture, 1.9–2.1% water uptake at 60% RH, 58–90 °C T_g	no	Titre loss: 1.6 to 2.0 log_{10} EID_{50}, 50% egg infectious dose) after spray-drying, max. 0.8 log_{10} EID_{50} after 6 months at 6 °C (conditioned room)	Huyge et al. (2012)
LZ58 strain	100 ml of a 15% w/w aqueous solution containing 60 w% mannitol and 40 w% BSA 1500 ml of a 1% w/w aqueous solution containing 80 w% mannitol and 20 w% BSA	4M8-TriX spray dryer (Procept), 25 kHz ultrasonic nozzle, T_{in}/T_{out}: 100 °C/65 °C, F: 2 mL/min Mobile Minor D-2000 pilot spray dryer (GEA Niro), two-fluid nozzle (1 mm), airflow: 113.9 L/min, T_{in}/T_{out}: 120 °C/65 °C, F: 12 mL/min	37 µm (Span 0.8) 6.5 µm (Span 1.7)	Respiratory (broiler hens in isolators)	n.a	Landman et al. (2015)

6.21.3 Conclusions on Spray-Dried Newcastle Disease Vaccines

In conclusion, this brief review shows that only a few laboratory and pilot-scale studies have been conducted on spray drying of NDV vaccines. Research activities mainly cover Ghent University (Belgium) and Utrecht University (The Netherlands). Further scale-up is needed. Spray drying allows particle size adjustment and optimization of a narrow particle size distribution for more efficient targeting of poultry's upper or lower respiratory tract. The issue of thermostability as a consumer-friendly and cost-effective spray-dried NDV vaccine powder would be an innovative alternative drying technology to freeze-drying that is faster and potentially more cost-effective. Dry powder formulations based on trehalose-BSA or mannitol-inositol-BSA increase shelf life to 10 months at 25 °C storage temperature.

6.22 Tularemia Vaccines

6.22.1 A Short Introduction to Tularemia

Tularemia is a rare but potentially serious zoonotic disease of the Northern hemisphere caused by the coccobacillus bacterium *Francisella tularensis* (*F. tularensis*) (Ellis et al. 2002; WHO 2007; Sjöstedt 2007; Oyston 2008; Nigrovic and Wingerter 2008; Rubin 2018; Cieślik et al. 2018; Maurin 2020). In many parts of the world, the disease is colloquially known as rabbit fever, hare fever, deerfly fever, lemming fever, meat cutter's disease, and market men's disease (Ellis et al. 2002; Oyston 2008; Nigrovic and Wingerter 2008; Rubin 2018).

F. tularensis is one of the most infectious pathogenic bacteria known (Conlan and Oyston 2007), transmitting to humans either through the bite of an infected arthropod (e.g., ticks, mosquitoes), contact with infected animals, or through inhalation of aerosolized particles (Nigrovic and Wingerter 2008). In addition, inhalation of contaminated air containing as few as 10 bacteria can cause the disease (Saslaw et al. 1961b, a). The average annual number of cases reported in the U.S. ranged between 90 and 154 from 2001 to 2010 (Nelson et al. 2013).

The clinical manifestations of tularemia fall into two main groups: ulceroglandular (approximately 75% of patients) and typhoidal (about 25%) (Nigrovic and Wingerter 2008). Without treatment, the clinical course could progress to respiratory failure, shock, and death. Therefore, immediate therapy with antibiotics like streptomycin, gentamicin, doxycycline, or ciprofloxacin is recommended (Dennis et al. 2001; WHO 2007; Rubin 2018). After introducing effective antibiotics, the overall case fatality has dropped to approximately 2% (Nigrovic and Wingerter 2008; Oyston 2009). Figure 6.61a shows painful skin ulceration on the thumb of a patient infected with tularemia and (b) a tularemia lesion on the dorsal skin of a hand with characteristic heaped-up edges typically between 0.4 and 3.0 cm in diameter (Nigrovic and Wingerter 2008).

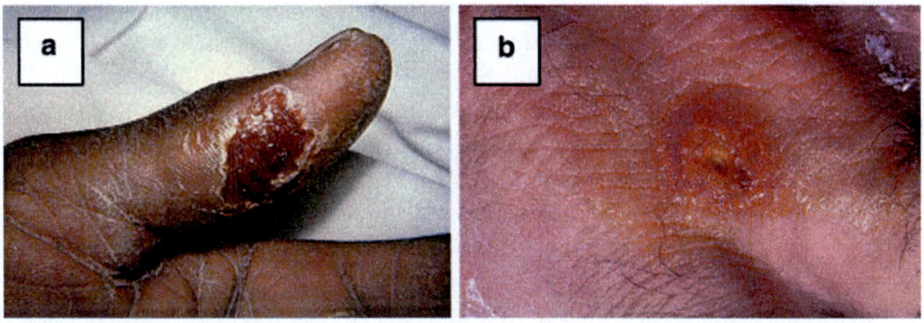

Fig. 6.61 **a** Thumb with a skin ulcer of tularemia. **b** A tularemia lesion on the dorsal skin of the right hand with characteristic heaped-up edges (Reproduced from Nigrovic and Wingerter (2008) with permission from Elsevier)

There is increasing concern about *F. tularensis* use as a powerful biological weapon because of its extreme infectivity, ease of dissemination, and substantial morbidity and mortality (Christopher et al. 1997; Klietmann and Ruoff 2001; Dennis et al. 2001; Nigrovic and Wingerter 2008; Cieślik et al. 2018). For example, estimates of casualties from a hypothetical biological attack with a release of 50 kg of bacteria by an aircraft along a 2-km line upwind of a population center of 500,000 could result in 30,000 deaths and 125,000 disabled people (Christopher et al. 1997). Therefore, the Centers for Disease Control and Prevention (CDC) classified *F. tularensis* as a class A bioterrorism agent (Nigrovic and Wingerter 2008).

6.22.2 Research on *F. Tularensis* Vaccines

As a result, there has been renewed research interest in developing a vaccine and other countermeasures against *F. tularensis* (Lu and Hickey 2007; Ohtake et al. 2011b; Marohn and Barry 2013). In addition, a readily available stockpiled vaccine would facilitate mass vaccination campaigns under threatening conditions (Ohtake et al. 2011b).

The history of *F. tularensis* vaccine development has been extensively reviewed (Conlan and Oyston 2007; Griffin et al. 2007; Cross et al. 2007; Oyston 2009; Marohn and Barry 2013). The first live-attenuated vaccine was developed in the former Soviet Union that partially protects against infection resulting from the respiratory and intracutaneous transmission of *F. tularensis* (Saslaw et al. 1961b, a). This vaccine has been used in limited settings with good results for protecting at-risk personnel (particularly laboratory personnel who routinely handle potentially infectious specimens). Attenuated live vaccines such as *F. tularensis* LVS (live vaccine strain) provide good, although not complete protection against high-dose aerosol challenges (Tärnvik 1989; Sandström 1994). In addition, serological tularaemia tests are varied and poorly standardized, so more accurate tests are needed to confirm the diagnosis in patients (Maurin 2020).

Currently, there is no licensed vaccine against tularemia (WHO 2007). However, several new effective vaccines have been characterized in animal models and are still under development (Marohn and Barry 2013). However, it remains critical to bridge the preclinical studies with animal models to human subjects due to the lack of reliable correlates of protection (Lyons and Wu 2007; De Pascalis et al. 2015).

Recent clinical studies with a new lot of LVS confirmed its immunogenicity in humans (El Sahly et al. 2009). The vaccine cell suspension was mixed with formulation buffer (10 mM potassium phosphate, 10% sucrose, and 1.3% gelatin), filled into 1-mL single-dose vials, and lyophilized. The final product was stored at -20 ± 10 °C. However, for long-term storage, this liquid vaccine must be stored frozen at −80 °C or below (Ohtake et al. 2011b).

More recently, Mulligan et al. (2017) conducted a phase 2 study with 228 subjects and compared a new batch of lyophilized LVS prepared according to cGMP with an existing vaccine. The reconstituted vaccine contained 10% sucrose, 1.9% gelatin, and 10 mM potassium phosphate. The new vaccine was safe, showed a rapid antibody response, and was well tolerated by the study population. After reconstitution, the vaccines were stored at 2–8 °C and used within 8 h.

In another study, Day et al. (2009) tested the long-term survival of lyophilized *F. tularensis* LVS in dehydrated infant food formula. Here, the vaccine titer decreased by > 3 $\log_{10}$ after 3 months of storage at ambient conditions, suggesting that *F. tularensis* may be more thermolabile than originally presumed. Thus, further development of vaccines against *F. tularensis* requires optimizing manufacturing and storage (Marohn and Barry 2013). Therefore, formulating a dried powder and bulk storage until a need for large-scale vaccination arises would be an economic strategy for a tularemia vaccine.

Ohtake et al. (2011b) demonstrated that *F. tularensis* LVS can be stabilized successfully in the dried state using foam drying with trehalose-gelatin-based formulations (e.g., 30% w/v trehalose, 5% w/v gelatin) (see Chap. 3, Sect. 3.6). The foam-dried *F. tularensis* LVS decreased less than 1 $\log_{10}$ in titer following 12 weeks of storage at 25 °C (Ohtake et al. 2011b), and there was no activity loss for at least 12 weeks under refrigerated conditions (4 °C). The reported stability is remarkable, but the scalability of the foam-drying process is yet to be demonstrated (Walters et al. 2014).

6.22.3 Research on Spray-Dried *F. Tularensis* Vaccine

To the author's best knowledge, there is only one published study of spray-dried *F. tularensis* vaccine research. As a proof-of-concept, (Kunda et al. 2016) used spray-drying to formulate a recombinant live attenuated Listeria monocytogenes (Lm) vaccine powder against *F. tularensis* bacteria. The Lm strain expresses *F. tularensis* immune protective antigen proteins, including IglC (intracellular growth locus subunit C), protecting mice against lethal *F. tularensis* LVS challenge administered intranasally (Jia et al. 2009). Non-reducing sugars such as trehalose, inositol, and mannitol were used to form hydrogen bonds with the phosphate

Table 6.26 Optimized spray drying conditions for live-attenuated bacteria Listeria monocytogenes (Lm) expressing Francisella tularensis antigen IglC (intracellular growth locus subunit C)

Process parameters	Optimal values
Spray dryer model	Büchi Mini Spray Dryer B-290 installed in a BioPROtect III Jr® Bio Safety Cabinet
Nozzle	Standard two-fluid nozzle (0.7 mm diameter)
Exhaust gas filters	PTFE membrane filter and 0.2 μm EMFLON filter
Formulation	2.05% w/v solution with Lm suspension containing 85.4% mannitol, 1.71% trehalose, 0.85% dextran, 7.85% L-leucine, and 4.27% inositol
Inlet/outlet temperature	130 °C/47 ± 2 °C
Drying gas flow rate	100% (aspirator rate setting)
Feed rate	6 mL/min (10% setting)
Spray gas flow rate	742 L/h (Nitrogen)
Feed solids concentration	21 g/100 mL (21% w/v)
Powder yield	74 ± 10% w/w
Residual moisture content	0.7% w/w (by TGA)
Particle size and morphology	2.2 ± 0.3 μm, smooth, spherical (by SEM)
Storage stability	> 1 year at room temperature (~23 °C), 6 months at 40 °C (bioassay)

Data from Kunda et al. (2016)

groups of the bacterial membrane and stabilize the bacteria during spray drying and storage. Mannitol provides low hygroscopy (Jin et al. 2010), trehalose has a high T_g (Ohtake et al. 2011b), dextran decreases crystallization and increases the T_g of dry powders (Tonnis et al. 2014a), inositol enhances bacterial survival during spray drying (Lin et al. 2011; Agarkhedkar et al. 2014), and leucine increases the dispersibility of dry powder through low hygroscopicity (Garcia-Contreras et al. 2008). Table 6.26 summarizes the optimized spray drying conditions.

As shown in Fig. 6.62, the spray dryer was housed in a BioPROtect III Jr® Bio Safety Cabinet to protect the dry powder vaccine from contamination and the operator's exposure to Lm. All further handling of the powders was performed in a biosafety cabinet. The spray drying parameters were 130 °C inlet and 47 ± 2 °C outlet temperature, 100% aspirator setting, 742 L/h nitrogen flow rate, and 6 mL/min feed rate. As a result, the dry Lm powders obtained a spray-drying yield of 74 ± 10% w/w. Furthermore, the geometric diameter of the Lm-loaded dry powders was approximately 2.2 ± 0.3 μm, an approximate particle size that can potentially be used for pulmonary immunization (Muttil et al. 2010b; Kunda et al. 2015a, b). The residual moisture content in the Lm powders was 0.7% (w/w). Figure 6.62 shows an SEM image of the spray-dried powder, consisting of smooth, spherical particles. After spray drying, complete encapsulation of Lm was achieved, although a few bacteria were observed outside some particles (indicated by yellow arrows).

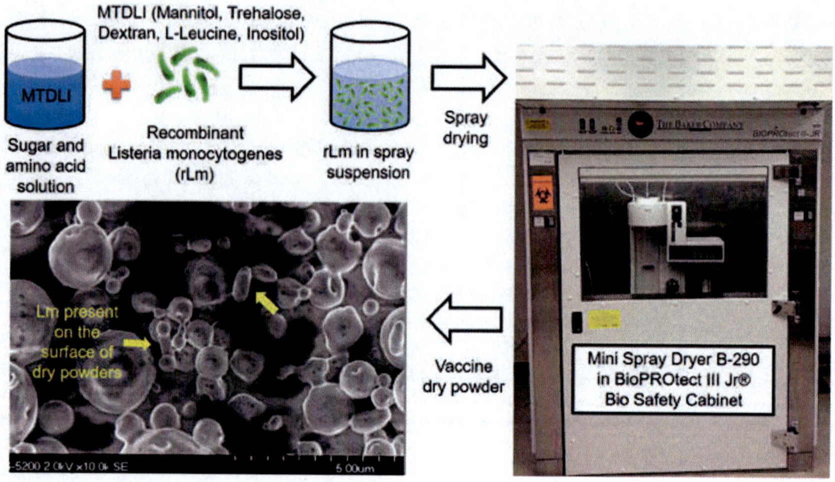

Fig. 6.62 The manufacturing process of spray-dried recombinant live attenuated Listeria monocytogenes (Lm) vaccine powder against *F. tularensis* (Pictures adapted from Kunda et al. (2016) with permission from Elsevier and courtesy of Prof. Pavan Muttil from the College of Pharmacy at the University of New Mexico in Albuquerque, USA)

Furthermore, the authors investigated the long-term stability of the Lm dry powders under different storage conditions. Powder samples stored under nitrogen purge (inert atmosphere) and with desiccant sachet and oxygen scavengers confined in the storage container exhibited a minimal loss of viability (approximately 0.5 $\log_{10}$) at 40 °C over 180 days, at both 15 and 75% relative humidity, as required by ICH (1995, 2003). At room temperature (around 23 °C), viability remained for more than a year.

Overall, it can be concluded that spray drying, modified atmosphere packaging, and improved storage conditions could be used for other live attenuated bacterial vaccines. In addition, the formulation in dry form with the thermal stability of 180 days at 40 °C is sufficient to administer the vaccines in remote regions without a reliable cold chain.

6.23 Pneumococcal Disease Vaccines

6.23.1 A Short Introduction to Pneumococcal Disease and Vaccines

Pneumococcal infections can lead to serious invasive diseases such as meningitis, septicemia, pneumonia, and milder but more common illnesses such as sinusitis and otitis media (O'Brien et al. 2016; WHO 2019e; Loughran et al. 2019) (Fig. 6.63). The causative agent is the bacteria Streptococcus pneumoniae, also known as

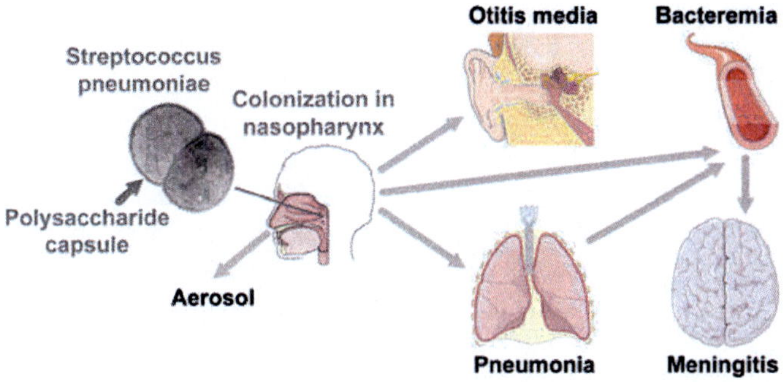

Fig. 6.63 Schematic representation of the spread and progression of *Streptococcus pneumoniae* infection. The bacteria are spread by aerosol from the nasopharynx to several different tissues. In children, the bacteria usually cause otitis media. The invasive disease usually begins in the lungs and spreads to the blood, with the most serious complication being meningitis (Adapted from Loughran et al. (2019) and O'Brien et al. (2016) with permission from Elsevier)

pneumococcus, colonizing the human nasopharynx and is transmitted mainly through respiratory droplets (i.e., aerosol). Infants and young children are the main reservoirs of this organism. *S. pneumoniae* bacteria are lancet-shaped, gram-positive, facultative anaerobic organisms (Gierke et al. 2021). Most pneumococci are encapsulated, and their surfaces are composed of complex polysaccharides.

Several vaccines are available to help prevent pneumococcal disease (Armstrong 2022). A 23-valent polysaccharide vaccine (PPSV23) was first approved in 1983, containing purified capsular polysaccharides from 23 of the most common S. pneumoniae serotypes and pneumococcal conjugate vaccines (PCV) have been available since 2009 (WHO 2019e). Today's pneumococcal vaccines on the market include:

- PPSV23 (Pneumovax23® from Merck Sharp & Dohme Corp.),
- PCV13 (Prevnar 13® from Wyeth Pharmaceuticals LLC, a subsidiary of Pfizer Inc.),
- PCV15 (Vaxneuvance® from Merck Sharp & Dohme Corp.),
- PCV20 (Prevnar 20® from Wyeth Pharmaceuticals LLC, a subsidiary of Pfizer Inc.), and
- PCV10 (Synflorix® from GSK Biologicals S.A.)

Moreover, several multi-valent vaccines are in Phase II and III clinical trials (Armstrong 2022). The WHO recommends the inclusion of PCVs in childhood immunization programs worldwide (WHO 2019e). Since 2017, *S. pneumoniae* has been one of the 12 priority pathogens of the WHO (Weiser et al. 2018). As of 2020, the global coverage rate of the third dose of PCV in 1-year-olds was 49%, with an increasing trend (WHO 2021g). The introduction and routine use of PCVs resulted in a significant decline in pneumococcal infections (CDC 2022).

Nevertheless, the vaccines do not protect against 100 documented serotypes (Gierke et al. 2021). Thus, there is still a possibility that someone will develop the pneumococcal disease after vaccination. A number of next-generation pneumococcal vaccine candidates are under development and in clinical trials, as reviewed by Masomian et al. (2020) and Oliveira et al. (2021). Current efforts are directed at increasing the number of serotypes covered by PCV or adding conserved pneumococcal proteins that induce serotype-independent immunity (Weiser et al. 2018).

In this regard, pneumococcal surface protein A (PspA) is considered one of the most prominent vaccine candidates and an alternative to polysaccharide-based vaccines (Tai 2006). In addition, PspA could be used to induce serotype-independent immunity at a low cost. The main function of PspA is to prevent the deposition of complement on the bacterium's surface, thereby inhibiting opsonization and phagocytosis (Anish et al. 2014; Kunda et al. 2015b; Gierke et al. 2021).

In addition, vaccine stability during shipment and distribution is crucial, especially in remote and rural places, since current pneumococcal vaccines are in a liquid dosage form and need to be stored between 2 and 8 °C (e.g., PPSV23 and PCV13) (WHO 2019e; Gierke et al. 2021). The Prevnar vaccine family (PCV suspensions for intramuscular injection) has a shelf life of 2 years at 2–8 °C and temperature stability of 4 days at 40 °C (Hill et al. 2016).

6.23.2 Research on Spray-Dried Pneumococcal Vaccines

Dry powder vaccine formulations offer improved antigen stability and integrity, do not necessarily require a cold chain, and can be administered non-invasively. These benefits have prompted several research groups to formulate a spray-dried vaccine against pneumococcal disease to improve its stability and investigate its feasibility in scale-up and implementation for pulmonary immunization (Anish et al. 2014; Kunda et al. 2015b; Rodrigues et al. 2018; Mehanny et al. 2022). Table 6.27 summarizes the most relevant research studies published on spray drying pneumococcal vaccines.

Anish et al. (2014) encapsulated recombinant PspA in polylactide (PLA, 45 kDa) particles by spray drying in a laboratory spray dryer (LU-227 Advanced, LabUltima, India) with a 0.5 mm two-fluid nozzle and twin-cyclone. Optimized process parameters were determined by a semi-factorial experimental design (Table 6.27). First, polymer particles were prepared as a water-in-oil-in-water (W1/O/W2) double emulsion by solvent evaporation. Then, either W1/O primary emulsions or W1/O/W2 secondary emulsions were spray-dried (Fig. 6.64).

The internal aqueous phase (W1) contained recombinant PspA, rat serum albumin, sodium bicarbonate, and sucrose, while the organic phase consisted of PLA in DCM (dichloromethane). The external aqueous phase contained PVA and sucrose as excipients. Spray drying of the W1/O/W2 secondary emulsion resulted in uniform particles of about 2–8 μm with narrow size distribution and smooth surface morphology (Fig. 6.64b). A high process yield of 79%, adequate antigen

Table 6.27 Research studies published on spray-dried pneumococcal vaccines

Vaccine type	Excipients, composition	Spray dryingprocess parameters	Particle size, moisture	Administration route (animal model)	References
Pneumococcal membrane vesicles (MVs)	Lactose (2.5% w/v), leucine (1% w/v), 2.5 mL of pneumococcal MVs suspension mixed with 17.5 mL of carrier solution, with or without fluorescein sodium	Nano Spray Dryer B-90 (Büchi), T_{in}/T_{out}: 90 °C/n.a., D: 110 L/min, 140 kHz, spraying: 40%, over-pressure 40 mbar	2.06 µm, corrugated surface, nanocrystalline/amorphous nature	Pulmonary (in-vitro), MMAD 2.3 µm, FPF 79.7%, emitted dose 88.8%	Mehanny et al. (2022)
PspA from clade 4 of family 2 (PspA4Pro)	Antigen-loaded polyvinyl alcohol (PVA) nanoparticles embedded in L-leucine microparticles (1:1.5 w/w ratio)	Mini Spray Dryer B-290 (Büchi), 0.7 mm two-fluid nozzle, HP cyclone, T_{in}/T_{out}: 100 °C/45–47 °C, F: 10%, A: 400 L/h, D: 100% aspirator	~2.0 µm, irregular corrugated surface, 56 ± 7% yield	Pulmonary (in-vitro, BALB/c mice), FPF 74.3% (< 4.46 µm), MMAD ~1.7 µm	Kunda et al. (2015b), Rodrigues et al. (2018)
Pneumococcal surface protein A (PspA)	W1/O/W2 emulsions, W1: PspA (5 mg/mL), rat serum albumin (RSA) (2.5% w/v), sodium bicarbonate ($NaHCO_3$) (2% w/v), sucrose (10% w/v), O: Poly-DL-lactide (PLA, 45 kDa, PURASORB) in DCM (dichloromethane) (50 mg/mL), W2: PVA (1% w/v), sucrose (10% w/v)	LU-227 laboratory spray drier (LabUltima), twin-cyclone, 0.5 mm two-fluid nozzle, T_{in}/T_{out}: 100 °C/n.a., F: 2 mL/min, D: 80% aspirator rate, A: 30 L/min (2 bar)	2–8 µm (Span 0.66), 79% yield, 3.7% w/w residual moisture content	Intramuscular (mice and rats), robust IgG responses	Anish et al. (2014)

Summarized from literature Anish et al. (2014), Kunda et al. (2015b), Rodrigues et al. (2018), Mehanny et al. (2022), *n.a.* not available

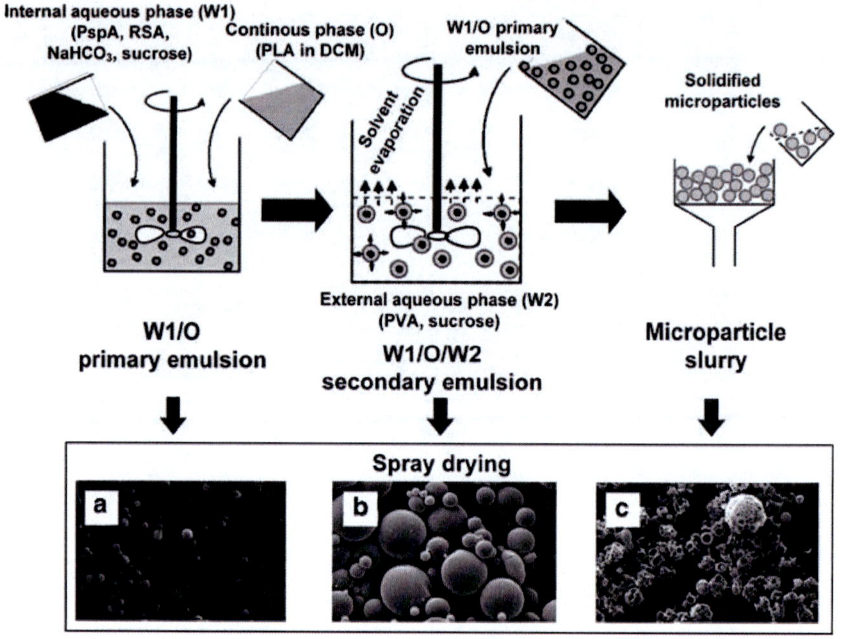

Fig. 6.64 Spray-dried polymer particles entrapping recombinant pneumococcal surface protein A (PspA) after various stages of the double emulsion evaporation method. **a** from the W1/O primary emulsion, **b** from the W1/O/W2 secondary emulsion, and **c** from spray drying microparticle slurry. DCM: dichloromethane, $NaHCO_3$: sodium bicarbonate, PLA: polylactide, PVA: polyvinyl alcohol, RSA: rat serum albumin (Adapted from Anish et al. (2014) with permission from Elsevier)

loading, and good flow properties was achieved. Porous particles were obtained by spray drying the slurry of preformed microparticles prepared by solvent evaporation in the double emulsion (Fig. 6.64c). The presence of PVA in the external aqueous phase ensured the sustained release of antigen protein from the particles. The optimized spray-dried vaccine formulation (suspended in sterile saline and injected intramuscularly) elicited robust IgG responses in animal models (Balb/c mice and Wistar rats). Finally, the authors concluded that spray drying might be an attractive method for large-scale production of biodegradable PLA microparticles, incorporating antigens for vaccine formulation development.

Kunda et al. (2015b) synthesized around 300 nm nanoparticles of poly(glycerol adipate-co-ω-pentadecalactone) (PGA-co-PDL) by an emulsion/evaporation process and adsorbed the antigen PspA from clade 4 of family 2 (PspA4Pro) onto the surface. Nanoparticle suspensions were spray-dried in L-leucine solution, and nanocomposite microparticles (NCMPs) were prepared for mucosal delivery targeting the lungs. A Mini Spray Dryer B-290 was applied at a 10% feed rate, 400 L/h atomizing airflow, 100% aspirator capacity, 100 °C inlet, and 45–47 °C outlet temperature. SEM pictures revealed approximately 2 μm NCMPs with a corrugated surface (Fig. 6.65). A reasonable process yield of 56 ± 7% was obtained.

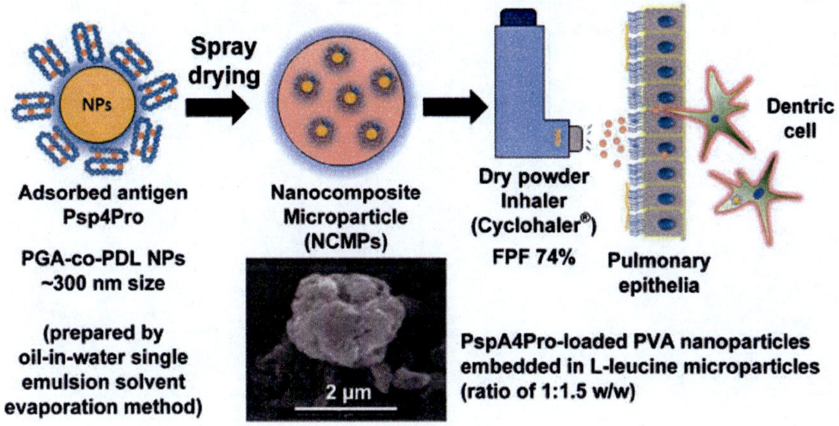

Fig. 6.65 Spray dried nanocomposite microparticles (NCMPs) for pulmonary delivery containing pneumococcal antigen (PspA4Pro) adsorbed to PGA-co-PDL nanoparticles (NPs, around 300 nm size, prepared by oil-in-water single emulsion solvent evaporation) and L-leucine, FPF: fine particle fraction (Adapted from Kunda et al. (2015b) with permission from Elsevier)

In addition, in-vitro lung deposition data of the spray-dried powders using a Cyclohaler® placed in a Next Generation Impactor (Copley) displayed a fine particle fraction (FPF < 4.46 μm) of 74% and an MMAD of 1.7 μm, suggesting a broncho-alveolar lung deposition facilitating the uptake of the nanoparticles by dendritic cells (Kunda et al. 2013). About 20 μg of PspA4Pro was adsorbed onto the surface per mg of PGA-co-PDL nanoparticles. Moreover, the released antigen from the dry powders maintained stability (primary and secondary structure), integrity, and activity. Further investigations were suggested to focus on determining the immunogenicity of the released PspA4Pro in animal models.

Rodrigues et al. (2018) continued the studies of Kunda et al. (2015b) with lung-targeted mucosal immunization of BALB/c mice. Spray-dried NP/NCMPs formulations containing 2 μg PspA4Pro were suspended in saline (50 μl/dose) and administered under anesthesia by nasal instillation into one nostril using a micropipette. Nasal instillation ensured that most of the inoculum reached the lungs of the mice. The mice received two doses of the formulations at a 15-day interval. Inoculation of this formulation induced anti-PspA4Pro IgG antibodies in serum and lungs. Serum IgG showed good binding efficiency against intact bacteria expressing PspA from clades 3, 4, and 5 of homologous pneumococcal family 2. However, no binding was observed to bacteria expressing PspA from clades 1 and 2 of family 1.

The authors declared that further improvements in the formulation were needed. Nonetheless, vaccination targeting the lungs is a promising alternative to induce local immune responses against respiratory pathogens, including *M. tuberculosis*, influenza virus, *S. pneumoniae*, and other diseases. In addition, the authors reiterated the advantage of dry powder vaccines for humans, as they could be transported without a cold chain requirement and delivered to the lungs with a powder inhaler or nebulizer after resuspension in liquid.

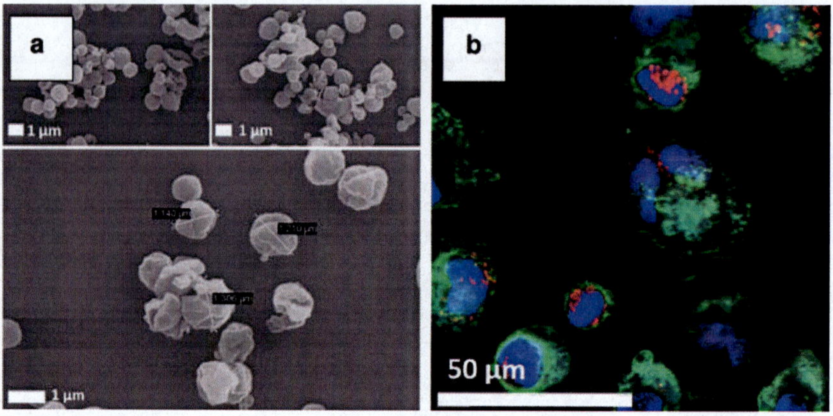

Fig. 6.66 **a** SEM images of pneumococcal MVs-loaded vaccine particles (1–2 µm) prepared with a Nano Spray Dryer B-90), **b** Confocal Laser scanning micrographs of fluorescently labeled pneumococcal MVs (red dots) in vaccine microparticles uptake into human macrophage-like cells after 8 h of incubation, cytoskeleton was labeled green and nuclei in blue (Adapted from Mehanny et al. (2022) with permission from Elsevier)

In addition to PspA, bacterial extracellular membrane vesicles (MVs) represent another potential candidate for developing cell-free vaccines against microbial infections. MVs harbor many microbial components, including virulence factors and antigens, and can therefore interact with the host immune system (Mehanny et al. 2021). Recently, Mehanny et al. (2022) presented a study on spray-dried pneumococcal MVs-loaded vaccines with lactose/leucine as inert carriers to improve their stability and flow properties. The focus was to investigate the feasibility of scale-up and translation for pulmonary immunization. A suspension of pneumococcal MVs (2.5 mL with 10^{12} particles/mL) was mixed with 17.5 mL of carrier solution (2.5% w/v lactose and 1% w/v leucine), with or without fluorescein sodium solution, followed by gentle magnetic stirring. The formulation was spray-dried using a Nano Spray Dryer B-90 (Flawil, Switzerland) at 90 °C inlet temperature, 140 kHz spray vibration frequency, 40% spray rate, and 110 L/min drying gas flow. The spray-dried vaccine particles were collected from the electrostatic particle collector with a plastic scraper and transferred to glass vials wrapped with aluminum foil for storage in a desiccator. Figure 6.66 shows SEM and confocal laser scanning micrographs of fluorescently labeled spray-dried pneumococcal MVs (red dots) from vaccine microparticles into human THP-1 macrophage-like cells (leukemia monocytic cells) after 8 h of incubation.

The optimized vaccine particles exhibited an average particle size of 1–2 µm, a corrugated surface, and a nanocrystalline nature. Their aerodynamic diameter of 2.34 µm, average emitted dose of 88.8%, and FPF of 79.7% indicated suitable flow characteristics for deep alveolar delivery using a next-generation impactor. In addition, confocal microscopy confirmed the successful encapsulation of pneumococcal MVs within the prepared microparticles. Almost 60% of fluorescently labeled MVs were taken up by the cells. Accordingly, the authors concluded that pneumococcal MVs in spray-dried vaccine microparticles are internalized into immune cells, deliver their antigen payload, and are expected to initiate an immune response. The authors also concluded that the optimized vaccine microparticles are suitable for further animal testing, scale-up, and clinical translation.

6.23.3 Conclusions on Spray-Dried Pneumococcal Vaccines

In short, this brief review presented new approaches to developing an effective pneumococcal vaccine by spray drying. Pneumococcal surface protein (PspA) and membrane vesicles (MVs) are promising cell-free vaccine candidates. Research is mainly focused on the formulation and design of spray-dried polymer particles. Encapsulation strategies include emulsion processes and polymeric nanoparticles in carrier materials. It has been reported that encapsulation unalters protein integrity, and robust immune responses are elicited in mice and rats. The process parameters of spray drying on a laboratory scale and particle design are optimized. The next steps are further animal testing, scale-up, and clinical translation. The aim is a pulmonary powder vaccine that can be administered to the lungs via an inhaler or nebulizer. Finally, thermostability studies are additionally needed to provide shelf-life information.

6.24 Meningococcal Disease Vaccines

6.24.1 A Short Introduction to Meningococcal Disease and Vaccines

Neisseria meningitidis, often referred to as meningococcus, is a bacterium that can cause meningitis and other forms of meningococcal disease such as life-threatening sepsis (Rosenstein et al. 2001; Stephens et al. 2007). Humans are the only host, and transmission occurs via aerosol droplets or contact with respiratory secretions (WHO 2020f). The bacterium is a round diplococcus and about 0.6–1.0 μm in size (Fig. 6.67). There are 13 serogroups of meningococci, six of which (A, B, C, W-135, X, and Y) are recognized as the main causes of disease (WHO 2022a).

Meningococcal meningitis occurs worldwide; however, recurring epidemics mainly affect sub-Saharan Africa, known as the "meningitis belt," which comprises 26 countries, from Senegal in the west to Ethiopia in the east (WHO 2020g) (Fig. 6.68).

Today, meningococcal meningitis is largely a vaccine-preventable disease, and several licensed vaccines are available for protection from the most common serogroups, including monovalent (A or C), quadrivalent (A, C, W, Y), or combination formulations (serogroup C, *Haemophilus influenzae* type b) (WHO 2020f, 2022a).

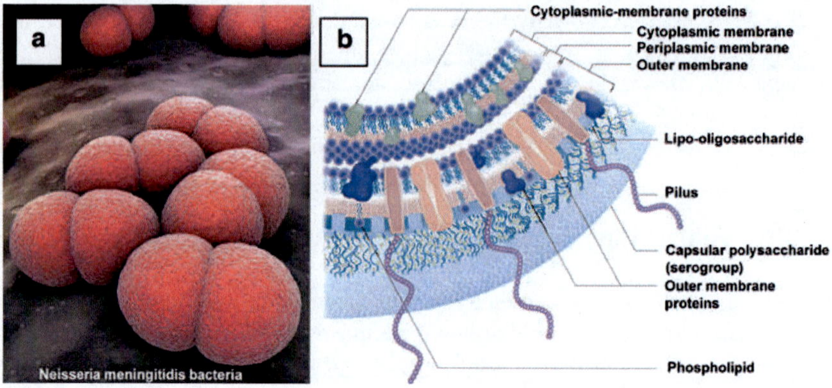

Fig. 6.67 a *Neisseria meningitidis* bacteria, the cause of meningococcal meningitis (Adapted from Encyclopædia Britannica (2022), ©Ilexx/Dreamstime.com), **b** Cross-sectional view of the meningococcus (Adapted from Stephens et al. (2007) with permission from Elsevier)

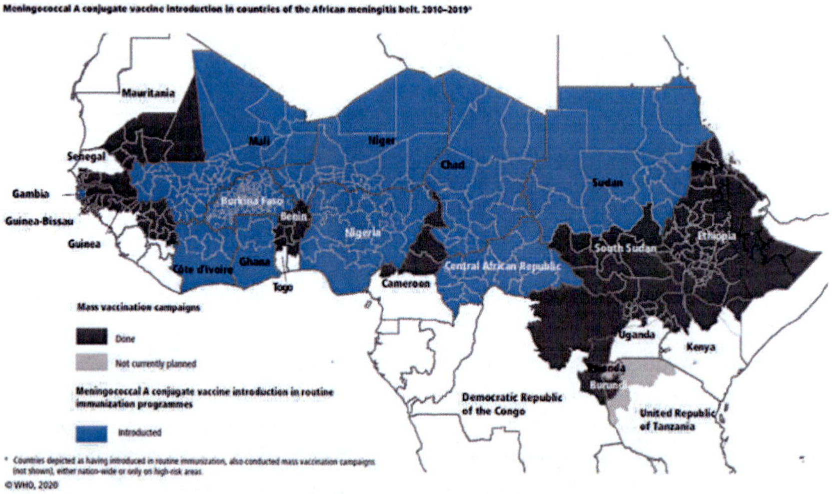

Fig. 6.68 Mass vaccination campaigns and Meningococcal A conjugate vaccine introduction in countries of the African meningitis belt, 2010–2019 (Adapted from WHO (2020g)

WHO recommends that countries with high (> 10 cases per 100,000 population/year) or intermediate cases (2–10 cases) conduct large-scale meningococcal vaccination programs (WHO 2011, 2022a). In countries where the disease occurs less frequently (< 2 cases), meningococcal vaccination is recommended for certain risk groups, e.g., laboratory personnel, travelers at risk of exposure, and persons with immunodeficiency. In addition, the International Coordinating Group (ICG) manages a global vaccine safety stockpile for vaccine provision for meningococcal meningitis. From 2009 to 2016, 20 million meningococcal vaccine doses were shipped to 13 countries for emergency response (WHO 2022b).

In the Meningitis Vaccine Project, a partnership between WHO and PATH, an affordable meningococcal A conjugate vaccine was developed (MenAfriVac™) (Okwo-Bele et al. 2015). Since its introduction in 2010, the cases and outbreaks of serogroup A across the "meningitis belt" have dramatically reduced (WHO 2022a). Before, major epidemics occurred every 7–14 years and resulted in excessive morbidity and mortality among children and young adults (WHO 2012b). However, the risk of outbreaks due to other serogroups continues.

In 2012, a lyophilized powder MenAfriVac™ vaccine was approved and prequalified by the WHO for use in a Controlled Temperature Chain (CTC) (WHO 2012b). Experience from mass vaccination campaigns in Chad (Lydon et al. 2014), Benin (Zipursky et al. 2014), and Togo (Mvundura et al. 2017) have shown that dry, heat-stable vaccine formulations under CTC reduce the need for a last-mile cold chain, which results in lower overall logistics, storage, and transportation costs (VIPS 2019), especially in hard-to-reach and resource-poor areas. Savings were also achieved by not having to purchase additional cold chain equipment for the vaccination campaign.

6.24.2 Research on Spray-Dried Meningococcal Vaccine

Research on spray drying of meningococcal vaccines is scarce, which is mainly attributed to the already available and licensed dry powder vaccine products in lyophilized form (WHO 2020f). Table 6.28 briefly summarizes the main findings of published laboratory tests on spray-dried meningococcal vaccines (Chen et al. 2010; Ubale et al. 2013).

Chen et al. (2010) developed spray-dried formulations of Neisseria meningitidis A (MenA) protein-polysaccharide conjugate vaccine in a pilot-scale Mobile Minor™ (GEA Niro, Denmark). The Serum Institute of India Ltd. provided the bulk vaccine as a freeze-dried formulation consisting of Neisseria meningitidis polysaccharide A (PSA), tetanus toxoid, Tris, mannitol, and sucrose. For the spray-drying studies, trehalose was added as the main stabilizer to achieve a glassy state, with sodium hydrogen phosphate (NaH_2PO_4) as a further excipient. N_2 was used as drying gas, and 82–97% process yields were achieved. The total solids concentration was 11 w%, and the spray drying parameters were 0.52 kg/h feed rate, 90–95 °C inlet, and 60–71 °C outlet temperature. The batch size of the feed mixture per experiment was approximately 1 L, sufficient to produce 5000–10,000 vaccine doses. The spray-dried powder vaccine had a particle size of 3.1–4.3 μm and a residual moisture content of 2.2–3.4 w%. Remarkably, the vaccine remained thermostable for 2 weeks up to 60 °C. In addition, the particle size did not change after storage. After single immunization with a reconstituted spray-dried formulation stored at 37 °C for 15 months, antibody responses in Balb/c mice were not significantly different from those of a standard liquid vaccine stored at 4 °C.

Table 6.28 Research published on spray-dried meningococcal vaccines

Vaccine type	Excipients, composition	Spray drying process conditions	Particle size, morphology, residual moisture	Route of administration, in-vitro, in-vivo	Shelf life, thermostability	References
Meningitis group A Protein-polysaccharide conjugate	Trehalose, Adjuvant: AlPO4, 11 w% solid concentration	Mobile Minor, 0.8 m chamber diameter, T_{in}/T_{out}: 90–95 °C/ 60–71 °C, D: N2 gas, F: 0.52 kg/h	3.1–4.3 um 2.2–3.4% RMC, 82–97% process yield	Balb/c mice, Intramuscular, 1 L batch (5000–10,000 doses)	> 20 weeks (40 °C), > 2 weeks (60 °C)	Chen et al. (2010)
Capsular polysaccharides (CPS)	BSA Adjuvant: unglycosylated meningococcal lipid A	Mini Spray Dryer B-191, 0.5 mm nozzle T_{in}/T_{out}: 110 °C/74 °C, D: 95% aspirator, F: 20 mL/h	70–75% process yield, 85–90% antigen content	In-vitro, macrophage cell culture	n.a	Bejugam et al. (2008), Uddin et al. (2009), Ubale et al. (2013)

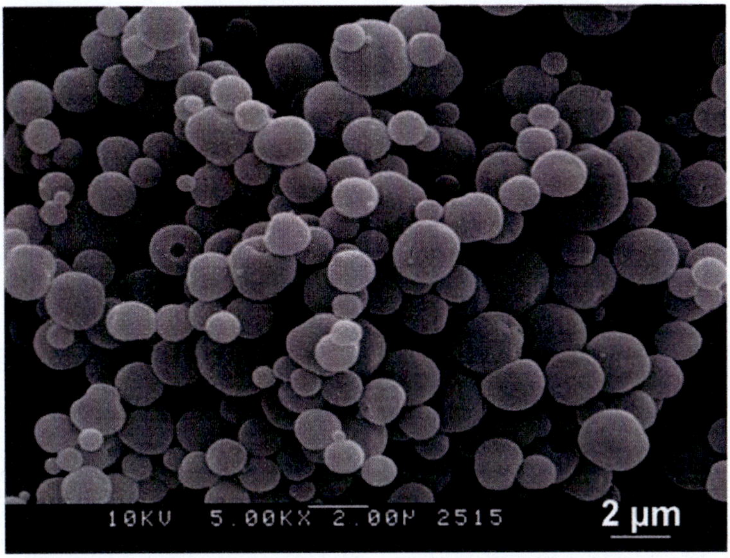

Fig. 6.69 Spray-dried albumin microparticles with encapsulated meningococcal capsular polysaccharides (CPS) (bar = 2 µm) (Adapted from Ubale et al. (2013) with permission from Taylor & Francis)

Capsular polysaccharides (CPS) are a major virulence factor in meningococcal infections and form the basis for serogroup designation and protective vaccines. In this context, Ubale et al. (2013) reported on a novel formulation of spray-dried microparticles loaded with meningococcal CPS. The authors encapsulated the vaccine in biodegradable albumin microparticles that mimic the chemical conjugation process of CPS with a protein carrier, enhancing antigen uptake via albumin receptors and eliciting a T-cell-dependent immune response (Fig. 6.69).

A 1% bovine serum albumin (BSA) solution was prepared and kept overnight for cross-linking with glutaraldehyde. The polysaccharide antigens and adjuvant (non-toxic meningococcal unglycosylated lipid A) were added to the solution and spray dried using a Mini Spray Dryer B-191 at a feed rate of 20 mL/h and 110 °C/74 °C inlet/outlet temperature (Bejugam et al. 2008; Uddin et al. 2009). The yield after spray drying was 70–75%, and about 90% of the CPS-loaded microparticles were between 1 and 2 µm in size and had a smooth surface and generally spherical shape (Fig. 6.69).

The micro/nanotechnology-based meningococcal vaccine formulation can be stored as a dry powder, allowing controlled antigen release. The albumin matrix encapsulation did not hamper CPS's immunostimulatory effect. Cytotoxicity studies showed that the vaccine-loaded microparticles were robustly taken up by macrophage cells and elicited strong innate immune responses that enhanced antigen presentation, which is a prerequisite for eliciting adaptive immune responses.

In summary, the reviewed studies demonstrate the feasibility of producing dry, thermostable, spray-dried meningococcal vaccines with advanced processing and formulation technology. Formulations containing trehalose withstand accelerated thermostability for up to 20 weeks at 40 °C and 2 weeks at 60 °C and promise to simplify the logistics of vaccine distribution out of the cold chain and may expand the vaccine coverage. As an alternative to lyophilization, spray drying offers advantages in particle size control, making the powder compatible with several alternative vaccine delivery routes, including aerosol and dry powder jet injection. Powder sizes of < 5 µm are suitable for pulmonary delivery or oral vaccination and uptake by Peyer's patch M cells. Vaccine encapsulation into biodegradable albumin as a protein matrix allows the powder to be filled into capsules for oral administration or other package types and varying dose quantities, providing further flexibility.

6.25 mRNA Vaccines

6.25.1 A Short Introduction to mRNA Vaccines

In response to the COVID-19 pandemic, current vaccine research is moving from first-generation live or inactivated whole viral and bacterial vaccines toward next-generation subunit, toxoid, plasmid DNA, and messenger (mRNA) vaccines. In particular, in vitro-transcribed mRNA vaccines open up a new era in vaccinology, offering capacity for rapid development, the potential for cost-effective manufacturing (e.g., large amounts in GMP quality), high potency, and safe administration, as described by Pascolo (2004, 2006, 2008, 2014, 2017), Brito et al. (2015), Pardi et al. (2018), Alameh et al. (2020), Kim et al. (2021), Rosa et al. (2021), Verbeke et al. (2021), Dolgin (2021), Chaudhary et al. (2021), Hussain et al. (2022). Various positive attributes of other vaccine platforms are combined in mRNA vaccines, including (Brito et al. 2015):

- Induction of both humoral and cell-mediated immunity (as live attenuated vaccines),
- focused immune responses (as subunit vaccines),
- safety and simplicity (as plasmid DNA vaccines), and
- robust, generic manufacturing methods (as plasmid DNA vaccines).

Figure 6.70 shows a representative schematic of mRNA vaccines and the mechanism of antigen expression in a cell (Maruggi et al. 2019). Both conventional and self-amplifying mRNA share the basic elements of encoding an antigen of interest (GOI) flanked by untranslated regions 5′ UTR and 3′ UTR, a cap, and a poly(A) tail of variable length. In addition, self-amplifying mRNA encodes viral nonstructural proteins (nsPs) required for intracellular RNA amplification and high antigen expression levels (Hekele et al. 2013; Brito et al. 2015; Ulmer et al. 2015; Maruggi et al. 2019; Blakney et al. 2021).

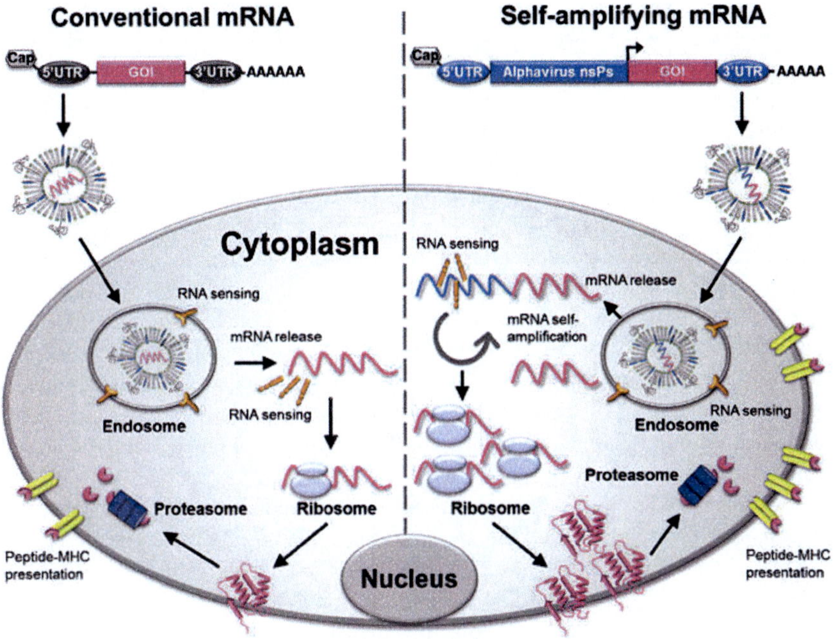

Fig. 6.70 Schematic representation of mRNA vaccines and mechanism of antigen expression (GOI: gene of interest; MHC: major histocompatibility complex, nsPs: nonstructural proteins) (Adapted from Maruggi et al. (2019) with permission from Elsevier under open access CC BY-NC-ND 4.0 license)

Compared to plasmid DNA and viral vectors that need to reach the nucleus, mRNA vaccines are active in the cell's cytoplasm and do not integrate the genome (Kallen and Theß 2014; Delany et al. 2014; Steinle et al. 2017; Liu 2019). Interestingly, whether they target tumors or infectious diseases, mRNA-based vaccines are formally considered gene therapy products or not (Hinz et al. 2017). Under current legislation in the EU, mRNA-based vaccines must be centrally approved in most cases.

Furthermore, in the cytoplasm, the translation of the genetic information encoded on the mRNA to produce the desired protein starts instantly (Guan and Rosenecker 2017). Finally, the mRNA is rapidly degraded by cellular processes in the cytoplasm so that no traces can be found after a few days to less than a week (Pascolo 2008, 2021a; Plews et al. 2010; Steinle et al. 2017; Uddin and Roni 2021). Furthermore, the in vivo half-life can be regulated through various modifications to the mRNA sequence (e.g., 3′ UTR and poly(A) tail length) and delivery vehicle (Pardi et al. 2018; Pascolo 2021a). Thus, the therapeutic time window of mRNA vaccines is well defined.

Another major benefit of mRNA vaccines is that they can be produced in vitro within days after obtaining gene sequence information, using completely synthetic manufacturing processes, making them suitable for rapid responses to newly emerging pathogens (Hekele et al. 2013; Brito et al. 2015; Ulmer et al. 2015; Richner et al. 2017; Pardi et al. 2018; Maruggi et al. 2019; Rosa et al. 2021). Moreover, the genetic code of mRNA vaccines can be rapidly modified to address new virus mutations (AboulFotouh et al. 2021).

Figure 6.71 shows a timeline of innovations that have contributed to developing mRNA vaccines and associated technologies (Blakney et al. 2021). As described by Dolgin (2021), hundreds of scientists had worked on mRNA vaccines for decades before the coronavirus pandemic brought the breakthrough.

In the use of mRNA for vaccination, a breakthrough was in 1993 when Martinon et al. (1993) demonstrated that in vitro produced mRNA encapsulated in liposomes could be used as a vaccine. A flu nucleoprotein-encoding mRNA-liposome formulation injected subcutaneously or intravenously triggered an anti-influenza response in mice. In 1995, Conry et al. (1995) showed that intradermal injections of carcinoembryonic antigen-encoding mRNA could induce antibodies in tumor-challenged mice.

Another breakthrough, although initially not intended at all for vaccines but purposed for the opposite, i.e., therapies, was provided by the fundamental studies of Karikó et al. (2005, 2008), which showed that modifications naturally found in human RNA, such as pseudouridine, thiouridine, and 5-methylcytidine, reduced its immunostimulatory potential. Replacing nucleotides like uridine with pseudouridine increased translational capacity and RNA stability compared to unmodified mRNA.

In 2003, synthetic mRNA was injected for the first time into humans (Probst et al. 2007; Weide et al. 2008, 2009). Dr. Steve Pascolo, cofounder of CureVac in 2000 and Chief Scientific Officer until 2006, volunteered as the first study subject and was injected by the dermatology Dr. Benjamin Weide in the University Clinic of Tuebingen with naked unmodified mRNA coding luciferase (bioluminescent) into the skin (Probst et al. 2007; Pascolo 2021b) (Fig. 6.72a). Sixteen hours after injection, 3-mm diameter punch biopsies were taken from the skin of the injection site. The in vivo production of luciferase was detected (Fig. 6.72b), and the emitted luminescence in the presence of luciferin substrate was above the background level (Fig. 6.72c). In contrast, punch biopsies outside the mRNA injected zone ("Mock") demonstrated no luciferase activity. The experiment was repeated three times. Those results demonstrated the first report on cellular uptake of unmodified synthetic mRNA in vivo and local protein expression after injection in human skin. It also demonstrated, as in mice, that injections of non-modified synthetic mRNA are safe. This way, the data obtained in mice could be extended to humans.

In 2020, the first synthetic COVID-19 spike-encoding mRNA vaccine was approved for emergency use (Polack et al. 2020; Baden et al. 2021). Notably, both COVID-19 vaccines from BioNTech/Pfizer and Moderna use mRNA in which pseudouridine replaces uridine (Nance and Meier 2021). This change is thought to prevent the immune system from reacting to the introduced mRNA and counterintuitively increase vaccine efficacy.

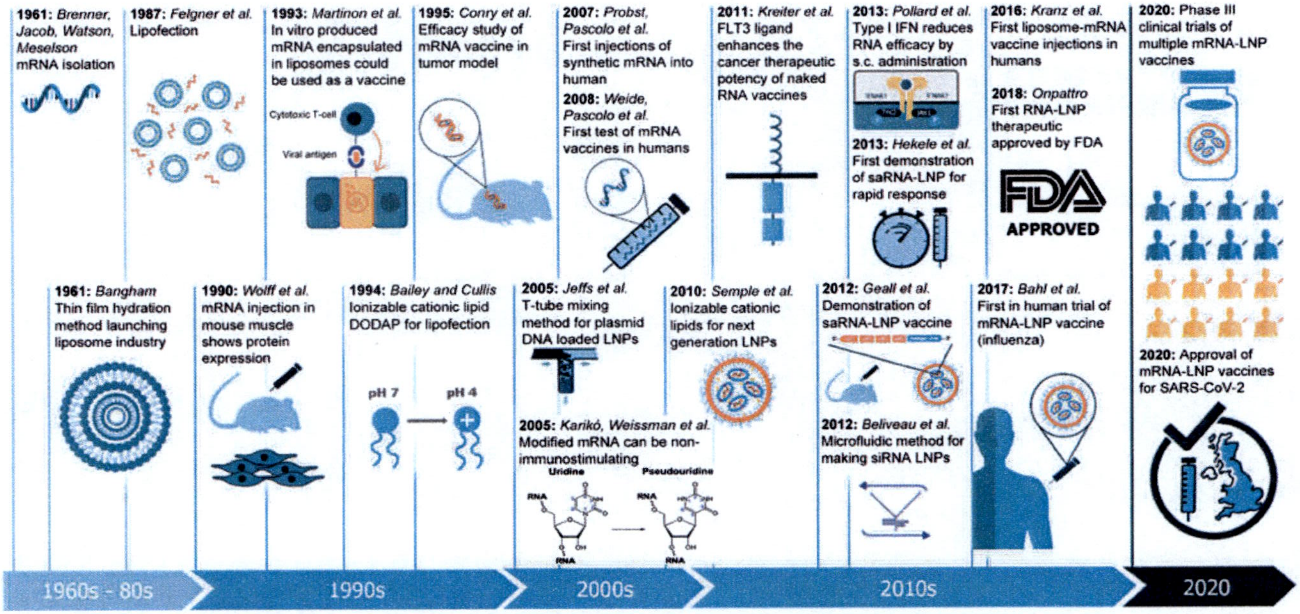

Fig. 6.71 Timeline of innovations that have contributed to the development of mRNA vaccines and associated technologies (Adapted and extended from Blakney et al. (2021), published by MDPI under open access CC BY 4.0 license)

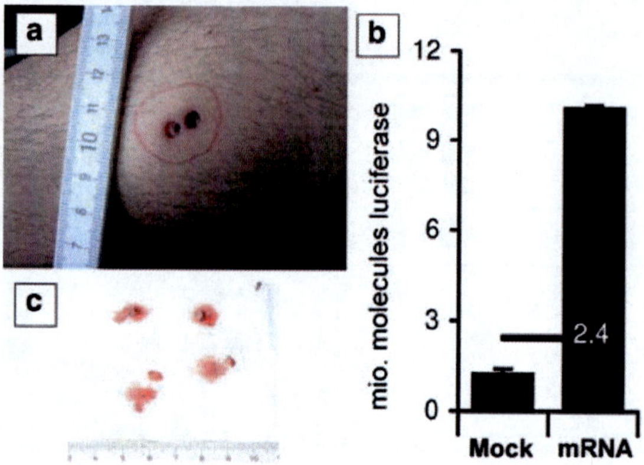

Fig. 6.72 The first injection of in vitro-transcribed mRNA coding luciferase into human skin (Probst et al. 2007). Dr. Steve Pascolo, cofounder of CureVac in 2000 and former Chief Scientific Officer until 2006, volunteered as the first study subject and got injected with 120 μg unmodified synthetic mRNA intradermally into his leg skin. **a** 3-mm punch biopsies are taken at the mRNA injection site (within a circle of approximately 2 cm). **b** Measurement of luciferase activity in punch biopsies that received mRNA and comparison with other biopsies made from skin distant from the injected site ("Mock"). **c** Biopsies from the mRNA area emitting luminescence above the background level in the presence of luciferin substrate (Figures adapted from Pascolo (2021b), published by MDPI under open access CC BY 4.0 license)

In addition, before COVID-19, various mRNA vaccines were already tested in preclinical and clinical trials for infectious diseases like influenza, zika, HIV, ebola, rabies, tuberculosis, chikungunya, malaria, genital herpes, toxoplasma gondii, cytomegalovirus, and others, as summarized in several reviews (Richner et al. 2017; Bahl et al. 2017; Hajj and Whitehead 2017; Evers et al. 2018; Kowalski et al. 2019; Maruggi et al. 2019; Liu 2019; Alameh et al. 2020; Gómez-Aguado et al. 2020; Buschmann et al. 2021; Blakney et al. 2021; Pascolo 2021a; Tenchov et al. 2021; Chaudhary et al. 2021; Hussain et al. 2022; Shi et al. 2022; Barbier et al. 2022).

A challenge of mRNA vaccines is to cross the cell membrane to reach the cytoplasm due to the negative charge of the molecule, its relatively large size (300–5000 kDa), and its degradability by ribonuclease (Zeng et al. 2020). Although injected naked mRNA can reach some cells' cytosol and be expressed (as in Fig. 6.72b), formulations in delivery systems can lead to enhanced uptake. mRNA can be delivered using different delivery systems, including lipid-based nanoparticles (LNPs), virus-like replicon particles, polymer-based delivery, cationic nanoemulsions, naked mRNA, and peptide-based nanoparticles (Hussain et al. 2022) (Fig. 6.73).

Among those various delivery systems, LNPs have become the most attractive and widely used for mRNA vaccines, as discussed in several reviews (Whitehead et al. 2009; Midoux and Pichon 2015; Reichmuth et al. 2016; Guan and Rosenecker 2017; Pardi et al. 2018; Kowalski et al. 2019; Samaridou et al. 2020; Zeng et al.

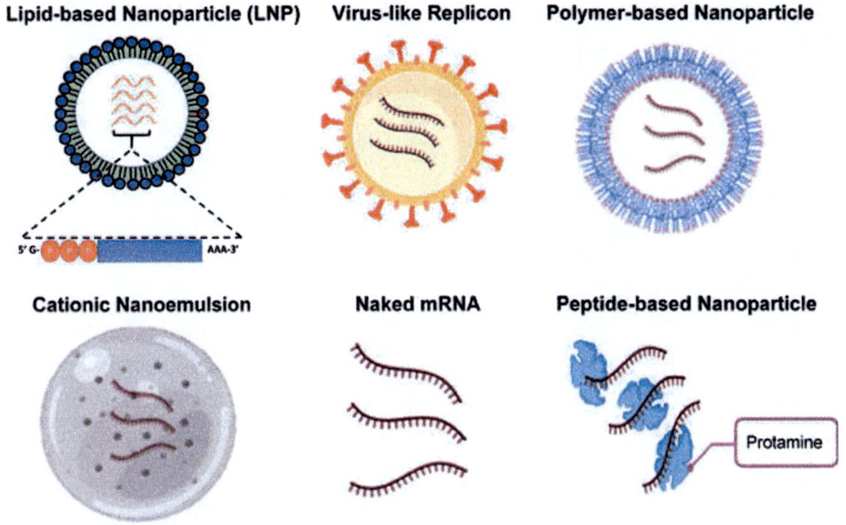

Fig. 6.73 Delivery systems for mRNA vaccines, including lipid-based nanoparticles (LNP), virus-like replicon particles, polymer-based nanoparticles, cationic nanoemulsion, naked mRNA, and peptide-based nanoparticles (Adapted from Hussain et al. (2022) with permission from Elsevier)

2020; Blakney et al. 2021; Rosa et al. 2021; Pilkington et al. 2021; Uddin and Roni 2021; Pascolo 2021b; Tenchov et al. 2021; Chaudhary et al. 2021; Hussain et al. 2022).

The main advantages of LNPs with encapsulated mRNA are protection from enzymatic degradation, efficient intracellular delivery of mRNA through receptor-mediated endocytosis, high reproducibility, and ease of scale-up (Kowalski et al. 2019; Samaridou et al. 2020; Zeng et al. 2020; Pascolo 2021b). Figure 6.74 shows a schematic of mRNA-loaded lipid nanoparticles (Verbeke et al. 2019).

The LNP formulations for the delivery of mRNA are typically composed of an ionizable cationic lipid and other "helper" lipids, such as a phospholipid, cholesterol, and polyethylene-glycol (PEG)-lipid, as reviewed by several authors (Midoux and Pichon 2015; Reichmuth et al. 2016; Verbeke et al. 2019; Ickenstein and Garidel 2019; Schoenmaker et al. 2021; Tenchov et al. 2021; Chaudhary et al. 2021; Hou et al. 2021; Shi et al. 2022). However, the choice and mixing ratio of the lipids affect the structural arrangement and functional properties. Therefore, an appropriate balance must be found between sufficient LNP stability in a biological context, mRNA cargo, and endosomal release of mRNA.

Microfluidic mixing is the current technique for the production of mRNA-containing LNPs by mixing an ethanol phase containing the lipids and an acidic aqueous buffer (e.g., acetic acid at a pH of 4) containing the mRNA (Zhigaltsev et al. 2012; Leung et al. 2015; Zhang et al. 2020; Buschmann et al. 2021; Schoenmaker et al. 2021; Pilkington et al. 2021; Maeki et al. 2022).

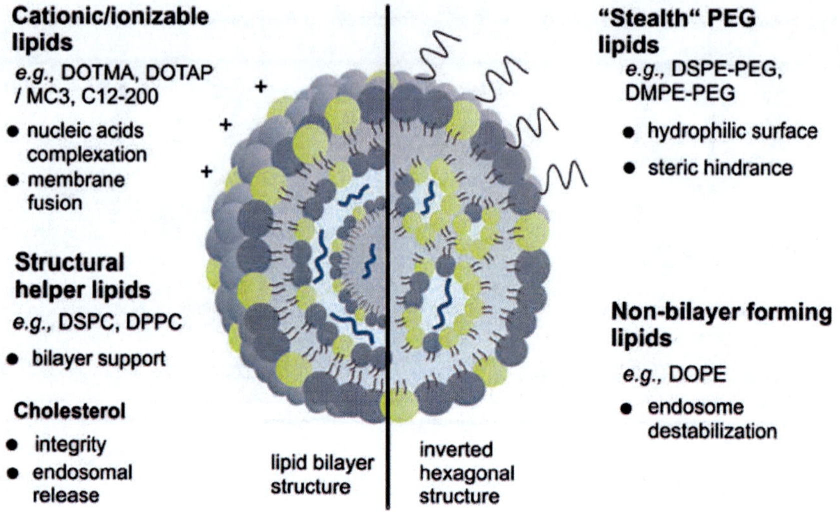

Fig. 6.74 Schematic of mRNA-LNP structures (Adapted from Verbeke et al. (2019) with permission from Elsevier). For abbreviations of lipids, see Midoux and Pichon (2015), Reichmuth et al. (2016), Ickenstein and Garidel (2019), Hou et al. (2021)

Figure 6.75a, b illustrates microfluidic mixing devices with staggered herring-bone (chaotic mixer) and baffle structures (Stroock et al. 2002; Zhigaltsev et al. 2012; Maeki et al. 2022) that enable rapid mixing of ethanol and aqueous phases and are easily fabricated with commonly used methods of planar lithography.

By diluting the ethanol phase below the level required to support lipid solubility, the lipids undergo a condensation process and form LNPs spontaneously while efficiently entrapping the mRNA.

Microfluidic mixing allows reproducible and scalable production of mRNA-containing LNP formulations that exhibit high encapsulation efficiency, homogeneous size distribution, and biocompatibility.

For intramuscular administration of mRNA vaccines, LNPs with a size between 75 and 95 nm are most suitable (Hassett et al. 2019; Schoenmaker et al. 2021; Maeki et al. 2022). For example, the mRNA-loaded LNPs of the COVID-19 vaccines nCoVsaRNA (Imperial College London) and ARCoV (Yunnan Walvax Biotechnology, Suzhou Abogen Biosciences) have an average hydrodynamic diameter of around 75 nm (McKay et al. 2020) and 89 nm (Zhang et al. 2020; Zhao et al. 2021), respectively.

The first small interfering RNA (siRNA) drug delivered by LNPs was Onpattro® (Alnylam Pharmaceuticals, Inc., USA), approved in the US and EU in 2018 (Akinc et al. 2019). It inhibits the synthesis of transthyretin protein in the liver, treating polyneuropathy caused by amyloidosis. The approval of Onpattro® has paved the way toward utilizing LNP delivery technology to deliver mRNA vaccines and other applications.

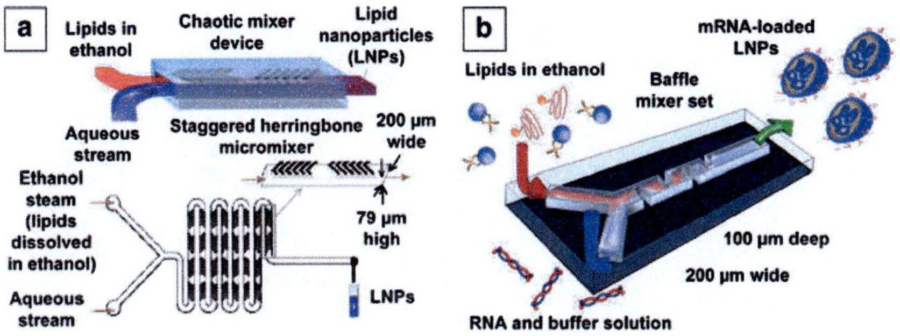

Fig. 6.75 Examples of microfluidic mixing devices for preparing mRNA-loaded LNP from lipid/ethanol and mRNA/buffer solutions. **a** Chaotic mixer with staggered herringbone mixer (Adapted from Zhigaltsev et al. (2012) with permission from the American Chemical Society), **b** T-shaped microfluid device with baffle mixer set (Adapted from Maeki et al. (2022) with permission from Elsevier)

In 2016, Kranz et al. (2016) reported the first utilization in humans of mRNA vaccines (non-modified mRNA) in liposomes. The first clinical trial of a modified mRNA vaccine in LNPs encoding hemagglutinin proteins against influenza was reported by Bahl et al. (2017). The mRNA-LNP vaccine technology generated high seroconversion rates, demonstrating robust prophylactic immunity in humans.

Furthermore, all leading mRNA COVID-19 vaccines are encapsulated in LNPs for delivery to the cells, which illustrates the successes achieved with LNP technology to stabilize mRNA (Blakney et al. 2021; Kim et al. 2021; Schoenmaker et al. 2021; Verbeke et al. 2021; Pilkington et al. 2021; Pascolo 2021b; Hussain et al. 2022). The efficacy of the COVID-19 mRNA vaccines developed by BioNTech/Pfizer and Moderna is about 95% (Polack et al. 2020; Baden et al. 2021).

As reported by Pascolo (2004, 2017, 2021b), mRNA molecules are stable in solution for weeks at room temperature as long as they are pure, i.e., in the absence of degrading enzymes like contaminating ribonucleases and in a neutral or acidic solution. However, even though mRNA is stable, a liposomal formulation may be unstable at room temperature and change in shape, size, or percentage of encapsulated mRNA by rounds of thawing/freezing (Pascolo 2021a). Figure 6.76 illustrates the main factors that impact mRNA-LNP stability, including water, pH, nucleases, oxygen, and storage condition, as discussed by Schoenmaker et al. (2021).

Indeed COVID-19 mRNA vaccines even require ultra-cold storage in freezers down to −80 and −60 °C to maintain efficacy (Crommelin et al. 2021b; Schoenmaker et al. 2021; Uddin and Roni 2021). To date, none of the approved COVID-19 mRNA vaccines has been able to be stored at room temperature for a long time (weeks) (Pascolo 2021a). Once thawed, the BioNTech/Pfizer vaccine can be stored for 5 days at 2–8 °C, while Moderna's vaccine for 30 days at 2–8 °C.

The development of thermostable mRNA vaccines is of great interest. The conversion to a dry powder form is one of the most commonly used methods for shipping and long-term storage (stockpiling).

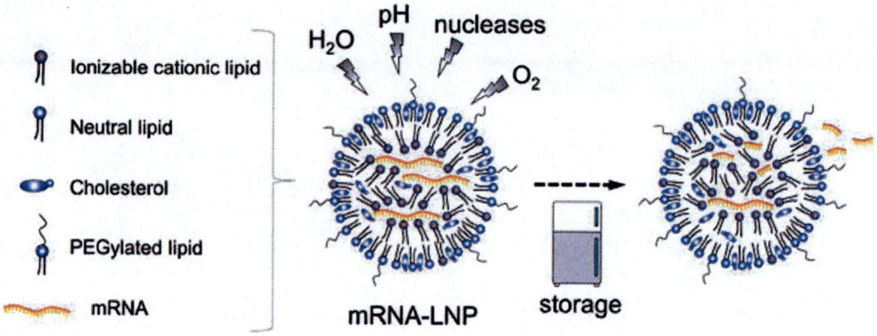

Fig. 6.76 Factors that impact mRNA-LNP stability (Adapted from Schoenmaker et al. (2021) with permission from Elsevier under open access CC-BY-NC-ND 4.0 license)

6.25.2 Research on Lyophilized mRNA Vaccines

Current literature suggests that lyophilization (freeze-drying) improves the shelf-life of mRNA vaccine formulations (Jones et al. 2007; Petsch et al. 2012; Kallen and Theß 2014; Stitz et al. 2017; Zhao et al. 2020b; Hong et al. 2021; Uddin and Roni 2021; Muramatsu et al. 2022; Gerhardt et al. 2022). However, few published studies have produced freeze-dried mRNA-LNP vaccines and demonstrated improved thermostability over time. Table 6.29 summarizes reported stability data.

Jones et al. (2007) showed that a freeze-dried self-replicating RNA in trehalose, encoding hepatitis C virus proteins, was stable for 10 months at 4 °C. Petsch et al. (2012) demonstrated that a lyophilized mRNA influenza vaccine was stable at 37 °C for 3 weeks. Vaccinated mice were fully protected, similar to siblings that received the frozen reference vaccine. Kallen and Theß (2014) emphasized that lyophilized mRNA-based vaccines were stable for several years or months at 25 °C and 40 °C, respectively. Stability was even demonstrated at 60 °C for several months. In a later publication, Stitz et al. (2017) freeze-dried a non-replicating mRNA-protamine complex encoding rabies virus glycoprotein based on CureVac's proprietary RNActive® technology (patents EP1392341, EP1857122, and WO2012019780A1) (Rauch et al. 2017). As determined by immunogenicity tests in mice, the lyophilized mRNA vaccine was stable for up to 3 months at 70 °C, up to 12 months at 25 °C, and after 20 cycles of oscillating temperatures between 4 and 56 °C simulating cold chain interruptions during transport. Unfortunately, the study does not provide information on the formulation or physicochemical characteristics before and after freeze-drying and storage.

In the field of mRNA-LNPs, Zhao et al. (2020b) generated lyophilized firefly luciferase-encoding mRNA-LNPs and demonstrated that the reconstituted vaccine maintained mRNA expression efficiency in mice as detected by bioluminescence imaging studies. Furthermore, the optimal storage condition for at least 3 months in liquid nitrogen was found by adding 5% (w/v) sucrose or trehalose as a lyoprotectant.

Table 6.29 Stability data of lyophilized mRNA vaccines published in the literature

Delivery system	Encoded antigen	Shelf life, thermostability	References
Self-replicating RNA in 10% (w/v) trehalose	Hepatitis C virus proteins	10 months at 4 °C	Jones et al. (2007)
mRNA (lyoprotectant n.a.)	Influenza A virus hemagglutinin	3 weeks at 37 °C	Petsch et al. (2012), Kallen and Theß (2014)
mRNA (lyoprotectant n.a.)	n.a	Several months at 60 °C	Kallen and Theß (2014)
mRNA-protamine complex	Rabies virus glycoprotein	12 months at 25 °C, 3 months at 70 °C	Stitz et al. (2017)
mRNA in LNP with 5% (w/v) sucrose or trehalose	Firefly luciferase	3 months in liquid nitrogen	Zhao et al. (2020b)
mRNA in LNP with 4% (w/v) sucrose	SARS-CoV-2 spike glycoprotein	1 week at 4 °C	Hong et al. (2021)
RNA in LNP with 10% (w/v) sucrose, mRNA in LNP with 20% (w/v) sucrose	Zika virus reporter protein, Ovalbumin	21 months at 4 °C, 8 months at 25 °C	Gerhardt et al. (2022)
mRNA in LNP with 10% (w/v) sucrose and 10% (w/v) maltose	Influenza virus hemagglutinin, firefly luciferase	24 weeks at 4 °C, 12 weeks at 25 °C	Muramatsu et al. (2022)

Hong et al. (2021) developed a lyophilized SARS-CoV-2 mRNA-LNP vaccine formulation with 4% (w/v) sucrose and showed that the reconstituted vaccine could induce strong immune responses in mice. Furthermore, mice sera successfully inhibited SARS-CoV-2 viral infection into Vero cells.

Gerhardt et al. (2022) lyophilized Zika self-amplifying RNA vaccine with nanostructured lipid carriers and 10% w/v sucrose, which retained the biophysical properties and protein expression ability in mice after 21 months of storage at 4 °C and 8 months at room temperature.

More recently, Muramatsu et al. (2022) demonstrated that the physicochemical properties of lyophilized nucleoside-modified mRNA-LNPs did not significantly change for 12 weeks after storage at room temperature (25 °C) and at least 24 weeks at 4 °C. Sucrose and maltose were used as excipients. Most importantly, comparative mouse studies showed that lyophilized firefly luciferase-encoding mRNA-LNPs maintained high expression efficiency after storage. In addition, no decrease in the immunogenicity of a lyophilized influenza virus hemagglutinin (HA)-encoding mRNA-LNP vaccine in mice was observed after 12 weeks of storage at 25 °C or for at least 24 weeks at 4 °C (Fig. 6.77a). Figure 6.77b shows

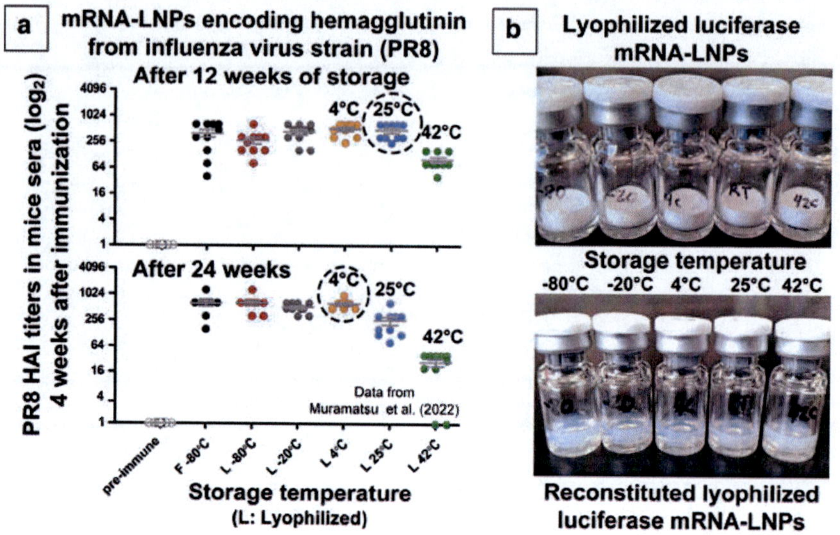

Fig. 6.77 **a** PR8 hemagglutination inhibition (HAI) titers of mice after immunization with frozen and lyophilized PR8 HA mRNA-LNPs stored at −80, −20, 4 °C, 25, and 42 °C for 12 and 24 weeks before reconstitution. Mice were intramuscularly immunized, and serum was collected 4 weeks post-immunization. Each symbol represents one animal. **b** Samples of lyophilized and reconstituted firefly luciferase-encoding mRNA-LNPs stored at −80, −20, 4, 25 °C (room temperature), and 42 °C for up to 24 weeks (Data and pictures from Muramatsu et al. (2022) with permission from Elsevier under open access CC-BY-NC-ND 4.0 license)

representative images of lyophilized and reconstituted firefly luciferase-encoding mRNA-LNPs. The samples were stored at −80, −20, 4, 25 °C (room temperature), and 42 °C for up to 24 weeks. All lyophilized mRNA-LNPs formed a uniform, dense, white cake, while the reconstituted samples acquired a clear, opalescent appearance with no visible solids.

Moreover, experiments by Ball et al. (2016) also showed that the lyoprotectants trehalose and sucrose at a concentration of about 20% (w/v) preserved LNP efficacy over three freeze–thaw cycles.

Finally, as reported by Dolgin (2020), Pfizer/BioNTech have clinical studies ongoing with lyophilized mRNA-LNP formulations of COVID-19 vaccines in humans. Recently, a Phase 3 clinical trial (clinicaltrials.gov, NCT04816669) was initiated to evaluate the safety, tolerability, and immunogenicity of a two-dose regimen of their approved COVID-19 vaccine (BNT162b2, Comirnaty®) comparing frozen liquid and lyophilized BNT162b2.

In addition, Arcturus Therapeutics, Inc. (San Diego, USA) is on track for Phase 3 with its lyophilized, nonmodified mRNA vaccine ARCT-021 (LUNAR-COV19) (Arcturus Therapeutics 2021; Cohen 2022). Unfortunately, room temperature stability data are not yet available.

Furthermore, U.S. researchers at the University of Massachusetts Lowell were recently awarded $930,000 to develop a freeze-drying technique for mRNA-based COVID-19 vaccines (Balfour 2022).

However, apart from challenges to preserving the integrity of mRNA-LNP through lyophilization, there are other hurdles such as the high energy consumption, the time-consuming freeze-drying process, and additional costs, e.g., those related to the necessary expansion of the lyophilization capacity worldwide in times of a pandemic (Schoenmaker et al. 2021).

Therefore, it is worthwhile to explore alternative drying techniques to improve the stability of mRNA vaccines, such as spray drying, spray freeze drying, thin-film freeze drying, foam drying, supercritical fluid drying, or evaporative drying onto dissolvable microneedles (Crommelin et al. 2021b, a; Schoenmaker et al. 2021; Uddin and Roni 2021). In particular, dry powder formulation of nucleic acids offers several additional advantages such as higher stability, longer shelf life and new dosage forms like inhalable formulations (Chow and Lam 2015; AboulFotouh et al. 2021; Heida et al. 2021).

6.25.3 Research on Spray-Dried mRNA Vaccines

So far, there is a lack of studies addressing the dry powder formulation of mRNA by spray drying. However, more research activity is recognized on spray drying of small interfering RNA (siRNA) (Jensen et al. 2010; Chow and Lam 2015; Chow et al. 2017; Agnoletti et al. 2017; Liang et al. 2018a, b; Miwata et al. 2018; Wu et al. 2019) or plasmid DNA (pDNA) (Bowey and Neufeld 2010; Munir et al. 2021).

In the area of mRNA vaccines, Qiu et al. (2019) developed an inhalable mRNA dry powder for pulmonary delivery using spray drying and spray freeze drying. A novel mRNA delivery vector was introduced comprising hydrophilic polyethylene glycol (PEG_{12}) bound to synthetic KL4 peptide ($PEG_{12}KL4$), resulting in improved solubility and transfection efficiency (Fig. 6.78).

The spray-dried formulations contained 10:1 w/w $PEG_{12}KL4$ peptide to mRNA ratio (0.1 and 0.5% w/w mRNA) and mannitol as an excipient (1.5 and 3% w/v). The powders were produced using a Mini Spray Dryer B-290 with high-performance cyclone and Dehumidifier B-296 in suction mode and closed loop (Büchi Labortechnik AG, Switzerland).

The process parameters were 80 °C/50 °C inlet/outlet temperature, 1.4 mL/min feed rate, 90% aspirator setting (approx. 35 m^3/h drying air flow), and 742 L/h atomization flow rate. Relatively low inlet temperature and atomization flow rate were selected to address thermal and shear stresses during processing. The yields ranged from 46 to 76%, and the morphology of the powders was spherical, with a particle size well below 5 µm (Figs. 6.78a and 6.79).

Both spray-dried and spray-freeze-dried powder exhibited desirable aerosol properties for inhalation even after 6 months of storage at room temperature. The

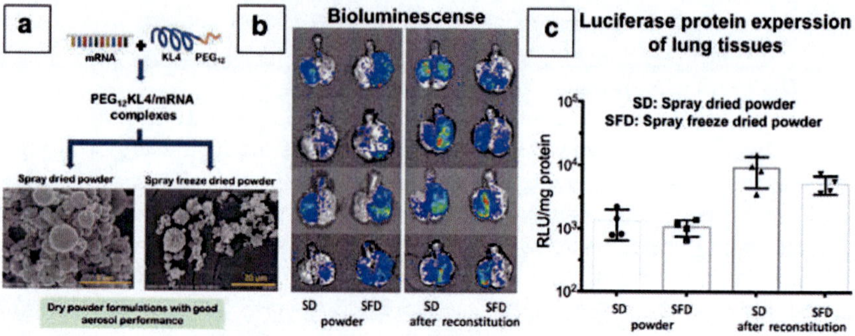

Fig. 6.78 **a** Novel mRNA delivery vector of hydrophilic polyethylene glycol (PEG_{12}) bound to synthetic KL4 peptide (PEG_{12}KL4). Spray freeze drying (SFD) and spray drying (SD) were used to prepare dry powder formulations with good aerosol properties. **b** Bioluminescence imaging of isolated lungs from BALB/c mice (around 20 g) after pulmonary (intracheal) administration of mRNA formulations as powder aerosol or reconstituted liquid aerosol. **c** Luciferase protein expression of lung tissue was measured 24 h after administration in relative light units (RLU) per mg protein (Figures adapted from Qiu et al. (2019) with permission from Elsevier and under open access CC-BY-NC-ND 4.0 license)

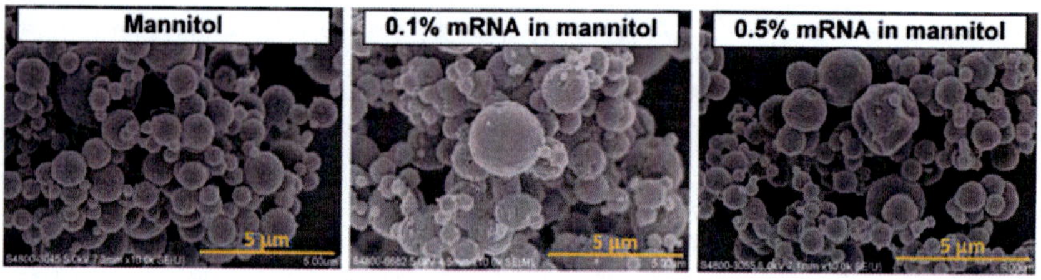

Fig. 6.79 Spray-dried formulations of PEG_{12}KL4/mRNA complexes encapsulated in mannitol (Adapted from Qiu et al. (2019) with permission from Elsevier and under open access CC-BY-NC-ND 4.0 license)

emitted fraction of the spray-dried powder was at least 75% for all formulations, while the fine particle fraction (aerodynamic diameter < 5.0 μm) was between 36 and 41%, respectively.

In-vitro transfection studies in cells showed that the mRNA remained intact and retained its biological activity after spray drying (Qiu et al. 2019). Flow cytometry and confocal microscopy showed that the presence of PEG promoted cellular uptake. When the PEG_{12}KL4/mRNA complexes were administered intratracheally either as a reconstituted liquid aerosol or powder aerosol into the lungs of mice, effective transfection and in vivo activity were observed (Fig. 6.78b). Both liquid and powder aerosol could express luciferase at 24 h post-administration, but the liquid aerosol displayed significantly higher transfection efficiency (Fig. 6.78c). A single dose of PEG_{12}KL4/mRNA at a 5 μg mRNA dose per mouse did not show any signs of inflammation and toxicity in the lungs of the animals.

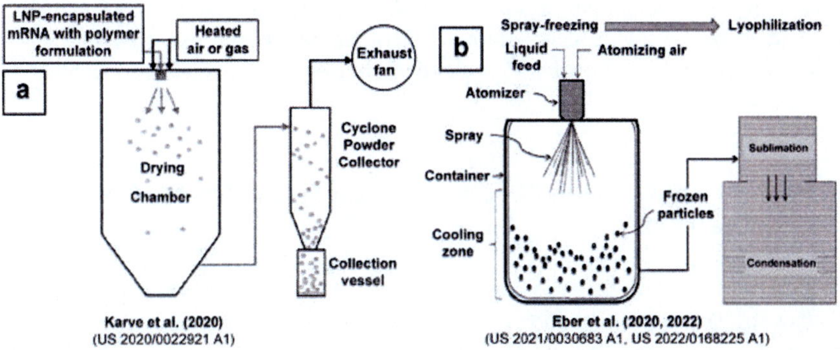

Fig. 6.80 **a** Patent from Translate Bio, Inc. (U.S.): dry powder formulations for messenger RNA, (Karve et al. 2020) (US 2020/0022921 A1), **b** Patent from CureVac AG (Germany): dry powder composition comprising long-chain RNA (Eber et al. 2021, 2022) (US 2021/0030683 A1, US 2022/0168225 A1)

Overall, the authors concluded that $PEG_{12}KL4$ had considerable potential to be developed as a non-viral vector for pulmonary delivery of mRNA vaccine in dry powder form. However, the thermostability profile was not reported.

In the patent literature, there are at least two inventions specifically addressing spray drying and spray freeze drying of mRNA vaccines (Fig. 6.80).

Karve et al. (2020) from Translate Bio, Inc. (U.S.) described in a patent (US 2020/0022921 A1) the spray drying of mRNA-LNPs using polymers (e.g., Eudragit or poly(lactide-co-glycolide PLGA) as stabilizers to ensure translational efficiency in vivo and stability during storage under various condition (Fig. 6.80a). The field of application was therapeutics, particularly dry powder formulation for delivery of mRNA expressing cystic fibrosis conductance regulator. However, the invention could be used to spray dry other mRNA formulations. Examples described include mRNA encoding firefly luciferase or argininosuccinate synthetase. For spray drying, an inlet/outlet temperature of 90 °C/46 °C to 50° C, an aspiration rate of 85%, and a pump rate of 25% were typically used. The mRNA retained integrity and was stable at or below room temperature for 6 months or longer.

Eber et al. (2021, 2022) from CureVac AG (Germany) presented an invention (US 2021/0030683 A1, US 2022/0168225 A1) directed to a storage-stable dry powder composition containing long-chain RNA molecules by spray freeze drying. Figure 6.80b shows a schematic diagram of the spray freeze-drying process in which liquid starting material is atomized and frozen in a cooling zone, and the frozen particles are subsequently freeze-dried in a freeze-drying step. Long-chain RNA preferably comprised at least 30 nucleotides, in contrast to shorter RNAs. In particular, the invention addressed the use of such dry powder in preparing pharmaceutical compositions and vaccines. Highlighted were the free-flowing nature of the powder (around 100 to about 200 μm particle size) and the potential for large-scale production, preferably in a continuous spray freeze-drying system with freeze-drying drum and flexible multi- or single-dose packaging of bulk powder.

Examples were protamine-formulated mRNA with trehalose and mRNA encoding the influenza A virus's hemagglutinin (HA) protein. Results showed that spray freeze drying enables powder formulations with extremely low water content (< 0.35 wt%).

6.25.4 Conclusions and Future View on Spray-Dried mRNA Vaccines

This section highlights the main benefits and challenges of mRNA vaccines, opening a new era in vaccinology. In particular, mRNA vaccines are characterized by low-cost production, the potential for high efficacy, safe administration, and rapid adaptation to a changing virus. However, as mRNA is degraded by nucleases and cannot easily cross the cell membrane due to its large size and negative charge, delivery is best achieved by encapsulation in vehicles, such as lipid nanoparticles (LNPs).

LNPs prepared by microfluid mixing has emerged as the most promising delivery carrier for mRNA, encapsulating the sensitive RNA molecule while allowing efficient delivery into cells where it can initiate expression of the desired disease-specific antigen protein. Notably, all leading mRNA COVID-19 vaccines utilize LNP formulation technology. However, a major challenge is the thermostability of LNPs, particularly the combination of mRNA-LNP as vaccines.

Conversion to a thermostable dry vaccine is a common method for long-term storage. Pfizer-BioNTech is currently conducting a Phase 3 clinical trial with lyophilized mRNA-LNP formulations of the COVID-19 vaccine BNT162b2 (Comirnaty®), and Arcturus Therapeutics, Inc. is aiming towards Phase 3 with the lyophilized, unmodified mRNA vaccine ARCT-021 (LUNAR-COV19).

Current literature shows that freeze-drying (or lyophilization) is used to improve the thermostability of mRNA vaccine formulations. Preclinical studies in mice demonstrated that reconstituted and injected mRNA vaccine stored at room temperature for several months retained their encoded antigen expression and immunogenicity. Trehalose and sucrose were mainly used as stabilizers immobilizing the mRNA-LNP vaccines in a glassy matrix. As a result, some vaccine manufacturers are already conducting clinical trials with lyophilized COVID-19 vaccines to test the vaccine's safety, tolerability, and immunogenicity in humans. Thus, freeze-dried mRNA vaccines may soon become commercially viable formats. On the other hand, freeze drying costs are higher than other drying technologies due to the high energy consumption, the time-consuming (batch) freeze-drying process, and additional expenses related to expanding freeze-drying capacities worldwide during a pandemic.

Therefore, it is worthwhile to explore alternative drying technologies to improve the stability of mRNA vaccines, such as spray drying, spray freeze drying, thin-film freeze drying, or evaporative drying of dissolvable microneedles.

To date, there are only a few scientific publications and patents reporting on the formulation of mRNA vaccines by spray drying. The publications deal mainly with dry powder formulations for pulmonary delivery. The studies demonstrate particle production capability and successful immunization by inhalation (intratracheal) in mice. Crucial is particle design with an appropriate size and low residual moisture content, allowing direct administration of the powder vaccine via the lungs. Immunization via the respiratory tract is attractive because it enables immunization without injection via the mucous membranes. To minimize process losses during spray drying, mild atomization at low pressure and drying temperatures is crucial. Typical outlet temperatures are around 50 °C to minimize thermal and shear stress. Another key requirement is to optimize the formulation composition for incorporating the mRNA-LNP vaccine into a protective matrix of glass-forming excipients such as trehalose, sucrose, maltose, and other suitable stabilizers.

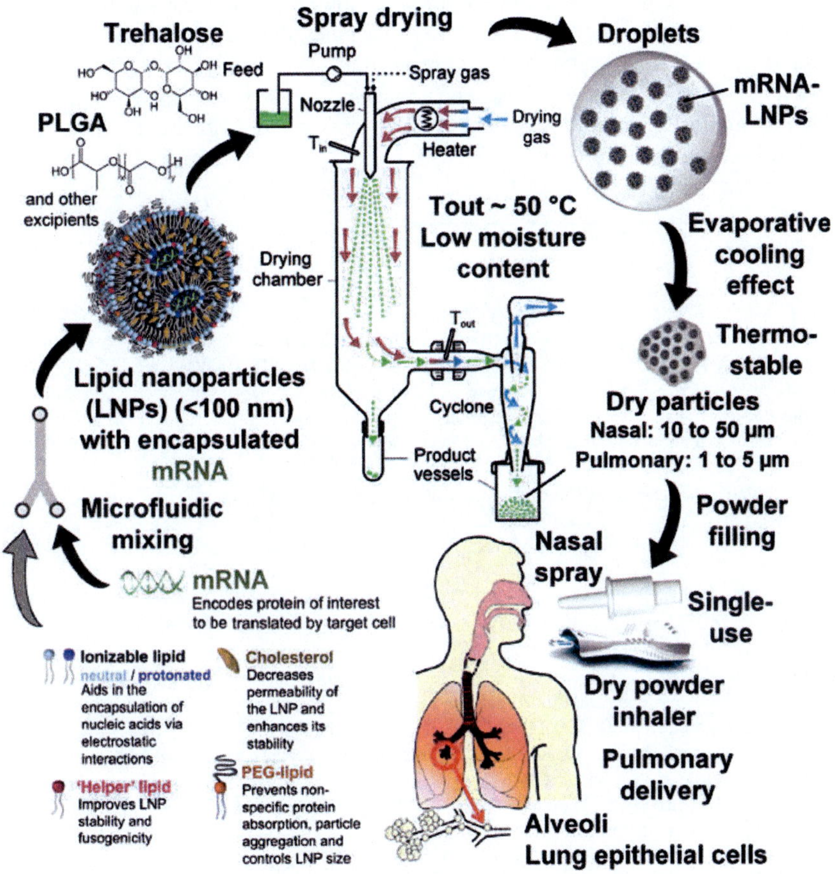

Fig. 6.81 Potential strategy for pulmonary or nasal delivery of spray-dried mRNA-LNP vaccines using single-use dry powder inhalers or nasal sprays. The LNPs formulated from lipids and mRNA by microfluidic mixing are encapsulated by spray drying in glass-stabilizing trehalose, biodegradable PLGA, and other excipients. By process optimization, thermostable powders can be produced at low outlet drying temperature (about 50 °C) and low residual moisture content with inhalable (1–5 µm) or nasal mucosa-adherent (10–50 µm) size

Overall, this brief review identified a research gap in spray-drying mRNA vaccine encapsulated in LNPs. As illustrated in Fig. 6.81, spray-dried powder form could be a viable technology for stabilizing mRNA-LNP vaccines in glass-forming carbohydrates (e.g., trehalose) or biodegradable polymers (e.g., PLGA) and other (mucoadhesive) excipients for pulmonary or nasal delivery using single-use dry powder inhalers or nasal spray devices. The particle size can be optimized between 1 and 5 μm for pulmonary delivery and about 10–50 μm for deposition on the nasal mucosa.

In the future, aseptic spray drying could open the door for other mRNA dry powder applications specific to a particular disease. However, bringing a spray-dried mRNA vaccine to market would require further investment in R&D and scale-up, starting with laboratory and pilot testing and preclinical studies.

In the long term, thermostable mRNA vaccines that can be stored for a few months at room temperature are highly desirable, especially for next-generation mRNA-LNP COVID-19 vaccines. Such vaccines would allow stockpiling, reduce distribution costs, and improve access in countries where cold chain infrastructure and maintenance are difficult.

References

Abdul-Fattah AM, Truong-Le V, Yee L et al (2007a) Drying-induced variations in physico-chemical properties of amorphous pharmaceuticals and their impact on stability II: stability of a vaccine. Pharm Res 24:715–727. https://doi.org/10.1007/s11095-006-9191-2

Abdul-Fattah AM, Truong-Le V, Yee L et al (2007b) Drying-induced variations in physico-chemical properties of amorphous pharmaceuticals and their impact on stability (I): stability of a monoclonal antibody. J Pharm Sci 96:1983–2008. https://doi.org/10.1002/jps.20859

Abdul-Fattah AM, Lechuga-Ballesteros D, Kalonia DS, Pikal MJ (2008) The impact of drying method and formulation on the physical properties and stability of methionyl human growth hormone in the amorphous solid state. J Pharm Sci 97:163–184. https://doi.org/10.1002/jps.21085

AboulFotouh K, Cui Z, Williams RO (2021) Next-generation COVID-19 vaccines should take efficiency of distribution into consideration. AAPS PharmSciTech 22:126. https://doi.org/10.1208/s12249-021-01974-3

Adhikari BB, Goodson JL, Chu SY et al (2016) Assessing the potential cost-effectiveness of microneedle patches in childhood measles vaccination programs: the case for further research and development. Drugs R D 16:327–338. https://doi.org/10.1007/s40268-016-0144-x

Adwar T, Lukešová D (2008) Evaluation of thermostable vaccines against newcastle disease in village chicken used in tropics and subtropics. Agric Trop Subtrop 41:74–79

Afkhami S, LeClair DA, Haddadi S et al (2017) Spray dried human and chimpanzee adenoviral-vectored vaccines are thermally stable and immunogenic in vivo. Vaccine 35:2916–2924. https://doi.org/10.1016/j.vaccine.2017.04.026

Agarkhedkar S, Kulkarni PS, Winston S et al (2014) Safety and immunogenicity of dry powder measles vaccine administered by inhalation: a randomized controlled Phase I clinical trial. Vaccine 32:6791–6797. https://doi.org/10.1016/j.vaccine.2014.09.071

Agnoletti M, Bohr A, Thanki K et al (2017) Inhalable siRNA-loaded nano-embedded microparticles engineered using microfluidics and spray drying. Eur J Pharm Biopharm 120:9–21. https://doi.org/10.1016/j.ejpb.2017.08.001

Ajmera A, Scherließ R (2014) Stabilisation of proteins via mixtures of amino acids during spray drying. Int J Pharm 463:98–107. https://doi.org/10.1016/j.ijpharm.2014.01.002

Akinc A, Maier MA, Manoharan M et al (2019) The Onpattro story and the clinical translation of nanomedicines containing nucleic acid-based drugs. Nat Nanotechnol 14:1084–1087. https://doi.org/10.1038/s41565-019-0591-y

Alameh M, Weissman D, Pardi N (2020) Messenger RNA-based vaccines against infectious diseases. Current topics in microbiology and immunology. Springer, Berlin, Heidelberg, pp 1–35

Alders R, Spradbrow P (2001) Controlling newcastle disease in village chickens: a field manual. https://www.aciar.gov.au/publication/books-and-manuals/MN082_controlling_ND_in_village_chickens

Alders R, Dos Anjos F, Bagnol B et al (2002) Controlling newcastle disease in village chickens: a training manual. In: ACIAR Monogr. No. 87. https://www.aciar.gov.au/publication/books-and-manuals/controlling-newcastle-disease-village-chickens-training-manual

Alders RG (2014) Making Newcastle disease vaccines available at village level. Vet Rec 174:502–503. https://doi.org/10.1136/vr.g3209

Alders RG, Bagnol B, Young MP (2010) Technically sound and sustainable Newcastle disease control in village chickens: lessons learnt over fifteen years. Worlds Poult Sci J 66:433–440. https://doi.org/10.1017/S0043933910000516

Ali M, Nelson AR, Lopez AL, Sack DA (2015) Updated global burden of cholera in endemic countries. PLoS Negl Trop Dis 9:e0003832. https://doi.org/10.1371/journal.pntd.0003832

Al-Khattawi A, Bayly A, Phillips A, Wilson D (2018) The design and scale-up of spray dried particle delivery systems. Expert Opin Drug Deliv 15:47–63. https://doi.org/10.1080/17425247.2017.1321634

Allison LMC, Mann GF, Perkins FT, Zuckerman AJ (1981) An accelerated stability test procedure for lyophilized measles vaccines. J Biol Stand 9:185–194. https://doi.org/10.1016/S0092-1157(81)80022-4

Altrichter J, Kemter K, Ströver R, Scholz M (2019) Exploring amino acid-based and stable spray-dried vaccinations. In: EBR—Eur. Biopharm. Rev. https://en.calameo.com/books/0061133855dbf0ab162d7

Al-Zoubi N, Gharaibeh S, Aljaberi A, Nikolakakis I (2021) Spray drying for direct compression of pharmaceuticals. Processes 9:267. https://doi.org/10.3390/pr9020267

Amacker M, Smardon C, Mason L et al (2020) New GMP manufacturing processes to obtain thermostable HIV-1 gp41 virosomes under solid forms for various mucosal vaccination routes. npj Vaccines 5:41. https://doi.org/10.1038/s41541-020-0190-9

Ameri M, Maa Y-F (2006) Spray drying of biopharmaceuticals: stability and process considerations. Dry Technol 24:763–768. https://doi.org/10.1080/03602550600685275

Amidi M, Pellikaan HC, Hirschberg H et al (2007) Diphtheria toxoid-containing microparticulate powder formulations for pulmonary vaccination: preparation, characterization and evaluation in guinea pigs. Vaccine 25:6818–6829. https://doi.org/10.1016/j.vaccine.2007.05.064

Amorij J-P (2007) The development of stable influenza vaccine powder formulations for new needle-free dosage forms. PhD thesis, University of Groningen, The Netherlands. https://www.researchgate.net/publication/50843621

Amorij J-P, Saluja V, Petersen AH et al (2007) Pulmonary delivery of an inulin-stabilized influenza subunit vaccine prepared by spray-freeze drying induces systemic, mucosal humoral as well as cell-mediated immune responses in BALB/c mice. Vaccine 25:8707–8717. https://doi.org/10.1016/j.vaccine.2007.10.035

Amorij J-P, Huckriede A, Wilschut J et al (2008) Development of Stable Influenza Vaccine Powder Formulations: Challenges and Possibilities. Pharm Res 25:1256–1273. https://doi.org/10.1007/s11095-008-9559-6

Anderson SD, Daviskas E, Brannan JD, Chan HK (2018) Repurposing excipients as active inhalation agents: The mannitol story. Adv Drug Deliv Rev 133:45–56. https://doi.org/10.1016/j.addr.2018.04.003

Angelidou A, Conti M-G, Diray-Arce J et al (2020) Licensed Bacille Calmette-Guérin (BCG) formulations differ markedly in bacterial viability, RNA content and innate immune activation. Vaccine 38:2229–2240. https://doi.org/10.1016/j.vaccine.2019.11.060

Angelo R, Rousseau K, Grant M et al (2009) Technosphere® insulin: Defining the role of technosphere particles at the cellular level. J Diabetes Sci Technol 3:545–554. https://doi.org/10.1177/193229680900300320

Anish C, Upadhyay AK, Sehgal D, Panda AK (2014) Influences of process and formulation parameters on powder flow properties and immunogenicity of spray dried polymer particles entrapping recombinant pneumococcal surface protein A. Int J Pharm 466:198–210. https://doi.org/10.1016/j.ijpharm.2014.03.025

Año G, Esquisabel A, Pastor M et al (2011) A new oral vaccine candidate based on the microencapsulation by spray-drying of inactivated *Vibrio cholerae*. Vaccine 29:5758–5764. https://doi.org/10.1016/j.vaccine.2011.05.098

Arcturus Therapeutics (2021) Arcturus—a clinical-stage mRNA therapeutics and vaccines company. In: Biopharma Deal. https://www.nature.com/articles/d43747-021-00073-3

Armstrong M (2022) GSK goes back to its vaccine bread and butter, May 31, 2022. In: Eval. Vantage. https://www.evaluate.com/vantage/articles/news/deals/gsk-goes-back-its-vaccine-bread-and-butter

Arpagaus C (2018) Pharmaceutical particle engineering via nano spray drying—process parameters and application examples on the laboratory-scale. Int J Med Nano Res 5. https://doi.org/10.23937/2378-3664.1410026

Arpagaus C (2020) Spray drying of vaccines: a review. In: IDS 2020, 22nd International drying symposium Worcester, Massachusetts, USA, June 28–July 1, 2020, pp 1–14

Arpagaus C, Collenberg A, Rütti D et al (2018) Nano spray drying for encapsulation of pharmaceuticals. Int J Pharm 546:194–214. https://doi.org/10.1016/j.ijpharm.2018.05.037

Ashford DA, Perkins B, Rotz LD (2001) Use of anthrax vaccine in the United States: recommendations of the advisory committee on immunization practices. J Toxicol Clin Toxicol 39:85–100. https://doi.org/10.1081/CLT-100102886

Augustin MA, Sanguansri L (2017) Microencapsulation technologies. In: Engineering foods for bioactives stability and delivery. Springer Science+Business Media, New York, pp 119–142

Aziz AB, Dembinski JL, Jahan Y (2020) Debate on Bacille Calmette-Guérin vaccination against COVID-19: Is it worth performing clinical trials? Biosaf Heal 2:113–114. https://doi.org/10.1016/j.bsheal.2020.07.001

Baden LR, El Sahly HM, Essink B et al (2021) Efficacy and safety of the mRNA-1273 SARS-CoV-2 vaccine. N Engl J Med 384:403–416. https://doi.org/10.1056/NEJMoa2035389

Badizadegan K, Goodson JL, Rota PA, Thompson KM (2020) The potential role of using vaccine patches to induce immunity: platform and pathways to innovation and commercialization. Expert Rev Vaccines 19:175–194. https://doi.org/10.1080/14760584.2020.1732215

Baehner F, Bogaerts H, Goodwin R (2016) Vaccines against norovirus: state of the art trials in children and adults. Clin Microbiol Infect 22:S136–S139. https://doi.org/10.1016/j.cmi.2015.12.023

Bahamondez-Canas TF, Cui Z (2018) Intranasal immunization with dry powder vaccines. Eur J Pharm Biopharm 122:167–175. https://doi.org/10.1016/j.ejpb.2017.11.001

Bahl K, Senn JJ, Yuzhakov O et al (2017) Preclinical and clinical demonstration of immunogenicity by mRNA vaccines against H10N8 and H7N9 Influenza Viruses. Mol Ther 25:1316–1327. https://doi.org/10.1016/j.ymthe.2017.03.035

Bailey CJ, Barnett AH (2007) Why is Exubera being withdrawn? BMJ 335:1156–1156. https://doi.org/10.1136/bmj.39409.507662.94

Balducci AG, Cagnani S, Sonvico F et al (2014) Pure insulin highly respirable powders for inhalation. Eur J Pharm Sci 51:110–117. https://doi.org/10.1016/j.ejps.2013.08.009

Balfour H (2022) Could Covid-19 vaccines be freeze dried? In: Eur. Pharm. Rev. https://www.europeanpharmaceuticalreview.com/news/167848/could-covid-19-vaccines-be-freeze-dried/

Ball R, Bajaj P, Whitehead K (2016) Achieving long-term stability of lipid nanoparticles: examining the effect of pH, temperature, and lyophilization. Int J Nanomedicine 12:305–315. https://doi.org/10.2147/IJN.S123062

Ball JP, Springer MJ, Ni Y et al (2017) Intranasal delivery of a bivalent norovirus vaccine formulated in an in situ gelling dry powder. PLoS ONE 12:e0177310. https://doi.org/10.1371/journal.pone.0177310

Barbier AJ, Jiang AY, Zhang P et al (2022) The clinical progress of mRNA vaccines and immunotherapies. Nat Biotechnol 40:840–854. https://doi.org/10.1038/s41587-022-01294-2

Barré-Sinoussi F, Ross AL, Delfraissy J-F (2013) Past, present and future: 30 years of HIV research. Nat Rev Microbiol 11:877–883. https://doi.org/10.1038/nrmicro3132

Bartsch SM, Lopman BA, Ozawa S et al (2016) Global economic burden of norovirus gastroenteritis. PLoS ONE 11:1–16. https://doi.org/10.1371/journal.pone.0151219

Baumann JM, Adam MS, Wood JD (2021) Engineering Advances in Spray Drying for Pharmaceuticals. Annu Rev Chem Biomol Eng 12:annurev-chembioeng-091720–034106. https://doi.org/10.1146/annurev-chembioeng-091720-034106

Bejugam NK, Uddin AN, Gayakwad SG, D'Souza MJ (2008) Formulation and evaluation of albumin microspheres and its enteric coating using a spray-dryer. J Microencapsul 25:577–583. https://doi.org/10.1080/02652040802476187

Beller K-D (2018) Single-dose powder inhalator and method for the production thereof, Patent EP 3223890 B1, Filed: July 29, 2015, Priority Date: November 26, 2014 (DE 102014017409), Assignee: Perlen Packaging AG (CH), Beller, Klaus-Dieter (DE). 1–25

Beller K-D (2021) Single-dose powder inhalator and method for the production thereof, Patent US 11110233 B2, PCT Filed: July 29, 2015, Assignee: Perlen Packaging AG (CH). 1–19

Bend Research (2013) Vaccine and biotherapeutic formulation development and manufacture using spray-drying. https://www.youtube.com/watch?v=rDP6ldHWcO0

Bennett JV, Fernandez de Castro J, Valdespino-Gomez JL et al (2002) Aerosolized measles and measles-rubella vaccines induce better measles antibody booster responses than injected vaccines: randomized trials in Mexican schoolchildren TT—Les vaccins antirougeoleux et antirougeoleux-antirubéoleux en aérosol induisent une. Bull World Health Organ 80:806–812

Berkenfeld K, Lamprecht A, McConville JT (2015) Devices for dry powder drug delivery to the lung. AAPS PharmSciTech 16:479–490. https://doi.org/10.1208/s12249-015-0317-x

Bhowmik T, D'Souza B, Uddin MN, D'Souza MJ (2012) Oral delivery of microparticles containing plasmid DNA encoding hepatitis-B surface antigen. J Drug Target 20:364–371. https://doi.org/10.3109/1061186X.2012.662686

Bhujbal SV, Mitra B, Jain U et al (2021) Pharmaceutical amorphous solid dispersion: a review of manufacturing strategies. Acta Pharm Sin B 11:2505–2536. https://doi.org/10.1016/j.apsb.2021.05.014

Bivas-Benita M, Zwier R, Junginger HE, Borchard G (2005) Non-invasive pulmonary aerosol delivery in mice by the endotracheal route. Eur J Pharm Biopharm 61:214–218. https://doi.org/10.1016/j.ejpb.2005.04.009

Blakney AK, Ip S, Geall AJ (2021) An update on self-amplifying mRNA vaccine development. Vaccines 9:97. https://doi.org/10.3390/vaccines9020097

Boehm G, Peyre M, Sesardic D et al (2002) On technological and immunological benefits of multivalent single-injection microsphere vaccines. Pharm Res 19:1330–1336. https://doi.org/10.1023/A:1020354809581

Boraey MA, Vehring R (2014) Diffusion controlled formation of microparticles. J Aerosol Sci 67:131–143. https://doi.org/10.1016/j.jaerosci.2013.10.002

Bot AI, Tarara TE, Smith DJ et al (2000) Novel lipid-based hollow-porous microparticles as a platform for immunoglobulin delivery to the respiratory tract. Pharm Res 17:275–283. https://doi.org/10.1023/A:1007544804864

Bot AI, Smith DJ, Bot S et al (2001) Receptor-mediated targeting of spray-dried lipid particles coformulated with immunoglobulin and loaded with a prototype vaccine. Pharm Res 18:971–979

Boumart Z, Hamdi J, Daouam S et al (2016) Thermal stability study of five Newcastle disease attenuated vaccine strains. Avian Dis 60:779–783. https://doi.org/10.1637/11426-042116-Reg.1

Bowey K, Neufeld RJ (2010) Systemic and mucosal delivery of drugs within polymeric microparticles produced by spray drying. BioDrugs 24:359–377. https://doi.org/10.2165/11539070-000000000-00000

Brachman PS, Friedlander AM, Grabenstein JD (2008) Anthrax vaccine. In: Vaccines. Elsevier, pp 111–126

Brandau DT, Jones LS, Wiethoff CM et al (2003) Thermal stability of vaccines. J Pharm Sci 92:218–231. https://doi.org/10.1002/jps.10296

Branswell H (2020) Why a decades-old TB vaccine is getting attention in the fight against Covid-19, April 14, 2020. Stat. https://www.statnews.com/2020/04/14/decades-old-tb-vaccine-attracts-attention-and-skepticism-as-a-potential-weapon-against-covid-19/

Braun LTJ, Jezek J, Peterson S et al (2009) Characterization of a thermostable hepatitis B vaccine formulation. Vaccine 27:4609–4614. https://doi.org/10.1016/j.vaccine.2009.05.069

Brito LA, Kommareddy S, Maione D et al (2015) Self-amplifying mRNA vaccines. In: Advances in genetics. Elsevier Ltd, pp 179–233

Broadhead J, Edmond Rouan SK, Rhodes CT (1992) The spray drying of pharmaceuticals. Drug Dev Ind Pharm 18:1169–1206. https://doi.org/10.3109/03639049209046327

Brough C, Williams RO (2013) Amorphous solid dispersions and nano-crystal technologies for poorly water-soluble drug delivery. Int J Pharm 453:157–166. https://doi.org/10.1016/j.ijpharm.2013.05.061

Bukreyev A, Huang Z, Yang L et al (2005) Recombinant Newcastle disease virus expressing a foreign viral antigen is attenuated and highly immunogenic in primates. J Virol 79:13275–13284. https://doi.org/10.1128/JVI.79.21.13275-13284.2005

Burger JL, Cape SP, Braun CS et al (2008) Stabilizing formulations for inhalable powders of live-attenuated measles virus vaccine. J Aerosol Med Pulm Drug Deliv 21:25–34. https://doi.org/10.1089/jamp.2007.0658

Burke RM, Tate JE, Kirkwood CD et al (2019) Current and new rotavirus vaccines. Curr Opin Infect Dis 32:435–444. https://doi.org/10.1097/QCO.0000000000000572

Burke RM, Mattison CP, Pindyck T et al (2021) Burden of Norovirus in the United States, as estimated based on administrative data: updates for medically attended illness and mortality, 2001–2015. Clin Infect Dis 73:e1–e8. https://doi.org/10.1093/cid/ciaa438

Buschmann MD, Carrasco MJ, Alishetty S et al (2021) Nanomaterial delivery systems for mRNA vaccines. Vaccines 9:65. https://doi.org/10.3390/vaccines9010065

Byron PR (1986) Prediction of drug residence times in regions of the human respiratory tract following aerosol inhalation. J Pharm Sci 75:433–438. https://doi.org/10.1002/jps.2600750502

Cal K, Sollohub K (2010) Spray drying technique. I: hardware and process parameters. J Pharm Sci 99:575–586. https://doi.org/10.1002/jps.21886

Calmette A (1931) Preventive vaccination against tuberculosis with BCG. In: Proceedings of the royal society of medicine. https://doi.org/10.1177/003591573102401109

Cape SP, Villa JA, Huang ETS et al (2008) Preparation of active proteins, vaccines and pharmaceuticals as fine powders using supercritical or near-critical fluids. Pharm Res 25:1967–1990. https://doi.org/10.1007/s11095-008-9575-6

Cape SP, McAdams DH, Manion JR et al (2010) Development of stable, immunogenic, and protective measles vaccines for needle-free administration. In: 2010 FIP pharmaceutical sciences world congress and AAPS annual meeting and exposition, 14–18 November 2010, New Orleans, LA, USA

Capua I, Alexander DJ (2009) Avian Influenza and Newcastle disease: a field and laboratory manual. Springer, Milan

Cargill P (1999) Vaccine administration in poultry. In Pract 21:323–328. https://doi.org/10.1136/inpract.21.6.323

Catalent (2021) Catalent news, January 13, 2011: catalent agrees to acquire dry powder inhaler spray drying, capsule manufacturing, and packaging capabilities from acorda therapeutics

Cates JE, Vinjé J, Parashar U, Hall AJ (2020) Recent advances in human norovirus research and implications for candidate vaccines. Expert Rev Vaccines 19:539–548. https://doi.org/10.1080/14760584.2020.1777860

CDC (2016) Guide to understanding anthrax. https://www.cdc.gov/anthrax/pdf/evergreen-pdfs/anthrax-evergreen-content-english.pdf

CDC (2019) Typhoid vaccines—what you need to know. Vaccine Inf Statement

CDC (2020) Human Papillomavirus (HPV) vaccination & cancer prevention. Centers for Disease Control and Prevention. https://www.cdc.gov/vaccines/vpd/hpv/index.html. Accessed 17 Oct 2021

CDC (2021) CDC vaccine price list. Centers for Disease and Control and Prevention (adult vaccine price list). https://www.cdc.gov/vaccines/programs/vfc/awardees/vaccine-management/price-list/. Accessed on 11 Feb 2021

CDC (2022) Pneumococcal vaccination: what everyone should know. https://www.cdc.gov/vaccines/vpd/pneumo/public/index.html

Celik M, Wendel SC (2005) Chapter 5: spray drying and pharmaceutical applications. In: Handbook of pharmaceutical granulation technology, pp 129–157

Chae J, Choi Y, Tanaka M, Choi J (2021) Inhalable nanoparticles delivery targeting alveolar macrophages for the treatment of pulmonary tuberculosis. J Biosci Bioeng. https://doi.org/10.1016/j.jbiosc.2021.08.009

Chaudhary N, Weissman D, Whitehead KA (2021) mRNA vaccines for infectious diseases: principles, delivery and clinical translation. Nat Rev Drug Discov 20:817–838. https://doi.org/10.1038/s41573-021-00283-5

Chen D, Kristensen D (2009) Opportunities and challenges of developing thermostable vaccines. Expert Rev Vaccines 8:547–557. https://doi.org/10.1586/erv.09.20

Chen D, Kapre S, Goel A et al (2010) Thermostable formulations of a hepatitis B vaccine and a meningitis A polysaccharide conjugate vaccine produced by a spray drying method. Vaccine 28:5093–5099. https://doi.org/10.1016/j.vaccine.2010.04.112

Chhabra P, de Graaf M, Parra GI et al (2019) Updated classification of norovirus genogroups and genotypes. J Gen Virol 100:1393–1406. https://doi.org/10.1099/JGV.0.001318

Chow M, Lam J (2015) Dry powder formulation of plasmid DNA and siRNA for inhalation. Curr Pharm Des 21:3854–3866. https://doi.org/10.2174/1381612821666150820105916

Chow MYT, Qiu Y, Lo FFK et al (2017) Inhaled powder formulation of naked siRNA using spray drying technology with L-leucine as dispersion enhancer. Int J Pharm 530:40–52. https://doi.org/10.1016/j.ijpharm.2017.07.013

Christopher GW, Cieslak TJ, Pavlin JA, Eitzen EM (1997) Biological warfare: a historical perspective. JAMA 278:412. https://doi.org/10.1001/jama.1997.03550050074036

Cieślik P, Knap J, Bielawska-Drózd A (2018) Francisella Tularensis—review. Postępy Mikrobiol—Adv Microbiol 57:58–67. https://doi.org/10.21307/PM-2018.57.1.058

Clarke DK, Hendry RM, Singh V et al (2016) Live virus vaccines based on a vesicular stomatitis virus (VSV) backbone: standardized template with key considerations for a risk/benefit assessment. Vaccine 34:6597–6609. https://doi.org/10.1016/j.vaccine.2016.06.071

Coalition Against Typhoid (2021) Currently available typhoid vaccines recommended by the WHO

Cohen J (2022) First self-copying mRNA vaccine proves itself in pandemic trial. Science (80–) 376:446–446. https://doi.org/10.1126/science.abq7232

Conlan JW, Oyston PCF (2007) Vaccines against *Francisella tularensis*. Ann N Y Acad Sci 1105:325–350. https://doi.org/10.1196/annals.1409.012

Conry RM, LoBuglio AF, Wright M et al (1995) Characterization of a messenger RNA polynucleotide vaccine vector. Cancer Res 55:1397–1400

Corbanie E (2007) Development of a dry powder vaccine for mass vaccination of poultry. Doctoral Thesis, Ghent University. https://biblio.ugent.be/publication/469294/file/1880031

Corbanie EA, Remon JP, Van Reeth K et al (2007) Spray drying of an attenuated live Newcastle disease vaccine virus intended for respiratory mass vaccination of poultry. Vaccine 25:8306–8317. https://doi.org/10.1016/j.vaccine.2007.09.049

Corbanie EA, Vervaet C, van Eck JHH et al (2008) Vaccination of broiler chickens with dispersed dry powder vaccines as an alternative for liquid spray and aerosol vaccination. Vaccine 26:4469–4476. https://doi.org/10.1016/j.vaccine.2008.06.055

CORDIS (2018) EU horizon 2020 Project, MAnufacturing process for Cold-chain Independent VIrosome-based VAccines (MACIVIVA). Grant agreement ID: 646122. https://cordis.europa.eu/project/id/646122

Crawford SE, Ramani S, Tate JE et al (2017) Rotavirus infection. Nat Rev Dis Prim 3:17083. https://doi.org/10.1038/nrdp.2017.83

Crommelin DJA, Anchordoquy TJ, Volkin DB et al (2021a) Addressing the cold reality of mRNA vaccine stability. J Pharm Sci 110:997–1001. https://doi.org/10.1016/j.xphs.2020.12.006

Crommelin DJA, Volkin DB, Hoogendoorn KH et al (2021b) The science is there: key considerations for stabilizing viral vector-based Covid-19 vaccines. J Pharm Sci 110:627–634. https://doi.org/10.1016/j.xphs.2020.11.015

Cross AS, Calia FM, Edelman R (2007) From rabbits to humans: the contributions of Dr. Theodore E. Woodward to Tularemia research. Clin Infect Dis 45:S61–S67. https://doi.org/10.1086/518150

Crowe LM, Crowe JH, Rudolph A et al (1985) Preservation of freeze-dried liposomes by trehalose. Arch Biochem Biophys 242:240–247. https://doi.org/10.1016/0003-9861(85)90498-9

Crowe JH, Hoekstra FA, Crowe LM (1992) Anhydrobiosis. Annu Rev Physiol 54:579–599. https://doi.org/10.1146/annurev.ph.54.030192.003051

Crowe LM, Reid DS, Crowe JH (1996) Is trehalose special for preserving dry biomaterials? Biophys J 71:2087–2093. https://doi.org/10.1016/S0006-3495(96)79407-9

Crowe JH, Carpenter JF, Crowe LM (1998a) The role of vitrification in anhydrobiosis. Annu Rev Physiol 60:73–103. https://doi.org/10.1146/annurev.physiol.60.1.73

Crowe JH, Clegg JS, Crowe LM (1998b) Anhydrobiosis: the water replacement hypothesis. In: Reid D (ed) The properties of water in foods ISOPOW 6. Springer, US, Boston, MA, pp 440–455

Crowe JH, Crowe LM, Oliver AE et al (2001) The trehalose myth revisited: introduction to a symposium on stabilization of cells in the dry state. Cryobiology 43:89–105. https://doi.org/10.1006/cryo.2001.2353

CTRI (2012) Clinical trials registry, India. CTRI number: CTRI/2012/02/002447 [registered on: 24/02/2012]. In: CTRI Website URL—http//ctri.nic.in. http://ctri.nic.in/Clinicaltrials/pdf_generate.php?trialid=46186&EncHid=&modid=&compid=%27,%2746186det%27

Cui F (2020) Strategies to increase timely uptake of hepatitis B vaccine birth dose. Lancet Glob Heal 8:e869–e870. https://doi.org/10.1016/S2214-109X(20)30257-6

Cutts FT, Clements CJ, Bennett JV (1997) Altenative routes of measles immunization: a review. Biologicals 25:323–338. https://doi.org/10.1006/biol.1997.0103

Danilov A (1973) The basis for vaccination against measles by aerosols of live vaccine. In: Hers JFP, Winkler KC (eds) Airborne transmission and airborne infection. Wiley, New York, pp 344–345

Das SC, Stewart PJ (2016) Understanding the respiratory delivery of high dose anti-tubercular drugs. In: Drug delivery systems for tuberculosis prevention and treatment. Wiley, Chichester, UK, pp 258–274

Day JB, Nguyen H, Sharma SK et al (2009) Effect of dehydrated storage on the survival of *Francisella tularensis* in infant formula. Food Microbiol 26:932–935. https://doi.org/10.1016/j.fm.2009.06.005

De Boer AH, Hagedoorn P (2015) The role of disposable inhalers in pulmonary drug delivery. Expert Opin Drug Deliv 12:143–157. https://doi.org/10.1517/17425247.2014.952626

De Boer AH, Hagedoorn P, Westerman EM et al (2006) Design and in vitro performance testing of multiple air classifier technology in a new disposable inhaler concept (Twincer®) for high powder doses. Eur J Pharm Sci 28:171–178. https://doi.org/10.1016/j.ejps.2005.11.013

De Pascalis R, Mittereder L, Chou AY et al (2015) *Francisella tularensis* vaccines elicit concurrent protective t- and B-Cell immune responses in BALB/cByJ Mice. PLoS ONE 10: e0126570. https://doi.org/10.1371/journal.pone.0126570

De Swart RL, LiCalsi C, Quirk AV et al (2007) Measles vaccination of macaques by dry powder inhalation. Vaccine 25:1183–1190. https://doi.org/10.1016/j.vaccine.2006.10.019

de Boer AH, Hagedoorn P, Hoppentocht M et al (2017) Dry powder inhalation: past, present and future. Expert Opin Drug Deliv 14:499–512. https://doi.org/10.1080/17425247.2016.1224846

de Martel C, Plummer M, Vignat J, Franceschi S (2017) Worldwide burden of cancer attributable to HPV by site, country and HPV type. Int J Cancer 141:664–670. https://doi.org/10.1002/ijc.30716

de Martel C, Georges D, Bray F et al (2020) Global burden of cancer attributable to infections in 2018: a worldwide incidence analysis. Lancet Glob Heal 8:e180–e190. https://doi.org/10.1016/S2214-109X(19)30488-7

Delany I, Rappuoli R, De Gregorio E (2014) Vaccines for the 21st century. EMBO Mol Med 6:708–720. https://doi.org/10.1002/emmm.201403876

Dennis DT, Inglesby TV, Henderson DA et al (2001) Tularemia as a biological weapon. JAMA 285:2763. https://doi.org/10.1001/jama.285.21.2763

Dilraj A, Cutts FT, Fernandez de Castro J et al (2000) Response to different measles vaccine strains given by aerosol and subcutaneous routes to schoolchildren: a randomised trial. Lancet 355:798–803. https://doi.org/10.1016/S0140-6736(99)95140-1

Dimitrov KM, Afonso CL, Yu Q, Miller PJ (2017) Newcastle disease vaccines—a solved problem or a continuous challenge? Vet Microbiol 206:126–136. https://doi.org/10.1016/j.vetmic.2016.12.019

DiNapoli JM, Kotelkin A, Yang L et al (2007) Newcastle disease virus, a host range-restricted virus, as a vaccine vector for intranasal immunization against emerging pathogens. Proc Natl Acad Sci 104:9788–9793. https://doi.org/10.1073/pnas.0703584104

Dolgin E (2020) COVID-19 vaccines poised for launch, but impact on pandemic unclear. Nat Biotechnol. https://doi.org/10.1038/d41587-020-00022-y

Dolgin E (2021) The tangled history of mRNA vaccines. Nature 597:318–324. https://doi.org/10.1038/d41586-021-02483-w

Donaldson B, Lateef Z, Walker GF et al (2018) Virus-like particle vaccines: immunology and formulation for clinical translation. Expert Rev Vaccines 17:833–849. https://doi.org/10.1080/14760584.2018.1516552

DoseOne LLC (2011) DoseOne®: a simple patient experience. www.dose-one.com

Eagle H (1959) Amino acid metabolism in mammalian cell cultures. Science (80–) 130:432–437. https://doi.org/10.1126/science.130.3373.432

Eber FJ, Yazdan Panah B, Sewing S et al (2021) Dry powder composition comprising long-chain RNA, Patent US 2021/0030683 A1, Filed: June 18, 2020. 1–42

Eber FJ, Yazdan Panah B, Sewing S et al (2022) Dry powder composition comprising long-chain RNA, Patent US US2022/168225A1, Filed: Feb 21, 2022. 1–42

Edens C, Collins ML, Goodson JL et al (2015) A microneedle patch containing measles vaccine is immunogenic in non-human primates. Vaccine 33:4712–4718. https://doi.org/10.1016/j.vaccine.2015.02.074

El Sahly HM, Atmar RL, Patel SM et al (2009) Safety, reactogenicity and immunogenicity of *Francisella tularensis* live vaccine strain in humans. Vaccine 27:4905–4911. https://doi.org/10.1016/j.vaccine.2009.06.036

Eldridge JH, Hammond CJ, Meulbroek JA et al (1990) Controlled vaccine release in the gut-associated lymphoid tissues. I. Orally administered biodegradable microspheres target the peyer's patches. J Control Release 11:205–214. https://doi.org/10.1016/0168-3659(90)90133-E

Eldridge JH, Staas JK, Meulbroek JA et al (1991) Biodegradable microspheres as a vaccine delivery system. Mol Immunol 28:287–294. https://doi.org/10.1016/0161-5890(91)90076-V

Ellis J, Oyston PCF, Green M, Titball RW (2002) Tularemia. Clin Microbiol Rev 15:631–646. https://doi.org/10.1128/CMR.15.4.631-646.2002

Emergent BioSolutions (2015) BioThrax® (anthrax vaccine adsorbed), Initial U.S. approval: 1970. https://www.fda.gov/media/71954/download

Emergent BioSolutions (2018) News release, 28 December 2018, Emergent biosolutions announces submission to FDA of application covering emergency use authorization for NuThrax™. https://investors.emergentbiosolutions.com/news-releases/news-release-details/emergent-biosolutions-announces-submission-fda-application

Emergent Travel Health Inc. (2020) Vivotif® (typhoid vaccine live oral Ty21a) package insert. https://vivotif.com/downloads/Vivotif_Prescribing_Information_2017.pdf

Encyclopædia Britannica (2022) Neisseria meningitidis; meningococcal meningitis. https://www.britannica.com/science/meningococcus#/media/1/375072/146173

Esposito S, Principi N (2020) Norovirus vaccine: priorities for future research and development. Front Immunol 11:1–9. https://doi.org/10.3389/fimmu.2020.01383

Evers MJW, Kulkarni JA, van der Meel R et al (2018) State-of-the-art design and rapid-mixing production techniques of lipid nanoparticles for nucleic acid delivery. Small Methods 2:1700375. https://doi.org/10.1002/smtd.201700375

Fathi A, Dahlke C, Addo MM (2019) Recombinant vesicular stomatitis virus vector vaccines for WHO blueprint priority pathogens. Hum Vaccines Immunother 15:2269–2285. https://doi.org/10.1080/21645515.2019.1649532

FDA (2013) Drug approval package, TOBI Podhaler (tobramycin inhalation powder) oral inhalation. Company: Novartis Pharmaceuticals Corporation, Application No.: 201688, Approval Date: 3/22/2013. https://www.accessdata.fda.gov/drugsatfda_docs/nda/2013/201688Orig1s000TOC.cfm

FDA (2022) Vaccines licensed for use in the United States, Content current as of 1/31/2022. https://www.fda.gov/vaccines-blood-biologics/vaccines/vaccines-licensed-use-united-states

Fernandes JV, Villax P (2019) Hovione technology—dry powder inhalers: towards, effective, affordable, sustainable respiratory healthcare, 8th November 2019. ONdrugDelivery. https://ondrugdelivery.com/dry-powder-inhalers-towards-effective-affordable-sustainable-respiratory-healthcare/

Fernando GJP, Hickling J, Jayashi Flores CM et al (2018) Safety, tolerability, acceptability and immunogenicity of an influenza vaccine delivered to human skin by a novel high-density microprojection array patch (Nanopatch™). Vaccine 36:3779–3788. https://doi.org/10.1016/j.vaccine.2018.05.053

Ferreira MPA, Martins JP, Hirvonen J, Santos HA (2020) Spray-drying for the formulation of oral drug delivery systems, chapter 9. In: Nanotechnology for oral drug delivery. Elsevier, pp 253–284

Foged C (2016) Thermostable subunit vaccines for pulmonary delivery: how close are we? Curr Pharm Des 22:2561–2576. https://doi.org/10.2174/1381612822666160202141603

Fourie P, Germishuizen W, Wong Y-L, Edwards D (2008) Spray drying TB vaccines for pulmonary administration. Expert Opin Biol Ther 8:857–863. https://doi.org/10.1517/14712598.8.7.857

Friebel C, Steckel H (2010) Single-use disposable dry powder inhalers for pulmonary drug delivery. Expert Opin Drug Deliv 7:1359–1372. https://doi.org/10.1517/17425247.2010.538379

Friesen DT, Shanker R, Crew M et al (2008) Hydroxypropyl methylcellulose acetate succinate-based spray-dried dispersions: an overview. Mol Pharm 5:1003–1019. https://doi.org/10.1021/mp8000793

Gala RP, Popescu C, Knipp GT et al (2017) Physicochemical and preclinical evaluation of a novel buccal measles vaccine. AAPS PharmSciTech 18:283–292. https://doi.org/10.1208/s12249-016-0566-3

Galazka A, Milstien J, Zaffran M (1999) The thermostability of vaccines. Global Programme for Vaccines and Immunization, WHO, Geneva, 1998. In: WHO/GPV/98.07. https://apps.who.int/iris/bitstream/handle/10665/64980/WHO_GPV_98.07.pdf

Galazka AM (1989) Expanded programme on immunization, stability of vaccines. WHO/EPI/GEN/89.8

Ganar K, Das M, Sinha S, Kumar S (2014) Newcastle disease virus: current status and our understanding. Virus Res 184:71–81. https://doi.org/10.1016/j.virusres.2014.02.016

Gander B, Merkle HP, Nguyen VP, Ho N-T (1995a) A new thermodynamic model to predict protein encapsulation efficiency in poly(lactide) microspheres. J Phys Chem 99:16144–16148. https://doi.org/10.1021/j100043a066

Gander B, Wehrli E, Alder R, Merkle HP (1995b) Quality improvement of spray-dried, protein-loaded d, l-PLA microspheres by appropriate polymer solvent selection. J Microencapsul 12:83–97. https://doi.org/10.3109/02652049509051129

Gander B, Johansen P, Nam-Trân H, Merkle HP (1996) Thermodynamic approach to protein microencapsulation into poly(D, L-lactide) by spray drying. Int J Pharm 129:51–61. https://doi.org/10.1016/0378-5173(95)04240-7

Ganem CF, Ganem J, Ganem S (2013) Medication inhaler, Patent US 8464712 B2, Filed: September 7, 2007, Applicant: Dose One LLC, Continuation-in-part of Patent US 7832399 B2. 1–39

Garcea RL, Meinerz NM, Dong M et al (2020) Single-administration, thermostable human papillomavirus vaccines prepared with atomic layer deposition technology. npj Vaccines 5:45. https://doi.org/10.1038/s41541-020-0195-4

Garcia-Contreras L, Wong Y-L, Muttil P et al (2008) Immunization by a bacterial aerosol. Proc Natl Acad Sci 105:4656–4660. https://doi.org/10.1073/pnas.0800043105

Gatto NM, Koralek DO, Bracken MB et al (2019) Lung cancer-related mortality with inhaled insulin or a comparator: follow-up study of patients previously enrolled in Exubera controlled clinical trials (FUSE) final results. Diabetes Care 42:1708–1715. https://doi.org/10.2337/dc18-2529

Geisbert TW, Feldmann H (2011) Recombinant vesicular stomatitis virus-based vaccines against Ebola and Marburg virus infections. J Infect Dis 204:S1075–S1081. https://doi.org/10.1093/infdis/jir349

Geller DE, Weers J, Heuerding S (2011) Development of an inhaled dry-powder formulation of tobramycin using PulmoSphere™ technology. J Aerosol Med Pulm Drug Deliv 24:175–182. https://doi.org/10.1089/jamp.2010.0855

Geng O (2018) SPX FLOW aseptic spray drying, 13 Mar 2018. https://file.vogel.com.cn/2018/08/31/5b8905c261000.pdf

Gerhardt A, Voigt E, Archer M et al (2022) A flexible, thermostable nanostructured lipid carrier platform for RNA vaccine delivery. Mol Ther Methods Clin Dev 25:205–214. https://doi.org/10.1016/j.omtm.2022.03.009

Gerke C, Frantz PN, Ramsauer K, Tangy F (2019) Measles-vectored vaccine approaches against viral infections: a focus on Chikungunya. Expert Rev Vaccines 18:393–403. https://doi.org/10.1080/14760584.2019.1562908

Ghaemmaghamian Z, Zarghami R, Walker G et al (2022) Stabilizing vaccines via drying: quality by design considerations. Adv Drug Deliv Rev 187:114313. https://doi.org/10.1016/j.addr.2022.114313

Gibani MM, Britto C, Pollard AJ (2018) Typhoid and paratyphoid fever. Curr Opin Infect Dis 31:440–448. https://doi.org/10.1097/QCO.0000000000000479

Gierke R, Wodi AP, Kobayashi M (2021) Chapter 17: pneumococcal disease. In: Epidemiology and prevention of vaccine-preventable diseases. The Pink book: course textbook, 14th edn

Giersing BK, Vekemans J, Nava S et al (2019) Report from the World Health Organization's third product development for vaccines advisory committee (PDVAC) meeting, Geneva, 8–10th June 2016. Vaccine 37:7315–7327. https://doi.org/10.1016/j.vaccine.2016.10.090

Gill D (2014) Heat stable oral rotavirus vaccine, MSD Wellcome Trust Hilleman Laboratories. In: 11th International rotavirus symposium, 4 September 2014, New Delhi, India, pp 1–15

Gill D (2018) Development of a heat-stable Rotavirus vaccine using innovative delivery technology, MSD Wellcome Trust Hilleman Laboratories. In: 13th International rotavirus symposium, 29–31 August 2018, Minsk, Belarus, pp 1–15

Gill D, Saigal N, Kale S et al (2017) Thermostable spray dried rotavirus vaccine formulation and process thereof, Patent US 2017/0173145 A1, PCT filed: July 17, 2015. 1–17

Gill D, Saigal N, Kale S et al (2020) Thermostable spray dried rotavirus vaccine formulation and process thereof, MSD Wellcome Trust Hilleman Laboratories Pvt. Ltd., Patent US10556000 B2, PCT Filed: July 17, 2015. 1–17

Glass RI, Parashar UD, Estes MK (2009) Norovirus gastroenteritis. N Engl J Med 361:1776–1785. https://doi.org/10.1056/NEJMra0804575

Glassman HN (1958) World incidence of anthrax in man. Public Heal Reports 73:22. https://doi.org/10.2307/4590029

Goldberg T, Wong E (2015) Afrezza (insulin human) inhalation powder. Pharm Therap 40(11), 735–741. https://www.ncbi.nlm.nih.gov/pmc/articles/PMC4634344/

Gomez M, Vehring R (2022) Spray drying and particle engineering in dosage form design for global vaccines. J Aerosol Med Pulm Drug Deliv 35:1–18. https://doi.org/10.1089/jamp.2021.0056

Gomez M, Johnson S, Gennaro ML (2000) Identification of secreted proteins of *Mycobacterium tuberculosis* by a bioinformatic approach. Infect Immun 68:2323–2327. https://doi.org/10.1128/IAI.68.4.2323-2327.2000

Gomez M, Archer M, Barona D et al (2021a) Microparticle encapsulation of a tuberculosis subunit vaccine candidate containing a nanoemulsion adjuvant via spray drying. Eur J Pharm Biopharm 163:23–37. https://doi.org/10.1016/j.ejpb.2021.03.007

Gomez M, McCollum J, Wang H et al (2021b) Development of a formulation platform for a spray-dried, inhalable tuberculosis vaccine candidate. Int J Pharm 593:120121. https://doi.org/10.1016/j.ijpharm.2020.120121

Gomez M, McCollum J, Wang H et al (2021c) Evaluation of the stability of a spray-dried tuberculosis vaccine candidate designed for dry powder respiratory delivery. Vaccine 39:5025–5036. https://doi.org/10.1016/j.vaccine.2021.07.002

Gomez M, Ahmed M, Das S et al (2022) Development and testing of a spray-dried tuberculosis vaccine candidate in a mouse model. Front Pharmacol 12:1–23. https://doi.org/10.3389/fphar.2021.799034

Gómez-Aguado I, Rodríguez-Castejón J, Vicente-Pascual M et al (2020) Nanomedicines to deliver mRNA: state of the art and future perspectives. Nanomaterials 10:364. https://doi.org/10.3390/nano10020364

Gonzalez-Cruz P, Gill HS (2021) Demystifying particle-based oral vaccines. Expert Opin Drug Deliv 18:1455–1472. https://doi.org/10.1080/17425247.2021.1946511

Goosney DL, Nemerow GR (2003) Adenovirus infection: taking the back roads to viral entry. Curr Biol 13:R99–R100. https://doi.org/10.1016/S0960-9822(03)00037-X

Greb E (2009) Is spray drying a viable alternative to lyophilization. PharmTech 1–2

Green SL (1993) Vesicular stomatitis in the horse. Vet Clin North Am Equine Pract 9:349–353. https://doi.org/10.1016/S0749-0739(17)30401-7

Green KY (2013) Chapter 20, Calicivirdae: the Noroviruses. In: Knipe DM, Howley PM (eds) Fields virology. Lippincott Williams & Wilkins, Philadelphia, PA, pp 582–608

Griffin KF, Oyston PCF, Titball RW (2007) Francisella tularensis vaccines. FEMS Immunol Med Microbiol 49:315–323. https://doi.org/10.1111/j.1574-695X.2007.00219.x

Guan S, Rosenecker J (2017) Nanotechnologies in delivery of mRNA therapeutics using nonviral vector-based delivery systems. Gene Ther 24:133–143. https://doi.org/10.1038/gt.2017.5

Guntur VP, Dhand R (2007) Inhaled insulin: extending the horizons of inhalation therapy. Respir Care 52(911–922):12p

Hajj KA, Whitehead KA (2017) Tools for translation: non-viral materials for therapeutic mRNA delivery. Nat Rev Mater 2:17056. https://doi.org/10.1038/natrevmats.2017.56

Hall AJ, Lopman BA, Payne DC et al (2013) Norovirus disease in the United States. Emerg Infect Dis 19:1198–1205. https://doi.org/10.3201/eid1908.130465

Hallowell BD, Parashar UD, Hall AJ (2019) Epidemiologic challenges in norovirus vaccine development. Hum Vaccin Immunother 15:1279–1283. https://doi.org/10.1080/21645515.2018.1553594

Hanks JH, Wallace RE (1949) Relation of oxygen and temperature in the preservation of tissues by refrigeration. Exp Biol Med 71:196–200. https://doi.org/10.3181/00379727-71-17131

HARRO Magazine (2020) Semi-automatic filling of dry powder. HARRO Magazine 10/2020. https://www.harro-magazine.com/en/10-2020/semi-automatic-filling-of-dry-powder/

Hassett KJ, Meinerz NM, Semmelmann F et al (2015a) Development of a highly thermostable, adjuvanted human papillomavirus vaccine. Eur J Pharm Biopharm 94:220–228. https://doi.org/10.1016/j.ejpb.2015.05.009

Hassett KJ, Vance DJ, Jain NK et al (2015b) Glassy-state stabilization of a dominant negative inhibitor anthrax vaccine containing aluminum hydroxide and glycopyranoside lipid A adjuvants. J Pharm Sci 104:627–639. https://doi.org/10.1002/jps.24295

Hassett KJ, Benenato KE, Jacquinet E et al (2019) Optimization of lipid nanoparticles for intramuscular administration of mRNA vaccines. Mol Ther Nucleic Acids 15:1–11. https://doi.org/10.1016/j.omtn.2019.01.013

Heida R, Hinrichs WLJ, Frijlink HW (2021) Inhaled vaccine delivery in the combat against respiratory viruses: a 2021 overview of recent developments and implications for COVID-19. Expert Rev Vaccines 00:1–18. https://doi.org/10.1080/14760584.2021.1903878

Heinemann L (2008) The failure of exubera: are we beating a dead horse? J Diabetes Sci Technol 2:518–529. https://doi.org/10.1177/193229680800200325

Heininger U (2012) Pertussis. Pediatr Infect Dis J 31:78–79. https://doi.org/10.1097/INF.0b013e31823b034e

Hekele A, Bertholet S, Archer J et al (2013) Rapidly produced SAM® vaccine against H7N9 influenza is immunogenic in mice. Emerg Microbes Infect 2:1–7. https://doi.org/10.1038/emi.2013.54

Herzog C, Hartmann K, Künzi V et al (2009) Eleven years of Inflexal® V-a virosomal adjuvanted influenza vaccine. Vaccine 27:4381–4387. https://doi.org/10.1016/j.vaccine.2009.05.029

Hester (2017) Hester Biosciences Limited, India, product catalogue, January 2017, THERMOSTABL ND. https://static1.squarespace.com/static/584173905016e1ca1b9c2115/t/5a28caf7085229b3c96fbf24/1512622899462/Product+Catalogue+2017+-+Webview.pdf

Hickey AJ, Garmise RJ (2009) Dry powder nasal vaccines as an alternative to needle-based delivery. Crit Rev Ther Drug Carr Syst 26:1–27. https://doi.org/10.1615/CritRevTherDrugCarrierSyst.v26.i1.10

Hickey AJ, Stewart IE, Jones A (2019) Why we need to deliver large amounts of powder to the lungs and the concurrent challenges. In: Drug delivery to the lungs (DDL2019), Edinburgh, Scotland, UK, 11–13 December, 2019, pp 1–4

Hill AB, Kilgore C, McGlynn M, Jones CH (2016) Improving global vaccine accessibility. Curr Opin Biotechnol 42:67–73. https://doi.org/10.1016/j.copbio.2016.03.002

Hinz T, Kallen K, Britten CM et al (2017) The European regulatory environment of RNA-based vaccines. In: Methods in molecular biology, pp 203–222

Hiremath GS, Omer SB (2005) A meta-analysis of studies comparing the respiratory route with the subcutaneous route of measles vaccine administration. Hum Vaccin 1:30–36. https://doi.org/10.4161/hv.1.1.1423

Hirobe S, Azukizawa H, Hanafusa T et al (2015) Clinical study and stability assessment of a novel transcutaneous influenza vaccination using a dissolving microneedle patch. Biomater 57:50–58. https://doi.org/10.1016/j.biomaterials.2015.04.007

Hoe S, Ivey JW, Boraey MA et al (2014) Use of a fundamental approach to spray-drying formulation design to facilitate the development of multi-component dry powder aerosols for respiratory drug delivery. Pharm Res 31:449–465. https://doi.org/10.1007/s11095-013-1174-5

Holman DH, Wang D, Woraratanadharm J, Dong JY (2009) Viral vectors. In: Vaccines for biodefense and emerging and neglected diseases. Elsevier, pp 77–91

HollyCon Italy (2021) What is PillHaler? https://www.hollyconitaly.com/en/technologies-solutions/

Hong HC, Kim KS, Park SA et al (2021) An mRNA vaccine against SARS-CoV-2: lyophilized, liposome-based vaccine candidate EG-COVID induces high levels of virus neutralizing antibodies. Prepr bioRxiv. https://doi.org/10.1101/2021.03.22.436375

Hoppentocht M, Akkerman OW, Hagedoorn P et al (2015) The cyclops for pulmonary delivery of aminoglycosides; a new member of the Twincer™ family. Eur J Pharm Biopharm 90:8–15. https://doi.org/10.1016/j.ejpb.2015.01.012

Hou X, Zaks T, Langer R, Dong Y (2021) Lipid nanoparticles for mRNA delivery. Nat Rev Mater 6:1078–1094. https://doi.org/10.1038/s41578-021-00358-0

Hovione (2021b) Large dose DPI—disposable DPI for high dose drug delivery (TwinMax, 8Shot). http://www.hovionetechnology.com/en/portfolio/product-family/large-dose-dpi/

Hovione (2021a) TwinCaps—World's largest selling disposable DPI. http://www.hovionetechnology.com/en/portfolio/product-family/twincaps/

Hsu DC, O'Connell RJ (2017) Progress in HIV vaccine development. Hum Vaccin Immunother 13:1018–1030. https://doi.org/10.1080/21645515.2016.1276138

Huang Y, Dai W-G (2014) Fundamental aspects of solid dispersion technology for poorly soluble drugs. Acta Pharm Sin B 4:18–25. https://doi.org/10.1016/j.apsb.2013.11.001

Huang J, Mikszta JA, Ferriter MS et al (2007) Intranasal administration of dry powder anthrax vaccine provides protection against lethal aerosol spore challenge. Hum Vaccin 3:90–93. https://doi.org/10.4161/hv.3.3.4011

Huber B, Wang JW, Roden RBS, Kirnbauer R (2021) RG1-VLP and other L2-based, broad-spectrum HPV vaccine candidates. J Clin Med 10:1044. https://doi.org/10.3390/jcm10051044

Hurt K, Bilton D (2012) Inhaled mannitol for the treatment of cystic fibrosis. Expert Rev Respir Med 6:19–26. https://doi.org/10.1586/ers.11.87

Hussain A, Yang H, Zhang M et al (2022) mRNA vaccines for COVID-19 and diverse diseases. J Control Release 345:314–333. https://doi.org/10.1016/j.jconrel.2022.03.032

Huyge K (2012) Dry powder vaccines for mass vaccination of poultry: New formulations, comparison with liquid vaccines and feasibility to decrease adverse vaccinal reactions. Doctoral thesis, Ghent University

Huyge K, Van Reeth K, De Beer T et al (2012) Suitability of differently formulated dry powder Newcastle disease vaccines for mass vaccination of poultry. Eur J Pharm Biopharm 80:649–656. https://doi.org/10.1016/j.ejpb.2011.11.018

Ibrahim M, Verma R, Garcia-Contreras L (2015) Inhalation drug delivery devices: technology update. Med Devices Evid Res 8:131. https://doi.org/10.2147/MDER.S48888

ICH (1995) ICH Quality Guideline Q5C: stability testing of biotechnological/biological product. https://www.ich.org/page/quality-guidelines

ICH (2003) ICH quality guideline Q1A(R2): stability testing of new drug substances and products. https://www.ich.org/page/quality-guidelines

Ickenstein LM, Garidel P (2019) Lipid-based nanoparticle formulations for small molecules and RNA drugs. Expert Opin Drug Deliv 16:1205–1226. https://doi.org/10.1080/17425247.2019.1669558

Iconovo (2021) Iconovo develops inhaled Covid-19 vaccine. ICOone News, 12 October 2021. https://iconovo.se/iconovo-develops-inhaled-covid-19-vaccine/

Inglesby TV, O'Toole T, Henderson DA et al (2002) Anthrax as a biological weapon, 2002. JAMA 287:2236. https://doi.org/10.1001/jama.287.17.2236

Ingvarsson PT, Schmidt ST, Christensen D et al (2013a) Designing CAF-adjuvanted dry powder vaccines: spray drying preserves the adjuvant activity of CAF01. J Control Release 167:256–264. https://doi.org/10.1016/j.jconrel.2013.01.031

Ingvarsson PT, Yang M, Mulvad H et al (2013b) Engineering of an inhalable DDA/TDB liposomal adjuvant: a quality-by-design approach towards optimization of the spray drying process. Pharm Res 30:2772–2784. https://doi.org/10.1007/s11095-013-1096-2

Iuliano AD, Roguski KM, Chang HH et al (2018) Estimates of global seasonal influenza-associated respiratory mortality: a modelling study. Lancet 391:1285–1300. https://doi.org/10.1016/S0140-6736(17)33293-2

IVEK (2012) Sonicair, atomizing systems, www.ivek.com. https://www.ivek.com/literature/sonicair.pdf

Ivers LC (2017) Vaccines plus water, sanitation, and hygiene interventions in the fight against cholera—authors' reply. Lancet Glob Heal 5:e395. https://doi.org/10.1016/S2214-109X(17)30086-4

Ivey J, Bhambri P, Lewis D et al (2016) Dried corticosteroid particle formation from evaporating monodisperse propellant solution droplets, poster. In: AAPS annual meeting and exposition, 13–17 Nov 2016, Denver, CO, USA

Ivey JW, Vehring R (2016) A methodology for efficient pharmaceutical spray drying process development. In: AAPS annual meeting and exposition, 13–17 Nov 2016, Denver, CO, USA

James C, Harfouche M, Welton NJ et al (2020) Herpes simplex virus: global infection prevalence and incidence estimates, 2016. Bull World Health Organ 98:315–329. https://doi.org/10.2471/BLT.19.237149

Jensen DMK, Cun D, Maltesen MJ et al (2010) Spray drying of siRNA-containing PLGA nanoparticles intended for inhalation. J Control Release 142:138–145. https://doi.org/10.1016/j.jconrel.2009.10.010

Jermain SV, Brough C, Williams RO (2018) Amorphous solid dispersions and nanocrystal technologies for poorly water-soluble drug delivery—an update. Int J Pharm 535:379–392. https://doi.org/10.1016/j.ijpharm.2017.10.051

Jernigan DB, Raghunathan PL, Bell BP et al (2002) Investigation of bioterrorism-related anthrax, United States, 2001: epidemiologic findings. Emerg Infect Dis 8:1019–1028. https://doi.org/10.3201/eid0810.020353

Jezek J, Chen D, Watson L et al (2009) A heat-stable hepatitis B vaccine formulation. Hum Vaccin 5:529–535. https://doi.org/10.4161/hv.5.8.8600

Jia Q, Lee B-Y, Clemens DL et al (2009) Recombinant attenuated Listeria monocytogenes vaccine expressing *Francisella tularensis* IglC induces protection in mice against aerosolized Type A F. tularensis. Vaccine 27:1216–1229. https://doi.org/10.1016/j.vaccine.2008.12.014

Jiang G, Joshi SB, Peek LJ et al (2006) Anthrax vaccine powder formulations for nasal mucosal delivery. J Pharm Sci 95:80–96. https://doi.org/10.1002/jps.20484

Jin T, Tsao E (2013) Powder vaccines for pulmonary delivery. In: Flower DR, Perrie Y (eds) Immunomic discovery of adjuvants and candidate subunit vaccines. Springer, New York, New York, NY, pp 291–301

Jin TH, Tsao E, Goudsmit J et al (2010) Stabilizing formulations for inhalable powders of an adenovirus 35-vectored tuberculosis (TB) vaccine (AERAS-402). Vaccine 28:4369–4375. https://doi.org/10.1016/j.vaccine.2010.04.059

Johansen P, Moon L, Tamber H et al (1999a) Immunogenicity of single-dose diphtheria vaccines based on PLA/PLGA microspheres in guinea pigs. Vaccine 18:209–215. https://doi.org/10.1016/S0264-410X(99)00191-7

Johansen P, Tamber H, Merkle HP, Gander B (1999b) Diphtheria and tetanus toxoid microencapsulation into conventional and end-group alkylated PLA/PLGAs. Eur J Pharm Biopharm 47:193–201. https://doi.org/10.1016/S0939-6411(98)00095-2

Johansen P, Merkle HP, Gander B (2000) Technological considerations related to the up-scaling of protein microencapsulation by spray-drying. Eur J Pharm Biopharm 50:413–417. https://doi.org/10.1016/S0939-6411(00)00123-5

Jones A, Miller RL (2012) Dose delivery device for inhalation, Patent US 8291901 B2, Filed: July 7, 2008, Assignee: Manta Devices LLC. 1–69

Jones A, Miller RL (2017) Dose delivery device for inhalation with plunger, Patent US 9713684 B2, Filed: December 13, 2013, Assignee: Manta Devices LLC. 1–69

Jones A, Miller RL (2018) Dose delivery device for Inhalation with first and second portions to open a dose chamber, Patent US 9919115 B2, Filed: April 10, 2014, Assignee: Manta Devices LLC. 1–69

Jones KL, Drane D, Gowans EJ (2007) Long-term storage of DNA-free RNA for use in vaccine studies. Biotechniques 43:675–681. https://doi.org/10.2144/000112593

Jones RM, Burke M, Dubose D et al (2017) Stability and pre-formulation development of a plant-produced anthrax vaccine candidate. Vaccine 35:5463–5470. https://doi.org/10.1016/j.vaccine.2016.12.009

Joyce JC, Carroll TD, Collins ML et al (2018) A microneedle patch for measles and rubella vaccination is immunogenic and protective in infant Rhesus Macaques. J Infect Dis 218:124–132. https://doi.org/10.1093/infdis/jiy139

Jyothi NVN, Prasanna PM, Sakarkar SN et al (2010) Microencapsulation techniques, factors influencing encapsulation efficiency. J Microencapsul 27:187–197. https://doi.org/10.3109/02652040903131301

Kahn A-L, Kristensen D, Rao R (2017) Extending supply chains and improving immunization coverage and equity through controlled temperature chain use of vaccines. Vaccine 35:2214–2216. https://doi.org/10.1016/j.vaccine.2016.10.091

Kallen K-J, Theß A (2014) A development that may evolve into a revolution in medicine: mRNA as the basis for novel, nucleotide-based vaccines and drugs. Ther Adv Vaccines 2:10–31. https://doi.org/10.1177/2051013613508729

Kanojia G (2018) Drying made easy: spray drying a promising technology for the production of stable vaccine and therapeutic protein formulations. PhD thesis, University of Groningen

Kanojia G, Willems G-J, Frijlink HW et al (2016) A design of experiment approach to predict product and process parameters for a spray dried influenza vaccine. Int J Pharm 511:1098–1111. https://doi.org/10.1016/j.ijpharm.2016.08.022

Kanojia G, ten Have R, Soema PC et al (2017) Developments in the formulation and delivery of spray dried vaccines. Hum Vaccin Immunother 13:2364–2378. https://doi.org/10.1080/21645515.2017.1356952

Kanojia G, Raeven RHM, van der Maas L et al (2018a) Development of a thermostable spray dried outer membrane vesicle pertussis vaccine for pulmonary immunization. J Control Release 286:167–178. https://doi.org/10.1016/j.jconrel.2018.07.035

Kanojia G, ten Have R, Brugmans D et al (2018b) The effect of formulation on spray dried Sabin inactivated polio vaccine. Eur J Pharm Biopharm 129:21–29. https://doi.org/10.1016/j.ejpb.2018.05.021

Karikó K, Buckstein M, Ni H, Weissman D (2005) Suppression of RNA recognition by toll-like receptors: the impact of nucleoside modification and the evolutionary origin of RNA. Immunity 23:165–175. https://doi.org/10.1016/j.immuni.2005.06.008

Karikó K, Muramatsu H, Welsh FA et al (2008) Incorporation of pseudouridine into mRNA yields superior nonimmunogenic vector with increased translational capacity and biological stability. Mol Ther 16:1833–1840. https://doi.org/10.1038/mt.2008.200

Karve S, DeRosa F, Heartlein M et al (2020) Dry powder formulations for messenger RNA, Patent US2020/0022921 A1, Filed: July 23, 2019. 1–66

Kawabata Y, Wada K, Nakatani M et al (2011) Formulation design for poorly water-soluble drugs based on biopharmaceutics classification system: basic approaches and practical applications. Int J Pharm 420:1–10. https://doi.org/10.1016/j.ijpharm.2011.08.032

Kim S-H, Samal S (2016) Newcastle disease virus as a vaccine vector for development of human and veterinary vaccines. Viruses 8:183. https://doi.org/10.3390/v8070183

Kim J, Eygeris Y, Gupta M, Sahay G (2021) Self-assembled mRNA vaccines. Adv Drug Deliv Rev 170:83–112. https://doi.org/10.1016/j.addr.2020.12.014

King DM, Spencer JA, Liang X et al (2007) Atomic layer deposition on particles using a fluidized bed reactor with in situ mass spectrometry. Surf Coatings Technol 201:9163–9171. https://doi.org/10.1016/j.surfcoat.2007.05.002

King DM, Liang X, Weimer AW (2012) Functionalization of fine particles using atomic and molecular layer deposition. Powder Technol 221:13–25. https://doi.org/10.1016/j.powtec.2011.12.020

Kirkwood CD, Ma L-F, Carey ME, Steele AD (2019) The rotavirus vaccine development pipeline. Vaccine 37:7328–7335. https://doi.org/10.1016/j.vaccine.2017.03.076

Kisich KO, Higgins MP, Park I et al (2011) Dry powder measles vaccine: particle deposition, virus replication, and immune response in cotton rats following inhalation. Vaccine 29:905–912. https://doi.org/10.1016/j.vaccine.2010.10.020

Klas SD, Petrie CR, Warwood SJ et al (2008) A single immunization with a dry powder anthrax vaccine protects rabbits against lethal aerosol challenge. Vaccine 26:5494–5502. https://doi.org/10.1016/j.vaccine.2008.07.062

Klietmann WF, Ruoff KL (2001) Bioterrorism: implications for the clinical microbiologist. Clin Microbiol Rev 14:364–381. https://doi.org/10.1128/CMR.14.2.364-381.2001

Kling J (2008) Inhaled insulin's last gasp? Nat Biotechnol 26:479–480. https://doi.org/10.1038/nbt0508-479

Kols A, Sherris J (2000) HPV vaccines: promise and challenges, July 2000. PATH

Kondakova OA, Nikitin NA, Evtushenko EA et al (2019) Vaccines against anthrax based on recombinant protective antigen: problems and solutions. Expert Rev Vaccines 18:813–828. https://doi.org/10.1080/14760584.2019.1643242

Konstan MW, Geller DE, Minić P et al (2011) Tobramycin inhalation powder for *P. aeruginosa* infection in cystic fibrosis: the EVOLVE trial. Pediatr Pulmonol 46:230–238. https://doi.org/10.1002/ppul.21356

Koster KL, Lei YP, Anderson M et al (2000) Effects of vitrified and nonvitrified sugars on phosphatidylcholine fluid-to-gel phase transitions. Biophys J 78:1932–1946. https://doi.org/10.1016/S0006-3495(00)76741-5

Kowalski PS, Rudra A, Miao L, Anderson DG (2019) Delivering the messenger: advances in technologies for therapeutic mRNA delivery. Mol Ther 27:710–728. https://doi.org/10.1016/j.ymthe.2019.02.012

Kraan H, van Herpen P, Kersten G, Amorij J-P (2014) Development of thermostable lyophilized inactivated polio vaccine. Pharm Res 31:2618–2629. https://doi.org/10.1007/s11095-014-1359-6

Kranz LM, Diken M, Haas H et al (2016) Systemic RNA delivery to dendritic cells exploits antiviral defence for cancer immunotherapy. Nature 534:396–401. https://doi.org/10.1038/nature18300

Krishnan R, Stuart PM (2021) Developments in vaccination for herpes simplex virus. Front Microbiol 12. https://doi.org/10.3389/fmicb.2021.798927

Kristensen D (2008) Progress in stabilizing vaccines. In: TECHNET consultation, Tunis, Tunisia, 3 Dec 2008, pp 1–20

Kristensen D (2012) Summary of stability data for investigational formulations of vaccines, 9 May 2012

Kukut Hatipoglu M, Hickey AJ, Garcia-Contreras L (2018) Pharmacokinetics and pharmacodynamics of high doses of inhaled dry powder drugs. Int J Pharm 549:306–316. https://doi.org/10.1016/j.ijpharm.2018.07.050

Kumru OS, Joshi SB, Smith DE et al (2014) Vaccine instability in the cold chain: mechanisms, analysis and formulation strategies. Biologicals 42:237–259. https://doi.org/10.1016/j.biologicals.2014.05.007

Kunda NK, Somavarapu S, Gordon SB et al (2013) Nanocarriers targeting dendritic cells for pulmonary vaccine delivery. Pharm Res 30:325–341. https://doi.org/10.1007/s11095-012-0891-5

Kunda NK, Alfagih IM, Dennison SR et al (2015a) Bovine serum albumin adsorbed PGA-CO-PDL nanocarriers for vaccine delivery via dry powder inhalation. Pharm Res 32:1341–1353. https://doi.org/10.1007/s11095-014-1538-5

Kunda NK, Alfagih IM, Miyaji EN et al (2015b) Pulmonary dry powder vaccine of pneumococcal antigen loaded nanoparticles. Int J Pharm 495:903–912. https://doi.org/10.1016/j.ijpharm.2015.09.034

Kunda NK, Wafula D, Tram M et al (2016) A stable live bacterial vaccine. Eur J Pharm Biopharm 103:109–117. https://doi.org/10.1016/j.ejpb.2016.03.027

Kunda NK, Peabody J, Zhai L et al (2019) Evaluation of the thermal stability and the protective efficacy of spray-dried HPV vaccine, Gardasil® 9. Hum Vaccin Immunother 15:1995–2002. https://doi.org/10.1080/21645515.2019.1593727

Labiris NR, Dolovich MB (2003) Pulmonary drug delivery. Part I: physiological factors affecting therapeutic effectiveness of aerosolized medications. Br J Clin Pharmacol 56:588–599. https://doi.org/10.1046/j.1365-2125.2003.01892.x

Lal M, Zhu C, McClurkan C et al (2014) Development of a low-dose fast-dissolving tablet formulation of Newcastle disease vaccine for low-cost backyard poultry immunisation. Vet Rec 174:504–504. https://doi.org/10.1136/vr.101926

Lam J, Vaughan S, Parkins MD (2013) Tobramycin inhalation powder (TIP): an efficient treatment strategy for the management of chronic *Pseudomonas aeruginosa* infection in cystic fibrosis. Clin Med Insights Circ Respir Pulm Med 7:CCRPM.S10592. https://doi.org/10.4137/CCRPM.S10592

Landman WJM, Huyge K, Remon JP et al (2015) Comparison of Newcastle disease vaccine administered as powder or liquid in relation to the serum antibody response and adverse vaccinal reactions in broilers. Avian Pathol 44:114–123. https://doi.org/10.1080/03079457.2015.1007920

Landman WJM, Vervaet C, Remon JP et al (2017) Primary Newcastle disease vaccination of broilers: comparison of the antibody seroresponse and adverse vaccinal reaction after eye–nose drop or coarse spray application, and implication of the results for a previously developed coarse dry powder vaccine. Avian Pathol 46:451–461. https://doi.org/10.1080/03079457.2017.1307941

Langford A, Ohtake S (2019) Spray drying of biotherapeutic compounds, chapter 43. In: Nema S, Ludwig JD (eds) Parenteral medications. Taylor & Francis Incorporated, pp 929–945

Langford A, Bhatnagar B, Walters R et al (2018) Drying technologies for biopharmaceutical applications: recent developments and future direction. Dry Technol 36:677–684. https://doi.org/10.1080/07373937.2017.1355318

Lastow O, Arvidsson L (2016) Single dose dry powder inhaler, Patent WO 2016/193379 A1, Filed: June 2, 2015, Applicant: Iconovo AB (SE). 1–29

Laube BL, Dolovich MB (2014) Aerosols and aerosol drug delivery systems. In: Middleton's allergy. Elsevier, pp 1066–1082

Laube BL, Janssens HM, de Jongh FHC et al (2011) What the pulmonary specialist should know about the new inhalation therapies. Eur Respir J 37:1308–1417. https://doi.org/10.1183/09031936.00166410

Lay J (2011) The man fighting measles one breath at a time. In: Color. Mag. Alumni Assoc. Univ. Color. Boulder, 1 March 2011. https://www.colorado.edu/coloradan/2011/03/01/man-fighting-measles-one-breath-time

LeClair DA, Cranston ED, Xing Z, Thompson MR (2016a) Evaluation of excipients for enhanced thermal stabilization of a human type 5 adenoviral vector through spray drying. Int J Pharm 506:289–301. https://doi.org/10.1016/j.ijpharm.2016.04.067

LeClair DA, Cranston ED, Xing Z, Thompson MR (2016b) Optimization of spray drying conditions for yield, particle size and biological activity of thermally stable viral vectors. Pharm Res 33:2763–2776. https://doi.org/10.1007/s11095-016-2003-4

LeClair DA, Li L, Rahman N et al (2019) Stabilization of HSV-2 viral vaccine candidate by spray drying. Int J Pharm 569:118615. https://doi.org/10.1016/j.ijpharm.2019.118615

Ledet GA, Graves RA, Bostanian LA, Mandal TK (2015) Spray-drying of biopharmaceuticals. In: Varshney D, Singh M (eds) Lyophilized biologics and vaccines. Springer, New York, New York, NY, pp 273–297

Lema D, Garcia A, De Sanctis JB (2014) HIV vaccines: a brief overview. Scand J Immunol 80:1–11. https://doi.org/10.1111/sji.12184

Leung AKK, Tam YYC, Chen S et al (2015) Microfluidic mixing: a general method for encapsulating macromolecules in lipid nanoparticle systems. J Phys Chem B 119:8698–8706. https://doi.org/10.1021/acs.jpcb.5b02891

Leung V, Mapletoft J, Zhang A et al (2019) Thermal stabilization of viral vaccines in low-cost sugar films. Sci Rep 9:7631. https://doi.org/10.1038/s41598-019-44020-w

Levine MM (2011) "IDEAL" vaccines for resource poor settings. Vaccine 29:D116–D125. https://doi.org/10.1016/j.vaccine.2011.11.090

Levine MM, Simon R (2018) Typhoid fever vaccines to prevent typhoid fever. https://www.sabin.org/sites/sabin.org/files/levinetyphoidfever_rev7.20.18.pdf

Levy ML, Carroll W, Izquierdo Alonso JL et al (2019) Understanding dry powder inhalers: key technical and patient preference attributes. Adv Ther 36:2547–2557. https://doi.org/10.1007/s12325-019-01066-6

Li J, Zhan L, Qin C (2021) The double-sided effects of *Mycobacterium bovis* bacillus Calmette–Guérin vaccine. npj Vaccines 6:14. https://doi.org/10.1038/s41541-020-00278-0

Liang X, Lynn AD, King DM et al (2009) Biocompatible interface films deposited within porous polymers by atomic layer deposition (ALD). ACS Appl Mater Interfaces 1:1988–1995. https://doi.org/10.1021/am9003667

Liang W, Chan AYL, Chow MYT et al (2018a) Spray freeze drying of small nucleic acids as inhaled powder for pulmonary delivery. Asian J Pharm Sci 13:163–172. https://doi.org/10.1016/j.ajps.2017.10.002

Liang W, Chow MYT, Chow SF et al (2018b) Using two-fluid nozzle for spray freeze drying to produce porous powder formulation of naked siRNA for inhalation. Int J Pharm 552:67–75. https://doi.org/10.1016/j.ijpharm.2018.09.045

LiCalsi C, Christensen T, Bennett JV et al (1999) Dry powder inhalation as a potential delivery method for vaccines. Vaccine 17:1796–1803. https://doi.org/10.1016/S0264-410X(98)00438-1

LiCalsi C, Maniaci MJ, Christensen T et al (2001) A powder formulation of measles vaccine for aerosol delivery. Vaccine 19:2629–2636. https://doi.org/10.1016/S0264-410X(00)00503-X

Lin JH, Weng CN, Liao CW et al (2003) Protective Effects of oral microencapsulated mycoplasma hyopneumoniae vaccine prepared by co-spray drying method. J Vet Med Sci 65:69–74. https://doi.org/10.1292/jvms.65.69

Lin W-H, Griffin DE, Rota PA et al (2011) Successful respiratory immunization with dry powder live-attenuated measles virus vaccine in rhesus macaques. Proc Natl Acad Sci 108:2987–2992. https://doi.org/10.1073/pnas.1017334108

Liu (2019) A comparison of plasmid DNA and mRNA as vaccine technologies. Vaccines 7:37. https://doi.org/10.3390/vaccines7020037

Looker KJ, Magaret AS, Turner KME et al (2015) Global estimates of prevalent and incident herpes simplex virus type 2 infections in 2012. PLoS ONE 10:e114989. https://doi.org/10.1371/journal.pone.0114989

Lopman B (2015) Global burden of norovirus and prospects for vaccine development. Atmar, Robert Baric, Ralph Estes, Mary Green, Kim Glas Roger Hall, Aron Iturriza-Gomara, Miren Kang, Cherry Lee, Bruce Parashar, Umesh Riddle, Mark Vinjé, Jan

Lopman BA, Steele D, Kirkwood CD, Parashar UD (2016) The vast and varied global burden of norovirus: prospects for prevention and control. PLoS Med 13:1–12. https://doi.org/10.1371/journal.pmed.1001999

Loughran AJ, Orihuela CJ, Tuomanen EI (2019) Streptococcus pneumoniae: invasion and inflammation. Microbiol Spectr 7:1–18. https://doi.org/10.1128/microbiolspec.GPP3-0004-2018

Lovalenti PM, Truong-Le V (2020) Foam drying. In: Ohtake S, Izutsu K-I, Lechuga-Ballesteros D (eds) Drying technologies for biotechnology and pharmaceutical applications. Wiley, pp 257–282

Lovalenti PM, Anderl J, Yee L et al (2016) Stabilization of live attenuated influenza vaccines by freeze drying, spray drying, and foam drying. Pharm Res 33:1144–1160. https://doi.org/10.1007/s11095-016-1860-1

Low N, Kraemer S, Schneider M, Restrepo AMH (2008) Immunogenicity and safety of aerosolized measles vaccine: systematic review and meta-analysis. Vaccine 26:383–398. https://doi.org/10.1016/j.vaccine.2007.11.010

Low N, Bavdekar A, Jeyaseelan L et al (2015) A Randomized, controlled trial of an aerosolized vaccine against measles. N Engl J Med 372:1519–1529. https://doi.org/10.1056/NEJMoa1407417

Lowinger M, Baumann J, Vodak DT, Moser J (2015) Practical considerations for spray dried formulation and process development. In: Discovering and developing molecules with optimal drug-like properties, pp 383–435

Lu D, Hickey AJ (2007) Pulmonary vaccine delivery. Expert Rev Vaccines 6:213–226. https://doi.org/10.1586/14760584.6.2.213

Lu D, Garcia-Contreras L, Xu D et al (2007) Poly (lactide-co-glycolide) microspheres in respirable sizes enhance an in vitro T cell response to recombinant *Mycobacterium tuberculosis* antigen 85B. Pharm Res 24:1834–1843. https://doi.org/10.1007/s11095-007-9302-8

Lu D, Garcia-Contreras L, Muttil P et al (2010) Pulmonary immunization using antigen 85-B polymeric microparticles to boost tuberculosis immunity. AAPS J 12:338–347. https://doi.org/10.1208/s12248-010-9193-1

Lucero Y, Vidal R, O'Ryan GM (2018) Norovirus vaccines under development. Vaccine 36:5435–5441. https://doi.org/10.1016/j.vaccine.2017.06.043

Ludwig-Begall LF, Mauroy A, Thiry E (2021) Noroviruses—the state of the art, nearly fifty years after their initial discovery. Viruses 13:1541. https://doi.org/10.3390/v13081541

Lydon P, Zipursky S, Tevi-Benissan C et al (2014) Economic benefits of keeping vaccines at ambient temperature during mass vaccination: the case of meningitis A vaccine in Chad. Bull World Health Organ 92:86–92. https://doi.org/10.2471/BLT.13.123471

Lyons CR, Wu TH (2007) Animal models of *Francisella tularensis* Infection. Ann N Y Acad Sci 1105:238–265. https://doi.org/10.1196/annals.1409.003

Maa YF, Nguyen PA, Sweeney T et al (1999) Protein inhalation powders: spray drying vs spray freeze drying. Pharm Res 16:249–254

Maa Y-F, Zhao L, Payne LG, Chen D (2003) Stabilization of alum-adjuvanted vaccine dry powder formulations: mechanism and application. J Pharm Sci 92:319–332. https://doi.org/10.1002/jps.10294

Maa Y, Ameri M, Shu C et al (2004) Influenza vaccine powder formulation development: spray-freeze-drying and stability evaluation. J Pharm Sci 93:1912–1923. https://doi.org/10.1002/jps.20104

Maa Y-F, Ameri M, Shu C et al (2007) Hepatitis-B surface antigen (HBsAg) powder formulation: process and stability assessment. Curr Drug Deliv 4:57–67. https://doi.org/10.2174/156720107779314758

Mack GS (2007) Pfizer dumps Exubera. Nat Biotechnol 25:1331–1332. https://doi.org/10.1038/nbt1207-1331

Madan M, Sikriwal D, Sharma G et al (2018) Rational design of heat stable lyophilized rotavirus vaccine formulations. Hum Vaccin Immunother 14:2132–2141. https://doi.org/10.1080/21645515.2018.1487499

Madani F, Hsein H, Busignies V, Tchoreloff P (2020) An overview on dosage forms and formulation strategies for vaccines and antibodies oral delivery. Pharm Dev Technol 25:133–148. https://doi.org/10.1080/10837450.2019.1689402

Maeki M, Uno S, Niwa A et al (2022) Microfluidic technologies and devices for lipid nanoparticle-based RNA delivery. J Control Release 344:80–96. https://doi.org/10.1016/j.jconrel.2022.02.017

Mann GF, Allison LMC, Lloyd JS et al (1983) Stability of further-attenuated measles vaccines. Clin Infect Dis 5:482–486. https://doi.org/10.1093/clinids/5.3.482

Manta Devices LLC (2017a) SOLO™ disposable dry powder inhaler. http://mantadevices.com/dry-powder-inhaler-development/disposable-dpi/

Manta Devices LLC (2017b) Torus™ large dose DPI. http://mantadevices.com/dry-powder-inhaler-development/large-dose-dpi/

Manufacturing Chemist (2016) A green light for cGMP aseptic spray drying, 21 Jan 2016. https://www.manufacturingchemist.com/news/article_page/A_green_light_for_cGMP_aseptic_spray_drying/114844

Marohn ME, Barry EM (2013) Live attenuated tularemia vaccines: recent developments and future goals. Vaccine 31:3485–3491. https://doi.org/10.1016/j.vaccine.2013.05.096

Martinon F, Krishnan S, Lenzen G et al (1993) Induction of virus-specific cytotoxic T lymphocytes in vivo by liposome-entrapped mRNA. Eur J Immunol 23:1719–1722. https://doi.org/10.1002/eji.1830230749

Maruggi G, Zhang C, Li J et al (2019) mRNA as a transformative technology for vaccine development to control infectious diseases. Mol Ther 27:757–772. https://doi.org/10.1016/j.ymthe.2019.01.020

Masomian M, Ahmad Z, Ti Gew L, Poh CL (2020) Development of next generation *Streptococcus pneumoniae* vaccines conferring broad protection. Vaccines 8:132. https://doi.org/10.3390/vaccines8010132

Mast J, Demeestere L (2009) Electron tomography of negatively stained complex viruses: application in their diagnosis. Diagn Pathol 4:5. https://doi.org/10.1186/1746-1596-4-5

Masters K (1976) Chapter 16: applications in the pharmaceutical-biochemical industry. In: Spray drying: an introduction to principles, operational practice and applications, 2nd edn. Wiley, New York

Masters K (1985) Spray drying handbook, 4th edn. George Godwin, London

Masters K (2002) Spray drying in practice. SprayDryConsult International ApS, Denmark

Masters K, Vestergaard I (1975) Aseptic and closed cycle spray drying in pharmaceuticals/bio-chemicals manufacture. Process Biochem 10:21–29

Mattison CP, Cardemil CV, Hall AJ (2018) Progress on norovirus vaccine research: public health considerations and future directions. Expert Rev Vaccines 17:773–784. https://doi.org/10.1080/14760584.2018.1510327

Mattoo S, Cherry JD (2005) Molecular pathogenesis, epidemiology, and clinical manifestations of respiratory infections due to *Bordetella pertussis* and other Bordetella subspecies. Clin Microbiol Rev 18:326–382. https://doi.org/10.1128/CMR.18.2.326-382.2005

Maurin M (2020) *Francisella tularensis*, Tularemia and serological diagnosis. Front Cell Infect Microbiol 10:1–20. https://doi.org/10.3389/fcimb.2020.512090

McAdams D, Chen D, Kristensen D (2012) Spray drying and vaccine stabilization. Expert Rev Vaccines 11:1211–1219. https://doi.org/10.1586/erv.12.101

McAdams DH (2011) Production, optimization and characterization of pharmaceutical and excipient powders produced by carbon dioxide-assisted nebulization with a bubble dryer (CAN-BD). PhD thesis, University of Colorado. https://scholar.colorado.edu/downloads/0z708w64z

McKay PF, Hu K, Blakney AK et al (2020) Self-amplifying RNA SARS-CoV-2 lipid nanoparticle vaccine candidate induces high neutralizing antibody titers in mice. Nat Commun 11:3523. https://doi.org/10.1038/s41467-020-17409-9

McKeage K (2015) Raplixa™: a review in improving surgical haemostasis. Clin Drug Investig 35:519–524. https://doi.org/10.1007/s40261-015-0307-5

Mehanny M, Lehr C-M, Fuhrmann G (2021) Extracellular vesicles as antigen carriers for novel vaccination avenues. Adv Drug Deliv Rev 173:164–180. https://doi.org/10.1016/j.addr.2021.03.016

Mehanny M, Boese A, Bornamehr B et al (2022) Spray-dried pneumococcal membrane vesicles are promising candidates for pulmonary immunization. Int J Pharm 121794. https://doi.org/10.1016/j.ijpharm.2022.121794

Melhem NM (2016) Norovirus vaccines: correlates of protection, challenges and limitations. Hum Vaccines Immunother 12:1653–1669. https://doi.org/10.1080/21645515.2015.1125054

Merck (2021) Merck. Gardasil-9, human papillomavirus 9-valent vaccine, recombinant, storage and handling. https://www.merckvaccines.com/gardasil9/storage-handling/. Accessed 23 Oct 2021

Midoux P, Pichon C (2015) Lipid-based mRNA vaccine delivery systems. Expert Rev Vaccines 14:221–234. https://doi.org/10.1586/14760584.2015.986104

Mikszta JA, Sullivan VJ, Dean C et al (2005) Protective immunization against inhalational anthrax: a comparison of minimally invasive delivery platforms. J Infect Dis 191:278–288. https://doi.org/10.1086/426865

Miller PJ, Koch G (2020) Newcastle disease, chapter 3. In: Swayne DE, Boulianne M, Louge CM et al (eds) Diseases of poultry, 14th edn. Wiley Blackwell in partnership with the American Association of Avian Pathologists (AAAP), Hoboken, NJ, USA, pp 112–129

Mischler R, Metcalfe IC (2002) Inflexal®V a trivalent virosome subunit influenza vaccine: production. Vaccine 20:B17–B23. https://doi.org/10.1016/S0264-410X(02)00512-1

Miwata K, Okamoto H, Nakashima T et al (2018) Intratracheal administration of siRNA dry powder targeting vascular endothelial growth factor inhibits lung tumor growth in mice. Mol Ther Nucleic Acids 12:698–706. https://doi.org/10.1016/j.omtn.2018.07.009

Molinari N-AM, Ortega-Sanchez IR, Messonnier ML et al (2007) The annual impact of seasonal influenza in the US: measuring disease burden and costs. Vaccine 25:5086–5096. https://doi.org/10.1016/j.vaccine.2007.03.046

Momin MAM, Tucker IG, Das SC (2018) High dose dry powder inhalers to overcome the challenges of tuberculosis treatment. Int J Pharm 550:398–417. https://doi.org/10.1016/j.ijpharm.2018.08.061

Morello M, Krone CL, Dickerson S et al (2009) Dry-powder pulmonary insufflation in the mouse for application to vaccine or drug studies. Tuberculosis 89:371–377. https://doi.org/10.1016/j.tube.2009.07.001

Mu Z, Haynes BF, Cain DW (2021) HIV mRNA vaccines—progress and future paths. Vaccines 9:134. https://doi.org/10.3390/vaccines9020134

Mulligan MJ, Stapleton JT, Keitel WA et al (2017) Tularemia vaccine: safety, reactogenicity, "Take" skin reactions, and antibody responses following vaccination with a new lot of the *Francisella tularensis* live vaccine strain—a phase 2 randomized clinical trial. Vaccine 35:4730–4737. https://doi.org/10.1016/j.vaccine.2017.07.024

Munir M, Jena L, Kett VL et al (2021) Spray drying: Inhalable powders for pulmonary gene therapy. Mater Sci Eng C 112601. https://doi.org/10.1016/j.msec.2021.112601

Muramatsu H, Lam K, Bajusz C et al (2022) Lyophilization provides long-term stability for a lipid nanoparticle-formulated, nucleoside-modified mRNA vaccine. Mol Ther 30:1941–1951. https://doi.org/10.1016/j.ymthe.2022.02.001

Muttil P, Prego C, Garcia-Contreras L et al (2010a) Immunization of Guinea Pigs with novel Hepatitis B Antigen as nanoparticle aggregate powders administered by the pulmonary route. AAPS J 12:330–337. https://doi.org/10.1208/s12248-010-9192-2

Muttil P, Pulliam B, Garcia-Contreras L et al (2010b) Pulmonary immunization of Guinea Pigs with Diphtheria CRM-197 antigen as nanoparticle aggregate dry powders enhance local and systemic immune responses. AAPS J 12:699–707. https://doi.org/10.1208/s12248-010-9229-6

Mvundura M, Lydon P, Gueye A et al (2017) An economic evaluation of the controlled temperature chain approach for vaccine logistics: evidence from a study conducted during a meningitis A vaccine campaign in Togo. Pan Afr Med J 27:27. https://doi.org/10.11604/pamj.supp.2017.27.3.12087

Mymetics (2022) COVID-19 virus vaccine candidate. https://www.mymetics.com/vaccine-pipeline/covid-19/

Naik SP, Zade JK, Sabale RN et al (2017) Stability of heat stable, live attenuated Rotavirus vaccine (ROTASIIL®). Vaccine 35:2962–2969. https://doi.org/10.1016/j.vaccine.2017.04.025

Nance KD, Meier JL (2021) Modifications in an emergency: the role of N1-methylpseudouridine in COVID-19 vaccines. ACS Cent Sci 7:748–756. https://doi.org/10.1021/acscentsci.1c00197

Nelson C, Kugeler K, Petersen J, Mead P (2013) Tularemia—United States, 2001–2010. MMWR Morb Mortal Wkly Rep 62:962–966

Neumiller JJ, Campbell RK (2010) Technosphere® Insulin. BioDrugs 24:165–172. https://doi.org/10.2165/11536700-000000000-00000

New RRC (2019) Formulation technologies for oral vaccines. Clin Exp Immunol 198:153–169. https://doi.org/10.1111/cei.13352

Ng'uni T, Chasara C, Ndhlovu ZM (2020) Major scientific hurdles in HIV vaccine development: historical perspective and future directions. Front Immunol 11:1–17. https://doi.org/10.3389/fimmu.2020.590780

Nieto K, Weghofer M, Sehr P et al (2012) Development of AAVLP(HPV16/31L2) particles as broadly protective HPV vaccine candidate. PLoS ONE 7:e39741. https://doi.org/10.1371/journal.pone.0039741

Nigrovic LE, Wingerter SL (2008) Tularemia. Infect Dis Clin North Am 22:489–504. https://doi.org/10.1016/j.idc.2008.03.004

O'Brien KL, Ramakrishnan M, Finn A, Malley R (2016) Pneumococcus, pneumococcal disease, and prevention. In: The vaccine book, 2nd edn. Elsevier, pp 225–243

Odevall L, Hong D, Digilio L et al (2018) The Euvichol story—development and licensure of a safe, effective and affordable oral cholera vaccine through global public private partnerships. Vaccine 36:6606–6614. https://doi.org/10.1016/j.vaccine.2018.09.026

Ohtake S, Martin RA, Yee L et al (2010) Heat-stable measles vaccine produced by spray drying. Vaccine 28:1275–1284. https://doi.org/10.1016/j.vaccine.2009.11.024

Ohtake S, Martin R, Saxena A et al (2011a) Room temperature stabilization of oral, live attenuated Salmonella enterica serovar Typhi-vectored vaccines. Vaccine 29:2761–2771. https://doi.org/10.1016/j.vaccine.2011.01.093

Ohtake S, Martin RA, Saxena A et al (2011b) Formulation and stabilization of *Francisella tularensis* live vaccine strain. J Pharm Sci 100:3076–3087. https://doi.org/10.1002/jps.22563

Ohtake S, Lechuga-Ballesteros D, Truong-Le V, Patzer EJ (2015) Strategies for heat-stable vaccines, chapter 10. In: Vaccine development and manufacturing. Wiley, pp 287–318

Okwo-Bele J-M, LaForce FM, Borrow R, Preziosi M-P (2015) Documenting the results of a successful partnership: a new meningococcal vaccine for Africa. Clin Infect Dis 61:S389–S390. https://doi.org/10.1093/cid/civ592

Oliveira GS, Oliveira MLS, Miyaji EN, Rodrigues TC (2021) Pneumococcal vaccines: past findings, present work, and future strategies. Vaccines 9:1338. https://doi.org/10.3390/vaccines9111338

Osman N, Goovaerts D, Sultan S et al (2021) Vaccine quality is a key factor to determine thermal stability of commercial newcastle disease (ND)vaccines. Vaccines 9:363. https://doi.org/10.3390/vaccines9040363

Oyston PCF (2008) Francisella tularensis: unravelling the secrets of an intracellular pathogen. J Med Microbiol 57:921–930. https://doi.org/10.1099/jmm.0.2008/000653-0

Oyston PCF (2009) Francisella tularensis vaccines. Vaccine 27:D48–D51. https://doi.org/10.1016/j.vaccine.2009.07.090

Öztürk AA, Arpagaus C (2021) Nano spray-dried drugs for oral administration: a review. Assay Drug Dev Technol 19:412–441. https://doi.org/10.1089/adt.2021.053

Perlen Packaging (2021) PERLAMED™—BLISTair single-use dry powder inhaler (DPI). https://www.perlenpackaging.com/en/products/perlamedtm-blistair

Paik J (2020) Levodopa inhalation powder: a review in Parkinson's disease. Drugs 80:821–828. https://doi.org/10.1007/s40265-020-01307-x

Pardi N, Hogan MJ, Porter FW, Weissman D (2018) mRNA vaccines—a new era in vaccinology. Nat Rev Drug Discov 17:261–279. https://doi.org/10.1038/nrd.2017.243

Pascolo S (2004) Messenger RNA-based vaccines. Expert Opin Biol Ther 4:1285–1294. https://doi.org/10.1517/14712598.4.8.1285

Pascolo S (2006) Vaccination with messenger RNA. In: Saltzman WM, Shen H, Brandsma JL (eds) DNA vaccines. Humana Press, New Jersey, pp 23–40

Pascolo S (2008) Vaccination with messenger RNA (mRNA). In: Toll-like receptors (TLRs) and innate immunity. Handbook of experimental pharmacology. Springer, Berlin, Heidelberg, pp 221–235

Pascolo S (2014) The messenger's great message for vaccination. Expert Rev Vaccines 14:153–156. https://doi.org/10.1586/14760584.2015.1000871

Pascolo S (2017) Messenger RNA: the inexpensive biopharmaceutical. J Multidiscip Eng Sci Technol. http://www.jmest.org/wp-content/uploads/JMESTN42352119.pdf

Pascolo S (2021a) Synthetic messenger RNA-based vaccines: from scorn to hype. Viruses 13:270. https://doi.org/10.3390/v13020270

Pascolo S (2021b) Vaccines against COVID-19: priority to mRNA-based formulations. Cells 10:2716. https://doi.org/10.3390/cells10102716

Pastor M, Esquisabel A, Talavera A et al (2013a) An approach to a cold chain free oral cholera vaccine: in vitro and in vivo characterization of *Vibrio cholerae* gastro-resistant microparticles. Int J Pharm 448:247–258. https://doi.org/10.1016/j.ijpharm.2013.02.057

Pastor M, Pedraz JL, Esquisabel A (2013b) The state-of-the-art of approved and under-development cholera vaccines. Vaccine 31:4069–4078. https://doi.org/10.1016/j.vaccine.2013.06.096

Pastor M, Esquisabel A, Marquínez I et al (2014) Cellulose acetate phthalate microparticles containing *Vibrio cholerae*: steps toward an oral cholera vaccine. J Drug Target 22:478–487. https://doi.org/10.3109/1061186X.2014.888071

Patel MM, Hall AJ, Vinjé J, Parashar UD (2009) Noroviruses: a comprehensive review. J Clin Virol 44:1–8. https://doi.org/10.1016/j.jcv.2008.10.009

PATH (2012) Summary of stability data for licensed vaccines, 29 Nov 2012, produced by Working in Tandem, Ltd. for PATH

Patton JS, Bukar JG, Eldon MA (2004) Clinical pharmacokinetics and pharmacodynamics of inhaled insulin. Clin Pharmacokinet 43:781–801. https://doi.org/10.2165/00003088-200443120-00002

Peabody J, Muttil P, Chackerian B, Tumban E (2017) Characterization of a spray-dried candidate HPV L2-VLP vaccine stored for multiple years at room temperature. Papillomavirus Res 3:116–120. https://doi.org/10.1016/j.pvr.2017.03.004

Peck DE, Reeves WK, Pelzel-McCluskey AM et al (2020) Management strategies for reducing the risk of equines contracting vesicular stomatitis virus (VSV) in the Western United States. J Equine Vet Sci 90:103026. https://doi.org/10.1016/j.jevs.2020.103026

Petit D, Tevi-Benissan C, Woodring J et al (2017) Countries' interest in a hepatitis B vaccine licensed for the controlled temperature chain; survey results from African and Western Pacific regions. Vaccine 35:6866–6871. https://doi.org/10.1016/j.vaccine.2017.10.025

Petsch B, Schnee M, Vogel AB et al (2012) Protective efficacy of in vitro synthesized, specific mRNA vaccines against influenza A virus infection. Nat Biotechnol 30:1210–1216. https://doi.org/10.1038/nbt.2436

Peyraud N, Zehrung D, Jarrahian C et al (2019) Potential use of microarray patches for vaccine delivery in low- and middle-income countries. Vaccine 37:4427–4434. https://doi.org/10.1016/j.vaccine.2019.03.035

Pfützner A, Forst T (2005) Pulmonary insulin delivery by means of the Technosphere™ drug carrier mechanism. Expert Opin Drug Deliv 2:1097–1106. https://doi.org/10.1517/17425247.2.6.1097

Pfützner A, Mann AE, Steiner SS (2002) Technosphere™/insulin—a new approach for effective delivery of human insulin via the pulmonary route. Diabetes Technol Ther 4:589–594. https://doi.org/10.1089/152091502320798204

Pharmaxis (2021) Orbital® dry powder inhaler. https://www.pharmaxis.com.au/product-pipeline/other-pipeline-assets/orbital-dry-powder-inhaler/

Pilkington EH, Suys EJA, Trevaskis NL et al (2021) From influenza to COVID-19: lipid nanoparticle mRNA vaccines at the frontiers of infectious diseases. Acta Biomater 131:16–40. https://doi.org/10.1016/j.actbio.2021.06.023

Pinto JT, Faulhammer E, Dieplinger J et al (2021) Progress in spray-drying of protein pharmaceuticals: literature analysis of trends in formulation and process attributes. Dry Technol 0:1–32. https://doi.org/10.1080/07373937.2021.1903032

Plews JR, Li J, Jones M et al (2010) Activation of pluripotency genes in human fibroblast cells by a novel mRNA based approach. PLoS ONE 5:e14397. https://doi.org/10.1371/journal.pone.0014397

Polack FP, Thomas SJ, Kitchin N et al (2020) Safety and efficacy of the BNT162b2 mRNA Covid-19 vaccine. N Engl J Med 383:2603–2615. https://doi.org/10.1056/NEJMoa2034577

Porton Biopharma Ltd (2015) Anthrax vaccine (Alum precipitated sterile filtrate). https://portonbiopharma.com/products/anthrax-vaccine/

Prausnitz MR (2017) Engineering microneedle patches for vaccination and drug delivery to skin. Annu Rev Chem Biomol Eng 8:177–200. https://doi.org/10.1146/annurev-chembioeng-060816-101514

Prausnitz MR, Goodson JL, Rota PA, Orenstein WA (2020) A microneedle patch for measles and rubella vaccination: a game changer for achieving elimination. Curr Opin Virol 41:68–76. https://doi.org/10.1016/j.coviro.2020.05.005

Preston KB, Randolph TW (2021) Stability of lyophilized and spray dried vaccine formulations. Adv Drug Deliv Rev 171:50–61. https://doi.org/10.1016/j.addr.2021.01.016

Price DN, Kunda NK, McBride AA, Muttil P (2016) Vaccine preparation: past, present, and future, chapter 4. In: Drug delivery systems for tuberculosis prevention and treatment. Wiley, Chichester, UK, pp 67–90

Price DN, Kunda NK, Ellis R, Muttil P (2020) Design and optimization of a temperature-stable dry powder BCG vaccine. Pharm Res 37:11. https://doi.org/10.1007/s11095-019-2739-8

Probst J, Weide B, Scheel B et al (2007) Spontaneous cellular uptake of exogenous messenger RNA in vivo is nucleic acid-specific, saturable and ion dependent. Gene Ther 14:1175–1180. https://doi.org/10.1038/sj.gt.3302964

Pulliam B, Sung JC, Edwards DA (2007) Design of nanoparticle-based dry powder pulmonary vaccines. Expert Opin Drug Deliv 4:651–663. https://doi.org/10.1517/17425247.4.6.651

PureIMS (2021) Cyclops platform inhalation technology. https://pureims.com/products

Putranto A, Chen XD (2019) Drying in biotechnology, 3rd edn. Elsevier

Qi W, Orgel S, Francon A et al (2018) Urea improves stability of inactivated polio vaccine serotype 3 during lyophilization and storage in dried formulations. J Pharm Sci 107:2070–2078. https://doi.org/10.1016/j.xphs.2018.04.019

Qiu Y, Man RCH, Liao Q et al (2019) Effective mRNA pulmonary delivery by dry powder formulation of PEGylated synthetic KL4 peptide. J Control Release 314:102–115. https://doi.org/10.1016/j.jconrel.2019.10.026

Quarta E, Chierici V, Flammini L et al (2020) Excipient-free pulmonary insulin dry powder: pharmacokinetic and pharmacodynamics profiles in rats. J Control Release 323:412–420. https://doi.org/10.1016/j.jconrel.2020.04.015

Raeven RHM, Brummelman J, La PJ et al (2018) Molecular and cellular signatures underlying superior immunity against *Bordetella pertussis* upon pulmonary vaccination. Mucosal Immunol 11:979–993. https://doi.org/10.1038/mi.2017.81

Rani S, Gogoi P, Kumar S (2014) Spectrum of Newcastle disease virus stability in gradients of temperature and pH. Biologicals 42:351–354. https://doi.org/10.1016/j.biologicals.2014.08.006

Rauch S, Lutz J, Kowalczyk A et al (2017) RNActive® technology: generation and testing of stable and immunogenic mRNA vaccines. In: Methods in molecular biology, pp 89–107

Ré M-I (2006) Formulating drug delivery systems by spray drying. Dry Technol 24:433–446. https://doi.org/10.1080/07373930600611877

Ré MI (1998) Microencapsulation by spray drying. Dry Technol 16:1195–1236. https://doi.org/10.1080/07373939808917460

Rouphael NG, Paine M, Mosley R et al (2017) The safety, immunogenicity, and acceptability of inactivated influenza vaccine delivered by microneedle patch (TIV-MNP 2015): a randomised,

partly blinded, placebo-controlled, phase 1 trial. Lancet 390:649–658. https://doi.org/10.1016/S0140-6736(17)30575-5
Royal Pas Reform (2020) Spray vaccination of day-old-chicks at the hatchery. https://www.pasreform.com/en/knowledge/43/spray-vaccination-of-day-old-chicks-at-the-hatchery
Reichmuth AM, Oberli MA, Jaklenec A et al (2016) mRNA vaccine delivery using lipid nanoparticles. Ther Deliv 7:319–334. https://doi.org/10.4155/tde-2016-0006
Rein DT, Breidenbach M, Curiel DT (2006) Current developments in adenovirus-based cancer gene therapy. Futur Oncol 2:137–143. https://doi.org/10.2217/14796694.2.1.137
Rexroad J, Wiethoff CM, Jones LS, Middaugh CR (2002) Lyophilization and the thermostability of vaccines. Cell Preserv Technol 1:91–104. https://doi.org/10.1089/153834402320882593
Richardson PC, Boss AH (2007) Technosphere® Insulin technology. Diabetes Technol Ther 9:65–72. https://doi.org/10.1089/dia.2007.0212
Richardson C, Bargatze RF, Goodwin R, Mendelman PM (2013) Norovirus virus-like particle vaccines for the prevention of acute gastroenteritis. Expert Rev Vaccines 12:155–167. https://doi.org/10.1586/erv.12.145
Richner JM, Himansu S, Dowd KA et al (2017) Modified mRNA vaccines protect against Zika virus infection. Cell 168:1114-1125.e10. https://doi.org/10.1016/j.cell.2017.02.017
Rodrigues TC, Oliveira MLS, Soares-Schanoski A et al (2018) Mucosal immunization with PspA (Pneumococcal surface protein A)-adsorbed nanoparticles targeting the lungs for protection against pneumococcal infection. PLoS ONE 13:e0191692. https://doi.org/10.1371/journal.pone.0191692
Rodríguez LL (2002) Emergence and re-emergence of vesicular stomatitis in the United States. Virus Res 85:211–219. https://doi.org/10.1016/S0168-1702(02)00026-6
Roediger EK, Kugathasan K, Zhang X et al (2008) Heterologous boosting of recombinant adenoviral prime immunization with a novel vesicular stomatitis virus–vectored tuberculosis vaccine. Mol Ther 16:1161–1169. https://doi.org/10.1038/mt.2008.59
Rosa SS, Prazeres DMF, Azevedo AM, Marques MPC (2021) mRNA vaccines manufacturing: challenges and bottlenecks. Vaccine 39:2190–2200. https://doi.org/10.1016/j.vaccine.2021.03.038
Rose JK, Clarke DK (2015) Rhabdoviruses as vaccine vectors: from initial development to clinical trials. In: Biology and pathogenesis of Rhabdo- and Filoviruses. WORLD SCIENTIFIC, pp 199–230
Rose NF, Marx PA, Luckay A et al (2001) An Effective AIDS vaccine based on live attenuated vesicular stomatitis virus recombinants. Cell 106:539–549. https://doi.org/10.1016/S0092-8674(01)00482-2
Rosenstein NE, Perkins BA, Stephens DS et al (2001) Meningococcal disease. N Engl J Med 344:1378–1388. https://doi.org/10.1056/NEJM200105033441807
Rossi I, Spagnoli G, Buttini F et al (2021) A respirable HPV-L2 dry-powder vaccine with GLA as amphiphilic lubricant and immune-adjuvant. J Control Release 340:209–220. https://doi.org/10.1016/j.jconrel.2021.11.002
Rota PA, Moss WJ, Takeda M et al (2016) Measles. Nat Rev Dis Prim 2:16049. https://doi.org/10.1038/nrdp.2016.49
Roth Y, Chapnik JS, Cole P (2003) Feasibility of aerosol vaccination in humans. Ann Otol Rhinol Laryngol 112:264–270. https://doi.org/10.1177/000348940311200313
Rubin LG (2018) *Francisella tularensis* (Tularemia). In: Principles and practice of pediatric infectious diseases, 5th edn. Elsevier, pp 923–925.e1
Sabin AB (1983) Immunization against measles by aerosol. Clin Infect Dis 5:514–523. https://doi.org/10.1093/clinids/5.3.514
Sabin AB (1991) Measles, killer of millions in developing countries: strategy for rapid elimination and continuing control. Eur J Epidemiol 7:1–22. https://doi.org/10.1007/BF00221337
Saboo S, Tumban E, Peabody J et al (2016) Optimized formulation of a thermostable spray-dried virus-like particle vaccine against Human Papillomavirus. Mol Pharm 13:1646–1655. https://doi.org/10.1021/acs.molpharmaceut.6b00072

Sadrzadeh N, Miller DP, Lechuga-Ballesteros D et al (2010) Solid-state stability of spray-dried insulin powder for inhalation: chemical kinetics and structural relaxation modeling of exubera above and below the glass transition temperature. J Pharm Sci 99:3698–3710. https://doi.org/10.1002/jps.21936

Saha A, Khan A, Salma U et al (2016) The oral cholera vaccine Shanchol™ when stored at elevated temperatures maintains the safety and immunogenicity profile in Bangladeshi participants. Vaccine 34:1551–1558. https://doi.org/10.1016/j.vaccine.2016.02.020

Salama AH (2020) Spray drying as an advantageous strategy for enhancing pharmaceuticals bioavailability. Drug Deliv Transl Res 10:1–12. https://doi.org/10.1007/s13346-019-00648-9

Saluja V, Amorij J-P, Kapteyn JC et al (2010) A comparison between spray drying and spray freeze drying to produce an influenza subunit vaccine powder for inhalation. J Control Release 144:127–133. https://doi.org/10.1016/j.jconrel.2010.02.025

Samaridou E, Heyes J, Lutwyche P (2020) Lipid nanoparticles for nucleic acid delivery: current perspectives. Adv Drug Deliv Rev 154–155:37–63. https://doi.org/10.1016/j.addr.2020.06.002

Samborska K, Poozesh S, Barańska A et al (2022) Innovations in spray drying process for food and pharma industries. J Food Eng 321. https://doi.org/10.1016/j.jfoodeng.2022.110960

Sandström G (1994) The tularaemia vaccine. J Chem Technol Biotechnol 59:315–320. https://doi.org/10.1002/jctb.280590402

Saslaw S, Eigelsbach HT, Wilson HE et al (1961a) Tularemia vaccine study, I. Intracutaneous challenge. Arch Intern Med 107:689–701. https://doi.org/10.1001/archinte.1961.03620050055006

Saslaw S, Eigelsbach HT, Wilson HE et al (1961b) Tularemia vaccine study, II. Respiratory challenge. Arch Intern Med 107:702. https://doi.org/10.1001/archinte.1961.03620050068007

Sawatdee S, Srichana T (2021) Single-use dry powder inhalers for pulmonary drug delivery. Handbook of lung targeted drug delivery systems. CRC Press, Boca Raton, pp 503–511

Schellenbacher C, Roden RBS, Kirnbauer R (2017) Developments in L2-based human papillomavirus (HPV) vaccines. Virus Res 231:166–175. https://doi.org/10.1016/j.virusres.2016.11.020

Schellenbacher C, Huber B, Shafti-Keramat S, Kirnbauer R (2021) HPV-Impfstoffe – zugelassene Vakzinen und experimenteller RG1-VLP-Impfstoff der nächsten Generation. hautnah 20:155–160. https://doi.org/10.1007/s12326-021-00453-7

Scher G, Schnell MJ (2020) Rhabdoviruses as vectors for vaccines and therapeutics. Curr Opin Virol 44:169–182. https://doi.org/10.1016/j.coviro.2020.09.003

Scherließ R, Ajmera A, Dennis M et al (2014) Induction of protective immunity against H1N1 influenza A(H1N1)pdm09 with spray-dried and electron-beam sterilised vaccines in non-human primates. Vaccine 32:2231–2240. https://doi.org/10.1016/j.vaccine.2014.01.077

Schiller JT, Müller M (2015) Next generation prophylactic human papillomavirus vaccines. Lancet Oncol 16:e217–e225. https://doi.org/10.1016/S1470-2045(14)71179-9

Schoenmaker L, Witzigmann D, Kulkarni JA et al (2021) mRNA-lipid nanoparticle COVID-19 vaccines: structure and stability. Int J Pharm 601:120586. https://doi.org/10.1016/j.ijpharm.2021.120586

Scholz M, Altenhofen W (2011) Stabilising biologics for vaccine production. Manuf Chem. https://manufacturingchemist.com/news/article_page/Stabilising_biologics_for_vaccine_production/63953

Schuster A, Haliburn C, Döring G, Goldman MH (2013) Safety, efficacy and convenience of colistimethate sodium dry powder for inhalation (Colobreathe DPI) in patients with cystic fibrosis: a randomised study. Thorax 68:344–350. https://doi.org/10.1136/thoraxjnl-2012-202059

Schwartzbach H (2011) Achieving aseptic drying with spray drying technologies. Pharm Technol Eur 23:90–92

Scott N, Palmer A, Morgan C et al (2018) Cost-effectiveness of the controlled temperature chain for the hepatitis B virus birth dose vaccine in various global settings: a modelling study. Lancet Glob Heal 6:e659–e667. https://doi.org/10.1016/S2214-109X(18)30219-5

Seaman CP, Morgan C, Howell J et al (2020) Use of controlled temperature chain and compact prefilled auto-disable devices to reach 2030 hepatitis B birth dose vaccination targets in LMICs: a modelling and cost-optimisation study. Lancet Glob Heal 8:e931–e941. https://doi.org/10.1016/S2214-109X(20)30231-X

Searles J, Mohan G (2010) Spray drying of biopharmaceuticals and vaccines. Formulation and process development strategies for manufacturing biopharmaceuticals. Wiley, Hoboken, NJ, USA, pp 739–761

Seville PC, Learoyd TP, Li H-Y et al (2007) Amino acid-modified spray-dried powders with enhanced aerosolisation properties for pulmonary drug delivery. Powder Technol 178:40–50. https://doi.org/10.1016/j.powtec.2007.03.046

Sharma A, Khamar D, Cullen S et al (2021) Innovative drying technologies for biopharmaceuticals. Int J Pharm 609:121115. https://doi.org/10.1016/j.ijpharm.2021.121115

Shastri PN, Kim M, Quan F et al (2012) Immunogenicity and protection of oral influenza vaccines formulated into microparticles. J Pharm Sci 101:3623–3635. https://doi.org/10.1002/jps.23220

Shattock RJ, Moore JP (2003) Inhibiting sexual transmission of HIV-1 infection. Nat Rev Microbiol 1:25–34. https://doi.org/10.1038/nrmicro729

Shi J, Huang M-W, Lu Z-D et al (2022) Delivery of mRNA for regulating functions of immune cells. J Control Release 345:494–511. https://doi.org/10.1016/j.jconrel.2022.03.033

Shin S, Desai SN, Sah BK, Clemens JD (2011) Oral vaccines against Cholera. Clin Infect Dis 52:1343–1349. https://doi.org/10.1093/cid/cir141

Sibum I, Hagedoorn P, de Boer AH et al (2018) Challenges for pulmonary delivery of high powder doses. Int J Pharm 548:325–336. https://doi.org/10.1016/j.ijpharm.2018.07.008

Sibum I, Hagedoorn P, Kluitman MPG et al (2019) Dispersibility and storage stability optimization of high dose isoniazid dry powder inhalation formulations with L-Leucine or Trileucine. Pharmaceutics 12:24. https://doi.org/10.3390/pharmaceutics12010024

Sibum I, Hagedoorn P, Botterman CO et al (2020) Automated filling equipment allows increase in the maximum dose to be filled in the Cyclops® high dose dry powder inhalation device while maintaining dispersibility. Pharmaceutics 12:645. https://doi.org/10.3390/pharmaceutics12070645

Sievers R (2015) Needle-free delivery of stable respirable powder vaccine, grand challenges in global health: 2005–2015 retrospective. https://drupal.grandchallenges.org/sites/default/files/files/robertsievers.pdf

Sievers RE, Quinn BP, Cape SP et al (2007) Near-critical fluid micronization of stabilized vaccines, antibiotics and anti-virals. J Supercrit Fluids 42:385–391. https://doi.org/10.1016/j.supflu.2007.03.001

Siew A (2016) Exploring the use of aseptic spray drying in the manufacture of biopharmaceutical injectables. Pharm Technol 40:24–27

Singh A, Van den Mooter G (2016) Spray drying formulation of amorphous solid dispersions. Adv Drug Deliv Rev 100:27–50. https://doi.org/10.1016/j.addr.2015.12.010

Sjöstedt A (2007) Tularemia: history, epidemiology, pathogen physiology, and clinical manifestations. Ann N Y Acad Sci 1105:1–29. https://doi.org/10.1196/annals.1409.009

Slack M, Esposito S, Haas H et al (2020) Haemophilus influenzae type b disease in the era of conjugate vaccines: critical factors for successful eradication. Expert Rev Vaccines 19:903–917. https://doi.org/10.1080/14760584.2020.1825948

Smith DJ, Bot S, Dellamary L, Bot A (2003) Evaluation of novel aerosol formulations designed for mucosal vaccination against influenza virus. Vaccine 21:2805–2812. https://doi.org/10.1016/S0264-410X(03)00224-X

Snyder HE, Lechuga-Ballesteros D (2008) Spray drying: theory and pharmaceutical applications, chapter 7. In: Augsburger LL, Hoag SW (eds) Pharmaceutical dosage forms: tablets, third edition, volume 1: unit operations and mechanical properties. Informa Healthcare USA, Inc., New York, NY, pp 227–260

Sollohub K, Cal K (2010) Spray drying technique: II. Current applications in pharmaceutical technology. J Pharm Sci 99:587–597. https://doi.org/10.1002/jps.21963

Son Y-J, Miller DP, Weers JG (2021) Optimizing spray-dried porous particles for high dose delivery with a portable dry powder inhaler. Pharmaceutics 13:1528. https://doi.org/10.3390/pharmaceutics13091528

Sou T, Meeusen EN, de Veer M et al (2011) New developments in dry powder pulmonary vaccine delivery. Trends Biotechnol 29:191–198. https://doi.org/10.1016/j.tibtech.2010.12.009

Sou T, Morton DAV, Williamson M et al (2015) Spray-dried influenza antigen with trehalose and leucine produces an aerosolizable powder vaccine formulation that induces strong systemic and mucosal immunity after pulmonary administration. J Aerosol Med Pulm Drug Deliv 28:361–371. https://doi.org/10.1089/jamp.2014.1176

Splino M, Patocka J, Prymula R, Chlibek R (2005) Anthrax vaccines. Ann Saudi Med 25:143–149. https://doi.org/10.5144/0256-4947.2005.143

Springer MJ, Ni Y, Finger-Baker I et al (2016) Preclinical dose-ranging studies of a novel dry powder norovirus vaccine formulation. Vaccine 34:1452–1458. https://doi.org/10.1016/j.vaccine.2016.01.064

Steiner S, Pfützner A, Wilson BR et al (2002) Technosphere™/insulin—proof of concept study with a new insulin formulation for pulmonary delivery. Exp Clin Endocrinol Diabetes 110:17–21. https://doi.org/10.1055/s-2002-19989

Steinle H, Behring A, Schlensak C et al (2017) Concise review: application of in vitro transcribed messenger RNA for cellular engineering and reprogramming: progress and challenges. Stem Cells 35:68–79. https://doi.org/10.1002/stem.2402

Stephens DS, Greenwood B, Brandtzaeg P (2007) Epidemic meningitis, meningococcaemia, and Neisseria meningitidis. Lancet 369:2196–2210. https://doi.org/10.1016/S0140-6736(07)61016-2

Stevenson CL, Bennett DB (2014) Development of the Exubera® Insulin pulmonary delivery system. In: das Neves J, Sarmento B (eds) Mucosal delivery of biopharmaceuticals. Springer US, Boston, MA, pp 461–481

Stitz L, Vogel A, Schnee M et al (2017) A thermostable messenger RNA based vaccine against rabies. PLoS Negl Trop Dis 11:e0006108. https://doi.org/10.1371/journal.pntd.0006108

Stocchi F, Vacca L, Stirpe P, Torti M (2018) Pharmacokinetic drug evaluation of CVT-301 for the treatment of Parkinson's disease. Expert Opin Drug Metab Toxicol 14:1189–1195. https://doi.org/10.1080/17425255.2018.1550483

Strebel PM, Papania MJ, Gastañaduy PA, Goodson JL (2018) Chapter 37: measles vaccines. In: Plotkin's vaccines. Elsevier, pp 579–618.e21

Stroock AD, Dertinger SKW, Ajdari A et al (2002) Chaotic mixer for microchannels. Science (80–) 295:647–651. https://doi.org/10.1126/science.1066238

Summary Group (2015) Developing aerosol vaccines for *Mycobacterium tuberculosis*: workshop proceedings. Vaccine 33:3038–3046. https://doi.org/10.1016/j.vaccine.2015.03.060

Sung JC, Pulliam BL, Edwards DA (2007) Nanoparticles for drug delivery to the lungs. Trends Biotechnol 25:563–570. https://doi.org/10.1016/j.tibtech.2007.09.005

Tai SS (2006) *Streptococcus pneumoniae* protein vaccine candidates: properties, activities and animal studies. Crit Rev Microbiol 32:139–153. https://doi.org/10.1080/10408410600822942

Takeda (2016) Takeda initiates world's first norovirus vaccine field trial, 20 June 2016. https://www.takeda.com/newsroom/newsreleases/2016/takeda-initiates-worlds-first-norovirus-vaccine-field-trial/

Tan M (2021) Norovirus vaccines: current clinical development and challenges. Pathogens 10:1–8. https://doi.org/10.3390/pathogens10121641

Tärnvik A (1989) Nature of protective immunity to *Francisella tularensis*. Rev Infect Dis https://www.jstor.org/stable/4454898

Tate JE, Burton AH, Boschi-Pinto C, Parashar UD (2016) Global, regional, and national estimates of rotavirus mortality in children 5 years of age, 2000–2013. Clin Infect Dis 62:S96–S105. https://doi.org/10.1093/cid/civ1013

Tawde SA, Chablani L, Akalkotkar A et al (2012) Formulation and evaluation of oral microparticulate ovarian cancer vaccines. Vaccine 30:5675–5681. https://doi.org/10.1016/j.vaccine.2012.05.073

Tawde SA, Chablani L, Akalkotkar A, D'Souza MJ (2016) Evaluation of microparticulate ovarian cancer vaccine via transdermal route of delivery. J Control Release 235:147–154. https://doi.org/10.1016/j.jconrel.2016.05.058

Tenchov R, Bird R, Curtze AE, Zhou Q (2021) Lipid nanoparticles—from liposomes to mRNA vaccine delivery, a landscape of research diversity and advancement. ACS Nano 15:16982–17015. https://doi.org/10.1021/acsnano.1c04996

Terskikh II, Danilov AI, Shchelchkov GI et al (1971) Theoretical substantiation and the effectiveness of immunization with aerosols of liquid anitmeasles vaccines [article in Russian]. Vestik Akad Meditsinskikh Nauk SSSR 26:84–90

Thakur A, Ingvarsson PT, Schmidt ST et al (2018) Immunological and physical evaluation of the multistage tuberculosis subunit vaccine candidate H56/CAF01 formulated as a spray-dried powder. Vaccine 36:3331–3339. https://doi.org/10.1016/j.vaccine.2018.04.055

Timoney P (2016) Vesicular stomatitis. Vet Rec 179:119–120. https://doi.org/10.1136/vr.i4075

Tlaxca JL, Ellis S, Remmele RL (2015) Live attenuated and inactivated viral vaccine formulation and nasal delivery: potential and challenges. Adv Drug Deliv Rev 93:56–78. https://doi.org/10.1016/j.addr.2014.10.002

Tomar J, Born PA, Frijlink HW, Hinrichs WLJ (2016) Dry influenza vaccines: towards a stable, effective and convenient alternative to conventional parenteral influenza vaccination. Expert Rev Vaccines 15:1431–1447. https://doi.org/10.1080/14760584.2016.1182869

Tomar J, Tonnis WF, Patil HP et al (2019) Pulmonary immunization: deposition site is of minor relevance for influenza vaccination but deep lung deposition is crucial for hepatitis B vaccination. Acta Pharm Sin B 9:1231–1240. https://doi.org/10.1016/j.apsb.2019.05.003

Toniolo SP, Afkhami S, Mahmood A et al (2019) Excipient selection for thermally stable enveloped and non-enveloped viral vaccine platforms in dry powders. Int J Pharm 561:66–73. https://doi.org/10.1016/j.ijpharm.2019.02.035

Toniolo SP, Afkhami S, D'Agostino MR et al (2020) Spray dried VSV-vectored vaccine is thermally stable and immunologically active in vivo. Sci Rep 10:13349. https://doi.org/10.1038/s41598-020-70325-2

Tonnis WF, Kersten GF, Frijlink HW et al (2012) Pulmonary vaccine delivery: a realistic approach? J Aerosol Med Pulm Drug Deliv 25:249–260. https://doi.org/10.1089/jamp.2011.0931

Tonnis WF, Lexmond AJ, Frijlink HW et al (2013) Devices and formulations for pulmonary vaccination. Expert Opin Drug Deliv 10:1383–1397. https://doi.org/10.1517/17425247.2013.810622

Tonnis WF, Amorij J-P, Vreeman MA et al (2014a) Improved storage stability and immunogenicity of hepatitis B vaccine after spray-freeze drying in presence of sugars. Eur J Pharm Sci 55:36–45. https://doi.org/10.1016/j.ejps.2014.01.005

Tonnis WF, Bagerman M, Weij M et al (2014b) A novel aerosol generator for homogenous distribution of powder over the lungs after pulmonary administration to small laboratory animals. Eur J Pharm Biopharm 88:1056–1063. https://doi.org/10.1016/j.ejpb.2014.10.011

Troeger C, Khalil IA, Rao PC et al (2018) Rotavirus vaccination and the global burden of Rotavirus diarrhea among children younger than 5 years. JAMA Pediatr 172:958. https://doi.org/10.1001/jamapediatrics.2018.1960

Trovato M, D'Apice L, Prisco A, De Berardinis P (2018) HIV vaccination: a roadmap among advancements and concerns. Int J Mol Sci 19:1241. https://doi.org/10.3390/ijms19041241

Truong VL, Abdul-Fattah AM (2010) The impact of formulation and drying processes on the characteristics and performance of biopharmaceutical powders. In: Jameel H (ed) Formulation and process development strategies for manufacturing biopharmaceuticals. Wiley, pp 555–585

Truong-Le V, Pham B (2007) Preservation of bioactive materials by spray drying, Patent US 7258873 B2, filed: Apr 10, 2003. 1–30

Truong-Le V, Scherer T (2008) high pressure spray-dry of bioactive materials, Patent US 7378110 B2, filed: Dec 16, 2003. 1–22

Truong-Le V, Ohtake S, Martin RA et al (2012) Sonic low pressure spray drying, Patent US 8268354 B2, Filed: 6 Nov 2008. 1–17

Truong-Le V, Ohtake S, Martin RA et al (2014) Sonic low pressure spray drying, Patent US 8673357 B2, Filed: Aug 14, 2012. 1–18

Tscheliessnig R, Zörnig M, Herzig EM et al (2012) Nano-coating protects biofunctional materials. Mater Today 15:394–404. https://doi.org/10.1016/S1369-7021(12)70166-9

Tumban E, Muttil P, Escobar CAA et al (2015) Preclinical refinements of a broadly protective VLP-based HPV vaccine targeting the minor capsid protein, L2. Vaccine 33:3346–3353. https://doi.org/10.1016/j.vaccine.2015.05.016

Tyne AS, Chan JGY, Shanahan ER et al (2013) TLR2-targeted secreted proteins from *Mycobacterium tuberculosis* are protective as powdered pulmonary vaccines. Vaccine 31:4322–4329. https://doi.org/10.1016/j.vaccine.2013.07.022

Ubale RV, D'souza MJ, Infield DT et al (2013) Formulation of meningococcal capsular polysaccharide vaccine-loaded microparticles with robust innate immune recognition. J Microencapsul 30:28–41. https://doi.org/10.3109/02652048.2012.692402

Uddin MN, Roni MA (2021) Challenges of storage and stability of mRNA-based COVID-19 vaccines. Vaccines 9:1033. https://doi.org/10.3390/vaccines9091033

Uddin AN, Bejugam NK, Gayakwad SG et al (2009) Oral delivery of gastro-resistant microencapsulated typhoid vaccine. J Drug Target 17:553–560. https://doi.org/10.1080/10611860903067301

Ulmer JB, Mansoura MK, Geall AJ (2015) Vaccines 'on demand': science fiction or a future reality. Expert Opin Drug Discov 10:101–106. https://doi.org/10.1517/17460441.2015.996128

UNAIDS (2022) Global HIV statistics: Fact Sheet 2022. https://www.unaids.org/sites/default/files/media_asset/UNAIDS_FactSheet_en.pdf

Usmani OS, Biddiscombe MF, Barnes PJ (2005) Regional lung deposition and bronchodilator response as a function of β 2-agonist particle size. Am J Respir Crit Care Med 172:1497–1504. https://doi.org/10.1164/rccm.200410-1414OC

Valdespino-Gómez JL, de Lourdes Garcia-Garcia M, Fernandez-de-Castro J et al (2006) Measles aerosol vaccination. In: Current topics in microbiology and immunology, pp 165–193

Van Boven M, Bouma A, Fabri THF et al (2008) Herd immunity to Newcastle disease virus in poultry by vaccination. Avian Pathol 37:1–5. https://doi.org/10.1080/03079450701772391

van den Pol AN (2013) Polio, still lurking in the shadows. J Neurosci 33:855–862. https://doi.org/10.1523/JNEUROSCI.2861-12.2013

Varmus H (2003) Grand challenges in global health. Science (80–) 302:398–399. https://doi.org/10.1126/science.1091769

Vaxart (2021) Vaxart announces first subject enrolled in phase 1b norovirus dose-ranging trial in elderly adults. Press Release, 7 May 2021. https://investors.vaxart.com/news-releases/news-release-details/vaxart-announces-first-subject-enrolled-phase-1bnorovirus-%0Adose

Vehring R (2008) Pharmaceutical particle engineering via spray drying. Pharm Res 25:999–1022. https://doi.org/10.1007/s11095-007-9475-1

Vehring R, Foss WR, Lechuga-Ballesteros D (2007) Particle formation in spray drying. J Aerosol Sci 38:728–746. https://doi.org/10.1016/j.jaerosci.2007.04.005

Vehring R, Snyder H, Lechuga-Ballesteros D (2020) Spray drying. In: Ohtake S, Izutsu K, Lechuga-Ballesteros D (eds) Drying technologies for biotechnology and pharmaceutical applications. Wiley, pp 179–216

Velasquez LS, Shira S, Berta AN et al (2011) Intranasal delivery of Norwalk virus-like particles formulated in an in situ gelling, dry powder vaccine. Vaccine 29:5221–5231. https://doi.org/10.1016/j.vaccine.2011.05.027

Verbeke R, Lentacker I, De Smedt SC, Dewitte H (2019) Three decades of messenger RNA vaccine development. Nano Today 28:100766. https://doi.org/10.1016/j.nantod.2019.100766

Verbeke R, Lentacker I, De Smedt SC, Dewitte H (2021) The dawn of mRNA vaccines: the COVID-19 case. J Control Release 333:511–520. https://doi.org/10.1016/j.jconrel.2021.03.043

VIPS (2019) VIPS Phase I executive summary: heat-stable/controlled temperature chain (CTC) qualified dry formulations (vaccine innovation prioritisation strategy). 1–7

VIPS (2020) VIPS Phase II executive summary: heat stable/controlled temperature chain (CTC) qualified liquid formulations (vaccine innovation prioritisation strategy), 1–16

Vodak DT, Dobry DE, Falk R et al (2012) Enhancing the immune response to intranasal vaccination with Norovirus VLPs—size matters. In: Drug delivery to the lungs (DDL 23), Edinburgh, pp 46–49

Wahl B, O'Brien KL, Greenbaum A et al (2018) Burden of *Streptococcus pneumoniae* and *Haemophilus influenzae* type b disease in children in the era of conjugate vaccines: global, regional, and national estimates for 2000–15. Lancet Glob Heal 6:e744–e757. https://doi.org/10.1016/S2214-109X(18)30247-X

Wain J, Hendriksen RS, Mikoleit ML et al (2015) Typhoid fever. Lancet 385:1136–1145. https://doi.org/10.1016/S0140-6736(13)62708-7

Walters RH, Bhatnagar B, Tchessalov S et al (2014) Next generation drying technologies for pharmaceutical applications. J Pharm Sci 103:2673–2695. https://doi.org/10.1002/jps.23998

Wang SH, Kirwan SM, Abraham SN et al (2012) Stable dry powder formulation for nasal delivery of anthrax vaccine. J Pharm Sci 101:31–47. https://doi.org/10.1002/jps.22742

Watts P, Smith A, Hinchcliffe M (2014) ChiSys® as a chitosan-based delivery platform for nasal vaccination. Mucosal delivery of biopharmaceuticals. Springer, US, Boston, MA, pp 499–516

Weers J, Tarara T (2014) The PulmoSphere™ platform for pulmonary drug delivery. Ther Deliv 5:277–295. https://doi.org/10.4155/tde.14.3

Weers JG, Tarara TE, Clark AR (2007) Design of fine particles for pulmonary drug delivery. Expert Opin Drug Deliv 4:297–313. https://doi.org/10.1517/17425247.4.3.297

Weide B, Carralot J-P, Reese A et al (2008) Results of the first phase I/II clinical vaccination trial with direct injection of mRNA. J Immunother 31:180–188. https://doi.org/10.1097/CJI.0b013e31815ce501

Weide B, Pascolo S, Scheel B et al (2009) Direct Injection of protamine-protected mRNA: results of a phase 1/2 vaccination trial in metastatic melanoma patients. J Immunother 32:498–507. https://doi.org/10.1097/CJI.0b013e3181a00068

Weimer AW (2019) Particle atomic layer deposition. J Nanoparticle Res 21:9. https://doi.org/10.1007/s11051-018-4442-9

Weiser JN, Ferreira DM, Paton JC (2018) Streptococcus pneumoniae: transmission, colonization and invasion. Nat Rev Microbiol 16:355–367. https://doi.org/10.1038/s41579-018-0001-8

Weiss S (2020) Does the BCG vaccine work agains coronavirus? We just don't know, 15 Apr 2020. WIRED. https://www.wired.co.uk/article/bcg-vaccine-coronavirus-uk

Wells CR, Pandey A, Parpia AS et al (2019) Ebola vaccination in the Democratic Republic of the Congo. Proc Natl Acad Sci 116:10178–10183. https://doi.org/10.1073/pnas.1817329116

Weniger BG, Papania MJ (2013) Alternative vaccine delivery methods. Vaccines 1200–1231. https://doi.org/10.1016/B978-1-4557-0090-5.00063-X

Westerman EM, de Boer AH, Le Brun PPH et al (2007a) Dry powder inhalation of colistin sulphomethate in healthy volunteers: a pilot study. Int J Pharm 335:41–45. https://doi.org/10.1016/j.ijpharm.2006.11.021

Westerman EM, De Boer AH, Le Brun PPH et al (2007b) Dry powder inhalation of colistin in cystic fibrosis patients: a single dose pilot study. J Cyst Fibros 6:284–292. https://doi.org/10.1016/j.jcf.2006.10.010

White S, Bennett DB, Cheu S et al (2005) EXUBERA®: pharmaceutical development of a novel product for pulmonary delivery of insulin. Diabetes Technol Ther 7:896–906. https://doi.org/10.1089/dia.2005.7.896

Whitehead KA, Langer R, Anderson DG (2009) Knocking down barriers: advances in siRNA delivery. Nat Rev Drug Discov 8:129–138. https://doi.org/10.1038/nrd2742

Whitley RJ, Roizman B (2001) Herpes simplex virus infections. Lancet 357:1513–1518. https://doi.org/10.1016/S0140-6736(00)04638-9

WHO (1970) Health aspects of chemical and biological weapons: a report of a WHO Group of Consultants, Geneva, Switzerland. https://apps.who.int/iris/bitstream/handle/10665/39444/24039.pdf

WHO (1994) Annex 3, Requirements for measles, mumps and rubella vaccines and combined vaccine (live). WHO Tech Rep Ser 840:100–176

WHO (2003) Background document: the diagnosis, treatment and prevention of typhoid fever. WHO/V&B/03.07, May 2003

WHO (2006) Temperature sensitivity of vaccines, WHO/IVB/06.10, August 2006. World Heal Organ Dep Immunization, Vaccines Biol 1–58. WHO/IVB/06.10

WHO (2007) WHO guidelines on Tularaemia, WHO/CDS/EPR/2007.7. http://apps.who.int/iris/bitstream/handle/10665/43793/9789241547376_eng.pdf

WHO (2008) Anthrax in humans and animals, 4th edn. https://www.who.int/csr/resources/publications/anthrax_webs.pdf

WHO (2011) Meningococcal vaccines: WHO position paper, November 2011 (WER8647). Wkly Epidemiol Rec. https://www.who.int/publications/i/item/WER8647

WHO (2012a) Vaccines against influenza WHO position paper—November 2012. Wkly Epidemiol Rec 461–476

WHO (2012b) Revolutionary meningitis vaccine breaks another barrier; first to gain approval to travel outside cold chain. Newsroom, 14 Nov 2012, Atalanta. https://www.who.int/immunization/newsroom/menafrivac_20121114/en/

WHO (2013a) Oral Cholera Vaccine stockpile for cholera emergency response, International Coordinating Group (ICG) on vaccine provision. https://www.who.int/publications/m/item/oral-cholera-vaccine-stockpile-for-cholera-emergency-response

WHO (2013b) Haemophilus influenzae type b (Hib) vaccination position paper—July 2013. Wkly Epidemiol Rec. https://apps.who.int/iris/handle/10665/242126

WHO (2015a) Preventing perinatal Hepatitis B virus transmission: a guide for introducing and strengthening Hepatitis B birth dose vaccination. http://www.who.int/iris/handle/10665/208278

WHO (2015b) Pertussis Vaccines: WHO position paper—August 2015. Wkly Epidemiol Rec. https://apps.who.int/iris/bitstream/handle/10665/242413/WER9035_433-458.PDF

WHO (2016a) A systematic review of monovalent Hepatitis B vaccine thermostability

WHO (2016b) Polio vaccines: WHO position paper—March, 2016. Wkly Epidemiol Rec. https://www.who.int/wer/2016/wer9112.pdf

WHO (2017a) WHO preferred product characteristics for next-generation influenza vaccines

WHO (2017b) Measles vaccines: WHO position paper—April 2017. Wkly Epidemiol Rec 92:205–228

WHO (2017c) Hepatitis B vaccines: WHO position paper—July 2017. https://apps.who.int/iris/bitstream/handle/10665/255873/WER9227-369-392.pdf

WHO (2017d) Controlled temperature chain: strategic roadmap for priority vaccines 2017–2020. WHO/IVB/17.20 (orginally distributed as WHO/IVB/17.17)

WHO (2017e) Human papillomavirus vaccines: WHO position paper, May 2017. Wkly Epidemiol Rec 19:241–268

WHO (2017f) Cholera vaccines: WHO position paper—August 2017. Wkly Epidemiol Rec. http://apps.who.int/iris/bitstream/10665/258764/1/WER9234-477-498.pdf

WHO (2017g) Background Paper to SAGE on typhoid vaccine policy recommendations, 24 Sept 2017. https://www.who.int/immunization/sage/meetings/2017/october/1_Typhoid_SAGE_background_paper_Final_v3B.pdf

WHO (2017h) Diphtheria vaccine: WHO position paper—August 2017. Wkly Epidemiol Rec 417–43s5

WHO (2017i) Tetanus vaccines: WHO position paper—February 2017. Wkly Epidemiol Rec. https://www.who.int/publications/i/item/tetanus-vaccines-who-position-paper-february-2017

WHO (2018a) Critically important antimicrobials for human medicine, 6th revision 2018, 1–52

WHO (2018b) Fact sheet influenza (seasonal), 6 Nov 2018

WHO (2018c) Global vaccine action plan, monitoring, evaluation & accountability, Secretariat annual report 2018

WHO (2018d) Typhoid vaccine: WHO position paper—March 2018. Wkly Epidemiol Rec 13:153–172

WHO (2018e) Rotavirus, vaccine-preventable diseases, surveillance standards, September 5, 2018

WHO (2019a) Model List of Essential Medicines, 21st edn. WHO/MVP/EMP/IAU/2019.06. 1–65

WHO (2019b) Measles key facts, 5 Dec 2019. https://www.who.int/news-room/fact-sheets/detail/measles

WHO (2019c) Poliomyelitis, key facts, 22 July 2019. https://www.who.int/news-room/fact-sheets/detail/poliomyelitis

WHO (2019d) WHO preferred product characteristics for herpes simplex virus vaccines. World Heal Organ. https://www.who.int/publications/i/item/9789241515580

WHO (2019e) Pneumococcal conjugate vaccines in infants and children under 5 years of age WHO position paper—February 2019. Wkly Epidemiol Rec. https://apps.who.int/iris/handle/10665/310970

WHO (2020a) Hepatitis B, Key facts, 27 July 2020. https://www.who.int/news-room/fact-sheets/detail/hepatitis-b

WHO (2020b) Tuberculosis, key facts, 14 Oct 2020. https://www.who.int/news-room/fact-sheets/detail/tuberculosis

WHO (2020c) Bacille Calmette-Guérin (BCG) vaccination and COVID-19: Scientific brief, 12 Apr 2020. Sci Br. https://www.who.int/news-room/commentaries/detail/bacille-calmette-guérin-(bcg)-vaccination-and-covid-19

WHO (2020d) Human papillomavirus (HPV) and cervical cancer. https://www.who.int/news-room/fact-sheets/detail/human-papillomavirus-(hpv)-and-cervical-cancer. Accessed 11 Nov 2020

WHO (2020e) WHO vaccine-preventable diseases: monitoring system 2020 global summary. https://www.who.int/teams/immunization-vaccines-and-biologicals/immunization-analysis-and-insights/surveillance/surveillance-for-vpds

WHO (2020f) The immunological basis for immunization series: module 15: meningococcal disease, Updated 2020. https://www.who.int/publications/i/item/9789240017481

WHO (2020g) Control of epidemic meningitis in countries in the African meningitis belt, 2019. Wkly Epidemiol Rec 14(15):133–144

WHO (2021a) Global tuberculosis report 2021. https://www.who.int/publications/i/item/9789240037021

WHO (2021b) WHO Prequalified Vaccines, online database: https://extranet.who.int/gavi/PQ_Web. Accessed on 20 Mar 2021

WHO (2021c) Cholera, key facts. https://www.who.int/news-room/fact-sheets/detail/cholera. Accessed 5 Feb 2021

WHO (2021d) Immunization coverage, key facts, 15 July 2021. https://www.who.int/news-room/fact-sheets/detail/immunization-coverage

WHO (2021e) Rotavirus vaccines: WHO position paper—July 2021. Wkly Epidemiol Rec. https://www.who.int/publications/i/item/weekly-epidemiological-record-vol.-28-2021-96-301-320

WHO (2021f) Global polio eradication initiative, annual report 2020 and semi-annual status update, January–June and July–December 2020. https://polioeradication.org/wp-content/uploads/2021/08/GPEI-2020-Annual-Report-ISBN-9789240030763.pdf

WHO (2021g) Pneumococcal conjugate (PCV3): immunization coverage estimates by WHO region. https://apps.who.int/gho/data/view.main.PCV3vREG

WHO (2022a) Meningococcal meningitis. https://www.who.int/teams/immunization-vaccines-and-biologicals/diseases/meningitis

WHO (2022b) International Coordinating Group (ICG) on vaccine provision, meningitis vaccine stockpiles. https://www.who.int/groups/icg/meningitis/stockpiles

WHO/UNICEF (2014) Vaccine vial monitor (VVM) assignments for different WHO-prequalified vaccines and their proper handling, July 2014. Inf Bull https://www.who.int/immunization/programmes_systems/service_delivery/EN_Information_Bulletin_VVM_assignments.pdf

WHO, UNICEF (2019) Measles-rubella microarray patch (MR-MAP) target product profile, June 2019. https://www.who.int/immunization/research/ppc-tpp/WHO_MR_MAP_TPP.pdf

WHO/UNICEF (2020) Progress towards global immunization goals—2019, Summary presentation of key indicators, updated July 2020, 1–28

Wimer-Mackin S, Hinchcliffe M, Petrie CR et al (2006) An intranasal vaccine targeting both the *Bacillus anthracis* toxin and bacterium provides protection against aerosol spore challenge in rabbits. Vaccine 24:3953–3963. https://doi.org/10.1016/j.vaccine.2006.02.024

Wong Y-L, Sampson S, Germishuizen WA et al (2007) Drying a tuberculosis vaccine without freezing. Proc Natl Acad Sci 104:2591–2595. https://doi.org/10.1073/pnas.0611430104

Wong J, Ricci M, Chan H-K (2016) Spray drying strategies to stop tuberculosis. In: Drug delivery systems for tuberculosis prevention and treatment. Wiley, Chichester, UK, pp 161–196

Wong G, Mendoza EJ, Plummer FA et al (2018) From bench to almost bedside: the long road to a licensed Ebola virus vaccine. Expert Opin Biol Ther 18:159–173. https://doi.org/10.1080/14712598.2018.1404572

Wong-Chew RM, Islas-Romero R, García-García MDL et al (2006) Immunogenicity of aerosol measles vaccine given as the primary measles immunization to nine-month-old Mexican children. Vaccine 24:683–690. https://doi.org/10.1016/j.vaccine.2005.08.045

World Bank; TAFS Forum (2011) World livestock disease atlas: a quantitative analysis of global animal health data (2006–2009). World Bank, Washington, DC and TAFS Forum, Bern. © World Bank. https://openknowledge.worldbank.org/handle/10986/27118

Wu Q, Wu H (2018) Disposable monodose inhaler for powdered medicaments, Patent WO2018051266A1, Prioritiy Data: September 19, 2016, Applicant: Hollycon Italy S.R.L. 1–23

Wu J, Wu L, Wan F et al (2019) Effect of thermal and shear stresses in the spray drying process on the stability of siRNA dry powders. Int J Pharm 566:32–39. https://doi.org/10.1016/j.ijpharm.2019.05.019

Xu H, Cai L, Hufnagel S, Cui Z (2021) Intranasal vaccine: Factors to consider in research and development. Int J Pharm 609:121180. https://doi.org/10.1016/j.ijpharm.2021.121180

Yadav R, Zhai L, Tumban E (2019) Virus-like particle-based L2 vaccines against HPVs: where are we today? Viruses 12:18. https://doi.org/10.3390/v12010018

Yang M (2011) Spray drying pharmaceuticals. Eur Pharm Rev. https://www.europeanpharmaceuticalreview.com/article/10406/spray-drying-pharmaceuticals/

Young PM (2018) High dose powders: a critical assessment of needs, formulation, and delivery technologies. Respir Drug Deliv 207–216

Young PM, Crapper J, Philips G et al (2014) Overcoming dose limitations using the orbital® multi-breath dry powder inhaler. J Aerosol Med Pulm Drug Deliv 27:138–147. https://doi.org/10.1089/jamp.2013.1080

Young PM, Salama RO, Zhu B et al (2015) Multi-breath dry powder inhaler for delivery of cohesive powders in the treatment of bronchiectasis. Drug Dev Ind Pharm 41:859–865. https://doi.org/10.3109/03639045.2014.909841

Yu CP, Nicolaides P, Soong TT (1979) Effect of random airway sizes on aerosol deposition. Am Ind Hyg Assoc J 40:999–1005. https://doi.org/10.1080/15298667991430613

Zehrung D (2017) Overview of emerging delivery technologies. In: 2017 WHO Prod. Dev. Vaccines Advis. Comm. Consult. 21–22 June 21–22, WHO, Geneva. https://www.who.int/immunization/research/meetings_workshops/PDVAC_2017_Delivery_Tech_Update_Zehrung_PATH.pdf

Zehrung D (2018) Novel primary containers and delivery technologies, Program leader, devices & tools at PATH. In: 13th international Rotavirus symposium, 29–31 Aug 2018, Minsk. Belarus. https://www.sabin.org/sites/sabin.org/files/darin_zehrung.pdf

Zemp F, Rajwani J, Mahoney DJ (2018) Rhabdoviruses as vaccine platforms for infectious disease and cancer. Biotechnol Genet Eng Rev 34:122–138. https://doi.org/10.1080/02648725.2018.1474320

Zeng C, Zhang C, Walker PG, Dong Y (2020) Formulation and delivery technologies for mRNA vaccines. Current topics in microbiology and immunology. Springer, Berlin, Heidelberg, pp 1–40

Zhai L, Tumban E (2016) Gardasil-9: a global survey of projected efficacy. Antiviral Res 130:101–109. https://doi.org/10.1016/j.antiviral.2016.03.016

Zhai L, Yadav R, Kunda NK et al (2019) Oral immunization with bacteriophage MS2-L2 VLPs protects against oral and genital infection with multiple HPV types associated with head & neck cancers and cervical cancer. Antiviral Res 166:56–65. https://doi.org/10.1016/j.antiviral.2019.03.012

Zhang M, Dong C, Xiong S (2017) Vesicular stomatitis virus-vectored multi-antigen tuberculosis vaccine limits bacterial proliferation in mice following a single intranasal dose. Front Cell Infect Microbiol 7:1–9. https://doi.org/10.3389/fcimb.2017.00034

Zhang N-N, Li X-F, Deng Y-Q et al (2020) A Thermostable mRNA Vaccine against COVID-19. Cell 182:1271-1283.e16. https://doi.org/10.1016/j.cell.2020.07.024

Zhao H, Zhou X, Zhou YH (2020a) Hepatitis B vaccine development and implementation. Hum Vaccines Immunother 16:1533–1544. https://doi.org/10.1080/21645515.2020.1732166

Zhao P, Hou X, Yan J et al (2020b) Long-term storage of lipid-like nanoparticles for mRNA delivery. Bioact Mater 5:358–363. https://doi.org/10.1016/j.bioactmat.2020.03.001

Zhao H, Wang T-C, Li X-F et al (2021) Long-term stability and protection efficacy of the RBD-targeting COVID-19 mRNA vaccine in nonhuman primates. Signal Transduct Target Ther 6:438. https://doi.org/10.1038/s41392-021-00861-4

Zhigaltsev IV, Belliveau N, Hafez I et al (2012) Bottom-up design and synthesis of limit size lipid nanoparticle systems with aqueous and triglyceride cores using millisecond microfluidic mixing. Langmuir 28:3633–3640. https://doi.org/10.1021/la204833h

Zhou QT, Tang P, Leung SSY et al (2014) Emerging inhalation aerosol devices and strategies: where are we headed? Adv Drug Deliv Rev 75:3–17. https://doi.org/10.1016/j.addr.2014.03.006

Zhu C, Shoji Y, McCray S et al (2014) Stabilization of HAC1 influenza vaccine by spray drying: formulation development and process scale-up. Pharm Res 31:3006–3018. https://doi.org/10.1007/s11095-014-1394-3

Zhu B, Young PM, Ong HX et al (2015) Tuning aerosol performance using the multibreath orbital® dry powder inhaler device: controlling delivery parameters and aerosol performance via modification of puck orifice geometry. J Pharm Sci 104:2169–2176. https://doi.org/10.1002/jps.24458

Zhu B, Padroni M, Colombo G et al (2016) The development of a single-use, capsule-free multi-breath tobramycin dry powder inhaler for the treatment of cystic fibrosis. Int J Pharm 514:392–398. https://doi.org/10.1016/j.ijpharm.2016.04.009

Ziaee A, Albadarin AB, Padrela L et al (2019) Spray drying of pharmaceuticals and biopharmaceuticals: critical parameters and experimental process optimization approaches. Eur J Pharm Sci 127:300–318. https://doi.org/10.1016/j.ejps.2018.10.026

Zipursky S, Djingarey MH, Lodjo J-C et al (2014) Benefits of using vaccines out of the cold chain: delivering Meningitis A vaccine in a controlled temperature chain during the mass immunization campaign in Benin. Vaccine 32:1431–1435. https://doi.org/10.1016/j.vaccine.2014.01.038

Zuniga A, Wang Z, Liniger M et al (2007) Attenuated measles virus as a vaccine vector. Vaccine 25:2974–2983. https://doi.org/10.1016/j.vaccine.2007.01.064

Zwerling A, Behr MA, Verma A et al (2011) The BCG World Atlas: a database of global BCG vaccination policies and practices. PLoS Med 8:e1001012. https://doi.org/10.1371/journal.pmed.1001012

Conclusions and Future Perspectives of Spray-Dried Vaccines

7

Abstract

This chapter offers a condensed summary of the main findings from the preceding Chaps. 1 to 6. In particular, this chapter provides answers to the following questions:

- What are the key motivating factors for spray drying vaccines?
- What are the main advantages of spray drying compared to other drying technologies?
- What are the challenges, barriers, and opportunities?
- What are spray dryer models used on a laboratory and commercial scale?
- What types of excipients and process conditions are suitable for improving stability?
- What is the application spectrum of spray-dried vaccines?
- What are the typical drying temperatures?
- What particle size and powder morphologies are produced by spray drying?
- What is the achievable shelf-life of spray-dried vaccines?
- By what routes of administration can spray-dried vaccine powders be delivered?
- How close to the market are spray-dried vaccine products?

The chapter concludes with an insight into the future perspectives for aseptic spray drying of vaccines.

7.1 Motivation of Spray Drying Vaccines

Vaccines are one of modern medicine's greatest achievements for humanity, providing an effective and cost-effective intervention to prevent deadly diseases (Ulmer et al. 2006; Bregu et al. 2011; Nabel 2013; Delany et al. 2014; Greenwood 2014;

C. Arpagaus, *Spray Drying of Vaccines*,
https://doi.org/10.1007/978-3-031-24323-3_7

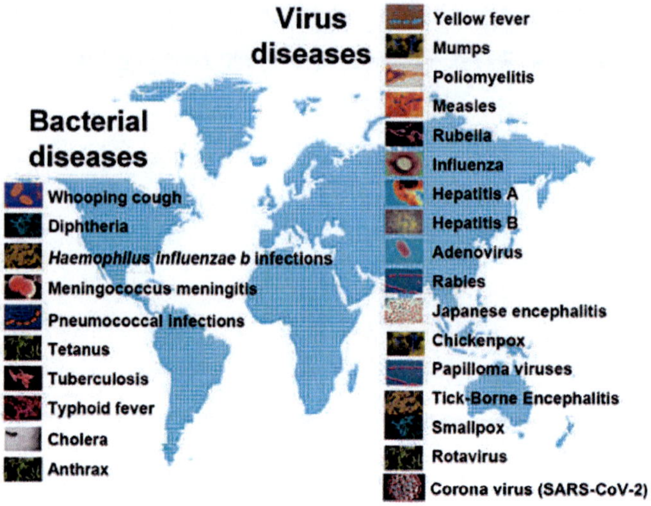

Fig. 7.1 Vaccines for humans against bacterial and viral infectious diseases (Extended from Kaddar (2013))

GAVI 2021a, b). Today, more than 70 vaccines are approved by the World Health Organization (WHO) to prevent at least 25 different bacterial and viral infectious diseases, including COVID-19 (SARS-CoV-2 virus) (Fig. 7.1) (Nabel 2013; WHO 2021a). In addition, more than 120 new vaccine products were under development (before the COVID-19 era) (Kaddar 2013; D'Amore and Yang 2019; GEA 2019). Thus, vaccines will continue to play an essential role in protecting human health in the twenty-first century (Rappuoli 2011; Rappuoli et al. 2011, 2021; Delany et al. 2014).

More than 100 million infants are vaccinated annually, saving at least 2 to 3 million lives per year worldwide (WHO/UNICEF/World Bank 2009; Delany et al. 2014; WHO 2020a). However, despite successes in global vaccination, more than 1.5 million people still die annually from vaccine-preventable diseases (Vanderslott et al. 2019; WHO 2020b; GAVI 2021a), mainly in low- and middle-income countries where distributing life-saving vaccines to people in poorly served remote villages, fragile states, and strife-torn regions is difficult (WHO/UNICEF/World Bank 2009).

With COVID-19, global attention on vaccination is at an all-time high. As of August 8, 2022, more than 582 million confirmed cases of COVID-19 infection have occurred, including more than 6.4 million deaths. A total of 12.3 million vaccine doses have been administered, according to the WHO (2022a).

From an economic perspective, the vaccine market is growing rapidly. Figure 7.2a shows the projected global vaccine market from 2000 to 2028 based on various market reports (Evaluate 2018; Globe Newswire 2018a; PR Newswire 2018; Research and Markets 2018, 2020; Markets and Markets 2020). It is

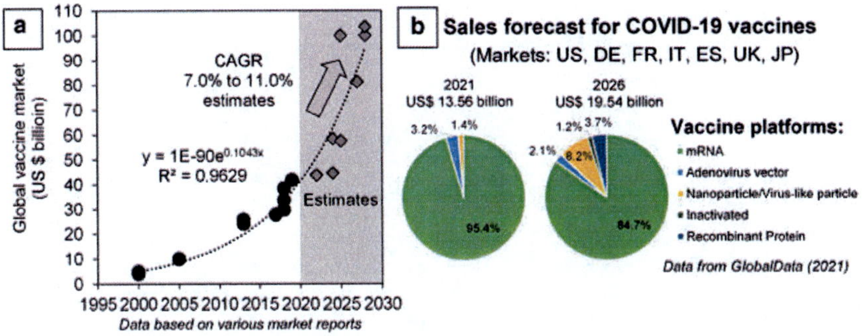

Fig. 7.2 **a** Global vaccine market from 2000 to 2028 in billion US$ based on various market reports (Douglas and Samant 2018; Evaluate 2018; Globe Newswire 2018; PR Newswire 2018; Research and Markets 2018, 2020; D'Amore and Yang 2019; GEA 2019; Markets and Markets 2020). **b** Sales forecast of COVID-19 vaccines until 2026 for the seven major markets (US, Germany, France, Italy, Spain, UK, and Japan) (Data adapted from GlobalData (2021))

estimated to rise to $100 billion by 2028 at a compound annual growth rate (CAGR) of 7–11% (Globe Newswire 2018). In particular, the sales forecast for COVID-19 vaccines is expected to grow from $13.6 billion in 2021 to $19.5 billion in 2026, at a CAGR of 7.6% in seven major markets (USA, Germany, France, Italy, Spain, UK, and Japan) (Global Data 2021) (Fig. 7.2b). COVID-19 vaccines based on mRNA technology will dominate in these countries. Chapter 1 provides more information on vaccine suppliers and cooperations to improve vaccine accessibility.

Approximately 72% of the 70 vaccines approved by the United States Food and Drug Administration (FDA) today (FDA 2022) are injected in liquid form intramuscularly, followed by subcutaneous injections (12%) into the fatty layer under the skin (hypodermis) (WHO 2021b) (Fig. 7.3). Others are injected percutaneously (3%) and intradermally (1%), targeting the dermis. In addition, some vaccines are administered orally (e.g., typhoid, rabies, and rotavirus) or intranasally (e.g., influenza) to mimic the natural route of infection. The dosage forms are mainly liquid suspensions (approx. 63%), lyophilized powder (26%), solutions (10%), or emulsions (1%).

The bulk of the 70 vaccines (approximately 87%) requires storage under refrigerated conditions from 2 to 8 °C. Only a few of the listed vaccines can be kept at room temperature (25 °C) for a few days, and none are stable long-term above refrigerated temperatures (Gomez and Vehring 2022), which is a key challenge. Several vaccines must be kept frozen at temperatures below −15 °C and even below −60 °C (ultracold storage) to ensure their efficacy when reaching the patients. Examples are COVID-19 mRNA or Ebola vaccines (Cavalier 2020; CureVac 2020a, b; Moderna 2020; Pfizer-BioNTech 2020; Crommelin et al. 2021; Schoenmaker et al. 2021).

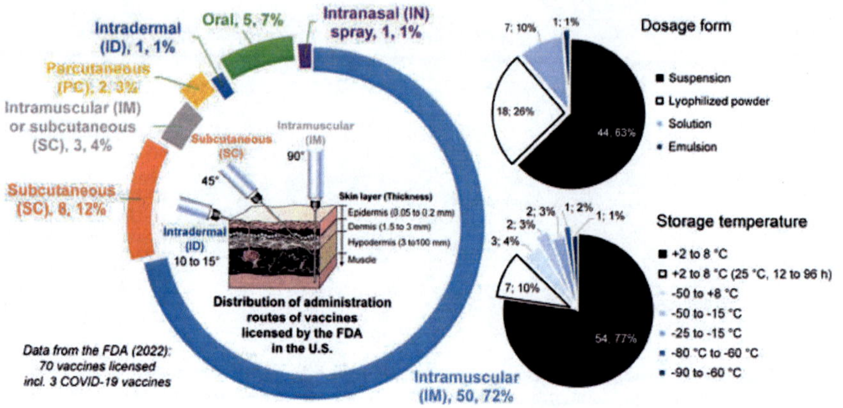

Fig. 7.3 Most vaccines are injected as liquids and need cold storage. Data from the FDA (2022), as of 31 January 2022, of 70 vaccines licensed, including 3 emergency use authorized COVID-19 vaccines

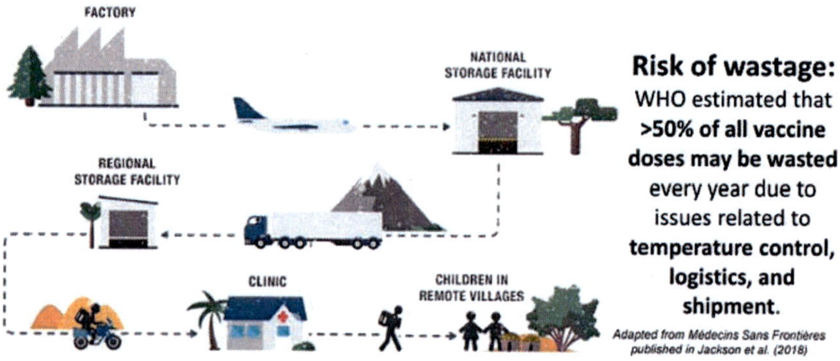

Fig. 7.4 The cold chain is a challenge. Most vaccines must be kept cold between + 2 to + 8 °C from the factory to the patients at the most remote places on earth (Adapted from Médecins Sans Frontières (MSF) published in Jackson et al. (2018)). As a result, over 50% of all vaccine doses may be wasted globally yearly because of issues related to temperature control, logistics, and shipment (WHO 2005)

The associated complex logistics of the cold chain for transport, distribution, and storage (Fig. 7.4) make mass vaccination campaigns costly, especially in regions with inadequate refrigeration infrastructure and limited financial resources, i.e., vulnerable low- and middle-income countries. It is estimated that maintaining the cold chain and refrigeration of vaccines from production to patient costs about $200 to 300 million each year (Lloyd 2000; Das 2004), which increases the cost of vaccination by 14% to 20% (Lloyd 2000; Manufacturing Chemist 2010). Other studies in low-income countries evaluated that cold chain costs account for up to 80% of the vaccination program (Kols and Sherris 2000).

Although there is strong support for strengthening the cold chain by upgrading the refrigeration facilities in many health centers with new equipment (UNICEF 2019), delivering vaccines anywhere and on time remains a major challenge in the twenty-first century (Cavalier et al. 2021). In high-income countries, the availability of a vaccine cold chain may not be as much of a problem, but in low-income countries, getting vaccines to a village or community can be a real challenge, especially with the introduction of newer and more expensive vaccines (Zaffran et al. 2013). A cold chain is only as strong as its weakest link (Kartoglu and Milstien 2014).

In addition, there is a risk of vaccine wastage due to improper storage. The WHO estimated that over 50% of all vaccine doses might be wasted globally every year because of issues related to temperature control, logistics, and shipment (WHO 2005). In addition, liquid injection requires trained health workers, sterility, and proper disposal of sharps.

Therefore, an important strategy for preventing a pandemic is stockpiling stable vaccines for rapid intervention in the event of a disease outbreak and reducing dependence on the cold chain. In addition, thermostable vaccines in liquid or powder form offer more convenience and may reduce transportation and distribution costs.

7.2 Drying Technologies for Vaccine Formulations

The most common method for stabilizing vaccines and increasing shelf-life without losing immunogenicity is conversion to dry powder using appropriate stabilizing excipients and drying technologies. More stable solid formulations are explained by the water replacement and glass vitrification hypotheses, providing lower mobility of the biopharmaceutical molecules and slower degradation kinetics (e.g., hydrolysis) (see discussion in Chap. 3). In addition, a thermostabilized dried formulation enables several benefits, including (Tlaxca et al. 2015; Geng 2018; Bahamondez-Canas and Cui 2018; Bhambhani and Antochshuk 2020):

- improved handling, storage, transport, and distribution (i.e., room temperature storage, lighter weight, less dependence on a cold chain),
- a logistical advantage for more effective stockpiling in the event of a pandemic, facilitating mass vaccination, and
- more flexible dosage forms and new routes of administration, such as pulmonary, intranasal, oral, intradermal, or intramuscular (after reconstitution) (see illustration in Fig. 7.5).

Several drying technologies are suitable for vaccine stabilization, including freeze-drying (lyophilization), spray drying, spray freeze drying, vacuum foam drying, supercritical fluid drying, air drying, and others (see Chap. 3).

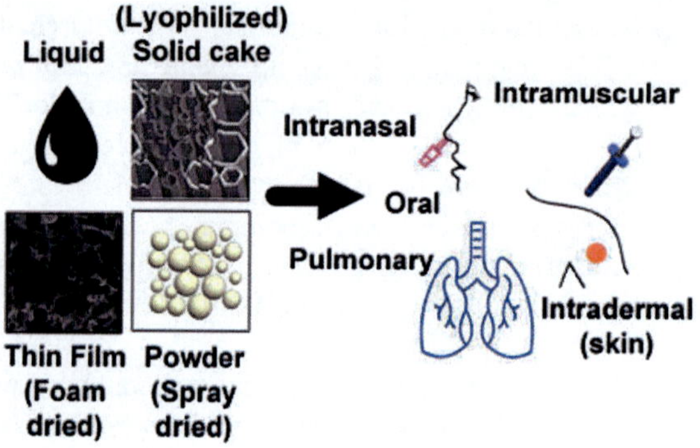

Fig. 7.5 The typical stabilized dosage forms of vaccines are liquid, thin film (foam-dried), solid cake (lyophilized), or powder (spray-dried). Besides intramuscular administration, a dry powder offers new delivery routes like pulmonary, intranasal, oral, or intradermal (skin) (Adapted from Tlaxca et al. (2015) with permission from Elsevier)

In addition, there are also other innovative drying methods for vaccines that enable new delivery forms such as rapidly soluble oral thin films, sugar pills, fiber membranes, or porous foams as attachments for syringes, powders dried at room temperature in laminar air, and microarray patches that enable vaccine delivery via the skin. A summary is given in Chap. 3, Sect. 3.9.

Each drying technology has its advantages and challenges that must be considered when selecting a drying method, as discussed by many authors (Amorij 2007; Amorij et al. 2008; Maltesen and van de Weert 2008; Abdul-Fattah and Truong 2010; Rosini et al. 2011; Walters et al. 2014; Searles and Cherian 2015; Ohtake et al. 2015; Langford et al. 2018a; Emami et al. 2018; Vass et al. 2019; Lovalenti and Truong-Le 2020; Thorat et al. 2020).

The most important decision criteria for selecting a drying technology are (Amorij 2007; Amorij et al. 2008; Arpagaus 2020):

- process load (e.g., atomization, freezing, heating),
- process speed,
- crystallization risk,
- simplicity of particle design,
- capacity,
- recovery,
- cost, and
- current (industrial) experience.

In addition, the choice of drying method is associated with the economics of drying and the intended route of administration.

So far, freeze-drying has established itself as a gold standard for drying vaccines. It generates a solid cake-like product in vials that can be aseptically sealed inside the freeze-drying chamber before removal, providing sterility and particle-free

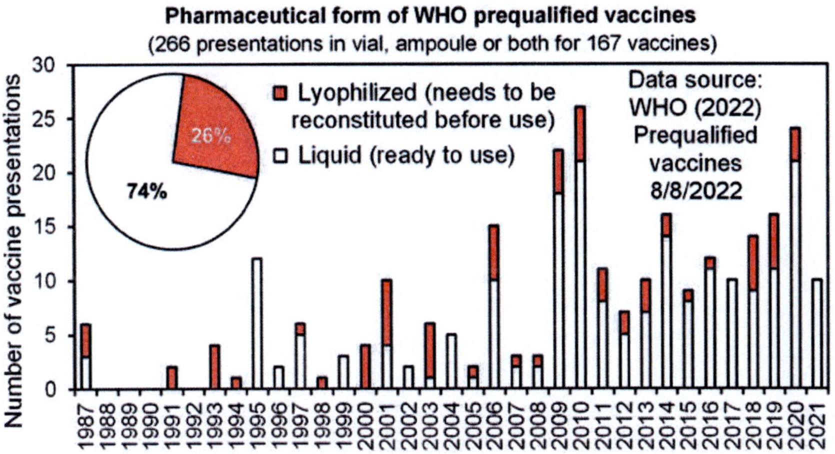

Fig. 7.6 The pharmaceutical form of WHO prequalified vaccines. 69 of 266 presentations (26%) are lyophilized, and 74% (190 of 266) are in liquid form (ready to use) (*Data source* WHO (2022b), https://extranet.who.int/gavi/PQ_Web, accessed on August 8, 2022)

quality attributes for injectable products (Abdul-Fattah et al. 2007a, b). However, before injection to the patient, the lyophilized vaccines have to be reconstituted with excipient diluent or liquid active components.

Figure 7.6 shows that 74% of the 266 listed WHO prequalified vaccine presentations are in liquid form (ready-to-use) and 26% (69 of 266) in a lyophilized form (in vials and/or ampoules or two sets of vials) (WHO 2022b). The lyophilized vaccine presentations cover 16 vaccine types, such as Bacille Calmette-Guérin (6 vaccine presentations), dengue treatment (1), diphtheria-tetanus-pertussis (6), Hemophilus influenzae (2), influenza (2 pandemics, 1 seasonal), Japanese encephalitis (3), measles (6), measles and rubella (7), measles-mumps-rubella (7), meningococcal (4), rabies (4), rotavirus (4), rubella (4), varicella (2), and yellow fever (10).

Unlike freeze drying, the favorable factors for spray drying include no freezing stress, process speed (fast drying), particle design possibility, scalability, and high capacity to overcome bottlenecks in production (Table 7.1). On the other hand, the main challenges are thermal stress, residual moisture content, aseptic processing, and powder filling (see more detailed discussion in Sect. 7.6).

Today, no spray-dried WHO prequalified vaccine is available on the market (WHO 2022b). However, there are numerous approved spray-dried pharmaceuticals like poorly soluble drugs, therapeutic peptides, proteins, enzymes, DNA/RNA, antibodies, blood products (e.g., plasma, serum), hormones, viruses, live cells, bacteria, vaccines, and others (Masters 2002; Celik and Wendel 2005; Guntur and Dhand 2007; Greb 2009; Konstan et al. 2011; Kawabata et al. 2011; Lam et al. 2013; Brough and Williams 2013; Huang and Dai 2014; Siew 2016; Geng 2018; Jermain et al. 2018; Anderson et al. 2018; Arpagaus 2018; Ziaee et al. 2019; Vehring et al. 2020; Baumann et al. 2021; Bhujbal et al. 2021; Sharma et al. 2021). Section 6.1 in Chap. 6 provides a summary.

Table 7.1 General comparison of drying technologies for vaccines

	Freeze drying	Spray drying	Spray freeze drying	Vacuum foam drying	Supercritical fluid drying
Atomization stress	+	−	−	+	−
Freezing stress	−	+	+	+	+
Thermal stress	+	−	+	±	±
Process speed	−	+	±	−	+
Crystallization risk	+	−	+	−	±
Particle design	−	+	+	−	+
Capacity	±	+	±	±	+
Industrial experience	+	±	−	+	−

+ favorable, − unfavorable (Adapted from Amorij et al. (2008), Arpagaus (2020) with permission from Springer Nature)

While most spray-dried pharmaceuticals are administered orally in the form of tablets and capsules (Al-Zoubi et al. 2021), there is a growing number of spray-dried inhalable drugs like Exubera®, Aridol/Bronchitol®, TOBI Podhaler®, or Inbrija®. These products pave the way for spray drying technology to become a viable alternative to established drying methods.

Although freeze-drying is well established in the vaccine industry, it is among the most time-consuming and expensive unit operations (Roser 1991; Colaço and Roser 1994; Domínguez 2011; Walters et al. 2014; Alexeenko and Topp 2020). For example, a lyophilization process may take 2 to 3 days per batch (Jin and Tsao 2013; Chan et al. 2017; GEA 2021). Even freeze dryer filling takes a few hours for 100,000 vials and a fill rate of 600 vials per minute (Gladd 2014). However, efforts are underway to develop continuous freeze-drying processes for individual vials that will allow direct monitoring of the drying history for each vial (De Meyer et al. 2015; Pisano et al. 2019; Ghaemmaghamian et al. 2022).

A cost comparison of different drying methods for foods shows that the costs of freeze-drying relative to spray drying are about 8.3 times and the operating costs 4.9 times higher (Table 7.2). In the flavor industry, freeze-drying may even be 30–50 times more costly than spray drying (Desobry et al. 1997). However, cost data must be treated cautiously as they depend on many aspects, such as location, utilities, building requirements, financing exchange rates, materials, and labor costs (Masters 2002).

For vaccines, cost factors such as labor, materials, consumables (e.g., process gases, electricity, cooling water), and equipment costs of freeze drying and spray drying can be assumed to be comparable (Jin and Tsao 2013).

In contrast to freeze-drying, where the vaccine solution is first filled into vials by sterile filtration and then dried to a powder cake in the vials (Fig. 7.7), aseptic spray-drying is followed by powder transfer to a powder filling line with sterile

Table 7.2 Costs of different drying methods for foods given as costs per kg of water evaporated per hour and relative to spray drying

Drying method	Fixed costs		Manufacturing costs	
	(US$/kg-h)	Relative to spray drying	(Cents/kg)	Relative to spray drying
Air drying	610	0.4	17	0.9
Fluid bed drying	1000	0.7	17	0.9
Drum drying	1040	0.8	23.4	1.2
Spray drying	1360	1.0	19.4	1.0
Vacuum drying	5860	4.3	49	2.5
Freeze-drying	11,228	8.3	95.2	4.9

The fixed costs include equipment purchase, installation, and engineering costs (Data from Roser (1991), Colaço and Roser (1994), Domínguez (2011))

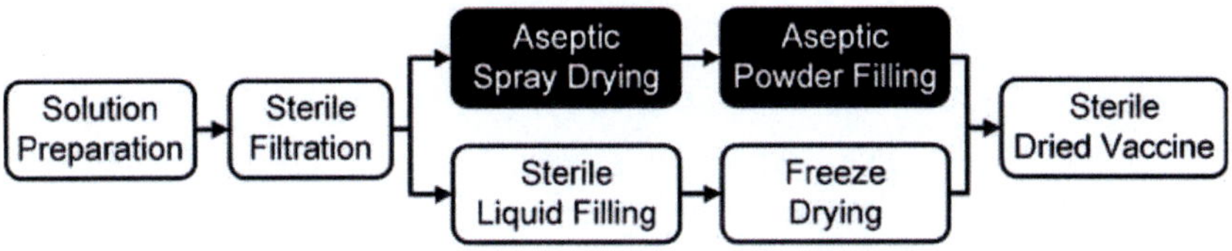

Fig. 7.7 Process steps of aseptic spray drying with aseptic powder filling (top) in contrast to sterile liquid filling and freeze-drying (bottom) (Adapted from Andersen (2016, 2020) with kind permission from the author)

containers and airlock systems, able to fill the spray-dried powder into various desired product formats such as vials, blisters, capsules, pouches, dual-chamber syringes, tubes, inhalers, and other medical containers (Schwartzbach 2011; Gladd 2014; Andersen 2016, 2020; Siew 2016; Luy and Stamato 2020; Vehring et al. 2020).

Powder filling is generally more expensive than liquid filling, so the overall costs balance out again (Andersen 2016, 2020). Thus, it can be assumed that the investment in an aseptic spray drying system with powder filling is comparable to freeze drying, although the investment costs for large-scale spray drying might be somewhat lower in terms of equipment, plant, and operating resources (Greb 2009; Gladd 2014).

7.3 Aseptic Spray Drying Considerations

In many cases, spray drying is the most mature and complementary alternative technology to freeze drying, as stated by several authors (Roser 2009; Searles and Mohan 2010; Schwartzbach 2011; McAdams et al. 2012; Breit and DuBose 2013; Walters et al. 2014; Ledet et al. 2015; Kanojia et al. 2017; Geng 2018; Langford et al. 2018a, b; Andersen 2020; Arpagaus 2020; Ohtake et al. 2020; Bhambhani and Antochshuk 2020; Thorat et al. 2020; Preston and Randolph 2021; Gomez and Vehring 2022; Ghaemmaghamian et al. 2022).

According to expert interviews, aseptic spray drying is expected to expand its position in the vaccine industry and gain market share as an alternative to traditional freeze drying, especially when large product volumes and particle technology capabilities are required. In particular, aseptic spray drying for vaccine production offers:

- Particle engineering by adjusting the process parameters of spray drying and the formulation (i.e., particle size, distribution, shape, morphology, porosity, density, chemical composition, moisture content, flowable powder, aerosolization properties),
- flexibility to powder dosage forms for alternative routes of administration (not requiring needles nor reconstitution) by pulmonary, intranasal, intradermal, oral delivery, or injection (after reconstitution),
- vaccine encapsulation and room temperature stability with appropriate excipients and stabilizers (although formulation efforts may be required),
- fast and continuous drying process (i.e., in seconds rather than days for a batch), and
- high product throughput (e.g., can handle high concentration solutions) and scalability to overcome bottlenecks and better match production capacity to product demand.

The concept of aseptic spray drying dates back to 1975 (Masters and Vestergaard 1975; Masters 1976) (see Chap. 4). Since then, the technology has been used for biopharmaceutical applications, such as peptides, proteins, antibiotics, hormones, allergens, and other products requiring aseptic processing (see applications in Chaps. 5 and 6).

Figure 7.8 illustrates a flow diagram of a typical aseptic open-cycle spray drying system for aqueous formulations, such as vaccines and other biological compounds, based on various literature sources (Masters and Vestergaard 1975; Masters 1976, 1985, 2002; Mujumdar and Alterman 1997; BETE 2005; Turok and Rasmussen 2010; Pakowski and Mujumdar 2014; Manufacturing Chemist 2016; GEA Niro 2017; Arpagaus 2020). A closed-cycle configuration using nitrogen as process gas may be used to dry certain vaccines under inert conditions to protect them from oxidation degradation (see Chap. 4).

The development of temperature-resistant high-efficiency particulate air (HEPA) filters (with bag-in/bag-out replacement principle) (Turok and Rasmussen 2010; Camfil 2020), sterile liquid filters (based on resistant membranes with 0.22 μm pore size), combined with contamination-free atomization, powder discharge systems, and careful design have made aseptic spray drying systems possible that comply with aseptic processing guidelines, such as the FDA Guidance for Industry (FDA 2004) and the ISPE Baseline Guide on Sterile Product Manufacturing Facilities (ISPE 2018).

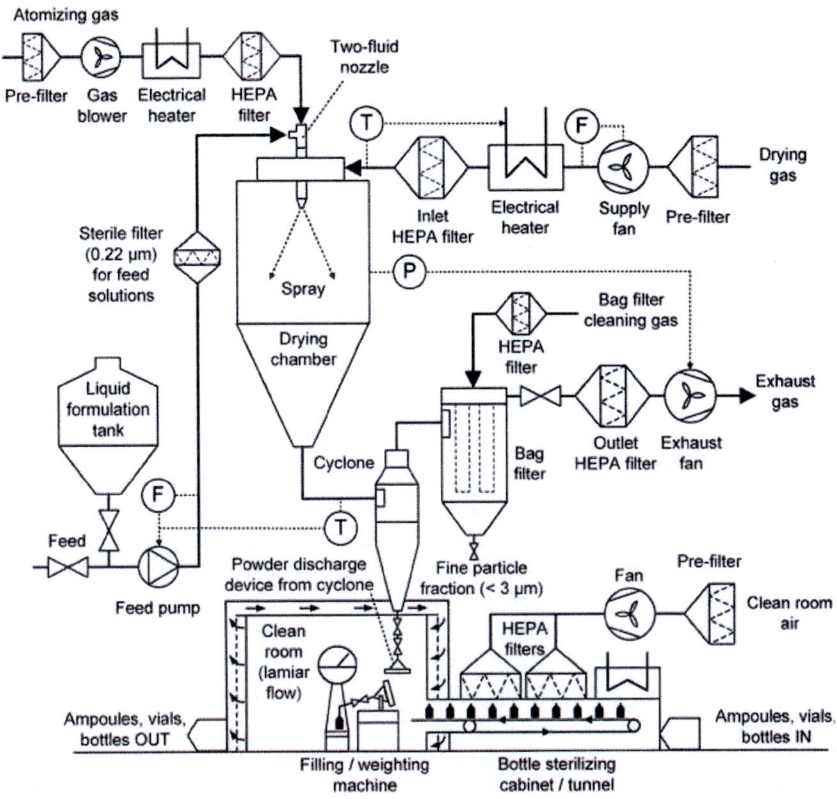

Fig. 7.8 Flow diagram of an aseptic open-cycle spray drying system with two-fluid nozzle, feed pump, sterile filter for feed solutions, prefilter for drying and atomization gas (air), supply and exhaust fans, electrical heater, high-efficiency particulate air (HEPA) filters, cyclone separator, powder discharge device from the cyclone (collection vessel), packaging area with clean room, filling/weighing machine, and sterilizing cabinet/tunnel (Image from the author Arpagaus (2020))

When building an aseptic spray dryer, particular attention has to be paid to quality standards such as ASME Bioprocessing Equipment (BPE) (ASME 2019) and cGMP regulations (McGuire 1998; FDA 2004; EMEA 2006; ICH 2010) which concern personnel, buildings, equipment, production controls, records, and other aspects of the vaccine manufacturing process (Artaud et al. 2019). In addition, filtration, draining, cold spots, steam supply, materials science, cleaning-in-place (CIP) and sterilization-in-place (SIP) equipment, and zoning of clean areas are all topics that need particular attention (WHO 2002; Mogavero 2014). For large-scale spray dryers where disassembly of the components is not feasible (in contrast to lab-scale units, see Chap. 4), appropriate semi- or fully-automatic CIP and SIP systems are required for plant cleaning and sterilization.

Basic considerations to achieving aseptic conditions in a spray drying process are (Schwartzbach 2011; Vehring et al. 2020):

- Initial CIP systems and cleaning procedures of product contact surfaces with detergents, acids, alkaline solutions, and purified water,
- SIP systems and pressure-monitored sterilization cycles with saturated steam of at least 121 °C for about 30 min,
- polished internal steel surfaces and short pipes to minimize surface contamination,
- maintaining sterility during the production of spray-dried powder, and
- aseptic removal of the dried powder from the drying system and powder filling.

For example, Fig. 7.9 illustrates the basic requirements for aseptic powder transfer from a spray dryer to a powder filling line via charge bottles. To prevent contamination, facility design needs to ensure that the area between a spray dryer and a powder filling line provides Grade A protection (Class 100, ISO 5) (FDA 2004). In addition, transport and loading procedures should afford the same protection conditions. The cleanroom area adjacent to an aseptic processing line is recommended to meet at least Grade C standards (Class 10,000, ISO 7) (FDA 2004). Monitoring the microbiological cleanliness should be frequent, and methods such as settle plates and volumetric air and surface sampling (e.g., swabs and contact plates) should be used.

Using highly contained docking solutions (e.g., split butterfly valve) enables the safe aseptic transfer of the vaccine powder from the cyclone exit to a collector container (GEA 2016; ChargePoint Technology 2019; Dunne 2020).

Table 7.3 lists several aseptic powder filling systems available on the market that allow for high throughputs in the milligram range of powder (1 to 50 mg filling weight) with robotics, helping the breakthrough of aseptic spray drying applications. Leading companies are IMA Life (Italy), ROMACO (Italy), Syntegon Technology GmbH (formerly Bosch Packaging Technology GmbH, Germany), NKP Pharma (India), Cazzoli (USA), Harro Höfliger (Germany), Dec Group (Switzerland), and 3P Innovation (UK).

For injectable vaccines, an aseptic spray drying system is a must-have criterion for maintaining sterile conditions during powder production and obtaining a dry powder suitable for clinical use without terminal sterilization (Siew 2016). On the other hand, oral, pulmonary, or nasal vaccines require controlled processing with low bioburden, and sterility in manufacturing is not a prerequisite.

The product range of aseptically designed cGMP spray dryers used for vaccine research and production is limited, as shown in Chap. 4. SPX Anhydro (US) (SPX FLOW Inc. 2017; Geng 2018) (Fig. 7.10) and GEA Niro (DK) play a pioneering role and have the most experience with spray drying of vaccines.

GEA Niro® ASEPTICSD™ spray dryers (GEA Niro 2017) or Anhydro™ MicraSpray® Aseptic spray dryers (SPX FLOW Inc. 2017) are designed for processing heat-sensitive biopharmaceutical products, such as peptides, proteins, vaccines, antibiotics, hormones, allergens, and others requiring aseptic processes.

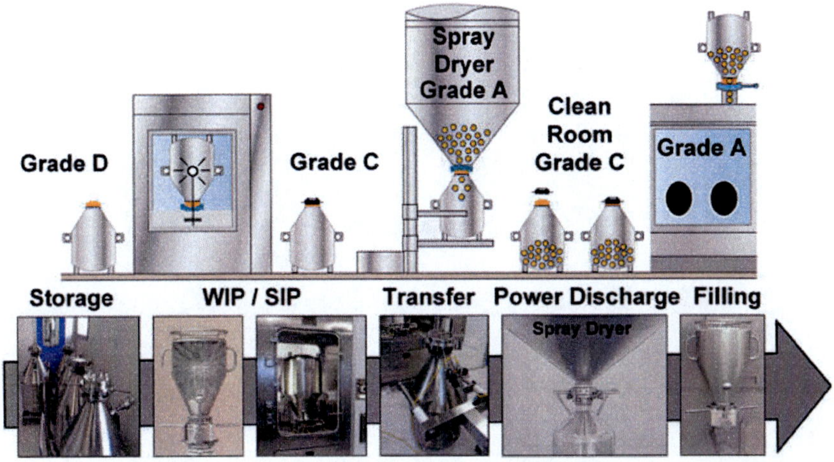

Fig. 7.9 Aseptic powder transfer from the spray dryer discharge point to a powder filling line via charge bottles. Empty bottles are stored in standard Grade D (Class 100,000, ISO 8) conditions, then washed (WIP, washing-in-place), sterilized (SIP, sterilization-in-place), and transferred to the spray dryer clean room in Grade C (Class 10,000, ISO 7) (Image courtesy of ChargePoint Technology Ltd., UK)

Basic requirements are initial plant cleaning and sterilization of product contact surfaces, maintaining sterility during the production of spray-dried powder, and aseptic extraction, filling, and packaging of the spray-dried powder (Turok and Rasmussen 2010; Schwartzbach 2011; Mogavero 2014; GEA Niro 2017; SPX FLOW Inc. 2017; Vehring et al. 2020).

In addition, GEA Niro offers several spray dryers for pharmaceutical application from pilot to production scale, the so-called Pharma-SD® (PSD™ types) and Mobile Minor™ models with nominal drying gas rates ranging from 100 to 5000 kg/h (GEA Process Engineering A/S 2020a). The Anhydro MicraSpray dryers MS 35 to MS 2500 from SPX FLOW Inc. (USA) provide drying gas flow rates from about 50 to 2500 kg/h, allowing water evaporation of up to 100 kg/h (SPX FLOW Inc. 2017) (see Chap. 4, Sect. 4.9).

In addition, there are several service providers and CDMOs (Contract Development and Manufacturing Organization), whereas Upperton (UK) and Nova Laboratories (UK) are particularly worth mentioning in the field of vaccines, and Hovione has a lot of experience in scaling up spray drying technology for inhalable drugs (see Chap. 4, Sect. 4.9).

Figure 7.11 shows the range of available spray dryers at Upperton, including a GEA Niro PharmaSD™ PSD-1, a GEA Niro SDMicro (GEA 2013), a ProCepT 4M8-TriX, and several Büchi Mini Spray Dryer B-290 models covering batch sizes from a few milligrams to several kilograms of powder.

Table 7.3 A selection of aseptic powder filling systems available on the market for small fill weights

Company (Country)	Aseptic powder filling systems	Min. fill weight (Accuracy, RSD)	Max. throughput (vials per hour)	Remarks	References
ROMACO (Italy)	Macofar MicroMaxX, MicroRobot 50	20 to 30 mg (± 2%)	Up to 24,000	Dual dosing disc, vacuum-pressure technology	ROMACO (2018), Cattenati (2019)
IMA Life (Italy)	SMART PWD, Microfill, MD	50 mg (± 2%)	7200, 9000 to 24,000	Double-dosing disks filling up to 4 vials	IMA Life (2012, 2020)
Syntegon (Germany)	AFG 5000	50 mg (± 7.5%)	9600 to 28,800	100% in-process control of dosing weights	Syntegon (2021)
NKP (India)	NKPF	50 mg (± 1%)	3600 to 14,400	Single to four doses	NKP (2020)
Cazzoli (USA)	PF	50 mg (± 1%)	Up to 18,000	2 to 16 dosing heads	Cazzoli (2019)
Harro Höfliger (Germany)	Microvibe SDM (auger dosing)	1 mg (< 3%)	600 to 4800	Dosing time < 2 s for ~10 mg	Höfliger (2021)
		50 mg (< 3%)	1200 to 4800	Dosing time < 2 s for 500 mg	
Dec Group (Switzerland)	µPTS	< 1 mg (± 2%)	900 to 3600	Single head/cavity can be parallelized	Dec Group (2021)
3P Innovation (UK)	Fill2Weight	1 mg (± 3%)	1800	Dosing time: around 2 s	3P Innovation (2020)

Fig. 7.10 Aseptic spray drying system MicraSpray MS-1250 from SPX Anhydro (Image courtesy of SPX FLOW Inc., USA)

Figure 7.12 illustrates the aseptic spray drying facility at Nova Laboratories Ltd. (UK) based on a GEA Niro ASEPTICSD™ size ASD-1, capable of producing vaccine batches of 5000 to 10,000 doses (GEA 2021; Nova Laboratories 2021). The plant was installed within a sterile Class 100 isolator (Grade A) (Roser 2009) located in a separate Grade D room (Roser 2005). Pharmaceutical-grade nitrogen is used as drying and atomization gas. The spray drying facility has undergone numerous validation cycles and is available for contract manufacturing (Nova Laboratories 2021). Pharmaceutical biologics can be spray dried continuously for up to 5 days to manufacture large kilogram quantities of powder (De Costa 2014).

7.4 Spray Dryer Models Used in Vaccine Research

Most laboratory research studies on spray-dried vaccines (about 76% according to Fig. 7.13) have been conducted on a Mini Spray Dryer B-290, B-191, or B-190 (Buchi Labortechnik AG, Switzerland). For the pilot scale, a Mobile Minor is used (GEA Process Engineering A/S 2020b). These units are similar to production-scale units in configuration and construction materials (e.g., stainless steel, CIP, pressure nozzles). In addition, a few researchers used custom-built spray dryers (Zhu et al. 2014; Jones et al. 2017; Gomez et al. 2021a, b, c; Gomez and Vehring 2022) or

Fig. 7.11 Spray dryer equipment at Upperton Ltd. (Nottingham, UK) for comprehensive services from proof-of-concept studies in the laboratory through drug development and manufacturing for clinical trials. **a** GEA Niro PharmaSDTM PSD-1 spray dryer, **b** GEA Niro SDMicro spray dryer, **c** Büchi Mini Spray Dryer B-290 Advanced, **d** ProCepT 4M8-TriX spray dryer (Images courtesy of Upperton Pharma Solutions, UK)

ultrasonic spray nozzles (Ohtake et al. 2010; Truong-Le et al. 2012, 2014; Lovalenti et al. 2016) to achieve particles with a narrow size distribution. Further options are using a high-performance cyclone, optimized for small powder quantities and particles < 5 µm, and a Dehumidifier B-296 for conditioning the inlet drying air humidity.

On a laboratory scale, offline sterilization in an autoclave or oven is feasible. The dismountable parts can be cleaned, sterilized, and then reassembled under sterile conditions inside a Class 100 (ISO 5) isolator system or a biosafety cabinet (Biosafety Level BL-2 or BL-3) (DiBerardinis et al. 2013) capable of maintaining sterility during spray drying (Gladd 2014).

Figure 7.14a shows a Mini Spray Dryer B-290 installed inside a BioPROTECT® III Jr. biosafety cabinet (The Baker Company 2010) and Fig. 7.14b inside an enclosure designed for active pharmaceutical ingredients (API) processing. The enclosure effectively contains compounds with Occupational Exposure Limits (OEL) of less than 1 µg/m^3 (Flow Sciences Inc. 2020). It also provides space for weighing and dissolving the samples in various solvents before spray drying. For example, the setup in Fig. 7.14a was used for the aseptic preparation of thermostable BCG dry powder vaccine (Price et al. 2020), virus-like particle vaccine against Human Papilloma Virus (HPV) (Tumban et al. 2015; Saboo et al. 2016),

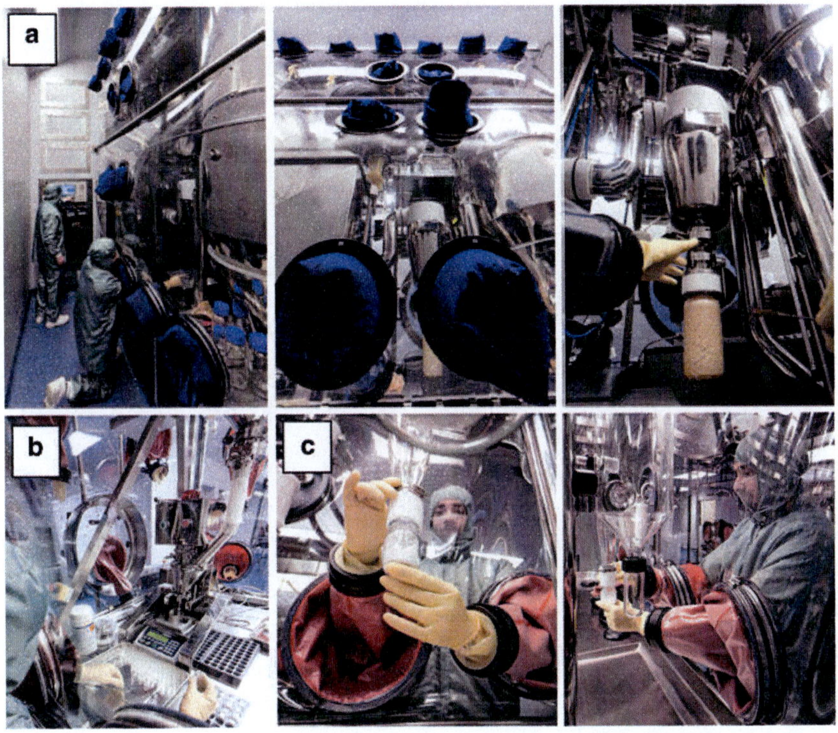

Fig. 7.12 **a** Aseptic spray dryer on a commercial scale with powder collection. **b** Aseptic powder filling within a sterile and low-humidity environment for manual filling into vials. **c** Small-scale spray dryer in enclosure (Images courtesy of Nova Laboratories Ltd., Leicester, UK)

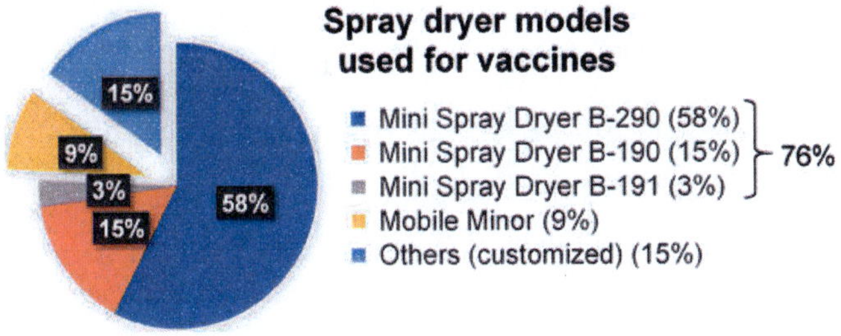

Fig. 7.13 Spray dryer models used for vaccines. Around 76% used Mini Spray Dryers B-290, B-190, and B-191

and live attenuated bacterial vaccine (Listeria monocytogenes) against tularemia (Kunda et al. 2016). In addition, there are also options for working with the Mini Spray Dryer B 290 under "low bioburden" conditions (see more details in Chap. 4, Sect. 4.8).

Fig. 7.14 **a** Buchi Mini Spray Dryer B-290 installed in a BioPROTECT® III Jr. containment enclosure to protect the operator and the dry powder vaccine from contamination (Images courtesy of Prof. Pavan Muttil from the University of New Mexico and The Baker Company, USA). **b** Enclosure for processing active pharmaceutical ingredients with a Mini Spray Dryer B-290. The enclosure effectively contains compounds with Occupational Exposure Limits of less than 1 μg/m^3. It provides space for weighing and dissolving the samples in various solvents (Image courtesy of Flow Sciences Inc., USA)

7.5 Scale-Up of Spray-Dried Vaccines

The transfer of the spray drying process from the laboratory to pilot and commercial scale has been demonstrated in some vaccine examples. Table 7.4 summarizes scale-up studies by vaccine type, spray dryer model, scale-up factor, and temperature conditions (Johansen et al. 2000; Jin et al. 2010; Chen et al. 2010; Jin and Tsao 2013; Bowen et al. 2013; Zhu et al. 2014; Gikanga et al. 2015; Tlaxca et al. 2015).

An important scale-up criterion is to keep the water vapor concentration at the spray dryer outlet constant across scales, as it correlates with the residual moisture content in the powder (Arpagaus and Schwartzbach 2008; Dobry et al. 2009, 2015; DuBose et al. 2013; Costa et al. 2014; Gaspar et al. 2014; Al-Khattawi et al. 2018). For this purpose, the outlet drying gas temperature is controlled by adjusting the inlet temperature, feed rate, and drying gas flow rate (see Chap. 3, Sect. 3.4).

For example, with a batch size of approximately 1 L (11.2 wt% solids in feed) and a high yield of > 90%, a pilot-scale Mobile Minor (GEA Niro) could produce sufficient powder for 5000 to 10,000 doses of hepatitis B subunit vaccine (Chen et al. 2010). In addition, storage stability based on bioassays for up to 2 years at 37 °C was achieved.

Table 7.4 Scale-up studies of spray-dried vaccines

Vaccine type	Spray dryer models	Scale-up factor	Inlet/outlet temperature (yield)	References
Influenza antigen HAC1	B-290 → BLD-1	100× (0.5 g → 50 g) 6.7× (36 mL/h → 240 mL/h)	100 °C/58 to 62 °C (> 90%)	Zhu et al. (2014)
Respiratory syncytial virus	B-190 → MS75	10× (1 mL/min → 10 mL/min)	n.a./50 to 60 °C (> 80%)	Tlaxca et al. (2015)
Tuberculosis	B-290 → MS35	4× (15 g/h → 60 g/h powder production)	65 to 125 °C/34 to 50 °C (70 to 85%)	Jin et al. (2010), Jin and Tsao (2013)
Tetanus toxoid	B-191	10× (1 g → 10 g powder batch)	45 °C/n.a. (ethyl formate) (yield n.a.)	Johansen et al. (2000)
Hepatitis B, Meningitis A	Mobile Minor	1-L batch providing 5000 to 10,000 vaccine doses	90 to 95 °C/ 60 to 71 °C (> 90%)	Chen et al. (2010)
Monoclonal antibodies	MS35 → MS150	3.3× (15 mL/min → 50 mL/min)	180 to 190 °C/ 80 to 90 °C (> 95%)	Bowen et al. (2013), Gikanga et al. (2015)

B-190/B-191/B-290: Mini Spray Dryers (Büchi), BLD-1: custom-made pilot, MS35/MS75/MS150: MicraSpray Dryer (SPX), Mobile Minor (GEA)

7.6 Design of Experiments (DOE) and Quality by Design (QbD)

Design of experiments (DOE) (Kanojia et al. 2017; Kanojia 2018) and Quality by Design (QbD) approaches (Baldinger et al. 2012; Gaspar et al. 2014; Andersen 2018; Scrafford 2020; Baumann et al. 2021) for spray drying of vaccines enable rapid determination of optimal process conditions and critical/non-critical parameters, thereby reducing costs and material consumption.

As discussed in Chap. 3, several studies have used the DOE approach to optimize the operating parameters of vaccine spray drying, including:

- Tuberculosis (Price et al. 2020),
- Poliovirus (Kanojia et al. 2018b),
- Influenza virus (Lovalenti et al. 2016; Kanojia et al. 2016),
- Rotavirus (Gill et al. 2017),
- Human type 5 adenoviral vector (LeClair et al. 2016a),
- Herpes simplex virus (HSV) (LeClair et al. 2019),
- Human papillomavirus (HPV) (Saboo et al. 2016; Peabody et al. 2017),
- Pneumococcus (Anish et al. 2014), and
- others described in (Arpagaus 2022).

The primary response parameters are particle size, residual moisture content, vaccine activity (in-vitro), immune response (in-vivo), and storage stability. Related or secondary parameters include powder yield, particle morphology, and glass transition temperature.

For example, Fig. 7.15 shows the predicted design space with optimal spray drying parameters for a trehalose-based whole inactivated influenza virus vaccine. The green zone ("Sweet Spot") provides inhalable particles (1 to 5 μm size), powder yield > 70%, outlet drying gas temperature < 100 °C, residual moisture content between 1 and 3 wt%, and no significant loss in vaccine antigenicity (in-vitro) during storage for 12 months at 2 to 8 °C (Kanojia et al. 2016). Moreover, the optimized influenza vaccine powder retained its antigenic stability even after storage at 60 °C for 3 months.

An Ishikawa (fishbone) diagram (Fig. 7.16) can serve as a basis for a risk analysis that allows a cross-functional team of experts to collaboratively identify potential sources of variability that may impact desired product quality. For example, certain factors can be kept constant, such as the designed spray drying equipment, water as a solvent, or the humidity of the inlet drying gas, while others, such as process and formulation parameters, have to be optimized depending on the vaccine to meet the targeted product profile. Possible factors influencing the outcome of a spray-dried dry powder vaccine include equipment design, process parameters, product characteristics, analytical methods, preparation method, formulation parameters, post-production, and plant factors.

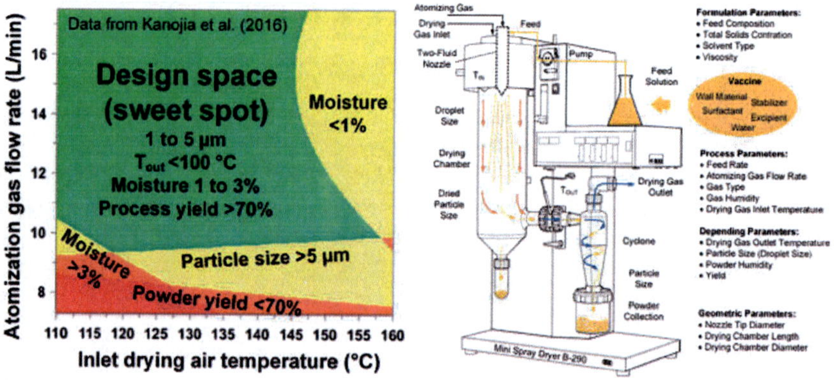

Fig. 7.15 Predicted design space for whole inactivated influenza virus vaccine stabilized with trehalose in a Mini Spray Dryer B-290 (Image courtesy of Büchi Labortechnik AG, Switzerland). Feed flow rate (1.0 mL/min) and feed concentration (100 mg/mL) were fixed. The design space criteria were 1 to 5 μm particle size, < 100 °C outlet temperature (T_{OUT}), 1 to 3 wt% residual moisture content, and > 70% process yield. In the green zone, all criteria are met (sweet spot) (Data adapted from Kanojia et al. (2016), open access under CC-BY 4.0 license)

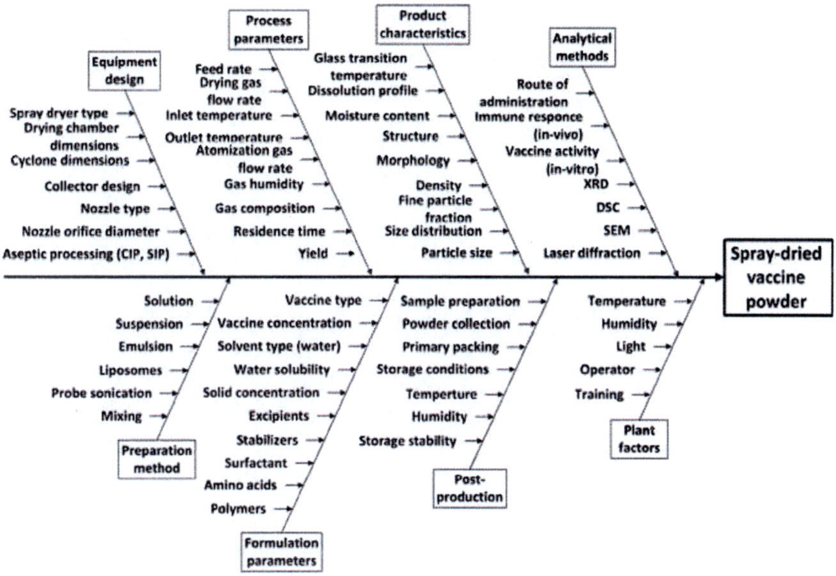

Fig. 7.16 Ishikawa (fishbone) diagram to visualize the factors affecting a spray-dried vaccine powder quality. Researchers can use the chart for risk analysis and evaluation of the impact of critical process parameters (CPPs) on the critical quality attributes (CQAs) (Adapted from various literature sources (ICH 2009; Baldinger et al. 2012; Ingvarsson et al. 2013; Karimi et al. 2016; Benjasirimongkol et al. 2019; Dormenval et al. 2019; Soni et al. 2020; Ignjatović et al. 2021))

7.7 Application Spectrum of Spray-Dried Vaccines

Over the past 25 years, a well-established body of literature has been published on the spray drying of vaccines. Chapter 6 provides a review of the applications structured by disease. Figure 7.17a shows the development of the number of publications based on the Web of Science and PubMed using the search terms ALL FIELDS [spray drying AND vaccine]. The search yielded 270 hits on the Web of Science and 189 hits in PubMed (accessed on 8 August 2022). Over time, there has been a steady increase in publications, indicating a lively research activity, with about 15 to 20 additional publications per year.

Table 7.5 summarizes the main research articles on spray-dried vaccines used prophylactically. As can be seen, the range of applications is quite broad and includes vaccines against various diseases:

- Tuberculosis (TB) (Wong et al. 2007; Lu et al. 2007, 2010; Garcia-Contreras et al. 2008; Fourie et al. 2008; Morello et al. 2009; Jin et al. 2010; Tyne et al. 2013; LeClair et al. 2016a, b; Afkhami et al. 2017; Thakur et al. 2018; Price et al. 2020; Gomez et al. 2021b, a, c, 2022; Gomez and Vehring 2022),

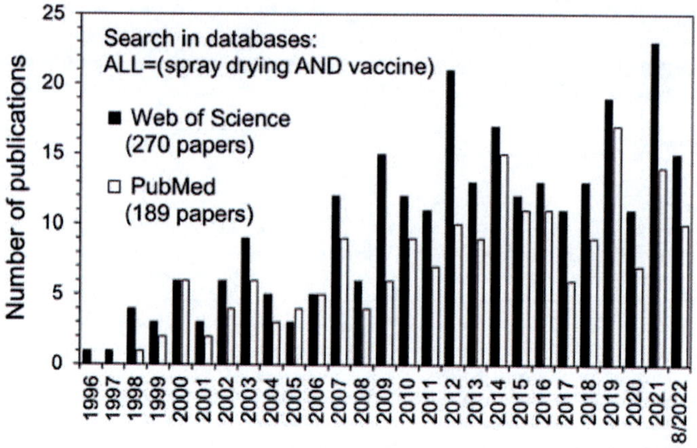

Fig. 7.17 Number of publications identified in the Web of Science and PubMed online databases using the search term ALL FIELDS [spray drying AND vaccine] (accessed on 8 August 2022)

- Influenza (Bot et al. 2000, 2001; Smith et al. 2003; Amorij 2007; Abdul-Fattah et al. 2007a, b; Amorij et al. 2008; Saluja et al. 2010; Shastri et al. 2012; Ajmera and Scherließ 2014; Scherließ et al. 2014; Zhu et al. 2014; Sou et al. 2015; Lovalenti et al. 2016; Kanojia et al. 2016),
- Diphteria, tetanus, pertussis, Hib (DTaP/Hib) (Gander et al. 1995b, a, 1996; Johansen et al. 1999a, b, 2000; Boehm et al. 2002; Muttil et al. 2010b; Kanojia et al. 2018a),
- measles (De Swart et al. 2007; Ohtake et al. 2010; Kisich et al. 2011; Lin et al. 2011; Agarkhedkar et al. 2014; Gala et al. 2017; Kanojia et al. 2017; Putranto and Chen 2019),
- Newcastle disease virus (NDV) (Corbanie 2007; Corbanie et al. 2007, 2008; Huyge 2012; Huyge et al. 2012; Landman et al. 2015, 2017),
- Hepatitis B (HepB) (Maa et al. 2003, 2007; Chen et al. 2010; Muttil et al. 2010a; Bhowmik et al. 2012; Tomar et al. 2019)
- Human papillomavirus (HPV) (Tumban et al. 2015; Saboo et al. 2016; Peabody et al. 2017; Kunda et al. 2019; Garcea et al. 2020; Rossi et al. 2021),
- Rotavirus (Gill 2014, 2018; Ohtake et al. 2015; Gill et al. 2017, 2020),
- Pneumococcal disease (Anish et al. 2014; Kunda et al. 2015; Rodrigues et al. 2018; Mehanny et al. 2022),
- Cholera (Año et al. 2011; Pastor et al. 2013b, 2014),
- Typhoid (Bejugam et al. 2008; Uddin et al. 2009; Ohtake et al. 2011),
- Vesicular stomatitis virus (VSV) (Toniolo et al. 2019, 2020),
- Meningococcal disease (Chen et al. 2010; Ubale et al. 2013),
- Norovirus (Velasquez et al. 2011; Vodak et al. 2012),
- Tularemia (Kunda et al. 2016),
- Anthrax (Jones et al. 2017),
- Poliovirus (Polio) (Kanojia et al. 2018b)

Table 7.5 Analyzed diseases and published research articles on spray-dried vaccines from the scientific literature. Search in Web of Science, PubMed, SCOPUS, and Google Scholar using the keywords [spray drying] AND [vaccine]

Disease	No. of articles	References on spray-dried vaccines
Tuberculosis (TB)	18	Wong et al. (2007), Lu et al. (2007, 2010), Garcia-Contreras et al. (2008), Fourie et al. (2008), Morello et al. (2009), Jin et al. (2010), Tyne et al. (2013), LeClair et al. (2016a, b), Afkhami et al. (2017), Thakur et al. (2018), Price et al. (2020), Gomez et al. (2021a, b, c, 2022), Gomez and Vehring (2022)
Influenza	14	Bot et al. (2000, 2001), Smith et al. (2003), Amorij (2007), Abdul-Fattah et al. (2007a, b), Amorij et al. (2008), Saluja et al. (2010), Shastri et al. (2012), Ajmera and Scherließ (2014), Scherließ et al. (2014), Zhu et al. (2014), Sou et al. (2015), Lovalenti et al. (2016), Kanojia et al. (2016)
Diphteria, tetanus, pertussis, Hib (DTaP/Hib)	9	Gander et al. (1995a, b, 1996), Johansen et al. (1999a, b, 2000), Boehm et al. (2002), Muttil et al. (2010b), Kanojia et al. (2018a)
Measles	8	De Swart et al. (2007), Ohtake et al. (2010), Kisich et al. (2011), Lin et al. (2011), Agarkhedkar et al. (2014), Gala et al. (2017), Kanojia et al. (2017), Putranto and Chen (2019)
Newcastle disease virus (NDV)	7	Corbanie (2007), Corbanie et al. (2007, 2008), Huyge (2012), Huyge et al. (2012), Landman et al. (2015, 2017)
Hepatitis B (HepB)	6	Maa et al. (2003, 2007), Chen et al. (2010), Muttil et al. (2010a), Bhowmik et al. (2012), Tomar et al. (2019)
Human papillomavirus (HPV)	6	Tumban et al. (2015), Saboo et al. (2016), Peabody et al. (2017), Kunda et al. (2019), Garcea et al. (2020), Rossi et al. (2021)
Rotavirus	5	Gill (2014, 2018), Ohtake et al. (2015), Gill et al. (2017, 2020)
Pneumococcal disease	4	Anish et al. (2014), Kunda et al. (2015), Rodrigues et al. (2018), Mehanny et al. (2022)
Cholera	3	Año et al. (2011), Pastor et al. (2013b, 2014)
Typhoid	3	Bejugam et al. (2008), Uddin et al. (2009), Ohtake et al. (2011)
Vesicular stomatitis virus (VSV)	2	Toniolo et al. (2019, 2020)
Meningococcal disease	2	Chen et al. (2010), Ubale et al. (2013)
Norovirus	2	Velasquez et al. (2011), Vodak et al. (2012)
Tularemia	1	Kunda et al. (2016)
Anthrax	1	Jones et al. (2017)
Poliovirus (Polio)	1	Kanojia et al. (2018b)
Herpes simplex virus (HSV)	1	LeClair et al. (2019)
Human immunodeficiency virus (HIV)	1	Amacker et al. (2020)
Total	76	Scientific articles

- Herpes simplex virus (HSV) (LeClair et al. 2019), and
- Human immunodeficiency virus (HIV) (Amacker et al. 2020).

According to the number of publications, TB, influenza, DTaP/Hib, measles, NDV, HepB, and HPV received the highest research activity. In addition, numerous reviews and book chapters have been published on vaccine stabilization by spray drying that are worth reading (Chen and Kristensen 2009; Searles and Mohan 2010; Kristensen 2012; McAdams et al. 2012; Kanojia et al. 2017; Kanojia 2018; Putranto and Chen 2019; Langford and Ohtake 2019; Ziaee et al. 2019; Arpagaus 2020; Vehring et al. 2020; Pinto et al. 2021; Preston and Randolph 2021; Gomez and Vehring 2022; Ghaemmaghamian et al. 2022).

On the other hand, for some diseases, there are only a few articles on spray-dried vaccines (e.g., VSV, meningococcal, norovirus, tularemia, anthrax, poliovirus, herpes simplex virus, HIV). Reasons may be that certain vaccines are already available in a freeze-dried form or that the research results have only recently been published (e.g., HIV) or are treated proprietarily.

Table 7.6 summarizes the most relevant findings structured by vaccine platform, excipients (formulation), process parameters, spray dryer model, drying temperatures, powder properties, route of administration, animal model, immunogenicity, and storage stability (shelf-life). More details on each disease have been reviewed and discussed in Chap. 6. It is noted that the major research activities are ongoing in various countries, especially in the USA, Canada, the Netherlands, the UK, India, Denmark, Germany, Australia, Belgium, South Africa, Spain, Cuba, Italy, and Switzerland.

Overall, the main motivation drivers for researching spray drying of vaccines are:

- improving stability in dry powder form and extending shelf life (e.g., storage at room temperature),
- particle technology for needle-free administration (non-invasive, no need for reconstitution) via the lungs (inhalable vaccine), nose (nasal spray), skin, and oral administration (as capsules or tablets),
- improving efficacy as an alternative to traditional freeze-dried vaccines,
- increasing the cost-effectiveness of bulk applications for mass vaccination (e.g., of poultry), and
- easier dosing without water or buffer for mixing components in multivalent vaccines and flexible packing containers.

7.8 Excipients Used for Improving Spray-Dried Vaccine Stability

A key element in stabilizing vaccines by spray drying and for long-term storage is a formulation with suitable excipients depending on the vaccine type, like subunit proteins, live attenuated viruses, live attenuated bacteria, vectors, virus-like particles, etc., and the route of administration (see Table 7.6).

Table 7.6 Overview of research studies with spray-dried vaccines (in alphabetic order)

Vaccines	Vaccine platform	Excipients, formulation	Spray dryer model, T_{in}/T_{out}	Particle size, residual moisture	Animal model, route of administration	Shelf-life/Stability (Storage conditions)	References (Country)
Anthrax	Recombinant (protein rPA, pp-dPA83)	Trehalose, hydrolyzed gelatin, salts, Alhydrogel, Tween 80	BLD-35 (Custom build lab-scale), 100 to 120 °C/58 °C	1 to 10 μm, 3.8 to 4.7 wt%	Mice (reconstituted, intramuscular)	3 months (40 °C/75% RH) (bioassay and immunogenicity)	Jones et al. (2017) (USA)
Cholera	Inactivated Vibrio cholerae bacteria	Eudragit® L30 D-55, alginic acid, Carbopol®, triethyl citrate	Mini Spray Dryer B-191, 80 °C/60 °C	7 to 9 μm, n.a	Rats (dry, oral)	12 months (25 °C/60% RH), 6 months (40 °C/75% RH) (antigenicity)	Pastor et al. (2013a) (Spain, Cuba)
Hepatitis B (HepB)	Recombinant (surface antigen HBsAg)	Trehalose, mannitol, NaH_2PO_4	Mobile Minor, 90 to 95/61 to 65 °C	5 to 10 μm, 2.4 to 4.1 wt%	Mice (reconstituted, intraperitoneal injection)	24 months (37 °C) (bioassay), 15 months (37 °C) (immunogenicity)	Chen et al. (2010) (USA, India, UK, Denmark)
Herpes simplex virus 2 (HSV-2)	Live-attenuated	Trehalose or sucrose	Mini Spray Dryer B-290 with HP cyclone, 105 °C/n.a	n.a 3.6 wt%	In vitro (*Vero cells*)	10 days at 45 °C (trehalose) (< 1.5 log activity loss)	LeClair et al. (2019) (Canada)
Human immune-deficiency virus (HIV)	Recombinant virosomes (HIV-1 gp41 protein and 3 M-052 adjuvant on surface)	Trehalose, sodium alginate	n.a., 85 to 90 °C/60 °C	> 9 to 10 μm, 3 wt% (nasal), about 5 μm, 2.5 wt% (oral)	Rats (nasal dry powder, oral dry powder)	3 months (40 °C/75% RH) (immunogenicity)	Amacker et al. (2020) (Switzerland, The Netherlands, UK, Germany)
Human papilloma virus (HPV)	Recombinant virus-like particles (bacteriophage L2 VLPs)	Mannitol/dextran and L-leucine/trehalose	Mini Spray Dryer B-290, 155 °C/55 °C	2.7 to 5.0 μm, 1.0 to 2.5 wt%	Rats (dry, oral, in gelatin capsules), Mice (reconstituted, intramuscular)	14 months (37 °C) (immunogenicity)	Saboo et al. (2016) (USA)
	Recombinant virus-like particles (bacteriophage L2 VLPs)	Mannitol, trehalose, dextran, L-leucine, inositol	Mini Spray Dryer B-290, 140 °C/47 ± 2 °C	2 to 4 μm, n.a	Mice (reconstituted, intramuscular)	34 months (25 °C), 7.5 months (25 °C), > 1 month (37 °C) (immunogenicity)	Tumban et al. (2015), Peabody et al. (2017) (USA)

(continued)

Table 7.6 (continued)

Vaccines	Vaccine platform	Excipients, formulation	Spray dryer model, T_{in}/T_{out}	Particle size, residual moisture	Animal model, route of administration	Shelf-life/Stability (Storage conditions)	References (Country)
	Recombinant (Gardasil® 9 (L1 VLPs))	L-Leucine, mannitol, trehalose, *myo*-inositol, dextran, PVP, Gardasil®9 liquid suspension	Mini Spray Dryer B-290, 145 ± 10 °C/ 47 ± 3 °C	n.a	Mice (reconstituted, intramuscular)	3 months (40 °C/60% RH) (immunogenicity, protected from a challenge)	Kunda et al. (2019) (USA)
	Recombinant (model HPV16 L1 capsomeres)	Histidine, trehalose, hydroxyethyl starch, NaCl, Tween 80	Mini Spray Dryer B-290 with HP cyclone, n.a	2 to 4 µm, 5 wt%	Mice (reconstituted, intramuscular)	1 month (50 °C) (immunogenicity)	Garcea et al. (2020) (USA)
	Recombinant (PfTrx-HPV-L2 antigen)	Mannitol, glucopyranosyl lipid A	Mini Spray Dryer B-290, 125 °C/n.a	1.2 to 5.3 µm, n.a	Mice (dry, intratracheal with insufflator)	5 months (25 °C)	Rossi et al. (2021) (Italy, Germany)
Influenza	Live (viral vector)	Trehalose and trehalose/dextran	Mini Spray Dryer B-290, 110 °C/n.a	n.a	In vitro (*bioassay*)	30 days (37 °C/< 10% RH) (~15% activity retention)	Toniolo et al. (2019) (Canada)
	Inactivated virus	Trehalose	Mini Spray Dryer B-290 with Dehumidifier B-296, 110 to 160 °C/48 to 91 °C	2.4 to 8.5 µm, 1.2 to 4.9 wt%	In vitro (*bioassay*)	3 months (60 °C) (86% to 100% activity preserved)	Kanojia et al. (2016) (The Netherlands)
	Inactivated virus	Trehalose, L-leucine	Mini Spray Dryer B-190, n.a./70 °C	< 5 µm (49.3%), 3.2 to 4.6 wt%	Rats (pulmonary liquid instillation, subcutaneous)	2 months (40 °C) (immunogenicity)	Sou et al. (2015) (Australia)
	Recombinant (HAC1 hemagglutinin protein)	Trehalose, hydrolyzed gelatin	Mini Spray Dryer B-290, 100 °C/58 to 62 °C, BLD-1 (pilot-scale), ~100 °C/58 °C	3.4 to 7.2 µm, ~ 4 wt%, 0.5 to 30 µm, 4.0 ± 0.5 wt%	Mice (reconstituted, intramuscular)	6 months (50 °C) (bioassay), 3 months (50 °C) (immunogenicity)	Zhu et al. (2014) (USA)

(continued)

Table 7.6 (continued)

Vaccines	Vaccine platform	Excipients, formulation	Spray dryer model, T_{in}/T_{out}	Particle size, residual moisture	Animal model, route of administration	Shelf-life/Stability (Storage conditions)	References (Country)
	Live-attenuated virus	Trehalose, sucrose, Pluronic F-68, sorbitol, histidine, ZnCl2	Mini Spray Dryer B-190 with ultrasonic nozzle, ~60 °C/~45 °C	n.a., 1 to 4 wt%	Ferrets (intranasal instillation)	> 1 log titer loss after 1 week (37 °C) and 4 weeks (25 °C)	Lovalenti et al. (2016) (USA)
	Live-attenuated virus	Sucrose, surfactant (Pluronic F68)	Mini Spray Dryer B-190 with a custom-made high-pressure nozzle, 40 °C/30 °C	n.a., 4 to 5 wt% (1 to 2 wt% after vacuum drying)	In vitro *(bioassay)*	20 weeks (25 °C), 1 week (37 °C) (bioactivity)	Abdul-Fattah et al. (2007a, b) (USA)
	Inactivated influenza A split virus	Trehalose, different amino acids (e.g., arginine, glycine), glycyrrhizinic acid, chitosan	Mini Spray Dryer B-290 with HP cyclone, 120 °C/50 to 55 °C	12.8 µm, 7 to 8 wt%	Macaques (reconstituted, intramuscular)	3 months (25 °C/60% RH) (immunogenicity)	Ajmera and Scherließ (2014), Scherließ et al. (2014) (Germany, UK)
	Recombinant (subunit protein HA)	Inulin	Mini Spray Dryer B-190, 130 °C/70 °C	1 to 6 µm, < 5 µm (37%), n.a	Mice (pulmonary dry powder insufflation)	> 3 years (20 °C/10% RH) (bioassay, immunogenicity)	Saluja et al. (2010) (The Netherlands)
Measles	Live attenuated virus	Trehalose, sucrose, human serum albumin, L-arginine	Mini Spray Dryer B-190 with ultrasonic nozzle, 60 °C/40 °C	5 to 25 µm, ~3 wt%	In vitro *(bioassay)*	8 weeks (37 °C) (< 1 log titer loss)	Ohtake et al. (2010), Truong-Le et al. (2012, 2014) (USA)
	Live attenuated virus	Myo-inositol, sorbitol, L-arginine	CAN-BD (CO_2 assisted nebulization with Bubble Dryer®), n.a/50 °C	< 5.8 µm (46%), 0.7 wt%	Cotton rats, rhesus macaques, human clinical study Phase I (pulmonary dry powder)	> 2 years (2 to 8 °C), 1 week (37 °C)	Kisich et al. (2011), Lin et al. (2011), Agarkhedkar et al. (2014) (USA, India)

(continued)

Table 7.6 (continued)

Vaccines	Vaccine platform	Excipients, formulation	Spray dryer model, T_{in}/T_{out}	Particle size, residual moisture	Animal model, route of administration	Shelf-life/Stability (Storage conditions)	References (Country)
Meningitis A (MenA)	Protein-polysaccharide conjugate	Trehalose, $AlPO_4$	Mobile Minor, 90 to 95 °C/60 to 71 °C	3.1 to 4.3 µm, 2.2 to 3.4 wt%	Mice (Intramuscular)	20 weeks (40 °C), 2 weeks (60 °C)	Chen et al. (2010) (USA, India, UK, Denmark)
Newcastle disease (NCD)	Clone 30 strain	Trehalose, PVP, BSA, mannitol	Mobile Minor, 160 °C/70 °C	28 µm and 7 µm, n.a	Broiler chickens (pulmonary in isolators)	10 months at 25 °C (bioassay)	Corbanie (2007), Corbanie et al. (2007, 2008) (Belgium, The Netherlands)
Poliomyelitis (Polio)	Sabin inactivated polio vaccine (sIPV)	Trehalose, monosodium glutamate, maltodextrin, arginine, cell culture medium M199, Pluronic	Mini Spray Dryer B-290 with HP cyclone and Dehumidifier B-296, 110 °C/48 to 53 °C	5 to 7 µm, 1 to 7 wt%	In vitro *(bioassay)*	1 week (40 °C)	Kanojia et al. (2018b) (The Netherlands)
Rotavirus	Live attenuated	Trehalose, maltodextrin, NaCl, glutamic acid, histidine, $CaCl_2$, and HEPES buffer	Mini Spray Dryer B-290 with HP cyclone and Dehumidifier B-296, 70 to 75 °C/43 to 48 °C	1 to 24 µm, < 2 wt%	In vitro *(bioassay)*	4 weeks (45 °C), 24 weeks (37 °C), 24 months (2 to 8 °C)	Gill et al. (2020) (India)
Tuberculosis (TB)	Live attenuated (Calmette-Guérin BCG bacteria)	L-leucine, mannitol, PVP, BSA, trehalose	Mini Spray Dryer B-290 in a biosafety cabinet, 150 °C/50 °C	2.0 to 5.3 µm, 0.6 to 1.5 wt%	Mice (reconstituted, intradermal)	1 year (25 °C/ 60% RH) (viability assay), 2 years (25 °C) (immunogenicity) (in protective packing)	Price et al. (2020) (USA)

(continued)

Table 7.6 (continued)

Vaccines	Vaccine platform	Excipients, formulation	Spray dryer model, T_{in}/T_{out}	Particle size, residual moisture	Animal model, route of administration	Shelf-life/Stability (Storage conditions)	References (Country)
	Live attenuated (BCG bacteria)	L-leucine	Mini Spray Dryer B-290 with HP cyclone, 100 to 125 °C/40 °C	2 to 3 μm, n.a	Guinea pigs and mice (pulmonary dry powder insufflation)	9 months (25 °C/20% RH) 4 months (25 °C/60% RH), (bacterial viability)	Wong et al. (2007), Garcia-Contreras et al. (2008), Fourie et al. (2008), Morello et al. (2009) (USA, South Africa)
	Live attenuated (human adenoviral AdHu5 vector)	Mannitol, dextran	Mini Spray Dryer B-290, 90 to 120/48 to 65 °C	< 10 μm	Mice (reconstituted, intramuscular)	3 months (20 °C/< 10% RH) (immunogenicity)	LeClair et al. (2016a, b), Afkhami et al. (2017) (Canada)
	Recombinant adenovirus (Ad35) vector (AERAS-402)	Mannitol, cyclodextrin, trehalose, dextran, L-leucine, sucrose, inositol, histidine, PVP, Tween 80	Mini Spray Dryer B-290, 65 to 125 °C/34 to 50 °C	1 to 5 μm, 1.45 wt%	In vitro (*bioassay*)	12 months (25 °C), 5 weeks (37 °C) (0.12 log virus titer loss)	Jin et al. (2010), Jin and Tsao (2013) (USA, The Netherlands)
	Recombinant (fusion protein ID93 and glucopyranosyl lipid adjuvant GLA in nanoemulsion)	Trehalose, trileucine, L-leucine, pullulan, squalene	Custom-designed research spray dryer, 65 to 70 °C/36 to 46 °C	1 to 20 μm, 1.8 to 2.6 wt%	In vitro (*bioassay*) Mice (reconstituted intramuscular, nasal/pulmonary dry powder)	26 months (40 °C) (chemical and physical stability), 1 year (50 °C) (45% antigen present)	Gomez et al. (2021a, b, c, 2022), Gomez and Vehring (2022) (Canada, USA)
Tularemia	Live attenuated bacteria (Listeria monocytogenes)	Mannitol, trehalose, dextran, leucine, inositol	Mini Spray Dryer B-290 in a biosafety cabinet 130 °C/47± 2 °C	2.2 ± 0.3 μm, 0.7 wt%	In vitro (*bioassay*)	1 year at 23 °C, 6 months at 40 °C/75% RH (bioassay) (stored in protective packaging)	Kunda et al. (2016) (USA)

(continued)

Table 7.6 (continued)

Vaccines	Vaccine platform	Excipients, formulation	Spray dryer model, T_{in}/T_{out}	Particle size, residual moisture	Animal model, route of administration	Shelf-life/Stability (Storage conditions)	References (Country)
Typhoid	Live attenuated Salmonella Typhi bacteria (Ty21a)	Sucrose	Mini Spray Dryer B-190, Sonicair nozzle, n.a./45 °C	n.a., 3.14 wt%	Mice (reconstituted, intramuscular)	4 weeks (25 °C)	Ohtake et al. (2011) (USA)
Vesicular stomatitis virus (VSV)	Live attenuated (viral vector)	Trehalose, dextran	Mini Spray Dryer B-290 with HP cyclone, 110 °C/n.a	n.a	Mice (reconstituted, intramuscular)	30 days (37 °C) (bioassay), 15 days (25 °C) (bioassay and immunogenicity)	Toniolo et al. (2019, 2020) (Canada)
Pertussis (Whooping cough)	Recombinant (outer membrane vesicles from B. pertussis B1917)	Trehalose	Mini Spray Dryer B-290 with HP cyclone and Dehumidifier B-296, 135 °C/n.a	5.1 µm, 3.3 wt%	Mice (pulmonary as reconstituted powder by aerosolizer)	4 weeks (40 and 65 °C) (bioassay)	Kanojia et al. (2018a) (The Netherlands)

Mini Spray Dryer B-290, B-191, B-190 from Büchi Labortechnik AG, Mobile Minor from GEA Niro, BCG: Bacillus Calmette-Guérin, BSA: bovine serum albumin, PVP: polyvinyl pyrrolidine, CAN-BD: CO_2 assisted nebulization with bubble dryer, rPA: recombinant protein antigen, T_{in}/T_{out}: inlet/outlet drying gas temperature, RH: relative humidity, VLP: virus-like particles, AdHu5: human adenovirus

Table 7.7 Excipients used for improving the stability of spray-dried vaccines

Category	Examples of excipients	Mechanism of action
Surfactants	Polysorbate 20 and 80, Triton X-100, Pluronic F68	Prevent protein aggregation at air–liquid interfaces, increase solubility, and surface coverage on droplets
Inorganic salts	NaCl, KCl, $MgCl_2$	Adjust tonicity, specific stabilization
Buffering agents	Phosphate buffer (KPO_4), succinate buffer, sodium borate	Maintain solution pH, modulation of antigen binding to aluminum salt adjuvants
Amino acids	L-glutamate, L-arginine, L-histidine, L-leucine, Trileucine, glycine, methionine	Specific interaction with proteins, charge shielding of protein surfaces, stabilizing the conformations of proteins, reducing aggregation, increasing dispersibility of dry powder, antioxidant, buffering agent
Sugars and polyols	Sucrose, trehalose, inulin, mannitol, dextran, maltodextrin, inositol, sorbitol, pullulan	Glassy state stabilization, water replacement, preferential hydration, low hygroscopicity, interfacial tension in lipid bilayers
Proteins	Human and bovine serum albumin, gelatin	Inhibition of protein aggregation by competing for surface coverage on the droplets and reducing the heterogeneity of protein distribution inside the powder particles
Antioxidants	Ethylenediaminetetraacetic acid (EDTA), citric acid, ascorbic acid, metabisulphate	Chelating agents, oxygen scavenger
Polymers	Eudragit, cellulose acetate phthalate, hydroxypropyl methylcellulose acetate succinate (HPMCAS), poly-lactic-co-glycolic acid (PLGA), polyvinylpyrrolidone (PVP)	Enteric coating polymers, encapsulating polymers

Table 7.7 gives some guidance on the main excipients used and their stabilization mechanism to improve the stability of spray-dried vaccines. The information is based on various literature (Maa et al. 1998; Adler and Lee 1999; Truong-Le and Pham 2007; Vehring 2008; Hasija et al. 2013; Ajmera and Scherließ 2014; Saboo et al. 2016; Lovalenti et al. 2016; Kunda et al. 2016; Kanojia et al. 2017; Ziaee et al. 2019; Arpagaus 2020; Vehring et al. 2020; Gomez et al. 2021b).

The range of excipients extends from simple and small buffer molecules to amino acids, sugars, proteins, and more complex polymers. Each ingredient plays an essential role in feed formulation and vaccine stabilization.

Surfactants (e.g., Polysorbate 20 and 80, Triton X-100, Pluronic F68) are added to prevent protein aggregation due to mechanical stresses during droplet atomization, compete for surface coverage of proteins on liquid droplets, and prevent adsorption to different surfaces.

Inorganic salts (e.g., NaCl, KCl, $MgCl_2$) are typically added to vaccine liquid formulation to adjust tonicity and thus prevent pain upon injection.

Buffering agents (e.g., phosphate buffer (KPO_4), succinate buffer, sodium borate) are added to maintain solution pH and stabilize antigens adsorbed to aluminum salts.

Amino acids (e.g., L-glutamate, L-arginine, L-histidine, L-leucine, trileucine, glycine, methionine) are well-known stabilizers for protein-based pharmaceuticals and vaccines and act by diverse mechanisms, including preferential hydration, prevention of aggregation, charge shielding of protein surfaces, direct binding to protein molecules, and increasing dispersibility (e.g., leucine provides low hygroscopy and increases the dispersibility of a dry powder inhalation).

Sugars and polyols (e.g., trehalose, sucrose, inulin, mannitol, dextran, maltodextrin, myo-inositol, sorbitol, pullulan) are rich in hydroxyl functional groups and give structure and protection to the powder particles (glass state stabilization), stabilize the preferential hydration (water replacement), conserve the 3D-conformation of the vaccine molecules by replacing the water by hydrogen-bond interaction and increase the activation energy for protein unfolding. Mannitol is also characterized by low hygroscopicity. Dextran reduces crystallization and raises the glass transition temperature of dry powders. Pullulan has good film-forming abilities and low oxygen permeability.

Human and bovine serum albumin and animal-derived gelatin are other frequent stabilizers of vaccines competing for surface coverage and reducing the heterogeneity of protein distribution in the powder particles.

Several antioxidants (e.g., ethylenediaminetetraacetic acid, citric acid, ascorbic acid, metabisulphate) are used as pharmaceutical excipients to prevent oxidation.

Encapsulating polymers (e.g., Eudragit, cellulose acetate phthalate, hydroxypropyl methylcellulose acetate succinate, poly-lactic-co-glycolic acid, polyvinylpyrrolidone) protect antigens in the enteric environment and control release for delivery via the oral route.

7.9 Spray Drying Temperatures for Vaccines

Spray dryers operating in the co-current mode are particularly suitable for gentle spray drying of heat-sensitive vaccines. Due to the intense heat and mass transfer and the evaporative cooling, the temperature of the droplets rises to the wet-bulb temperature (i.e., the temperature of the saturated surface) (see Chap. 3, Sect. 3.4), and the particles approach the gas temperature at the spray dryer outlet (T_{out}), which decreases with lower inlet temperature (T_{in}), higher liquid feed rate, and lower drying gas flow rate. Thus, T_{out} can be considered as the maximum temperature load the vaccine may experience. As shown in Fig. 7.18, the average T_{out} for spray drying vaccine formulations is around 53 ± 11 °C. In addition, there is a linear correlation between T_{in} and ($T_{in} - T_{out}$), as shown in Fig. 7.19 and described in Eq. 7.1

$$T_{in} = 50.188 + 1.0515(T_{in} - T_{out}) \text{ with } R^2 = 0.8942 \quad (7.1)$$

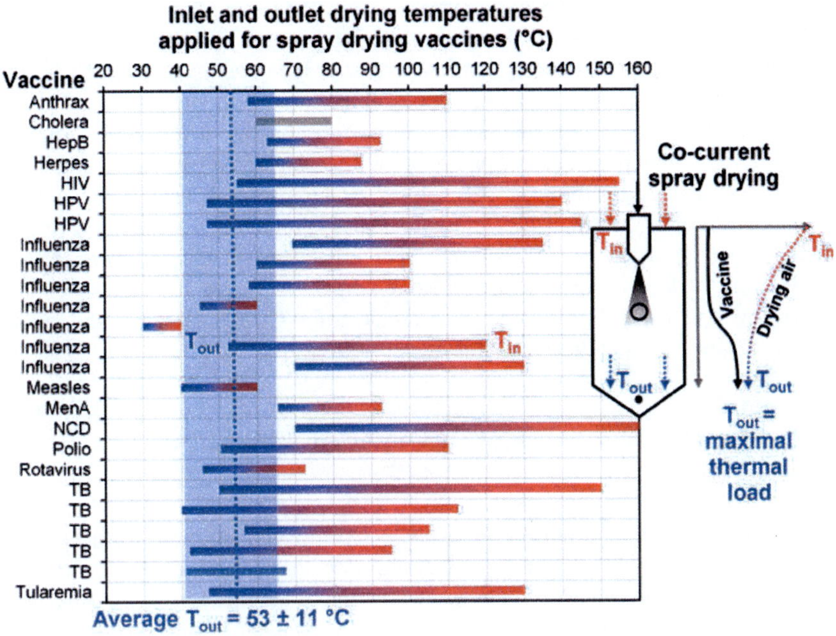

Fig. 7.18 Inlet and outlet drying temperatures applied for spray drying vaccines. In co-current mode, the outlet temperature (T_{out}) is the maximal temperature the vaccine may experience. Average $T_{out} = 53 \pm 11$ °C

Fig. 7.19 Linear correlation between inlet drying temperature (T_{in}) and temperature difference between inlet and outlet ($T_{in} - T_{out}$) for spray drying vaccines. The average $T_{out} = 53 \pm 11$ °C (between 42 °C and 64°) is marked as a grey area

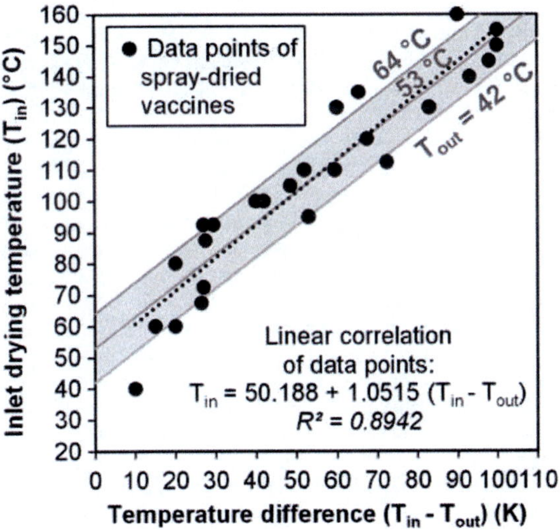

A final powder product with small residual moisture content is obtained at a given inlet temperature when decreasing the temperature difference between the inlet and the outlet gas. The residual moisture content of the reviewed vaccine studies is typically between 1 and 5 wt% (Table 7.6).

In addition, appropriate moisture-protective packaging is normally used to maintain the integrity of the spray-dried vaccine powder during storage. Researchers applied desiccants (e.g., silica gel or gypsum) to control the relative humidity in powder storage packaging or sealed aluminum pouches or Pharma-Keep® packets (Wong et al. 2007; Kunda et al. 2019; Price et al. 2020; Gomez et al. 2021a). An increase in the residual moisture content of sugars (e.g., trehalose, see Chap. 3) would lower the glass transition temperature (Chen et al. 2000) and reduce the effectiveness of stabilization (Grasmeijer et al. 2013).

7.10 Particle Size and Morphologies of Spray-Dried Vaccines

The particle size and morphology are other decisive quality attributes for the dosage form and the application. They depend strongly on the material properties and the spray drying process conditions. For inhalation, the target aerodynamic particle size needs to be between 1 and 5 μm (Byron 1986; Patton et al. 2004; Kanojia et al. 2017). Larger particles are suitable for nasal, skin, or oral administration.

Fig. 7.20 SEM pictures of spray-dried vaccine powders. **a** Anthrax vaccine containing trehalose, hydrolyzed gelatin, PBS buffer salts, antigen, and Tween 80 after 3 months of storage at 40 °C (Adapted from Jones et al. (2017) with permission from Elsevier). **b** Inactivated V. cholerae vaccine encapsulated in Eudragit® L30D55 and alginate after 6 months of storage at 40 °C (Adapted from Pastor et al. (2013a) with permission from Elsevier). **c** Hepatitis B surface antigen (HBsAg) vaccine in inulin (1:250 weight ratio) (Adapted from Tomar et al. (2019), open access under BY 4.0 license). **d** Whole inactivated influenza virus stabilized in trehalose (Adapted from Kanojia et al. (2016), open access under CC-BY 4.0 license). **e** Recombinant hemagglutinin (HA) influenza vaccine formulated with trehalose and hydrolyzed gelatin after 2 months of storage at 50 °C (Adapted from Zhu et al. (2014) with permission from Springer Nature). **f** Influenza subunit vaccine (HA protein) stabilized with inulin in PBS buffer (1:200 w/w HA:inulin) (Adapted from Saluja et al. (2010) with permission from Elsevier). **g** Human Papilloma Virus (HPV) L2 VLP-based vaccine stabilized with mannitol, trehalose, dextran, inositol, and L-leucine (Adapted from Tumban et al. (2015) with permission from Elsevier). **h** Live-attenuated measles virus vaccine encapsulated in bovine serum albumin crosslinked with glutaraldehyde (Adapted from Gala et al. (2017) with permission from Springer Nature). **i** Meningococcal capsular polysaccharides encapsulated in albumin microparticles (Adapted from Ubale et al. (2013) with permission from Taylor & Francis). **j** Newcastle disease vaccine formulated with mannitol-inositol-BSA (Adapted from Huyge et al. (2012) with permission from Elsevier). **k** Recombinant live-attenuated Listeria monocytogenes vaccine against F. tularensis (Adapted from Kunda et al. (2016) with permission from Elsevier). **l** Pertussis vaccine based on outer membrane vesicles (OMVs from B. pertussis B1917) stabilized in trehalose (Adapted from Kanojia et al. (2018a) with permission from Elsevier). **m** Bacille Calmette-Guérin (BCG) vaccine (heat-inactivated bacteria) encapsulated in L-leucine, mannitol, PVP, BSA, and trehalose (Adapted from Price et al. (2020) with permission from Springer Nature). **n** Tuberculosis subunit vaccine containing antigen ID93 and adjuvant system GLA-SE in trehalose after 3 months of storage at 40 °C (Adapted from Gomez et al. (2021a) with permission from Elsevier). **o** Tuberculosis subunit vaccine containing antigen ID93 and adjuvant system GLA-SE in trehalose after 26 months of storage at 40 °C. The cracked particle indicates a higher concentration of encapsulated nanoemulsion droplets near the surface wall (Adapted from Gomez et al. (2021a) with permission from Elsevier)

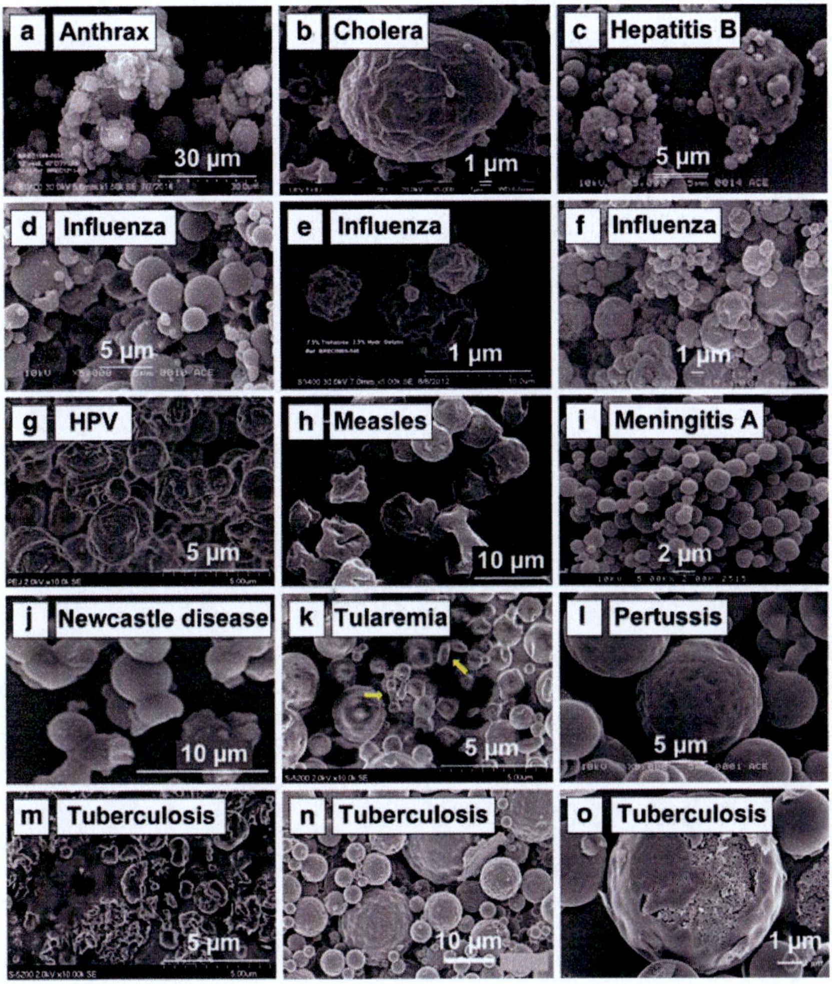

Figure 7.20 shows some SEM images of spray-dried vaccines against anthrax (Jones et al. 2017), cholera (Pastor et al. 2013a), Hepatitis B (Tomar et al. 2019), influenza (Kanojia et al. 2016), (Zhu et al. 2014) (Saluja et al. 2010), HPV (Tumban et al. 2015), measles (Gala et al. 2017), meningitis A (Ubale et al. 2013), Newcastle disease (Huyge et al. 2012), tularemia (Kunda et al. 2016), pertussis (Kanojia et al. 2018a), and tuberculosis (Price et al. 2020; Gomez et al. 2021a). In addition, the vaccines were stabilized in:

- Trehalose (Zhu et al. 2014; Kanojia et al. 2016, 2018a; Jones et al. 2017; Price et al. 2020; Gomez et al. 2021a),
- Mannitol (Huyge et al. 2012; Tumban et al. 2015; Price et al. 2020),

- Dextran (Tumban et al. 2015),
- Inositol (Huyge et al. 2012; Tumban et al. 2015),
- Inulin (Saluja et al. 2010; Tomar et al. 2019),
- L-leucine (Tumban et al. 2015; Price et al. 2020),
- Hydrolyzed gelatin (Zhu et al. 2014; Jones et al. 2017),
- Bovine serum albumin (Huyge et al. 2012; Ubale et al. 2013; Gala et al. 2017; Price et al. 2020),
- Eudragit (Pastor et al. 2013a),
- Alginate (Pastor et al. 2013a), and
- ether excipient combinations.

The particles have a spherical or shrunken (raisin-like) shape and range in size from about 1 to 20 µm (see Table 7.6), with the upper limit of particle size determined mainly by the limited drying chamber size of the laboratory-scale spray dryers used. The photos generally illustrate the particle engineering capabilities of spray drying.

7.11 Long-Term Storage Stability of Spray-Dried Vaccines

The WHO set the tolerance of a thermostable vaccine in a controlled temperature chain (CTC) to 40 °C for at least 3 days, 25 °C for 1 month, and 8 °C for more than 2 years (WHO 2016). In addition, there are different Vaccine Vial Monitor (VVM) categories with the number of days required for a VVM to reach its end-point when exposed to 37, 25, and 5 °C (Table 7.8). For example, high thermal stability (VVM30) requires stability of 30 days (1 month) at 37 °C, 193 days (around 6.3 months) at 25 °C, and over 4 years at 5 °C (WHO/UNICEF 2014).

Figure 7.21 shows the thermal stability data (shelf life) of the spray-dried vaccine formulations from Table 7.6 as a function of storage temperature and time. The data demonstrate remarkable long-term stability for several spray-dried vaccines, exceeding the "high thermal stability" criteria of VVM30. This is particularly true for vaccines in dry powder form against influenza, HPV, TB, and HepB, with storage stability of over 24 months at room temperature. In addition, most reported studies show improved storage stability in the order of several months when stored at 20 to 60 °C (Truong-Le and Pham 2007; Chen et al. 2010; Agarkhedkar et al. 2014; Zhu et al. 2014; Ohtake et al. 2015; Sou et al. 2015; Kanojia et al. 2016).

In addition, most data also fulfill the ICH quality guidelines for vaccine storage stability testing in protective packaging (ICH 1995, 2003):

- at least 12 months at 25 °C/60% RH or 30 °C/65% RH for long-term studies,
- at least 6 months at 30 °C/65% RH for intermediate studies,
- at least 6 months at 40 °C/75 RH for accelerated studies.

Table 7.8 Tolerance for thermostable vaccines in a CTC (WHO 2016) and WHO specified Vaccine Vial Monitor (VVM) categories with time to reach its endpoint when exposed to 37, 25, and 5 °C (WHO/UNICEF 2014)

	40 °C	25 °C	8 °C
WHO tolerance for CTC	3 days	1 month	> 2 years
VVM categories	37 °C	25 °C	5 °C
VVM30 (High stability)	30 days	193 days	> 4 years
VVM14 (Medium stability)	14 days	90 days	> 3 years
VVM7 (Moderate stability)	7 days	45 days	> 2 years
VVM2 (Least stability)	2 days	n.a.	225 days

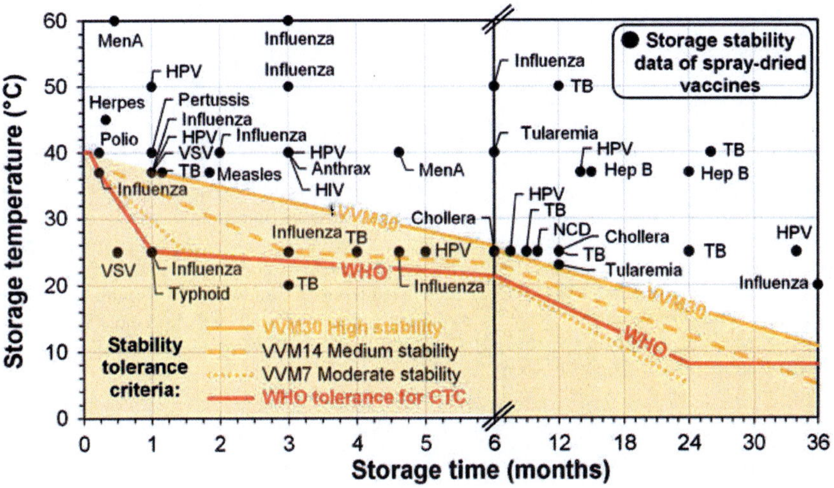

Fig. 7.21 Storage stability data of spray-dried vaccines (data from Table 7.6) compared to the tolerance for thermostable vaccines in a CTC (WHO 2016) (see criteria in Table 7.8, here linearized between endpoints) and the WHO-specified Vaccine Vial Monitor (VVM) categories (WHO/UNICEF 2014)

The encouraging results on improved long-term stability enable stockpiling vaccines in powder form at room temperature and could facilitate mass vaccination.

Figure 7.22 shows three spray-dried vaccine examples compared to their liquid counterparts. Figure 7.22a presents storage data of live-attenuated B/Harbin influenza virus vaccine formulated with 5% sucrose, 2% trehalose, 2% arginine, 0.2% Pluronic F68, 50 mM pH 7.2 KPO_4 buffer, 10 mM methionine, and 2 mM EDTA (ethylenediaminetetraacetic acid) (Truong-Le and Pham 2007; Ohtake et al. 2015). Using a fluorescent biological activity assay, a slope of – 0.087 log loss per month was measured at 25 °C storage temperature. Figure 7.22b illustrates the potency of a spray-dried live rotavirus vaccine formulation compared to a liquid formulation, showing improved room temperature stability at 25 °C (Ohtake et al. 2015). Further

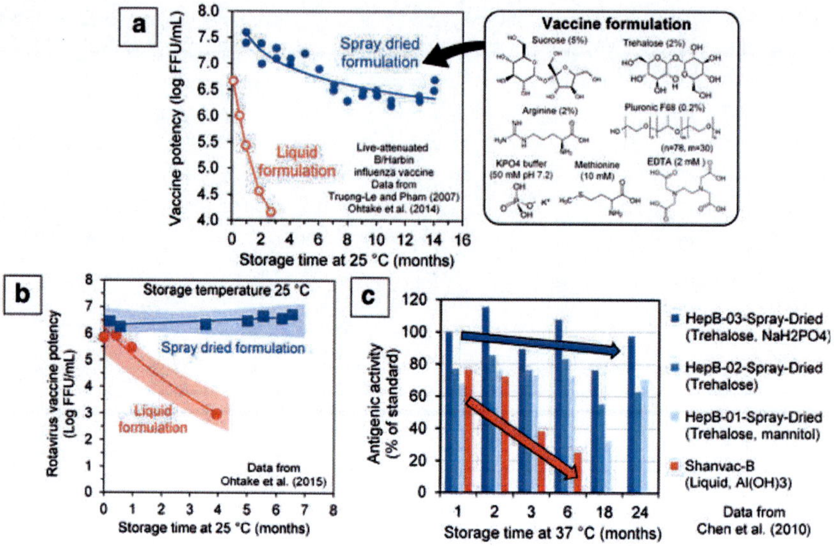

Fig. 7.22 Examples of spray-dried vaccine formulations with improved long-term stability compared to their liquid counterparts. **a** Live-attenuated B/Harbin influenza virus vaccine formulation stored at 25 °C (Slope = − 0.087 log loss per month) (Data adapted from Truong-Le and Pham (2007), Ohtake et al. (2015) with permission from John Wiley and Sons), **b** Live rotavirus vaccine formulation at 25 °C (Data adapted from Ohtake et al. (2015) with permission from John Wiley and Sons), **c** Hepatitis B vaccine formulation with trehalose and NaH_2PO_4 at 37 °C in comparison to liquid Shanvac-B vaccine stored at 4 °C (Data adapted from Chen et al. (2010) with permission from Elsevier)

development involves incorporating the spray-dried vaccine powder into a rapidly dissolving oral thin film, which is mucoadhesive and expected to ease administration and improve patient compliance. Figure 7.22c shows the antigenic activity of spray-dried hepatitis B (HepB) vaccine formulations at 37 °C storage compared to the original liquid Shanvac-B vaccine (Chen et al. 2010). The spray-dried formulations were stable at 37 °C for more than 24 months while maintaining immunogenicity in mice. The composition of the spray-dried HepB-03 formulation was 12.4 mg vaccine, 85 mg trehalose, and 15 mg NaH_2PO_4 per dose.

7.12 Administration Routes of Spray-Dried Vaccines

Although needle vaccination is the standard way for vaccine administration, it has several limitations and risks, including needle-stick injuries, unsafe injection practices (WHO 2010), potential disease transmission through needle reuse, sharps disposal costs, and the need for health care professionals to administer the dose.

The growing scientific interest in particle engineering and nano/micro-technologies contributes to the advancement of spray drying and needle-free approaches to vaccination. In addition, spray drying enables new delivery routes to target where infection occurs, e.g., as a powder for inhalation, intranasal administration (Gomez et al. 2022; Gomez and Vehring 2022), oral capsule, thin film, or microarray patch on the skin. The number of applications described underscores the versatility of spray-drying technology for developing thermostable vaccines. Therefore, an increase in clinical trials can be expected in the coming years.

Spray-dried vaccine powder can be packaged in single or multiple doses in a variety of product formats, including vials, blisters, capsules, inhalers, coated microneedles, dissolving films (for buccal administration) (Gala et al. 2017), and other medical containers (Schwartzbach 2011; Gladd 2014; Andersen 2016, 2020; Siew 2016; Luy and Stamato 2020; Vehring et al. 2020).

Other options include compressed tablets dissolved in water before vaccination or dual chambers of powder and diluent with a breakable seal for oral administration, or dual-chamber devices for integrated reconstitution of vaccine powder and diluent and injection with a needle (Gill 2014, 2018; Ghaemmaghamian et al. 2022).

Pulmonary and intranasal administration seems to be the most promising route for the delivery of spray-dried vaccines (Gomez and Vehring 2022). First, administration eliminates the need for a needle (Weers et al. 2007; Tonnis et al. 2012), improving patient compliance and avoiding sharps disposal, making administration more user-friendly (Sou et al. 2015). Second, aerosol administration of powder vaccines stimulates strong mucosal and systemic immunity comparable to liquid vaccines (Jin and Tsao 2013).

For pulmonary administration of spray-dried vaccines, the powders can be packaged in:

- blister packages (e.g., single doses containing a few mg of powder) (LiCalsi et al. 1999; Vehring et al. 2007; Cape et al. 2010; Searles and Mohan 2010; McAdams et al. 2012),
- capsules (Weniger 2011; Weniger and Papania 2013), and
- dry powder inhalers (DPI) (Friebel and Steckel 2010; De Boer and Hagedoorn 2015; Kanojia et al. 2017; Sawatdee and Srichana 2021)

Figure 7.23 gives an overview of single-use DPIs available on the market. DPIs are available as simple, cheap, highly compact, and disposable devices in a single-use format suitable for storage, distribution, and administration (Sou et al. 2011; de Boer et al. 2017; Hellfritzsch and Scherließ 2019). More details on single-use DPIs are discussed in Chap. 6 (Sect. 6.3) and various specific articles (Friebel and Steckel 2010; Tonnis et al. 2013; De Boer and Hagedoorn 2015; Berkenfeld et al. 2015; Sibum et al. 2018; Sawatdee and Srichana 2021).

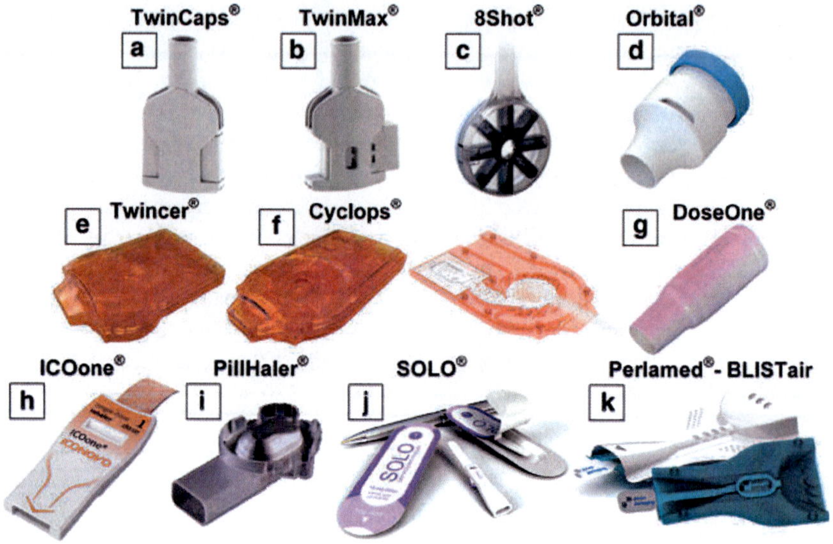

Fig. 7.23 An overview of single-use dry powder inhalers. **a** TwinCaps®, **b** TwinMax®, and **c** 8Shot® by Hovione (Portugal) are available for licensing and supply, **d** Orbital® by Pharmaxis (Australia), **e** Twincer® and **f** Cyclops® developed at the University of Groningen (The Netherlands) and produced by PureIMS, **g** DoseOne® by DoseOne LLC (USA), **h** ICOone® by Iconovo (Sweden), **i** PillHaler® by HollyCon Italy S.r.l., **j** SOLO® inhaler by Manta Devices LLC (USA), and **k** PERLAMED™-BLISTair by Perlen Packaging AG (Switzerland) (Images courtesy of all manufacturers, Twincer® and Cyclops® adapted from Sibum et al. (2018), open access under CC-BY 4.0 license)

7.13 Challenges and Proposed Solutions for Spray-Dried Vaccines

Despite the many advantages of spray-drying vaccines and years of intensive research, no spray-dried vaccine product has yet been market approved according to the WHO list of prequalified vaccines (WHO 2021a). The status to date is that several preclinical studies have been conducted with animal models (e.g., mice, guinea pigs, rhesus macaques) (Amorij et al. 2007; Chen and Kristensen 2009; Saluja et al. 2010; Sou et al. 2015; Kanojia et al. 2017; Vehring et al. 2020; Preston and Randolph 2021; Arpagaus 2022; Gomez and Vehring 2022), but there are limited data from clinical trials.

One of the most advanced studies is an initial Phase I clinical trial of a spray-dried measles vaccine for inhalation tested in adult males (Agarkhedkar et al. 2014). All subjects tolerated the vaccine well, it was safe, and the immunogenicity profiles were comparable to the licensed lyophilized measles vaccine administered subcutaneously after reconstitution. Thus, the question arises as to why spray-dried vaccines are not yet on the market.

Table 7.9 summarizes the remaining challenges for aseptic spray drying of vaccines and proposes solutions on the way to market based on various studies (Ståhl et al. 2002; Vehring et al. 2007, 2020; Snyder and Lechuga-Ballesteros 2008; Vehring 2008; Fourie et al. 2008; Roser 2009; Greb 2009; Germishuizen et al. 2010; Abdul-Fattah and Truong 2010; Schwartzbach 2011; McAdams et al. 2012; Breit and DuBose 2013; Walters et al. 2014; Upperton 2015; Sou et al. 2015; LeClair et al. 2016a; Geng 2018; Arpagaus 2018, 2020; Langford and Ohtake 2019; Grasmeijer et al. 2019; Andersen 2020; Morgan et al. 2020; Pinto et al. 2021; Roth and Scherließ 2021; Preston and Randolph 2021).

First, freeze-drying is already well established as a drying technology in the vaccine industry. In addition, the vaccine market is highly regulated, so that vaccine manufacturers are reluctant to introduce new production methods and drying technologies. Therefore, spray drying is still considered a new technology in the vaccine field, and there is no precedent case. The market introduction is moving forward, but the regulatory review sounds like a lengthy approval process.

Then, the scalability of a spray-drying process to produce millions of doses must be demonstrated. Even though the concepts have been known for many years, establishing and maintaining an aseptic spray-drying process for vaccine powder is challenging. Today, only a limited number of manufacturers offer aseptic spray drying technology and services. More market players and suppliers are needed for broadening applications.

Next, powder filling after spray drying in milligram quantities with high dosing accuracy and different dosage forms (e.g., blisters, vials, inhalers, etc.) remains a challenge, requiring high throughput rates and aseptic handling, transportation, and sealing. Some powder filling systems are available on the market, as presented in Chap. 4 (Sect. 4.6.3), but extensive validation and expertise are needed.

Table 7.9 Benefits (+) and challenges (−) of aseptic spray drying vaccines to reach the market

Benefits	Challenges
+ Improved thermal storage stability (shelf life) + The fast and continuous drying process + High capacity and scalability + Particle engineering (e.g., tunable particle size, shape, morphology, free-flowing powder, aerosolization properties, vaccine encapsulation) + Enables alternative administration routes (e.g., pulmonary, intranasal, skin, oral, etc.) + Many research studies on a laboratory scale + No need for a vacuum + No freezing step required + Industrial experience with pharmaceuticals + Mature technology in the food industries (knowledge and technology transfer)	− Freeze drying is well established for vaccine stabilization − Lack of commercialized spray-dried vaccines (limited clinical trials, lack of regulatory precedent, no spray-dried vaccine listed in the WHO database) − Lack of manufacturers offering aseptic spray dryer equipment for aseptic processing and services (lack of existing aseptic facilities) − Aseptic powder filling of milligrams as an additional step − Formulation efforts can take a long time − Thermal and shear stresses during spray drying (e.g., shear forces during atomization, air–liquid interface) − Residual moisture content (potential secondary drying step needed) − Limited yield (50% to > 70%) in laboratory scale, > 95% in industrial-scale

After that, formulation science plays an important role in selecting suitable excipients and adjuvants to improve the vaccine's chemical and physical stability and potency. However, developing an optimal formulation for a spray-dried vaccine remains a major challenge because it can be a lengthy development process, and most decisions are based on real-time stability testing that can take 12 months or longer. Therefore, it is suggested to include spray-drying technology as a process in the early stages of vaccine development.

Afterward, during spray drying, high gas temperatures and atomization stress can affect the immunogenicity of the vaccine. In this respect, the evaporative cooling effect limits the thermal load that is controlled by the gas outlet temperature.

Furthermore, a compromise must be found between the required outlet temperature for efficient drying and achieving a low residual moisture content. Therefore, an additional drying step may be necessary, e.g., in a vacuum or microwave dryer. Here, the effects of secondary drying on the microstructure of the vaccine formulation, the mechanical relaxation of the powders, and the integrity of the vaccine are topics for a thorough investigation in the future.

Finally, the yield is not 100% because the spray-dried product comes into contact with the wall of the spray dryer during processing. Powder losses, e.g., wall deposits and filter losses, can be an issue in vaccine applications. Therefore, the cleaning procedures are of critical importance. The yield is limited to about 50 to 65% at the laboratory scale but can increase to more than 70% by using a high-performance cyclone (see Chap. 4) or an electrostatic collector like in the Nano Spray Dryer B-90. Pilot spray dryers routinely recover more than 95% of the spray-dried product, and industrial-scale spray dryers achieve as much as 98 to 99% under optimized conditions.

7.14 Future Perspectives of Spray-Dried Vaccines

Vaccine research is experiencing a rapid evolution from the first-generation live or inactivated whole viral and bacterial vaccines to next-generation subunit, toxoid, DNA, and mRNA vaccine platforms. Stability has become an urgent requirement for the biopharmaceutical industry and is becoming increasingly important as public health becomes more globalized, as demonstrated by the example of the COVID-19 pandemic. Of particular interest are thermostable vaccines that can be administered needle-free, i.e., by means other than injection, like via the lungs, nose, oral, and skin.

Converting a vaccine into a thermostable dry powder is advantageous because it reduces the resource burden associated with the cold chain and increases flexibility in dosage form and administration via multiple routes. In particular, drying technologies that enable the stabilization of vaccines are on the rise, as dry powder vaccine formulations can be stored and transported without refrigeration and have the advantage of a longer shelf-life. In addition to freeze-drying, spray-drying processes may also play a role in the future development of thermostable vaccines.

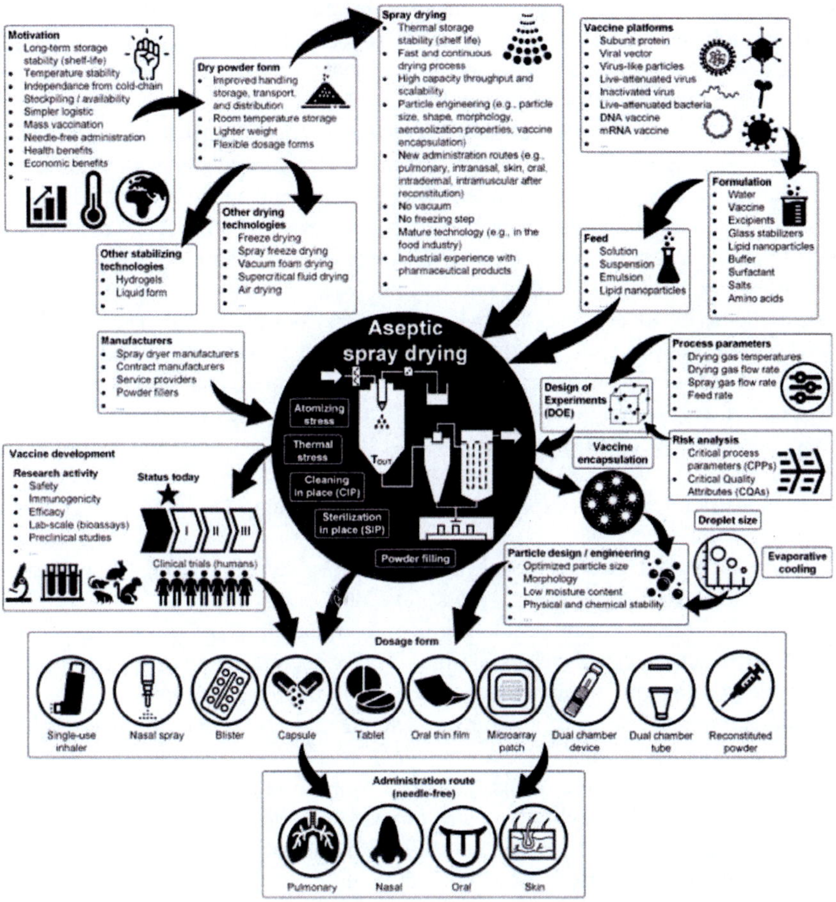

Fig. 7.24 The multi-faceted relationships that shape aseptic spray drying of vaccines and serve as a potential basis for discussion among stakeholders in the vaccine market

Spray drying is an inherently fast and continuous process that can eliminate bottlenecks in vaccine production and match capacity to demand, providing economic and public health benefits (e.g., during a pandemic).

Figure 7.24 illustrates the multi-faceted relationships that shape aseptic spray drying of vaccines, from motivation, vaccine development, formulation, and process parameters, to dosage form and route of administration, that shape the discussion around aseptic spray drying of vaccines and serve as a potential basis for discussion among stakeholders in the vaccine market.

Aseptic spray dryers equipped with sterile filters and cleaning-in-place and steam sterilization processes are used in the biopharmaceutical industry for injectable drugs or blood products. Much of the aseptic technology platform has been developed, and some manufacturers offer aseptic spray dryers suitable for various applications, including vaccines. Therefore, aseptic spray drying ensuring sterility and cleaning-in-place is expected to further expand its position in the vaccine industry as an alternative to conventional freeze drying. However, only a few manufacturers offer aseptic spray drying technology and services. More players would need to enter the market to enable wider application and improve collaboration. The advances in spray drying of vaccines deserve more attention from manufacturers and other stakeholders. Biopharma companies, spray dryer manufacturers, powder fillers, service providers, researchers, and other stakeholders must join forces to invest further in aseptic spray drying technology.

Numerous feasibility studies on various vaccine candidates show a lively research activity to improve the stability of spray-dried powder vaccines. Spray drying technology is well suited for stabilizing viral vectors, attenuated and inactivated whole or subunit vaccines, and the critical process parameters have been determined. Advancing the development of vaccines by spray drying in the coming years is a multi-faceted task that covers several areas:

- Demonstrating further experimental evidence of long-term stability at room temperature.
- Addressing and solving legal and regulatory concerns and approving a spray-dried vaccine powder as a first precedent case.
- Improving preclinical models to mimic human anatomy and physiology.
- Implementing risk analysis and design of experiments minimizing the impact of applied drying and atomization stresses on the final vaccine product.
- Formulation screening of a variety of excipients to determine the optimal lead formulations.
- Improving the final quality of spray-dried powders through optimal formulation for each vaccine, which is closely related to good stability and low hygroscopicity.
- Extending particle engineering capabilities by adjusting spray drying and formulation parameters to improve product quality.
- Expanding the flexibility of powder dosage forms for alternative routes of administration.
- Developing economical and efficient disposable inhalers and nasal sprays that enable mass vaccination campaigns at low cost.

Finally, a major research gap has been identified in the stabilization of mRNA vaccines, heralding a new era of vaccine platforms. Thus, future R&D studies should pay special attention to spray-drying mRNA vaccines encapsulated in lipid nanoparticles suitable for pulmonary or nasal administration with disposable dry powder inhalers or nasal sprays. This area includes developing guidelines for spray drying mRNA vaccines and translating preclinical results to clinical trials demonstrating safety, immunogenicity, and efficacy in humans. It is also a matter of overcoming the mental block that high drying temperatures would not work for heat-sensitive products in spray drying. The next few years will show the extent to which spray drying can be an alternative to conventional drying methods for vaccines.

References

3P Innovation (2020) Fill2Weight: aseptic powder filling, product brochure. https://www.3pinnovation.com/machines/powder-filling/fill2weight

Abdul-Fattah AM, Truong VL (2010) Drying process methods for biopharmaceutical products: an overview. In: Jameel H (ed) Formulation and process development strategies for manufacturing biopharmaceuticals. John Wiley & Sons Inc., Hoboken, NJ, USA, pp 705–738

Abdul-Fattah AM, Truong-Le V, Yee L et al (2007a) Drying-induced variations in physico-chemical properties of amorphous pharmaceuticals and their impact on stability II: stability of a vaccine. Pharm Res 24:715–727. https://doi.org/10.1007/s11095-006-9191-2

Abdul-Fattah AM, Kalonia DS, Pikal MJ (2007b) The challenge of drying method selection for protein pharmaceuticals: product quality implications. J Pharm Sci 96:1886–1916. https://doi.org/10.1002/jps.20842

Adler M, Lee G (1999) Stability and surface activity of lactate dehydrogenase in spray-dried trehalose. J Pharm Sci 88:199–208. https://doi.org/10.1021/js980321x

Afkhami S, LeClair DA, Haddadi S et al (2017) Spray dried human and chimpanzee adenoviral-vectored vaccines are thermally stable and immunogenic in vivo. Vaccine 35:2916–2924. https://doi.org/10.1016/j.vaccine.2017.04.026

Agarkhedkar S, Kulkarni PS, Winston S et al (2014) Safety and immunogenicity of dry powder measles vaccine administered by inhalation: a randomized controlled Phase I clinical trial. Vaccine 32:6791–6797. https://doi.org/10.1016/j.vaccine.2014.09.071

Ajmera A, Scherließ R (2014) Stabilisation of proteins via mixtures of amino acids during spray drying. Int J Pharm 463:98–107. https://doi.org/10.1016/j.ijpharm.2014.01.002

Alexeenko A, Topp E (2020) Future directions. In: Ohtake S, Izutsu K-I, Lechuga-Ballesteros D (eds) Drying technologies for biotechnology and pharmaceutical applications. Wiley, pp 355–372

Al-Khattawi A, Bayly A, Phillips A, Wilson D (2018) The design and scale-up of spray dried particle delivery systems. Expert Opin Drug Deliv 15:47–63. https://doi.org/10.1080/17425247.2017.1321634

Al-Zoubi N, Gharaibeh S, Aljaberi A, Nikolakakis I (2021) Spray drying for direct compression of pharmaceuticals. Processes 9:267. https://doi.org/10.3390/pr9020267

Amacker M, Smardon C, Mason L et al (2020) New GMP manufacturing processes to obtain thermostable HIV-1 gp41 virosomes under solid forms for various mucosal vaccination routes. NPJ Vaccines 5:41. https://doi.org/10.1038/s41541-020-0190-9

Amorij J-P (2007) The development of stable influenza vaccine powder formulations for new needle-free dosage forms. PhD Thesis, University of Groningen, The Netherlands. https://www.researchgate.net/publication/50843621

Amorij J-P, Saluja V, Petersen AH et al (2007) Pulmonary delivery of an inulin-stabilized influenza subunit vaccine prepared by spray-freeze drying induces systemic, mucosal humoral as well as cell-mediated immune responses in BALB/c mice. Vaccine 25:8707–8717. https://doi.org/10.1016/j.vaccine.2007.10.035

Amorij J-P, Huckriede A, Wilschut J et al (2008) Development of stable influenza vaccine powder formulations: challenges and possibilities. Pharm Res 25:1256–1273. https://doi.org/10.1007/s11095-008-9559-6

Andersen SK (2016) Challenges in spray drying process and formulation development. In: Lyophilisation Europe, 5 July 2016, London, UK, pp 1–31

Andersen SK (2018) Integration of quality-by-design into qualification and validation of spray drying processes. In: ECA training course, 10–12 April 2018, Lisbon, Portugal, pp 1–18

Andersen SK (2020) New drying processes for biopharmaceuticals. Bright, 5 May 2020. https://www.brighttalk.com/webcast/10519/373808/new-drying-processes-for-biopharmaceuticals

Anderson SD, Daviskas E, Brannan JD, Chan HK (2018) Repurposing excipients as active inhalation agents: the mannitol story. Adv Drug Deliv Rev 133:45–56. https://doi.org/10.1016/j.addr.2018.04.003

Anish C, Upadhyay AK, Sehgal D, Panda AK (2014) Influences of process and formulation parameters on powder flow properties and immunogenicity of spray dried polymer particles entrapping recombinant pneumococcal surface protein A. Int J Pharm 466:198–210. https://doi.org/10.1016/j.ijpharm.2014.03.025

Año G, Esquisabel A, Pastor M et al (2011) A new oral vaccine candidate based on the microencapsulation by spray-drying of inactivated Vibrio cholerae. Vaccine 29:5758–5764. https://doi.org/10.1016/j.vaccine.2011.05.098

Arpagaus C (2018) Pharmaceutical particle engineering via nano spray drying—process parameters and application examples on the laboratory-scale. Int J Med Nano Res 5. https://doi.org/10.23937/2378-3664.1410026

Arpagaus C (2020) Spray drying of vaccines: a review. In: IDS 2020, 22nd international drying symposium Worcester, Massachusetts, USA, June 28–July 1, 2020, pp 1–14

Arpagaus C (2022) Spray drying of vaccines: from laboratory research to industrial applications. Springer Nature

Arpagaus C, Schwartzbach H (2008) Scale-up from the Büchi Mini Spray Dryer B-290 to the Niro MOBILE MINOR, best@buchi Information Bulletin, Number 52/2008

Artaud C, Kara L, Launay O (2019) Vaccine development: from preclinical studies to phase 1/2 clinical trials. In: Methods in molecular biology, pp 165–176

ASME (2019) Bioprocessing equipment, ASME BPE-2019, The American Society of Mechanical Engineers. ISBN 978-0-7918-7225-3, pp 1–380

Bahamondez-Canas TF, Cui Z (2018) Intranasal immunization with dry powder vaccines. Eur J Pharm Biopharm 122:167–175. https://doi.org/10.1016/j.ejpb.2017.11.001

Baldinger A, Clerdent L, Rantanen J et al (2012) Quality by design approach in the optimization of the spray-drying process. Pharm Dev Technol 17:389–397. https://doi.org/10.3109/10837450.2010.550623

Baumann JM, Adam MS, Wood JD (2021) Engineering advances in spray drying for pharmaceuticals. Annu Rev Chem Biomol Eng 12:annurev-chembioeng-091720-034106. https://doi.org/10.1146/annurev-chembioeng-091720-034106

Bejugam NK, Uddin AN, Gayakwad SG, D'Souza MJ (2008) Formulation and evaluation of albumin microspheres and its enteric coating using a spray-dryer. J Microencapsul 25:577–583. https://doi.org/10.1080/02652040802476187

Benjasirimongkol P, Piriyaprasarth S, Moribe K, Sriamornsak P (2019) Use of risk assessment and plackett-burman design for developing resveratrol spray-dried emulsions: a quality-by-design approach. AAPS PharmSciTech 20:14. https://doi.org/10.1208/s12249-018-1220-z

Berkenfeld K, Lamprecht A, McConville JT (2015) Devices for dry powder drug delivery to the lung. AAPS PharmSciTech 16:479–490. https://doi.org/10.1208/s12249-015-0317-x

BETE (2005) Spray dry manual. BETE Fog Nozzle, Inc

Bhambhani A, Antochshuk V (2020) Vaccines and microorganisms. In: Ohtake S, Izutsu K-I, Lechuga-Ballesteros D (eds) Drying technologies for biotechnology and pharmaceutical applications. Wiley, pp 121–136

Bhowmik T, D'Souza B, Uddin MN, D'Souza MJ (2012) Oral delivery of microparticles containing plasmid DNA encoding hepatitis-B surface antigen. J Drug Target 20:364–371. https://doi.org/10.3109/1061186X.2012.662686

Bhujbal SV, Mitra B, Jain U et al (2021) Pharmaceutical amorphous solid dispersion: a review of manufacturing strategies. Acta Pharm Sin B 11:2505–2536. https://doi.org/10.1016/j.apsb.2021.05.014

Boehm G, Peyre M, Sesardic D et al (2002) On technological and immunological benefits of multivalent single-injection microsphere vaccines. Pharm Res 19:1330–1336. https://doi.org/10.1023/A:1020354809581

Bot AI, Tarara TE, Smith DJ et al (2000) Novel lipid-based hollow-porous microparticles as a platform for immunoglobulin delivery to the respiratory tract. Pharm Res 17:275–283. https://doi.org/10.1023/A:1007544804864

Bot AI, Smith DJ, Bot S et al (2001) Receptor-mediated targeting of spray-dried lipid particles coformulated with immunoglobulin and loaded with a prototype vaccine. Pharm Res 18:971–979

Bowen M, Turok R, Maa Y-F (2013) Spray drying of monoclonal antibodies: investigating powder-based biologic drug substance bulk storage. Dry Technol 31:1441–1450. https://doi.org/10.1080/07373937.2013.796968

Bregu M, Draper SJ, Hill AVS, Greenwood BM (2011) Accelerating vaccine development and deployment: report of a Royal Society satellite meeting. Philos Trans R Soc B Biol Sci 366:2841–2849. https://doi.org/10.1098/rstb.2011.0100

Breit J, DuBose D (2013) Spray-dry manufacture of vaccine formulations. Bioprocess Int. https://bioprocessintl.com/manufacturing/monoclonal-antibodies/spray-dry-manufacture-of-vaccine-formulations-347400/

Brough C, Williams RO (2013) Amorphous solid dispersions and nano-crystal technologies for poorly water-soluble drug delivery. Int J Pharm 453:157–166. https://doi.org/10.1016/j.ijpharm.2013.05.061

Byron PR (1986) Prediction of drug residence times in regions of the human respiratory tract following aerosol inhalation. J Pharm Sci 75:433–438. https://doi.org/10.1002/jps.2600750502

Camfil (2020) Termikfil 2000, High temperature HEPA filter. Product datasheet. https://www.camfil.com/product/documents/dam/30482/Product-Documentation-Termikfil-2000.pdf

Cape SP, McAdams DH, Manion JR et al (2010) Development of stable, immunogenic, and protective measles vaccines for needle-free administration. In: 2010 FIP pharmaceutical sciences world congress and AAPS annual meeting and exposition, 14–18 Nov 2010, New Orleans, LA, USA

Cattenati C (2019) Business case on low powder dosages, Romaco s.r.l., 20th Sept 2019

Cavalier G (2020) The cold chain: a logistical challenge at the heart of the distribution of the Covid-19 vaccine. https://iifiir.org/en/news, https://iifiir.org/en/news/the-cold-chain-a-logistical-challenge-at-the-heart-of-the-distribution-of-the-covid-19-vaccine

Cavalier G, Fertel C, De Paoli F et al (2021) Cold chain technology brief: vaccines

Cazzoli (2019) PF—powder filler series, data sheet. https://www.cozzoli.com/wp-content/uploads/2019/01/PF-Data-Sheet.pdf

Celik M, Wendel SC (2005) Chapter 5: Spray drying and pharmaceutical applications. In: Handbook of pharmaceutical granulation technology, pp 129–157

Chan MY, Dutill TS, Kramer RM (2017) Lyophilization of adjuvanted vaccines: methods for formulation of a thermostable freeze-dried product. In: Methods in molecular biology, pp 215–226

ChargePoint Technology (2019) AseptiSafe contained aseptic transfer valves, Brochure 04/2019, pp 1–12

Chen D, Kristensen D (2009) Opportunities and challenges of developing thermostable vaccines. Expert Rev Vaccines 8:547–557. https://doi.org/10.1586/erv.09.20

Chen T, Fowler A, Toner M (2000) Literature review: supplemented phase diagram of the trehalose-water binary mixture. Cryobiology 40:277–282. https://doi.org/10.1006/cryo.2000.2244

Chen D, Kapre S, Goel A et al (2010) Thermostable formulations of a hepatitis B vaccine and a meningitis A polysaccharide conjugate vaccine produced by a spray drying method. Vaccine 28:5093–5099. https://doi.org/10.1016/j.vaccine.2010.04.112

Colaço CALS, Roser B (1994) Trehalose-a multifunctional additive for food preservation. Food packaging and preservation. Springer, US, Boston, MA, pp 123–140

Corbanie E (2007) Development of a dry powder vaccine for mass vaccination of poultry. Doctoral Thesis, Ghent University. https://biblio.ugent.be/publication/469294/file/1880031

Corbanie EA, Remon JP, Van Reeth K et al (2007) Spray drying of an attenuated live Newcastle disease vaccine virus intended for respiratory mass vaccination of poultry. Vaccine 25:8306–8317. https://doi.org/10.1016/j.vaccine.2007.09.049

Corbanie EA, Vervaet C, van Eck JHH et al (2008) Vaccination of broiler chickens with dispersed dry powder vaccines as an alternative for liquid spray and aerosol vaccination. Vaccine 26:4469–4476. https://doi.org/10.1016/j.vaccine.2008.06.055

Costa E, Neves F, Andrade G, Winters C (2014) Scale-up & QbD approaches for spray-dried inhalation formulations. ONdrugDelivery. https://www.ondrugdelivery.com/scale-qbd-approaches-spray-dried-inhalation-formulations/

Crommelin DJA, Volkin DB, Hoogendoorn KH et al (2021) The science is there: key considerations for stabilizing viral vector-based Covid-19 vaccines. J Pharm Sci 110:627–634. https://doi.org/10.1016/j.xphs.2020.11.015

CureVac (2020a) Clinical trial protocol. COVID-19: a phase 2b/3, randomized, observer-blinded, placebo-controlled, multicenter clinical study evaluating the efficacy and safety of investigational SARS-CoV-2 mRNA vaccine CVnCoV in adults 18 years of age and older, 11 Nov, pp 1–152

CureVac (2020b) CureVac's COVID-19 vaccine candidate, CVnCoV, suitable for standard fridge temperature logistics, Press Release, 12 Nov 2020

D'Amore T, Yang Y (2019) Advances and challenges in vaccine development and manufacture. www.bioprocessintl.com, https://bioprocessintl.com/manufacturing/vaccines/advances-and-challenges-in-vaccine-development-and-manufacture

Das P (2004) Revolutionary vaccine technology breaks the cold chain. Lancet Infect Dis 4:719. https://doi.org/10.1016/S1473-3099(04)01222-8

de Boer AH, Hagedoorn P (2015) The role of disposable inhalers in pulmonary drug delivery. Expert Opin Drug Deliv 12:143–157. https://doi.org/10.1517/17425247.2014.952626

de Boer AH, Hagedoorn P, Hoppentocht M et al (2017) Dry powder inhalation: past, present and future. Expert Opin Drug Deliv 14:499–512. https://doi.org/10.1080/17425247.2016.1224846

de Costa S (2014) Advanced delivery devices—new drug delivery & stabilization platforms gaining global interest. Drug Dev Technol. https://drug-dev.com/advanced-delivery-devices-new-drug-delivery-stabilization-platforms-gaining-global-interest/

de Meyer L, Van Bockstal P-J, Corver J et al (2015) Evaluation of spin freezing versus conventional freezing as part of a continuous pharmaceutical freeze-drying concept for unit doses. Int J Pharm 496:75–85. https://doi.org/10.1016/j.ijpharm.2015.05.025

de Swart RL, LiCalsi C, Quirk AV et al (2007) Measles vaccination of macaques by dry powder inhalation. Vaccine 25:1183–1190. https://doi.org/10.1016/j.vaccine.2006.10.019

Dec Group (2021) Microdosing μPTS. Product datasheet. https://www.dec-group.net/micropts

Delany I, Rappuoli R, De Gregorio E (2014) Vaccines for the 21st century. EMBO Mol Med 6:708–720. https://doi.org/10.1002/emmm.201403876

Desobry SA, Netto FM, Labuza TP (1997) Comparison of spray-drying, drum-drying and freeze-drying for β-Carotene encapsulation and preservation. J Food Sci 62:1158–1162. https://doi.org/10.1111/j.1365-2621.1997.tb12235.x

DiBerardinis LJ, Baum JS, First MW et al (2013) Guidelines for laboratory design: health, safety, and environmental considerations. John Wiley & Sons

Dobry DE, Settell DM, Baumann JM et al (2009) A model-based methodology for spray-drying process development. J Pharm Innov 4:133–142. https://doi.org/10.1007/s12247-009-9064-4

Dobry DE, Settell DM, Baumann JM (2015) Chapter 9: Spray drying and scale-up. In: Newman A (ed) Pharmaceutical amorphous solid dispersions. John Wiley & Sons Inc., Hoboken, NJ, USA, pp 1–26

Domínguez JM (2011) Drying. In: Comprehensive biotechnology. Elsevier, pp 727–735

Dormenval C, Lokras A, Cano-Garcia G et al (2019) Identification of factors of importance for spray drying of small interfering RNA-loaded lipidoid-polymer hybrid nanoparticles for inhalation. Pharm Res 36:142. https://doi.org/10.1007/s11095-019-2663-y

Douglas RG, Samant VB (2018) The vaccine industry. In: Plotkin's vaccines. Elsevier, pp 41–50.e1

DuBose D, Settell D, Baumann J (2013) Spray-dried dispersions: efficient scale-up strategy for spray-dried amorphous dispersions. Drug Dev Technol 13:54–62

Dunne C (2020) Aseptic spray dryer application. ChargePoint Technology, www.thechargepoint.com, pp 1–10

Emami F, Vatanara A, Park E, Na D (2018) Drying technologies for the stability and bioavailability of biopharmaceuticals. Pharmaceutics 10:131. https://doi.org/10.3390/pharmaceutics10030131

EMEA (2006) IHC topic Q7 good manufacturing practice for active pharmaceutical ingredients, step 5, pp 1–48

Evaluate (2018) Evaluate pharma world preview 2018, Outlook to 2024

FDA (2004) Guidance for industry, sterile drug products produced by aseptic processing—current good manufacturing practice

FDA (2022) Vaccines licensed for use in the United States, content current as of 1/31/2022. https://www.fda.gov/vaccines-blood-biologics/vaccines/vaccines-licensed-use-united-states

Flow Sciences Inc. (2020) Buchi B-290 mini-spray dryer enclosure for API processing. https://flowsciences.com/buchi-b-290-mini-spray-dryer-enclosure-for-api-processing

Fourie P, Germishuizen W, Wong Y-L, Edwards D (2008) Spray drying TB vaccines for pulmonary administration. Expert Opin Biol Ther 8:857–863. https://doi.org/10.1517/14712598.8.7.857

Friebel C, Steckel H (2010) Single-use disposable dry powder inhalers for pulmonary drug delivery. Expert Opin Drug Deliv 7:1359–1372. https://doi.org/10.1517/17425247.2010.538379

Gala RP, Popescu C, Knipp GT et al (2017) Physicochemical and preclinical evaluation of a novel buccal measles vaccine. AAPS PharmSciTech 18:283–292. https://doi.org/10.1208/s12249-016-0566-3

Gander B, Merkle HP, Nguyen VP, Ho N-T (1995a) A new thermodynamic model to predict protein encapsulation efficiency in Poly(lactide) microspheres. J Phys Chem 99:16144–16148. https://doi.org/10.1021/j100043a066

Gander B, Wehrli E, Alder R, Merkle HP (1995b) Quality improvement of spray-dried, protein-loaded d, l-PLA microspheres by appropriate polymer solvent selection. J Microencapsul 12:83–97. https://doi.org/10.3109/02652049509051129

Gander B, Johansen P, Nam-Trân H, Merkle HP (1996) Thermodynamic approach to protein microencapsulation into poly(D, L-lactide) by spray drying. Int J Pharm 129:51–61. https://doi.org/10.1016/0378-5173(95)04240-7

Garcea RL, Meinerz NM, Dong M et al (2020) Single-administration, thermostable human papillomavirus vaccines prepared with atomic layer deposition technology. NPJ Vaccines 5:45. https://doi.org/10.1038/s41541-020-0195-4

Garcia-Contreras L, Wong Y-L, Muttil P et al (2008) Immunization by a bacterial aerosol. Proc Natl Acad Sci 105:4656–4660. https://doi.org/10.1073/pnas.0800043105

Gaspar F, Vicente J, Neves F, Authelin J-R (2014) Chapter 8: spray drying: scale-up and manufacturing. In: Shah N et al (ed) Amorphous solid dispersions, advances in delivery science and technology. Controlled Release Society, pp 261–302

GAVI (2021a) Facts & figures, updated 22 Feb 2021. https://www.gavi.org/sites/default/files/document/2021a/Gavi-Facts-and-figures-February-21.pdf

GAVI (2021b) Annual progress report, year 4 of our five-year strategy. https://www.gavi.org/sites/default/files/programmes-impact/our-impact/apr/Gavi-Progress-Report-2019_1.pdf

GEA (2013) GEA Niro Spray Dryer SDMicro, Ideal for research and development, pp 1–4

GEA (2016) Containment technology for solid dosage processing, pp 1–20

GEA (2019) High quality vaccine production with GEA technology

GEA (2021) Spray dried vaccines, customer case. https://www.gea.com/en/customer-cases/spray-dried-vaccines.jsp. Accessed 28 Jan 2021

GEA Niro (2017) Aseptic spray dryers: an effective method for sterile solid drug manufacture, BNA 971/GB 1011 CBC

GEA Process Engineering A/S (2020a) GEA pharmaceutical spray drying, brochure BTC 0012-EN 052020, pp 1–16

GEA Process Engineering A/S (2020b) GEA spray drying: small-scale solutions for R&D and production, brochure BTC0100-EN 0520

Geng O (2018) SPX FLOW aseptic spray drying, 13 March 2018. https://file.vogel.com.cn/2018/08/31/5b8905c261000.pdf

Germishuizen WA, Venter L, Khosa A et al (2010) Establishing a platform for spray drying inhalable vaccines in South Africa. Eur Cells Mater 19:16

Ghaemmaghamian Z, Zarghami R, Walker G et al (2022) Stabilizing vaccines via drying: quality by design considerations. Adv Drug Deliv Rev 187:114313. https://doi.org/10.1016/j.addr.2022.114313

Gikanga B, Turok R, Hui A et al (2015) Manufacturing of high-concentration monoclonal antibody formulations via spray drying–the road to manufacturing scale. PDA J Pharm Sci Technol 69:59–73. https://doi.org/10.5731/pdajpst.2015.01003

Gill D (2014) Heat stable oral rotavirus vaccine, MSD Wellcome Trust Hilleman Laboratories. In: 11th International rotavirus symposium, 4th Sept 2014, New Delhi, India, pp 1–15

Gill D (2018) Development of a heat-stable Rotavirus vaccine using innovative delivery technology, MSD Wellcome Trust Hilleman Laboratories. In: 13th International rotavirus symposium, 29–31 Aug 2018, Minsk, Belarus, pp 1–15

Gill D, Saigal N, Kale S et al (2017) Thermostable spray dried rotavirus vaccine formulation and process thereof. Patent US 2017/0173145 A1, PCT filed: 17 July 2015, pp 1–17

Gill D, Saigal N, Kale S et al (2020) Thermostable spray dried rotavirus vaccine formulation and process thereof, MSD Wellcome Trust Hilleman Laboratories Pvt. Ltd. Patent US10556000 B2, PCT filed: 17 July 2015, pp 1–17

Gladd T (2014) Using spray drying for flu vaccines: what challenges do we need to overcome? 14 April 2014. www.bioprocessonline.com, https://www.bioprocessonline.com/doc/using-spray-drying-for-flu-vaccines-what-challenges-do-we-need-to-overcome-0001

Global Data (2021) COVID-19 vaccine market set to reach $19.5 bn by 2026. https://www.globaldata.com/covid-19-vaccine-market-set-reach-19-5bn-2026/. Accessed 14 July 2021

Globe Newswire (2018) Global vaccine market analysis & forecast (2018–2028): a $103.57 billion opportunity, growing at a CAGR of 11.02%

Gomez M, Vehring R (2022) Spray drying and particle engineering in dosage form design for global vaccines. J Aerosol Med Pulm Drug Deliv 35:1–18. https://doi.org/10.1089/jamp.2021.0056

Gomez M, Archer M, Barona D et al (2021a) Microparticle encapsulation of a tuberculosis subunit vaccine candidate containing a nanoemulsion adjuvant via spray drying. Eur J Pharm Biopharm 163:23–37. https://doi.org/10.1016/j.ejpb.2021.03.007

Gomez M, McCollum J, Wang H et al (2021b) Development of a formulation platform for a spray-dried, inhalable tuberculosis vaccine candidate. Int J Pharm 593:120121. https://doi.org/10.1016/j.ijpharm.2020.120121

Gomez M, McCollum J, Wang H et al (2021c) Evaluation of the stability of a spray-dried tuberculosis vaccine candidate designed for dry powder respiratory delivery. Vaccine 39:5025–5036. https://doi.org/10.1016/j.vaccine.2021.07.002

Gomez M, Ahmed M, Das S et al (2022) Development and testing of a spray-dried tuberculosis vaccine candidate in a mouse model. Front Pharmacol 12:1–23. https://doi.org/10.3389/fphar.2021.799034

Grasmeijer N, Stankovic M, de Waard H et al (2013) Unraveling protein stabilization mechanisms: vitrification and water replacement in a glass transition temperature controlled system. Biochim Biophys Acta—Proteins Proteomics 1834:763–769. https://doi.org/10.1016/j.bbapap.2013.01.020

Grasmeijer N, Tiraboschi V, Woerdenbag HJ et al (2019) Identifying critical process steps to protein stability during spray drying using a vibrating mesh or a two-fluid nozzle. Eur J Pharm Sci 128:152–157. https://doi.org/10.1016/j.ejps.2018.11.027

Greb E (2009) Is spray drying a viable alternative to lyophilization. PharmTech, pp 1–2

Greenwood B (2014) The contribution of vaccination to global health: past, present and future. Philos Trans R Soc B Biol Sci 369:20130433. https://doi.org/10.1098/rstb.2013.0433

Guntur VP, Dhand R (2007) Inhaled insulin: extending the horizons of inhalation therapy. Respir Care 52(911–922):12p

Hasija M, Li L, Rahman N, Ausar SF (2013) Forced degradation studies: an essential tool for the formulation development of vaccines. Vaccine Dev Ther 11.https://doi.org/10.2147/VDT.S41998

Hellfritzsch M, Scherließ R (2019) Mucosal vaccination via the respiratory tract. Pharmaceutics 11:1–24. https://doi.org/10.3390/pharmaceutics11080375

Höfliger H (2021) Dosing & filling: sophisticated processes, individual solutions, machines. https://www.hoefliger.com/en/machines-technology/dosing-filling

Huang Y, Dai W-G (2014) Fundamental aspects of solid dispersion technology for poorly soluble drugs. Acta Pharm Sin B 4:18–25. https://doi.org/10.1016/j.apsb.2013.11.001

Huyge K (2012) Dry powder vaccines for mass vaccination of poultry: new formulations, comparison with liquid vaccines and feasibility to decrease adverse vaccinal reactions. Doctoral Thesis, Ghent University

Huyge K, Van Reeth K, De Beer T et al (2012) Suitability of differently formulated dry powder Newcastle disease vaccines for mass vaccination of poultry. Eur J Pharm Biopharm 80:649–656. https://doi.org/10.1016/j.ejpb.2011.11.018

ICH (1995) ICH quality guideline Q5C: stability testing of biotechnological/biological product. https://www.ich.org/page/quality-guidelines

ICH (2003) ICH quality guideline Q1A(R2): stability testing of new drug substances and products. https://www.ich.org/page/quality-guidelines

ICH (2009) ICH quality guideline Q8(R2) Pharmaceutical development. https://www.ich.org/page/quality-guidelines

ICH (2010) ICH quality guideline Q7: good manufacturing practice guide for active pharmaceutical ingredients. https://www.ich.org/page/quality-guidelines

Ignjatović J, Đuriš J, Cvijić S et al (2021) Development of solid lipid microparticles by melt-emulsification/spray-drying processes as carriers for pulmonary drug delivery. Eur J Pharm Sci 156:105588. https://doi.org/10.1016/j.ejps.2020.105588

IMA Life (2012) Microdosing series: powder microdosers for aseptic environments, Brochure February 2012-991ZS021101, pp 1–12

IMA Life (2020) SMART PWD powder filling & stoppering machine. https://ima.it/pharma/machine/smart-pwd/

Ingvarsson PT, Yang M, Mulvad H et al (2013) Engineering of an inhalable DDA/TDB liposomal adjuvant: a quality-by-design approach towards optimization of the spray drying process. Pharm Res 30:2772–2784. https://doi.org/10.1007/s11095-013-1096-2

ISPE (2018) ISPE baseline guide: sterile product manufacturing facilities

Jackson T, Peterson G, Naunton M et al (2018) Over half of vaccines are wasted globally for these simple reasons. World Economic Forum. https://www.weforum.org/agenda/2018/07/the-biggest-hurdle-to-universal-vaccination-might-just-be-a-fridge

Jermain SV, Brough C, Williams RO (2018) Amorphous solid dispersions and nanocrystal technologies for poorly water-soluble drug delivery—an update. Int J Pharm 535:379–392. https://doi.org/10.1016/j.ijpharm.2017.10.051

Jin T, Tsao E (2013) Powder vaccines for pulmonary delivery. In: Flower DR, Perrie Y (eds) Immunomic discovery of adjuvants and candidate subunit vaccines. Springer, New York, New York, NY, pp 291–301

Jin TH, Tsao E, Goudsmit J et al (2010) Stabilizing formulations for inhalable powders of an adenovirus 35-vectored tuberculosis (TB) vaccine (AERAS-402). Vaccine 28:4369–4375. https://doi.org/10.1016/j.vaccine.2010.04.059

Johansen P, Moon L, Tamber H et al (1999a) Immunogenicity of single-dose diphtheria vaccines based on PLA/PLGA microspheres in guinea pigs. Vaccine 18:209–215. https://doi.org/10.1016/S0264-410X(99)00191-7

Johansen P, Tamber H, Merkle HP, Gander B (1999b) Diphtheria and tetanus toxoid microencapsulation into conventional and end-group alkylated PLA/PLGAs. Eur J Pharm Biopharm 47:193–201. https://doi.org/10.1016/S0939-6411(98)00095-2

Johansen P, Merkle HP, Gander B (2000) Technological considerations related to the up-scaling of protein microencapsulation by spray-drying. Eur J Pharm Biopharm 50:413–417. https://doi.org/10.1016/S0939-6411(00)00123-5

Jones RM, Burke M, Dubose D et al (2017) Stability and pre-formulation development of a plant-produced anthrax vaccine candidate. Vaccine 35:5463–5470. https://doi.org/10.1016/j.vaccine.2016.12.009

Kaddar M (2013) Global vaccine market features and trends. In: WHO, IVB, Geneva. https://www.who.int/immunization/programmes_systems/procurement/market/world_vaccine_market_trends.pdf

Kanojia G (2018) Drying made easy: spray drying a promising technology for the production of stable vaccine and therapeutic protein formulations, PhD thesis, University of Groningen

Kanojia G, Willems G-J, Frijlink HW et al (2016) A design of experiment approach to predict product and process parameters for a spray dried influenza vaccine. Int J Pharm 511:1098–1111. https://doi.org/10.1016/j.ijpharm.2016.08.022

Kanojia G, ten Have R, Soema PC et al (2017) Developments in the formulation and delivery of spray dried vaccines. Hum Vaccin Immunother 13:2364–2378. https://doi.org/10.1080/21645515.2017.1356952

Kanojia G, Raeven RHM, van der Maas L et al (2018a) Development of a thermostable spray dried outer membrane vesicle pertussis vaccine for pulmonary immunization. J Control Release 286:167–178. https://doi.org/10.1016/j.jconrel.2018.07.035

Kanojia G, ten Have R, Brugmans D et al (2018b) The effect of formulation on spray dried Sabin inactivated polio vaccine. Eur J Pharm Biopharm 129:21–29. https://doi.org/10.1016/j.ejpb.2018.05.021

Karimi K, Pallagi E, Szabó-Révész P et al (2016) Development of a microparticle-based dry powder inhalation formulation of ciprofloxacin hydrochloride applying the quality by design approach. Drug Des Devel Ther 10:3331–3343. https://doi.org/10.2147/DDDT.S116443

Kartoglu U, Milstien J (2014) Tools and approaches to ensure quality of vaccines throughout the cold chain. Expert Rev Vaccines 13:843–854. https://doi.org/10.1586/14760584.2014.923761

Kawabata Y, Wada K, Nakatani M et al (2011) Formulation design for poorly water-soluble drugs based on biopharmaceutics classification system: basic approaches and practical applications. Int J Pharm 420:1–10. https://doi.org/10.1016/j.ijpharm.2011.08.032

Kisich KO, Higgins MP, Park I et al (2011) Dry powder measles vaccine: particle deposition, virus replication, and immune response in cotton rats following inhalation. Vaccine 29:905–912. https://doi.org/10.1016/j.vaccine.2010.10.020

Kols A, Sherris J (2000) HPV vaccines: promise and challenges, July 2000. PATH

Konstan MW, Geller DE, Minić P et al (2011) Tobramycin inhalation powder for P. aeruginosa infection in cystic fibrosis: the EVOLVE trial. Pediatr Pulmonol 46:230–238. https://doi.org/10.1002/ppul.21356

Kristensen D (2012) Summary of stability data for investigational formulations of vaccines, 9 May 2012

Kunda NK, Alfagih IM, Miyaji EN et al (2015) Pulmonary dry powder vaccine of pneumococcal antigen loaded nanoparticles. Int J Pharm 495:903–912. https://doi.org/10.1016/j.ijpharm.2015.09.034

Kunda NK, Wafula D, Tram M et al (2016) A stable live bacterial vaccine. Eur J Pharm Biopharm 103:109–117. https://doi.org/10.1016/j.ejpb.2016.03.027

Kunda NK, Peabody J, Zhai L et al (2019) Evaluation of the thermal stability and the protective efficacy of spray-dried HPV vaccine, Gardasil® 9. Hum Vaccin Immunother 15:1995–2002. https://doi.org/10.1080/21645515.2019.1593727

Lam J, Vaughan S, Parkins MD (2013) Tobramycin inhalation powder (TIP): an efficient treatment strategy for the management of chronic Pseudomonas Aeruginosa infection in cystic fibrosis. Clin Med Insights Circ Respir Pulm Med 7:CCRPM.S10592. https://doi.org/10.4137/CCRPM.S10592

Landman WJM, Huyge K, Remon JP et al (2015) Comparison of Newcastle disease vaccine administered as powder or liquid in relation to the serum antibody response and adverse vaccinal reactions in broilers. Avian Pathol 44:114–123. https://doi.org/10.1080/03079457.2015.1007920

Landman WJM, Vervaet C, Remon JP et al (2017) Primary Newcastle disease vaccination of broilers: comparison of the antibody seroresponse and adverse vaccinal reaction after eye–nose drop or coarse spray application, and implication of the results for a previously developed coarse dry powder vaccine. Avian Pathol 46:451–461. https://doi.org/10.1080/03079457.2017.1307941

Langford A, Ohtake S (2019) Spray drying of biotherapeutic compounds. In: Nema S, Ludwig JD (eds) Parenteral medications, Ch 43. Taylor & Francis Incorporated, pp 929–945

Langford A, Bhatnagar B, Walters R et al (2018a) Drying of biopharmaceuticals: recent developments, new technologies and future direction. Japan J Food Eng 19:15–24. https://doi.org/10.11301/jsfe.18514

Langford A, Bhatnagar B, Walters R et al (2018b) Drying technologies for biopharmaceutical applications: recent developments and future direction. Dry Technol 36:677–684. https://doi.org/10.1080/07373937.2017.1355318

LeClair DA, Cranston ED, Xing Z, Thompson MR (2016a) Optimization of spray drying conditions for yield, particle size and biological activity of thermally stable viral vectors. Pharm Res 33:2763–2776. https://doi.org/10.1007/s11095-016-2003-4

LeClair DA, Cranston ED, Xing Z, Thompson MR (2016b) Evaluation of excipients for enhanced thermal stabilization of a human type 5 adenoviral vector through spray drying. Int J Pharm 506:289–301. https://doi.org/10.1016/j.ijpharm.2016.04.067

LeClair DA, Li L, Rahman N et al (2019) Stabilization of HSV-2 viral vaccine candidate by spray drying. Int J Pharm 569:118615. https://doi.org/10.1016/j.ijpharm.2019.118615

Ledet GA, Graves RA, Bostanian LA, Mandal TK (2015) Spray-drying of biopharmaceuticals. In: Varshney D, Singh M (eds) Lyophilized biologics and vaccines. Springer, New York, New York, NY, pp 273–297

LiCalsi C, Christensen T, Bennett JV et al (1999) Dry powder inhalation as a potential delivery method for vaccines. Vaccine 17:1796–1803. https://doi.org/10.1016/S0264-410X(98)00438-1

Lin W-H, Griffin DE, Rota PA et al (2011) Successful respiratory immunization with dry powder live-attenuated measles virus vaccine in rhesus macaques. Proc Natl Acad Sci 108:2987–2992. https://doi.org/10.1073/pnas.1017334108

Lloyd J (2000) Technologies for vaccine delivery in the 21st century, WHO/V&B/00.35, Dec 2000

Lovalenti PM, Truong-Le V (2020) Foam drying. In: Ohtake S, Izutsu K-I, Lechuga-Ballesteros D (eds) Drying technologies for biotechnology and pharmaceutical applications. Wiley, pp 257–282

Lovalenti PM, Anderl J, Yee L et al (2016) Stabilization of live attenuated influenza vaccines by freeze drying, spray drying, and foam drying. Pharm Res 33:1144–1160. https://doi.org/10.1007/s11095-016-1860-1

Lu D, Garcia-Contreras L, Xu D et al (2007) Poly (Lactide-co-Glycolide) microspheres in respirable sizes enhance an in vitro T Cell response to recombinant mycobacterium tuberculosis antigen 85B. Pharm Res 24:1834–1843. https://doi.org/10.1007/s11095-007-9302-8

Lu D, Garcia-Contreras L, Muttil P et al (2010) Pulmonary immunization using antigen 85-B polymeric microparticles to boost tuberculosis immunity. AAPS J 12:338–347. https://doi.org/10.1208/s12248-010-9193-1

Luy B, Stamato H (2020) Spray freeze drying. In: Ohtake S, Izutsu K-I, Lechuga-Ballesteros D (eds) Drying technologies for biotechnology and pharmaceutical applications. Wiley, pp 217–237

Maa Y-F, Nguyen P-AT, Hsu SW (1998) Spray-drying of air-liquid interface sensitive recombinant human growth hormone. J Pharm Sci 87:152–159. https://doi.org/10.1021/js970308x

Maa Y-F, Zhao L, Payne LG, Chen D (2003) Stabilization of alum-adjuvanted vaccine dry powder formulations: mechanism and application. J Pharm Sci 92:319–332. https://doi.org/10.1002/jps.10294

Maa Y-F, Ameri M, Shu C et al (2007) Hepatitis-B surface antigen (HBsAg) powder formulation: process and stability assessment. Curr Drug Deliv 4:57–67. https://doi.org/10.2174/156720107779314758

Maltesen MJ, van de Weert M (2008) Drying methods for protein pharmaceuticals. Drug Discov Today Technol 5:e81–e88. https://doi.org/10.1016/j.ddtec.2008.11.001

Manufacturing Chemist (2010) No need for the cooler, 6 Sept 2010. www.manufacturingchemist.com

Manufacturing Chemist (2016) A green light for cGMP aseptic spray drying, 21 Jan 2016. https://www.manufacturingchemist.com/news/article_page/A_green_light_for_cGMP_aseptic_spray_drying/114844

Markets and Markets (2020) Vaccines market—global forecast to 2024, published Jan 2020

Masters K (1976) Chapter 16: Applications in the pharmaceutical-biochemical industry. In: Spray drying: an introduction to principles, operational practice and applications, 2nd ed. John Wiley & Sons, New York

Masters K (1985) Spray drying handbook, 4th edn. George Godwin, London

Masters K (2002) Spray drying in practice. SprayDryConsult International ApS, Denmark

Masters K, Vestergaard I (1975) Aseptic and closed cycle spray drying in pharmaceuticals/biochemicals manufacture. Process Biochem 10:21–29

McAdams D, Chen D, Kristensen D (2012) Spray drying and vaccine stabilization. Expert Rev Vaccines 11:1211–1219. https://doi.org/10.1586/erv.12.101

McGuire MT (1998) WHO good manufacturing practices for pharmaceutical products: main principles. who-GMP-Annex 2 170–170 p.

Mehanny M, Boese A, Bornamehr B et al (2022) Spray-dried pneumococcal membrane vesicles are promising candidates for pulmonary immunization. Int J Pharm 121794.https://doi.org/10.1016/j.ijpharm.2022.121794

Moderna (2020) Fact sheet for healthcare providers administring vaccine (vaccination providers) emergency use authorization (EUA) of the Moderna COVID-19 vaccine to prevent Corona Virus disease 2019 (COVID-19) in individuals 18 years of age and older, pp 1–5

Mogavero M (2014) Insight into the aseptic spray drying process, 17 Dec 2014. www.powderbulksolids.com. pp 4–7

Morello M, Krone CL, Dickerson S et al (2009) Dry-powder pulmonary insufflation in the mouse for application to vaccine or drug studies. Tuberculosis 89:371–377. https://doi.org/10.1016/j.tube.2009.07.001

Morgan BA, Manser M, Jeyanathan M et al (2020) Effect of shear stresses on adenovirus activity and aggregation during atomization to produce thermally stable vaccines by spray drying. ACS Biomater Sci Eng 6:4304–4313. https://doi.org/10.1021/acsbiomaterials.0c00317

Mujumdar AS, Alterman DS (1997) Chapter 10: Drying in the pharmaceutical and biotechnology fields. Handbook of downstream processing. Chapman & Hall, London, UK, pp 235–260

Muttil P, Prego C, Garcia-Contreras L et al (2010a) Immunization of guinea pigs with novel Hepatitis B antigen as nanoparticle aggregate powders administered by the pulmonary route. AAPS J 12:330–337. https://doi.org/10.1208/s12248-010-9192-2

Muttil P, Pulliam B, Garcia-Contreras L et al (2010b) Pulmonary immunization of guinea pigs with Diphtheria CRM-197 antigen as nanoparticle aggregate dry powders enhance local and systemic immune responses. AAPS J 12:699–707. https://doi.org/10.1208/s12248-010-9229-6

Nabel GJ (2013) Designing tomorrow's vaccines. N Engl J Med 368:551–560. https://doi.org/10.1056/NEJMra1204186

NKP (2020) Product catalogue 2019–2020, N.K.P. Pharma Pvt. Ltd. https://www.nkppharma.com

Nova Laboratories (2021) Stabilization technologies. https://www.novalabs.co.uk/stabilisation-technologies. Accessed 28 Jan 2021

Ohtake S, Martin RA, Yee L et al (2010) Heat-stable measles vaccine produced by spray drying. Vaccine 28:1275–1284. https://doi.org/10.1016/j.vaccine.2009.11.024

Ohtake S, Martin R, Saxena A et al (2011) Room temperature stabilization of oral, live attenuated *Salmonella enterica serovar* Typhi-vectored vaccines. Vaccine 29:2761–2771. https://doi.org/10.1016/j.vaccine.2011.01.093

Ohtake S, Lechuga-Ballesteros D, Truong-Le V, Patzer EJ (2015) Strategies for heat-stable vaccines (Ch 10). In: Vaccine development and manufacturing. Wiley, pp 287–318

Ohtake S, Izutsu K, Lechuga-Ballesteros D (2020) Drying technologies for biotechnology and pharmaceutical applications. Wiley, VCH

Pakowski Z, Mujumdar A (2014) Drying of pharmaceutical products. In: Mujumdar AS (ed) Handbook of industrial drying, 4th ed. CRC Press

Pastor M, Esquisabel A, Talavera A et al (2013a) An approach to a cold chain free oral cholera vaccine: in vitro and in vivo characterization of Vibrio cholerae gastro-resistant microparticles. Int J Pharm 448:247–258. https://doi.org/10.1016/j.ijpharm.2013.02.057

Pastor M, Pedraz JL, Esquisabel A (2013b) The state-of-the-art of approved and under-development cholera vaccines. Vaccine 31:4069–4078. https://doi.org/10.1016/j.vaccine.2013.06.096

Pastor M, Esquisabel A, Marquínez I et al (2014) Cellulose acetate phthalate microparticles containing Vibrio cholerae : steps toward an oral cholera vaccine. J Drug Target 22:478–487. https://doi.org/10.3109/1061186X.2014.888071

Patton JS, Bukar JG, Eldon MA (2004) Clinical pharmacokinetics and pharmacodynamics of inhaled insulin. Clin Pharmacokinet 43:781–801. https://doi.org/10.2165/00003088-200443120-00002

Peabody J, Muttil P, Chackerian B, Tumban E (2017) Characterization of a spray-dried candidate HPV L2-VLP vaccine stored for multiple years at room temperature. Papillomavirus Res 3:116–120. https://doi.org/10.1016/j.pvr.2017.03.004

Pfizer-BioNTech (2020) Fact sheet for healthcare providers administring vaccine, emergency use authorization (EUA) of the Pfizer-BioNTech COVID-19 vaccine to prevent coronavirus disease 2019 (COVID-19) in individuals 16 years of age and older, pp 1–6

Pinto JT, Faulhammer E, Dieplinger J et al (2021) Progress in spray-drying of protein pharmaceuticals: literature analysis of trends in formulation and process attributes. Dry Technol 0:1–32. https://doi.org/10.1080/07373937.2021.1903032

Pisano R, Arsiccio A, Capozzi LC, Trout BL (2019) Achieving continuous manufacturing in lyophilization: technologies and approaches. Eur J Pharm Biopharm 142:265–279. https://doi.org/10.1016/j.ejpb.2019.06.027

PR Newswire (2018) Global vaccines market forecasts 2018–2025: strong pipeline, rising adoption of vaccination, increasing government focus on immunization programs & technological progress in vaccine administration, 12 Dec 2018

Preston KB, Randolph TW (2021) Stability of lyophilized and spray dried vaccine formulations. Adv Drug Deliv Rev 171:50–61. https://doi.org/10.1016/j.addr.2021.01.016

Price DN, Kunda NK, Ellis R, Muttil P (2020) Design and optimization of a temperature-stable dry powder BCG vaccine. Pharm Res 37:11. https://doi.org/10.1007/s11095-019-2739-8

Putranto A, Chen XD (2019) Drying in biotechnology, 3rd ed. Elsevier

Rappuoli R (2011) Twenty-first century vaccines. Philos Trans R Soc B Biol Sci 366:2756–2758. https://doi.org/10.1098/rstb.2011.0075

Rappuoli R, Mandl CW, Black S, De Gregorio E (2011) Vaccines for the twenty-first century society. Nat Rev Immunol 11:865–872. https://doi.org/10.1038/nri3085

Rappuoli R, De Gregorio E, Del Giudice G et al (2021) Vaccinology in the post−COVID-19 era. Proc Natl Acad Sci 118:e2020368118. https://doi.org/10.1073/pnas.2020368118

Research and Markets (2018) Vaccines—global market outlook (2018–2027)

Research and Markets (2020) Vaccines global market report 2020, December 2019

Rodrigues TC, Oliveira MLS, Soares-Schanoski A et al (2018) Mucosal immunization with PspA (Pneumococcal surface protein A)-adsorbed nanoparticles targeting the lungs for protection against pneumococcal infection. PLoS ONE 13:e0191692. https://doi.org/10.1371/journal.pone.0191692

ROMACO (2018) Aseptic powder dosing solutions, Brochure 2018/06. https://www.romaco.com/files/Dokumente/delivering-solutions/romaco_aseptic_powder_filling-EN.pdf

Roser B (1991) Trehalose, a new approach to premium dried foods. Trends Food Sci Technol 2:166–169. https://doi.org/10.1016/0924-2244(91)90671-5

Roser B (2005) Sterile spray drying for stable liquid 21st century pharmaceuticals. Innov Pharm Technol. http://iptonline.com/articles/public/IPT192005_p50+nonprintable.pdf

Roser B (2009) Aseptic spray drying for stable liquid 21st century pharmaceuticals. Cambridge Biostability Ltd, UK. https://freelancecopy.files.wordpress.com/2009/09/cb-aseptic-spray-drying.pdf

Rosini R, Barocchi MA, Rappuoli R (2011) Microbial vaccine design: the reverse vaccinology approach. Development of vaccines. John Wiley & Sons Inc., Hoboken, NJ, USA, pp 1–18

Rossi I, Spagnoli G, Buttini F et al (2021) A respirable HPV-L2 dry-powder vaccine with GLA as amphiphilic lubricant and immune-adjuvant. J Control Release 340:209–220. https://doi.org/10.1016/j.jconrel.2021.11.002

Roth F, Scherließ R (2021) Protein DPI production: discriminating destabilising influences during spray drying. In: Drug delivery to lungs. https://ddl-conference.com/wp-content/uploads/2021/11/03.Roth_.pdf

Saboo S, Tumban E, Peabody J et al (2016) Optimized formulation of a thermostable spray-dried virus-like particle vaccine against human Papillomavirus. Mol Pharm 13:1646–1655. https://doi.org/10.1021/acs.molpharmaceut.6b00072

Saluja V, Amorij J-P, Kapteyn JC et al (2010) A comparison between spray drying and spray freeze drying to produce an influenza subunit vaccine powder for inhalation. J Control Release 144:127–133. https://doi.org/10.1016/j.jconrel.2010.02.025

Sawatdee S, Srichana T (2021) Single-use dry powder inhalers for pulmonary drug delivery. Handbook of lung targeted drug delivery systems. CRC Press, Boca Raton, pp 503–511

Scherließ R, Ajmera A, Dennis M et al (2014) Induction of protective immunity against H1N1 influenza A(H1N1)pdm09 with spray-dried and electron-beam sterilised vaccines in non-human primates. Vaccine 32:2231–2240. https://doi.org/10.1016/j.vaccine.2014.01.077

Schoenmaker L, Witzigmann D, Kulkarni JA et al (2021) mRNA-lipid nanoparticle COVID-19 vaccines: structure and stability. Int J Pharm 601:120586. https://doi.org/10.1016/j.ijpharm.2021.120586

Schwartzbach H (2011) Achieving aseptic drying with spray drying technologies. Pharm Technol Eur 23:90–92

Scrafford J (2020) An integrated approach to spray drying: from scale-up to manufacturing. Catalent, Pharm Technol Spec Spons Sect 1–6

Searles J, Cherian M (2015) Alternatives to vial lyophilization. Lyophilized biologics and vaccines. Springer, New York, New York, NY, pp 257–271

Searles J, Mohan G (2010) Spray drying of biopharmaceuticals and vaccines. Formulation and process development strategies for manufacturing biopharmaceuticals. John Wiley & Sons Inc., Hoboken, NJ, USA, pp 739–761

Sharma A, Khamar D, Cullen S et al (2021) Innovative drying technologies for biopharmaceuticals. Int J Pharm 609:121115. https://doi.org/10.1016/j.ijpharm.2021.121115

Shastri PN, Kim M, Quan F et al (2012) Immunogenicity and protection of oral influenza vaccines formulated into microparticles. J Pharm Sci 101:3623–3635. https://doi.org/10.1002/jps.23220

Sibum I, Hagedoorn P, de Boer AH et al (2018) Challenges for pulmonary delivery of high powder doses. Int J Pharm 548:325–336. https://doi.org/10.1016/j.ijpharm.2018.07.008

Siew A (2016) Exploring the use of aseptic spray drying in the manufacture of biopharmaceutical injectables. Pharm Technol 40:24–27

Smith DJ, Bot S, Dellamary L, Bot A (2003) Evaluation of novel aerosol formulations designed for mucosal vaccination against influenza virus. Vaccine 21:2805–2812. https://doi.org/10.1016/S0264-410X(03)00224-X

Snyder HE, Lechuga-Ballesteros D (2008) Spray drying: theory and pharmaceutical applications. In: Augsburger LL, Hoag SW (eds) Pharmaceutical dosage forms: tablets, 3rd ed, Ch 7, vol 1: Unit operations and mechanical properties. Informa Healthcare USA, Inc., New York, NY, pp 227–260

Soni G, Yadav KS, Gupta MK (2020) QbD based approach for formulation development of spray dried microparticles of erlotinib hydrochloride for sustained release. J Drug Deliv Sci Technol 57:101684. https://doi.org/10.1016/j.jddst.2020.101684

Sou T, Meeusen EN, de Veer M et al (2011) New developments in dry powder pulmonary vaccine delivery. Trends Biotechnol 29:191–198. https://doi.org/10.1016/j.tibtech.2010.12.009

Sou T, Morton DAV, Williamson M et al (2015) Spray-dried influenza antigen with trehalose and leucine produces an aerosolizable powder vaccine formulation that induces strong systemic and mucosal immunity after pulmonary administration. J Aerosol Med Pulm Drug Deliv 28:361–371. https://doi.org/10.1089/jamp.2014.1176

SPX FLOW Inc. (2017) Anhydro spray drying for the pharmaceutical industry, Brochure ANH-116-GB, Issued 12/2017, pp 1–8

Ståhl K, Claesson M, Lilliehorn P et al (2002) The effect of process variables on the degradation and physical properties of spray dried insulin intended for inhalation. Int J Pharm 233:227–237. https://doi.org/10.1016/S0378-5173(01)00945-0

Syntegon (2021) AFG 5000—filling and sealing machines for pharmaceutical powder and antibiotics. https://www.syntegon.com/products/afg-5000-filling-and-sealing-machines-pharmaceutical-powder-and-antibiotics?t=1644&tg=1587&vpa=&kp=1559&prod=1508&ind

Thakur A, Ingvarsson PT, Schmidt ST et al (2018) Immunological and physical evaluation of the multistage tuberculosis subunit vaccine candidate H56/CAF01 formulated as a spray-dried powder. Vaccine 36:3331–3339. https://doi.org/10.1016/j.vaccine.2018.04.055

The Baker Company (2010) BioPROtect® III Class II style walk-in clean air and containment enclosures. Datasheet, 134-043–10, 11/10, pp 1–6

Thorat BN, Sett A, Mujumdar AS (2020) Drying of vaccines and biomolecules. Dry Technol 0:1–23. https://doi.org/10.1080/07373937.2020.1825293

Tlaxca JL, Ellis S, Remmele RL (2015) Live attenuated and inactivated viral vaccine formulation and nasal delivery: potential and challenges. Adv Drug Deliv Rev 93:56–78. https://doi.org/10.1016/j.addr.2014.10.002

Tomar J, Tonnis WF, Patil HP et al (2019) Pulmonary immunization: deposition site is of minor relevance for influenza vaccination but deep lung deposition is crucial for hepatitis B vaccination. Acta Pharm Sin B 9:1231–1240. https://doi.org/10.1016/j.apsb.2019.05.003

Toniolo SP, Afkhami S, Mahmood A et al (2019) Excipient selection for thermally stable enveloped and non-enveloped viral vaccine platforms in dry powders. Int J Pharm 561:66–73. https://doi.org/10.1016/j.ijpharm.2019.02.035

Toniolo SP, Afkhami S, D'Agostino MR et al (2020) Spray dried VSV-vectored vaccine is thermally stable and immunologically active in vivo. Sci Rep 10:13349. https://doi.org/10.1038/s41598-020-70325-2

Tonnis WF, Kersten GF, Frijlink HW et al (2012) Pulmonary vaccine delivery: a realistic approach? J Aerosol Med Pulm Drug Deliv 25:249–260. https://doi.org/10.1089/jamp.2011.0931

Tonnis WF, Lexmond AJ, Frijlink HW et al (2013) Devices and formulations for pulmonary vaccination. Expert Opin Drug Deliv 10:1383–1397. https://doi.org/10.1517/17425247.2013.810622

Truong-Le V, Pham B (2007) Preservation of bioactive materials by spray drying, Patent US 7258873 B2, filed: 10 April, 2003, pp 1–30

Truong-Le V, Ohtake S, Martin RA et al (2012) Sonic low pressure spray drying, Patent US 8268354 B2, filed: 6 Nov, 2008, pp 1–17

Truong-Le V, Ohtake S, Martin RA et al (2014) Sonic low pressure spray drying. Patent US 8673357 B2, filed: 14 Aug, 2012, pp 1–18

Tumban E, Muttil P, Escobar CAA et al (2015) Preclinical refinements of a broadly protective VLP-based HPV vaccine targeting the minor capsid protein, L2. Vaccine 33:3346–3353. https://doi.org/10.1016/j.vaccine.2015.05.016

Turok RH, Rasmussen SS (2010) Process gas filtration, pp 1–7

Tyne AS, Chan JGY, Shanahan ER et al (2013) TLR2-targeted secreted proteins from Mycobacterium tuberculosis are protective as powdered pulmonary vaccines. Vaccine 31:4322–4329. https://doi.org/10.1016/j.vaccine.2013.07.022

Ubale RV, D'souza MJ, Infield DT et al (2013) Formulation of meningococcal capsular polysaccharide vaccine-loaded microparticles with robust innate immune recognition. J Microencapsul 30:28–41. https://doi.org/10.3109/02652048.2012.692402

Uddin AN, Bejugam NK, Gayakwad SG et al (2009) Oral delivery of gastro-resistant microencapsulated typhoid vaccine. J Drug Target 17:553–560. https://doi.org/10.1080/10611860903067301

Ulmer JB, Valley U, Rappuoli R (2006) Vaccine manufacturing: challenges and solutions. Nat Biotechnol 24:1377–1383. https://doi.org/10.1038/nbt1261

UNICEF (2019) Supply annual report 2019, pp 1–65

Upperton (2015) Spray drying—an alternative to freeze drying? Youtube-Video, 23 Feb, 2015. https://www.youtube.com/watch?v=uz2Hf_6u-N8

Vanderslott S, Dadonaite B, Roser M (2019) Vaccination, published online at OurWorldInData.org. Our World Data. https://ourworldindata.org/vaccination

Vass P, Démuth B, Hirsch E et al (2019) Drying technology strategies for colon-targeted oral delivery of biopharmaceuticals. J Control Release 296:162–178. https://doi.org/10.1016/j.jconrel.2019.01.023

Vehring R (2008) Pharmaceutical particle engineering via spray drying. Pharm Res 25:999–1022. https://doi.org/10.1007/s11095-007-9475-1

Vehring R, Foss WR, Lechuga-Ballesteros D (2007) Particle formation in spray drying. J Aerosol Sci 38:728–746. https://doi.org/10.1016/j.jaerosci.2007.04.005

Vehring R, Snyder H, Lechuga-Ballesteros D (2020) Spray drying. In: Ohtake S, Izutsu K, Lechuga-Ballesteros D (eds) Drying technologies for biotechnology and pharmaceutical applications. Wiley, pp 179–216

Velasquez LS, Shira S, Berta AN et al (2011) Intranasal delivery of Norwalk virus-like particles formulated in an in situ gelling, dry powder vaccine. Vaccine 29:5221–5231. https://doi.org/10.1016/j.vaccine.2011.05.027

Vodak DT, Dobry DE, Falk R et al (2012) Enhancing the immune response to intranasal vaccination with Norovirus VLPs—size matters. In: Drug delivery to the lungs (DDL 23), Edinburgh, pp 46–49

Walters RH, Bhatnagar B, Tchessalov S et al (2014) Next generation drying technologies for pharmaceutical applications. J Pharm Sci 103:2673–2695. https://doi.org/10.1002/jps.23998

Weers JG, Tarara TE, Clark AR (2007) Design of fine particles for pulmonary drug delivery. Expert Opin Drug Deliv 4:297–313. https://doi.org/10.1517/17425247.4.3.297

Weniger BG (2011) Research on new vaccine delivery methods: focus on influenza. In: Second WHO consultation on the global action plan for influenza vaccines (GAP-II), 12–14 July 2011, Geneva, Switzerland. https://www.who.int/influenza_vaccines_plan/resources/weniger.pdf

Weniger BG, Papania MJ (2013) Alternative vaccine delivery methods. Vaccines 1200–1231. https://doi.org/10.1016/B978-1-4557-0090-5.00063-X

WHO/UNICEF/World Bank (2009) State of the world's vaccines and immunization, 3rd ed. World Health Organization, Geneva. https://apps.who.int/iris/bitstream/handle/10665/44169/9789241563864_eng.pdf

WHO/UNICEF (2014) Vaccine vial monitor (VVM) assignments for different WHO-prequalified vaccines and their proper handling, July 2014. Inf Bull. https://www.who.int/immunization/programmes_systems/service_delivery/EN_Information_Bulletin_VVM_assignments.pdf
WHO (2002) WHO expert committee on specifications for pharmaceutical preparations, annex 6: good manufacturing practices for sterile products. WHO Tech Rep Ser No 902:76–93
WHO (2005) Monitoring vaccine wastage at country level. Guidelines for programme managers, WHO/V&B/03.18.Rev.1
WHO (2010) WHO best practices for injections and related procedures toolkit
WHO (2016) The controlled temperature chain (CTC): frequently asked questions. https://www.who.int/publications/m/item/controlled-temperature-chain-(ctc)-faqs
WHO (2020a) Immunization coverage: key facts, 15 July 2020. https://www.who.int/en/news-room/fact-sheets/detail/immunization-coverage
WHO (2020b) Tuberculosis, key facts, 14 Oct 2020. https://www.who.int/news-room/fact-sheets/detail/tuberculosis
WHO (2021a) WHO prequalified vaccines, online database: https://extranet.who.int/gavi/PQ_Web . Accessed on 20 March 2021
WHO (2021b) Vaccine safety basics e-learning course, Module 2: types of vaccines and adverse reactions. Route of administration. https://vaccine-safety-training.org/route-of-administration.html
WHO (2022a) WHO Coronavirus (COVID-19) dashboard, as of 8 August, 2022. https://covid19.who.int/. Accessed 8 Aug 2022
WHO (2022b) WHO prequalified vaccines, online database. https://extranet.who.int/pqweb/vaccines/prequalified-vaccines. Accessed 8 Aug 2022
Wong Y-L, Sampson S, Germishuizen WA et al (2007) Drying a tuberculosis vaccine without freezing. Proc Natl Acad Sci 104:2591–2595. https://doi.org/10.1073/pnas.0611430104
Zaffran M, Vandelaer J, Kristensen D et al (2013) The imperative for stronger vaccine supply and logistics systems. Vaccine 31:B73–B80. https://doi.org/10.1016/j.vaccine.2012.11.036
Zhu C, Shoji Y, McCray S et al (2014) Stabilization of HAC1 influenza vaccine by spray drying: formulation development and process scale-up. Pharm Res 31:3006–3018. https://doi.org/10.1007/s11095-014-1394-3
Ziaee A, Albadarin AB, Padrela L et al (2019) Spray drying of pharmaceuticals and biopharmaceuticals: critical parameters and experimental process optimization approaches. Eur J Pharm Sci 127:300–318. https://doi.org/10.1016/j.ejps.2018.10.026